D0909212

THE CHEMICAL REVOLUTION

P. J. de Loutherbourg, R.A. (1740-1812)

THE IRON WORKS AT COALBROOK DALE

Here Abraham Darby first smelted iron ore with coke, and revolutionised iron production. In time, scenes like this provided new sources of inspiration for such artists as J. S. Cotman and J. M. W. Turner

THE CHEMICAL REVOLUTION

A Contribution to Social Technology

by

ARCHIBALD CLOW

and

NAN L. CLOW

"Like a good citizen Lavoisier turned his thoughts to political economy."

Essay Index Reprint Series

BOOKS FOR LIBRARIES PRESS

FREEPORT, NEW YORK

First Published 1952
Reprinted 1970

INTERNATIONAL STANDARD BOOK NUMBER:
0-8369-1909-2

LIBRARY OF CONGRESS CATALOG CARD NUMBER:
79-134069

PRINTED IN THE UNITED STATES OF AMERICA

CONTENTS

LIST OF ILLUSTRATIONS

DIAGRAMS

INTRODUCTION

THIS book is a synthesis. It is an attempt to create something new out of hitherto unexplored ground common to economic history and chemical technology. It is addressed to practitioners in both fields, in the full consciousness that the new compound, while having specific beneficial properties to some, may conceivably prove toxic to others. To effect the synthesis and do justice to the theme it has been necessary to sift original sources of contemporary material, for the most part not readily available. At the same time facts that are already familiar to economic historians have been re-examined from a point of view that has been neglected. It is not the last word on the subject but rather the first.

The evolution of chemical industry must be regarded as an important concomitant of the Industrial Revolution, but, to a far greater extent than would be true of mechanical invention of the period with which our book deals, a satisfactory appraisal of the technological nexus of chemical manufacture during the same period presupposes an intimate understanding of the nature of the technical processes involved. Also, in greater measure than would be true of mechanical innovation of the period, the technical issues then facing the chemical manufacturer are now obscured by the gulf that separates us from his concepts, and by the nomenclature current at the time. To see clearly why and how an innovation took place, when and where it did take place, implies something more than a knowledge of the social need that prompted the solution of a particular technical problem. It also calls for a knowledge of the class of raw materials suitable for the solution of the problem and the circumstances that made them available.

Just how the history of the Industrial Revolution reveals its dependence on prior theoretical knowledge may be illustrated by citing from a discourse which Wm. Thos. Brande (1788–1866) gave at the Royal Institution in 1819 when the tide of innovation was in full flow.

> Among the useful arts, it is difficult to select one that is not very immediately dependent upon chemical principles, and in reverting to the history of these arts, we shall find ourselves obliged to confess that their leading improvements have derived from the same source. It would be trite and tedious to enumerate all that

> chemistry has done for the arts of bleaching, dyeing, calico-printing, and tanning; in the arts of pottery, of glass, and porcelain, or in the apparently more remote operations of the brewer and distiller: but it may not be useless to inform you, that the discovery of the present way of making oil of vitriol, of preparing vinegar from wood, of extracting pure acid from the lemon; that the abstruse and apparently abstract enquiries into the propagation and effects of heat, are so many sources whence these improvements have been derived, and whence individuals, often ignorant of their origin, have enriched themselves, and benefited the community.[1]

To reinterpret what we already know of the history of chemical manufacture, and of such industries as made use of chemical knowledge during this period, in their relation to advancing theoretical knowledge, no less than to explore sources of new relevant information, will be the task to which we shall apply ourselves. If, therefore, our narrative contains many facts with which economic historians are already familiar it is also our hope that the student of economic history will be stimulated to see their significance in a new way. Clearly a proper understanding of the developments with which this book deals calls for a closer cooperation between the trained historian and the scientific worker in the laboratory. We have written it with this in mind.

Inversely, therefore, if on one hand the scientific practitioner whose acquaintance with economic history is slight finds the technical detail unnecessary, the economic historian on the other will be familiar with the social changes brought about by coextensive mechanical technology.

The magnitude of the task before us set limits to the scope of our investigations. Neither in point of time nor in location have these been chosen arbitrarily. In the first place a galaxy of Scotsmen[2] gave scientific direction to the Industrial Revolution, particularly to its hitherto neglected non-mechanical aspects. The vital contributions of these men were made for the most part between 1750 and 1830. Nor is that all. We have been fortunate in having certain foundations on which to build. The economic history of the Industrial Revolution in Scotland has been worked out by Henry Hamilton: David Bremner has given a useful if somewhat limited account of specific Scottish industries. But of

[1] *Quarterly Journal*, 1819, 7, 205.

[2] Joseph Black, Francis Home, James Watt, James Keir, Patrick Copland, the Earl of Dundonald, James Hutton, William Murdock, George and Charles Macintosh, Charles Tennant, James B. Neilson, David Mushet, George and Cuthbert Gordon.

greater significance than these secondary works of reference, invaluable though they be, John Sinclair's *Statistical Account of Scotland* (1791–9), and the *New Statistical Account* (1845) which it inspired, provide mines of contemporary raw material for the social technologist. There are no statistical accounts for England, and the English *General Views of the Agriculture* of the various counties merely parallel the corresponding publications for Scotland. Therefore without apology, we have throughout made special, though by no means exclusive, reference to Scotland, and to the north of England with which it is on occasion linked economically. Other regional social technologies must await the cooperative syncretism of a group of historians and technologists familiar with local peculiarities and having access to available sources of material which diligent search would no doubt unearth.

With the exception of the opening chapter, in which we thought it advisable to discuss technological preparations for what is usually designated *the* Industrial Revolution, our point of departure is approximately 1750: the time at which sheer mass of material to be handled enforced a stop, roughly 1830. The first chemical works for the manufacture of sulphuric acid in Scotland was established in 1749 at Prestonpans. In 1828 James Beaumont Neilson started heating the blast supplied to iron-furnaces and so transformed that industry. Between these two focal points the pattern of social technology which persisted for many generations was worked out. Agriculture and agricultural products have been brought down a little later to allow for the characteristic time-lag always operative there.

J. U. Nef has pointed out in his *Rise of the British Coal Industry* that the main non-mechanical problems which faced seventeenth and eighteenth century technologists arose out of finding solutions for shortage of timber, and that efforts to bring about a transition from an economy based on wood and water to an economy based on coal and iron constituted a powerful determinant to the course of industrial evolution. Our indebtedness to Nef's painstaking analysis cannot be overstated.

Since it is convenient to have terms of reference to cover the various stages through which technics evolved we have adopted Lewis Mumford's extension of a terminology originated by Patrick Geddes. The following condensed citation from Mumford's *Technics and Civilization* illustrates what we mean by them:

> . . . one can divide the development of the machine and the machine civilization into three successive but *overlapping and interpenetrating phases:* eotechnic, palaeotechnic, neotechnic. . . .

> Speaking in terms of power and characteristic materials, the eotechnic phase is a water-and-wood complex: the palaeotechnic phase is a coal-and-iron complex, and the neotechnic phase is an electricity-and-alloy complex. . . . The dawn-age of our modern technics stretches roughly from the year 1000 to 1750. It did not, of course, come suddenly to an end in the middle of the eighteenth century. A new movement appeared in industrial society which had been gathering headway almost unnoticed from the fifteenth century on: after 1750 industry passed into a new phase, with a different source of power, different materials, different objectives.

In the middle, palaeotechnic, phase, chemistry abandoned remnants of alchemical tradition and evolved a rational theory of combustion, a fitting accompaniment to Dr. Joseph Black's theory of latent heat which James Watt applied with such success in his separate condenser. Combustion and energy transformation were inseparable from steam-engine economy. It is with the influence of chemistry on social conditions during this phase that we are in the main concerned. Since Mumford's *Technics and Civilization* is perhaps the only study that regards society from a standpoint comparable with our own, his influence on the evolution of our thesis is not unnaturally discernible. We acknowledge this inspiration wholeheartedly.

In conclusion let it be said that, in the interests of historical and geographical accuracy, where we have written *England* we do not mean the *British Isles*, a nicety not infrequently neglected by those from whom we have had occasion to cite.

Chapter I
MINERALS AND MANUFACTURES

The scientific details which now terrify the adult manufacturer will be mere trifles to his children when they shall be taught at school, a little more Mathematics and a little less Latin, a little more Chemistry, and a little less Greek.

DUMAS.

PART ONE
Encouragement of Trade and Manufactures

TO understand and appreciate the significant impact of technics on eighteenth-century social life, during that period of flux which was originally characterized as the Industrial Revolution, it is advisable first of all to examine the technological anticipations and preparations of the eotechnic age. For the 'complicated intermingling of economic, social, and intellectual forces, which underlie the great inventions', a term of wide connotation is necessary. While labour, assisted by an unprecedented concentration of capital, was being regimented in mill and factory, important technological changes in processes and methods of manufacture were being effected. If one views the changes from outside the traditional ambit of mechanical technology, even cursory examination shows that the phenomenon called *the* Industrial Revolution—what Witt Bowden describes as 'the initial decisive change in England by which, in the technique of manufacturing and allied industries, power-operated machinery took the place of prominence formerly held by the hand and the tool'—was in fact no revolution, but a slow, evolutionary change.[1]

Moreover, the 'decisive change' of the eighteenth century was not the first awakening of English industry. Both Lewis Mumford in his *Technics and Civilization* and J. U. Nef in his masterly study of the *Rise of the British Coal Industry* emphasize that technological development in England began to gather impetus about the time of Henry VIII's dissolution of the monasteries, and exploded into a first industrial revolution coincident with the founding of the Royal Society of London for Improving Natural Knowledge (1660). 'Seldom has there been a more perfect marriage between pure science and practical achievement than in Restoration

[1] W. Bowden, *Industrial Society in England towards the End of the Eighteenth Century*, p. vii.

England, through the efforts of the group of men who founded and carried on the Royal Society.'[1] The galaxy included Boyle, Newton, Leibniz, Evelyn, Pepys, Wren, Petty, Southwell, Lowther, Brereton, Papin, Huygens, and Becher. As the seventeenth century came to a close this efflorescence spent itself. It was only when the next century was well advanced that the spirit of invention and enterprise revived in England again, during the so-called, or in Nef's terminology, the second, Industrial Revolution. 'This second revolution multiplied, vulgarized, and spread the methods and goods produced by the first: above all, it was directed towards the quantification of life, and its success could be gauged only in terms of the multiplication table.'[2]

In the interim, Scotland made vital contributions to social technology. The focus of practical experiment and intellectual expansion moved north during the period which intervenes between the two southern 'revolutions'. It is with developments during this alteration of focus that we are mainly concerned in this preparatory chapter.[3]

The long era of consolidation, or if you will stagnation, which England went through between the founding of the London (1660) and Edinburgh Royal Societies (1783) may be gauged by the number of English patents of invention granted between 1660 and 1790.

1660–69	..	31	1730–39	..	56
1670–79	..	51	1740–49	..	82
1680–89	..	53	1750–59	..	92
1690–99	..	102	1760–69	..	205
1700–09	..	22	1770–79	..	294
1710–19	..	38	1780–89	..	477[4]
1720–29	..	89			

When represented graphically these figures leave no doubt as to the profound change which took place in the second half of the eighteenth century.

This method of representing statistical data, it might be mentioned, was invented outright by William Playfair (1759–1823), younger brother of John Playfair, mathematician and geologist, and Founder Fellow of the Edinburgh Royal Society. William became connected with the Boulton and Watt concern at Birmingham.[5]

One cannot now decide with certainty the cause of the two

[1] J. U. Nef, *Rise of the British Coal Industry*, Vol. 1, p. 252.

[2] L. Mumford, *Technics and Civilization*, p. 151.

[3] J. U. Nef, *Progress of Technology, 1540–1640; Economic History Review*, 1934–6, *5*, 3.

[4] W. Bowden, *op. cit.*, p. 12.

[5] H. H. Funkhouser and H. M. Walker, *Economic History*, *1934–7*, *3*, 103.

peaks which characterize the earlier part of the curve, nor can one assess their influence in Scotland, since the founding of the Royal Society of London coincides with measures by the Scots Parliament for encouraging trade and manufactures. It may be that some of its Acts resulted from the stimulus given by the initial activities of the Royal Society whose list of queries, proposals, experiments, and observations, clearly illustrates the

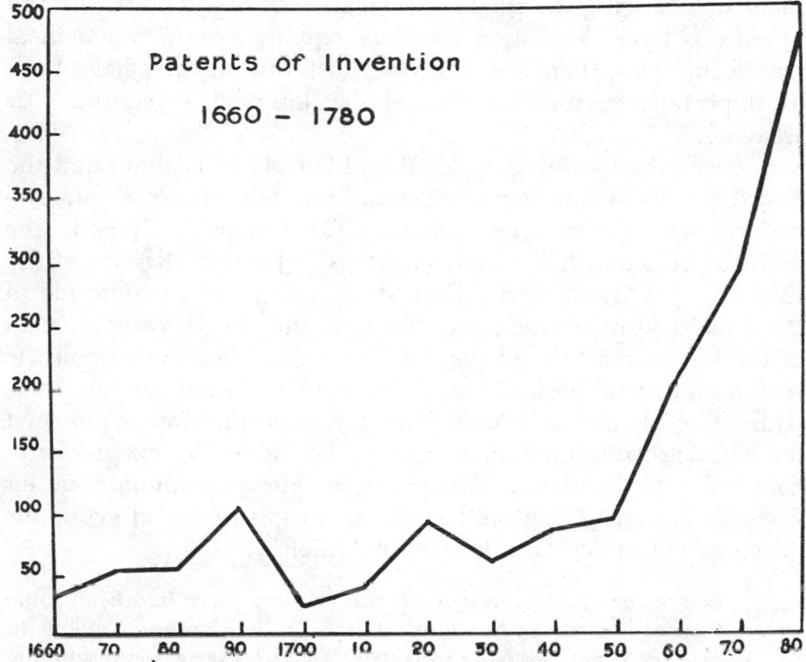

growing impact of science upon technological advance. Trade and industry were the points of departure of many of the Fellows' initial investigations, e.g. in 1667 they proposed to collect material for histories of English mines, ores, lead-refining, iron-making, saffron, colours, dyeing, making copperas, alum and nitre, making salt from sea-water, potash, paper-making, brewing beer and ale, pitch, tar, and so on.[1]

The new phase which developed in the latter half of the eighteenth century must be regarded as the result of a quickening of pace rather than from the introduction of new techniques of invention.

[1] T. Sprat, *History of the Royal Society*, p. 155.

> The eighteenth century witnesses a closer *rapprochement* between science and technology. On the one hand, men of science took a more active interest in practical problems; on the other hand, practical craftsmen or technicians showed a new interest in the scientific aspects of their work.[1]

Few of the new inventions were mechanical, despite the fascination which mathematical and mechanical invention has for the historian of science. 'While tools and machines transform the environment by changing the shape and location of objects, utensils and apparatus have been used to effect equally necessary chemical transformations. Tanning, brewing, distilling, dyeing have been as important in man's technical development as smithing or weaving.'[2]

Between the founding of the Royal Society of London and the Royal Society of Edinburgh, Scotland suffered a series of cataclysmal changes: the failure of Paterson's Darien project (1699), the political union with England (1707), the Jacobite Risings of the Fifteen (1715), and of the Forty-five (1745). The magnitude of these political milestones, together with the alleged consequences of the Union, and the exaggerated attention bestowed upon the textile and metallurgical industries, have obscured the fact that, while English genius rested from the scientific *élan* of the first industrial revolution, there was a considerable extension of eotechnics in Scotland. For example, Henry Hamilton in his *Economic History of Scotland* has the following to say of economic development in Scotland before the Union:

> Before the end of the seventeenth century there had been some industrial activity in Scotland. The Scots Parliament, anxious to encourage manufacturing industries, passed a series of Acts giving extensive privileges to people who would invest capital and skill in industrial enterprises. As a consequence many companies were founded for prosecuting different branches of the textile industries, as well as various other trades. Two of these industries are of particular interest from our point of view, and curiously enough they received the smallest measure of State assistance and yet enjoyed the greatest permanent success. One was sugar-boiling, which marks the beginnings of the sugar-refining industry, and the other soap-boiling, which might be regarded as the seed from which the chemical industry has sprung. Both were started in 1667, and both are still with us.[3]

[1] A. Wolf, *History of Science, Technology, and Philosophy, XVIIIth Century*, p. 499.
[2] L. Mumford, *op. cit.*, p. 11.
[3] H. Hamilton, *Industrial Revolution in Scotland*, p. 2.

The principal Act referred to by Hamilton was the *Act for Encouraging Trade and Manufactures* of 1661, the provisions of which were extended by subsequent Acts.[1] The consequence of these Acts was that some fifty undertakings became 'privileged manufactories', i.e. virtually protected private monopolies. Many of them were joint-stock companies with a total capital so large that it could not have been found by the earlier craftsmen. Even quite primitive establishments at that time required a capital outlay often amounting to many thousands of pounds, i.e. sums quite incommensurate with individual resources.[2] 'From the beginning factory production made direct demands for capital far above the small advances necessary to provide the old-style handicraft worker with tools or keep him alive. The freedom to operate independent workshops and factories, to use machines and profit by them, went to those who had command of capital; while the feudal families, with their command over the land, often had a monopoly over such natural resources as were found in the earth, and often retained an interest in glass-making, coal-mining, and iron-works right down to modern times.'[3]

While the sixteenth century saw the introduction into England of the manufacture of alum, copperas, cannon, gunpowder, paper, sugar, and saltpetre, it was not till the later decades of the seventeenth century that the majority of such manufactures were established in Scotland. This is illustrated by the monopoly granted to Coln. Ludovick Leslie and Coln. James Scott for

> . . . making saltpetre, salt upon salt, potashes, tanning of skins and hydes without bark, pitch, tarr, whiteyron, yron threid, making of yron with coale, castle-soap, raising of water and weights out of pits, improving of ground, making of plenchs and salt-pans, and making of anything in cristall alswell for ornament as use.[4]

The monopoly was to last for nineteen years from 1661 and the only condition attached was that the inventors 'discover and make experiment of their inventions'. Its extraordinary scope illustrates the virgin field awaiting inventors in Scotland. The granting of such a monopoly does not necessarily infer, however, that there were no manufactures of the kind already established, since a monopoly was not infrequently granted to more than one party. Be that as it may, varied industries were established in Scotland

[1] *Acts of the Parliament of Scotland*, Vol. 7, p. 255; Vol. 8, p. 348.
[2] Th. Keith, *Commercial Relations between England and Scotland, 1603–1707*, p. 80.
[3] L. Mumford, *op. cit.*, p. 25.
[4] *Acts of the Parliament of Scotland*, Vol. 7, p. 47.

between the regal and political Unions. They included soap and sugar boiling, distilling, glass and paper making, tile making, industries related to mining, iron and steel, and ancillaries to textile manufacture.[1] Their establishment was, as in England, often the consequence of imported continental technical skill. As Mumford points out, 'all the critical instruments of modern technology—the clock, the printing-press, the water-mill, the magnetic compass, the loom, the lathe, gunpowder, paper, to say nothing of mathematics and chemistry and mechanics—existed in other cultures' before they took root in England or Scotland.[2]

Since frequent reference will be made in the ensuing pages to the granting of monopoly, patent, or privilege, it will not be inappropriate to mention briefly the evolution of patent law, bearing in mind that Scotland had a separate system till the nineteenth century. The following summary of the development of the idea of a patent is taken from H. T. Wood's *Industrial England in the Eighteenth Century*.

From early times sovereigns granted monopolies for manufacturing or trading in various articles. This was a recognized method of establishing new trades and industries. The earliest recorded in England is in the reign of Edward III. There was no control over the sovereign, and it frequently happened that grants were made for inventions which were not new, and also that more than one grant was made for the same manufacture or trade to different persons. This led to the passing of the Statute of Monopolies in 1624 which transferred the power of granting Letters Patent for inventions from the sovereign to the State. The Statute contained a clause excepting 'any manner of new manufactures within this realm'. The clause (sect. 6) fixed the term of the grant at fourteen years, and this period has continued to the present day. Upon it has been built up our whole patent system. For long after the passing of the Statute of Monopolies Scotland and Ireland retained their own patent laws. Scottish patents were registered under the Great Seal at Edinburgh till the beginning of the nineteenth century at least.

Many early inventions were introductions from abroad, and grants were given to anyone who could show that his project was not already in operation. Patentees were in the habit of referring to their travel, study, and observation abroad, and this would seem to imply an acknowledgment that technical achievement had been pushed farther on the Continent than at home. In the

[1] W. R. Scott, *Scottish Industrial Undertakings before the Union*, *Scottish Historical Review*, 1903, *1*, 407 *et seq.*

[2] L. Mumford, *op. cit.*, p. 41.

majority of the early grants no specification or description of the invention was recorded. The exact date at which it became obligatory to disclose fully the patentee's secret is doubtful, but it was some time during Anne's reign. This stipulation appears in many grants about 1725, and in all of them a few years later. This change began at the end of the first quarter of the eighteenth, and was completed by the middle of that century. Mechanical invention was a stimulus to accurate registration of inventions.[1]

The history of eotechnics in Scotland is not dramatic. Little is known of the personnel taking part; less even of the undertakings established. No names shine out against a European background as did those of Cullen, Black, and Home in the late part of the eighteenth century. Yet it was on the solid foundation of eotechnic economy that Scottish intellect reared a culture comparable for many generations with anything to be found on the Continent of Europe, with the possible exception of Paris. To the social technology of the eotechnic age we must now apply ourselves.

Undoubtedly the most important factor in the evolution of seventeenth-century social technology was the replacement of wood by substitute materials, e.g. coal. As early as 1610, Sir William Slingsby divided processes being affected by the transition into two classes: (*a*) 'bear, dies, allom, sea-salt', where coal could be substituted without difficulty; (*b*) 'malt, brede, brycke, tyles, pottes', and the melting of bell metal, copper, brass, iron, lead, and glass, where successful substitution was awaited.[2] Replacement of wood by coal at this time was less a matter of choice than sheer necessity. Timber famine was a dominant note in eotechnic Scotland as elsewhere.

> To such desperate straits were the inhabitants of some parts of Scotland driven by 1621 'that nomberis of thame bothe to burgh and land hes bene constrayned not onlie to cutt doun and distroy thair policie and planting, bot thair movable tymmer worke, to make fyre of it. . . . and in mony placeis the trade of brewing and baiking for want of fyre is neglectit and cassen up.'[3]

Scarcity of fuel led Edinburgh brewers to cut heather for fuel to fire their brewing coppers, and so great were the quantities used that Parliament passed an Act forbidding storage within the burgh on account of fire hazards.[4] Sir William Brereton (1604–61) on his journey to Glasgow in 1635 remarked the absence of trees in the Lowlands:

[1] See also E. W. Hulme, *Law Quarterly Review*, 1897, *13*, 313; 1900, *16*, 44.
[2] J. U. Nef, *Rise of the British Coal Industry*, Vol. 1, p. 215.
[3] *Ibid.*, Vol. 1, p. 159. [4] *Acts of the Parliament of Scotland*, Vol. 4, p. 627.

> There is very little or no timber in any of the south or west parts of this kingdom, much less than in England. I have diligently observed, but cannot find any timber in riding near one hundred miles; all the country (is) poor and barren, save where it is helped by lime or sea-weeds.[1]

Scarcity persisted for many generations: a highly significant deprivation. A century later, Thomas Pennant observed that further into the Highlands, on the south bank of Loch 'Raynach', 'trees are now grown so scarce as not to admit of exploitation'.[2] In the *Statistical Account* also will be found instances of impeded economic development arising from lack of fuel. Iron could not be smelted at Monymusk; there was an instance of decrease in population, at Udny; while in Glenorchy, we are told, young and old were obliged to remain in bed throughout the winter on account of the cold which could not be assuaged through lack of fuel.[3] So late as 1790 the inhabitants of the Hebrides almost perished from cold and hunger, being compelled to burn their scanty household furniture as firewood.[4]

But throughout the seventeenth century new forest trees were introduced into Scotland: lime in 1664, silver fir in 1682, maple in 1690. In 1727 larch was first planted, but it was not till the latter half of the century that active steps were taken to augment the supply of timber by systematic planting. Then Clerk of Penicuik planted 300,000 trees, Grant of Monymusk softened the barren prospects of Aberdeenshire, the Duke of Atholl covered no less than 16,000 acres with larches. He even had a special gun devised to fire seeds into inaccessible crevices in barren crags. Indeed afforestation by landowners was carried out in the Highlands with great vigour from 1780–1820, but was often violently resisted by the people.[5]

The complete dependence of European civilization on wood till comparatively recently is now somewhat difficult to appreciate. J. U. Nef has listed many operations which in the seventeenth century depended on wood as a raw material and we have drawn from him freely in compiling the following uses.[6]

Gunpowder, for example, was made by mixing sulphur (S) and nitre (KNO_3) with charcoal (C). From the bark of trees a sap was

[1] W. Brereton, *Travels in Scotland, etc., 1634–5*, p. 114.

[2] T. Pennant, *Tour in Scotland in 1769*, p. 88.

[3] *Statistical Account*, Vol. 3, pp. 24, 49, 68; Vol. 4, p. 159; Vol. 8, p. 358.

[4] MS. 642, f. 21, *National Library of Scotland*.

[5] H. Hamilton, *Economic History of Scotland*, p. 45; H. G. Graham, *Social Life in Scotland in the Eighteenth Century*, p. 73.

[6] J. U. Nef, *op. cit.*, Vol. 1, p. 190 *et seq.*

extracted which was used for making pitch and tar to caulk the hulls of ships. From wood ashes potash (K_2CO_3) was extracted. It was essential to the makers of glass, soap, and saltpetre. All industries relied on wood as fuel. Wooden ships carried metal fitments for which metal had to be smelted, forged, and shaped. In the textile trade also, wood was needed, in the preparation of alum to mordant the cloth, in making the soap to clean it, and in the actual process of dyeing. The manufacture of cauldrons in which dyeing was carried out consumed considerable amounts of wood as fuel. Fuel was needed by all metal workers, and in industries depending on boiling or firing, making starch, refining sugar, boiling soap, baking, firing pottery, tiles, bricks, tobacco pipes, etc. Also for drying malt and hops and for tanning skins. Lime-burners were other extensive consumers. Nef continues:

> Wood can be made to grow almost anywhere and most societies depending on wood fuel can therefore obtain their supplies of timber in large measure near the place of consumption. But coal resources are normally localized in a limited number of regions, and the adoption of coal in place of wood necessarily involves a greater economic interdependence between districts, because the number of persons who are obliged to secure fuel from a distance is immensely increased.[1]

We may divide trades using coal under two headings: (*a*) those in which coal was used before the sixteenth century, or where substitution for wood presented no serious technical problems; and (*b*) trades where the substitution did present problems. Considering the position which the timber famine occupies in the history of technics, we propose to make this classification the basis of the discussion that follows.

In the Middle Ages the principal users of coal were, according to Nef, lime-burners and smiths, i.e. workers in all metals whether iron, lead, silver, gold, copper, or tin. In Scotland there are early records of the purchase of coal for arms manufacture and melting down lead for bullets and shot.[2] As we shall see in the chapter on agriculture, from the point of view of chemical technology, the single substance which did most to alter the condition of the land during the improving movement was lime. Of the small quantities burned at first no records survive, but we know that in 1678 application of lime was the usual way of 'gooding' the land. In November of that year there was a civil process between Sir R. Hepburn of East Lothian and a tenant concerning lime.[3]

[1] J. U. Nef, *op. cit.*, Vol. 1, p. 256. [2] *Ibid.*, Vol. 1, p. 201.
[3] R. Chambers, *Domestic Annals of Scotland*, Vol. 2, p. 398.

Generally it may be said that minerals, particularly coal, and the operations required to win them, occupy a fundamental position in social technology.

> Quarrying and mining are the prime extractive occupations: without stones and metals with sharp edges and resistant surfaces neither weapons nor tools could have passed beyond a very crude shape and a limited effectiveness—however ingeniously wood, shell and bone may have been used by primitive man before he had mastered stone.[1]

In the eotechnic age the story of civilization is the story of the progress of mine engineering and the exact science which supported it, and 'nowhere, perhaps, is the connexion between mineral wealth and the early growth of manufactures more plain than in the case of Scotland'.[2]

In the seventeenth century the increasing depth of mines is reflected in the enquiries issued by the Royal Society regarding the wonders and curiosities to be observed in *deep mines*.[3] To maintain the deepening mines free from water, the fire engine evolved in the first instance as a pump. 'Between the first known attempt in 1631 "to raise water from coal pitts by fire" and the erection in 1712 at a Staffordshire colliery of a Newcomen steam engine, which made the raising of water from a mine "by fire" a commercial success, there is a long and intricate story, still only partially known, of efforts to embody an increasing amount of scientific experience in a practical form.'[4] Viewed from the standpoint of its evolution, rotative motion was a late mutation.

Although impeded by lack of facilities for proper draining, and primitive methods of working in general, mining in Scotland increased in importance throughout the seventeenth century. As the timber famine increased in intensity mineral fuel was substituted for wood. Thus we find that coal was substituted for wood on the Forth littoral at an early date. The salt-pans there became an important absorber of small coal, or culms, which could not be sold to other consumers. Indeed coal had been in common use as far south as the Wear and Tees salt-producing area, at least from the beginning of the sixteenth century—a hundred years earlier than in Worcester and Cheshire which were much farther from coal mines. Other trades in which substitution of coal for wood presented no major technical problem were alum and copperas

[1] L. Mumford, *Technics and Civilization*, p. 65.
[2] J. U. Nef, *op. cit.*, Vol. 1, p. 232.
[3] T. Sprat, *History of the Royal Society*, p. 156.
[4] J. U. Nef, *op. cit.*, Vol. 1, p. 243.

making, and in the production of saltpetre and gunpowder. By the beginning of the eighteenth century coal was used for sugar and soap-boiling and in the making of starch and candles. Dyers and brewers also used coal, though in those trades there was a prejudice that the taste and smell of the products of its combustion clung to materials prepared with it.

Before going on to discuss problems involved in substitution of coal in other processes we give short historical summaries of the establishment in Scotland of those trades in which no particular problem was involved. They are salt-boiling, alum and copperas making, gunpowder, sugar-boiling, soap-boiling, brewing, etc. The reader who wishes to avoid the detail will find a table of events during the eotechnic age at page 32.

PART TWO

Primary Manufactures

SALT. Passing reference has already been made to salt. According to Nef, increase in the scale of manufacture was in no industry more impressive than in making salt by the evaporation of sea-water. From 1550–1700 the production of ships, salt, and glass in Great Britain increased much more rapidly than did population. The early economy of salt concerns its use for the preservation of meat to supply food throughout the winter and for use at sea, in addition to its culinary use. Its history, during the industrial expansion which both England and Scotland underwent before the eighteenth century, reveals how from mere domestic beginnings certain substances became the key raw materials of vast industries. By 1630 even, one-half of all Scottish shipping was employed in carrying salt and coal.[1]

From the sixteenth century the Scottish Parliament encouraged the manufacture of salt by a series of Acts dated 1561, 1573, 1673, 1681.[2] As a result, along the shores of the Firth of Forth wherever there was coal there were groups of salt-pans, as at, for instance, Kennetpans, Grangepans, Bonnardpans, and Prestonpans, where the manufacture of salt from sea-water was a highly important industry till the removal of the salt duties.

Prior to the sixteenth century little coal was used in salt-boiling, and it is probable that on the Firth of Forth the manufacture of salt antedates the extensive use of coal, there having been salt-pans there since the reign of David I (1124–53). This conjecture is

[1] J. V. Nef, *op. cit.*, Vol. 1, p. 224 *et seq.*
[2] *Pamphlet on Salt*, General Register House, Edinburgh.

based on the fact that small coal or dross, which was used in firing salt-pans, was called at one time *panwood*. Early in the sixteenth century it became usual to find salt-pans associated with coal-mining activities as on the Forth. There were others similarly associated, in Scotland, on the Ayrshire coast (Saltcoats), and for a time at Brora (Sutherland); and in England at the mouths of the rivers Tyne and Tees. So acute was the timber crisis that by 1700 coal was almost the only fuel used in the evaporation of brine. By this time Nef estimates, from records in the General Register House, Edinburgh, the annual consumption of coal by salt-works in the Forth area had reached a hundred and fifty thousand tons, half, in fact, of the total consumed in both England and Scotland.[1]

GUNPOWDER. Prior to the general preparations for war which heralded the ascent of Charles I to the throne (1625), England was equipped with powder-mills, probably from about 1554. This made the Privy Council of Scotland reflect on the inconvenience of being wholly dependent upon foreign countries for supplies of gunpowder. Consequently they entrusted Sir James Baillie of Lochend with the task of inducing Englishmen to come and settle in Scotland with a view to the setting up of gunpowder-mills.[2]

J. U. Nef asserts[3] that there were powder-mills both in England and in Scotland before the reign of Charles I. None the less it seems that Sir James Baillie did not succeed in his mission, for in 1690 James Gordon, a London merchant, was applying to the Scottish Parliament for a privilege for himself and partners to establish a gunpowder manufactory.[4] The outcome of Gordon's petition is uncertain, but there was no such factory in Scotland five years later when Sir Alexander Hope of Kerse and his co-partners asked encouragement to establish the manufacture of powder and alum in Scotland.[5] Figures are not available, but Scott informs us that 'a great stock of money' went into these concerns. Coal was used to dry the powder when it came from the mill.

ALUM AND COPPERAS. The powder production side of Sir Alexander Hope's venture prospered. The powder is said to have

[1] *Accounts of the Torry Salt and Coal Works, 1679*, cited by Nef.
[2] R. Chambers, *Domestic Annals of Scotland*, Vol. 2, p. 11.
[3] J. U. Nef, *op. cit.*, Vol. 1, p. 185.
[4] *Acts of the Parliament of Scotland*, Vol. 9, App. p. 42; R. W. Cochrane-Patrick, *Mediaeval Scotland*, p. 64.
[5] W. R. Scott, *Joint Stock Companies*, Vol. 3, p. 193; *Acts of the Parliament of Scotland*, Vol. 9, p. 420.

been of the highest quality. The alum, on the other hand, was a complete failure. Alum is mentioned several times during the eotechnic era in Scotland, but never as a successful manufacture. The same is true of the north of England, where in 1612 William Philips of Newcastle announced the mining of alum-stone in conjunction with coal.[1] Alum-making consumed large quantities of coal as fuel. In Scotland, 'a gift of making starch and alum' was granted to Timothie Langley in June 1663,[2] and again in 1672 Lockhart of Cleghorne petitioned to be allowed to establish the manufacture of alum.[3] Some time about 1750 the manufacture was prosecuted with greater success, but it was not till the end of the eighteenth century that Charles Macintosh really established the manufacture of alum in Scotland. As was so often the case, the technical knowledge required, or the inspiration to development, came from abroad, in this case from Sweden.[4]

Since there is also no evidence of the manufacture of copperas from pyrites in Scotland till the eighteenth century, it is probable that Scottish dyers were supplied from the Tyne area. Sibbald, however, recognized that it would be a profitable manufacture to introduce into Scotland.

> This is a good commodity to be exported at great profit, and is of use for our dyers and chirurgeons, and others too. We have in sundry places of the country the earth and stones it is made of: and we want not old iron (which is so wormd as it can be useful for nothing else). The vessels it is made in are of lead too.[5]

SUGAR. Two industries, both considerable consumers of coal, and employing somewhat similar techniques—separation from aqueous solution—were established in Scotland prior to the political union. These were sugar-boiling and soap-boiling. Both involved boiling solutions in open iron pans over open fires, and later on it was in the former of these industries that the first fortunes in business were made in the west of Scotland.[6]

Our first clue to sugar-boiling is a concession of the sole right of sugar-boiling to a partnership of Scottish and London merchants about 1628.[7] Not till the rise of the *Western Sugar House* in Glasgow did the profits to be reaped from this technique become

[1] J. U. Nef, *op. cit.*, Vol. 1, p. 227.
[2] *Miscellaneous Law Tracts*, 25.3.4, p. 50, National Library of Scotland.
[3] *Acts of the Parliament of Scotland*, Vol. 8, App. p. 23b.
[4] G. Macintosh, *A Memoir of Charles Macintosh*, p. 44.
[5] Sibbald, Catalogue and Discourses, (1701), N.L.S., MS. 33.5.16, p. 34.
[6] J. Gibson, *History of Glasgow*, p. 246.
[7] *Privy Council Register*, Vol. 12, p. 91.

apparent. In 1667 a partnership was set up by four persons[1] who acquired a small apartment for sugar-boiling under the guidance of a Dutchman as master-boiler.[2] So successful was the venture that they soon built a great stone tenement with convenient office space for their work. This was the *Western Sugar House.*

In two years a further partnery of five persons[3] set up the *Eastern Sugar House*. According to W. R. Scott the capital of one of these houses amounted to £10,000 sterling by 1684. It is said that they reduced the price of sugar for sale to 8*s*. Scots per pound.

That considerable technical skill was necessary for running these *sugaries* is illustrated by the difficulties which followed the death of Peter Gemmill himself. It is recorded that his wife, to whom he left his share, was unable to take part in the management, which required 'great skill and pains'. Further evidence of advancing technical knowledge is the association of rum manufacture with sugar-boiling. A proposal to make *wine* out of sugar canes had been put before the Royal Society by a Dr. Goddard.[4] By rum distilling it was possible to utilize on the spot by-product molasses which would otherwise have gone to waste, or would have had to be exported. Indeed the waste molasses was originally exported to Holland, but finding themselves injured by this trade, the Dutch forbade its importation. This was why Scottish manufacturers turned to the making of rum, though its production in Scotland did not go entirely without opposition, and a short-lived Act was passed in 1695 in deference to the malt spirit interests, forbidding production for export.[5] Also, a sugar manufactory at Leith was only allowed to make and sell eighteen tuns of rum per annum and 'provided it was not made for use within the kingdom'.[6]

An attempt to establish the production of sugar and rum was made by Robert Douglas, of Leith, but despite his established success with manufacture of earthenware, soap, and starch, which he could trade for raw sugar, the Clyde area continued to dominate the sugar trade. There Robert and James Montgomery established the *South Sugar House* in 1696.[7]

The last development in sugar-boiling as far as we are concerned

[1] Peter Gemmill, Frederick Hamilton, John Caldwell, and Robert Cumings, along with whom John Graham of Dougalston, Richard Graham, William Anderston, John Stark, William Craig, and James Craig were also associated.

[2] J. M'Ure, *Glasghu Facies*, p. 871.

[3] John Cross, James Peaddie, John Like, George Bogle and Robert Cross.

[4] T. Sprat, *op. cit.*, p. 193.

[5] *Acts of the Parliament of Scotland*, Vol. 9, p. 462; Vol. 10, p. 34.

[6] *Ibid.*, Vol. 9, p. 491.

[7] J. M'Ure, *op. cit.*, p. 283.

here was the erection of the *King Street Sugar House* or *Distillery Manufactory*, as it was sometimes called. To this end an Act in favour of Matthew and Daniel Campbell was passed in 1700. Here distilling was associated with sugar-boiling, a frequent and usual combination.[1] The number of partners was soon increased to six; and in 1736 the work was purchased by 'Robert M'Nair, Jean Holmes and Co.'.

Sugar-refining is still associated with this quarter of Scotland, and its location is an 'exception to the general rule that few industrial companies founded in seventeenth-century Scotland survived the removal of protection after the Union.'[2]

SOAP. Till early in the seventeenth century, importation from Flanders sufficed for the Scottish consumption of soap.

In 1619 Nathaniel Uddart had set up a soap manufactory in Leith, and two years later petitioned the Privy Council to prohibit the importation of all foreign soap. He claimed that he could supply the entire demands of the country, and his petition was granted on condition that the maximum price for green soap should not exceed £24 per barrel and white soap £32 per barrel (sterling or Scots not indicated).

Robert Chambers in his *Domestic Annals of Scotland* reckoned that the quantity of soap produced at Uddart's time was in the neighbourhood of 400,880 lb.[3] When Uddart's monopoly was coming to an end in 1634 Parliament made a new concession to Patrick Mauld of Panmure for 'the sole and full licence to make and cause to be made, within the said kingdom, soap for washing clothes, of all such colours and quantity as they shall think good'. Mauld's concession also covered the collection and production of the necessary raw materials. In fact he got as well 'licence to fish and trade in the country and seas of Greenland and in the isles and other parts adjacent thereto, and that for provision of the said soap-works with oils and other materials necessary thereto'. He also had the right 'to make potasses (alkaline carbonates) of all sorts, of such wood within the said kingdom, as is most fit for that purpose, and that can be most conveniently spared, . . . and of all sorts of fern and other vegetable things'.[4]

When the earlier Act which prohibited importation of foreign soap lapsed, further Acts to the same end were passed in 1661 and 1669.[5] This led to the formation of a company whose capital was

[1] J. M. Hutcheson, *Notes on the Sugar Industry of the United Kingdom, passim.*
[2] W. R. Scott, *op. cit.*, Vol. 3, p. 137.
[3] R. Chambers, *op. cit.*, Vol. 1, p. 508.
[4] *Ibid.*, Vol. 2, p. 81.
[5] *Acts of the Parliament of Scotland*, Vol. 7, pp. 88, 203–4, 560.

£11,700 sterling. Each of nine partners[1] subscribed £1,300.[2] The manufactory which they operated was known as the *Glasgow Soaperie*. The company fitted out a ship of 700 tons burden to supply them with blubber, which was boiled down at Greenock. The resultant oil was converted into soap at the Soaperie in Glasgow. In the *Glasgow Courant* in 1715 they were advertising:

> Anyone who wants good black or speckled soap may be served by Robert Luke, Manager of the Soaperie of Glasgow, at reasonable rates.[3]

So it also survived the Union and prospered.

The difficulties which faced these early capitalist enterprises may be gauged by their having to transport from Holland the bricks with which to build their furnaces.[4] The whaling side of the venture involved the Company in heavy losses, and it is well nigh certain that supplies of oil did not always come readily to the hands of the soap-boilers. We get evidence of such a shortage of oil in 1690 when Robert Douglas, the Leith soap-boiler, bought the carcasses of *mere-swine* (porpoises) for oil.[5] In spite of these difficulties, soap-boiling was firmly established as a Scottish eotechnic industry by 1700.[6] Even at a later date whale-fishing from Scottish ports was said to be 'a poor concern'.[7]

FERMENTATION. Not the least of the crafts which partially succumbed to capitalist enterprise during the seventeenth century was brewing. When Sir William Brereton visited Edinburgh in 1635 he described the state of the manufacture in these words:

> I took notice here of that common brewhouse which supplieth the whole city with beer and ale, and observed there the greatest, vastest leads, boiling keeres, cisterns and combs (wooden tubs), that ever I saw: the leads to cool the liquor in were as large as the whole house, which was as long as my court.[8]

Brewing was a voracious consumer of fuel, and early felt the effects of timber shortage. In the north of England, coked coal

[1] The partners were Sir George Maxwell of Pollock, Sir John Bell of Hamilton's Farm, John Campbel of Woodside, John Graham of Dougalston, John Henderson of Downhill, John Luke of Claythorn, Ninian Anderson, James Colquhoun and John Anderson. (J. M'Ure, *Glasghu Facies*, p. 874.)

[2] W. R. Scott, *op. cit.*, Vol. 3, p. 131.

[3] *New Statistical Account*, Vol. 6, p. 135.

[4] J. M'Ure, *op. cit.*, p. 874.

[5] R. Chambers, *op. cit.*, Vol. 3, p. 24.

[6] Sibbald, *loc. cit.*

[7] J. Sinclair, *Statistical Account*, Vol. 5, pp. 480 and 580; J. Cleland, *Rise and Progress of Glasgow*, p. 24.

[8] W. Brereton, *Travels in Holland . . . Scotland, etc., 1634–5*, Chetham Society, Vol. 1, p. 104.

provided artificial warmth for malting, but Scottish maltsters preferred peat. Everywhere, use of new fuel necessitated recourse to kilns of new type, like the one invented by Eustatius Rogle in 1599.[1] But in 1704 Patrick Smith announced the secret of drying malt *by any kind of fuel*. His special claim was that it was dried in his kiln 'so as to receive no impression from the smoke thereof, and that in a more short and less expensive manner than hath hitherto been known in the kingdom'. The drink brewed from it was to be 'as clear as white wine, free from all bad tincture, more relishing and pleasant to the taste, and altogether more agreeable to human health than the ale hath been heretofore known in the kingdom'.[2]

Even in malting, however, the substitution of coal for wood, as we have just seen, was not effected entirely without difficulty, and this was general in a wide variety of other technical processes. Where ore had to be mixed with fuel to be reduced, or where products of combustion could contaminate grain, for example, the substitution of coal was a complex problem: (*a*) either the coal had to be freed from noxious products (coking), or (*b*) improved furnaces and kilns had to be invented to afford protection from the products of combustion. It was only gradually that such inventions came into use.

Nef states that the first unmistakable reference to the removal of noxious substances in coal by coking occurs in 1620, when a patent was granted for the process. Between 1660 and 1700 there are indications that 'the principle of a second burning' was well understood. It was on the success of such experiments that Darby's successful smelting of iron ore depended. In this case a modified furnace was not enough.[3] And so we come to coal itself.

COAL. The history of the successful utilization of coal is the history of advancing technology in the eighteenth century. Nef points out that the formation of few joint-stock enterprises in the coal trade may be the reason why so many economic historians overlook the scope of business organizations in connection with coal-mining.[4] There were, however, combines of other types. Mines were divided into *parts*: each man got a *part* of the coal mined and arranged for its sale. Later on, the *parts* were leased or sold in shares with a consequent increase in the number of part-

[1] *Acts of the Parliament of Scotland*, Vol. 4, pp. 187–8.

[2] R. Chambers, *op. cit.*, Vol. 3, p. 303.

[3] See J. Houghton, *Husbandry and Trade*, Vol. 1, pp. 185 and 205; Vol. 2, p. 81: J. U. Nef, *op. cit.*, Vol. 1, p. 248.

[4] J. U. Nef, *op. cit.*, Vol. 1, p. 42 *et seq.*

ners, an arrangement not unlike that which operated in the Stanneries in Cornwall.[1]

> Coal had long been used as a fuel in a limited scale in Scotland. As early as the twelfth century power was granted to the monks of Holyrood to work coal in the lands of Carriden, near Bo'ness, and in the beginning of the thirteenth century the coal in the estate of Tranent was made over by charter to the monks of Newbattle. At the end of it, that on the estate of Pittencrief was granted to the monks of Dunfermline. It seems to have been used largely in connection with salt-works, but the growing scarcity of wood led to its more general use in hall and kitchen as well as forge. Its value as fuel had become so generally recognized that Acts were passed in this century by the Scottish Parliament prohibiting its export. In the seventeenth an export duty was substituted for this prohibition.[2]

At the end of the seventeenth century the average Scottish mine probably employed about one hundred men. In all Scotland there were probably some three thousand miners producing several hundred thousand tons of coal per annum. The capital involved and the number of operatives employed must have already exceeded that of all the other industries referred to in this chapter. Contiguous occurrence of iron ore, coal, and limestone in Scotland might have given her world preeminence in ferrous metallurgy if the developments which took place at Carron during the sixties of the eighteenth century had taken place a century earlier; but such innovations were the product of new scientific knowledge in the era of Drs. William Cullen (1710–90) and Francis Home (1719–1813) of Edinburgh. When it did develop the face of Scotland changed: coal made dirt a necessary accompaniment of chemical and metallurgical industry.

While coal deposits abounded along the Firth of Forth and surface outcrops were sometimes accessible by *ingean e'es*, it often had to be mined at a depth of sixteen to twenty fathoms.[3] Only towards the end of the seventeenth century did progress of invention make it possible to recover by means of pumps such mines as had been worked out of their accessible coal or drowned due to the ingress of water from underground streams. In 1693 Marmaduke Hudson, the inventor of a draining engine, was granted the exclusive right of using the engine he had invented, for nineteen years, together with the privilege of a manufactory.[4] Eight years

[1] G. R. Lewis, *The Stanneries, passim.*

[2] J. Mackinnon, *Social and Economic History*, p. 17.

[3] W. Brereton, *op. cit.*, p. 112.

[4] *Acts of the Parliament of Scotland*, Vol. 9, p. 323; App. p. 91.

later James Gregory, Professor of Mathematics in the University of Edinburgh, petitioned for an Act for the exclusive use of a machine invented by him for raising water.[1] The great significance of the transition to coal fuel in bringing palaeotechnics into being, as a result of its utilization when timber supplies dwindled following the ravages of eotechnics, cannot be underrated.

> Even before 1700 coal seems to have had a decided, though not always easily discerned influence on the development of industry. Coal being cheaper near mines, if other things were equal, manufactures were more likely to thrive in mining districts. Where the cost of coal was small compared with other costs, it mattered less.[2]

During the seventeenth century attempts were made to improve the reduction of metallic ores (iron, copper, tin, and lead), and the subsequent manufacture of them into hardware, such as pins, needles, scissors, scythes, tobacco-boxes, and English knives. The Royal Society of London had recommendations to make upon the smelting of lead with pit coal, about refining lead and tin ores, and hardening steel 'so as to cut porphyry with it'.[3] These improvements required some knowledge of the chemical reactions of ores.

> Something more important than gold came out of the researches of the alchemists: the retort and the furnace and the alembic: the habit of manipulation by crushing, grinding, firing, distilling, dissolving—valuable apparatus for real experiments, valuable methods for real science.[4]
>
> If the use of metals came at a relatively late date in technics, the reason is not far to seek. Metals, to begin with, usually exist as compounds in ores. The extraction of metals, unlike the cutting down of trees or the digging of flint, requires high temperatures over considerable periods. Even after the metals are extracted they are hard to work: only by being softened do the metals respond: where there is metal there must be fire.[5]

So we find that it is only towards the end of the seventeenth century that working in metals becomes of importance in Scotland. In 1699 the Privy Council passed an Act in favour of some English merchants who sought to establish the manufacture of hardware in Glasgow;[6] and in 1701 the Scots Parliament passed an Act in

[1] *Acts of the Parliament of Scotland*, Vol. 10, pp. 240 and 267.
[2] J. U. Nef, *op. cit.*, Vol. 1, p. 225.
[3] T. Sprat, *op. cit.*, pp. 191 and 197.
[4] L. Mumford, *op. cit.*, p. 40.
[5] *Ibid.*, pp. 68 and 69.
[6] R. Chambers, *op. cit.*, Vol. 3, p. 127.

favour of a co-partnery for the smelting of minerals.[1] A certain amount of japanning was also in progress since Messrs. Le Blanc and Scott opposed the petition of Sarah Dalrymple asking for the privilege of a manufactory for her establishment.[2] Yet metal industries remained in a very subordinate position not only in Scotland but throughout all the countries of Europe. Of the groups in existence to-day, textiles and even leather were more important. According to A. P. Usher:

> The leather industry was probably more important than metals in England and France, and though in Germany the metals were in all probability a greater factor in general industrial development, we have no grounds for supposing that metals outranked leather, even in Germany. The relative position of the industries in 1700 represents the culmination of the general factors in industrial development that became notable in the twelfth and thirteenth centuries.[3]

By the establishment of a variety of capitalist enterprises, Scotland made a bold bid in the last decades of the seventeenth century to obliterate her lag behind other countries.

> The merchant accumulated capital by widening the scale of his operations, quickening his turnover, and discovering new territories for exploitation: the inventor carried on a parallel process by exploiting new methods of production and devising new things to be produced.[4]

Albeit coal had not replaced charcoal in the manufacture of iron, mining already occupied an important place in what became the industrial belt of Scotland, an area which has always supported a large fraction of her total population. Even in 1603 Scotland was exporting salt and coal and importing grain into this part of the country.[5] Nef's estimate is that the rise of the Scottish coal industry to a place of great importance in the economic life of the country during the period 1550–1700 is scarcely less impressive than that of the north of England. Contemporary Englishmen referred to Durham and Northumberland as the Black Indies, a term which would have been equally apt for the part of Scotland bordering on the Firth of Forth.[6]

> Coal was in general use as a fuel among glassmakers, brewers, distillers, sugar-bakers, soap-boilers, smiths, dyers, brick-makers,

[1] *Acts of the Parliament of Scotland*, Vol. 10, App. p. 99.
[2] *Ibid.*, Vol. 11, p. 249.
[3] A. P. Usher, *Industrial History of England*, p. 254.
[4] L. Mumford, *op. cit.*, p. 26.
[5] J. U. Nef, *op. cit.*, Vol. 1, p. 258.
[6] *Ibid.*, Vol. 1, p. 42.

lime-burners, founders and calico printers. But in the meanwhile a more significant use had been found for coal: Dud Dudley, at the beginning of the seventeenth century, sought to substitute coal for charcoal in the production of iron: this aim was successfully accomplished by a Quaker, Abraham Darby, in 1709. By that invention the high-powered blast furnace became possible; but the method itself did not make its way to Coalbrookdale in Shropshire, to Scotland and the North of England until the 1760's. The next development in the making of cast-iron awaited the introduction of a pump which should deliver to the furnace a more effective blast of air; this came with the invention of Watt's steam pump, and the demand for more iron, which followed, in turn increased the demand for coal.[1]

IRON. Iron had a long eotechnic history in Scotland before it was smelted with coal. It was originally produced in small charcoal-fired furnaces or *bloomeries*, which could be built without great outlay of capital. They had an output amounting to a hundredweight or so per day, but the range of iron articles in use was very limited. Steel, for example, was used solely for cutting edges.[2] The equipment of a bloomery consisted of little more than the forge itself and the rudest of tools. The existence in Scotland of upwards of a hundred such bloomeries has been demonstrated by W. Ivison Macadam.[3] By chemical and mineralogical analysis Macadam was able to deduce the degree of technical skill forthcoming in their operation. Though many of the bloomeries constructed on open hill-sides or in waterless valleys must have relied solely upon natural draught, ample evidence from the nature of still extant slag-heaps shows that in some of them an artificial blast was produced with the aid of water, and used in a furnace of the more advanced Catalan design. In the most advanced bloomeries water-wheels supplied mechanical power as well as air blast.

Technical accomplishment in the Highlands must have been of a high order if we can believe the following comments by the Rev. Donald McNicoll, minister of Lismore, Argyllshire:

> The smelting and working of iron was well understood, and constantly practised, over all the Highlands and Islands from time immemorial. Instead of improving in that art, we have fallen off exceedingly of late years, and at present make little or none. Tradition bears that it was made in the *blomary* way; that is, by laying it under the hammers, in order to make it malleable with the same heat that melted it in the furnace.

[1] L. Mumford, *op. cit.*, p. 156.
[2] J. Campbell, *Political Survey of Britain*, Vol. 2, p. 43.
[3] *Pro. Soc. Antiq.* 1886, *22*, 89.

This is in all probability an overstatement, but it gives one some idea of the impression left by iron-working in the Highlands at the end of the eighteenth century.[1] These Highland furnaces were based entirely on Scottish timber resources. We shall see that the next step forward depended on the solution of the problem of smelting iron with coal, a palaeotechnic development which did not penetrate Scotland till 1760.

Though Scotland was not important as a copper producer, the non-ferrous branches of the metallurgical industries also participated in the process of rationalisation. In 1695, Nicholas Dupin, who was already associated with the manufacture of paper and linen, having got leases of mineral properties from several persons, founded the *Company for working Mines and Minerals in the Kingdom of Scotland*. We may infer that the minerals included copper ores, because he later petitioned Parliament for permission to make coin out of the copper of the nation. The petition was not granted, though the scarcity of coin makes it appear an opportune suggestion.[2]

Thus concentration of ownership went on both in mining and metallurgy. Labour was gradually drawn from subsistence employment into enterprises which supplied needs arising from rising standards of living. At least one large-scale enterprise for supplying the needs of Scottish craftsmen, the Leith Wool-card Manufactory (1663), was established as a result of the Act of 1661. Its size may be estimated from the claim that it made when its monopoly was opposed. It then supported sixty families.

About 1680 the coppersmiths and founders of Edinburgh got an Englishman, William Smith, to 'give them his insight into the airt of casting in brass'.

Early in the eighteenth century, following the Rising of 1715, a single organization got into its control more land, mines, and minerals, than any organization which preceded it. It was usually known as the *York Buildings Company*. The Company's influence on Scottish economic development cannot be neglected, so we shall now say something about its activities.

An institution which had a considerable effect on Scottish manufacture (quite apart from the mass of litigation in which it became involved serving almost as a text-book of law) was the *Company of Undertakers for Raising the Thames Water in York Buildings*. The Company had its origin in letters-patent granted by Charles II on 7th May, 1675, to Ralph Bucknall and Ralph Wayne. Till

[1] *Remarks on Dr. Samuel Johnson's Journal of Journey to the Hebrides, 1774.*
[2] W. R. Scott, *op. cit.*, Vol. 3, p. 187.

1719 it pursued its appointed business, but in that year it entered on a new phase of activity by the purchase of estates in Scotland which had been forfeited as a result of the Rising of the 'Fifteen. These purchases made the Company the greatest landowner in Scotland, with holdings in many counties. As the Company could not work the estates, mines, and minerals themselves, they let them out in the ordinary way. Some of their tenants are of interest. On the estate of Belhelvie was George Fordyce, merchant in Aberdeen, whose sons were Alexander, the banker, Sir William, the physician, and David, Professor of Philosophy in the University of Aberdeen. Another tenant was Sir Archibald Grant of Monymusk. He is described in earlier life as 'of a singularly speculative turn, and connected with various mercantile projects, many of them not of the most reputable character'. He speculated in York stock for several years, and 'if not the evil genius of the Company, was certainly mixed up in several transactions which proved disastrous to it'.

From the conditions of lease to James Stark of the estate of Kilsyth in 1721, there are indications that the Company also had improvement of their lands in mind. Stark was bound to plant two trees for every one cut down for necessary repairs: an unusual attempt to offset the ravages of eotechnics. But cutting timber, the Company found, was a quicker route to financial gain, and they used it to their utmost.

In the north of Scotland there were then large tracks of virgin forest which had been unaffected by the timber shortage of more populous areas. In 1726 Colonel Horsey, Governor of the Company, was recommended to acquire some of this timber, in particular the great pine-woods of Abernethy, belonging to the Grants of Strathspey. As a result the Company purchased 60,000 firs for some £7,000. Cutting was begun at once, and continued over fifteen years. When a specimen cargo was sent to Deptford, it turned out that the timber was not, as had been thought, suitable for main masts of His Majesty's ships. As an agreement had been made, however, and the Company were in possession of the timber, large-scale operations were undertaken. At the end of four years, debits exceeded credits by £27,913 19*s*. 10½*d*. Their activities brought them no great reputation on Speyside. The parish minister described them as 'the most profuse and profligate set that ever were heard of in this corner'. Unable to keep up their payments to the Laird of Rothiemurchus, a settlement was arrived at, and another contract entered into with him for the supply of wood for charring. Setting up iron furnaces to use the charcoal followed. These will be discussed elsewhere,[1] as also the Company's attempts at salt-boiling and glass-making.[2]

LEAD MINES. Other than deposits of coal and iron, the only extensive metallic minerals in Scotland were the lead ores in the Clyde and Nith districts. For centuries these have occupied an important if neglected place in the economy of Scotland. In 1561 John Acheson, master-cunyer, and John Ashlowan, burgess of Edinburgh, entered into an arrangement with Queen Mary to work the lead mines of Glengonner and Wanlockhead and 'carry as much as 20,000 stone-weight of the ore to Flanders'. One of the conditions of the lease was that they should deliver 45 oz. fine silver for every thousand-stone-weight of ore exported. How long these partners operated we do not know, but in 1565 John, Earl of Atholl, was granted a licence within 'the nether lead hole' of Glengonner and Wanlock. He offered the State a better bargain, viz. 50 oz. of silver instead of 45, per thousand-stone-weight. A similar licence was granted to James Carmichael, James Lindsay, and Andrew Stevenson for any part of the realm excepting Glengonner and Wanlockhead.

In the last decade of the century the anti-papist activities of James VI put a considerable strain on the regal purse. By September 1594 the royal debt amounted to £14,598, much of it due to Thomas Foulis, an Edinburgh goldsmith. On consideration of this obligation, the king granted Foulis a lease of the gold, silver, and lead mines of Crawford Muir and Glengonner for twenty-one years. Although the gold and silver did not amount to much, Foulis made a success of mining the Lanarkshire lead, and he was showing a profit by 1597. Smelting must almost certainly have been in the first instance with peat, the nearest supplies of coal being many miles distant, and it is unlikely that the wild moors of upper Lanarkshire could have yielded sufficient timber for making charcoal. Moreover, it was only towards the end of the seventeenth century that smelting with coal proved successful.[3]

Following a variety of activities the York Buildings Company, already loser in timber, iron, salt, glass, and coal, engaged in further operations in mines and minerals in Scotland, this time copper and lead. Colonel Horsey in 1729 entered into a nineteen-years lease with the Earl of Hopetoun to work all mines and minerals, for gold, silver, copper, lead, tin, iron, and all other minerals and metals in the lands of Ballencrieff, Bathgate,

[1] Ch. xvi. [2] D. Murray, *York Buildings Company, passim.*

[3] J. U. Nef, *op. cit.*, Vol. 1, p. 224: R. Chambers, *op. cit.*, Vol. 1, pp. 18–19, 253–4, 289.

Drumcross, Knock of Drumcross, Hilderston, Torphichen, and Tortarven, in the Sheriffdom of Linlithgow. While this was in progress, Horsey was considering purchasing shares in lead mines near Fort William 'in which General Wade is concerned'. These were the property of Sir Alexander Murray of Stanhope, who had purchased the Ardnamurchan peninsula on the strength of the lead ore it was supposed to contain. Murray in 1724 had granted a lease to the following partnership, Thomas, Duke of Norfolk, Sir Gervas Clifton, Sir Robert Clifton, Sir Robert Sutton, Sir Archibald Grant, Marshal Wade, William Neilson, merchant of Edinburgh, Richard Graham and Peter Murdoch, merchants in Glasgow. It is worth noting the personnel common to both undertakings.

These partners, by methods which need not be gone into here, transferred their lease of the Ardnamurchan mines to the York Buildings Company on 31st July, 1730. For a time all appeared to be going well.

> Additional land was leased to secure a supply of timber. Shiploads of material and stores arrived at Strontian almost daily—bricks and castings, coals and billet wood, framed houses put together in London, deals from Norway, oatmeal and malt; improved furnaces, smelting mills and hearths were constructed. As many as five hundred workmen are said to have been employed, for whom a village, known as New York, was built.[1]

Expenditure by the Company in doing all this amounted to some £40,000 in the first two or three years. Their return did not suffice to meet their wages bill. In fact, they smelted only two hundred and forty-four tons of lead in four years. So another venture of the York Buildings Company was a failure. The mines were abandoned and the workmen dismissed at Christmas, 1740. Much litigation followed.

The subsequent history of the Company is not relevant to the present study. At last, in 1829, after an existence of more than a hundred and fifty years, an Act was brought in bringing about the dissolution of the Company and providing for sale of its property and division of proceeds amongst stockholders of the Company.

David Murray sums up this phase of Scottish development by saying:

> The design of purchasing the forfeited estates was a magnificent one, and if wisely carried out might have resulted in much

[1] D. Murray, *op. cit.*, p. 74.

benefit to Scotland, and great profit to the Company. The conduct of the Company's business often showed considerable ingenuity, but most of its schemes were wanting in honesty, and it seems strange that one generation after another of directors and officials should all have been inoculated with the evil principles which sprung into life in the great Bubble year.[1]

PART THREE
Technical Innovations

There remains to be discussed only those instances in which the alternative to coking was adopted, viz., the redesign of furnaces, etc. A reverberatory furnace in which the substance under treatment comes into contact with the products of combustion, but not with the fuel, was patented by Viscount Grandison in 1678, but it was not till 1696 that a really satisfactory one was devised. This furnace facilitated the smelting of lead and copper, and the making of brass with pit coal. Lead was more easily desilvered (it had been tried in Scotland as early as 1613), and in the Philosophical Transactions of the Royal Society there is an improved furnace for the making of copperas.[2] In the glass industry the open crucible in which the *metal* was melted was covered in, and only a small hole left for the glass blower to insert his tool. Glass-making in eotechnic Scotland merits further discussion.

GLASS. In most operations technical problems of considerable magnitude and far-reaching consequences were involved when it became necessary to change from wood to coal fuel. This is well illustrated by the glass industry.

Glass-making in Scotland started with the granting of a licence to Sir Jerome Bowes, Elizabeth's ambassador to the Court at Edinburgh. He mismanaged his licence, but was soon followed by other glass enthusiasts, there being attractive quantities of the raw materials, sand, kelp, and timber, in Scotland.

The next to get a gift of making glass was Sir George Hay of Nethercliff, the Clerk Registrar, afterwards Earl of Kinnoull and Lord Clarendon. In 1610 he was granted 'the sole and onlie privilege of making of glasses' within the kingdom for a thirty-one year period. This was confirmed by Parliament in 1612.[3] Hay established the manufacture of glass in Scotland at Wemyss in Fife on the property of Lord Elcho, probably under the supervision of a Venetian who was brought in to act as master. After

[1] D. Murray, *op. cit.*, p. 101. [2] *Phil. Trans.*, 1678, *12*, 1056.

[3] *Miscellaneous Law Tracts*, 25.3.4, National Library of Scotland; *Acts of the Parliament of Scotland*, Vol. 4, p. 515.

only a few years the Scottish Privy Council became alarmed at the depletion of the forests, and advocated a return to lattice windows and to the 'ancient manner of drinking out of stone'.[1] Hay's outlay is said to have been large, but in return his manufactory succeeded in producing *braid glass* (i.e. for windows) 'measuring three-quarters of a Scots ell and a nail in length, and the breadth at the head was an ell wanting half a nail', which glass took the place of that formerly imported from Danzig.[2] With the availability of glass for glazing windows, another change came in its train. In 1621 an Act ordered that Edinburgh houses should be roofed with 'slate, lead, or tiles'. A similar provision regarding the houses in Glasgow, Aberdeen, Dundee, and Stirling came into force in 1681.[3]

Although there are numerous reports that Hay was disappointed in the return from his venture, its output undoubtedly contributed considerably to the national economy. It was of sufficient importance to be bought up by Sir Robert Mansell, Bowes's successor in the English glass monopoly. Mansell may have closed down the Scottish works, for there was a petition in 1620 from Scottish glaziers to establish a proposed glassworks so that they would not in future be compelled to buy Sir Robert Mansell's glass, which was 'scarce, bad and brittle'. Because of the lack of Scottish manufacture they were forced to buy at full price what the London undertakers did not want.

It was contemporary with these foundations that the transition to coal took place. Nef states that 'shortly before 1612 glass-making was transformed by the discovery of a method of closing the clay crucible in which the potash and sand were melted down, in order to make it possible to substitute a coal for a wood fire'. This change was of great consequence, since the lowered temperature inside the pot caused the glass-makers to add lead oxide, and so a separate species of glass was evolved.[4]

Scottish coal was regarded as particularly suitable for glass-making, and its export was fostered by James VI. In 1619 it was ruled in England that Scottish glass came under the heading *foreign* in the English Act of 1615 which forbade the importation of foreign glass. As a result the Scottish Privy Council threatened to prohibit the export of Scottish coal to London and Newcastle glass-makers. This threat quickly resulted in a new Act permitting the importation of Scottish glass into England. While the prohibi-

[1] J. A. Fleming, *Scottish and Jacobite Glass*, p. 5.
[2] R. Chambers, *op. cit.*, Vol. 1, p. 428.
[3] *Acts of the Parliament of Scotland*, Vol. 4, p. 626; Vol. 8, p. 357.
[4] Ch. xiv.

tion lasted Lady Mansell, who managed her husband's English glass-works in his absence, experimented with Newcastle coal, with evident success.[1]

From Wemyss in Fife, but still keeping to the vicinity of the coal-fields, the manufacture of glass moved to Leith where in 1661 another of the Hays, Charles, 'brother german to William now Earl of Kinnoull' began the manufacture of glass.[2] The manufacture of glass flourished in Leith till its products rivalled those of English firms, both in quality and quantity. 'The making of glass at Leith has long been achieved there, and is carried on to great perfection and profit.' One of the attractions of Leith was that it was a continental entrepôt. Importation and distribution of wine naturally suggested the manufacture of bottles, which conceivably were made in Leith earlier than the date of Charles Hay's establishment.[3]

In August 1661 the traveller John Ray gave an account of having seen glass being made at Prestonpans, where local kelp and sand were calcined in crucibles made locally, from the same clay as was used for making tobacco pipes.[4] This may be a confusion of localities, but the parish of Prestonpans did become the seat of a manufactory of glass. In 1696 William Morison of Prestongrange was granted the privileges of a manufactory to set up glass-works at Aichison's or New Haven, for making all sorts of glass, such as bottles, drinking glasses, window glass, etc.[5] When the York Buildings Company took over the estate of the Earl of Winton they acquired the coal-works and salt-pans in Tranent parish. Here they installed one of the first fire-engines in Scotland, and built a wooden railway connecting coal-pits with the salt-works at Preston and the harbour of Port Seton. At the latter they undertook the manufacture of glass, again without profit. David Murray gives the following quotation from the *Edinburgh Evening Courant*, 9th February, 1730:

> At the Glasshouse at Port-Setton there is to be sold window glass of several sorts. . . . Also, all sorts of flint or chrystal glass, consisting of drinking glasses, all sorts, decanters, lamps, gelly glasses, mustard boxes, salvers, and vials, etc., glasses for alchymists, and bell-glasses for gardeners.[6]

[1] J. U. Nef, *op. cit.*, Vol. 1, p. 182.
[2] *Miscellaneous Law Tracts*, 25.3.4, p. 42, National Library of Scotland.
[3] J. A. Fleming, *op. cit.*, p. 110.
[4] J. Ray, *Itinerary* p. 104, quoted by Hume Brown, *Early Travellers in Scotland.*
[5] *Acts of the Parliament of Scotland*, Vol. 10, p. 180; R. Chambers, *op. cit.*, Vol. 3, p. 154; *Statistical Account*, Vol. 17, p. 67, *n.*
[6] D. Murray, *York Buildings Company*, p. 65.

It seems likely that the Company was acting as agent only for some of the kinds of glass.

Still another east-coast co-partnery, that of David, Lord Elcho and Methil, which revived the manufacture of glass at Wemyss (1698), had obtained the privileges of a manufactory before the making of glass in the Clyde area was begun.[1]

In 1701 James Montgomery, already mentioned in connection with the South Sugar House, Glasgow, stated that for the previous ten months he and his partners had been erecting glass-works in Glasgow at a large outlay. In Montgomery's scheme there was more technical coordination for the utilization of by-products. He had discovered that fern-ashes (as a substitute for potash made from wood-ashes) were a most useful material in glass-making, and ferns abounded near Glasgow. In the Highlands there were also large quantities of timber to yield wood-ash which could be employed in the making of soap. Montgomery proposed therefore to manufacture glass and soap,[2] stating that it was 'the policy and interest of all nations to improve industry, especially in societies, who design improvement of any part of the natural product of the country'.

The Glasgow collieries supplied fuel to the sugar, soap, and glass works. The consumption of coal in the manufacture of glass varied according to the nature of the raw materials employed and the product produced. Window glass and bottle glass consumed the greatest quantities. 'Broad glass spendeth both more coales and asse (ash) quantitie for quantitie than drinking glass dothe.' Had it not been for the readily available Scottish coal and increasing demand for the comfort which cheap glass confers, glass-making could not have flourished as a capitalistic enterprise, with its large and costly glass-houses.

POTTERY. Along with glass, Scottish pottery-making expanded rapidly in the last decade of the seventeenth century. Before this it does not seem to have attained any importance. Wooden platters and pewter mugs were the order of the day. About 1667 the Royal Society of London was making recommendations that trials should be made with English *earths* (clays) to ascertain if they would yield products comparable with those of China, and also that clays for tiles and bricks should be studied. This possibly stimulated development in Scotland.[3]

As early as 1595 a small pot-works was producing water jugs and tubs for butter at Blackfaulds from red clays dug round

[1] *Acts of the Parliament of Scotland*, Vol. 10, p. 179.

[2] *Ibid.*, Vol. 10, p. 209, App. p. 49.

[3] T. Sprat, *op. cit.*, p. 191.

Shawfield, Glasgow, and primitive pottery was also made on the island of Lewis.[1] Apart from these however there are no further records of pottery industry till the petition of James Colquhoun in 1688 for permission to establish a pottery. His petition was granted.[2]

The quantity of capital necessary for the setting up of a pottery is illustrated by the efforts of Robert Douglas of the Leith Whale Fishing Company to establish a pottery in Leith. In 1695 he was granted permission to erect a kiln, etc., at Leith. This he did, but the expense was too great for him to bear alone. Being unable to obtain partners, he could not bring the works into active production.[3] This being so William Montgomery, with a partner, made arrangements for setting up a 'pot-house and all conveniences for making laim, purselane and earthenware', and was given the privilege of a manufactory with the exclusive right of manufacture for fifteen years. Douglas opposed the granting of the monopoly, but it was upheld by the Privy Council.[4] As has already been mentioned, sand and kelp were fused at Prestonpans in crucibles made from local clay, and a clause in the Act in favour of Montgomery and Linn mentioned 'that the manufacture of the coarse earthenware commonly in use not to be affected by the Act'. So there may have been other pot-works turning out coarse earthenware in Scotland of which there is now no record.[5]

PAPER. Although not part of the coal nexus, among the new capitalist enterprises of this period, those which made paper figure prominently.

About 1589–90 the Lords of Council granted to Peter Groot Heres, a German, and his co-partners, a licence to manufacture paper for nine years without competition or tax. Nothing more is heard of his venture.[6]

Almost a century elapses before the setting-up of the Dalry Mills on the Water of Leith in 1675. A second mill was established by a company of Edinburgh merchants and printers at Duddingston Loch, prior to the revocation of the Edict of Nantes, in 1679. In the same year Nicholas de Champ, a 'Norman', came with two of his countrymen and started a mill at Colinton (Colington), also on the Water of Leith. Shortly afterwards he formed a co-partnery with a company of Glasgow merchants who had built a mill at Woodside. He then built a mill on the Cart at Newlands. John

[1] J. A. Fleming, *Scottish Pottery*, p. 132; *New Statistical Account*, Vol. 14, p. 134.
[2] J. A. Fleming, *op. cit.*, p. 236.
[3] W. R. Scott, *op. cit.*, Vol. 3, p. 195.
[4] J. A. Fleming, *op. cit.*, p. 150.
[5] *Acts of the Parliament of Scotland*, Vol. 9, p. 122.
[6] R. Chambers, *Domestic Annals of Scotland*, Vol. 1, p. 194; Vol. 2, p. 398.

Hall, his first apprentice, married his daughter, and erected another and larger mill at Milnholm (Millhome).[1]

A further mill was established by Peter Bruce at Restalrig about 1685. It remained in his hands only for a short time, and in less than a decade it was transferred to James Hamilton of Little Earnock. These mills confined their activities to production of coarse grey and blue papers.[2]

Lack of technical skill or sufficient capital prevented the production of white writing paper. Only with the incorporation of Nicolas Dupin, Denis Manes and Company, under the name of *The Scots White Paper Manufactory* about 1694 was a relatively large capitalist enterprise set up in Scotland at Yester, Haddingtonshire.[3] The raw material for the paper was linen rags.[4] On 19th August, 1694 the articles of partnership were signed. They prescribed the internal management of the undertaking and fixed the terms for a new issue of shares: 'the capital already paid in, together with that now offered for subscription amounting to £5,000 sterling', but subscribers were hard to come by, and the promoters were forced to take up the shares themselves.[5] According to D. Bremner, the articles of the Company are in the British Museum, and are the earliest documents extant on paper manufacture in Scotland.[6]

By 1696 the Company was in production and was producing enough paper to supply the whole country. The paper found a ready market, albeit the directorate complained that the Government continued to use paper produced abroad. Such was the scale of their activities that before long the Company had exhausted readily available supplies of raw material. The result was that they asked for an Act to prevent candle-makers using rags for wicks. The candle-makers of Edinburgh vigorously opposed the plea as was to be expected. After 1699, we hear nothing further concerning the Company. It must be presumed that it ceased to manufacture before 1705.[7]

By the time that the *Scots White Paper Manufactory* had come to an end, paper manufacturers had begun in another well-watered east coast district, the valley of the Don, near Aberdeen. Patrick Sandilands started the manufacture of paper at Gordon's Mills

[1] J. M'Ure, *Glasghu Facies*, p. 1224; A Brown, *History of Glasgow*, Vol. 2, p. 211. [2] *Acts of the Parliament of Scotland*, Vol. 9, p. 340.

[3] Ibid., Vol. 9, p. 429.

[4] R. Chambers, *op. cit.*, Vol. 3, p. 88.

[5] W. R. Scott, *op. cit.*, Vol. 3, p. 183

[6] D. Bremner, *The Industries of Scotland*, p. 322.

[7] W. R. Scott, *op. cit.*, Vol. 3, pp. 183–5.

on the Don about 1696.[1] The only other establishment to which reference will be made here was, like that of Patrick Sandilands at Aberdeen, the first in a district with a continuous history of paper manufacture, viz., the mill at Valleyfield, Penicuik, set up in 1709 by And. Anderson, printer to Queen Anne. It is still in operation (under the title of Alex. Cowan and Sons). The subsequent social and technical history of paper will be found in Ch. XIII.

SCOTTISH TECHNICS TO 1765

1400 Salt-pans of Forth mostly fired with wood.
1424 Scottish deposits of lead ore declared *mines royal.*
1500 Coal in use in Forth salt-pans.
Production of salt, glass, and ships in Great Britain increasing.
1513 Discovery of Leadhills mines.
1561 First Act of Parliament to encourage salt-boiling.
Glengonner and Wanlockhead Lead Mines (John Acheson).
1565 Glengonner and Wanlockhead Lead Mines (Earl of Atholl).
1567 Discovery of coal at Brora, Sutherland.
1590 Paper Mill (Groot Heres).
1594 Lead mines leased to Thomas Foulis.
1595 Pottery at Blackfaulds (Glasgow).
1600 Coal in use at Worcester and Cheshire salt-pans.
1603 REGAL UNION under JAMES VI and I.
1610 Sir William Slingsby on effects of timber famine.
Glass-making established at Wemyss (Sir George Hay).
1612 Coal used in glass-making; glass pots covered in.
1613 Attempt to desilver lead in Scotland.
1614 Salt-pans at Brora.
1619 First certain reference to coking coal.
Soap-boiling established at Leith (Nathaniel Uddart).
1620 Petition for more glass-works in Scotland.
1621 Houses in Edinburgh ordered to be roofed with slate, lead, or tiles.
1624 Statute of Monopolies (England).
1628 Sugar-boiling partnership.
1630 Half Scottish shipping engaged in salt and coal trade.
1634 Soap-boiling: potash-making (Patrick Mauld of Panmure).

[1] P. Morgan, *Annals of Woodside*, p. 71.

1635 Coal mines in Forth area 16–20 fathoms deep.

Capitalist brewery in Edinburgh.

c. 1660 First English Industrial Revolution.

1661 Act for Encouraging Trade and Manufactures.

Act for planting and inclosing of Ground.

Glass manufacture at Leith (Charles Hay).

Country-houses still without glass windows.

Monopoly to Ludovick Leslie and James Scott.

1662 Royal Society of London receives Charter.

1663 Manufacture of alum and starch (Timothie Langley).

Leith Wool-card Manufactory.

1667 Western Sugar House (Glasgow).

1668 Iron manufacture at Red Smiddy (Loch Maree) ends.

1669 Eastern Sugar House (Glasgow).

Act anent inclosing of Ground.

c. 1670 Soap manufacture established (Glasgow Soaperie).

1672 Manufacture of alum (Lockhart of Cleghorne).

1675 York Buildings Company receives Charter.

Paper manufacture at Dalry Mill.

1678 Lead mine at Gilfinnan.

1679 Paper Manufacture at Colinton (Nicholas de Champ), Duddingston, etc.

1680 Sugar manufacture established at Leith.

1680 Brass casting (Edinburgh).

Wanlockhead Lead Mines (Sir James Stampfield).

1681 Houses in Glasgow, Dundee, and Aberdeen ordered to be roofed with slate, tiles, etc.

1685 Paper manufacture at Restalrig (Peter Bruce).

Act in favour of Planters and Inclosers of Ground.

1686 Expansion of salt industry at Saltcoats.

1688 Pottery at Glasgow (James Colquhoun).

1690 Coal only fuel used in Forth salt-pans: consumption 150,000 tons.

Petition to manufacture gunpowder (James Gordon).

Calico-printing in England.

1691 First reference to water-proofing (Sutton and Hager).

1693 Draining engine patent (Marmaduke Hudson).

1694 Scots White Paper Manufactory at Yester, Haddington (Nicholas Dupin).

1695 Petition for manufacture of gunpowder (Alex. Hope of Kerse).
Pottery established at Leith (Wm. Montgomery).
Company for working Mines and Minerals in the Kingdom of Scotland (Nicholas Dupin).
1696 Contract between English Linen Company, and Convention of Royal Burghs.
South Sugar House (Glasgow).
Paper manufacture established on Donside (Patrick Sandilands).
Successful reverberatory furnace devised.
Glass manufacture established at Aichison's or New Haven (Wm. Morison).
1698 Glass manufacture revived at Wemyss (Lord Elcho).
1699 Hardware manufacture established at Glasgow.
Darien colonization scheme.
1700 3,000 coal miners in Scotland.
King Street or Distillery Sugar House (Glasgow).
Consumption of sugar in Scotland: 3 lb. per person per annum.
1701 Mineral smelting co-partnery.
Glass and soap at Glasgow (James Montgomery).
1702 Mine-draining engine (James Gregory, Edinburgh).
1703 Cloth manufacture at Gordon's Mills on River Don: French bleachers.
1707 Transfer of Scottish administration to Westminster.
1709 Reduction of iron ore with coke at Coalbrookdale (A. Darby).
Manufacture of paper at Valleyfield Mill, Penicuik.
1715 Jacobite Rising.
Dalquhurn Bleach field.
1719 Purchase of land in Scotland by York Buildings Company.
Scotland forced to relinquish rights to exemption from British sugar duties.
1720 Manufacture of Kelp started on Firth of Forth.
Calico Act: linen printing favoured.
1721 Friendly Mining Company (Wanlockhead).
1722 Manufacture of Kelp started in Orkney.
1723 Honourable The Society of Improvers in the Knowledge of Agriculture.

1724 Lease of Ardnamurchan Lead Mines to company of adventurers.

1726 Timber purchased at Abernethy (York Buildings Company).

Chemistry taught (Edinburgh University).

1727 Act for the better Regulation of the Linen and Hempen Manufacture in Scotland.

Act for encouraging Fisheries and other Manufactures and Improvements in Scotland.

1728 Establishment of Board of Trustees for Manufactures.

Galloway Bleachfield (A. Adair).

Iron manufacture at Abernethy (York Buildings Company).

1729 Iron manufacture at Invergarry.

Printing of linen introduced (Gorgie Bleachfield).

Lease of Hopetoun Lead Mines (York Buildings Co.).

1730 Manufacture of Kelp in Hebrides.

Scottish Kelp on sale in Newcastle.

Manufacture of bottles at Glasgow (Glasgow Bottlehouse Company).

Iron manufacture at Bonawe.

Lease of Ardnamurchan Lead Mines (York Buildings Co.).

Manufacture of glass at Port Seton (York Buildings Co.).

1731 Society for improving medical knowledge.

1732 Pitkerro Bleachfield (Richard Holden).

1734 Board of Trustees request exemption from duty for bleaching materials.

1735 Spinning-wheel being introduced in Lowlands.

1736 Manchester Act (Calico printing).

Gin Act (Duty on Spirits).

c. 1736 Soap-boiling at Leith (Stevens and Company).

1738 E. Chambers's Cyclopædia.

1739 Philosophical Society (Edinburgh).

1740 Harlem Dye Company, Glasgow (Ayton).

1742 Calico-printing at Pollokshaws (H. Ingram).

1745 Second Jacobite Rising.

1746 British Linen Company.

1747 Chemistry taught (Glasgow University).

Leadhills Mining Co. and Scots Mining Co.

1749 Manufacture of Sulphuric Acid at Prestonpans (Roebuck and Garbett).

Bleaching on River Don (Leys, Still and Co.).
Pottery at Delftfield, Glasgow.

c. 1750 Dovehill Dyeing Manufactory, Glasgow (Ayton, etc.).

1750 Glass manufacture at Alloa.
Paper manufacture at Culter, Aberdeenshire (B. Smith).
Soap manufacture at Leith (W. St. Clair of Roslin).
Fifty per cent. rise in price of potashes.

1751 Export of sulphuric acid to Continent begins.

1754 Manufacture of Pottery at Prestonpans (Gordon and Watson).
Edinburgh Society for Encouraging Arts, Science, Manufactures and Agriculture.

1755 English Bone China made in Scotland.
Award to Professor William Cullen for bleaching research.
Research on blue dyeing (Board of Trustees).

1756 Manufacture of sal ammoniac (Hutton and Davie).
Research on bleaching (Professor Francis Home).

1758 Cudbear manufacture started at Leith (Gordon and Alexander).

1759 Manufacture of Iron at Carron (Roebuck, Garbett, and Cadell).
Cobalt from Alva for Prestonpans Porcelain Works.

1763 Memorial on potash production (William Cullen).
Bleaching machinery introduced (S. Read).

1765 Manufacture of Sugar at Greenock (M. Kuhll and Glasgow partners).

Note.—This Calendar brings the history of Scottish technics down to the establishment of the first chemical works, and the first dye manufactory; to the smelting of iron ore with coal; and to the introduction of machinery in the bleaching industry. Subsequent developments will be dealt with under the relevant chapter headings.

PART FOUR

Rapprochement

The introduction of the manufactures so far discussed into Scotland during the eotechnic age shows that Scotland was ripening for industrial capitalism before her merger with England. At the beginning of the seventeenth century, nine-tenths of Scottish exports had been either raw materials or manufactured articles in the unfinished state: by the end of the century the foundations of an economy had been laid which in the ensuing years enabled her to supply personnel which helped to bring about the second English Industrial Revolution.[1] No claim is made that enterprises were extensive in scope, but if the claims of their promoters be allowed, there is every indication that they were sufficient to meet the requirements of the Scottish home market and more. The Scots Parliament did its utmost to foster them, but with the translation of the legislature to Westminster, its protective policy came to an end.

To achieve an impartial estimate of the effect of Union with England is wellnigh impossible. For many years the much-lauded access to the English plantations benefited only Glasgow. While English competition must be reckoned with, it must not be forgotten that special provision was made for some of the Scottish industries, particularly salt, sugar, distilling, etc. If there was commercial depression, and no doubt the economic historian has evidence to support this thesis, there was certainly no intellectual bankruptcy. Following the foundation of several important Societies for improving manufactures, agriculture, etc., we come soon to the founding of the first chemical works in Scotland, the Prestonpans Vitriol Company, founded in 1749 by Dr. John Roebuck and Mr. Samuel Garbett. By that time direct application of the available resources of theoretical science to advancement of technology introduces a new agency of social change. At this point we can turn for a moment to examine the intellectual, as opposed to the commercial, life of Scotland.

In the first half of the eighteenth century, the Scottish Universities woke up from their state of stagnation, long before the corresponding revival of Oxford or Cambridge. 1690–1725 was a period of 'dreary stagnation of all intellectual life and destitution of scholarship in Scotland' according to H. G. Graham.[2] True, the early death of James Gregory (1675) and the translation of his nephew, David Gregory, to Oxford in 1692 robbed Scotland of two of her principal luminaries; 'but in the first half of the century natural philosophy had a distinguished exponent in Colin Maclaurin in Edinburgh, and geometry in Robert Simson in Glasgow; and in course of time in every department—especially in philosophy, science, medicine—the Universities were abreast of or in advance of their age'.

In Goldsmith's work, *The Present State of Polite Learning*,[3] there

[1] J. U. Nef, *op. cit.*, Vol. 1, p. 233.
[2] H. G. Graham, *Social Life in Scotland*, p. 449.
[3] Cunningham Edition, Vol. 2, p. 40.

is recorded the following anecdote which illustrates how the Scottish Universities took the place of Paris, Leyden, and Utrecht in the intellectual life of the country.

> Happening once in conversation with Gaubius at Leyden to mention the College of Edinburgh, he began by complaining that all the English students who formerly came to his University went entirely there. . . . He concluded by asking if the professors of Edinburgh were rich? I replied that the salary of a professor there seldom amounted to more than thirty pounds a year. 'Poor men,' said he, 'I heartily wish they were better provided for; until they are rich we can have no expectation of English students at Leyden'.[1]

Edinburgh, next only to Paris perhaps, became the intellectual centre of Europe. Students were attracted from England, Ireland, and abroad. In 1700 there were three hundred students and eight professors at Edinburgh; in 1800 six hundred students and twenty-one professors. The remuneration to be gained in the Universities was pitifully small. Colin Maclaurin gave up a salary of £50 per annum to travel with a young gentleman; David Hume became guardian to a lunatic; and Adam Smith found that to be a travelling companion was more lucrative than a Chair in Glasgow.[2]

Speaking of the part played by the Universities in the Industrial Revolution, by which he presumably means Oxford and Cambridge, Witt Bowden says:

> Men of science and learning in the faculties of the Universities and the Gentlemen of the Royal Society were interested in improvements in the technique of pure science, but any idea of the application of their knowledge in devising of inventions useful in the ordinary affairs of life was rarely entertained.[3]

This generalisation is so sweeping that it must be taken to indicate Bowden's unfamiliarity with the early projects of the Royal Society of London, and with the leadership of the English Industrial Revolution which came from personnel in the Scottish Universities.

While Oxford and Cambridge antedate the Scottish foundations, (St. Andrews, 1411; Glasgow, 1453; Aberdeen, 1492), the northern Universities are in towns which represent the foci of

[1] Joseph Black's salary at Glasgow in 1764 was £50 as professor of medicine, plus £20 for teaching chemistry, with the addition of fees. When Cullen was appointed to the chair of medicine in Edinburgh in 1756, he received no salary, only fees. (Graham, *op. cit.*, p. 471.)

[2] H. G. Graham, *op. cit.*, pp. 66, 464, *et seq.*

[3] Witt Bowden, *Industrial England in the Eighteenth Century*, p. 11.

population in Scotland, where, even at the beginning of the eighteenth century, there was the greatest measure of commerce and industry. Moreover, even at the end of the century, the only educational establishments in England to give adequate scientific training were the dissenting academies where Joseph Priestley and John Dalton taught. (The Artillery Schools in France played a parallel rôle.) This may explain the more ready adoption, in Scotland, of disciplines outside the humanities. Thus Chemistry was taught in the Scottish Universities from 1726 in Edinburgh, 1747 in Glasgow, 1793 King's College, Aberdeen, and *c.* 1817 Marischal College, Aberdeen.

At the same time other forces were at work in Scotland for the dissemination of scientific humanism among all sections of the public. Their beginnings are described by John Ramsay, author of *Scotland and Scotsmen in the Eighteenth Century:*

> Soon after the extinction of the rebellion of 1715, a number of promising young men began to distinguish themselves in science or polite literature. In order to improve themselves and counteract conceit, which is never more apt to spring up than in rich minds unaccustomed to contradiction, societies were instituted wherein at stated times, literary subjects were canvassed with freedom and impartiality: ingenious paradoxes were started and assailed with equal ingenuity. There the members used to submit their first essays in composition to the friendly censure of associates, which helped to lop away luxuriances and check presumption.... The most eminent of them was the Rankenian Club (so called after the master of a tavern) formed in 1717. It consisted, among others, of Dr. Wallace (1696–1771), Dr. William Wishart (1660–1729) Principal of the University, Mr. Colin Maclaurin (1698–1746) Professor of Mathematics, Sir Andrew Mitchell, and Dr. Young, a physician, all afterwards first-rate men.

Possibly the earliest to crystallize, so to speak, was the *Rankenian Club*. This was followed in 1731 by a *Society for improving medical knowledge by collecting and publishing Essays and Observations on Various Subjects*. In 1739 Colin Maclaurin (1698–1746), Professor of Mathematics in Aberdeen and Edinburgh, proposed the widening of the Society's aims to include philosophical and literary subjects. Its title was changed to the *Society for improving Arts and Sciences*, or more generally the *Philosophical Society of Edinburgh*.[1] The Jacobite Rising of 1745 upset the regular meetings of the Society, but in 1754 the first volume of its transactions was published under the title of *Essays and Observations, Physical and*

[1] H. Arnot, *History of Edinburgh*, pp. 328 and 549.

Literary. Succeeding volumes were published in 1756 and 1771. For a time after that the Society languished, till Henry Home, Lord Kames, one of Scotland's most enlightened agricultural improvers, was appointed president in 1777. Here we have the Invisible College of Scotland—precursor of the *Royal Society of Edinburgh*. In 1783 the Royal Society got its charter, the Philosophical Society being dissolved.[1]

While over a century elapsed between the founding of the Royal Societies of London and Edinburgh, activities were engaged in during the century, in particular the founding in Edinburgh of a number of Societies which played a part equivalent to the elder Society. These must be reckoned as potent forces in forging the intellectual life of Scotland. Although it was fashionable to be a natural philospher, the legacy of the fad of the moment was the brilliant personnel (Roebuck, Watt, Murdock, Keir, Dundonald, *et al.*) of the English Industrial Revolution.

The earliest Society whose activities were national in breadth was *The Honourable the Society of Improvers in the Knowledge of Agriculture*. Its secretary, and the moving spirit behind it, was Robert Maxwell of Arkland (1695–1765), who prepared the *Select Transactions* of the Society for publication in 1743. By experiments undertaken at his own expense, Maxwell ruined himself, and Lady Arkland was reduced to keeping a small shop in the High Street of Edinburgh. Unlike the Edinburgh Philosophical Society, the Society of Improvers did not survive the Jacobite Rising, but Maxwell continued in the interests of agricultural improvement, delivering lectures, and publishing his *Practical Husbandman* in 1751.

The first part of the *Select Transactions* contains, in the main, letters enquiring about the improvement of soil and the answers furnished by the Society to the queries. As one would expect, the manures advocated are confined to lime and the usual biological material, although Scot of Scotstarvat also mentions the use of sea-ware. There are enquiries about the horse-hoeing husbandry of Jethro Tull, and a reply from Tull himself. From agriculture, the Society extended to manufactures.

> The Society, therefore, after they had laid a Foundation upon Husbandry, the Life and Support of all Arts and Sciences, and the Source of all solid Riches, took the languishing State of the Manufactures very early under their Consideration; with a Design not only to prevent the then shameful Extent of Import, but, if possible, to make Provision for Export.[2]

[1] *General Index to the Transactions of the Royal Society of Edinburgh. Passim.*
[2] R. Maxwell, *Select Transactions*, p. 309.

They attempted to prevent the use of lime in bleaching. From Lady Saltoun they learned the bleaching secrets of Haarlem. Soap was another of their considerations.[1]

Shortly after the Society of Improvers came *The Honourable the Board of Trustees for Manufactures in Scotland*, in 1727. While their activities were restricted in scope, they also played an important part in the development of eotechnic economy in Scotland. The function of the Board of Trustees was to administer the *Equivalent*, i.e. funds received in perpetuity as arranged at the time of the political union of England and Scotland in 1707.

> The £2,000 annually granted for seven years by the terms of the Treaty of Union, for the purpose of developing manufactures and fisheries, were allowed to accumulate without practical application to this object.[2]

The Board was created to bring this state of affairs to an end. The linen industry of Scotland was the counterpart of the woollen industry of England, so the Board early turned its attention to fostering the linen industry and the ancillaries to it.

To do this they called on the scientific personnel in the Universities. Dr. William Cullen investigated home supplies of ashes. In addition, the Trustees subsidized the laying down of a great number of bleach-fields. It also stimulated the first great revolution in the art of bleaching by advising the substitution of sulphuric acid (H_2SO_4) for the sour milk (containing lactic acid) formerly used. Other activities included investigation of dye-plant cultivation in Scotland. Dr. Francis Home (1719–1813) analysed samples of dyes submitted for the reward of premiums. One may well gauge the important position which the Trustees occupied by the large number of volumes which were dedicated to them, on such diverse subjects as soap and agriculture. For the greater part of the eighteenth century the Board of Trustees discharged a useful function to Scottish manufactures.

The Honourable the Board of Trustees is often confused with another association, this one without official status: *The Edinburgh Society for the Encouragement of Arts, Science, Manufactures, and Agriculture in Scotland.* It was founded by Allan Ramsay, the painter, in 1754, and for short is referred to as the *Select Society.*

This Society, which held its meetings in the Library of the Faculty of Advocates, consisted at first of only fifteen members, but it became so popular, or perhaps fashionable, that in a few

[1] R. Maxwell, *Secret Transactions*, p. 315, *et seq.*

[2] J. Mackinnon, *Social and Industrial History of Scotland*, p. 11.

years it numbered three hundred, and included all the literati, and many of the gentry, lawyers, clergy, and physicians of the capital of Scotland. They met weekly on Friday evening and debated questions of social economy, politics and commerce. The ponderous Hume, the absent-minded Adam Smith, and Lord Kames *et al.*, were members.

In the course of time the Select Society took up the idea of encouraging commerce, agriculture, and art in Scotland. On the whole their exertions were well rewarded. A subscription list was opened and Lords Hopetoun, Morton, Marchmont, etc., promised to support the scheme.[1]

The career of the *Select Society* can be followed in detail in *The Scots Magazine* from the year 1755 onwards. First we have an account of the foundation of the Society[2], and the objects which it considers merit attention, with comments like the following:

(*a*) To the shame of the country it is supplied with paper from countries which use not half the quantity of linen that is here consumed.
(*b*) Whisky . . . is still capable of great improvement in the quality and taste.

In 1757 we have an account of the success of the Society[3], and the award of a gold medal to Professor Francis Home for his researches on vegetation and the principles of agriculture. In that year no less than ninety-two premiums were offered. In the year following, awards were made for marl, madder, woad, collecting rags for paper, sealing-wax, glue, soap, curing smoking chimneys, etc.[4] In 1758 new premiums were offered, now increased to 138 items. They are more extensive in scope and include, among other things, the collection of silver, gold, lead, mercury, antimony, cobalt, and zinc ores in Scotland. There were premiums for printing linen from copper plates, and for the best imitation for delft.[5] Further expansion took place, but in 1764, for the first time, there were definite signs of an impending decline.[6] The objective of the Society had altered, and it became a *Society for Promoting the Reading and Speaking of the English Language in Scotland*. An extract from a contemporary Edinburgh paper is worth quoting, since the Society may have found it cheaper to learn to speak English than promote manufactures.

[1] Rob. Brydale, *Art in Scotland, passim*, Henry Graham, *Scottish Men of Letters of the Eighteenth Century, passim; Scots Magazine*, 1755, *17*, 129.
[2] *The Scots Magazine*, p. 126. [3] *Ibid.*, 1756, *18*, 48.
[4] *Ibid.*, 1757, *19*, 49. [5] *Ibid.*, 1758, *20*, 211. [6] *Ibid.*, 1764, *26*, 229.

> The Edinburgh Society expected that the manifest utility of the undertaking would of itself have interested the public in its favour; and therefore they did not importunately solicit subscriptions. They now find that they have been too sanguine in their expectations. The number of contributions, instead of increasing, diminishes, and many gentlemen who have not recalled their subscriptions do neglect to make their annual payments.

They add their conclusion 'that in Scotland, every disinterested plan of public utility, is slighted as soon as it loses the charm of novelty'.[1] We cannot neglect, however, that throughout its short career it was strongly biased in favour of premiums for agricultural subjects and so directed attention to that subject.

To those familiar with the academic approach to chemistry through the Daltonian theory and the progressive quantification which culminated in Mendeléeff's generalization, it will already be evident that the sociological approach to the history of chemistry is quite different. Community need and social utility determine the inclusion of a topic in what follows. Yet the scientist familiar with theoretical science will be surprised by the many social relations of discoveries which have traditionally been made to appear as if they were developed in a vacuum of inutility. The economic historian will similarly find the mechanical nexus and metallurgy here relegated to places of seeming unimportance. The alteration in focus has been deliberate. Chemistry has a utility outside the academic laboratory; social technology ought to have a vital place in the studies of those who profess familiarity with sociology.

The crucial determinant in the evolution of chemical industry has been the existence of the necessary raw materials and sufficient theoretical knowledge for their utilization. Raw materials were made the basis of a classification of industries, predominantly chemical in nature, devised by the apologist of the Industrial Revolution, Andrew Ure (1778–1857), in his *Philosophy of Manufactures* (1833). For all his shortcomings, Ure's scheme anticipated to a remarkable degree the needs of a social technologist, and as such merits inclusion in our discussion, since it gave a picture of industrial-revolution chemistry the balance of which was disturbed by the disymmetric development of certain branches of chemistry, particularly organic chemistry, after the analytical examination of coal tar. Ure's classification serves both by way of introduction and index to the raw materials of eotechnic and

[1] *The Scots Magazine*, 1764, 26, 142.

palaeotechnic economy discussed in the following chapters. Other raw materials were included by him, but those outwith the discussion have been omitted for simplicity.

I. MANUFACTURES OPERATING ON MINERAL OR INORGANIC MATERIALS

A. *Metallic Ores:*

(1) Iron as metal and as compounds.
(2) Lead as metal and as compounds, e.g. in lead paint.
(3) Cobalt used as pottery colour.
(4) Manganese a raw material for bleachers, etc.
(5) Chromium used by calico-printers, etc.

B. *Minerals and Clays:*

(1) Limestone for agriculture, bleaching, building, etc.
(2) Argillaceous material for pottery, porcelain, bricks, etc.
(3) Silicious material, i.e. sand for glass-making, building, etc.

C. *Combustible minerals:*

(1) Coal fuel, raw material for gas manufacture, which in turn yields valuable new raw materials—ammonia, tar, ammonium chloride, etc.
(2) Sulphur, principally employed in sulphuric acid manufacture.

D. *Saline substances:*

(1) Salt (sodium chloride): which with the industrial application of chemical science became the centre of a whole complex of other industries—alkalis, chemical bleaching of textiles, paper, etc.
(2) Alum (either potassium, sodium, or ammonium, aluminium sulphate): has been inseparably connected with the art of dyeing throughout historic times.
(3) Potash (potassium carbonate): the first alkali available to eotechnic industry.
(4) Soda (sodium carbonate): The conversion of salt into soda was the basic operation upon which the chemical industry of the industrial revolution was founded, and marks the transition from eotechnic to palæotechnic economy in that field. Its successful industrialization had a profound effect on glass, soap, and other related industries.
(5) Sal-ammoniac (ammonium chloride): At one time a separate industry using soot as a raw material.
(6) Saltpetre (potassium nitrate): One of the principal constituents of gunpowder, and the raw material for nitric acid.

II. MANUFACTURES OPERATING ON VEGETABLE MATTER

The more complicated nature of these materials makes the industries using them less subject to scientific exactitude, but accumulated craft knowledge during the eotechnic period brought about considerable industrial developments even in Ure's time.

(1) Sugar-boiling.
(2) Oils for vegetable soaps.
(3) Rubber and waterproofing.
(4) Dyeing and calico-printing.
(5) Brewing, distilling, and vinegar manufacture.
(6) Destructive distillation. (Pyroligneous acid, etc.)
(7) Pharmaceutical chemistry.
(8) Agriculture in so far as it can be called a chemical art.

III. MANUFACTURES OPERATING ON ANIMAL MATTER

(1) Tanning.
(2) Soap from animal matter.
(3) Dyeing with animal pigments.
(4) Preservation of food.
(5) Sal-ammoniac from animal matter.
(6) Prussian Blue.

The mass of data which we had to analyse and interpret will be grouped under the following heads: (*a*) primary chemical substances such as salt, potash, etc.; (*b*) the problem of the synthesis of soda; (*c*) the first substance to be manufactured on a large scale, sulphuric acid, and an offshoot of its availability, aeronautics; (*d*) the textile complex—bleaching, dyeing, calico-printing, mordant production, waterproofing, paper; (*e*) furnace products —glass and pottery; (*f*) metals—iron and lead; (*g*) coal as a chemical raw material—the gas and tar industries, matches; (*h*) agriculture and industries based on its products—sugar, brewing and distilling, food preservation. While each chapter is a unit in itself, the complete analysis of the industrial revolution set-up requires an integration of them all, and the consideration of coextensive mechanical technology.

Chapter II

THE ECONOMY OF COMMON SALT

Distillation of salt from sea-water had long been associated with the coal industry.

Twentieth century economic historian.

REFERENCE was made in Chapter I to expansion in the manufacture of common salt, or sodium chloride (NaCl), which took place in England and Scotland during what J. U. Nef calls the first industrial revolution, i.e. the era of eotechnic development in the British Isles. Between the Union of the Crowns (1603) and the Union of the Parliaments (1707) the Scottish salt industry continued to expand, and salt figures prominently in the 1707 treaty itself. During the century that followed the political union of north and south Britain, the widely dispersed salt industry underwent a series of alterations in locus of operation and in methods of production. It also became subject to fiscal charges which had a profound effect upon the whole course of the evolution of chemical industry in Britain. Neither in England nor in Scotland were the fiscal charges and methods of production the same. Thus although primarily interested in the Scottish part of this basic industry, it is impossible to discuss the Scottish set-up apart from its English counterpart.

Our interest in salt arises from its use as a chemical raw material. While this did not become of major importance till the development of palaeotechnics, even in the eotechnic era salt was put to uses other than for preparation and preservation of food and skins. It found use as a flux in glass-making and in metal smelting. It was the first pottery glaze, and in consequence of this North Staffordshire potteries stimulated the development of salt-boiling at Lawton on the edge of the Pennines. It was one of the earliest medicines. The Rev. R. Dacre wrote a book on the medicinal use of salt.

As a preservative, the crude and variable salt of the early eighteenth century was by no means constant in its antiseptic properties. It was not a pure compound but a mixture of salts, the principal one being sodium chloride.[1]

[1] An average type of sea-water contains 3.5 per cent. of solids in solution; 2.7 per cent. of these are sodium salts; 0.07 per cent. potassium salts; 0.14 per cent. calcium salts; 0.59 per cent. magnesium salts.

According to J. Walker, coarsely crystalline *bay* salt, produced by the natural evaporation of brine in areas with favourable climatic conditions, e.g. the Mediterranean, was the best that could be obtained in Britain for fish preservation. Scottish salt was slightly alkaline and bitter to taste in consequence of the presence of vitriolated or muriated magnesia ($MgSO_4$ or $MgCl_2$), while bay salt, it was said, was sharp and sweet to taste. Accordingly, the coarsely crystalline bay salt was thought to be a more powerful antiseptic and better adapted to fish preservation. On the other hand, Scottish salt was highly esteemed by some London fishmongers and by the Victualling Office. Aberdeen pork, famous for its excellence during many generations, was cured with Scottish salt.[1]

These differences in salt from various localities interested contemporary scientists. Both Henry and Dalton of Manchester examined home and foreign salt chemically. They concluded that the small crystals of Cheshire salt resulted from the purity of the local brine, combined with the rapid boiling which was possible if St. Helens coal were used.

Salt was commonly thought to be the cause of scurvy during the eighteenth century. William Cullen, for one, adhered to this view. It was thought that perhaps the purification of the salt was at fault. To this end:

> . . . a certain Thomas Lowndes, a native of Middlewich, brought to the notice of the Admiralty a process for making and purifying salt. Most of the salt used in England at that time (*c.* 1750) was 'Bay-salt', made in France and Northern Spain by the evaporation of sea-water by the sun. Lowndes gave an unpleasant account of its manufacture, and alleges that it was often contaminated with filth arising from 'putrefying human bodies and dead fish'. Having seen the method which the Dutch used to refine salt, and observing that their salt herrings kept much better than those preserved with 'Bay-salt', he suggested that he should be allowed to submit samples of his purified product for the approval of the College of Physicians before asking the Admiralty to use it in the preparation of food for the Navy.[2]

As palaeotechnic economy evolved, salt became indispensable in many manufacturing processes. According to Samuel Parkes, who contributed to Dacre's *Testimonies*.

[1] J. Walker, *An Essay on the Natural, Commercial, and Economic History of the Herring*, Trans. Highland Society, 1803, *2*, 297 n.; *Remarks* (*anonymous*) *on the System of Laws in Scotland relative to Salt*, (Edin., 1794) p. 25. (General Register House, Edinburgh).

[2] J. C. Drummond, *The Englishman's Food*, p. 311.

> It is impossible to ascertain what number of new manufactures would rise up in consequence of the repeal of the salt duties, because I apprehend that when these duties shall be repealed, such an impetus will be given to the manufacturing interests in general, that every manufacturer and tradesman in the kingdom will immediately set himself to inquire and contrive what new way he can employ an article of such general application.[1]

Two circumstances determined the locus of the salt industry. Where salt was prepared by evaporation of sea-water, the presence of easily wrought supplies of coal in the neighbourhood was essential. To this category the whole of the Scottish industry belonged. Where salt was obtained from mineral deposits or brine springs, it proved an easier proposition to transport fuel to the salt-producing area than to transport concentrated brine. Occasionally, when there was a considerable local demand for salt, small units developed and remained in operation despite competition from more advantageously placed producers.

In Scotland there were three centres of production, (*a*) the Firth of Forth, (*b*) Ayrshire, (*c*) Brora in Sutherland. On the Firth of Forth, the production of salt began at least as early as the reign of David I (1124–53). Indeed, salt production in that area probably antedates the extensive use of coal. At any rate, small coal or dross, which was used in firing the salt-pans from the sixteenth century onwards, was, in the seventeenth century, still referred to as *panwood* in the Scottish salt works. There is ample evidence, too, that the Lowland forests were ravished in eotechnic times, and it is not unlikely that they were partly felled to fire salt-pans on the Firth of Forth.[2] Like coal-mining, salt-making became one of the chief occupations of the inhabitants in the Forth area. For seventy miles, from Musselburgh to Pittenweem, pans were clustered on the littoral, and clouds of dirty brownish smoke marked the uneconomic combustion of cheap coal in eotechnic furnaces. Sir W. Brereton, who visited Edinburgh in 1635, gives us a contemporary picture of the scene:

> About six or seven miles from the city (Edinbro.) I saw and took notice of divers salt-works in poor houses erected upon the sea-coast. I went into one of them and observed iron pans eighteen feet long, nine feet broad; these, larger pans and houses than those at the Shieldes: an infinite innumerable number of salt-works here are erected upon the shore; all make salt from sea-water.[3]

[1] R. Dacre, *Testimonies in favour of Salt*, p. 198.

[2] *Statistical Account of Scotland*, Vol. 18, p. 437; H. M. Cadell, *The Story of the Forth*, p. 320.

[3] W. Brereton, *Travels in Scotland, etc.*, 1634–5, p. 98.

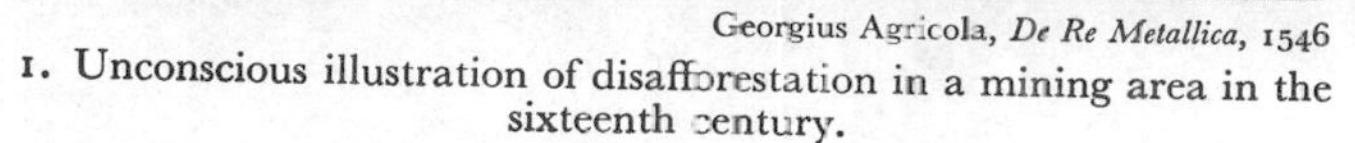

Georgius Agricola, *De Re Metallica*, 1546

1. Unconscious illustration of disafforestation in a mining area in the sixteenth century.

2. A distillery in the year 1729, illustrating an important method of purification.

Diderot and d'Alembert's *Encyclopédie*

3. Salt-production on the Normandy coast, illustrating solar evaporation and wood-fired salt-pans and basket filters.

J. M'Ure, *Glasghu Facies, A View of the City of Glasgow*

4. The Eastern Sugar House, Glasgow, founded 1669.

Speaking of the visitors to the isolated Scottish hamlets of the eighteenth century, George Robertson (1750–1832) says,

> The tinker, or horner, would now and then be calling; and the salt-man, more regularly than any, with a horse-loading from Salt Preston, in woollen bags, stuffed into panniers, from whence a regular supply of salt was got at least once a month.[1]

The intimate connexion of the villages of the area with salt-boiling is illustrated by the changes that the name Prestonpans underwent. It was formerly Salt-Priest's Town, then Salt-Preston as above, ultimately Prestonpans.[2]

It is not necessary to recount here in detail the gradual expansion of the Scottish industry. In the year of the Hanoverian Succession (1714), there were 164 coal-fired salt-pans in Scotland, and an additional small number which were fired with peat. Salt-boiling helped to forge a link in the evolution of palaeotechnic economy when in 1719 the York Buildings Company, as one of their many attempted undertakings in Scotland, constructed a railway of wood between one and two miles long in order to facilitate the conveyance of coal from the pits to the salt-pans at Port-Seton.[3] In 1798 there were 98 pans in the Forth area, or 106 on the east coast, if we include the littoral extending to Peterhead. Although the number of pans had declined, they had greatly increased in size. The greater part of the salt produced was exported to Holland.[4]

Short mention might also be made of the two other districts in which salt was the chief manufacture of the inhabitants. At Brora in Sutherland availability of coal encouraged the production of salt. The coal deposit was discovered before 1567 by John, tenth Earl of Sutherland, but remained undeveloped for several decades. Salt-pans were erected by Lady Jean Gordon, divorcee of the Bothwell who married Mary, Queen of Scots. They were further developed in 1614 by her son, John. The first Duke of Sutherland spent £16,000 on the coal works and £2,337 on the salt works, which, between 1814 and 1828, are said to have produced 20,000 tons of salt. The Brora salt was a valuable asset to the Highlands where, however, supply exceeded demand. The excess was exported to England and elsewhere.[5]

[1] J. G. Fyfe, *Scottish Diaries and Memoirs*, 1746–1843, p. 280.

[2] J. U. Nef, *The Rise of the British Coal Industry*, Vol. 2, p. 174.

[3] R. Chambers, *Domestic Annals of Scotland*, Vol. 3, p. 472.

[4] Sir W. Brereton, *op. cit.*, p. 112; Ed. Hughes, *Studies in Administration and Finance*, (1558–1825), p. 413. This work contains a valuable analysis of economic and fiscal aspects of salt production, and has been considerably drawn upon for the present chapter.

[5] Chambers, *op. cit.*, Vol. 1, p. 301.

The third centre of salt production in Scotland was at Saltcoats on the Ayrshire coast. It was originally a collection of clay-built cots, inhabited by poor persons who manufactured salt in small pans and kettles; and it thence obtained the name Saltcotes. But it possessed only a fitful prosperity; and, about the year 1660, it had dwindled away to only four houses. In 1686, however, Robert Cuninghame, built several large salt-pans at Saltcoats, placed the manufacture of salt on an entirely new and advantageous footing, constructed a harbour, etc. From then on the salt manufacture continued to flourish till the repeal of the salt duty in 1827. Magnesia works, started in connexion with the salt-pans in 1802, were the earliest establishment of their kind in Scotland. They were operated by a Mr. Burns, who sent most of his product to the large drug houses of London and New York.[1]

Production of salt on the west coast did not meet demand. In *The Case of the Panmasters* it was remarked that the opening of the Forth and Clyde Canal would ease the situation, but although the Canal was opened in 1790, salt continued to remain scarce in western districts. We hear of one Nathaniel Aiken of Wigtown trying to get salt from Fife in the summer of 1793. He was successful in getting one cargo from John Cadell of Cockenzie, but that was all.[2]

In England the main centre of production was initially the Tyne. At first in control of much of the supply of London, it suffered a series of reverses for various reasons. When, during the Commonwealth (1654), Cromwell repealed duties on the importation of Scottish salt into England, many pans at Shields were given up owing to Scottish competition which was stimulated.[3] The Tyne was again the victim of further external influences beyond its control when, towards the end of the seventeenth century, the English salt industry underwent a double revolution. Firstly, in 1692, new brine pits were sunk at Droitwich by Robert Steynor, and in the course of the next forty years, thirty-eight pans had come into operation. Secondly, rock-salt had been discovered in 1670 by William Marbury (or Madburry) on his land in Cheshire. Marbury obtained a patent to drain brine pits in 1682.[4]

It was some time, however, about a quarter of a century in fact, before the effect of this discovery was felt by the ancient *wiches* in the neighbourhood. When Collins wrote his *Salt and*

[1] *New Statistical Account*, Vol. 5 (1) p. 460.

[2] *National Library of Scotland*, MS. 1058, *f.* 31.

[3] J. U. Nef, *op. cit.*, Vol. 1, p. 174.

[4] *Cal. S. P. Dom.*, 1682, 282.

Fishery (1681) there were twenty-seven at Nantwich, twenty-four being town works, three being at Osterton in the possession of Sir Thomas Delves of Doddington. This lapse in time allowed expansion at Droitwich to precede development in Cheshire, with the result that, before the end of the century, price-cutting by Droitwich pan-masters was determining the pace in Cheshire.[1]

The discovery of rock-salt in Cheshire had two effects. It contributed to the decline of the industry on the Tyne, already lanquishing under alleged Scottish competition. It also contributed to the rise of Liverpool as a port in the early eighteenth century. 'Salt is Liverpool's chief article of commerce' said the Marquis de Bentivoglio. Before the end of the century return cargoes coming in to Liverpool consisted of Baltic and American timber which was used to prop up the south Lancashire and Staffordshire coal-pits from which coal was obtained for the salt industry.

The remaining centres of production in England ministered mainly to specific local needs. At the edge of the Pennines the industry at Lawton survived because of its nearness to the Derby and North Staffordshire coal fields from which it obtained fuel, and its proximity to a market in the shape of the Potteries.[2] The Lymington industry was near the naval depots at Portsmouth and Plymouth, and also enjoyed transport facilities to counties east of Devon and south of the Thames.[3] With the decline of the Tyne, Lymington gained ascendancy in the London market.[4]

It is apparent then, that salt was manufactured from three different raw materials: (*a*) sea-water; (*b*) brine springs; (*c*) deposits of mineral rock-salt. The implications of these differences has had a profound influence on the continuity and prosperity of the industry in different localities where it was prosecuted. Where the source of the salt was sea-water (as was the case at all the Scottish salt-pans) it was evaporated in trenches along the sea-shore if climatic conditions were considered sufficiently favourable, or concentrated by boiling down the sea-water in lead, or later iron, pans, fired with wood, peat, or coal. Where brine springs existed, these were utilized, because as a general rule they yielded a crude solution of salt more concentrated than sea-water. The use of brine springs in England developed contemporaneously with the Scottish sea-salt industry, the fortunes of salt-making in the two countries having had reciprocal consequences. In the following table the yield of salt from sea-water and from brine springs is

[1] Hughes, *op. cit.*, p. 379. [2] *Ibid.*, p. 388. [3] *Ibid.*, p. 410.

[4] Archibald Cochrane, ninth Earl of Dundonald, *The Present State of the Manufacture of Salt Explained* (1785), p. 1.

COMMON SALT
Sodium Chloride (NaCl)

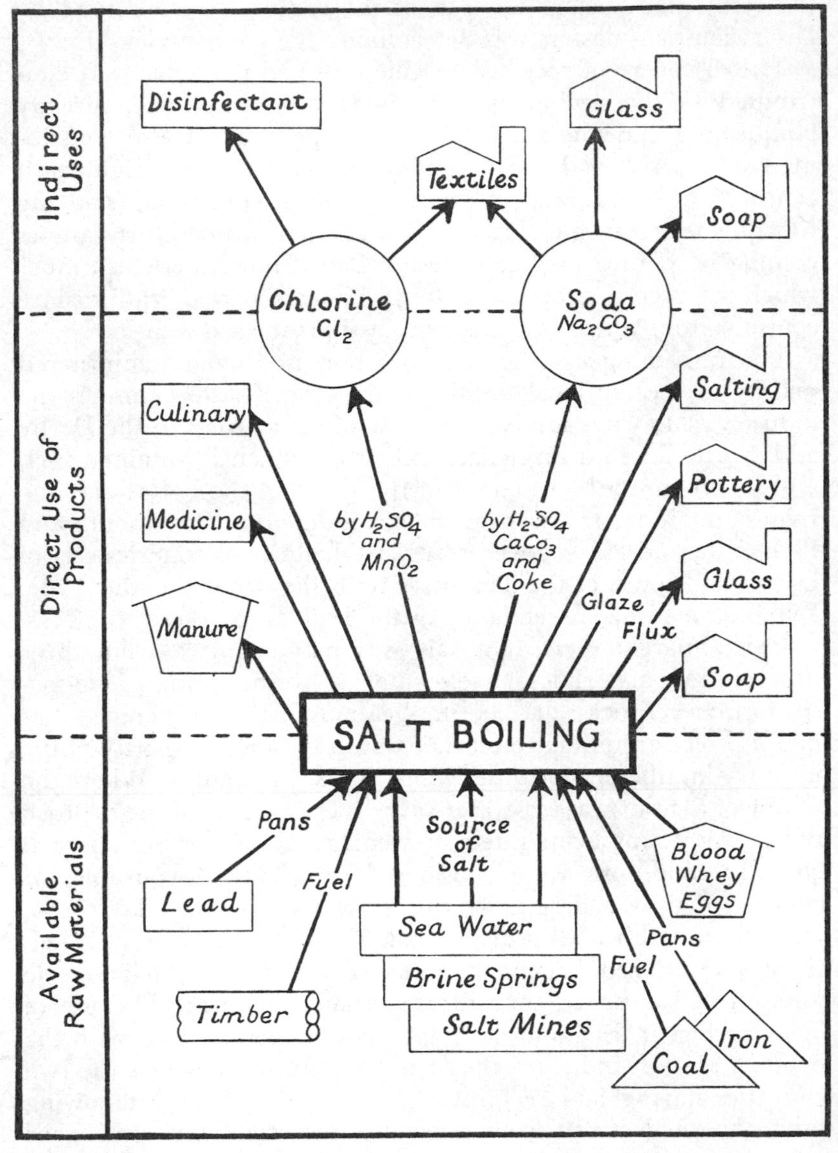

compared. The figures are taken from Lord Dundonald's *The Present State of the Manufacture of Salt Explained.*

Sea-water from the Firth of Forth yielded per hundred tons:

Of salt	2 T.	17 C.	0 Q.
Of water	97 T.	2 C.	3 Q.

while saturated brine from rock-salt contained in the same weight of solution:

Of salt	23 T.	0 C.	1 Q.
Of water	76 T.	19 C.	2 Q.

The economic advantages of rock-salt deposits are obvious, and it is clear why Cheshire ultimately became the main centre of production in Great Britain. It is worth noting in passing that, even there, sea-water continued to be used as the solvent in the recrystallization and purification of the salt from the deposits.

It is impossible to say when coal replaced wood and peat for the firing of salt-pans in the Forth and Tyne regions, but from the middle of the sixteenth century producers there were able to compete with continental *bay* salt on the London market due to their having access to supplies of cheap coal. The major advance in the technology of salt-boiling was the replacement, as a prelude to palaeotechnic operation, of lead by iron pans. This was necessitated by the higher temperatures resulting from the use of coal as fuel. In England it was the substitution of wood by coal in the sixteenth century that led to the convergence of salt-boiling on the Tyne, where, as in the Forth area, easily wrought coal abounded. In the case of the Tyne there was the additional advantage of associated brine springs. In Cheshire lead pans were still in use in the sixteenth century, and the fuel was timber. Hall says that iron pans were introduced at Nantwich in 1632, but Brereton, who visited the Tyne in 1635 and saw iron pans in use there, implies that they were still unknown in Cheshire.[1] At Lymington, iron replaced lead at the beginning of the reign of Charles I (1625) and even as early as 1489 iron pans had been used by Lionel Bell at South Shields. Brereton describes the South Shields pans which he saw in 1635 as made in sections and riveted together.

As was usually the case, the transition to palaeotechnic economy altered the balance of advantage between the different centres of production. Lymington and Portsea profited by their proximity to the surviving eotechnic iron industry in the Weald. Droitwich

[1] Hall, *The Civil War in Cheshire*, Lancashire and Cheshire Record Society, Vol. 19, quoted by Hughes; Brereton, *op. cit.*, p. 86.

was in all probability supplied from the Forest of Dean, as, in the days of lead pans, Cheshire had been supplied with pans made of Derbyshire lead. Pan-making was reckoned a skilled operation.[1] In Scotland, the several partners of the Carron Iron Works had interests in salt-boiling, and one of the early provisions made there was equipment for the fabrication of salt-pans.

When the substitution of iron for lead pans and coal for wood fuel was complete, there was not much room for revolutionary changes in the technics of the salt industry. The process of simple evaporation then established continued till modern times, though possible modifications were mooted from time to time.[2] One such came from Richard Watson, Bishop of Llandaff, who on the basis of careful meteorological observation suggested atmospheric evaporation of brine dispersed over suspended cloths.[3] Other schemes are due to Lord Dundonald.[4] According to James Leslie, magnesia manufacturer near Musselburgh:

> Of all the schemes yet thought of for refining the common salt, that of Lord Dundonald is both ingenious and strictly chemical, and it is therefore a little surprising that it is not practised universally.[5]

The method consisted of washing the impurities out of rock-salt with brine, but, while 'strictly chemical' it was found to be impracticable because the rock-salt as well as the impurities was washed away. Dundonald speaks of having communicated his method to the Secretary of the Royal Society of Edinburgh, but we have not seen it.[6] Somewhat later in the same social context in which James Watt evolved his steam engine and Joseph Black discovered the latent heat of steam; James Smith of Newton-upon-Ayr applied the heat capacity of steam to heat sea-water. It is said that by so doing coal consumption was more than halved, and a purer salt obtained as well. We hear no more of this idea. The profligate nature of palaeotechnics was yet to be realized.[7]

At the time of its original discovery, the mining branch of the salt industry had a great advantage over other branches on account of the concentrated form in which the raw material was obtained. The advent of the improved steam engine lessened this advantage,

[1] Hughes, *op. cit.*, p. 405. [2] *Ibid.*, p. 427.
[3] Richard Watson, *Chemical Essays*, Vol. 2, p. 57.
[4] *Scots Magazine*, 1804, *66*, 760. [5] *Trans. Highland Society*, 1804, *4*, 417.
[6] Dundonald, *The Present State*, p. 13.
[7] Sir J. Sinclair, *General Report of the Agricultural State and Political Circumstances of Scotland*. Appendix, Vol. 2, p. 300.

however, and finally tilted the balance back in favour of the older branches of the industry. Boulton and Watt were in touch with the salt industry in several districts and it ultimately became general practice to pump concentrated brine, rather than mine rock-salt. A Boulton and Watt engine was installed at Thurlwood, Cheshire, which threw 20,000 gallons of brine in twelve hours for the expenditure of three hundredweights of coal, costing nine pence.

As the salt industry expanded it called for larger pans and so stimulated the production of iron.[1] In the Restoration period Cheshire pans were about a yard square and six inches deep. Those at Droitwich were similar, whether made of lead or iron. By 1733 Cheshire pans had increased to 12–14 feet by 7–10 feet and 13–17 inches deep, while the capacity of some Northwich pans is given as between seven and eight hundred *ale gallons*. Pans subsequently grew to 24 by 15 feet, remaining about the same depth; ultimately they increased to between three and four times that size. In 1794 official surveys mention pans forty feet long and twenty-seven wide. It is not infrequently stated that Scottish pans increased but little. This does not agree with the observation that in 1765 the colliery and salt-works at Bo'ness were 'very great'. The works were operated by Dr. John Roebuck, founder of the vitriol works at Prestonpans and co-partner in the Carron Iron Works. The dimensions of the Bo'ness pan are given as fifty-five feet by thirty-two feet. Ten thousand tons of coal were consumed by it per annum. Little wonder atmospheric pollution was one of the distant symbols of salt-making. Such a pan required the outlay of three hundred pounds in capital and cost between thirty and forty pounds every two to three years to maintain it in working order. There were, on an average, two men employed to a pan, with a few extras to carry the salt to the 'granneries'.[2]

> Nothing ever exhibited such an idea of the infernall regions as this horrid furnace and the poor miserable naked wretches attending it.[3]

Economic conditions of labour in the Scottish salt-works were indeed ghastly. Prior to an Act of 1735, the wages of Scottish salters were paid in the commodity which they produced. In 1755 the well-known Act of Parliament 15, Geo. III, c. 28 opens 'Whereas many colliers, coal bearers, and salters in Scotland are in a state of slavery or bondage, bound to the collieries and

[1] *Phil. Trans. Roy. Soc.*, 1669, *4*, 1060.

[2] *Statistical Account*, Vols. 17, p. 65; and 12, p. 514.

[3] MS. Notes added to *Pennant's Tour* (Chester, 1771) in the Library of King's College, Aberdeen, at p. 86.

saltworks where they work for life, and are sold with the mines. . . .' Some of this was female labour into the bargain. 'The girls hair were snakes' says an observer, perhaps carried away by the semblance to the infernal regions. The salters were only effectually emancipated in 1799 by the Act 39 Geo. III, c. 56. One of the first salt-boilers and coal-owners to become sufficiently enlightened to abandon the employment of female labour in coal-mines was the author of *The Present State*, the Earl of Dundonald.[1]

According to Sir John Sinclair, an ordinary Scottish pan produced 3,000 lb. of salt per week, consuming fourteen tons of coal in the process. The coal cost from half-a-crown to four shillings per ton. This is a much smaller consumption than that given for Dr. Roebuck's pan, so it is likely that it was indeed exceptionally large.[2] About three tons of salt were made at each boiling. Sinclair's figure for coal consumption is, however, probably rather high, since it indicates that ten tons of coal were consumed for each ton of salt produced. Other authorities give six tons. It is important, however, to notice that five tons of salt could be made with three tons of coal at Larne, where rock-salt dissolved in sea-water was evaporated with Ballycastle, Ayrshire or Whitehaven coal.[3] Dundonald gives the ratio of the cost of production from sea-water on the Forth to production from rock-salt as eight to one.

The process of evaporation of sea-water and the purification of the concentrated solution has been described by H. M. Cadell, a descendant of one of the founder partners with Dr. Roebuck in the Carron Works, and himself one of the last pan-masters to operate on the Forth.

> The pans were rectangular in shape, ten to fifteen feet long, and one and a half feet deep, and were set upon long fires originally without bars. The brine was evaporated until sufficiently concentrated when a bucket of bullock's blood was thrown in. On mixing with the liquor it changed colour and by a sudden transformation the seething white liquor became immediately overspread with a thick variegated and brown bubbling scum. The albumen in the blood which had coagulated and risen to the surface carrying all the impurities with it was skimmed off, leaving the boiling liquid clear as crystal and devoid of all suspended matter. A salt pan had to be evaporated eight times before enough salt separated, requiring three and a half days firing. Creech or calcium sulphate separated first and then the salt crystals next and was drawn to the side and shuffled out

[1] T. S. Ashton and J. Sykes, *The Coal Industry of the Eighteenth Century*, p. 171.

[2] *General Report*, App., Vol. 2, p. 229.

[3] *Remarks (anonymous) on the System of Laws, etc.*, p. 25.

leaving behind a little 'bittern' containing iodine and bromine in combination, as well as a little magnesia.[1]

Prior to the introduction of bullock's blood, sweet whey had been used. This was essentially the method used in Holland where the Dutch were most particular about the purification of their salt. To a hundred pounds of salt dissolved in about three parts of water they added about two gallons of sweet whey. When the liquid was boiled a scum of lime or magnesia salts came to the surface, from which it was removed.[2] At many of the English salt pans whites of eggs were used, but at Newcastle, as in Scotland, blood was the coagulant.[3] After the salt was removed from the pan a solution called *bittern* was left. This was usually allowed to go to waste, but according to a writer in the *Statistical Account*, it was sometimes bought by *chymists*. A. J. Balard discovered a new chemical element, bromine, in it in 1826.

The process of salt-boiling was carried on continuously throughout the week, but was suspended at the week-end, when the pans were allowed to cool. The large crystals separating from the cooling solution were known in Scotland as *Sunday crystals*, and in England as *shivery salt*.

At the various foci of salt production, whether on the Forth or Tyne, at Droitwich, or in Cheshire, what were virtually different methods of production were in operation in relation to the available raw materials. As has been mentioned, in Scotland the raw material was a dilute solution of salt from the sea: in England either a concentrated solution from inland brine springs or deposits of rock-salt. This occasioned, *ipso facto*, marked difference in production costs. This must be borne in mind, together with the fact that subsequently industries developed in Scotland, on the Tyne, and in Cheshire and Lancashire, which used salt as an industrial raw material, when one is trying to obtain an understanding of the complicated fiscal charges to which this commodity was subject at different times.

The eighth Article of the Treaty of Union between England and Scotland gave Scotland exemption from the existing English salt taxes for seven years after the Union, and perpetual exemption from an additional duty of 2*s.* 4*d.* per bushel of fifty-six pounds which had been in force in England since 1697–8. The consequence of this Article in the Treaty was that, far from establishing free

[1] H. M. Cadell, *The Story of the Forth*, p. 322.

[2] J. Walker, *An Essay, etc.*, *Trans. Highland Society*, 1803, *2*, 380.

[3] W. Brownrigg, *The Art of Making Common Salt*, p. 57; for a biographical memoir of Brownrigg see *Annals of Philosophy*, 1817, *10*, 388 and 401.

trade between the two countries as is usually assumed, the Act of Union created a permanent inequality between them with regard to fiscal burdens. The inequality was in favour of Scotland only if we take no account of the different cost of production of salt from sea-water and from rock-salt. No sooner had the seven years allowed in the Act elapsed than an excise of 1*s*. was put on all salt made in Scotland, but since no extra duty was put on Scottish salt imported into England, for the next two generations the north of England was supplied with Scottish salt. Tyneside interests continually grumbled against the Scots, despite the existence in England of more powerful and advantageously placed competitors, at Droitwich, and especially in Cheshire, but Walpole maintained the *status quo* in the belief that the Scots were too poor to pay the full duty imposed in England. Whichever centre was to blame, it is admittedly true that most of the Tyne pan-owners failed before the end of the eighteenth century.[1] As late as the time of Ricardo (1772–1823) soap and glass manufacturers on the Tyne deplored the Scottish competition. An increasing ferment of discontent led to the institution in 1780 of an import duty of 7*d*. per bushel on all Scottish salt imported into England, but even this did not revivify Tyneside industry, which continued in its languishing condition. The ratio of the duties in the two countries remained at three to ten, i.e. 1*s*. 6*d*. on salt made and consumed in Scotland, and 5*s*. in England. These duties were high on a commodity which was continually revealing itself as a valuable industrial raw material, and agitation for repeal and readjustment came from widely different quarters. The salt laws operated not only against manufactures, but were an impediment to the West Coast fishings. At Loch Alsh fish caught could not be cured, and in the small isles, although a drawback was allowed, the number of forms to be filled up was a nuisance. To get even a small quantity of salt it had to be entered at the custom house, as had also the fish after it had been salted. All the unshipping and unpacking had to be gone through with fish even for domestic consumption.[2] Lord Selkirk, representing naval interests, expressed himself forcibly in a letter to Henry Dundas:

> For Godsake take off the Salt Duties—Think of the fishery—Consider what a Nursery for hardy Seamen it would be: and at this hour when it is found so difficult to Man the Navy, that consideration would have weight. . . . It has been said, if the Duty is taken off then Salt will be smuggled from Scotland to England,

[1] Hughes, *op. cit.*, p. 408 *et seq.*

[2] *Statistical Account*, Vol. 11, p. 428, and Vol. 17, p. 291.

> Probably some would be smuggled: but at the worst, that would only be transferring some of these illicit Profits from the Irish smugglers to the Scottish ones.[1]

The next development was a suggestion mooted *c.* 1793 to admit cheaply produced English rock-salt into Scotland. By this time two vital advances in the evolution of technics had been made. The eotechnic Scottish linen bleaching industry had passed into the palaeotechnic phase as the result of the discovery of chlorine by Scheele in 1774, during that expansion in pneumatic chemistry which saw the discovery of sulphur dioxide (1770), hydrochloric acid gas and nitric oxide (1772), ammonia and chlorine (1774), and sulphuretted hydrogen (1777). The bleaching properties of chlorine were discovered a few years later, and this property commercialized in Scotland through the activities of Patrick Copland, and James Watt, whose father-in-law was a Glasgow bleacher.[2]

The other advance was the solution of a problem which had engaged scientific minds throughout the greater part of the second half of the eighteenth century, the synthesis of soda ($Na_2CO_3.10H_2O$). As we shall see in Ch. IV, the Earl of Dundonald was associated with the operation of this synthetic process, and the suggestion to admit rock-salt from England into Scotland probably came from him. Some of the contemporary arguments are set forth in *Reasons for allowing English Salt to be brought to Scotland at the Scotch Duty*—1793.[3] In a pamphlet published during the final fight for a complete abolition of the Salt duties, Samuel Parkes enumerates the advantages accruing from cheap salt as an industrial raw material:

> The manufacture of sal-ammoniac will then be revived; the manufacture of soap in several districts in the interior of the kingdom will be increased and improved; several kinds of glass will be rendered cheaper; muriatic acid, or spirit of salt, which is an article of great consumption with all fancy dyers and calico printers, will be furnished at half its present price. Crystalline soda will be reduced in the same proportion. . . . All kinds of stone pottery . . . might be sold at reduced prices.[4]

At this time the possibility of taxing the solution left after crystallization of the salt was under consideration. Lockhart, Collector of Customs at Port-Glasgow, asked Dr. Joseph Black

[1] National Library of Scotland, MS. 640, ff. 176–7.
[2] Ch. IX. [3] National Library of Scotland, MS. 640, f. 182.
[4] R. Dacre, *op. cit.*, p. 185 *et seq.*

for a definition of *bittern*. Black replied 'the residue after crystallizing salt', but to Lockhart's further question as to what duty would have to be put on it to be equivalent to the duty on salt, he declined to answer on the ground that he did not want to take the responsibility. All that he would do was determine the density of a sample of bittern sent by Lockhart.[1]

Although himself a Scot, Dundonald was one of the new class of manufacturers anxious for cheap raw materials irrespective of the effect on the prosperity of the long established pan-masters in his country. He argued that, in any case, Scottish masters were being driven out of the important West of Scotland market by Cheshire salt, smuggled in from Ireland where it had been refined. Salt was undoubtedly scarce in the west. Nathaniel Aiken of Wigtown, having failed to get more than one cargo from the Forth, spoke of getting a sloop from Liverpool for himself and others. He did not mind the duty provided that he could get the salt.[2]

Their vested interests threatened, the Forth pan-masters began pamphleteering in an attempt to counter Dundonald's activities. Among other arguments they mention, not entirely to their own credit, that:

> . . . all the workmen in the coal-mines and salt-pans, being now fit for no other employment, would starve as the small coal or 'culm' was useless for anything other than salt-boiling and brick and lime burning.

The salt-makers of Edinburgh were also determined to oppose the importation of rock-salt from England. If unsuccessful they were to apply to Parliament for compensation, and appointed a committee to guard their interests. One of the committee was William Cadell of Banton.[3] An associate of the Cadells, Dr. John Roebuck, referred to above, also entered the lists. In a letter dated 30th December, 1785, he reviews the whole situation. According to his estimate about 300,000 bushels of salt were manufactured annually, principally at the collieries bordering on the Firth of Forth, only about one-tenth being made elsewhere.

> The whole of this salt is made from sea-water by evaporating it in Iron Pans by means of small Coal or Culm which is refuse arising from the getting of Big or Chow Coal, and which is of no value except for this purpose and the burning of lime. And if it

[1] W. Ramsay, *Life and Letters of Joseph Black*, p. 108.
[2] National Library of Scotland, MS. 1058, f. 31.
[3] National Library of Scotland, MS. 640, f. 190.

> is not consumed for this purpose it would be left in the Pits as useless, which would, of course, greatly increase the Price of Big and Chow Coals.
>
> There are no Vessels solely employed in carrying Coals from the Firth of Forth to London as there are from Newcastle and Sunderland. The price of the Coals and the length of the Voyage would make it a losing trade and the small proportion of Scotch Coal which is sent to London may be considered almost as back Carriage, and if the price of the Coal were increased its quantity would still be less to the injury of the Navigation of this part of the kingdom.
>
> The Collieries in the Firth of Forth are in a great measure supported by the sale of Salt. If they were deprived of this trade they would be losing advantage to the great injury of the undertakings who have laid out considerable pains in full confidence that no alteration would be made in the Salt Laws which should permit Salt to be imported into this Country to their detriment. But if the Duty on Salt is taken off and an importation permitted into Scotland, the Proprietors of Scottish Works must be ruined as Salt from Liverpool can be imported cheaper than they can make it. For a Salt Pan of the same dimensions will make eight times the quantity of Salt by dissolving Rock Salt in Sea Water than can be made by boiling Sea-Water.[1]

These various appeals went unheard, and when the Salt Acts of 1798 were introduced they completely forbade, in deference to the Newcastle interests, the importation by land of Scottish salt into England, and decreased the ratio of the duties in the two countries (previously 3: 10) so that the duty on Scottish salt was 6*s*. 6*d*., and on English 10*s*. per bushel. An additional duty amounting to 3*s*. 4*d*. per bushel was, however, imposed on Scottish salt imported into England by sea. That is, in reality the duties on English and Scottish salt were identical, despite the fact that costs of production remained widely different.[2]

A drawback was allowed for salt used in industry, e.g. as a raw material in chemical bleaching. It is doubtful if it was equal in England and Scotland, as the Glasgow Chamber of Commerce, the first in Britain, founded in 1783, complained that it was to the prejudice and loss of Scottish manufacturers engaged in bleaching.[3]

The new burden so suddenly placed on Scottish salt regardless of manufacturing costs was so catastrophic in effect that it had to be repealed almost immediately.[4] The new Act of 1799 reduced

[1] National Library of Scotland MS. 640, f. 166.
[2] 38 Geo. III, c. 89 § 86.
[3] Hughes, *op. cit.*, p. 418.
[4] 38 Geo. III, c. 77.

the duty by two shillings, but increased to six shillings the import duty on salt exported from Scotland to England. Huskisson now explained the reduction as taking into account the fact that Scottish salt was made from sea-water, and required large quantities of coal to evaporate it.

The advent of war with Napoleon in the early years of the nineteenth century forced the salt duty up to fifteen shillings per bushel, but the additional duty on Scottish salt imported into England was always adjusted to make the total duty on Scottish salt the same as that which had to be paid in England. This operated to the fatal detriment of the industry in Scotland. Salt in Glasgow cost £1 2*s.* 3*d.* per bushel. High prices fostered smuggling. The disparity in the duties and the restrictive Act of 1702, which prevented the refining of English salt except at certain refineries, coupled with the free export of rock-salt to Ireland, led to smuggled refined salt being sold in the West of Scotland at a price less than the cost of producing even crude Scottish salt. The war, too, caused a scarcity of salt throughout Scotland, so it is assumed that in times of peace there must have been an extensive smuggling trade, probably from Spain, Portugal, France and Holland, in addition to Ireland.[1]

Finally, in 1822, the pressure of the palaeotechnic industrialists became so great that it was proposed to repeal the Salt Acts altogether. On this question the Tyne salt interests remained strangely silent, but the soap and glass manufacturers there, still harbouring their traditional grudge against the Scots, opposed the repeal, fearing that the Scots would be left in an advantageous position, though it is difficult to see why. In Scotland the voice of the Orkney kelp-producers was raised in protest. They deplored the removal of the tax, which, by reducing the price of salt, would enable glass manufacturers to use it instead of kelp, at that time the cheaper raw material.[2] Although Jeffrey, the editor of the *Edinburgh Review*, tried to stir up public opinion in favour of repeal—and it must be borne in mind that the people of Scotland were paying more for the salt than for the meat which it savoured—when it came to the vote in the House, only five of the Scottish Members supported the motion, one of these with reservations.

But the repeal of the Salt Act passed into law, and with it the decline and collapse of the Scottish salt industry set in. The following decade saw the number of salt-pans in Scotland fall from a hundred and sixty-four to only fifteen in 1836. The

[1] *Remarks (anonymous) on the System of Laws, etc.*, p. 10.
[2] National Library of Scotland, MS. 640, f. 216.

survivors were all on the Forth, with the exceptions of one at Montrose and one at Saltcoats.

The advantages of the privileged position enjoyed since eotechnic times passed from the owners of coal-mine and salt-pan to the new manufacturing classes whose prosperity was being built up, not on a basis of material inheritance, but upon the application of the growing corpus of exact science then expanding with unprecedented velocity. The repeal of the Salt Acts had profound social and economic effects. One of these, the effect on industrial hygiene and industrial disease, had been used by Parkes as an argument in favour of repeal:

> There is a considerable consumption of salt in glazing some kinds of earthenware. And I submit, whether, if the duty were taken off salt, it would not in many districts be substituted for the poisonous preparations of lead which are now so abundantly employed in glazing many of our culinary utensils. . . . If the use of common salt were to be much extended in the potteries the health of the workmen in that district would be restored, whereas at present they are all subject to paralysis, from the circumstance of their hands being constantly immersed in so deleterious a mixture as that containing a large portion of the solution of oxide of lead.[1]

Another direction of possible exploration was in the use of salt as a fertilizer. As early as 1598, James VI granted letters patent for the use of salt in agriculture to Archibald Napier, Ninth of Merchison (1576–1645), son of John Napier (1550–1617), the author of *Merifici Logarithmorum Canonis Discriptio* (1614).[2] When agrarian revolution had further developed, one Thomas Liveings obtained a patent allowing him to make *new compound manure* from foul salt. He failed to follow up his privilege because he was unable to obtain salt duty free.[3] In 1818 Parliament did go so far as to pass an Act reducing the duties on salt to be used for agricultural purposes. This stimulated the Highland Society to offer premiums for essays describing the results of experiments with salt as a manure. Experimental results were submitted by Andrew Robertson and others indicating that salt is of doubtful utility. Despite this, Richard Watson, Bishop of Llandaff, tells us that no less than 3,000 tons were sold at Northwich for use as a fertilizer.[4]

The removal of the salt duties had a double-edged effect upon Scottish economy. The kelp industry, dependent upon the high

[1] R. Dacre, *op. cit.*, p. 206.
[2] R. Chambers, *Domestic Annals of Scotland*, Vol. 1, p. 300.
[3] Hughes, *op. cit.*, p. 428. [4] *Trans. Highland Society*, 1820, 5, XXVIII.

duties on salt and imported *barilla* (impure alkali made by burning *Salsola Soda* of the Mediterranean), had developed and flourished exceedingly round the coasts and islands of Scotland. While its decline has been variously attributed to the reduction of the duty on barilla, etc., even if barilla had been excluded altogether the kelp industry would not have continued in a flourishing state in the absence of salt duties, the removal of which made the synthesis of alkali on a commercial scale a feasible proposition. It had been argued against the duties on salt:

> There can be no doubt but that a great number of manufactories of mineral alkali will be established in every part of these kingdoms, immediately the repeal of the duties on common salt . . . and I do believe that in a very short time the importation of foreign barilla will entirely cease.[1]

These comments were indeed prophetic. The repeal of the salt duties helped the rise of the synthetic alkali industry, and this in turn dealt the death blow to the manufacture of kelp, which must now be studied.

[1] R. Dacre, *op. cit.*, p. 199.

Cinerum Clavellatorum Præparatio..

M. B. Valentine, 1704

5. Preparation of potash for making glass, soap, etc.

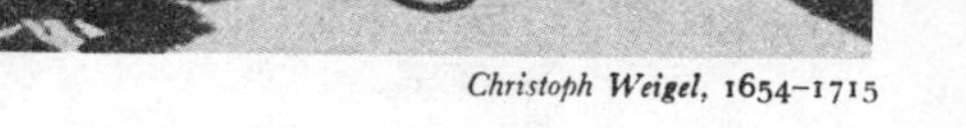

Christoph Weigel, 1654–1715

6. The Soap-boiler.

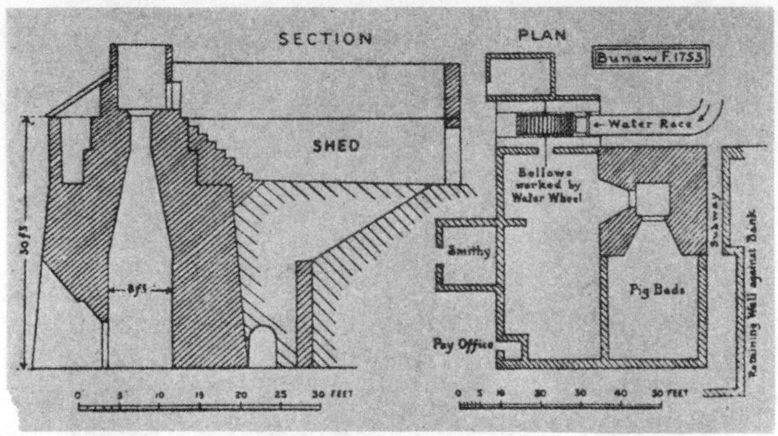

7. Bonawe Charcoal Furnace at Taynuilt, Argyllshire. Started by an Irish company in 1730, it was worked by Richard Ford from 1753 to 1874.

8. Covered-in glass-pots. The alteration in the shape of glass-pots was dictated by the change in fuel from timber to coal. This brought other changes in its train which led to the evolution of a distinct species of glass, viz. flint-glass.

Chapter III

THE TRADE IN ASHES AND KELP

> I now have two hen-houses at the kelp shore, and derive the same amount of income in a very much easier way from the hens.
>
> Scottish crofter on kelping.

DURING the evolution of eotechnic economy wood was a basic necessity to an extent now difficult to visualize. The derivative of its destructive distillation, charcoal, was an essential constitutent of the only known eotechnic explosive, gunpowder; pitch and tar were indispensable for shipbuilding; potash, originally got by lixiviating vegetable or common ashes and evaporating the solution in iron pots, was essential, first of all to manufacturers of soap, glass, alum, and saltpetre (KNO_3), and later to the bleacher, and farmer.

> As the woods here decay so the glass-houses remove and follow the woods with small change . . .

related a Worcestershire man in the 1690's. Where it was impossible for industry to follow the supply of raw materials, these were first reduced to as small bulk as possible and then transported. Hence the conversion of crude vegetable ashes into *potash*. 'It was more usual to transport potashes in place of common ashes', says James Dunbar in *Smegmatologia, or The Art of Making Potashes, Soap, and Bleaching of Linen.*[1] Five hogsheads of the former produced one of the latter. Regarding the use of ashes in agriculture, in some districts of England farmers carted *turf* from the commons free of charge if only they were given the residual ash in exchange.[2]

Apart from the industries mentioned above, mining and metal working made great inroads into supplies of timber drawn from forests already suffering depletion by eotechnics. Thus, from an early date, home supplies of potash had to be supplemented by the importation of alkaline materials in one or other of several forms, either weed or wood ashes in the crude, or potash and pearlash in the refined, form. But timber was not the only material from which imported supplies originated. On the Mediterranean littoral

[1] J. Dunbar, *Smegmatologia*, p. 7.

[2] Witt Bowden, *Industrial Society in England towards the End of the Eighteenth Century*, p. 225.

crude alkali (barilla) was made by burning *Salsola soda*, particularly on the eastern shores of Spain where it was cultivated in the *huerta* of Murthia.[1] 'Natural soda' or *trona* ($Na_2CO_3.NaHCO_3.2H_2O$) occurs as a mineral deposit in Egypt. While Spain was the main producer of barilla, the coniferous forests of the cool, temperate regions of continental Europe were the reservoir from which Britain's insular industries drew supplies of wood ashes. These figured largely in the trade of Riga, Danzig, and Königsberg.

As the timber famine increased in intensity throughout the first half of the eighteenth century it became more difficult to get supplies of alkali.

> We are, in Britain, destitute of woods, that can be employed for producing potashes. We must depend for this commodity upon foreign nations, in the northern parts of Europe and America.[2]

Between 1740 and 1750 the price increased by fifty per cent., that is from £16 to £25 per ton of ashes. When a Parliamentary Committee, which was set up to enquire into the rise, cross-examined a soap-boiler about the advance in price, he replied that the merchants explained it as arising from the woods being burned, but he himself believed it was on account of unlimited orders which French, Dutch, and Flemish merchants had to buy east-country ashes.[3]

> The several kinds of Potashes . . . are purchased by their (i.e. Dutch) merchants from sixty to seventy per cent. cheaper than they cost our Bleachers here,

wrote Patrick Lindsay in *The Interest of Scotland Considered*.[4] The demand for ashes in Holland was already greater than in Great Britain, and Francis Home (1719–1813), Professor of *Materia Medica* in the University of Edinburgh, who made important contributions to the art of bleaching, mentions an attempted monopoly by two Dutch merchants, who bought up ashes and sold them at two or three times the orginal price.[5] Moreover, monopolistic dealing was easily achieved because, in the case of Russian ashes, they could only be imported by those with contracts granted directly by the Czarina.

It was possible, however, to counter the difficulties arising from

[1] The culture of Salsola, as established in the environs of Narbonne, is described in the *Annales de Chimie*.

[2] J. Walker, *An Essay on Kelp*, *Trans. Highland Soc.* 1799, *1*, 3.

[3] *Journal of the House of Commons*, 1751, 25, 239.

[4] P. Lindsay, *The Interest of Scotland Considered*, p. 174.

[5] Francis Home, *Experiments on Bleaching*, p. 16.

scarcity and high prices in at least three ways: (*a*) organized importation; (*b*) the production of wood-ash at home; or (*c*) the production of ash from other biological materials, e.g. seaweed. Synthesis, it is true, was a fourth possibility, but that requires separate consideration.

It was probably on account of the progressive rise in prices given above that the British Linen Company, incorporated in 1746, took steps to assist Scottish bleachers by importing wood-ashes and distributing them on credit.[1] Another body, the *Board of Trustees for Fisheries, Manufactures, and Improvements in Scotland* was importing ashes at the same time. In 1751[2] they imported direct to:

Aberdeen	from	Christiansand	4,974 lb.
Leith	,,	Danzig	2,022 lb.
Montrose	,,	Riga	1,100 lb.
Stranraer	,,	Donnochadie	140 lb.
			8,236 lb.

In the following years the quantities imported by the Board of Trustees greatly increased.[3]

On reviewing the situation, the Parliamentary Committee mentioned above decided that to ease the situation the Acts imposing duties on ashes imported from His Majesty's colonies should be repealed. Stimulated by this, a committee of the *Society for the Promotion of Arts, Manufactures, and Commerce*, sitting under the chairmanship of Benjamin Franklin (1706–90), who visited London in 1757[4], was successful in promoting the production of potash in the British colonies, then consisting principally of North America and the West Indies. Supplies remained short, however, and as Francis Home pointed out:

> . . . it costs, I am told, Great Britain and Ireland £300,000 for ashes every year. . . . Our manufacture could not have subsisted during the late war with Spain (1717) unless the order of the King and Council had passed, allowing the importation of Spanish ashes.

Spanish barilla was the form of alkali most sought after by eighteenth-century manufacturers, and many generations before

[1] S. H. Higgins, *A History of Bleaching*, p 19.

[2] *Board of Trustees Papers*, General Register House, Edinburgh, 1751, Vol. 96: *Administration*.

[3] *Board of Trustees Papers, An Account of the Ashes imported into Scotland from 5th July 1753 to 5th July 1758.*

[4] On visit to Scotland he made the acquaintance of David Hume, whose teachings greatly influenced Franklin's outlook.

that it had become an important item of commerce. Here is an early description of the rigours of arranging for its importation:

> I am now, thanks to God, come to Alicant in Spain, for I am to send thence a commodity cald Barillia for making crystal glass, and I have treated with Signor Andriotti for a good round parcel of it to the value of £2,000. This Barillia is a strange kind of vegetable. . . . I think earthy shrub that bears berries like barberries. When ripe they dig it up by the roots and stack it in cocks like hay to dry. When dry they place the shrubs in a track and set fire to them. The pit is closed and when, after some days, it is opened, the Barillia juice is found turned into a blue hardstone.[1]

The long series of wars between Britain and Spain made barilla supplies precarious, with the result that steps were taken to augment domestic supplies. James Dunbar's *Smegmatologia* gives directions for making ashes from a variety of material available locally in Scotland, viz., 'brechans, nettles, thistles, hemlock, juniper bushes, whins, etc.' Other vegetables, according to Dunbar, are not worth using on account of the small quantity of residue which they leave on combustion.

Attempts were made to increase the quantity of available ashes in other directions. Dr. William Cullen (1712–90), Professor of Chemistry in the University of Edinburgh, spent a long vacation on an annexed estate in the Highlands experimenting on the production of ashes. Although the York Buildings Company were operating in the Highlands the conifers of the remote parts of Scotland had been uninfluenced by the demands of eotechnics, and Cullen showed that ashes could be made in Scotland from already felled timber at 3*d.* per lb., i.e. as cheaply as imported ashes. Cullen also favourably regarded ferns as an alternative source, and calculated that fern-ashes could be made at 1½*d.* per lb., since women could handle ferns while men were required to fell timber.

Although the timber resources of the Highlands might appear to be considerable, one must set against that the quantity of ash obtained from a given weight of timber. R. Warington gives some numerical data:

> The timber of freely-growing trees contains but 0.2–0.4 of ash constituents on 100 of dry matter. In seeds free from husk the ash is generally2–5 per cent. of the dry matter. In the straw of cereals 4–7 per cent. In roots and tubers 4–8 per cent. In hay 5–9 per cent. It is in leaves, and especially in old leaves, that the greatest

[1] Described in 1618 by James Howell : quoted by H. J. Powell, *Glass-making in England*, p. 33.

proportion of ash is found; in the leaves of root crops the ash will amount to 10–25 per cent. of the dry matter.[1]

This led Cullen to investigate the possiblities of getting ash from seaweed, but he concluded that only scarcity of timber would lead to the use of seaweed ashes since it was no cheaper than that obtained from other sources.[2]

> Alkaline or Bleaching Salt is also produced from every tree, shrub, weed, or vegetable that grows—An impure Salt is also produced from Sea Ware or Tangle, which when burned into an ash is called Kelp—very fit for bottle-making but very improper for bleaching. It is the Child of Sea-Water and is, of course, highly impregnated with Salt which no act of the chemist has hitherto been able to separate from the Alkaline salts.[3]

As timber became scarcer, kelp became increasingly important in the economy of Scotland, till making it became the dominant occupation in the Highlands and Islands.

Several circumstances contributed to control the price at which alkali could be had in the form of home-produced kelp. The original price offered for Orkney kelp at Newcastle was of the order of £3 per ton, but poor quality, not to mention wilful adulteration with sand and stones, led to a progressive fall in price till it stood at £1 in the middle 'forties. By 1764 Hebridean kelp was selling at £3, and Sir George Colebrook's attempted corner in alum had an effect on the price of kelp, which was inflated for a time during the 'sixties. The loss of imported wood-ash during the American war had a similar effect, and prices are said to have reached £8 per ton.[4] From the end of the American War to the beginning of the nineteenth century kelp prices followed fairly closely the price of imported barilla and potash.[5]

Up to the end of the American war records of kelp prices are fragmentary, and it is impossible to make a year-to-year comparison with the price of barilla and potash. Suffice it to say that till 1780 the price of kelp maintained an upward trend, modified from time to time by political events. Robert Jameson, in his *Observations on Kelp*,[6] gives the following average prices for the ten- and twenty-year periods:

[1] R. Warington, *Chemistry of the Farm*, p. 2.

[2] *Board of Trustees Papers. Register of Theses on Technical Subjects: Dr. Cullen's Memorial.*

[3] *Memorial relating to Ash-burning, etc. for the perusal of the Right Hon. the Lord Advocate for Scotland: by Ebenezer McCulloch.* National Library of Scotland MS. 642, ff. 1–10.

[4] *Statistical Account*, Vol. 7, p. 539 and Vol. 13, p. 305.

[5] T. Tooke, *History of Prices*, Vol. 2, p. 397.

[6] Trans. Highland Soc., 1799, *1*, 44.

1740–60	£2 5s. per ton
1760–70	£4 4s. "
1770–80	£5 "
1780–90	£6 "

After 1780 kelp prices are more complete, and the question of price will be considered again later.

In the Records of Scotland the earliest reference to the manufacture of kelp is in the papers of the Town Council of Anstruther in Fife, where in 1694 an Englishman offered the Council £4 for the privilege of cutting seaweed and making kelp from it. Since one of the bailies protested that it might prove prejudicial to the health of the inhabitants, it is probable that nothing came of the overtures. What Anstruther lost in manufactures, it certainly made up in sense of civic responsibility.[1]

According to J. Walker (1731–1803), who was professor of Natural History in Edinburgh, production was started tentatively on the Forth about 1720 and in the Orkneys in 1722. It was introduced there on the Island of Stronsay by James Fea, heritable proprietor of Whitehall, who got a man, Meldrum, from Fraserburgh to come and demonstrate the burning.[2] Kelp from this source was being retailed to bottle-makers in Newcastle several years before 1730, though this is often given as the date of its introduction into the Orkneys.[3]

In the Hebrides kelp-burning was begun by Macleod of Uist about the same time, i.e. *c.* 1730. He is said to have got the art from Ireland. This may refer to one Rory Macdonald being invited to North Uist from Ireland about 1735 to carry out experiments for Hugh Macdonald of Ballistar.[4] From the amounts of kelp exported, the manufacture cannot have been of much consequence till about 1746, although ten years earlier Dunbar, curiously enough, had written that the burning of kelp was so well known that it did not need description.[5]

In certain districts, e.g. the Orkneys, there was definite opposition to kelping which contained, as we shall see, more than the germ of sound sense, even if the arguments brought against it belong rather to fancy than sound logic.[6]

[1] *Statistical Account*, Vol. 3, p. 78; cf. John Ray, *Itinery*, 1661.

[2] *Statistical Account*, Vol. 15, p. 395.

[3] J. Walker, *Essay on Kelp, Trans. Highland Soc.*, 1799, *1*, 3.

[4] *Statistical Account*, Vol. 13, p. 305; Rob. Jameson, *Observations on Kelp, Trans. Highland Soc.*, 1799, *1*, 43. [5] J. Dunbar, *op. cit.*, p. 2.

[6] Highland Society Library: *Miscellaneous Pamphlet No.* 7, *passim.* Note: This pamphlet is almost identical with the conclusion of the section on Manufactures in J. Sinclair's *Analysis of the Statistical Account of Scotland.* (pp. 342–4.)

The inhabitants of Orkney contended that:

> They could have no doubt of its driving the fish from the coast, and it would therefore ruin the fishing; they were certain it would destroy both the corn and the grass, and they were much afraid, that it might even prevent their women from having children.[1]

The latter was perhaps merely wishful thinking. Kelping undoubtedly 'produced consequences obvious to the most superficial observer', and D. F. Macdonald has related the increase in population in the islands of Scotland to the prosecution of kelping. Here is his argument:

> One of the chief causes of excessive population was the kelp industry. The landlords had encouraged more people to settle on their estates than the soil alone could possibly maintain, so that they might have enough workers. Statistics clearly prove the effects of the kelp manufacture in increasing population. It was chiefly carried on in the islands, and, to a small extent, along the west coast of the mainland. Between 1755 and 1831 the population of the northern and north-western counties increased by 48 per cent.; whereas that of Lewis, Harris, and Uist increased in the same period by 139 per cent. The kelp manufacture was much more important in Uist than in Harris or Lewis, and the increase in the population of Uist was proportionately greater. The increase in South Uist was particularly striking, being no less than 211 per cent., while in Harris it was only 98 per cent.
>
> This increase of population was not accompanied by any increase in the productivity of the soil. On the contrary the demands of the kelp industry meant that the soil was to a certain extent neglected. But the landlords made huge profits. One of them, Lord Macdonald, had an average income from kelp of £20,000 a year, and Macdonald of Clanranald made £18,000 a year. The profits of the workers were small, but they were usually sufficient to pay the rent.

Percentage Increase in Population (1755–1831)[2]

(1)	Northern and North-Western counties	48 per cent.
(2)	Island of Harris	98 ,,
(3)	Island of Lewis	127 ,,
(4)	North Uist	141 ,,
(5)	South Uist	211 ,,

We must realize, then, that during the latter third of the eighteenth century, kelping, an operation then general from the Mull

[1] *Statistical Account*, Vol. 7, p. 539.

[2] D. F. Macdonald: *Scotland's Shifting Population, 1770–1850*, p. 144.

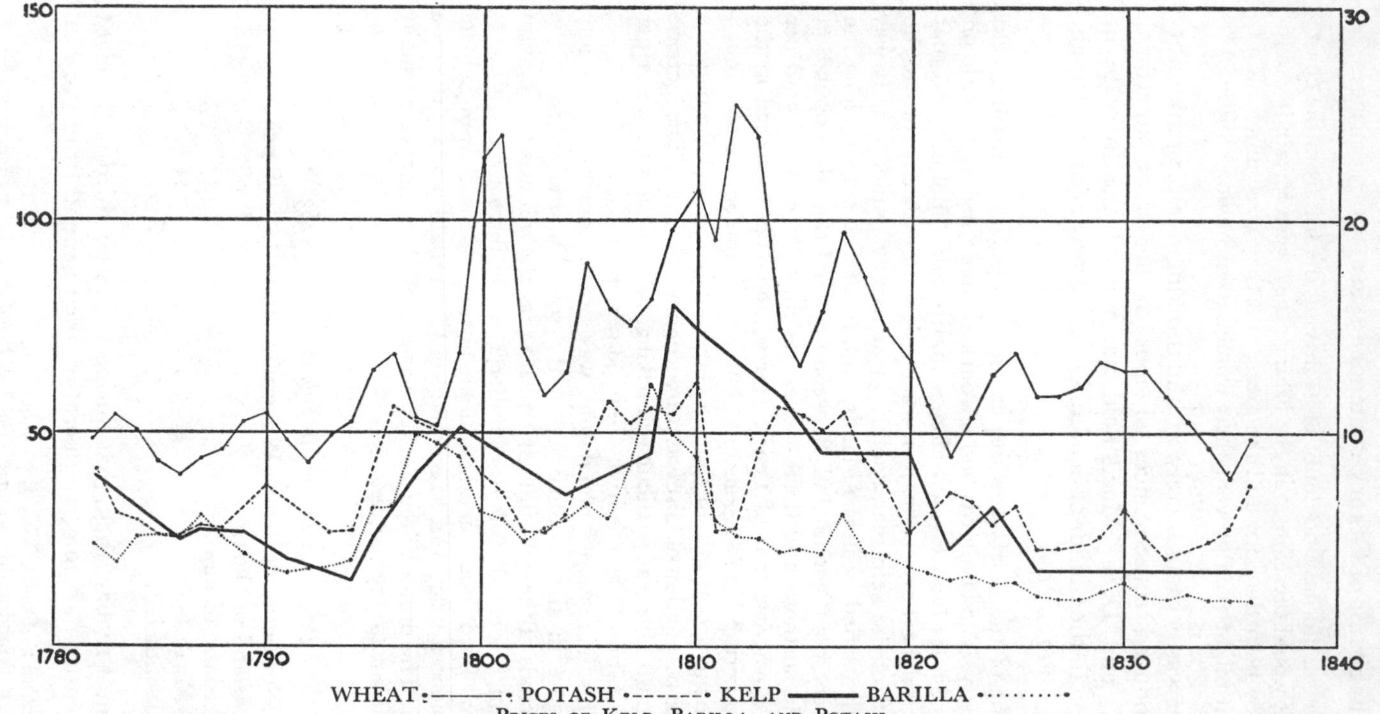

PRICES OF KELP, BARILLA, AND POTASH.

The above graph for the price of kelp is based as far as possible upon prices actually paid, not upon estimates or averages. The principal year to year source of prices was the Lochbuie and Mey Papers in General Register House, Edinburgh, and the Session Papers, Signet Library, Edinburgh. Other sources taken into account were the Old and New Statistical Accounts, the General Views of the Agriculture of the Highland Counties, Etc. From 1780–1800 more or less year to year prices were obtained: 1800–1808 is somewhat conjectural, as is also 1812–1818. The other graphs give the prices of barilla and potash from Tooke's History of Prices. For comparison the price of wheat is included.

of Kintyre to Cape Wrath, became a factor of considerable importance in the economics of the Highlands and Islands of Scotland.

Sir John Sinclair's *Statistical Account of Scotland*, for which the material was being collected in the 1790's, is full of references to kelping, and it is instructive to survey rapidly the activities on the shores of Scotland as described therein. Kelping was carried on in practically every littoral parish in Scotland, where, according to Walker, the four following plants supplied practically the whole of the raw materials. They were variously called:

(*a*) *Fucus vesiculosus:* common sea wrack, sea oak or black tang.
(*b*) *Fucus nodosus:* knotted sea wrack, bell wrack or yellow tang.
(*c*) *Fucus serratus:* serrated sea wrack or ware.
(*d*) *Fucus digitatus:* tangle.

In Berwickshire, where winter storms cast large quantities of ware upon the shores, the coarsest parts were used for manuring the land, and only the finer parts in making kelp.[1] In general the ware which was cast upon the shore was used for manure, the kelp being made exclusively from cut seaweed. This was the practice all the way round the coast to Forfarshire.[2] Where cutting was practised, it was carried out at two to three year intervals. Seemingly this gave the best results.[3] On the steep, rocky shores of Forfar and Kincardine, where little space was left by the receding tide, the collecting of seaweed for kelp was extremely difficult, but at Nigg, just outside Aberdeen, the quantities produced began to increase again.[4] Here, in 1791, kelping was carried on by thirty-three women under an overseer.[5] The shores of Aberdeen and Banff produced about a hundred tons between them, and it was from Aberdeenshire that the Orcadians learned the art of kelping.[6] From the Moray Firth area there are few records of kelping, but farther north again it flourished, the *redware* or *bellware* being cut and burned every two years.[7]

When one comes round to the islands in the west, kelp becomes the staple industry. Such was the vigour with which kelping was carried out that arable land degenerated owing to diversion of what had been the manure of the islands.[8] From the island of

[1] *Statistical Account,* Vol. 3, p. 119; Vol. 12, p. 46.

[2] *Ibid.*, Vol. 1, p. 112; Vol. 3, p. 194; Vol. 4, pp. 331 and 375; Vol. 14, p. 552.

[3] *Ibid.*, Vol. 2, p. 428; Vol. 3, p. 78.

[4] *Ibid.*, Vol. 6, p. 198; Vol. 7, p. 346; Vol. 12, p. 217.

[5] *Ibid.*, Vol. 7, p. 207.

[6] *Ibid.*, Vol. 1, p. 471; Vol. 6, p. 2; Vol. 11, p. 411; Vol. 15, p. 395; Vol. 16, p. 552.

[7] *Ibid.*, Vol. 3, p. 577; Vol. 6, p. 318.

[8] *Ibid.*, Vol. 10, p. 359; Vol. 13, p. 294; Vol. 16, pp. 155 and 223.

Barra alone, two hundred tons of kelp were exported annually to Leith and Liverpool. The precipitous shores of Argyll were not conducive to kelping, but from the south-west of Scotland, e.g. Kirkcudbright, considerable quantities of kelp were sold for making soap and bottles.[1] Dr. Richard Pococke speaks of the salting of sea fowl with kelp ashes in the islands around Skye.[2]

However, it was from the Orkney Islands that the largest quantities of kelp came, and particularly from Sanday, which usually produced about one-quarter of the whole output of these islands.

The great progress of the bleaching industry in Ireland first gave rise to the manufacture of kelp in that kingdom. From there it was transferred to the Hebrides, as explained above. In Great Britain kelp served as raw material for a wide range of manufactures. Demand for kelp for manufacture of glass and soap at Newcastle, and of alum at Whitby, where Lord Normanby's alum works used kelp from Orkney till as late as 1845,[3] seems to have encouraged the making of it on the Firth of Forth. Kelp found a ready market in Newcastle and the east of England, where crown glass manufacturers had a particular preference for supplies from Orkney. Liverpool manufacturers, on the other hand, preferred Highland kelp.[4] It is certain that the Firth of Forth glass- and soap-makers initiated production there, small quantities of glass having been made with local kelp and sand at Prestonpans by the time John Ray visited that district in 1661.[5] By the end of the eighteenth century, Dumbarton glass-works alone was consuming 1,200 tons per annum.[6]

It is now necessary to consider the economic and social effects of the kelp trade. First, it must be admitted that it is very difficult to arrive at a definite estimate of the quantity of kelp produced on the northern shores and at the numbers employed in making it. Kelping was a seasonal occupation and estimates vary greatly. As a result of 'pretty extensive observation and correspondence', James Walker in his *Essay* arrived at the following figures for the annual production over the period 1764–72[7]:

Hebrides (excluding Orkney)	1,890 tons
Orkneys	3,000 ,, [8]

[1] *Statistical Account*, Vol. 6, p. 177; Vol. 11, p. 13.
[2] R. Pococke, *Tours in Scotland*, p. 93.
[3] *New Statistical Account*, Vol. 15 (2), p. 91.
[4] *Statistical Account*, Vol. 7, p. 454.
[5] John Ray, *Itinery*, 1661; G. Chalmers, *Caledonia*, Vol. 2, p. 196.
[6] *Ibid.*, Vol. 3, p. 897.
[7] J. Walker, *Trans. Highland Soc.*, 1799, *1*, 6.
[8] Probably an over-estimate.

In 1814 Sir John Sinclair in his *General Report of the Agricultural State and Political Circumstances of Scotland* tells us that the quantity made in the Hebrides was 5,000 tons and that about an equal quantity was made on all the other shores including Orkney. Thus his total in the *General Report* is about 10,000 tons.[1] Walker ventures the additional information that a man made about a ton and a half of kelp in the three-month working season from which we can deduce that the number employed was less than 10,000.[2] Sinclair's estimate is 10,000. If we take the price of kelp as £10, on an average the sum received annually as a result of the kelp trade would be £100,000. Sinclair estimates the cost of production at one-third of this.

The *General View of the Agriculture of the Hebrides* (1811) gives figures that show a considerable but not unreasonable increase on those given by Walker and in the *General Report*, i.e. for the Hebrides alone:

> The average quantity of kelp made in the Hebrides for the past ten years, is from 5,000 to 5,500 tons; and the number of persons employed in it amounted to nearly 8,000. The price has been sometimes as high as *twenty guineas* a ton; but, upon the whole, £16 is a fair average. The money made by the district by this manufacture, therefore, has amounted to from £80,000 to £88,000 per annum.[3]

Similar corroborative evidence is found in John MacCulloch's *Description of the Western Islands*, published in 1819. John MacCulloch was a geologist of repute who made extensive explorations of the Scottish Highlands from 1811 onwards. His evidence is that the output of kelp from the Western Islands varied from 5,000 to 6,000 tons per annum. Regarding price and labour charges he says:

> The variations in the price of this article, resulting from the varying competition of foreign commerce, are very considerable; and as the total expense of manufacture has been estimated on the average at £5 a ton, a considerable deduction must, in calculating the profit, be made from the market price, which at the time of my last visit was £10. That price has occasionally varied even to £20, causing differences of serious amount in the value of these estates.

When we come to Sinclair's *Analysis of the Statistical Account* published in 1831, the figures given are of another order altogether. It seems worth while to cite *in extenso:*

[1] J. Sinclair, *General Report*, App. 2, p. 309.
[2] J. Walker, *op. cit.*, 1799, *1*, 19.. [3] J. Macdonald, *op. cit.*, p. 533.

In the space of fifty years, the proprietors of these remote islands (the Orkneys), where the seasons are often inclement, and the crops precarious, received, in addition to their former produce, £370,000. Such a sum, poured into the country where money was scarce, and where there was hardly any other kind of manufacture, soon produced consequences, obvious to the most superficial observer.

It is calculated that, in the Orkney Islands alone, 20,000 persons now depend on the manufacture of kelp for their subsistence. The introduction of this branch of industry has likewise prevented the ruinous effects which must otherwise have resulted from the failure of the corn crops, during six or seven years of scarcity, with which the Orkneys were afflicted.

In a memorial to the Lords of the Treasury, by the proprietors of kelp in the Western Islands, dated 30th December, 1822, it is urged that the making of kelp is the principal source of industry to which these islands are adapted: that no fewer than 80,000 individuals are personally engaged in it, besides at least an equal number indirectly supported by their labour: that to these must be added the seamen of perhaps 200 vessels, freighted with it to different parts of the kingdom.

The total quantity of kelp, produced in a favourable season, over the whole kingdom, is estimated at 20,000 tons; and on an average of 23 years, ending in 1822, the price was found to be £10 9*s*. 7*d*. per ton. The total value of this manufacture, therefore, may be stated at £200,000.[1]

Thus, there is a two to one ratio in the estimates of the annual value, but a ten to one ratio in the estimates of the numbers employed.

In the trade some hundred and fifty to two hundred vessels were employed transporting the kelp to Leith, Liverpool, Newcastle, etc.[2]

Several circumstances contributed to control the cost of producing kelp, for example the cost of labour, difficulties in getting plants cut, and particularly price and availability of imported barilla. Labour costs varied from district to district. Women kelpers were paid 8*d*. per day at Nigg, Kincardine.[3] Both women and boys in Orkney received 7*s*. 6*d*. per month, or 2*d*. per tide.[4] On the shores at Inverness, labour cost 35*s*. to 50*s*. per ton of kelp produced, £4 in the Hebrides, and up to £5 10*s*. in Ayrshire.[5] These figures are probably high. Walker says that in

[1] *Analysis of the Statistical Account*, Vol. 1, p. 344.
[2] *Miscellaneous Pamphlet No. 7, passim.*
[3] *Statistical Account*, Vol. 6, p. 198. [4] *Ibid.*, Vol. 7, p. 588.
[5] Sinclair, *General Report*, Appendix, Vol. 2, p. 309.

1764, on North Uist where the kelp was very easily prepared, it was manufactured for a guinea a ton, and for 30*s.* on other islands.

The method whereby production was organized varied from place to place. In Orkney two men rented a track of shore and paid so much per ton of kelp produced, hiring labour to gather and burn weed at 8*s.* to 15*s.* per month and 3–4 stones of meal.[1] In Shetland, shores were generally let to local farmers who made the kelp and received from the proprietors 40*s.* per ton for it.[2] On the estate of Lord Reay in Sutherland, a Peterhead Company had a rent of the kelp shores. Small tenants near the shores manufactured kelp at 30*s.* per ton delivered, but found that they could ill afford the time for such activities, and one has to bear in mind the considerable margin between the price paid by the kelp proprietor for getting kelp burned and the price he secured from the purchasing manufacturer.[3]

> How profitable kelp must have been to the proprietors can be judged from the fact that the tenants, who were bound by a condition of their tenure to make the kelp, were receiving for their labour about £2 5*s.* a ton at times when the kelp was selling at from £16 to £22 a ton, practically the entire difference, less freightage, representing profit to the estate. Moreover, on the basis of this profitable employment, the proprietors were actually able to raise the rents, thus making a double gain.[4]

Indeed it is small wonder that the inhabitants of some islands were slow to set themselves to the task of supplying raw materials for emergent palaeotechnics. According to Pennant:

> About Troup Head, some kelp is made; and the adventurers pay the Lord of the Manor £50 per annum for the liberty to collect the materials.[5]

In Berwickshire, manufacture of kelp on the shores took 'the commodious place of the former salt works'.[6] In Wigtown, Sir J. Maxwell let his shores for £100 per annum, and kelping greatly increased the value of his estates.[7] The writer of the *Statistical Account* for Sanday says, 'An estate which seventy years ago was worth £40 sterling a year is now worth £300 yearly'.[8] The shores

[1] Walker, *op. cit.*, p. 9. [2] *Statistical Account*, Vol. 7, p. 588.
[3] *Ibid.*, Vol. 6, p. 286. [4] J. Sinclair, *General Report*, App. 2, p. 309.
[5] Thos. Pennant, *Tour in Scotland in 1769*, p. 123.
[6] G. Chalmers, *op. cit.*, Vol. 2, p. 317.
[7] *Statistical Account*, Vol. 17, p. 562. [8] *Ibid.*, Vol. 7, p. 454.

ASHES and KELP

Potash (K_2CO_3) and Soda (Na_2CO_3)

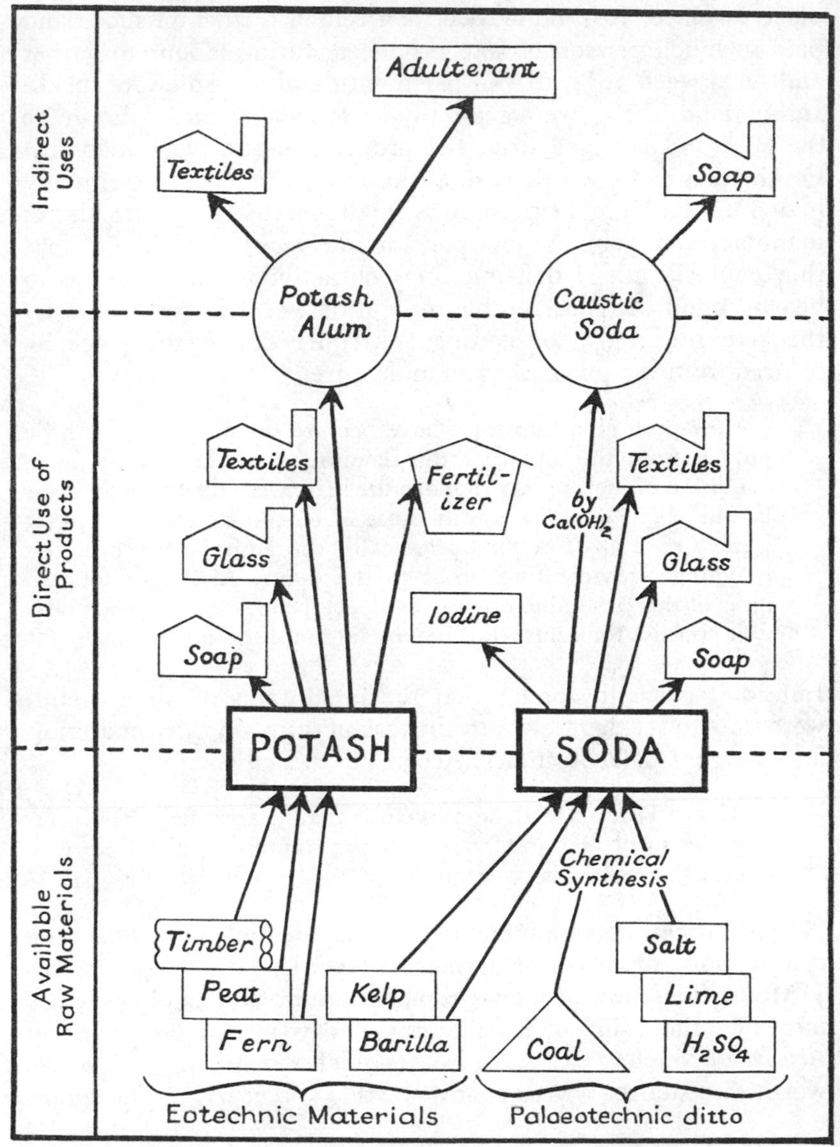

of North Uist were let during the Napoleonic Wars for £7,000 per annum, when barilla was unobtainable.[1]

Throughout the eighteenth century the price of kelp maintained an upward trend, modified from time to time by political events.[2] Before 1730, the original Orkney kelp had been sold in Newcastle for £1 5*s*., but by 1764, Hebridean kelp was fetching twice that figure, and during the American War the price reached £8, falling again to £6 10*s*. at Newcastle in 1788.[3] In 1791 the price rose again, and was £11 12*s*. per ton in 1798, the kelp boom reaching its height in the opening years of the nineteenth century. These probably represent peak rather than average prices, as the average price for the twenty-three years ending 1822 was £10 9*s*. 7*d*.[4]

When estimating the quality of kelp, the early merchants had to rely upon rule-of-thumb methods. 'It is estimated to be of good quality', says Angus Beaton, 'when on breaking a piece, it is found to be hard, solid, and resembling good indigo, that is, when it has some reddish and light blue shades running through it. When it has none of its peculiar salt taste, it is unfit for making ley, though it may be of use to glass makers'.[5] Advances in analytical chemistry during the eighteenth century soon revealed that such methods were useless. Dr. Richard Kirwan (1733–1812), the celebrated Irish chemist, was the first to ascertain the quantity of alkali in kelp with any degree of accuracy.[6] Dr. Joseph Black, successor to Cullen in the Chair in Edinburgh, devised a suitable method of analysis which revealed that high prices were frequently given for kelp of inferior quality. This is shown in the following table[7]:

Source	*Price per Ton*	*Per cent. Alkali*
Colonsay	£12 12 0	6.19
,,	£12 0 0	4.84
Mull	£11 11 0	0.81
Barra	£11 0 0	6.94
,, (driven ware)	£8 0 0	1.11
Loch Maddy	£8 10 0	1.21
,,	£9 0 0	1.41
Shetland	£8 5 0	3.09

[1] J. R. McCulloch, *Commercial Dictionary*, Vol. 1, p. 738.
[2] *Statistical Account*, Vol. 7, p. 539.
[3] *Ibid*., Vol. 13, p. 305; R. Jameson, *Observations on Kelp*, *Trans. Highland Society*, 1799, *1*, p. 44.
[4] *Miscellaneous Pamphlet* No. 7, *passim*.
[5] A. Beaton, *Trans. Highland Society*, 1799, *1*, 39.
[6] Richard Kirwan, *Trans. Roy. Irish Acad.*, 1789, *3*, 3.
[7] A. Fyfe, *Trans. Highland Society*, 1820, *5*, 29.

Kirwan also compared the percentage of alkali in kelp and barilla:

Barilla from Alicante	. .	23·5	per cent.	alkali
,, ,, Teneriffe	. . . .	8·5	,,	,,
Kelp from Scotland	. .	2·5–5	,,	,,
,, ,, Norway	. . .	2·75	,,	,,

Even allowing for the considerable difference in the alkali content of the two barillas, British consumers found it to be to their advantage to procure barilla from Teneriffe rather than Alicante when the Spanish resorted to adulterating their barilla in times of high prices.[1] When one examines the figures for the Scottish and Norway kelp, they hardly support the contention of the Orkney kelpers who complained of the good quality and low price of Norway kelp introduced into Scotland and sold to the glass-makers at less than £5 per ton.[2]

When a complete analysis of kelp and barilla became possible, price could be worked out in terms of saleable articles which barilla and kelp contain.[3] In this way Parkes[4] arrived at £28 5*s*. 9*d*. for barilla, which was then selling at £30 in Bristol, £14 19*s*. 2*d*. and £11 0*s*. 2*d*. for kelp from Skye and Lewis respectively, and £12 1*s*. 7*d*. for Irish kelp, the latter figure giving the lie to the remarks of the Orkney kelpers, that Irish kelp wasn't worth taxing, its quality was so low.[5]

What appears to be an early use of volumetric analysis (i.e. analysis carried out in solution with solutions of known concentration, a method which in industry has progressively replaced other methods on account of the speed with which it can be carried out) is contained in a paper by Charles Tennant on estimating barilla with a solution of sulphuric acid, using an extract of radishes as an indicator. The acid was added to a solution of the barilla from a tube graduated by five-grain divisions. A table to give the percentage of soda directly from the amount of acid added is included.[6]

Chemical analysis also enables one to comprehend the uses to which the component parts of kelp had been put. 'Kelp is a lixivial salt,' says Walker, 'and is always mixed with other salts, sea salt, Glauber's salt, magnesia salts, etc.'[7] Put in the language of our

[1] S. Parkes, *Trans. Highland Society*, 1820, *5*, 66.

[2] Letter from Kirkwall to Henry Dundas, 13th April, 1791. N. L. S., MS. 640, f. 65.

[3] A. Fyfe, *op. cit.*, 1820, *5*, 10.

[4] S. Parkes, *op. cit.*, 1820, *5*, 100.

[5] N. L. S., MS. 640, f. 65 quoted above.

[6] Charles Tennant, *Annals of Philosophy*, 1817, *10*, 114.

[7] J. Walker, *op. cit.*, 1799, *1*, 10.

9. View of Salt Works at Joppa. Some of the many other salt-pans on the Firth of Forth can be seen in the background.

10. Interior of salt-mine in Cheshire, illustrating the enormous economic advantage of mining salt over evaporation of sea-water.

Kay's *Edinburgh Portraits*

11. A Hawker of Salt. Calling daily at the salt-pans at Joppa, she purchased salt and hawked it at sixpence a *caup*.

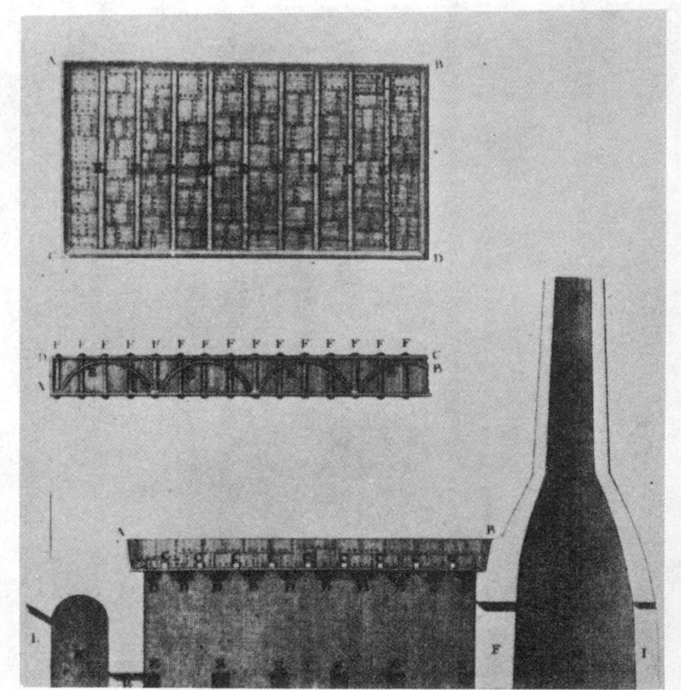

G. Jars, *Voyages métallurgiques*, 1774

12. Plan and elevation of salt-pans. the middle section is of the salt-pan at Kinneil.

own time, this amounts to saying that kelp is a mixture of Na_2CO_3, $NaCl$, Na_2SO_4, $MgSO_4$, $MgCl_2$, etc. These additions to the soda may, however, be harmless or even desirable. The 25 per cent. of sodium chloride ($NaCl$) which many samples of kelp contained was not regarded as impurity. It was essential in any case for the manufacture of soap and could be obtained from kelp much cheaper, on account of the salt duties, than if bought as salt. Similarly, glass-makers were not exercised by the presence of salt unless a very high percentage was present, and in any case the salt which rose to the top of their pots was skimmed off and resold as *sandiver*, fetching £8 to £10 per ton. Even the insoluble residue of kelp was valued as a manure.[1] It was, however, only the soluble part of kelp which was of value to bleachers and soapers. The use of kelp in the former art was introduced into Scotland by an Irishman, R. Holden, who was assisted by the *Honourable the Board of Trustees* to start a bleachfield at Dundee.[2]

As used by soapers, the kelp was reduced to powder, dissolved in water, and then rendered caustic by the action of lime (calcium hydroxide, $Ca(OH)_2$), after which it was well stirred and filtered.

$$\underset{\text{kelp}}{Na_2CO_3} + \underset{\text{lime}}{Ca(OH)_2} = \underset{\text{caustic soda}}{2\,NaOH} + \underset{\text{chalk}}{CaCO_3}$$

Thus the value of kelp to the soaper lay principally in the quantity of alkali it contained, but the value rose more steeply than in direct proportion to the quantity of alkali contained, since a greater quantity of alkali was obtained from rich kelp than from a poor one at much less expense in labour and utensils.

According to Dunbar[3] kelp and hemlock ashes were more suitable for making glass than soap. Since the glass-maker valued the kelp quite differently from the soaper, the admixture of the different neutral salts, which he used as a powerful flux, constituted its value to him.[4] Thus barilla was found to be no better than kelp, as the extra alkali was only offset by increased price.[5] Usually the entire kelp was used for glass, but sometimes the waste of the soap-makers was employed instead of the raw kelp. It was broken into small pieces, the stony matter picked out, and the remainder powdered by large stone rollers. The sand and kelp were mixed and calcined to remove the carbon and sulphur, care

[1] A. Fyfe, *op. cit.*, 1820, 5, 52.
[2] S. H. Higgins, *A History of Bleaching*, p. 17.
[3] J. Dunbar, *op. cit.*, p. 2.
[4] A. Fyfe, *op. cit.*, 1820, 5, 30.
[5] S. Parkes, *op. cit.*, 1820, 5, 106.

being taken not to overheat. It was then cooled, the resulting mass being known as *frit*. To convert it into glass the frit was heated, for two days, in 20-cwt. quantities, till it became clear under a scum of *glass gall* or *sandiver*.

Even these do not exhaust the uses to which the kelp was put. In addition to sodium salts, kelp contains about 3 per cent. of potassium chloride (KCl). Parkes points out that if it had been expedient to form nitre beds for production of saltpetre (potassium nitrate, KNO_3) in Great Britain, potash obtained from kelp would have been a source of great national importance. The principal users of the potash in kelp were the alum-makers, who bought evaporated lye from the soap-makers. In the manufacture of alum ($K_2SO_4.Al_2(SO_4)_3.24H_2O$ or $(NH_4)_2SO_4.Al_2(SO_4)_3.24\,H_2O$), either potash or ammonia had to be added to the aluminous liquor. Previously sal enixum (potassium sulphate, K_2SO_4) from the Aqua-fortis (nitric acid) works was used, but when this became too dear, the residue from kelp was found to answer even better, as it also precipitated the iron present as impurity and made the alum better for the dyers, etc.[1]

One use of kelp remains for discussion. Very little of the seaware which was cast upon the shores by storms was converted into kelp. Much of it was used as a manure. This, rather than natural richness of soil, says a writer in the *Statistical Account of Haddington*,[2] was responsible for the high price of land in the Lothians. Recommendations for seaweed as a manure came from the Lothians, Fife, Sutherland, Shetland, Argyll[3] and the islands in the west. Barley was the crop for which ware (i.e. seaweed in general) was principally used,[4] with which it appears to have produced excellent results. With potatoes, results seem to have been less striking.[5] The ease with which ware could be used as a manure tended to make the farmers in the Orkneys sow a grain crop year in year out, and this was presumed to have much impeded the introduction of the improved methods of agriculture.[6] Although there was a general opinion that effects of ware only lasted for a season,[7] it was used in some districts in preference to dung from compost heaps,[8] because it acted speedily and gave 'clean abundant crops'.

[1] A. Fyfe, *op. cit.*, p. 111. [2] *Statistical Account*, Vol. 1, p. 122.

[3] *Ibid.*, Vol. 3, p. 194; Fife, *ibid.*, Vol. 3, p. 78; Sutherland, *ibid.*, Vol. 3, p. 577; Shetland, *ibid.*, Vol. 1, p. 390; Argyll, *ibid.*, Vol. 14, p. 181.

[4] *Ibid.*, Vol. 1, p. 471; Vol. 2, p. 548. [5] *Ibid.*, Vol. 8, p. 93.

[6] *New Statistical Account*, Vol. 15 (2), p. 162.

[7] *Statistical Account*, Vol. 2, p. 428.

[8] *New Statistical Account*, Vol. 15 (2), p. 145.

The trade in kelp brought large sums of money, something like £120,000 per annum, into the Highlands and Islands of Scotland. Forty thousand pounds were expended in producing the kelp, leaving a profit of £80,000.[1] In a fifty-year period the Orkneys are said to have received no less than £370,000 sterling.[2] From a superficial examination one gets the impression from contemporary literature that the introduction of the art of kelping was a great boon to the unindustrialized Highlands and Islands. Yet now and then a discordant note creeps through, as here:

> While we trace with pleasure the advantages which it has produced, candour forbids us to conceal its disadvantages. Owing to the kelp manufacture, every species of provisions has greatly increased in price, which makes it difficult for those to live who have only fixed incomes: wages are much higher: agriculture, which in every country is the first and most necessary of all arts, is greatly neglected; and a style of living has been introduced among the proprietors, which their lands can by no means support, and which, if ever this manufacture should fail, must bring certain ruin upon them, their tenants, and their families.[3]

The actual gain per kelper, however, did not amount to more than a few pounds per annum, but even this small amount of ready money is said to have been readily accepted to meet steadily increasing rents. The increase in rents was phenomenal, e.g. in '. . . Orkney, where within these seven years past, small farms about £40 rent, have risen to £300 a year. . . .'[4]

Sinclair comments thus on the value of such a small sum of money to the kelpers:

> It may appear surprising, how that sum of money could be of such importance to 80,000 people; but it is to be considered, that it is acquired by the labour of a few weeks—that it is the only means, by which the inhabitants of these countries can convert their labour into money;—that they usually have small farms which furnish them with the necessities of life;—that the small sum which they derive from kelp enables them to pay their rents and supply themselves with any superfluities, for which they may have occasion; and that since even so small a sum as £8 to £10 per annum is sufficient, in their simple state of life and habits of frugal management, to enable a cottager or small farmer to

[1] Sir J. Sinclair, *General Report*, Appendix, Vol. 2, p. 301.

[2] *Miscellaneous Pamphlet No. 7, passim.*

[3] *Statistical Account*, Vol. 8, p. 539.

[4] R. Jameson, *Mineralogy of Shetland Isles and Arran, with an Appendix on Kelp, etc.*, p. 187; *New Statistical Account*, Vol. 15 (2), p. 91.

maintain himself and a family of four children, the deduction of £2 or £3 from an income so limited, would be distressing in the extreme.[1]

It was however not always possible to extract even this small sum from the inhabitants, and in some districts labour at kelping was taken in lieu of rent, the proprietor disposing of the kelp himself.[2] The occupation of kelper, too, appears to have been no sinecure. According to a writer in Lewis:

> The toil in cutting, drying, and burning the ware and watching the pot night and day till the ware is converted into boiling lava is terrible, and would require extraordinary wages.[3]

Any benefit which is mentioned is usually qualified by corresponding disadvantages such as increased cost of food.[4] In a case which was brought by a Peterhead Company against a parish minister who rented a small farm, the Court of Session decided that the kelp on the shores of the farm belonged to the Company. The result was that the farm was deprived of manure.[5] The high prices offered for kelp were a great temptation to burn ware instead of applying it to the land. Sometimes only the *overplus* was used for kelping, but very often it was argued that it was more profitable to sell the bulk of it.[6] This was doubtless true in some cases, but even in Barra kelping was described as a great obstacle to improving land.[7]

According to D. F. Macdonald, 'In the eighteenth and nineteenth centuries the landlords in the north-west and in the Hebrides were bitterly opposed to the emigration of their crofters. They exerted their influence to have passed the Passenger Acts of 1803, imposing restrictions on the emigration trade . . . which raised the cost of transport and prevented many from leaving the Highlands.'[8]

Even when the boom was at its height, it was not infrequently pointed out that the basis of the industry was factitious.[9] Vast quantities of kelp were exported from the islands, whose soil remained barren and unable to produce sufficient sustenance for the inhabitants, a state of affairs which could in all likelihood have been rectified by an application of the sea-ware to the island

[1] Sir J. Sinclair, *Analysis of the Statistical Account*, p. 344.
[2] *New Statistical Account*, Vol. 14 (1), p. 353. [3] *Ibid.*, Vol. 14 (1), p. 134.
[4] *Statistical Account*, Vol. 7, p. 539; Vol. 10, p. 359.
[5] *Ibid.*, Vol. 6, p. 286. [6] *Ibid.*, Vol. 4, p. 131; Vol. 7, p. 127.
[7] *New Statistical Account*, Vol. 14 (1) p. 213.
[8] D. F. Macdonald, *Scotland's Shifting Population*, p. 145.
[9] *Statistical Account*, Vol. 15, p. 395.

soil. This was occasionally realized, and in at least one island on the west coast the proprietor forbade all kelp-making so as to preserve the sea-ware as manure for the land.[1]

Perhaps the severest contemporary critic of the economic policy of kelping is James Macdonald, who wrote in his *General View of the Agriculture of the Hebrides* that:

> . . . the great body of seafaring Hebrideans are now metamorphosed into slavish kelpers; that the sea weeds, which constitute the only attainable manure for improving the soil of one half of the isles are lost to the ground, and perverted to a manufacture which indeed brings some money in the meantime to the proprietors, but which in the long run, by impeding all agricultural improvements, must prevent the rapid advance which these estates would otherwise make, and must redound to the severe loss of the landlords; that to sacrifice the soil to the precarious supply of a commodity which fluctuates yearly in price, and for which some cheap substitute may soon be invented, must eventually ruin both landlords and tenants; in short, that this most laborious and slavish mode of employment, destroys the bodies and minds of the people, without essentially benefiting the estates themselves.[2]

In the *New Statistical Account* it says that:

> Although this art of manufacture was a source from whence the Highland proprietors derived great benefit while it sold at high prices and employed the people at remunerative wages, yet it has turned out in the end ruinous to both proprietors and tenants, whose dependence has been much placed upon it.[3]

From the Orkneys comes the same opinion:

> We only remark that proprietors evidently placed more balance than they ought to have done upon the manufacture of that precarious art, and thereby neglected the improvement of their lands, which would have offered them a more permanent benefit.[4]

By the middle of July 1814 the boom had passed. It was stated at a general meeting of the Highland Society that the price of kelp had fallen so much as to offer no adequate inducement for manufacturing it.[5] There may have been two causes for this. One was the Treaty of Paris, which assured the re-opening of the barilla trade. The other was the appearance on the market of synthetic alkali made from sea-salt. More will be said about this in Ch. IV. The Highland Society strove to stem collapse by

[1] J. Walker, *op. cit.*, p. 16. [2] J. Macdonald, *op. cit.*, p. 140.
[3] *New Statistical Account*, Vol. 14 (1), p. 213. [4] *Ibid.*, Vol. 15 (2), p. 32.
[5] *Trans. Highland Soc.*, 1816, *4*, XIV.

offering premiums for improvements in the manufacture of kelp.

> The Society were sufficiently aware of the true principles of political economy, to make them cautious of interfering, as had sometimes been proposed, with the importation of the foreign article, which is the rival of kelp, the barilla; but they conceive it to be an object equally legitimate and useful, to endeavour to raise the value of the native kelp, so as to make this country more independent of that foreign importation than it had hitherto been, and therefore appointed a committee . . . who circulated a set of queries.

Awards of silver plate were actually made to Dr. Andrew Fyfe, Jr., of Edinburgh, and Samuel Parkes of Gorwell Street, London, for the results of their experiments.[1] Strangely enough there is only one reference to improvement in the art of kelping as a result of the Society's activities.[2]

Not that the proprietors had gone unwarned of the possible depression which might overtake the industry. Writing in 1804, Dr. Patrick Neill, of Edinburgh, warned them that:

> . . . should a cheap process for extracting soda from sea salt happen to be discovered or should the market for kelp on any other account unexpectedly fail, the landowners of Orkney will find when too late the great imprudence of thus neglecting the cultivation and improvement of their lands.[3]

Neill's words were prophetic. The conversion of sea-salt into soda was accomplished, and as a result the kelp industry rapidly declined and passed away. That the inhabitants of the Highlands did not by any means have a clear idea of the causes of the failure of their staple trade is revealed by a study of the *New Statistical Account*. As often as not the blame is laid on the government for removing the import duty on barilla and *other substances*. The reference to other substances is significant.

> Before barilla was allowed to enter our market duty free, and thereby exclude the kelp, there were no less than 150 tons annually manufactured in this parish, and of course it, in common with the Highlands in general have felt, and do still feel, the loss; for kelp cannot now be sold so as to bring the wages of the workers, as the former consumers of kelp prefer to work from barilla and other substances admitted duty free. In previous years, this manufacture employed and gave bread to many thousands in the Highlands and Islands, and the price it drew brought money to

[1] *Trans. Highland Soc.*, 1820, 5, XLVI.
[2] *New Statistical Account*, Vol. 7 (2), p. 57.
[3] Patrick Neill, *Tour through Orkney and Shetland.*

> the country, and this being again circulated through the kingdom at large kept that money at home, which now goes to enrich the foreigner at the poor Highlander's expense; a measure of policy which cannot be too strongly condemned—for whether it arose from ignorance on the part of the Government, or from any other cause, the Highlands have, since the admission, duty free, of barilla and other substances, presented scenes of much distress, bankruptcy, and poverty.[1]

Occasionally the economic effect of the successful conversion of sea-salt into soda was understood. This was clearly in the mind of the reporter from South Uist who wrote that:

> . . . because of the removing of the duty on salt and sulphur, the income of these islands is reduced from £15,000 to £5,000 per annum, and the manufacture of kelp is only continued to pay the rents.[2]

Here there is no reference to the removal of the tax on barilla, or to competition arising from importation, but to the removal of the tax on salt and sulphur, these being the materials which enabled a synthetic product to be made and substituted for natural ashes which had been obtained from plants for many centuries before. Another change, the use of sulphate of soda in the manufacture of glass, seems to be hinted at in another report:

> The use of kelp in the manufacture of glass has been superceded by Spanish barilla, brought in at low duty, and still more, it is said, by new chemical discoveries in the art of glass-making.[3]

How the inhabitants of the Highlands and Islands clung to kelping is revealed in the following letter of J. A. Banks, dated December 1817:

> The land rent of this county is in some cases inadequate for the payment of our feu-duties to Lord Dundas, who as Tacksman or Donatory of the Crown holds the rents, which by their (passing?) from the Crown into the hands of an Individual have by degrees been raised from a sum equal to what in other parts of Scotland is paid to the Crown as Superior, to what in fact now exceeds in many cases the total produce of the land . . . and the Landholder would thus be obliged to relinquish his property were it not for the profits of the Kelp burned on the shores. . . . By Kelp and Kelp alone this remote county is emerging from poverty, if this fails, decided and inevitable ruin must fall on every class of its inhabitants, and I do not hesitate to assert that in ten years, this

[1] *New Statistical Account*, Vol. 7 (2), p. 309.
[2] *Ibid.*, Vol. 14 (1), p. 194. [3] *Ibid.*, Vol. 14 (1), p. 92.

country now in an improving and comparatively with former times a flourishing condition, one of the best nurseries for seamen to the British Navy and also furnishing a yearly increasing revenue to the Government, would be a desolate and uninhabited waste.[1]

Yet occasionally a preference for kelp was expressed many years after the collapse of the trade. For example, Richard Baxter and Co. of Whitehaven wrote to Mull in 1835 informing the proprietor that:

> . . . since our last kelp from you we have been trying a variety of alkalis—home made of various kinds and Barillas. We still think that a portion of kelp assists the others and if you continue to manufacture shall be glad to have from you on what terms you will ship us a cargo of your very best kelp. . . . You must name your lowest price as British Alkalis from great competition in manufactory are offered at very low rates.

In 1812, when the Scottish kelp industry was entering its phase of final depression, B. Courtois (1777–1838), a manufacturer of saltpetre near Paris, found that the copper vessels in which he decomposed the product of the *nitre beds* (calcium nitrate, $Ca(NO_3)_2$) with an aqueous extract of kelp, were subject to severe and rapid corrosion.

$$\underset{\text{Calcium nitrate}}{Ca(NO_3)_2} + \underset{\text{potash}}{K_2CO_3} = \underset{\text{nitre}}{2KNO_3} + \underset{\text{chalk.}}{CaCO_3}$$

He traced the cause of the corrosion to an unidentified substance in the kelp lye. On concentrating the aqueous solution he obtained in succession, potassium sulphate (K_2SO_4), sodium sulphate (Na_2SO_4), sodium chloride ($NaCl$), and sodium carbonate (Na_2CO_3). In the residue he discovered a hitherto unknown element, *iodine*.

> This substance, lately discovered as a compound part of kelp, is most easily procured by evaporating the solution of kelp to dryness, and pouring sulphuric acid in it, and applying heat; violet-coloured vapours are given off, which condense in dark, shining, needle-formed crystals, on the cool part of the apparatus. I have endeavoured to procure this substance from other marine plants, and from sea water, but without success. The only other body I have been able to obtain it from is sponge.[2]

[1] *Melville Papers,* N. L. S. 640, f. 216,

[2] A. Fyfe, *Trans. Highland Soc.*, 1820, 5, 14.

This helped to retain kelp as a marketable commodity, chemical manufacturers purchasing kelp as a raw material for the production of iodine, or alternatively using spent lye from the soap works. Andrew Ure of Glasgow, where iodine was manufactured for the first time in 1841, devised a method of extracting it from the brown oily residue from the soap-boilers. By 1846 twenty manufacturers there were handling iodine.[1]

Small quantities of kelp were thus made at intervals throughout the nineteenth century, but the availability of synthetic alkali removed for all time the basis of the supposed period of prosperity which the Highlands were alleged to have enjoyed. Seaweed may again become a commodity of industrial importance, as a source of alginates, not of alkali, but the economics of the trade will perforce be vastly different. No longer will twopence per tide meet the cost of collection, and all to the good, for the real effect of the kelp industry upon the economics of the Highlands and Islands of Scotland in the years following the break-up of the ancient traditions in the Celtic fringe to European civilization cannot be better summed up than in the words of W. C. Mackenzie, who wrote:

> The boom in kelp reached its height at the commencement of the nineteenth century. Landlords were making large profits which were quickly dissipated by their extravagance. Agriculture and fishing had been neglected in the feverish haste to turn kelp into money, and when the bubble burst the problem of disposing of a starving population faced the impoverished proprietors who committed the blunder of striving to maintain at the increased level which they had reached, the rents of the struggling tenantry. While the boom lasted, emigration was discouraged, in order to secure a sufficiency of labour; but when it terminated, the unfortunate peasantry, who had shared but sparingly in their master's prosperity, were involved in ruin by their adversity. They were unable to pay their rents, and a renewed exodus to America took place.[2]

NOTE.—*Kelp in the Twentieth Century.*

The situation regarding kelp in 1941 was briefly as follows. According to information from the Ministry of Agriculture in North Uist, the 1914–18 war gave the surviving remnant of the kelp industry a considerable fillip when, owing to interrupted access to the Stassfurt deposits, potash, in addition to the iodine, became of value. The amounts produced were, South Uist 700

[1] A. McLean, Editor, *Local Industries of Glasgow and the West of Scotland.*
[2] W. C. Mackenzie, *A Short History of the Highlands and Islands*, p. 312.

tons, Barra only 20. Benbecula was only beginning to contribute. The maximum annual yield was reckoned at 1,000 tons. It paid to burn kelp at £5–6 per ton, and prices sometimes reached £9 per ton. So it will be seen that even up to that late date the art of kelping had not entirely died out. The economic crisis of 1933 again put an end to it for the time being.

The purchasers of the kelp were (1) The British Chemical Company Limited, Whitecrook Chemical Works, Clydebank; (2) Messrs. H. C. Fairlie and Company Limited, Camelon Chemical Works, Falkirk; (3) Scottish Alkali and Acid Company, Longford Chemical Works, Kilwinning; (4) Milnquarten Chemical Company, Bonnybridge.

Recent developments have been in the hands of the Scottish Seaweed Research Association, founded in 1945, whose staff have engaged in an extensive ecological survey of sources of seaweed in the Highlands and Islands with a view to its utilization as a modern chemical raw material.

Chapter IV

THE PALAEOTECHNIC TRANSITION

> Leblanc died in poverty in a French hospital, but it is satisfactory to know that statues were erected to his memory at Rome and in Greece.
>
> R. C. CLAPHAM.

THE successful commercialization of a process to convert common salt (sodium chloride, NaCl) into the alkaline substance, soda (sodium carbonate, Na_2CO_3), signalizes the founding of what is usually called the heavy chemical industry. This was secondary to, and dependent on, supplies of cheap oil of vitriol (sulphuric acid, H_2SO_4) which became available during the second half of the eighteenth century in quantities sufficient for commercial operation.[1] Nevertheless, the synthesis of soda may justly be regarded as representing in essence the transition from eotechnic to palaeotechnic economy. The principal reason for this is its connection with, among others, the textile, glass, and soap industries. Supplies of soda, obtained independent of biological material, converted primitive eotechnic soap-boiling into a fully-fledged capitalist industry. Soap and soda supplied to the finishing trades detergents with which they could keep pace with increased tempo in other branches of textile operation consequent to the introduction of power-operated units. By-products from the soda synthesis supplied raw materials to revolutionize the bleaching trade[2], and made a wider variety of rags, already growing scarce, usable by the paper-maker.[3]

> The works of an alkali manufacturer tended to become larger and more complicated; he began to make soda, using common salt and sulphuric acid and other raw materials. After a time he started to make his own sulphuric acid by burning sulphur or pyrites; if he used pyrites, it was probably a mixed sulphide of copper and iron, and it was comparatively easy to make copper sulphate and ferrous sulphate from the roasted pyrites. The process of making sodium sulphate produced large quantities of hydrochloric acid, and, as nitric acid was required in the manufacture of sulphuric acid, the alkali manufacturer easily developed into a manufacturer of hydrochloric, nitric, and sulphuric acids, and various salts of sodium, copper, and iron. It was a very

[1] Ch. VI. [2] Ch. IX. [3] Ch. XIII.

common development for the alkali manufacturer to use the chlorine he recovered so as to make bleaching-powder, and in these ways he became a maker of calcium chloride and bleaching-powder, and, as demands for them grew, he made other salts of sodium and calcium required in large quantities. The manufacture of all these 'heavy' chemicals became in this way an involved process, in which one part was dependent on the others and almost every effort to prevent waste involved the manufacture of some new product.[1]

INTEGRATION DIAGRAM FOR THE ALKALI TRADE

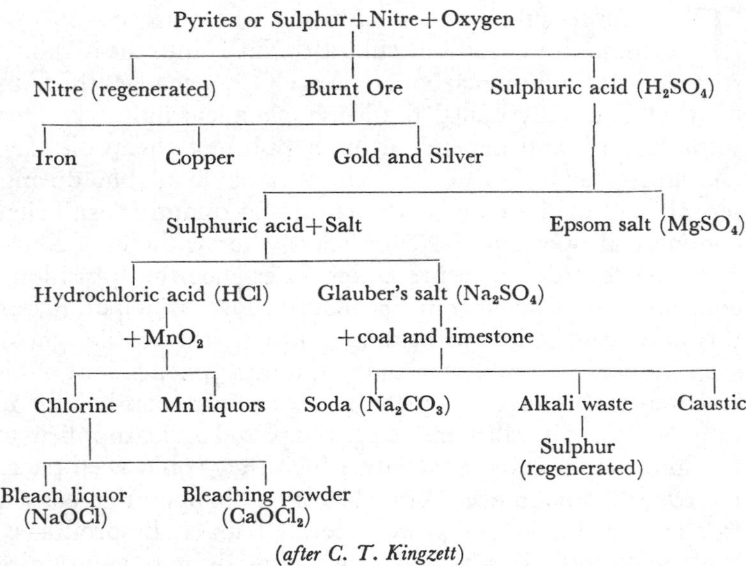

(*after C. T. Kingzett*)

Till the synthesis of the first synthetic dyestuff by William Perkin in 1856, *alkali* industry and *chemical* industry were virtually synonymous. The influence of these developments upon advance in industrial technology in the early nineteenth century was described by the illustrious German scientist, Justus Liebig:

> The manufacture of soda from common culinary salt may be regarded as the foundation of all our improvements in the domestic arts; and we may take it as affording an excellent illustration of the dependence of the various branches of human industry and commerce upon each other, and their relation to chemistry.[2]

The focal point of this development is usually taken as the founding in 1823, at the time the Salt Acts were being repealed at

[1] S. Miall, *History of British Chemical Industry*, p. 4.
[2] J. Liebig, *Familiar Letters on Chemistry*, p. 21.

the instigation of Huskisson, of a factory operating the Leblanc patent for the production of synthetic alkali by James Muspratt (1793–1885) in the neighbourhood of Liverpool.

This, however, cannot explain the fall in the price of kelp mentioned in Ch. III, which spread alarm and anxiety throughout the kelping districts of Scotland about the year 1814.

It is the purpose of the present chapter to trace the efforts which were made to solve the salt-to-soda problem between the founding, by Dr. John Roebuck, of the first sulphuric acid factory in Scotland in 1749, and the usually accepted founding of the alkali industry, i.e. the setting up of Muspratt's factory in 1823.

Details of the trade in ashes were given in the preceding chapter. The salient points were as follows. Prior to the second phase of the industrial revolution in the late eighteenth century, forests of Eastern Europe were the source of a considerable supply of *potashes* for the glass, soap, and textile industries. The eotechnic iron industry, dependent on wood as fuel, and the revolutionary state of Europe, made both home and imported ashes scarce. Thus, accompanying the process of industrial revolution, there was continual social impetus towards the production of alkalis, either from alternatives, or by synthesis. In Scotland the *Board of Trustees for Fisheries, Manufactures, and Improvements*, offered premiums for home production of suitable ashes, a problem that would have been solved by Dr. William Cullen (1712–90), Professor of Chemistry in the University of Edinburgh, had only home supplies of timber been adequate. The failure of biological materials as a possible source turned attention to salt as an alternative.

The personnel involved in the building up of an industry for the synthesis of alkali from salt falls into two distinct groups, without, so far as is known, much interconnexion: (*a*) members and associates of the Lunar Society in Birmingham; (*b*) Archibald Cochrane, Ninth Earl of Dundonald, in association with Lord Dundas and various manufacturers on the Tyne.

The first of the Birmingham pioneers to commercialize Scottish science was Dr. John Roebuck, whose initial contacts with Scotland are described by J. Lord as follows:

> Roebuck was born at Sheffield, in 1718. The son of a cutler, he was educated at Edinburgh and Leyden, and settled in Birmingham to practise as a physician. While there he met Samuel Garbett, the first of the great Birmingham men who were to have such an influence on the economic developments of the century. In conjunction with Samuel Garbett, Roebuck established a large

laboratory, in which he discovered more economical and improved processes for refining and working gold and silver, and made the important discovery of a commercial method of preparing sulphuric acid in leaden vessels, at a quarter the cost of the glass-retort method that had previously been in use.

The two partners established a manufactory of sulphuric acid at Prestonpans, in East Lothian, in 1749; the consumption of the article increased rapidly, and the undertaking was exceedingly profitable. They, therefore, added to their manufacture, and established a pottery for making white and brown ware, which was also successful.[1]

It is probable that Roebuck, while at Prestonpans, became interested in the production of synthetic soda. Associated with him were Dr. Joseph Black (1728–99), Professor of Chemistry in the University of Edinburgh, and James Watt (1736–1819), then experimenting with his separate condenser for Newcomen's engine. It has been suggested that this attempted synthesis may have been the cause of bringing Watt and Roebuck together through the intermediacy of Dr. Black, whom Watt had known when they were both on the staff of Glasgow University.[2] As we shall see later, Roebuck gave Watt accommodation for experimental work in his house at Kinneil, and was so convinced of the ultimate success of Watt's invention that in 1768 he repaid the £1,200 which Watt owed to Dr. Black. By this time Roebuck and Garbett had extended their activities by starting Carron Iron Works.[3]

Roebuck's relations with Watt are illustrated by the following citation:

The question of the coal supply now received the attention of Roebuck; he seems to have determined from the beginning to control all his raw materials. He, therefore, leased the coal-mines belonging to the Duke of Hamilton, at Borrowstoneness, as well as the salt-pans that were connected with them.

Roebuck discovered that in order to keep the pits clear of water, the most powerful pumping apparatus was necessary; Newcomen's engine was found to be insufficient, and it was with satisfaction that Roebuck heard of Watt's invention from their mutual friend, Dr. Black. Watt had been pursuing his invention for years, and, always lacking capital, had by 1768 borrowed £1,200 from Dr. Black.

Soon after the invention of the separate condenser, Watt was in

[1] J. Lord, *Capital and Steam Power*, p. 76; Watt to Small, 28th January, 1769, Assay Office Library, Birmingham.

[2] H. W. Dickinson, *James Watt*, p. 30. [3] Ch. XVI.

communication with Roebuck, and he kept him informed of the progress of his experiments.

Roebuck wished Watt to construct a large engine at Kinneil, and put his invention to a practical test, with the assistance of the skilled workmen he had brought from Carron. Watt, however, was busy making enough money to keep his family by surveying. However, in 1768, the engine was working, and Watt paid his long-promised visit to Kinneil, and an arrangement was made with Roebuck.

Roebuck, who was at this time at the height of his career and looked to Watt's engine to assure his coal-mine of permanent success, agreed to discharge his debts, obtain a patent, and pay for the expenses of the experiments, in exchange for a two-thirds share in the invention.[1]

When Roebuck became involved in financial disaster, Matthew Boulton (1728–1809) gained control of the patent for the separate condenser steam engine. It was about the same time that Watt and Roebuck were experimenting on the conversion of salt to soda. The method tried is thought to have been decomposition of sea-salt with lime, but details are not known. There is, moreover, no evidence of production on a commercial scale, although T. S. Ashton speaks of Roebuck acquiring interests in coal-mines and *soda-works* at Borrowstoneness.[2] One factor which would have undoubtedly made commercial operation difficult was the high duties on salt.[3]

The next extension of the problem, this time accompanied by production of soda at least experimentally, comes from another of the same coterie of industrial pioneers who helped to effect industrial revolution in England through the application of theoretical Scottish science. It was made by James Keir (1735–1820), nephew of Alexander Lind of Gorgie, Sheriff of the county, and of George Lind, Lord Provost of Edinburgh.[4] Keir was educated at Edinburgh High School and at Edinburgh University, along with Erasmus Darwin. After a career in the Army, he married and settled in Birmingham in 1770. He engaged in the manufacture of glass at Stourbridge 1775–8, in partnership with John Taylor, Skey of Bewdley, *et al.*[5] The glass-house with which he became connected was Rogers Amblecote Glass House. The Rogers came from Wales in the seventeenth century, but the

[1] J. Lord, *op. cit.*, p. 77; Boulton to Roebuck, 9th July, 1768, A. O. L., Birmingham.

[2] T. S. Ashton, *Iron and Steel in the Industrial Revolution*, p. 50.

[3] Ch. II.

[4] S. Timmins, *James Keir, F.R.S., 1735–1820.*, *Trans. Birmingham and Midland Institute*, 1891, *24*, 1.

[5] H. W. Dickinson, *Matthew Boulton*, p. 100.

greatest of the family was Thomas Rogers, who became Sheriff of Worcester in 1750. In 1768 his glass-house was offered for lease, and in 1775 James Keir became connected with it. He is believed to have left in 1778 to go to Birmingham, but the existence in 1789 of a firm, Scott, Keir, Jones and Co., is recorded.[1] In addition to making glass, they made red and white lead for other glass-houses, and for the potteries. They made the white lead by the action of carbon dioxide on lead chloride ($PbCl_2$).[2] Darwin introduced Keir to Wedgwood and Boulton, and for some time he was connected with the latter's firm. He is not known to have had any business connections with Wedgwood. Watt referred to him as 'a mighty chemist, and a very agreeable man'.

In 1771, he entered caveat at the Patents Office against the grant of a patent for making mineral alkali or soda from common salt, because of his interest in the business. It is in consequence of a Parliamentary Committee being set up to examine the basis of the proceedings instituted by Keir in support of his claims, that we learn of much of the activity towards the solution of alkali synthesis.[3] The immediate cause of a Committee being appointed was a petition by Alexander Fordyce of Harley Street, and Dr. George Fordyce (1736–1802), Professor and Lecturer in Chemistry at St. Thomas' Hospital:

> . . . alledging that he is possessed of the Principle of Chemistry, hitherto unknown, by the skilful Application of which the Alkali and Acid contained in Sea Salt and Rock Salt are separated. . . .

and asking for a drawback on the salt used in the process. This petition called forth immediate protest from Keir; Samuel Garbett, Roebuck's partner, already referred to; James Watt, on behalf of Dr. Black and himself; and several others. Among those examined by the Committee were James Watt and his partner at Birmingham, Matthew Boulton.

Watt, while making no mention of his own activities, declared that Keir, with whom he became acquainted in 1768, had been in possession of the process prior to 1769, and would have established a factory to work it, in Ireland one infers, if he had not been otherwise engaged in manufacturing pursuits in England. The high duty on the salt necessary as a raw material was a further deterrent. Boulton confirmed Watt's evidence, placing the date a little later (1773), and affirming Keir's intention of

[1] F. Buckley, *Notes on the Glasshouses of Stourbridge, 1700–1830.*, *Journal Soc. Glass Tech.*, 1927, *11*, 106. [2] A. Moiliet, *James Keir, F.R.S.*, p. 76.

[3] *Journal of the House of Commons*, 1778–80, *37*, pp. 891, 912, etc.

From an engraving in the possession of Dr. Lily Newton

13. The Island of St. Kilda, showing gathered heaps of sea-weed and a man with a long-handled cutting-knife.

14. Burning sea-weed for the production of kelp on the coast of Brittany.

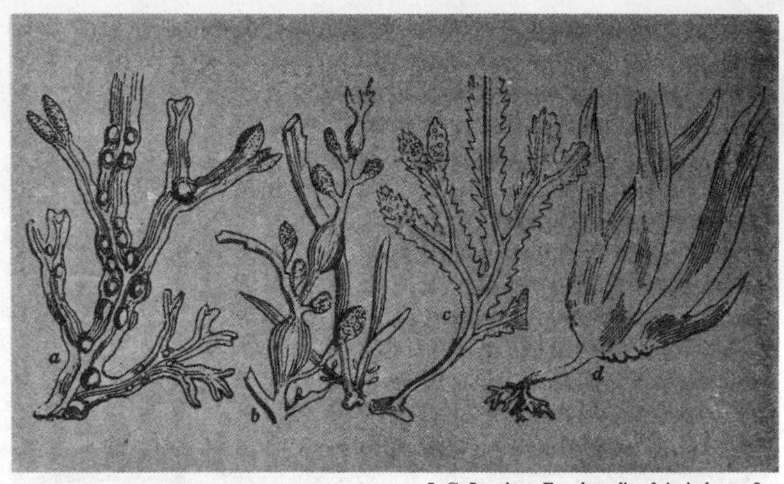

J. C. Loudon, *Encyclopaedia of Agriculture*, 1831

15. Sea-weeds burned for the production of kelp:
(a) *Fucus vesiculosus* (b) *Fucus nodosus* (c) *Fucus serratus* (d) *Fucus digitatus*

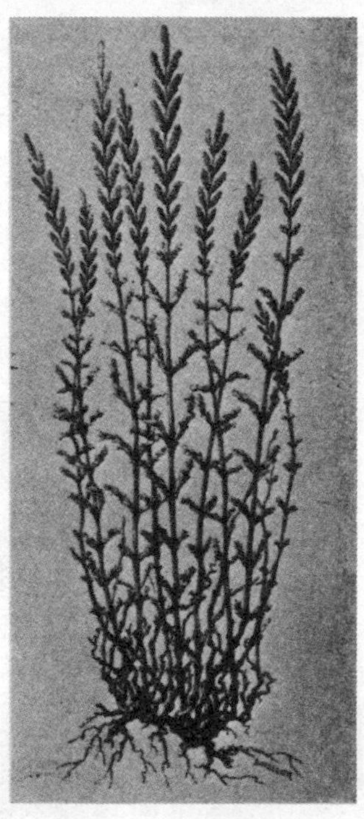

16. *Salsola soda*, and *Salicornia*, the herbs that were burned on the Mediterranean coasts for the production of *barilla*.

establishing a manufactory. Other witnesses evidenced Keir's actual synthesis of soda from common salt, and Samuel Moore, Secretary of the Society of Arts and Science, affirmed its quality and suitability for making glass and soap. Boulton, too, had an interest, if somewhat obscure, in alkalis. In the autumn of 1783 he set out on a visit to Scotland, going *via* Newcastle, where he visited the coal mines, and on to Edinburgh where he met Black and Robison. Here alkalis came under discussion. So he wrote:

> I talked with Dr. Black and another chemical friend respecting my plan for saving alkali at such bleach-grounds as our fire-engines are used at instead of water-wheels: the Doctor did not start any objections, but, on the contrary, much approved it.[1]

Further opposition to the Fordyces came from one Peter Theodore de Bruges, who declared that already he had expended £2,000 on equipping a works to work his process, and that he was in possession of 700 tons of necessary raw materials. Opposition triumphed, and it was decreed:

> . . . that no Person shall obtain a Patent for an Invention by which an Alkaline Salt may be extracted from Common Salt, or from Sea Water, or from the Water of Salt Springs, without notice being given to Mr. Henry Davidson, in Behalf of Messrs Archibald and James Keir.

The Committee were impressed with the evidence put before it regarding the possibilities of converting sea-salt into soda, and thought 'if it could be carried into such general effect as to become a substitute for the articles above mentioned (imported potashes and barilla), it would not only employ a considerable number of hands in a work perfectly new, but would prove a national saving and an additional source of public wealth'. A Bill followed on 2nd May, 1781, reducing the duties on *foul* salt to be used in making marine acid (hydrochloric acid, HCl), and fixed fossil or mineral alkali (sodium carbonate, Na_2CO_3). This concession, and the remark by Richard Watson in the next year, that the making of alkali in Great Britain was a matter of *great national concern*, give us an index of the importance attached to it.[2] Quite a number of patents were taken out about this time: Shannon's, in which Glauber's salt (sodium sulphate, $Na_2SO_4.10H_2O$) was heated with charcoal and lime in 1779 (British Patent No. 1223); Dr. Brian Higgins's, in which salt was treated with sulphuric acid and the

[1] S. Smiles, *Lives of Boulton and Watt*, p. 329.
[2] Richard Watson, *Chemical Essays*, Vol. 2, p. 354.

sodium sulphate reduced by charcoal, and soda produced by adding oxides of heavy metals, in 1781 (British Patent No. 1302). There is also a patent taken out by John Collison in 1782.

On the 7th February, 1782, Messrs. John Collison and Co., Southwark, sent Black a sample of their soda and asked him to analyse it, telling him that 'we have now entered into the manufacture of alkaline salts from the conviction that we can make them of a quality much better than any that were ever imported'. They enclosed a copy of their patent, which is quoted in full by Sir William Ramsay in his *Life of Black*. Their process is virtually the same as that of Leblanc. Black's report was that the alkali was 'very strong and powerfull. It contains more alkali than the best Alicant barilla in the proportion of 68 to 44, and more than the best kelp in the proportion 68 to 10. There is no need to use lime in drawing the leys from it, as it is already in a caustic state'. What became of Collison's works is not known.[1]

The Firth of Forth again became the focus of activity. When B. Faujas de St. Fond, the geologist and traveller, visited Edinburgh in 1784, he met a Dr. Swediaur, a European authority on venereal diseases, who told him that he had purchased an estate at Prestonpans—the site of Dr. Roebuck's vitriol works—where he intended to establish a manufactory of sea salt with the intention of separating the mineral alkali (sodium carbonate) from it. This venture did not live long however. By 1795 it had failed.[2]

Once again James Keir was more successful. In 1780 he joined forces with Alexander Blair, a brother officer in the army, at Tipton, Staffs., and among the activities in which they engaged were the making of alkali and soap, the alkali being made from common salt by the use of sulphuric acid.[3] The firm was Keir, Blair and Playfair.[4]

The method which they used was to pass a solution of sodium sulphate slowly through a sludge of lime.

$$Na_2SO_4 + Ca(OH)_2 = CaSO_4 + 2NaOH$$

The process was carried out slowly and it will be noticed that

[1] W. Ramsay, *Life and Letters of Joseph Black*, p. 67.

[2] B. Faujas de St. Fond, *Travels through England and Scotland, 1784*, Vol. 1, p. 173; G. Chalmers, *Caledonia*, Vol. 2, p. 497; *Statistical Account*, Vols. 10, p. 86, and 18, p. 437. The Assay Office, Birmingham, contains correspondence between Swediaur and Boulton.

[3] R. B. Prosser, *Birmingham Inventors and Inventions, passim.*

[4] *Local Notes and Queries from the Birmingham Journal and Daily Post*, 1867–70, Birmingham Reference Library.

caustic alkali was produced directly. A. Moilliet gives some side-lights on the ideas of Keir:

> Mr. Keir found that—by presenting the salts in an exceedingly weak solution, and by calling in the aid of a chemical agent (for which he always professed the greatest respect, and the functions of which in natural operations were, he thought, greatly underrated), *Time*—the rule of election was reversed. By passing a weak solution *slowly* through a thick body of lime, the sulphates were decomposed; the sulphuric acid meeting with the lime and leaving the alkalis disengaged. The liberated alkali had then only to be brought into a concentrated form for sale.[1]

Keir has an article on fossil alkali in the *Transactions of the Society for the Encouragement of Arts, Manufactures and Commerce*, but it is only concerned with the analysis of alkali imported from Bombay.[2]

The following citation from Ramsay's *Life of Black* amplifies the evidence of Keir's success:

> On January 15th, 1785, Black was consulted about the establishment of works for the manufacture of 'fossil alkali', or soda, by Mr. James Gerard, surgeon, in Liverpool. Gerard says that as the duty on salt is now lowered, it is a favourable time to consider such an undertaking, and Liverpool is a suitable place for such a manufacture; he proposes to make sal ammoniac, sulphuric acid, Glauber's salt, and alkali. He adds: 'I have copies of Dr. Higgins's, Fordyce's and Collison's patents, if you wish to consult them. The last is artful and intricate. I am told it answers, as also a work carried out at Birmingham by Keir, Boulton and Watt.'
>
> Black's answer is not preserved; but on June 7th Gerard wrote: 'I am concerned and disappointed to find that I must not expect your assistance in the alkali pursuit.' He does not reveal his own process; he merely says that he gets as much alkali from 4 oz. calcined Glauber's salt as will convert 3 oz. of tallow into good soap; he also gets sal ammoniac as a by-product. Can this be a foreshadowing of the ammonia-soda process?[3]

Like the other members of the Lunar Society, Keir's outlook was fresh and receptive of new ideas. Of geological disputation he wrote to James Watt:

> I am little disposed to pay much respect to the Edinburgh Wernerians, as I consider they are occupied more with the pedantry of systems and names, than with the substance of science.[4]

[1] A. Moilliet, *James Keir*, p. 75.
[2] 1788, Vol. 6, 2nd September.
[3] W. Ramsay, *op. cit.*, p. 69.
[4] J. Keir to J. Watt, 21st October, 1809.

The first volume of his *Dictionary of Chemistry* reveals in its introduction that he was one of the first British chemists to accept the antiphlogiston teaching of Lavoisier, and it was from the *Dictionary* that Wedgwood got his information about *terra ponderoso* (barium sulphate, $BaSO_4$) which was to effect a marked improvement in some of his wares. James Watt, Jr., and John Taylor of Manchester helped Keir with the *Dictionary*. Taylor undertook to supply articles on dyeing and calico-printing, while Watt, whose knowledge of German was useful to Keir, contributed on bleaching.[1]

The second chain of evidence that alkali was synthesized industrially in Great Britain before the opening of Muspratt's works now falls to be examined. It is associated with the names of Lords Dundas and Dundonald. Writing in 1785, Dundonald says:

> Attempts have been made in this country to save the annual drain of money by substituting fossil alkali or barilla of home manufacture; but for the encouragement of a manufacture of that sort in its infancy, it was judged necessary that the salt made use of in the course of the process should be exempt from duty.[2]

When his impecunious condition drove the Earl of Dundonald from Scotland to stay with some friends on the Tyne, he found in Newcastle two manufacturers, William Losh and Thomas Doubleday, experimenting independently on the conversion of salt into alkali. Dundonald brought them together and sent Losh to Paris in 1791 to acquire a knowledge of chemistry.[3] By this date one of the early alkali pioneers had turned up again. Dr. Francis Swediaur had evidently kept up his interest in the manufacture, since he wrote Andrew Smith at Boulton and Watt's on 24th March, 1791, asking if 'there is any worth in Lord Dundonald's discovery of obtaining mineral alkali from sea-salt'. Boulton replies on the 7th April, that 'neither Dr. Withering nor I know anything of Lord Dundonald's process of making mineral alkali from sea-salt, nor do we believe it'.

When Losh had returned from Paris, experiments in the production of soda were begun at Bell's Close, near Scotswood, and in consequence of their success a chemical works was started at Walker about 1796, the original partners being Lords Dundonald

[1] J. Keir to J. Watt, 13th December, 1789, 20th December, 1789, and 10th February, 1790, B.R.L.

[2] *The Present State of the Manufacture of Salt explained*, p. 82.

[3] *Newcastle: Chemical Manufactures in the District*, British Association Report, 1863, p. 701.

and Dundas, Aubone and John Surtees, John and William Losh, Thomas Doubleday apparently withdrawing. Miss H. Landell, sister of D. Landell of Landell and Chambers, Newcastle-on-Tyne, writing to Matthew Boulton on January 26th, 1796, says:

> Lord Dundonald has lived as a recluse in Newcastle for many months, and has at last exercised his chemical abilities to advantage, and will probably make a large fortune by his substitute for Barilla. Our glass manufacturers are contracting with him and have little doubt of his success.[1]

According to C. T. Kingzett[2] Losh utilized Scheele's discovery that common salt is decomposed by lead oxide with the formation of soda. The yellow lead compound which is formed in the process was sold by the firm as a pigment under the name of Turner's Patent Yellow. R. C. Clapham (*vide infra*) thought that they disposed of their by-products directly as lead and silver.

On 30th May, 1795, Dundonald had filed a patent which gives an account of the operations to be undertaken by the Walker Chemical Company.[3] The patent describes:

> . . . methods of obtaining mineral or fossil alkali or soda and vegetable alkali or pearl or pot-ash from neutral salts composed of these alkalis and an acid, or from solutions of these salts that in process previous to the engagement of these alkalis, or in processes connected therewith, several articles are disengaged or formed which may be collected and applied to several valuable purposes.

Then follow in detail possibilities envisaged by Dundonald, which include making Glauber's salt ($Na_2SO_4.10H_2O$) and decomposing it into sodium sulphide (Na_2S), from which soda could be obtained either in a mild (Na_2CO_3) or caustic ($NaOH$) state. Mention is also made of by-products which might be used in agriculture. Indeed, further evidence regarding the synthesis of alkali by methods similar to the Leblanc patent is contained in Dundonald's *Treatise on the Connection of Agriculture with Chemistry*:

> The disengagement and separation of this acid (HCl), from the alkaline basis with which it is united in sea or rock salt, may be accomplished by various methods; one only has as yet been discovered and effected at an expense which can admit a manufactory of alkaline salts being established on an extensive scale. The

[1] H. Landell to Matthew Boulton, 26th January, 1796.
[2] C. T. Kingzett, *History of the Alkali Trade*, p. 73.
[3] *Register of the Great Seal*, 1795, Vol. 20, Nos. 584 and 591, General Register House, Edinburgh.

accomplishment of this most desirable object, by a cheap and easy process, must appear, with respect to certain useful arts, as well as to the application of it to agriculture, to be one of the most important discoveries to which chemistry could have lent its aid.[1]

A further passage suggests that the preparation of sodium sulphate played a part in the process:

Methods of manufacturing, at a cheap rate, the most efficacious of these salts (Na_2SO_4), have been discovered, and farmers will soon be enabled to make the necessary experiments, and to satisfy themselves that such salts possess the powers ascribed to them. The price will be regulated by the duty that may be charged on sea-salt, and on the bitter refuse liquor of the salt works, whence these articles are to be made.[2]

There is more in the above patent than the manufacture of alkali by the action of sulphuric acid upon common salt. The salt tax then stood at £36 per ton, and constituted a serious impediment to commercialization of salt as an industrial raw material. Dundonald therefore turned his attention to making alkali from salt '*disengaged in a previous process*', i.e. waste or foul salt upon which a duty would not be levied, according to the above Act. Various expedients were tried to avoid using salt *per se*. They obtained brine from a salt spring in a coal-mine, and evaporated it with sulphuric acid, thus getting sodium sulphate instead of common salt.[3]

$$2NaCl + H_2SO_4 = Na_2SO_4 + 2HCl$$

R. C. Clapham, in his *Account of the Commencement of the Soda Manufacture on the Tyne*, gives the following additional data concerning the Walker company:

The works conducted at Bell's Close were removed in 1797 to Walker; a lease having been signed on 29th September for twenty-one years, the object being to obtain a more eligible site, and to make use of a salt spring recently discovered there. A coal pit, called the King's Pit, had been sunk in 1788 by the late Mr. Barnes, and an engine erected by the well-known engineers, Boulton and Watt, in the same year to work the coal.

In the shaft of this pit, the spring was discovered, and on the 24th June, 1798, a lease was signed with the Corporation of Newcastle, for its exclusive use, and an agreement entered into with the owners of the Walker Colliery to pump the brine. A four-inch lead pipe was put down on the outside of the metal tubbing

[1] Dundonald, *Treatise on Agriculture*, p. 60.
[2] *Ibid.*, p. 121. [3] *British Association Report*, 1863, p. 701

of the pit to the salt spring, and the engine did the double work of drawing the coal and pumping the brine.

Through the influence of the Earl of Dundonald with the Government, an order in Council was given that the salt made at Walker might be used in the manufacture of soda, free of duty, with a stipulation that soot or ground coal must be put in the pans, to prevent its use for other purposes; and an exciseman lived on the premises to watch proceedings. This peculiar privilege was found to be a great boon in establishing the soda trade at Walker.

The quantity of salt produced at first was not more than two or three tons weekly; but was afterwards increased to eight tons; and, as far as can be ascertained, this was the most that was made, which shows to what a limited extent chemical operations were then conducted.[1]

That may be so, but 8 tons per week is 400 per annum, and reckoning Scottish kelp as containing on an average 3 per cent. alkali, the Walker production is equivalent to 12,000 tons of kelp.

In 1808 a second Tyneside firm, Messrs. Doubleday and Easterby, of Bill Quay, began the manufacture of soda. This time the raw material was the waste salt from soap-boilers, chiefly a mixture of common and Glauber's salt ($NaCl$, $Na_2SO_4.10H_2O$), which they bought from Messrs. Jamieson and other soap-boilers of Leith at £7 per ton. They, also, avoided the salt tax by evaporating brine from a spring with sulphuric acid, thereby producing sodium sulphate directly. Initially Doubleday and Easterby bought their sulphuric acid, but later got plans from Charles Tennant of St. Rollox[2] and built sulphuric acid plant, *c.* 1812–1815, the first on the Tyne, handing on the plans in turn to Messrs. Cookson when they began manufacture of alkali at South Shields.

Another way of getting raw materials was to buy nitre cake ($KHSO^2$) from the nitric acid makers of Glasgow, etc.

It cost £14 per ton, and consisted, at that time, of bi-sulphate of potash, and was treated as under: 10 cwts. of nitre cake was ground with 1 cwt. of salt, it was then furnaced, 5 cwts. of salt was added, and the whole was mixed with 7½ cwts. of coals and 8 cwts. of slaked lime, and was balled. The ball was digested in water, and very pure chloride of potassium was separated, and the carbonate of soda was crystallised. In 1810, a parcel of 5 tons of Irish nitre cake, and 3 tons of salt, produced 42 cwts. muriates, and 64 cwts. British barilla (*soda*), and left a profit of £10 per ton on the latter.

[1] *Transactions Newcastle-upon-Tyne Chemical Society*, 1871, *1*, 32. [2] Ch. VI.

The Tyneside alkali manufacturers did business with a variety of associated interests. The potassium chloride which was a by-product of various processes was sold at £12–20 per ton to Macintosh of the Hurlet Alum Works near Glasgow, or to Lord Normanby for his works at Whitby. Both he and Macintosh in turn sent waste products from the alum works, particularly *alum slam* (a basic aluminium sulphate), to the alkali manufacturers. Cooksons, the glass manufacturers, who made their glass from kelp, disposed of sodium and potassium sulphates and chlorides which collected on the tops of the glass pots under the name of *sandiver*. Losh paid £8 per ton for it.[1]

Losh went to Paris again in 1815, and introduced further improvements on his return. From this nucleus, the solution of the alkali problem through the commercialization of sulphuric acid, the vast British alkali industry developed. Losh may justly be regarded as the father of the alkali trade of Great Britain. In 1816, synthetic soda was selling at £60 per ton.[2]

Several other Tyneside firms had been founded before the manufacture of alkali was begun in Lancashire. Those include Dr. Hutchinson at Felling, *c.* 1810, and Messrs. Cookson at Jarrow in 1823.[3] It is probably these that are referred to by Clapham in the following citations:

> In 1821, Mr. Losh erected acid chambers at Walker for the purpose of decomposing salt, and manufacturing soda on the French plan; and Mr. Charles Tennant commenced similar operations in Glasgow in 1819, so that these appear to be the first two alkali works commenced in this country on the French plan. The chambers erected at Walker were about 25 feet long by ten feet wide and ten feet high; and two were put up at first, in the lower part of the works, by a plumber from Glasgow. In the following year two large chambers were erected, and the four chambers were all that were put up for some years. The price of the sulphur at that time was £7 per ton (the duty being £15 per ton, which was remitted), and nitrate of potash was £32 per ton.
>
> .. .. ..
>
> On the repeal of the duty on salt, in 1823, a great development in soda works shortly followed. Next to Mr. Losh came Mr. T. H. Leighton, who commenced at Blyth in 1822, and used salt made from sea-water, and decomposed by copperas. He afterwards purchased sulphuric acid of Doubleday and Easterby, at Bill Quay, and from Tennant and Company, at Glasgow, which cost £18 10*s*. per ton.

[1] R. C. Clapham, *op. cit.*, p. 35. [2] *British Association Report*, 1863, p. 701.
[3] W. G. Armstrong, *Industrial Resources of the Three Northern Rivers*, p. 160.

Cookson and Company (now Jarrow Chemical Company) began, in 1823, by the erection of chambers and furnaces in the middle of South Shields; they removed a few years afterwards to Jarrow Slake, where very extensive works are now carried on.

John Allen commenced at Felling Shore, in 1827; A. Clapham at Friar's Goose, in 1829; C. Attwood (now Allhusen and Sons) at South Shore, in 1830; and H. L. Pattison and Company at Felling in 1834.[1]

Contemporary developments of considerable importance were taking place in Scotland. The first of these involved Dr. Joseph Black. In July 1789 he received from Mr. Samuel Birnie of Glasgow an *Estimate for a factory for marine alkali*, and although diffident about associating with new projects, became financially involved in the venture.

Birnie states that the cost of lead from ore, including wages, coals, etc., would be £59 15*s*. per ton. The lead was to be made into 'calx' (oxide) and ground, mixed with salt (at £6 a ton, including duty), costing with labour, coal, etc., £26 12*s*.; the sum is £86 7*s*. The sale of the silver from the lead at 5*s*. 9*d*. an ounce would bring in £60 7*s*. 6*d*., and there were produced four tons of mineral alkali, selling at £23 a ton, together making £152 7*s*. 6*d*.; deducting the cost, £86 7*s*., there is a weekly profit of £66 0*s*. 6*d*.[2]

In 1795, Lord Dundas bought an old candle-works at Burnfoot of Dalmuir, between the Clyde and the Canal, with the intention of making black-ashes for use in his alum works in Yorkshire.[3] Dundonald then persuaded Dundas to adopt the process for which he, Dundonald, had obtained the patent. Dundas installed as manager a Dr. Adair, who proved unsatisfactory. Next, Mr. Alves, foreman to Dundonald, tried to get things going, but with no greater success. Even Dundonald himself took charge for a time. These trials cost Dundas £10,000, and in despair he was about to convert the works into a manufactory of sal-ammoniac (ammonium chloride, NH_4Cl), then being made in Scotland with great profit by a medical graduate of Edinburgh, Dr. James Hutton, better known as the father of modern geology, when an alternative process was offered him.

The new process proved satisfactory, and a contract was entered into with Walt. Logan, who was to put up the capital, and Duncan Yule, the inventor, who was to act as sole manger for seven years, to receive a house, £150, and 5 per cent. of the

[1] R. C. Clapham, *Newcastle Chemical Society Trans.*, 1871, *1*, 38 *et seq.*
[2] W. Ramsay, *op. cit.*, p. 103.
[3] Ch. XII.

profits, under forfeit of £5,000 if he revealed the secret of the process. The profits up to 1809 were reckoned at £18,000, of which £10,000 went to Dundas. The workmen, of whom there were about twenty, were paid two shillings per day, eked out by the provision of houses and fuel by the proprietor.[1] The following information is given in the *New Statistical Account:*

> Soda works were first established here by the father of the present Lord Dundas, under the superintendence of the late Earl of Dundonald, who, much to his own honour, and his country's advantage, passed a considerable time at this place in scientific pursuits. His Lordship was the first who introduced the making of soda on a large scale, by using soap-maker's leys, which formerly used to be run off to waste. At the same time, the extended demand interested the scientific world, and induced the adoption of other modes for the production of this article.
>
> At these works, are made weekly about 30 tons sulphuric acid, which is wholly employed in decomposing common salt towards the manufacture of bleaching powder, or the chloride of lime. The residuum of the operation is converted into the well-known soda of commerce, used in great quantities for domestic purposes, and by the soap-makers, bleachers, dyers, potters, etc.
>
> The number of hands employed is from 80 to 100, all male adults, who work from six o'clock till six o'clock, and, as the operations go on night and day, one part relieves the other. The average wages are about 11*s.* per week.
>
> As the works have not been established a sufficient length of time, it is impossible to speak decisively of the effects on the health and longlivity of the workmen employed.[2]

These facts, and the evidence presented to a Parliamentary Committee[3] are of considerable interest, as they reveal the early activities of Scottish chemists. Within a few years of the starting of manufactories at Dalmuir and Newcastle, alkali was being produced in Scotland at Rutherglen Bridge, Port Dundas, Camlachie[4], Govan, Greenock, Leith, and particularly Queensferry, where there were important soap-works to create demand.[5] The manufacture of '*barilla, black ashes and soda*' was also carried out by 'Mr. Smith at Pencross Castle, Mr. Mitchell in Ayr,

[1] *Session Papers*, Signet Library, 241, No. 25; A. Whyte and D. Macfarlan, *General View of the Agriculture of Dumbarton*, p. 275.

[2] *New Statistical Account*, Vol. 8 (1), p. 28.

[3] *Journal of the House of Commons*, Vol. 37, pp. 891 and 912; Vol. 38, p. 323 *et seq.*

[4] P. 114.

[5] Sir J. Sinclair, *General Report on the Agricultural State and Political Circumstances of Scotland*, Appendix, 2, p. 301.

Mr. Burns in Saltcoats, and others'.[1] Other alkali manufacturers mentioned are, R. & J. Garroway, Netherfield Works, and Alexander Hope, Jr., both of Glasgow, and William Henderson and Co. at Irvine.[2]

These firms became purveyors of alkali; but before synthetic alkali ever came on the market, many soap-makers were carrying out what appears to have been virtually Leblanc's patent in order to regenerate their waste salt. Thus, according to Parkes:

> The sulphate of soda from barilla or kelp, when properly heated in a reverberatory furnace, with calcareous matter, will become converted into sulphuret of soda; an article very suitable for making that kind of soap (yellow) which is consumed by the poor. By a simple process the remaining alkali would be as fit for the best white as any they can procure from barilla.[3]

This is virtually Leblanc on a small scale.

Fyfe, writing in 1816 when he presented his recommendations to the Highland Society, says:

> At a time when the kelp manufacture is threatened with a total overthrow from the introduction of alkali matter made by the decomposition of sea salt, the cause of its inferiority to barilla with respect to the quantity of alkali it contains has at last been found out.

Dealing with his ideas for augmenting the soda obtained from kelp, he goes on:

> Should the plan now proposed have a fair trial and meet with the Society's approbation and ultimately succeed, I will consider myself amply repaid for the time and labour I have bestowed, if my efforts shall be the means of warding off the ruin now impending over the staple commodity of the Highlands and Islands of Scotland, by which one great source of sustenance must ultimately be withdrawn from the poorer part of their population, and the income of many of their proprietors be considerably reduced.[4]

From the foregoing there is much evidence to uphold our thesis that decline in the price of kelp in 1814 resulted from the activities of chemical works in the neighbourhood of Glasgow and on the Tyne whose operations appear to have been considerable before the founding in 1823 of the first alkali works actually operating the Leblanc patent.

[1] W. Aiton, *General View of the Agriculture of Ayr*, p. 607.
[2] A. McLean, *Local Industries of Glasgow and the West of Scotland*, p. 165.
[3] *Trans. Highland Society*, 1820, *5*, 104. [4] *Trans. Highland Soc.* 1824, *6*, 587.

When the scarcity of ashes throughout Europe became so acute during the revolutionary wars, the Académie des Sciences in 1775 offered a prize of 2,400 livres for a method of making alkali from non-vegetable sources. This stimulated Nicolas Leblanc to submit about 1790 the process which now bears his name. Although the process is a household word to chemists, few of them are familiar with the story of Leblanc's struggle, so we include the following citation from J. Fenwick Allen's *Some Founders of the Chemical Industry*.

> In the year 1753, at Issoudun, a town of some 10,000 inhabitants, about nineteen miles south of Orleans, where various manufactures were carried on, and where there was a college, was born Nicolas Leblanc. He was trained in an apothecary's shop, and having studied pharmacy, passed on to surgery. He must have been a man of considerable ability, for he was appointed surgeon to the Duke of Orleans, and he was the author of various scientific works. Incited to the research by the prize offered by the Académie, he devoted his attention to the treatment of common salt. Many others were also at work. A Benedictine Father, Malherbe, had his process of lixiviating the fused mixture of sulphate of soda, iron and charcoal. Guyton de Morveau and Carnay made their mixture of common salt and lime, and were so sanguine of their invention as to erect works at Croisac, in Picardy, in 1782. De la Métherie heated sulphate of soda and coal in close retorts, sulphurous acid was given off, which was condensed in leaden chambers. Athénas decomposed the salt with copperas, and decomposed the sulphate formed by employing Malherbe's method. Finally, in 1787, Nicolas Leblanc projected his process of decomposing common salt with oil of vitriol, condensing the muriatic acid in ammonia water: then after the sulphate had been well heated, it was mixed with half its weight of chalk and quarter its weight of charcoal, these were intimately ground together and heated in a crucible: then the contents were powdered and lixiviated; the soda evaporated and dried in hot air. The patent was obtained on the 25th September, 1791, and the works of "La Franciade" were built by Leblanc and Dizé at St. Denis—the Duke of Orleans having found the capital of 200,000 livres.
>
> When the Duke was executed on the 6th November, 1793, the works were confiscated. Leblanc was not awarded the prize, for a special commission appointed to examine the various processes, considered those of Malherbe and Athénas the most likely to prove successful, but judged none of the competitors worthy of the reward.
>
> On the death of his patron and the stoppage of his works, Leblanc was thrown into most distressing poverty. For several

> years he struggled on, and in the year 1806, the works that had been confiscated were returned to him by the Emperor Napoleon; but the property was useless to him as he had no capital, and with the burden of years and sorrow pressing upon him, he had to seek refuge in the workhouse, where he died by his own hand.
>
> The year Leblanc died, works were established by Payen at Dieuze, to carry out his process, and twelve months afterwards, plate-glass made with soda instead of potash was exhibited by the S. Gobain Plate Glass Company.[1]

Leblanc himself, together with Dizé and Pelletier *et al.*, described the manufacture of soda in a valuable memoir in the *Annales de Chimie*.[2]

A year or two before his execution, the Duke of Orleans addressed a query to Black through W. Fullerton as to the most expeditious method of extracting *sel de soude* from *sel de cuisine*, stating that he had been told that the method of obtaining pure mineral alkali was understood in Great Britain. Black's reply to Fullerton is interesting (1790).

> I had the honour to receive your letter of the 27th February, but do not know of any manufactory as yet established in this country for preparing an alkali from common salt. A person here called Birnie has lately got a patent for an invention or process for that purpose, and has built a house in which it is to be carried on or tried, but has not yet begun to work. The nature of his process will appear from the specification of his patent which is on record. In another work in which vitriolic acid is made from sulphur, an alkali is prepared from the vitriolated tartar (sulfate de potasse) which is produced in both these processes. But it is plain that such alkali is the vegetable alkali and not the fossil, neither is it of a good quality.[3]

Black must have known of the activities of Fordyce, Keir, and Collison. If they had not all ceased to operate, then it seems that Black must have shared with Watt, Priestley *et. al.*, sympathy with the Jacobins, and declined to reveal the state of alkali manufacture in Britain to the Duke's agent. That he was familiar with the political opinions of the Birmingham philosophers is certain, since he paid a visit to James Watt there in October 1788.

Cheshire, in Great Britain, with its vast resources of salt, was the natural focus upon which an extensive alkali industry should converge. The founder of the industry in this area was James

[1] J. Fenwick Allen, *Some Founders of the Chemical Industry*, p. 78. See also S. Miall, *A History of the British Chemical Industry*, p. 3.

[2] 1797, *19*, 58.

[3] Quoted by W. Ramsay, *op. cit.*, p. 105.

Muspratt (1793–1886). Indeed, the date at which he started operating Leblanc's patent in the neighbourhood of Liverpool, is usually taken as the foundation of the alkali industry in this country. As early as 1816, Muspratt, in conjunction with Abbott, had been engaged in a small chemical manufactory in Dublin, where he made yellow prussiate of potash (potassium ferrocyanide, $K_4Fe(CN)_6$). There he met Josiah Christopher Gamble (1776–1848), his partner to be.

The Gambles were an Ayrshire family, who left Scotland during the religious disturbances of the reign of James VI. They were staunch Presbyterians, and following the tradition of so many Scottish families, sent their son, J. C. Gamble, to Glasgow University, where he graduated M.A. in 1797. One of the classes which he attended was Dr. Cleghorn's class in chemistry. During his course he was brought into contact with chemical inventions by Charles Tennant of St. Rollox Chemical Works. Gamble practised his intended profession of Presbyterian pastor for a short time, but chemistry proving a greater attraction, he forsook the Church in 1812 and established a bleaching works at Monaghan in Ireland. Here he operated Tennant's processes for the preparation of bleaching materials[1], purchasing the necessary sulphuric acid from Tennant in Glasgow. After several years he moved nearer Dublin, and as a result came into contact with Muspratt.

In 1822 Muspratt left Ireland, and established himself at Vauxhall Road, Liverpool, in the following year, with the intention of manufacturing soda. His acquaintance with chemistry was limited, but with the aid of one or two treatises such as Nicholson's *Dictionary*, he acquired an understanding of the theory on which Leblanc's manufacture was based and from calculations of the quantities of materials used and products obtained he concluded that it would be a lucrative operation. Thus we see that the quantitative atomic theory introduced by the Manchester Schoolmaster, John Dalton (1766–1844), in 1807–8, soon enabled the manufacturer to cost his processes with increased accuracy. The opposite scheme drawn up by J. Mactear of St. Rollax for Kingzett[2] shows the quantities of the principal products of the heavy chemical industry, starting with one hundred parts of iron pyrites (FeS).

In 1828, Gamble followed Muspratt's example, left Ireland, and joined forces with him in Liverpool.[3]

[1] Ch. IX. [2] C. T. Kingzett, *op. cit.*, p. 247.
[3] J. Fenwick Allen, *op. cit.*, p. 43.

ALKALI RAW MATERIALS, PRODUCTS AND BY-PRODUCTS

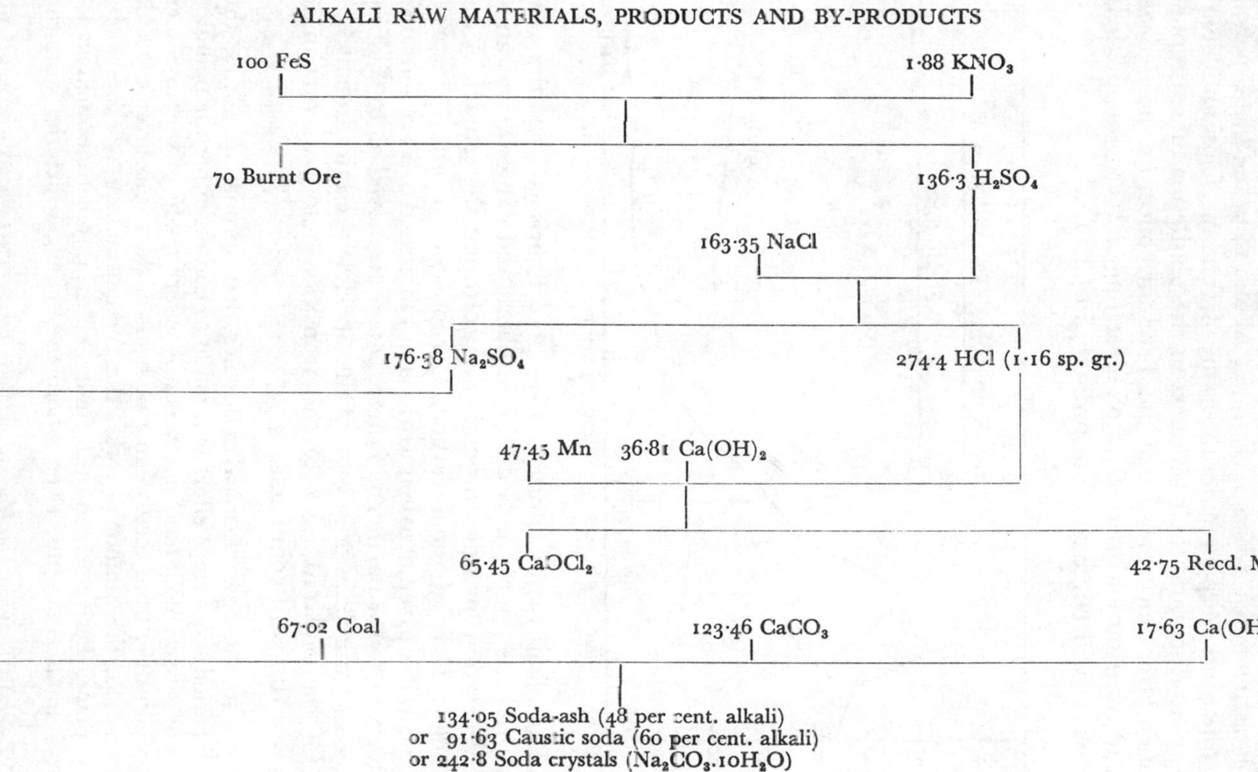

In 1825 manufacture of soda ash was begun at the St. Rollox works of Charles Tennant, in Glasgow, later to become one of the greatest chemical works in Europe. About 1830 the firm opened an office in Liverpool under the name of Tennant, Clow and Company, to distribute their products to the textile manufacturers of Lancashire. Since the rise of the Tennant concern depended on their contribution to the art of bleaching, an account of their activities will be given more properly in Ch. IX.[1]

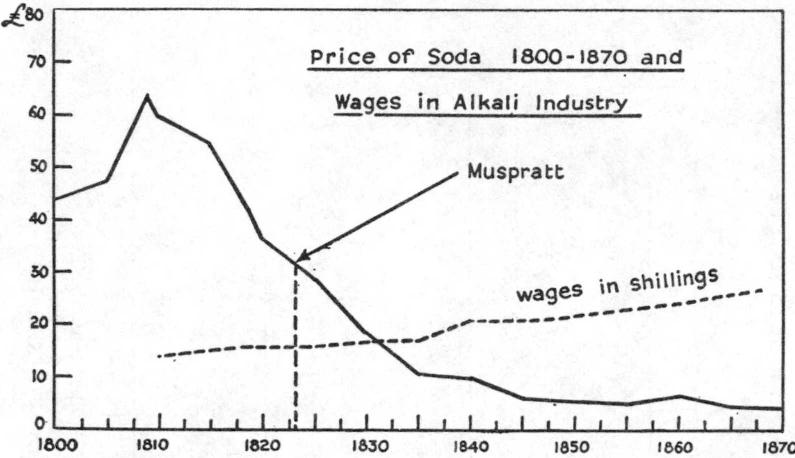

The alkali manufactory at Liverpool flourished, but the acid by-products poured into the atmosphere and caused devastation to vegetation for miles around, even crossing the estuary of the Mersey. This led the Liverpool Corporation to order Muspratt and Gamble to remove their works. They migrated to St. Helens and, after a two-year partnership there, Muspratt moved on to Widnes, leaving Gamble in possession of the St. Helens establishment.[2]

Of the outward signs of the transition into palaeotechnics, Lewis Mumford comments:

> In the new chemical industries that sprung up during this period, no serious effort was made to control either air pollution or stream pollution, nor was any effort made to separate such industries from the dwelling-quarters of the town. From the soda works, the ammonia-works, the cement-making works, the gas plant, there emerged dust, fumes, effluvia, sometimes noxious for human organisms. Even where the chemical factories were not

[1] E. W. D. Tennant, *One Hundred and Forty Years of the Tennant Companies*, p. 61.
[2] J. A. Picton, *Memorials of Liverpool*, Vol. 2, p. 77: quoted by N. F. Newbury in *Annals of Science*, 1938, *3*, 138.

17. Soda ſurnace for the conversion of common salt into soda. (For details, see below).

18. Section of soda furnace. Salt and sulphuric acid were brought together in *A*. After several hours, during which hydrochloric acid gas was given off, the mass was pushed into *C* and subsequently into *D* to drive off all the acid vapours.

19. Black-ash vats. The sodium sulphate from the decomposing furnaces was heated with chalk and coke, and the resulting black-ash extracted with water. The solution was evaporated in lead pans to give the soda-ash of commerce.

20. Soda crystallizing house at Chance's Alkali Works, Oldbury. For the manufacture of plate-glass and soda crystals the soda-ash had to be purified and crystallized in large cast-iron pans, as illustrated. The mother liquor yielded a residue containing 30 per cent. alkali which was used for the manufacture of common glass and soap.

conspicuously present, the railroad distributed smut and dirt: the reek of coal was the very incense of the new industrialism.[1]

Even by 1786, pollution arising from steam-engines was so common in Manchester that it caused Dr. Robert Percival (1756–1839), a prominent physician, to question James Watt concerning methods of abating the smoke nuisance.[2] It was due to the enlightened Dr. Percival that the first Factory Act, the Health and Morals of Apprentices Act (1802), was brought in by Sir Robert Peel, himself a large cotton-mill owner. Apparently, however, Watt in this direction was not without social conscience, since an earlier letter of his to De Luc mentions experiments on a furnace 'to be able to be quit of the abominable smoke which attends fire-engines'.[3] His interest in the abatement of the smoke nuisance possible arose through his connection with the erection of Albion Mills—an undertaking for which Professor Robison of Edinburgh recommended 'one of his experienced pupils', John Rennie, who later became one of the most celebrated structural engineers of his time. From a letter of David Landell's we learn that certain of his friends had erected a glass-house at Newcastle, and were in need of a remedy for the nuisance created by the great quantity of smoke. Landell approached Boulton, since he had heard that he and Watt had accomplished this at the Albion Mills.[4] But:

> . . . disregarding Benjamin Franklin's happy suggestion that coal smoke, being unburnt carbon, should be utilized a second time in the furnace, the new manufacturers erected steam-engines and factory chimneys without any effort to conserve energy by burning up thoroughly the products of the first combustion; nor did they at first attempt to utilize the by-products of the coke-ovens or burn up the gases produced in the blast-furnaces. . . .
>
> The hydrochloric evolved by the Leblanc process for manufacturing sodium carbonate was wasted until an Act of the British Parliament in 1863, incited by the corrosive action of the gas on the surrounding vegetation and metal work, compelled its conservation.[5]

Excessive production of corrosive by-products was a handicap to the operators of the Leblanc process right from its inception, and contemporaneous with the evolution of the industry, research

[1] L. Mumford, *Technics and Civilization*, p. 169.
[2] *Boulton and Watt MSS. Letter-Book* (*Office*), 1786–8, 109 and 110.
[3] *Watt to De Luc*, 10th September, 1785.
[4] M. Boulton to H. Holland, August 30th, 1801; D. Landell to M. Boulton, June 3rd, 1789, A.O.L.B.
[5] L. Mumford, *op. cit.*, p. 167.

was prosecuted to discover an alternative process free from these disadvantages. The one which eventually gained the field, and replaced the Leblanc patent, viz. the *ammonia-soda process*, was tried out experimentally by Thom at Camlachie as early as 1836, but successful commercial operation did not ensue for several decades.[1]

Returning to Gamble, we find that he carried on at St. Helens for some time without a partner. In 1835 he interested the brothers Joseph and James Crosfield, soap-boilers of Warrington, in the purchase of a defunct sulphuric acid works adjacent to his own alkali works. This laid the basis of an association between Gamble and the Crosfields. In the following year Simon Crosfield, a Liverpool tobacco manufacturer, joined the firm, which then became Gamble and Crosfield. So capital acquired in the West India trade was applied to the establishment of heavy industry in Lancashire in a way similar to that in which fortunes made by trade in sugar and tobacco in Glasgow helped to develop Lanark, Renfrew, and Dumbartonshire.[2]

There are further connexions between the Crosfield concern and Scottish academic science. James Shanks, who became one of their works managers, was born in Johnstone, Renfrewshire, on 24th April, 1800, of Fifeshire engineering stock.

> James remained at home, assisting his father and learning his business until he was of age, he then proceeded to Glasgow University, and in due time obtained his diploma. Returning to Johnstone, he there entered on his professional career, and, for two or three years, was the young doctor of that small manufacturing town. During his medical studies at Glasgow, Dr. Andrew Ure was the professor of chemistry, under whom it was his privilege to be placed.
>
> This was the master under whom Shanks studied, an enthusiastic teacher, accurate, painstaking, laborious, a man of broad culture and wide views, an author who occupies a first rank in technological literature, and whose name we associate with Payen, Knapp, Wagner, Watts, and Muspratt. Shanks forsook medicine for chemistry and set up as a maker of alum and potassium chromate in a small chemical works at Paisley, subsequently moving to Worcester and then to Newcastle. In association with Gossage he went to St. Helens to undertake constructional work for Gamble and Crosfields. J. C. Gamble, when in failing health, retained Shanks as a manager. He became a partner of Crosfields, Bros. and Company, and so continued for the rest of his life.[3]

[1] S. Miall, *op. cit.*, p. 12.
[2] J. Fenwick Allen, *op. cit.*, p. 51.
[3] *Ibid.*, p. 202.

The Wm. Gossage referred to in the preceding passage introduced the first social improvement in the Leblanc process in 1835. By the introduction of a tower of his invention, some of the acid fumes (HCl) were prevented from escaping into the atmosphere.

In conclusion it is interesting to look at the attitude which the early alkali manufacturers took to the social movements of their time, particularly in the present age of cartels and trusts. Both Muspratt and Gamble were active supporters of the Liberal cause. Both were associated with the anti-Corn-Law movement, indeed Muspratt was one of its earliest supporters, and stuck to his free trade principles even at the expense of his own interest. Gamble favoured the abolition of customs and the substitution of a levy on all incomes. The same is true of James Shanks. He was a radical of his day, associating with the demand for vote by ballot, extension of the franchise, and the suppression of patronage.

Chapter V
SOAP

> Scientists are the people who are changing the world and they don't know it.
>
> BALFOUR.

AT the present time the soap and glass industries absorb half the world's soda production between them. Throughout the evolution of technics both have at all times been intimately connected with the alkali trade, first in the form of ashes and kelp, latterly with synthetic alkali. When synthetic alkali became a commercial product, soap-making became still more closely connected with rising chemical industry. The *rapprochement* has become progressively closer and closer till soap-making may with justification be considered an integral part of the great industrial chemical set-up. Thus, though the soap industry throughout the palaeotechnic transition does not lend itself to a specific study, it does illustrate in an admirable way the reciprocal effect of techniques upon one another. Its commercialization too was not without social consequence. Speaking of the steady *general* drop in the death-rate throughout the nineteenth century, Lewis Mumford says *à propos* soap:

> Instead of giving credit for the steady advance to the industrial revolution, one should give due credit to quite another department—the increase of the food supply, which provided a better diet and helped raise resistance to disease. Still another factor may have had a part: the wider use of soap made possible through the increased amounts of available fats. The use of soap in personal hygiene may have extended from washing the nipples of the nursing mother to the child in her care: finally it passed from the feminine to the masculine half of society. This increased use of soap is not easily measured in trade schedules; for soap was originally a commercial monopoly, and as such, a luxury article: ordinary soap was produced and consumed within the household. The spread of the soap-and-water habit might well account for the lowering of infant mortality rates before the nineteenth century; even as the dearth of water and soap might account in part for the deplorable infant death-rates of the palaeotechnic town.[1]

[1] Lewis Mumford, *The Culture of Cities*, p. 172.

Few eotechnic crafts persisted so long as that of the domestic soap-producer, or primitive industrial *soaper*. Stephen Miall summarises the operations.

> Soap-making is an old art, centuries old—the art of boiling together oils and fats with alkali, and of then adding common salt which throws out the soap into a 'curd' that can be formed into bars, tablets, and the like. In the early days any kind of oil or fat, kitchen refuse, whale oil, or palm oil, was used, and barilla, an impure alkali containing both soda and potash and obtained by burning seaweed.[1]

The lag behind other ancillaries to the textile trade which is a characteristic of the soap industry—and it must be borne in mind that pressure for adequate supplies of soap came from industrial and not domestic users—can be explained in two ways. Firstly, although soap was made and consumed locally with raw materials available on the spot, e.g. tallow and kelp, the limited supply of these was a measure of the cleanliness of the inhabitants of Scotland at the time. Secondly, in contrast to salt-boiling (as an example of the purification of a substance already present in the solution from which it is produced) soap-making requires interaction of at least two materials in a chemical reaction. Till this was understood, little progress could be made in scientific soap-making, because to understand completely the process, both what is put in, *reactants*, and what is taken out, *resultants*, had to be isolated and identified. Fats, oils, etc., are complicated biological materials, consequently the history of soap-making from the seventeenth to the early nineteenth century is one of gradual approach of empirical operation towards scientific truth.

In the seventeenth century a comparative study of methods used throughout the country was begun with a view to obtaining information on the variety of soaps made at different places. In 1667 common methods of making soap then in use were examined by Sir Theodore de Vaux for the Royal Society. The results of his work are supposed to have been lost by Hooke, but may possibly be in the British Museum.[2] It is now difficult to ascertain whether the activities of the Royal Society had any effect upon contemporary soap-making, but the end of the seventeenth century saw the founding of the first *soaperies* in Scotland of which we have any detailed knowledge.[3] They were almost certainly

[1] S. Miall, *History of the British Chemical Industry*, p. 226.

[2] T. Birch, *History of the Royal Society*, Vol. 1, p. 406; F. W. Gibbs, *Saponis Artificium, Chemistry and Industry*, 1938, 57, 877.

[3] Dundee is said to have had even earlier soap-works. *Statistical Account*, Vol. 7, p. 218.

small affairs, and it is likely that soap-making remained largely a domestic craft till well through the eighteenth century. The early soaperies have been discussed in Ch. I, where it was pointed out that their success is difficult to assess on account of the political unrest of the period in which they were founded.[1]

The wide variety of materials which, when treated with alkali, will yield soap, did not help to speed up development towards industrial operation, because data discovered empirically in one culture, and as such satisfactory, did not necessarily transfer successfully to another. The raw materials and the soaps made from them included French soap made from olive oil and barilla, and white soap, also made from barilla, but with tallow instead of olive oil. In this country kelp was a cheaper source of alkali, so when the soap was made on a large scale, kelp replaced barilla. Mottled soap was made from *kitchen stuff*, presumably fatty waste, and barilla. If rosin were added to the tallow and barilla soap, a yellow soap resulted, the idea of the rosin being to cheapen the product, not improve its quality. By using potash along with whale or fish oil, a soft soap was obtained. In Russia, from which much of the tallow used in this country was obtained, rancid butter was converted into soap. As new oils, e.g. hemp, palm, linseed, and turnip, were brought into commerce, the variety of soaps was increased. Lord Dundonald suggested that hemp might be cultivated on the peat mosses of Scotland as a source of oil:

> Were hemp cultivated on an extensive scale in this country, the expressed oil from the seeds might advantageously be applied to the manufacture of soap, of a superior quality to that which is now made from tallow; for which purpose large quantities are annually imported from Russia, and other countries.[2]

Since the principles of soap formation were not understood, glue, peat, wool, etc., were experimented with as possible raw materials, unsuccessfully, as may be imagined. Mitchell of Edinburgh patented the use of offal. Sir John Dalrymple, in the name of John Crookes, obtained a patent for making soap from the muscles, fins, etc., of fish. The Board of Trustees in Edinburgh got Jameson of Leith to examine the soap produced, but he reported that the processes described were incorrect, and tests proved that fish soaps were practically as expensive, and their cleaning power certainly less, than ordinary soap.[3]

[1] J. Sinclair, *General Report*, Appendix, 2, p. 298; *Phil. Journal*, 1797, *1*, 40.
[2] Dundonald, *A Treatise on the Connexion of Agriculture with Chemistry*, p. 182.
[3] *Phil. Journal*, 1800, *3*, 108.

While Hippocrates and Galen knew that alkalis were rendered less caustic by treating with fats, we can now write:

$$\left.\begin{array}{l}\text{Fat} \;..\\ \text{Oil} \;..\\ \text{Wax} \;..\end{array}\right\} + \text{Alkali} = \text{Soap} + \text{Glycerine}$$

but it took a century and a half of industrial soap-making, followed by scientific investigation of the highest order, before this could be written down. Only then was it understood why some substances can be converted into soap, while others cannot. In addition to attempting the conversion of impossible crude products into soap, extraneous materials, unnecessary for the real soap-forming reaction, were added by soap-makers to improve the quality of their products. They were advocated by N. Lemery in the early eighteenth century because they gave a whiter and drier soap. Throughout the century variety was increased till it included rosin, talc, silica, starch, salt, borax, and Glauber's salt.

> Summarizing the history of the trade, we find that its line of development follows closely that of other crafts and trades, although in its initial stages it is unique. Few of the discoveries of ancient medicine have proved to be of such subsequent importance as the observation that alkaline substances were rendered less potent when mixed with fats and oils. Soap-like ointments found a considerable application in the treatment of skin diseases, and in toilet preparations for dyeing the hair. Later it was found that clothes were not harmed by being washed with soap: and soap received medical approval for personal use in the bath. Next came the discovery that unworked wool was greatly improved and cleansed by treatment with soap, and soon we find that the craft of the 'soaper' had become established in several European countries. In Southern Europe the discovery was made that caustic alkali reacts with oil better than the ordinary alkali or nitrum. When soap-making appeared as a localized art, the method consisted in boiling olive-oil (in the Mediterranean countries) or animal fats (in Northern Europe) with an extract of plant-ashes and lime. This was the main method throughout the whole of the Middle Ages right up to the seventeenth century.[1]

In 1731 Peter Shaw (1694–1763), a Fellow of the Royal Society, wrote his *Chemical Lectures for the Improvement of Arts, Trades, etc.*, in which he gave indications for more detailed investigations of the art of soap-making. He included suggestions for tests of purity, an index that the presence of extraneous matter had

[1] F. W. Gibbs, *The History of the Manufacture of Soap, Annals of Science*, 1939, *4*, 189.

SOAP
(Sodium and Potassium salts of fatty acids)

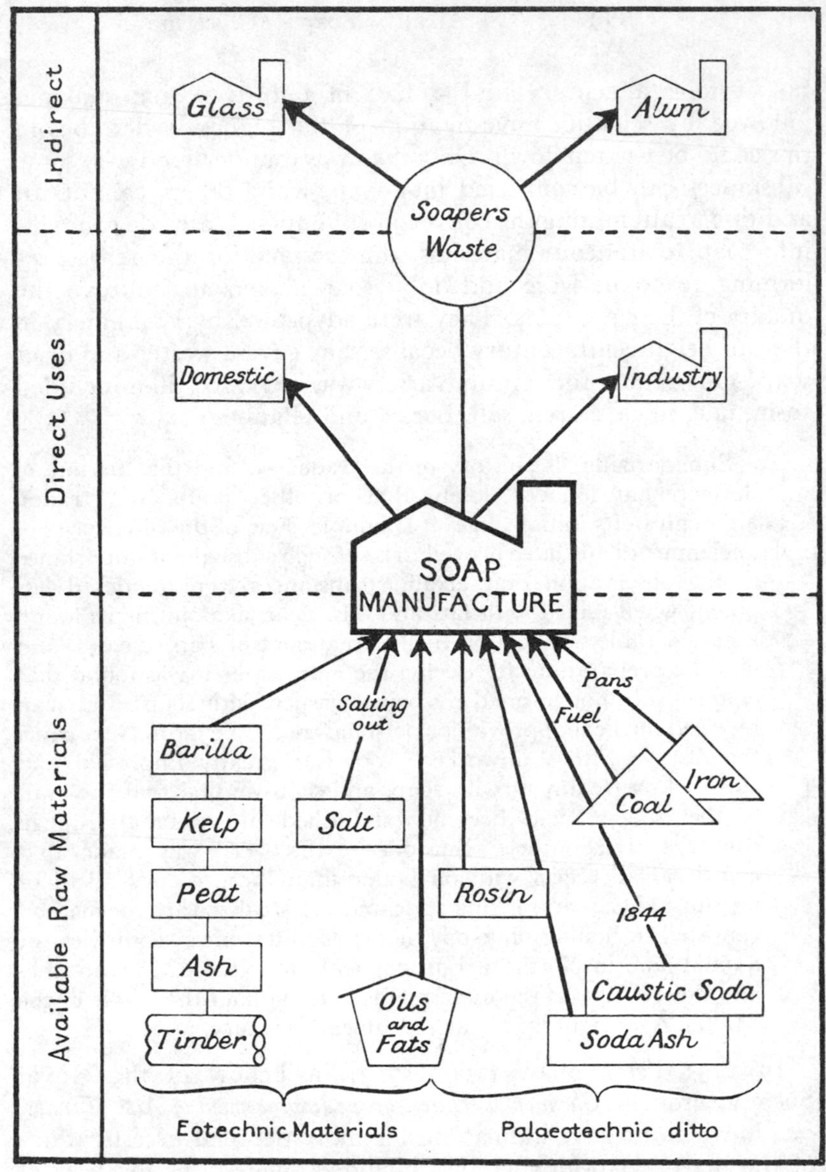

reduced the quality of soap below what might be reasonably expected. How small the consumption of soap was at this time may be gauged by figures given by J. U. Nef for annual consumption in the seventeen-thirties, viz. five to ten thousand tons.[1]

Shaw's lectures were followed in 1736 by James Dunbar's *Smegmatologia, or the Art of making Potashes, Soap, and Bleaching Linen.* The inclusion of bleaching signalizes the close connection which already subsisted between soap-making and the finishing of textiles, but soap-making still continued as a domestic craft, and the finishing trades went short.

This is substantiated by the inclusion of directions for soap-making in the *Select Transactions of the Honourable the Society of Improvers*, published in 1743. The information given consists of data concerning raw material production, viz. ash-burning, and the use of the resulting potash to convert, by boiling, tallow or other fat into soap. In less populous areas such locally available materials were employed and, according to Dunbar, Stevens and Company of Leith made soap from all-Scottish materials. Much tallow, however, was imported. This is important from the point of view of the ultimate locus of the soap industry after it had become commercialized. Thus we find soap-works in coastal towns; on the Clyde with its whaling connections; at Leith, an entrepôt for many continental wares, in particular, tallow from Russia. About 1785, 94,487 cwts. of tallow were imported into Scotland from St. Petersburg.[2] Soap-makers were not the only users of tallow. Soap-maker and candle-maker competed for tallow as a raw material. About the middle of the eighteenth century the demand for tallow was so great that there was imminent danger of a shortage of candles, and it looked for a time as though the use of tallow would have to be controlled. How the demand for soap increased relative to that for candles is illustrated by the following figures for the production at Leith:

In 1763,	1,500,000 lb. candles	against	500,000 lb. soap,
while in 1790,	3,000,000 ,, ,,	,,	6,000,000 ,, ,, [3]

Based on little more information than that contained in such works as Dunbar's *Smegmatologia*, a considerable soap industry developed in Scotland before science made any impression on the craft. Mention has been made of Stevens and Company of Leith, who were operating before 1736. Next there was William St. Clair

[1] J. U. Nef, *Rise of the British Coal Industry*, Vol. 1, p. 186.
[2] C. Wallace to H. Dundas, 21st November, 1786, N.L.S. MS. 640, f. 3–18.
[3] *Statistical Account*, Vol. 6, p. 602.

of Roslin, who began the manufacture of soap at Leith in 1750, and candles in 1770. It is thought that the soaps made in these works were of the soft variety, since David Neilson is credited with having introduced hard soap in 1770. Output of soap from Leith increased from half a million pounds in 1763, to six million in 1790. Candles over the same period increased from 1.5 to 3 million pounds.

The situation at Leith in the opening years of the nineteenth century is described in the following citation from the *General View of the Agriculture of Mid-Lothian*:

> There are two hard soap manufactories in Leith; the one under the firm of Anderson and Cundell, which employs from seven to eight men, whose wages are from 9*s*. 6*d*. to 10*s*. 6*d*. weekly, and which uses from 30 to 40 tons of coal, and pays from £3,000 to £5,000 of soap duty per year. The other soap house is under the firm of Jamison and Paton, and is very completely fitted up for an extensive trade, it being one of the largest soap-houses in the kingdom, having seven hard soap pans, some of which are 14 feet deep, and 14 feet wide, of boiling contents; and from 44 to 46 vats, etc., from 7 to 8 feet wide, and from 2 feet 6 inches to 2 feet 9 inches deep, most of which are cast iron, with the same number of receivers below ground, of the same metal. In this house are likewise employed six pans, or boilers, for evaporating waste or spent lees, which are very large, with two air furnaces; here is also burnt all the lime used in this manufactory, in a kiln on a new construction, which seems to be a very great improvement on the quality of the lime wanted for that purpose, but of the nature of which I am not fully informed. The quantity of coal used here is from 1,400 to 1,500 tons annually, and there are employed from 40 to 50 labourers, whose wages, each, are from 9*s*. 6*d*. to 12*s*. per week: from 21,000 to 22,000 pounds Sterling of duty is paid per annum. This house began business in 1777, and has also a very extensive soap manufactory in London, which commenced in the year 1789.[1]

As the industrial revolution progressed the east coast failed to maintain its position as a producer of soap. David Loch, who had been operating before 1769, failed in 1788.[2] Yet Thomas and Laurence Jameson, soap-boilers in Leith, were in a position to command supplies of kelp by offering £15 per ton for a hundred tons of best Highland kelp in 1810. This was £1 more than any sold in Scotland that year.[3] The writer of the

[1] G. Robertson, *General View of Mid-Lothian*, p. 186.

[2] Signet Library, *Session Papers*, 172, 38. [3] *Ibid*., 282, 6.

Statistical Account of Angus analysed the causes of the decline at Dundee:

> Soap was some years ago manufactured to a considerable amount, but this business now declines, and last year only yielded of duty to Government the sum of £1,828 19*s.* 3¼*d.* It is thought that this business will not only be abandoned here but that it will soon be totally lost to Scotland. The supposed causes are either regulations of excise, partial to England, or superior rigours in carrying the common regulations into execution. It is possible, however, that the real cause may be foolish attempts to undersell their richer English neighbours.[1]

The west of Scotland suffered a similar decline, but the art of soap-making was by no means eclipsed. The industry in Renfrew is described in the following extract:

> This art, which was known at Dundee so far back as the sixteenth century, has been carried on at Paisley to considerable extent since the year 1764,[2] which is said to have been the date of its commencement. The manufacture of soft soap was begun in 1776; and in 1781 there were five companies engaged in soap-making, who brought to market hard and soft soap to the value of £30,000 per annum. In the year 1791 there was the same number of houses in this business, one of them with works established at the royal burgh of Renfrew, but the manufacture was carried on to far greater extent than formerly. The duties paid to Government by two of them were £7,500, and, as the sum paid by the other three must have been of still greater amount, it is probable that the whole excise duties from these works might be about £16,000 per annum: but this manufacture has been declining for some years. The same soap-work is still carried on at Renfrew, and there are three works at Paisley, but the duties payable by all of them united do not exceed £13,000 per annum: so that this trade has declined considerably in this as well as in other places. The manufacture of candles, of excellent quality, was formerly carried on at Paisley to considerable extent; and, though the business is still prosecuted, it has also declined. The substituting of oil in place of candles in cotton mills and work-shops, and sometimes in private families, is supposed to be partly the cause. The houses engaged in this business are four in Paisley and one at Kilbarchan: the amount of their excise duties may be computed at £1,100 per annum.[3]

[1] *Statistical Account*, Vol. 7, p. 218.

[2] Wilson and Corse were Paisley soap-boilers (*Session Papers*, 367, 12 S.L.; G. Chalmers, *Caledonia*, Vol. 3, p. 815).

[3] J. Wilson, *General View of Renfrew*, p. 263.

Considerable quantities of soap were made at other centres in Scotland. Queensferry was important. Here manufacture of brown soap was brought it is said, to perfection, and made in four works. The industry was founded in 1770 and flourished till 1783, when there was a general set-back to soap-making in Scotland. At one time the contribution to the exchequer from this district was £8,000 to £10,000 per annum.[1] Production in Fife amounted to 250,000 lb. soap and 180,000 lb. candles, upon which a duty of

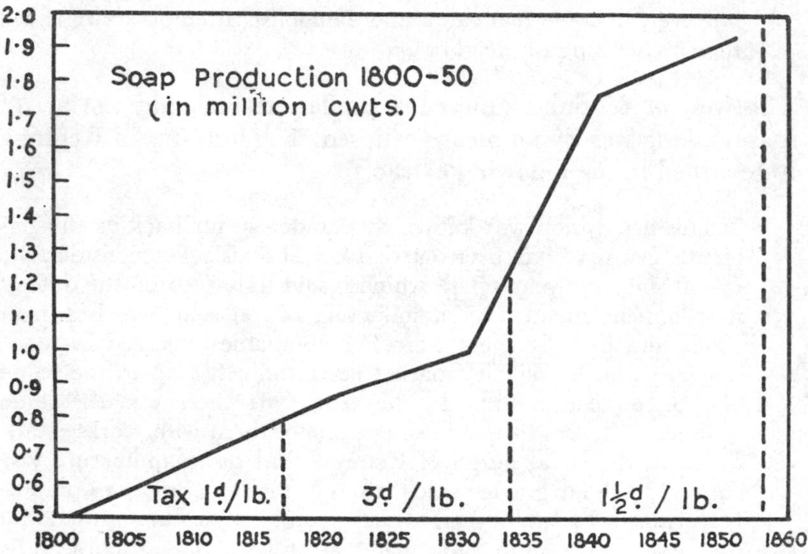

£3,000 was paid.[2] Yellow and white soap were made at Banff in a works 'fitted up in a very complete manner'. The same manufactory also turned out excellent candles.[3] The saponaceous requirements of the Highlands were supplied by a single soap-boiler at Inverness.[4] Miall gives a short account of a Glasgow firm which expanded from soap-boiling into general chemical manufacture:

> The firm of John and James White, Limited, of Glasgow, has a long history. It started as a small soap-works in 1810, and John and James White began in 1820 to make bichromate of potash, and in 1884 to make bichromate of soda. Later they and their

[1] *Statistical Account*, Vol. 17, p. 489.
[2] J. Thomson, *General View of the Agriculture of Fife*, p. 310.
[3] D. Souter, *General View, Banff*, p. 305.
[4] *Statistical Account*, Vol. 9, p. 618.

successors made considerable quantities of sulphuric acid, hydrochloric acid, and alkali. In 1919 the firm was converted into a private limited company, and it now makes, in addition to the products already mentioned, bichromate of ammonia, oxide of chromium, and chromic acid.[1]

For comparison, and also because of interest in the firm concerned, it is worth noting that the first *soaperie* in the Newcastle area was begun in 1770 by Lamb and Waldie at Westgate, whence they removed to The Close. In 1775 the works was purchased by Thomas Doubleday, and continued under the name of Doubleday and Easterby till 1841. This firm, it will be remembered, was connected with the efforts of Lord Dundonald to effect the conversion of salt into soda.[2] Among the first soap-boilers to appreciate the utility of science was George Crosfield.

On 11th May, 1814, George Crosfield, of Lancaster, of the Society of Friends, noted in his journal: 'I left Lancaster on my way to Warrington. The object of my journey was to view some premises near Bankey suitable for a Soapery, which business our son Joseph seems to have a strong inclination to.' That was the genesis of the great soap-making firm of Joseph Crosfield and Sons, Limited. The purchase was completed in June of that year. At first the business did not prosper, but by 1818 we learn from the same journal: 'Joseph appears in high spirits, his business has been profitable for about a year, by which he has regained his previous heavy loss'.[3]

Towards the end of the eighteenth century, scientists turned their attention to the study of the reaction by which soap is produced. In 1780, C. L. Berthollet published his *Sur la Combination des Huiles avec les Terres*, a scientific contribution of fundamental importance.[4] Only three years later, C. W. Scheele published his paper *On the Peculiar Saccharine Principle in Expressed Oils and Fats*. Scheele's researches revealed that glycerine (an alcohol, $C_3H_5(OH)_3$) is liberated in the same reaction as that which produces soap. Moreover, as it could be got from olive, almond, and rape oil, as well as from butter and lard, glycerine became, so to speak, the common factor on the *resultants* side of the soap-producing equation.

During the eighteenth century the procedure was improved both mechanically and by an attempt to test materials before and

[1] S. Miall, *op. cit.*, p. 35.

[2] W. G. Armstrong, *The Industrial Resources of the Three Northern Rivers, Tyne, Wear, and Tees*, p. 171.

[3] S. Miall, *op. cit.*, p. 230.

[4] *Journal de Physique*, 1785, *26*, 114.

during the actual preparation. But rule-of-thumb technique alone was possible in the absence of real chemical knowledge. At this time the salt-like nature of soap was known, and the fact that some change in the nature of the fats had taken place on saponification; and Scheele had discovered that glycerine was an essential constituent of certain fats and oils.[1]

Scheele was not able to contribute anything positive to conceptions of the constitution of soap itself. Thus, at the end of the eighteenth century, Laurens, who was connected with one of the main centres of soap-production in Europe, Marseilles, could still say of soap-makers:

> It is only from chemistry that they can expect to attain any improvement of the processes used for obtaining the article that they throw into trade.

There was no English work on soap-making. Indeed the only comprehensive account then extant appears to have been the *Rapport sur la fabrication des Savons*, which in 1797 was produced, in the heat of the French Revolution, by *Citoyens* D'Arcet, Lelièvre, and Pelletier.[2]

The first two decades of the next century saw the situation completely changed by the researches of M. E. Chevreul (1786–1889). These researches, published from 1813 onwards, placed soap-making on a sure quantitative basis, and technics was placed under one of its greatest debts to chemistry. Chevreul's first paper on the subject was read to the *Institute* on 5th July, 1813. Others followed rapidly in the succeeding years. His collected works, under the title *Recherches sur les corps gras d'origine animale* (Paris, 1823) is still the corner-stone of scientific soap-making. F. W. Gibbs, in his *History of the Manufacture of Soap*, says:

> It is difficult to overestimate the value of Chevreul's researches, for they enabled manufacturers for the first time to have an insight into the events which they endeavoured with varying degrees of success to control. He had only two facts to work on—Geoffroy had shown that the 'fat' that had combined with the alkali was different from the original fat, and Scheele had shown that Glycerine was an essential constituent of all oils and fats. Chevreul's great discovery was that *the laws of Constant and Multiple Proportions held with fats, oils, and waxes, as with inorganic compounds.*[3]

Chevreul began his researches in 1811, and announced the first of his discoveries in 1813. This was that fatty bodies were of

[1] F. W. Gibbs, *Annals of Science*, 1939, *4*, 190.
[2] *Annales de Chimie*, 1797, *19*, pp. 254–349.
[3] *Annals of Science*, 1939, *4*, 169.

a compound nature, consisting of an animal base and acid. The acid which Chevreul discovered he called margaric acid (from *margaron*, a pearl). In November 1813 he announced the discovery of another acid called oleic acid (*oleum* = oil), which he discovered in lard. The base with which fatty acids are combined in fats is no other than the glycerine isolated by Scheele. So we may define soap as a salt (stearate) of a fatty acid. We can write the equation for formation of soda soap:

$$\text{Fat (Glyceryl stearate)} + \text{Caustic Soda} = \text{Soap} + \text{Glycerine}$$
$$C_3H_5(\text{stearate})_3 + 3NaOH = 3Na(\text{stearate}) + C_3H_5(OH)_3$$

This equation is said to represent the *saponification* (or hydrolysis) of fat. In general, sodium soaps are hard, and potassium soaps soft. The *Parisian Society for the Encouragement of Manufactures* made Chevreul a gift of 12,000 francs for his researches. He was persuaded to enter industry, which he did in conjunction with Gay Lussac in 1825, but in this field he was not successful.[1]

Soon after these researches were made available, a rapid increase in soap production in Great Britain took place. True, the tax on soap was halved in 1833, but this is not likely to have been the cause of the increase, since a trebling of the tax in 1816 had no effect on the increase of production. The increase, however, takes place in the decade following the large-scale commercialization of soda production, when domestic demand was becoming significant. It is only when the squalor of palaeotechnic towns was reaching its height that the graph illustrating the quantity of soap produced turns upwards. The time-lag behind the same upward turn that is found in the case of coal is not without significance. It must be borne in mind that any scientific advance will be still-born if the social context of the discovery is inappropriate. Had Chevreul's researches been made half-a-century earlier, palaeotechnic economy would not have been sufficiently developed to benefit from them, because the large-scale commercialization of soap-making had to await two things: (*a*) the solution of the problem of the synthesis of soda from common salt in quantities adequate to augment (and ultimately replace) alkali which was initially obtained from kelp in the second half of the eighteenth century; (*b*) removal of salt duties to make this an economic proposition. At the time of Leblanc's discovery there were already established many soap-works.

The nineteenth century saw the introduction of steam-power, which enabled the manufacture to take place on a much larger

[1] G. F. Wilson, *Journal, Royal Society of Arts*, 1852.

scale, more efficiently, and under more readily controlled conditions. Leblanc showed how soda could be manufactured cheaply from sea-salt, a method which was considerably developed in this country by Muspratt. The researches of Chevreul followed, placing soap-manufacture on a scientific basis for the first time; they showed how chemistry could be applied to all the processes of saponification and with a much greater degree of control over the produce obtained from a great diversity of raw materials.[1]

Of the increasing interconnectedness of industries based on exact scientific knowledge, Stephen Miall sums up the situation concerning the alkali plexus:

> The manufacture of soap requires large quantities of soda, and soda, as we have seen, requires large quantities of sulphuric acid, so, as the use of soap spread to all parts of the civilized world, the demand for sulphuric acid grew. The manufacture of alum, which was carried on in a small way in this country for two or three centuries, also required sulphuric acid, and, as the demand for alum by the textile industry grew, so a further demand for sulphuric acid arose. All these developments helped the alkali manufacturer.[2]

The duty on salt impeded progress for a time, but when it was repealed, and the price of salt reduced to a minimum, the price of soda became dependent upon that of sulphuric acid.

Soap itself was subject to excise duty, and soap-boilers to irksome regulations. The following is a summary of them, taken from Chitty's *Commercial Law*.[3]

> Like bleachers, soap-makers were required to take out a licence which cost them £4. They were required to provide wooden covers for all coppers in which hard soap was boiled. These were sealed by the officer whenever soap was left in them. The furnace door, cover, and ash-hole door had to be locked and sealed at all times except when in use. Soap must not be conveyed by pipe or other *private conveyances*. Notice had to be given of the intention to take soap out of the coppers. The frames used in the making of hard soap had to be of regulation size, marked and numbered at the expense of the soap-maker. The making of yellow or mottled soap was regulated by 59 Geo. III, c. 90. No lees fit for the manufacture of soap might be manufactured for sale. Barilla might not be ground or pounded for sale. Quantities of soap in excess of 28 lb. could only be removed in boxes with the word *soap* printed on them in two-inch letters, and every waggon or

[1] F. W. Gibbs, *Annals of Science*, 1939, *4*, 190.
[2] S. Miall, *op. cit.*, p. 23. [3] Vol. 2, p. 418.

21. Kinneil House, home of Dr. John Roebuck, near Bo'ness, in an outhouse of which James Watt experimented with his first separate condenser engine.

22. St. Rollox Chemical Works, Glasgow, at one time the greatest chemical works in Europe.

Diderot and d'Alembert's *Encyclopédie*

23. The Pan-room of a Marseilles soap factory.

J. M. W. Turner, R.A. (1775–1851)

24. The Hawkes Soapworks, Blackfriars.

cart used in transporting such boxes had to have the word *soap* in three-inch letters on some conspicuous part of the same.

It is little wonder that Andrew Ure voiced the demand for a more logical fiscal policy:

> There is a second manufacture, of great interest to the health of the people of this country, which stands much in need of the minister's untrammelling hand; namely, that of soap; a manufacture for which, in the expert point of view, this country possesses peculiar facilities, in the boundless productiveness of its alkali works. It is remarkable, that the consumption of this necessary of civilized life has not increased with the increasing wealth and commerce of the country; that the quantity used by the manufacturers of silk, linen, and wool, has not increased in proportion to the progress of these manufacturers; that the export trade is in a ruinous state; that the duty is from 40 to 50 per cent. on the prime cost; that one-fourth (£240,119) of the whole amount collected is returned in drawbacks; and, finally, that another large proportion of that amount is wasted for the Treasury in the costs of collection, in maintaining an immense army of officers required to watch and stand guard over each trader.[1]

The changes for which Ure saw need were gradually effected during the course of the nineteenth century, and it was then that the great increase in the large-scale production of soap took place. The next technical change was substitution of caustic soda (sodium hydroxide, NaOH) for ordinary soda (sodium carbonate, Na_2CO_3), but since this did not take place till 1844, it is outside the scope of the present study.

[1] Andrew Ure, *Recent Improvements in Arts, Manufactures and Mines.* Preface, p. vii.

Chapter VI

VITRIOL IN THE INDUSTRIAL REVOLUTION

The chemical industry is the most polygamous of all industries.

WRITING in 1843, by which time the manufacture of sulphuric acid had been established in Great Britain for over a century, Justus Liebig in his *Familiar Letters on Chemistry remarks:* 'We may fairly judge of the commercial prosperity of a country from the amount of sulphuric acid it consumes'.

Prior to Liebig's generalization, during the first century of its commercialization, sulphuric acid, by contributing to the evolution of palaeotechnic economy, brought about more than one revolution in social technology. Despite this, it is wellnigh impossible to find a contemporary and extended account of the uses to which this fundamental reagent was put during the first hundred years of its production as an industrial chemical.

Early references to sulphuric acid appear in the works of the eotechnic alchemists, Valentine, Paracelsus, Agricola, and Glauber, during the period when alchemy was disintegrating, and mythological accounts were being replaced by accurate descriptions of the preparation and properties of substances which were to become fundamental *materia chemica* in the new economy. According to Samuel Parkes, the first correct account of sulphuric acid appears in the works of Gerard Dornaeus, published at Frankfurt in 1570, at which time it was made by distilling green vitriol (ferrous sulphate, $FeSO_4.7H_2O$).[1]

In Great Britain there was an abundant supply of green vitriol, derived by atmospheric oxidation of pyrites (iron sulphide, FeS_2),[2] on the island of Sheppey, and early in the seventeenth century a Fleming received a letter from King James VI agreeing to a proposal to start making '*brinston, vitreall, and allome*', and have a monopoly for thirteen years.[3] While this may refer to green or some other crystalline vitriol (i.e. sulphate), it probably refers to sulphuric acid (H_2SO_4). This early connection between sulphuric

[1] S. Parkes, *Chemical Essays*, Vol. 2, p. 377. [2] *Ibid.*, Vol. 2, p. 378.
[3] R. Chambers, *Domestic Annals of Scotland*, Vol. 1, p. 439.

acid and green vitriol is worthy of more than passing attention, as is also a complaint of clothiers about 1650:

> . . . that as oil of vitriol, used for bleaching, paid an excise on importation, the excise on all kinds of draperies should be abolished. . . .

because it signalizes, right from the beginning, the intimate connexion between sulphuric acid and the finishing processes of the textile trade. Green vitriol ($FeSO_4.7H_2O$), for example, was used as a *mordant* as early as 1576.[1] In response to the above complaint, the excise on *old* and *new* draperies was removed in 1654, and that on silks, linen, alum, and copperas (another name for green vitriol), two years later.[2]

By the beginning of the eighteenth century, since not more than sixty-four lb. of sulphuric acid could be obtained from six cwt. of green vitriol, the ever-increasing demand had focused attention upon processes using other substances as raw materials. A new epoch began with the use of sulphur. It lasted for over a century, until economic repercussions arising out of the Neapolitan sulphur monopoly again caused an alteration in the raw material. Sulphur alone, however, does not yield sulphuric acid. The accumulation of a considerable corpus of knowledge concerning its reactions, principally on the Continent, was necessary before the manufacture of sulphuric acid *per campanam* could be established. This process, using nitre (potassium nitrate, KNO_3) in addition to sulphur, is believed to have been introduced into England by Cornelius Drebbel (1572–1634), but may not have been operated without interruption.[3]

The focus of our first detailed knowledge of the manufacture of sulphuric acid in England is Joshua Ward (1685–1761), quack doctor, vendor of analeptic pills, and inventor of the noted white drops, who, returning in 1733 from France, where he had gone to evade justice in connexion with a Parliamentary offence, began manufacturing *oil of vitriol made by the bell* at Twickenham in 1736.[4] Whether Ward, or his partner, John White, was the chemist is uncertain, but there is little doubt, as will be seen by

[1] *Repertory of Arts*, Vol. 13, p. 269.

[2] B. M., Additional MS., 33051 f. 188, quoted by Ed. Hughes, *Studies in Administration and Finance* (1558–1825).

[3] G. Lunge, *The Manufacture of Sulphuric Acid and Alkali*, (Revised Edition, 1923), Vol. 1, p. 5. See also J. M. Jaeger, *Cornelius Drebbel and his Contemporaries*.

[4] *Gentleman's Magazine*, 1734–66, many references.

the citation of the following passage, that White supervised the chemical end of the venture.[1]

> Mr. White is the ingenious chymist who carries on the Great Vitriol Works at Twickenham for Mr. Ward, and was employed by him in other chymical preparations.

Although no local details of their operations can now be found, they appear to have been considerable in extent, since it is suggested that the cause of their removal to Richmond, which took place about 1740, was due to the offence caused to the genteel noses of Twickenham by the presence of the vitriol works.[2] Although Ward and White designated their product *oil of vitriol by the bell* to distinguish it from the acid made from green vitriol, neither was the inventor of the process. It was in contemporary operation in France and Holland, and was described in detail with illustration in Nicolas Lefèvre's *Compleat Body of Chymistry* (English Translation, 1670). Indeed it is probable that Ward may have gained first-hand knowledge of its operation while resident in France. Two points about Ward's activities are, however, worthy of notice: (*a*) the size of the glass vessels which he used—forty to fifty gallons—implies the coexistence of considerable technical skill among eighteenth-century glass-blowers; (*b*) his reduction of the price of sulphuric acid from 1*s*. 6*d*. to 2*s*. 6*d*. per oz. to 1*s*. 6*d*. to 2*s*. 6*d*. per lb., paved the way for sulphuric acid ceasing to be merely an apothecary's nostrum and becoming the chemical which prepared the way for palaeotechnics.

The empiricism characteristic of early eighteenth-century *materia medica* found in sulphuric acid a potent addition to its pharmacopœia, and applied it in medicine as a refrigerant and for stopping excessive perspiration, and in the preparation of the stomatic water of Rabel. For practical pharmacy it was important in connexion with the preparation of other derivatives, the most important of which was indubitably the *sal mirabile* of Glauber, (sodium sulphate, $Na_2SO_4.10H_2O$), the fashionable aperient of the eighteenth century. The very potency of sulphuric acid, however, may have been the cause of its rapid fall from favour, for even by 1783 Dr. Francis Home (1719–1813), first Professor of *Materia Medica* in Edinburgh indicated that he did not hold sulphuric acid in high regard.[3] It is therefore to its industrial, rather than to its pharmaceutical, applications that one must

[1] Jas. Mactear, *Proc. Phil. Soc.*, Glasgow, 1881, *13*, 409.
[2] H. W. Dickinson, *Trans. Newcomen Soc.*, 1937, *18*, 43.
[3] F. Home, *Clinical Histories and Cases*, p. 188.

turn for the impetus which stimulated subsequent rapid expansion in production.

Throughout the literature of the early Industrial Revolution we find sporadic references to the uses of sulphuric acid; more often, however, all that is said is that the uses of sulphuric acid are so well known that they do not need enumeration. Some of these uses can be gleaned from Parkes's *Chemical Essays*, without, however, much indication of the time of their introduction, excepting that it must be prior to the publication of the *Essays* in 1815. Of particular significance in view of the locus of the second sulphuric acid manufactory in England (Birmingham), was its use by hatters, paper-makers, tanners, tinplate-makers, brass-founders, button-makers, japanners and gilders, most of whom used it for metal pickling or cleaning. It was also in demand by refiners of precious metals for stripping precious metal from the base metal which it covered, the *stripping liquor*, sulphuric acid with nitre (KNO_3) dissolved in it, which has the property of dissolving the silver without attacking copper, probably having been in use for many years before James Keir (1735–1820) sent his communication about it to the Royal Society (1790). That the second persons to take up the manufacture of sulphuric acid in England should be consulting chemists, refiners and recoverers of gold and silver, and the site of their works Birmingham, need occasion no surprise. They were John Roebuck (1718–94) and Samuel Garbett (1717–1805) who, in 1746, began making sulphuric acid as an adjunct to their refinery at Steelhouse Lane, Birmingham. This is a pivotal event in eighteenth-century economic history.

This event is of two-fold significance. Considering the state of the roads in England in the middle of the eighteenth century, sulphuric acid must have been a load only transported with considerable risk. The setting up of Roebuck and Garbett's factory brought the actual manufacture of vitriol into the district where most would be used. Further, profiting by an observation by Glauber that lead is not attacked by sulphuric acid, they abandoned the glass *bell*, and built their plant of lead. The complete substitution of lead for glass removed the limits set by the size and fragility of glass vessels, and made possible a further reduction in the cost of sulphuric acid. So in the words of Liebig:

> The demand for sulphuric acid was increasing yearly and with every improvement in manufacture the price fell, and with every fall in price the sale increased.[1]

[1] J. Liebig, *Familiar Letters on Chemistry*, p. 24.

SULPHURIC ACID

Oil of Vitriol ($H_2 SO_4$)

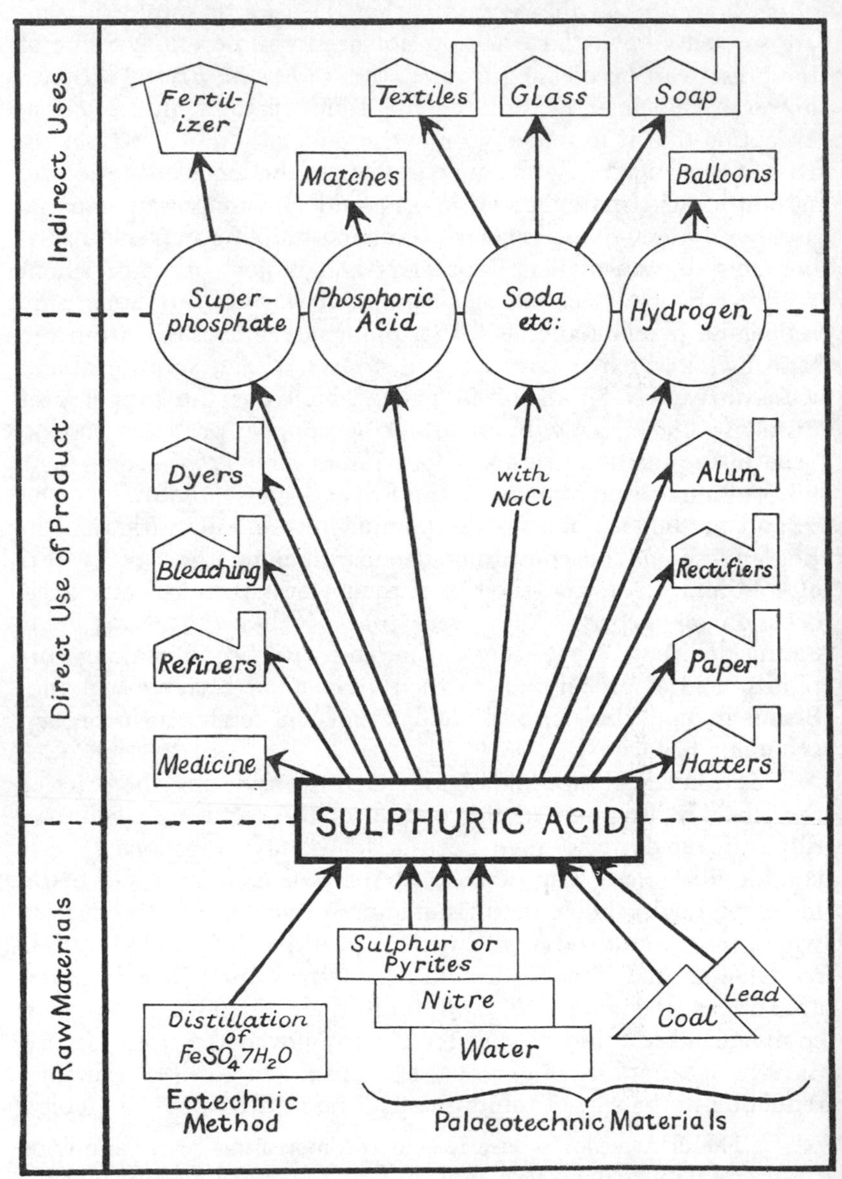

The fortunes of the Steelhouse Lane manufactory need not be followed in detail here. It continued to operate for over a century, becoming, in 1773, the sole property of Garbett, on whose failure in 1783 it went to James Alston, who took in Armitage as a partner. Garbett set up afresh as a refiner and acid manufacturer at the Lodge, Birmingham Heath. The firm finally became James Alston and Sons, only disappearing from the Birmingham Directory after 1847.[1] Alston had been Garbett's clerk, and took over when James Farquharson, a partner taken in in 1776, proved to be dishonest in his transactions.[2] Much of the output of this Midland manufactory was absorbed in the preparation of nitric acid (HNO_3), which was in great demand for dissolving copper, and for recovering silver and gold from waste gilt and plated metal.

In 1749, Roebuck and Garbett established a second vitriol works, this time at Prestonpans, on the Firth of Forth a few miles east of Edinburgh, in a district where the manufactures of salt, pottery and glass, were already established. It is invariably stated that Roebuck and Garbett used lead vessels, but the following notice, quoted by J. A. Fleming in his *Jacobean Glass*, suggests that they might have had occasion to work in glass at the beginning:

> A bottle made at Leith on 7th January, 1747 held ninety-four gallons. Later another was made to contain a hundred and five gallons. In Manchester Magazine of 15th January, 1751 we may read: 'Leith, December 28, 1750. A globular bottle has been lately blown here by Thomas Symmer, principal director of the glass works, South Leith, at the desire of several gentlemen undertakers of a private work at Prestonpans containing two hogsheads, and being measured, the dimensions are forty by forty-two inches'.[3]

We know of no other undertaking at Prestonpans to which this could refer. A suggested explanation of the move to Prestonpans is Ward's known opposition to his commercial rivals, but once again it may have been transport. In the newly established stage coach Glasgow was still twelve hours distant from Edinburgh. Be that as it may, Ward reacted somewhat violently to the setting up of the Steelhouse Lane establishment, and replied by patenting his process on 23rd June, 1749, the patent (No. 644) covering England and Wales for a period of fourteen years. A project to

[1] R. B. Prosser, *Birmingham Inventors and Inventions*, p. 16.

[2] Signed statement by J. Alston, March 1777, Assay Office Library, Birmingham.

[3] J. A. Fleming, *Jacobean Glass*, p. 112.

include Scotland did not materialise. A more probable explanation can be given in terms of Dr. Roebuck's activities before he ever settled in medical practice in Birmingham in 1745.

The following facts emerge from the only known biography of Roebuck.[1] From the Academy at Northampton, where he received a nonconformist education, he went to Edinburgh University to study medicine, and particularly chemistry, a subject which was attracting considerable attention in Scotland at that time. Edinburgh was then one of the foremost intellectual centres of Europe. From Edinburgh, Roebuck went to Leyden, the premier school of medicine in Europe, where he obtained his diploma in 1743. While in Edinburgh it is almost certain that he came into contact with the various movements, and the personnel associated therewith, which were giving expression to desire for improvements in agriculture, arts, and manufactures in Scotland.

For a decade before he took up residence in Edinburgh, one such body, the *Board of Trustees for Fisheries, Manufactures, and Improvements in Scotland*, had been doing its utmost to effect improvement in the finishing of Scottish linen by subsidizing the laying down of bleachfields, and by fostering research into new methods of bleaching. It is not possible to ascertain from the Minutes of the Trustees, now in General Register House, Edinburgh, when research began, but on 13th June, 1755, and 15th April, 1756, premiums were paid to Dr. William Cullen (1712–1790), appointed Professor of Chemistry in Edinburgh in 1756, and to Dr. Francis Home (1719–1813), for 'ingenious and useful experiments in the Art of Bleaching'. The results of Home's experiments were incorporated in a Memoir, *The Art of Bleaching*, published at the instigation of the Board of Trustees in Edinburgh in 1756. The main innovation which Home introduced was the substitution of sulphuric acid for sour milk, till then the only acid substance available as a *sour* for bleachers. Until the introduction of chlorine as the bleaching agent towards the end of the eighteenth century, the method of bleaching used was to boil with *ashes* (alkali), and then with a *sour* (acid), followed by exposure to sunlight, the process being repeated till the requisite degree of whiteness was attained.[2] Though Home was the first to publish an account of the use of sulphuric acid as a *sour*, Roebuck's biographer, Jardine, says:

> It is well known to several of Dr. Roebuck's chemical friends that he had tried it (i.e. sulphuric acid), found it effective, and

[1] G. Jardine, *Trans. Roy. Soc. Edin.*, 1796, *4*, 65.
[2] F. Home, *The Art of Bleaching*, *passim*.

> had frequently recommended it to bleachers before the date of that publication (i.e. 1756).[1]

Moreover, we know that sulphuric acid was in use in Scotland before it was manufactured by Roebuck. Andrew Brown in his *History of Glasgow* says:

> Before the year 1750, the bleachfields erected in Scotland, under the patronage of the Trustees for Fisheries and Manufactures, were supplied with what oil of vitriol they wanted in the process from England and Holland at the price of 16*d*. per lb.[2]

Thus there is circumstantial evidence that, while the use of lead freed the manufacture of sulphuric acid from many of its encumbrances, transport restrictions still prevailed, and it was to overcome these and extend the use of sulphuric acid to the linen industry of Scotland (and perhaps even Ireland), that Roebuck and Garbett settled upon Prestonpans for their manufactory. One thing can be said with certainty, it was not on account of locally available raw materials, as they imported their sulphur from Leghorn, and bought their nitre from the East India Company in London. Even the bricks of which the original factory was built were brought from Stourbridge.[3] P. S. Bebbington tells us:

> Garbett's enterprising spirit and able management rewarded them well, providing them with considerable capital for their future projects; despite his cleverness, Roebuck was not a business man. Articles were drawn up between the two, for carrying on the partnership at Prestonpans and Birmingham over a period of forty years.[4]

Pharmacy as well as bleaching, however, was an outlet for the Prestonpans vitriol. In a letter dated, Edinburgh, 15th January, 1754, from Dr. Joseph Black to Dr. William Cullen, occurs the following:

> This town (Edinburgh) is chiefly provided with genuine Glauber's salt, made at Prestonpans—one way of disposing of their oil of vitriol. . . .

but it is unlikely that the ailments of urban life, even in eighteenth century Edinburgh, could have absorbed more than a small fraction of the Prestonpans output.[5]

[1] Jardine, *op. cit.*, p. 79. [2] A. Brown, *History of Glasgow*, Vol. 2, p. 250.
[3] *Statistical Account*, Vol. 17, p. 67.
[4] P. S. Bebbington, *Samuel Garbett, 1717–1803*. Thesis, Birmingham Reference Library; Jottings of Samuel Garbett's case before Ilay Campbell, August 1783: Assay Office Library, Birmingham.
[5] J. Thomson, *Life of William Cullen*, Appendix F., p. 578.

From the *Statistical Account of Scotland*, we gather the following details of the vitriol works as they were in 1796:

> A manufactory of oil of vitriol, aquafortis, and spirit of salt is carried on here . . . of late it has extended to white ashes and Glauber salts. More than fifty men are employed, some by day and others by night. They are bound under indentures for twenty-one years, during which time they are paid weekly 6*s.* for stated wages, with a proportional allowance for extra work.

Then follow details of the raw materials consumed, the prices of products, etc. As mentioned above, the sulphur came from Leghorn, the nitre from the East India Company's sales in London, while sixty tons of local coal were consumed per week. The sulphuric acid sold at 3½*d.* per lb., aquafortis at 7½*d.*–10*d.* per lb., spirit of salt at 6*d.* per lb., Glauber's salt at 12*s.* per cwt., white ashes at £1 8*s.* per cwt., and at the time when the above *Account* was compiled, the Company also retailed manganese.[1]

The export activities of the firm can be followed from the Customs House Returns for Prestonpans, the first recorded exports being:

> 14th March, 1750/1: In the Huckster of Hamburgh, for Hamburgh 20 Large bottles containing 1,990 pounds wt. of Oyle of Vitriol. British manufacture to be exported Duty free. . . .

and two months later:

> 14th May, 1751: In the Buccleugh of Fisherraw, for Dunkirk 16 Large bottles containing 1,665 pounds wt. of Oyle of Vitriol. 3 Large bottles containing 203 pounds wt. of Aqua Fortis.

Two years elapse before another export is recorded, but the quantity has greatly increased:

> 16th Feb. 1753: In the Hope of Leith for Rotterdam 100 bottles containing 11,179 pounds wt. oil of Vitriol.

Cargoes follow to Bremen, Copenhagen, and Campvere, the average cargo by 1756 being about twenty-five thousand pounds weight contained in two hundred bottles. Import returns show that they imported sulphur—*not French*—through Campvere in seven-ton lots.[2] After 1764, Roebuck and Garbett are not mentioned in the Customs House Returns, and export was effected through their manager, Pat. Downey.

What was happening at this period is still obscure, but it seems

[1] *Statistical Account*, Vol. 17, p. 67.

[2] *Customs House Returns: Prestonpans*, in General Register House, Edinburgh.

evident that Roebuck and Garbett were not pulling together, as the following letter from Garbett shows:

> I have made a contact with K[ingscote] and W[alker][1] and Coney and Gascoigne, for seven thousand bottles of oil of vitriol annually; and we are not to sell to any body else (not even in Scotland) after Christmas next, when the whole sales are to commence on their account at 3½*d.* per lb. and six months credit as usual, for twenty-one years, expirable on two years notice. I don't let Doctor Roebuck know this until I come to Scotland.[2]

What this means is not plain, but it is certain that Bebbington draws the wrong conclusion from it. He says:

> The trade in acid was not good, and it must have been quite a relief to him to arrange a contract with one business house for the whole of their output, for this at least gave a certain market.

We have no evidence that sales were bad. An attempt by the authors to trace any of the firm's books or papers still extant was unfruitful.

For more than twenty years Roebuck and Garbett relied upon secrecy to protect them from rivals who wished to exploit their process. When the traveller B. Faujas de St. Fond visited Scotland in 1784, he found the Prestonpans factory surrounded by high walls. All operations were kept secret, and no strangers admitted.[3] The care with which they guarded the secrets of their manufacture is illustrated by the following citation from a letter of Dr. Joseph Black, who was one of Roebuck's most intimate scientific friends. A certain James Thenley desired information on the method of manufacture. Black replied:

> I am intimately acquainted with Dr. Roebuck, the contriver and principal proprietor of the work at Prestonpans, but he never dropt the smallest hint to me upon the subject, and as I know that they are at great pains to keep it a secret as being a very lucrative business, I never presumed to shew any curiosity with regard to it. I am therefore perfectly ignorant of the method they use.[4]

During those twenty years, however, Roebuck and Garbett were the unwitting suppliers of information to industrial rivals, who bribed their workmen, and offered jobs to dismissed or absconding employees from both Birmingham and Prestonpans.

[1] Garbett simply gives the initials *K and W.* There was a firm of sulphuric acid manufacturers of the name of Kingscote and Walker founded in 1772.

[2] S. Garbett to C. Gascoigne, 2nd November, 1776: quoted by Bebbington, p. 16.

[3] H. M. Cadell, *The Story of the Forth*, p. 146.

[4] Quoted W. Ramsay, *Life and Letters of Joseph Black*, p. 54.

One such, of the name of Samuel Falconbridge, was in the employment of Roebuck, in 1749, but was dismissed, and made for his home town, Bridgnorth. Being idle, he was confined in the Bridewell at Warwick, but a Mr. Rhodes, an eminent seed-crusher of Bridgnorth, paid for his liberty and was persuaded by Falconbridge to start a sulphuric acid works.[1] This happened about 1756. Some time later the same workman appears to have moved to Mr. Skey (1726–1800), a drysalter of Bewdley, also on the Severn, who had just begun to make sulphuric acid at Dowles, in chambers erected under the supervision of a workman from Prestonpans whom he met by accident.[2] Samuel Skey, Son and Company, Bewdley, are mentioned in the Boulton and Watt correspondence in the Assay Office Library, Birmingham, in 1786.

In consequence of these developments, and the presence of suspicious characters in the neighbourhood of Prestonpans, Roebuck and Garbett, on 9th August, 1771, registered a patent in fulfilment of a promise in letters-patent granted by the King to work their process in Scotland for fourteen years, their particular invention being the use of lead vessels for all operations. The parchment recording the patent can still be seen in the General Register House, Edinburgh.[3]

Roebuck tells how persons 'who kept their names and their business a profound secret' were apt to appear. One of these was Neil Macbrayne, who came in disguise to Prestonpans in June and July 1770, and entertained some of the Vitriol Works employees in his rooms. From one, whom he bribed with four and a half guineas, he discovered that lead vessels were used. Andrew Brown, in his *History of Glasgow*, indicates that the said Neil Macbrayne, whom he describes as 'a true friend to the freedom of the arts', was again active in 1771. According to Brown, he took a few workmen to Langlone, Prestonpans, 'and in the course of a summer's batheing the mystery came out in the process'. Then Matthew Machen erected a small works near Govan coal works.[4] At first Roebuck and Garbett did not comprehend the purport of these visits, but when it became clear that they were the activities of rivals, they applied for the protection of a patent.

Towards the end of 1772, Roebuck and Garbett, in terms of their patent, brought an action against Messrs. William and Andrew Stirling, merchants in Glasgow, who were known to be

[1] *Session Papers*, f. 166, No. 18.
[2] S. Parkes, *Chemical Essays*, Vol. 2, p. 399.
[3] Specification of Patents and Drawings, 1767–87.
[4] A. Brown, *op. cit.*, Vol. 2, p. 251.

erecting buildings near Glasgow with the intention of making sulphuric acid in lead vessels. On being challenged, the validity of the patent was called in question by the Stirlings.[1]

According to the respondents, they carried on 'an extensive Manufactory of Printing and Whitening Linen, in the neighbourhood of Glasgow', and 'consumed great quantities of the acid Spirit of Oil of Vitriol, in that Manufacture'. They had therefore, decided to erect oil of vitriol works of their own, and asked 'Mr. Kant, Chymist in London' to find them a manager. Kant suggested Mr. Copland, younger son of Mr. Copland of Collieston, who 'having been bred a Surgeon and Chymist, had applied himself to investigate the Methods of making Oil of Vitriol and discovered the use of Lead vessels'. Copland was offered a partnership, but refused, offering, however, to show them how to make oil of vitriol at his father's house in Galloway.[2] The Stirlings brought forward various arguments, and pointed out that another manufactory, in addition to their own, had been erected at great expense in the neighbourhood of the City of Edinburgh by Messrs. Steel, Gladstanes and Company (at Burrowmuirhead?).[3] They also made ammonium chloride.[4]

The Scottish Courts gave as their decision, that the patent was bad, on the ground that Roebuck and Garbett had practised the method for twenty years, and that the method was in general use elsewhere in Great Britain. The case was then taken to the House of Lords, where it was argued that: (*a*) the substitution of lead for glass was no new discovery, but only a slight variation; (*b*) that it was no *new* discovery, since it had been in use for twenty years; (*c*) the use of vessels of lead was by then known to various people in both England and Scotland. Lastly, it was pointed out that while the patent was in favour of Roebuck and Garbett, it was signed by Roebuck only, and therefore, should be declared invalid. Witnesses from Bridgnorth and Bewdley were called, one, the wife of Samuel Falconbridge, who had been dismissed from Prestonpans, and had communicated what he knew to the rival firm. The Lords upheld the decision of the Scottish Court, declaring that a patent obtained for an invention in Scotland was invalidated by proof of previous use in England.[5] By the time this decision was given, Roebuck had been forced to sever his connection with the firm, having become involved in the financial

[1] *Session Papers*, f. 166, 18.
[2] *Ibid.*, f. 31, 20.
[3] *Ibid.*, f. 166, 18.
[4] *Ibid.*, 307, 1.
[5] *Journal of the House of Lords*, 1774, *34*, 76 and 217; *Session Papers*, f. 31, 20; Paton's *Reports of Appeal Cases*, p. 346.

crisis of 1773. The Trustees sold the Prestonpans Works to Patrick Downey in 1782 for £11,000.[1]

During this period, Roebuck was actively engaged in assisting James Watt with the experiments on the separate condenser for the steam-engine. Lord describes the context in which Roebuck's share in the steam-engine patent passed to Matthew Boulton:

> The situation in 1769 was that Watt was engaged in constructing 'an engine of an 18-in. cylinder and 5-ft. stroke at Kinneil', and Watt says that 'Doctor Roebuck's colliery is in a very thriving condition, and daily improving', though 'some people in Birmingham have an interest in doing all in their power to lessen his character and credit'. This must have been the beginning of the end: Watt's engine was not available for pumping the colliery, and Roebuck began to be in financial difficulties. However, some of the other undertakings went well, for even in 1771, Watt says, 'our pottery does very well, tho' we make damned bad ware'. However flourishing the undertakings were in 1769, in that year an agreement was finally made by Roebuck with Boulton and his friend, Small, under which they were to take a third share, and, in exchange, to pay half the expenses.
>
> Roebuck's difficulties went on increasing, and lack of capital made him frightened of delays.
>
> Almost every private banker in Scotland failed during this period. This was an unfortunate time for Roebuck to be in difficulties. For a restriction of credit was the only method of restoring financial stability.[2]

As far as Roebuck's affairs were concerned, in 1776 he owed the firm £3,597 6*s.* 7½*d.* To balance this, he transferred some of his Carron Works shares to Garbett. The vitriol works were to belong solely to Garbett, but Roebuck was still to get a share in the profits. The Steelhouse Lane factory was also involved in the change. At the same date James Farquharson was admitted a partner there, and the firm became known as Samuel Garbett and Company. In 1772, Farquharson became bankrupt, and the name was changed to simply Samuel Garbett.[3] In 1782 Garbett also was overtaken by financial disaster.

These reverses of fortune and the overruling of the patent do not seem to have reacted adversely upon the production of the Prestonpans Vitriol Works. When B. Faujas de St. Fond visited

[1] P. S. Bebbington, *op. cit.*, p. 31.

[2] J. Lord, *Capital and Steam Power*, p. 84; Watt to Small, 28th April, 1769, 12th December, 1769, 24th December, 1771: Assay Office Library, Birmingham.

[3] *Session Papers*, f. 29, 31.

Roebuck in 1784, the works were at that date still the greatest manufactory of oil of vitriol in Great Britain.

> Roebuck lived on in obscurity till 1794, and though occasionally envious of the success of Boulton, deserves credit for his share in helping to lay the foundations of that success.[1]

It is now necessary to trace the contemporary extension of the manufacture of sulphuric acid elsewhere. We have referred to the offshoots from Birmingham and Prestonpans, which were set up at Bridgnorth and Bewdley, and the rival establishments at Govan (which may have been connected with the Stirlings), and near Edinburgh (Messrs. Steel, Gladstanes and Company). The amount of sulphuric acid being made by these firms was considerable, and its cheapness a direct impetus to extended application. The transition from ashes and sour milk as bleaching agents can be followed in the orders placed with London drysalters by Messrs. John and Nathaniel Philips and Company, tape-manufacturers of Tean, Staffs. In 1753–4 they ordered weed and wood ashes, 'a cask of Danzig ashes to look at', alum and tartar, etc., while a decade later, 1765–9, their orders are much more comprehensive, including in addition to a big variety of ashes, sulphuric acid, probably bought from Roebuck and Garbett in Birmingham.[2] The increased tempo effected by substituting sulphuric acid for sour milk was early appreciated by the finishing trades. According to Home:

> . . . the milk takes five days to perform its task, but the vitriol sours do it in as many hours, nay, perhaps in as many minutes.

Other branches of the finishing trades also made use of the cheapened acid, the dyers for dissolving indigo, calico printers for preparing *sours*, and in the preparation of citric acid, of which they used a great deal. Combrune mentions its application in fining *grey beer*, and it was also used by rectifiers and distillers of *ardent spirits* to remove fusel oil, and in the refining of rape-seed oil.[3] Jennings, in the *Bath Agricultural Papers*, speaks of it being used to prevent smut on wheat. This wide range of applications, and there were doubtless many others, stimulated a rapid development subsequent to the founding of the factories just mentioned. The quantities of sulphuric acid used during 1792–4 in the various

[1] J. Lord *op. cit.*, p. 88.

[2] Wadsworth and Mann, *Cotton Trade and Industry in Lancashire*, 1600–1780, p. 296.

[3] M. Combrune, *Theory and Practice of Brewing*, p. 274; *Mem., Manch. Lit. and Phil. Soc.*, 1798, *5*, 270.

enterprises with which Watt and Boulton were connected are known through the preservation of a statement in Boulton's writing in the Assay Office, Birmingham, which reveals that the annual consumption for 1792–4 at the Soho Foundry and its associates was 13,333 lb.

Although without definite records before 1792, it is possible that Benjamin Rowson founded a vitriol works in Bradford in 1750. In 1802 the works passed to Broadbent, and still continues under the name of the Leather Chemical Company Limited. Following this, among firms whose date of founding is recorded are two London firms; the first, Messrs. Kingscote and Walker, who began at Battersea with seventy-one chambers in 1772, but only survived for a short time; and secondly, Messrs. Thomas Farmer and Company, founded in 1778, who by contrast survived for over a century. In 1783 a nephew of Walker founded an offshoot of the former of these firms, the firm of Walker, Baker and Singleton, at Pitsworth Moor, near Manchester. Thereafter the pace of development is so rapid that detailed data for individual firms cannot be given.

By 1797 there were six to eight factories in the vicinity of Glasgow, probably those at Woodside, Napier's Hall, Port Dundas and Carntyne, while by 1814 there were also vitriol works at Carriden in West Lothian, and at Boroughmuir, near Edinburgh.

The vitriol works 'allowed to be the second of its kind in Scotland' was at Burntisland. It was founded between 1780 and 1790 by William Muir, who later went to Leith, leaving Alexander Pitcairn and other partners to carry on. By Candlemas 1790, Pitcairn was in sole possession, the other partners having died or withdrawn their shares. This works was extensive, being reputed to contain three hundred and sixty chambers. The cess (tax) paid by Pitcairn alone amounted to one twenty-fourth of the cess paid by the whole town. Between 1790 and 1805 he must have taken in fresh partners, since an advertisement in the *Edinburgh Gazette* (19th–23rd April, 1805) announces that the Burntisland Vitriol Company has been dissolved and will be continued by Alexander Pitcairn alone. Its subsequent history has not been traced, but the *New Statistical Account* (1845)[1] states:

> The vitriol works are long suspended, and the premises converted into cottages for summer visitors.

[1] *Session Papers*, 183, No. 30; J. Sinclair, *General Report*, Appendix 2, p. 307; *New Statistical Account*, Vol. 9, p. 425; J. Thomson, *General View of the Agriculture of Fife*, p. 310.

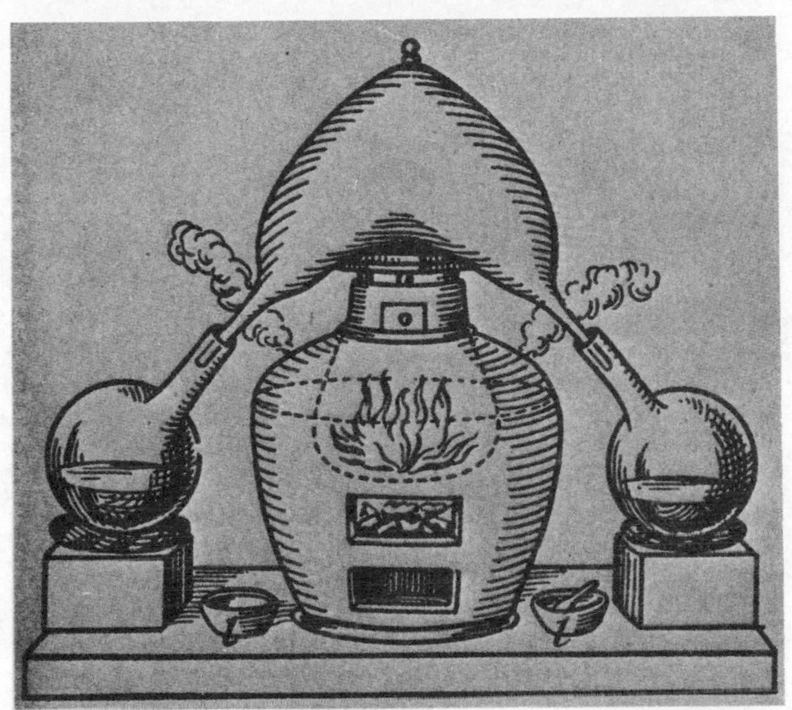

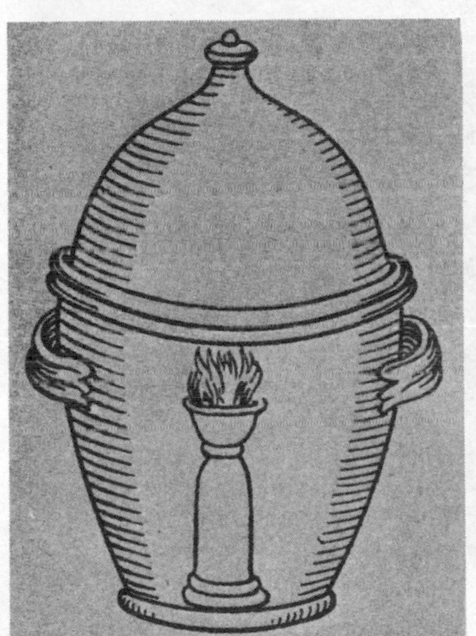

25. Vitriol manufacture *per campanam*. Various devices for the production of sulphuric acid before the introduction of lead apparatus by Dr. John Roebuck in 1746. (*Right and above from* Lefèvre *or* Le Febure; *left from* Lemery).

Engraving in the possession of Dr. Charles Singer

26. Manufacture of sulphuric acid in the first half of the eighteenth century.

27. Platinum still for concentrating sulphuric acid (*c.* 1840). The first of these was installed by Doubleday and Easterby at Bill Quay on the Tyne, at a cost of £700.

During this period two firms, intimately connected with the art of bleaching, had taken up the manufacture of sulphuric acid on their own account; first, in 1799, Messrs. Bealy of Radcliffe, Manchester, and second, in 1803, Messrs. Tennant, Knox and Company. Lead chambers were erected at St. Rollox, Glasgow, and sulphuric acid produced the following year. Messrs. Bealy had six chambers.

> They were ten feet square and twelve feet high, with a roof like a cottage, each house having a door, usually of mahogany, and a valve on top for ventilation between burnings. The floor was constructed to hold water to a depth of eight to nine inches. The 1 lb. charge consisting of a mixture of seven to eight parts of sulphur with one part of nitre was introduced upon two trays. This was lit and the doors shut for upwards of an hour until the combustion had taken place. Three hours from the time of lighting were allowed for condensation to take place, after which the door and valve were opened to 'sweeten' the chamber. This operation was repeated every four hours day and night for about six weeks, after which the acid was withdrawn and concentrated in lead vessels.

By 1820, in addition to the above factories in Scotland, there were twenty-four in England distributed, according to Mactear, as follows: London 7; Staffordshire 2; Bristol 2; Birmingham 4; Leeds 1; Halifax 1; Rotherham 1; Newcastle 1; Bolton 2; Manchester 2; Whitehaven 1. The rapid expansion which these figures illustrate is evidence of continually increasing pressure for larger and larger quantities of sulphuric acid, whose commercialization was in turn leading to the foundation of the synthetic alkali industry already discussed.[1]

To the second eighteenth-century revolution in the art of bleaching we attribute a further impetus to the expansion in sulphuric acid manufacture. Increased tempo in textiles put considerable pressure on cloth-finishers to accelerate their processes. Improvements were first effected by the substitution of sulphuric acid for sour milk, as mentioned earlier in this chapter, followed by the introduction of a new bleaching reagent, chlorine. Dry, and therefore readily transportable, bleaching powder ($CaOCl_2$) was patented in the name of Charles Tennant in 1798, Tennant already having an earlier patent for the absorption of chlorine to give a liquid bleach. There is little doubt, however, but that the working out of the process for its manufacture was the work of Charles Macintosh, then a partner of the firm.[2] In

[1] Ch. IV. [2] Ch. IX.

passing, it might be mentioned that Macintosh had, at one time, travelled on the Continent on behalf of the Prestonpans Vitriol Company.

Originally a bleacher at Darnley, near Glasgow, Charles Tennant moved to St. Rollox to manufacture bleaching powder, employing there, in 1814, from twelve to twenty men.[1] From this beginning developed one of the greatest chemical factories operating palaeotechnic chemical processes. It covered no less than ten acres by 1830, with plant engaged in the production of sulphuric acid, bleaching powder, alkalis, soap, etc. To begin with, Tennant bought sulphuric acid from the Prestonpans Vitriol Company, and from Halifax (Messrs. Norris and Sons), but in 1803 erected chambers of his own at a cost of about £50 per chamber, burning 1,000 lb. of sulphur per week. The rapid progress which this enterprise made may be attributed to the rapidity with which its promoters adopted any new technological improvements. From the beginning they burned their nitre (KNO_3) and sulphur external to the chambers proper, and four years later a floor was added to burn sulphur residues which could be bought at the low price of £5 per ton from other less progressive manufacturers. De la Follie next suggested blowing steam into the chambers instead of using water, and by 1813–14 this was adopted at St. Rollox. The annual production of sulphuric acid in Great Britain was then in the neighbourhood of three thousand tons. From Tennant, plans of sulphuric acid plant passed to Messrs. Doubleday and Easterby, Newcastle, who erected the first lead chambers at Bill Quay on the Tyne. The same firm at a cost of £700 erected the first platinum retort for rectifying acid.[2] This was probably similar to the vessel referred to by Parkes in the *Chemical Catechism*:[3]

> Some years ago I saw a vessel of platina constructed for the purpose of rectifying sulphuric acid. It holds 32 gallons, and cost a few hundred pounds; but the advantages which result from its employment are fully adequate to the expense.

Despite the large capital required for this single unit, two further retorts were soon added. The cheapness of the abundant supply of sulphuric acid which was now forthcoming began to lessen the cost of erecting the lead chambers themselves as hydrogen became correspondingly cheap and was used by lead-burners

[1] J. Sinclair, *General Report*, Appendix 2, p. 313.
[2] J. C. Stevenson and R. C. Chapman, *British Association Report*, 1863, p. 701.
[3] S. Parkes, *op. cit.*, p. 371 *n.*

in the jointing of the lead sheets from which the chambers were made.

In order to be nearer the bleachfields of Forfar and Fife, in 1836 Charles Tennant and Company of St. Rollox erected a small sulphuric acid works at Carnoustie, Forfarshire, output starting in 1838. With the introduction of artificial fertilizers, they installed a bone-crushing plant in 1846, and supplied the surrounding farmers with superphosphate and fertilizers. The works have been modernized, and now form part of Scottish Agricultural Industries, Limited.

Technological improvements brought about a gradual reduction in the price of sulphuric acid until the selling-price remained fairly constant at 3½*d.* per lb., or £30 per ton. The prices of the raw materials varied somewhat according to local conditions; sulphur from £7 to £16 per ton, nitre from £36 to £64 per ton. Although the ratio of nitre to sulphur used was small, being in the neighbourhood of one to ten, its high price had a marked effect upon the total cost of production, which worked out, on an average, at 2½*d.* per lb. or £22 per ton. A further reduction in the cost of production was effected, therefore, by the substitution of Chile saltpetre ($NaNO_3$), for nitre (KNO_3). The first cargo of the former was sold in England in 1831 at 30–40*s.* per ton,[1] thus releasing the nitre for the manufacture of gunpowder, and saving the Government hundreds of thousands of pounds annually.[2] Potash salts were also released by the economic production of sulphuric acid, because soda salts, made from brine or common salt, replaced them in the manufacture of glass and soap. Thus the potash, formerly used in these industries, but at the same time a most valuable and important fertilizer, was returned to the land.

Sulphuric acid manufacture, therefore, had an important influence on the cost of production of bleached and printed cotton goods, on soap, glass, and alkalis, and as Great Britain exchanged these commodities for such colonial products as raw cotton, silk, indigo, etc., its fundamental position in the transition to palaeotechnic economy cannot be overrated. Not only did it foster commercial ascendancy, it also ministered to social amenity at home. About 1802, William Murdock, still another member of the firm of Boulton and Watt, at Soho, Birmingham, brought gas-light within the region of practical operation. In a decade the logical complement to gas, viz., phosphorus matches, had been

[1] M. B. Donald, *Annals of Science*, 1936, *1*, 29.
[2] J. Liebig, *Familiar Letters*, p. 25.

commercialized as a result of the economic production of sulphuric acid.[1]

Towards the end of the period under discussion, sulphuric acid became a political influence of considerable magnitude. As already mentioned, the salt taxes, which during the Napoleonic Wars had risen to £40 per ton, making salt dearer than beef, constituted a serious impediment to the development of a synthetic alkali industry. Continued pressure led to the repeal of these and other fiscal duties in 1825. Some of the incidental effects of this can be gathered from the following citation from *A Memoir on Charles Macintosh, F.R.S.*, by George Macintosh.

> Since 1823, importers of sulphur, sulphuric acid and alkali makers, soap and glass makers, makers of factitious alum, in Great Britain have reaped millions at the public cost by the repeal of the duties on salt, sulphur, barilla, and tallow. They killed the kelp trade and almost killed the manufacture of alum and copperas, and by a lamentable fatality they crushed the trade in potashes, which was the branch to which emigrants, ruined in the Highland kelp trade, betook themselves in trans-atlantic colonies.[2]

Charles Macintosh himself had experienced the influence of the commercialization of sulphuric acid upon his alum works,[3] and although belonging to the new manufacturing classes himself, loyalties to the class from which he sprang made him feel acutely the distress which the new manufactures had brought in their train.

Sulphuric acid next passed into the sphere of international affairs.[4] The source from which British sulphuric acid manufacturers obtained their sulphur was Sicily. John Tennant of St. Rollox and James Muspratt entered into a joint venture and purchased some sulphur mines in Sicily. This caused a feeling of jealousy in the Neapolitans, and in 1838 a contract was drawn up between the King of the Two Sicilies and the house of Taix Aycard and Company of Marseilles in, it is said, an attempt to stabilize prices, which fluctuated due to speculation. The above contract gave a monopoly to Taix Aycard and Company, and was to reduce the annual production of sulphur from 900,000 to 600,000 cantars, while an export duty of £4 7*s.* was to be imposed. This caused the price of sulphur to rise from £5 10*s.* to £15 per

[1] Ch. XX. [2] G. Macintosh *op. cit.*, p. 181. [3] Ch. XII.

[4] *Documents concerning the Sulphur Monopoly*, consisting of the Parliamentary Inquiry into the conduct of the Foreign Secretary. London, 1841.

ton. The British Government regarded this as a breach of the Treaty of Commerce and Navigation, signed in London on 26th September, 1816.[1] In the words of Liebig:

> As the price of sulphur has such an important influence on the cost of production of so many manufactured goods, we can understand why the English Government should have resolved to resort to war with Naples in order to abolish the sulphur monopoly.

He goes on:

> Nothing could be more opposed to the true interests of Sicily than such a monopoly; indeed, had it been maintained a few years it is highly probable that sulphur, the source of her wealth, would have been rendered perfectly valueless to her.

This forecast was indeed true. The Neapolitan sulphur monopoly had a profound influence upon the subsequent technique of sulphuric acid manufacturers in Great Britain. According to R. C. Clapham:

> The effect of this was an immediate advance to £13 or £14 per ton, and a great depression and stagnation in the alkali trade. Under these circumstances, the attention of chemists was directed to pyrites, as a substitute for sulphur, which was found in abundance in Cornwall and Ireland. This mineral was first imported into the Tyne in 1840; and, in June of that year, was used by Mr. John Allen, at Felling Shore, and in September by Messrs. H. L. Pattison and Company at Felling.[2]

Attention was also turned, for example, to recovering sulphuric acid used in soda synthesis, some fifteen patents directed to this end being taken out during the operation of the monopoly, i.e. 1838–42. Interest also revived in the employment of alternative raw materials. Thomas Hills of Bromley-by-Bow had attempted the use of pyrites (iron sulphide, FeS_2) as early as 1818, but without success. Just preceding the sulphur monopoly, Messrs. Perret et Fils of Chessy took out a French patent for the manufacture of sulphuric acid from pyrites whose ultimate end was to be used as a raw material for copper-smelters. With the rise in the cost of sulphur occasioned by the monopoly, Thomas Farmer and Company, Kennington Common, began to use pyrites in 1839, Charles Tennant at St. Rollox in 1840. The decade following saw

[1] E. W. D. Tennant, *One hundred and forty Years of the Tennant Companies*, p. 61; J. F. Allen, *Some Founders of the Chemical Industry*, *passim*.

[2] *Newcastle-upon-Tyne Chemical Society*, 1871, *1*, 39.

an almost complete revolution in the raw material used, substantiating in full Liebig's dictum that:

> In commerce and industry every imprudence carries with it its own punishment, every oppression immediately and sensibly recoils upon the heads of those from whom it emanates.

Rock sulphur was, however, used at the Carnoustie Works of Charles Tennant and Company until 1870, when they made important extensions to the plant, and changed over to pyrites.

The use of pyrites instead of sulphur coincides with the end of the first century of sulphuric acid as a heavy chemical, and more or less with the complete establishment of a palaeotechnic economy in Great Britain. At this time, too, sulphuric acid began to be absorbed for the production of artificial fertilizers, and with the development of this new phase its further expansion belongs to the history of the agrarian revolution.

Chapter VII
BALLOONS

Men who are more familiar with books than with affairs are apt to over-estimate the influence of philosophers.

RUSSELL.

A CONSEQUENCE of the commercialization of sulphuric acid by Dr. John Roebuck was the development of aerial navigation. Some idea of the large quantities of acid required for the inflation of balloons will be gathered from the following data given for James Sadler's balloon which was exhibited in Manchester in 1812. The balloon had a diameter of 55 feet, a capacity of 87,114 cubic feet, and to fill two-thirds of the envelope required '4,119 lb. iron and a commensurate quantity of vitriol.'[1] James Sadler, a native of Oxford, who achieved great fame as an aeronaut, began his career as a laboratory boy in the Old Ashmolean Museum. It was in all probability there that he first became acquainted with hydrogen. Despite his spectacular success as an aeronaut, he returned to chemical pursuits in 1785, and continued Inspector of Chemistry to the Admiralty till 1807, when that post was suppressed without compensation.

An early suggestion of man's conquest of the sky in relation to advancing knowledge of physical conditions is contained in Joseph Galien's *L'Art de Navigeur dans les Airs* (1755), where it suggested that man might sail the skies in a huge ark filled with air from the upper regions, which Galien believed would be less dense than air at sea level. The same ideas were again put forward in 1766, after the determination of the density of hydrogen by Cavendish, as a result of which it was discovered that hydrogen produced by the action of metals on sulphuric acid, was, under identical conditions, less dense than air. Cavendish's experiments are recorded in the first part of his *Three Papers, containing Experiments on Factitious Air*, read before the Royal Society.[2] Hodgson says of Cavendish:

> A foremost place in the chemistry of aerostation may be claimed for Henry Cavendish (1731–1810). Though the existence of inflammable air from metals (hydrogen) had been recognised long

[1] T. E. Hodgson, *History of Aeronautics*, p. 151.
[2] H. Cavendish, *Phil. Trans.*, 1766, *56*, 144.

before the days of Cavendish, he was the first, not only to demonstrate the exact nature of it, but also the first to afford an approximate estimate of its density as compared with air, wherein lies the essence of his connexion with ballooning. It was on reading the results of Cavendish's experiments on the specific gravity of hydrogen, that Black suggested for the first time the aerostatic principle of the balloon.[1]

Thus the idea of filling balloons with hydrogen rendered cheap by Roebuck's acid was due to Roebuck's contemporary, Joseph Black, Professor of Chemistry in Edinburgh. Hodgson tells us:

It was Cavendish's announcement as to the specific gravity of hydrogen, made in 1766—nearly twenty years before the balloon became a *fait accompli*—that led Joseph Black (1728–99) to make the suggestion by reason of which he is most nearly associated with the invention of ballons.

His own account is to be found in two letters on the subject of aerostation, written in October and November 1784 in reply to Dr. James Lind (1736–1812), the Scottish scientist and inventor of an early type of anemometer, who was at this time living in Windsor as physician to the Royal Household. Lind had written asking Black to furnish him with particulars as to his experiments, and the name of 'the person that first discovered the true specific gravity of inflammable air', for use by Cavallo in his *History of Aerostation*, on which the latter was at this time engaged. Black replied (on November 13, 1784), that in 1766, after reading the paper in which Cavendish gave the results of his experiments on the specific gravity of hydrogen, it occurred to him 'as an obvious consequence of Mr. Cavendish's discovery, that if a sufficiently thin and light bladder were filled with inflammable air, the bladder and contained air would necessarily form a mass lighter than atmospheric air, and which would rise in it'.[2]

This was effected in 1782 by Tiberius Cavallo (1749–1809), who filled soap bubbles with hydrogen and found that they rose to the ceiling of the room. A full-scale experiment was, however, first realized, not with hydrogen, but by employing air, rendered less dense by heating, to fill the envelope. The experimenters were Joseph Michel Montgolfier (1740–1810) and Etienne Jacques Montgolfier (1745–1799), paper-makers at Annonay, near Lyons. They carried out a public trial on the 5th June, 1783, at Annonay. The balloon, made of linen, was filled with hot air, and rose to a height of six thousand feet. While the actual dis-

[1] T. E. Hodgson, *op. cit.*, p. 85.

[2] *Ibid.*, p. 88; quoting W. Ramsay, *Life and Letters of Joseph Black*, M.D., p. 77.

covery is attributed to Etienne Jacques Montgolfier, the younger of the two brothers, later research has shown that it is to Joseph Michel, the elder, that the honour is primarily due.[1]

Black, in his second letter, says:

> As you speak of the *birth* of aerostatic experiments, I beg leave to communicate to you more fully my thoughts on that subject. In the first place, although what I have already informed you of is strictly true, I by no means set up my claim for merit in the invention of machines for general flights and excursions. The experiment with the bladder, which I proposed as a striking example of Mr. Cavendish's discovery, was so very obvious that any person might have thought of it; but I certainly never thought of making large artificial bladders, and making these lift heavy weights, and carry men up into the air. I have not the least suspicion that this was thought of anywhere before we began to hear of its being attempted in France, and I do not doubt that what has been published in the newspapers is perfectly true, viz. that Mons. Mongolfier [*sic*] had sometimes before conceived the idea of flying up into the air by means of a very large bag or balloon of common air, simply rarifyed by the application of Fire of Flame.
>
> The idea being founded upon a principle which has long been known, and which has no connexion with Mr. Cavendish's discovery, it is only surprising that Mons. Mongolfier should not have put it sooner in practice. I suppose therefore, that though he might have formed the Project a long time before, he never was roused into an operation for making the trial until others began to think of flying by means of inflammable air. Who first thought of the method I cannot tell, for I confess I did not read the history of these Experiments; they never interested me in the least.[2]

The following extract from Sir John Sinclair's Correspondence throws further light on Dr. Joseph Black's influence on the Montgolfier brothers. Sinclair says:

> Towards the conclusion of the year 1785, some circumstances occurred, which induced me to take a short excursion from London to Paris, and accidentally I went in company with three distinguished foreigners, namely, *Argand*, so well known for his improvements in the art of making lamps; *Reveillon*, the greatest manufacturer of paper hangings then known (having about five hundred workmen in his employment); and *Montgolfier*, so celebrated for his discovery of balloons. I was able to obtain much information from the conversation of these intelligent men; and

[1] Hodgson, *op. cit.*, p. 95. [2] W. Ramsay, *op. cit.*, p. 79.

I remember, in particular, that the latter (M. Montgolfier), gave an account of the origin of his discovery of which the following is the substance.

Montgolfier said that he and his brother were paper manufacturers in Languedoc, but he had always felt a strong attachment to chemical and mathematical inquiries. They were thence led to procure all the information they could regarding those subjects. It seems that Montgolfier and his brother had talked over the possibility of being able to ascend themselves, or to send up large bodies from the earth, at a very early period, without, however, having made any experiments to prove whether the idea was practicable or not; but happening to read an account of some experiments made by Dr. Black, which explained the nature of the various kinds of airs or gases, and, in particular, their difference in point of weight, he immediately said to his brother, 'The possibility of effecting what we talked of some time ago, seems to be proved by a foreign chemist: let us try some experiments to ascertain its practicability'. The progress of the discovery afterwards is well-known. It is doubtful whether the attempt would ever have been made, had not the brothers been paper manufacturers as well as chemists; but the point which should be generally known is this, that had it not been for Dr. Black's discoveries, no experiment would probably have been tried by the two Montgolfiers. This I can assert upon the evidence of the elder Montgolfier, who was one of the most candid and able men I have met with, and who always mentioned Dr. Black with the respect to which he was so peculiarly entitled.[1]

An account of Black's experiment, referred to above, is given by Thomas Thomson (1773–1852), Black's successor in Glasgow University.

Soon after the appearance of Mr. Cavendish's paper on hydrogen gas, in which he made an approximation to the specific gravity of that body, showing that it was at least ten times lighter than atmospheric air, Dr. Black invited a party of his friends to supper, informing them that he had a curiosity to show them. Dr. Hutton, Mr. Clarke of Elden [*sic*], and Sir George Clarke of Pennicuik [*sic*] were of that number. When the company invited had assembled, he took them into a room. He had the allantois of a calf filled with hydrogen gas, and upon setting it at liberty, it immediately ascended and adhered to the ceiling. The phenomenon was easily accounted for: it was taken for granted that a small black thread had been attached to the allantois, that this

[1] Sir J. Sinclair's *Correspondence*, Vol. 11, p. 433.

> thread passed through the ceiling, and that some one in the apartment above, by pulling the thread, elevated it to the ceiling, and kept it in this position. This explanation was so probable, that it was acceded to by the whole company; though, like many other plausible theories, it turned out wholly unfounded; for when the allantois was brought down, no thread was found attached to it.[1]

Only a few weeks after the Montgolfiers' experiments, the first trial of a hydrogen balloon was put in operation by the brothers J. A. C. and Robert Charles. J. A. C. Charles was a physicist and worked under the stimulus of B. Faujas de St. Fond (1741–1819), the geologist. St. Fond was keenly interested in aerial navigation, studied hydrogen and the varying density of the different layers of the atmosphere, and was in charge of the arrangement for the Charles ascent. He also opened a subscription list to meet the cost of the experiment. About 1783–4 he published *Description des Expériences de la Machine Aérostatique de MM. de Montgolfier* (Paris) 1783–4, two 8vo volumes.[2]

The balloon of the brothers Charles was made of silk treated with rubber, an interesting precursor of textile waterproofing by Charles Macintosh in the thirties of the next century. The balloon was released in the Champs de Mars in Paris on 27th August, 1783. The experiment was repeated by the Montgolfiers on 19th September.

Human flight was next achieved. On the 15th October in the same year, J. F. Pilâtre de Rozier, a young scientist (1756–85) ascended eighty feet in a captive Montgolfier balloon from the Faubourg Saint-Antoine in Paris. A free flight was undertaken by the same aeronaut on 21st November, still using hot air, but human ascent by J. A. C. and Robert Charles in a hydrogen balloon followed on 1st December of that year.

In November 1783, the first balloon to go up from British soil was released by an Italian refugee, Count Francesco Zambeccari, from Cheapside, London. The balloon was made of oiled silk, gilded to make it less permeable.

Interest in balloons soon increased to great intensity. Aimé Argand (died 1803), the inventor of the Argand lamp referred to by Sinclair, demonstrated the ascent of a hydrogen balloon on 26th November, 1783, to George III and the Court at Windsor. The King had already displayed an interest in balloons, and had written to Sir Joseph Banks offering to subsidize experiments if

[1] Ramsay, *op. cit.*, p. 82; Hodgson, *op. cit.*, p. 91.

[2] A. Geikie, *Memoir of St. Fond*, p. 17.

they were undertaken by the Royal Society. The reply from this august body was that in their opinion 'no good whatsoever' could result from such experiments. The first human flight over Britain followed. Different authorities say it was made by different people. According to Geikie, in his *Memoir of Faujas de St. Fond*, the first Englishman to go up in a balloon was Dr. John Sheldon (1752–1808). In concert with a Col. Gardner, A.D.C. to Sir William Howe in America, he made a balloon of varnished linen in Lord Foley's garden in Portland Place, and filled it with hot air.[1] The actual details are obscure.[2]

The other person said to have made the first flight was James Tytler (1747–1804). Tytler was a native of Forfarshire and had studied medicine under Dr. Black at Edinburgh, and later became the editor of the Second Edition of the *Encyclopaedia Britannica*.[3] In August, 1784, Tytler ascended from Comely Gardens, Edinburgh, in a balloon filled with hot air, and came down later at Restalrig.[4]

By this time a celebrated French chemist, Guyton de Morveau, had made a flight to a height of over ten thousand feet, accompanied by the Abbé Bertrand, during which they engaged in barometric and thermometric observation. Thus the balloon early became a tool of the meteorologists.

During Sheldon's ascent with Blanchard from Little Chelsea on 16th October, 1784, Cavendish made observations on altitudes and balloon velocity in conjunction with Nevil Maskelyne (1732–1811), Astronomer Royal, and Dr. William Heberden (1710–1801), physician. The observations were printed in Blanchard's *Journal* and are the first of the kind made in England.[5]

The inflation of the balloon was carried out under the direction of Aimé Argand, who will be noticed in other connexions.[6] The balloon was made of varnished silk, the pieces of which were fixed together by passing a hot iron over them. Leakage was prevented by further sewing the seams and covering them with a riband. The varnish consisted of elastic gum dissolved in five times its weight of spirits of turpentine, 1 oz. of which he boiled in 8 oz. of drying linseed oil.[7]

The next ascents were made from a number of English and

[1] A. Geikie, *Memoir of St. Fond*, p. 37, *n.* [2] Hodgson, *op. cit.*, Ch. IV, *passim.*

[3] The Third Edition of the *Encyclopaedia Britannica* is the first to have an account of *Aerostation* (1797). In the Supplement to the Fourth, Fifth and Sixth Editions, there is a rewritten account by Sir John Leslie, under the title *Aeronautics.* (Vol. 1, p. 651.)

[4] J. Tytler, *Short History of the Edinburgh Fire Balloon*, printed in V. Lunardi's *Account of Five Aerial Voyages in Scotland* (1786). *See also Statistical Account*, Vol. 5, p. 391. [5] Hodgson, *op. cit.*, p. 87. [6] Ch. XIX. [7] Hodgson, *op. cit.*, p. 163.

Scottish towns by an Italian, Vincenzo Lunardi (1757–1806), using a hydrogen-filled envelope.[1]

Another chemist, the Aberdeen-born Dr. George Fordyce (1736–1802), is associated with the introduction of balloons into England. Fordyce offered both scientific advice and practical assistance to Lunardi, on the inflation of the balloon for his first ascent in London. According to Hodgson, 'Fordyce's first attempt to inflate a balloon was not wholly successful, though how far this was due to his inexperience or the carelessness of his helpers cannot be known. It is said that when at midnight on 14th September he left the Artillery Ground to snatch a few hours sleep, the workmen he left in charge of the gas-making apparatus got drunk, so that Fordyce on returning at four o'clock in the morning, found nothing further had been accomplished—a contretemps which might have had disastrous consequences, and which actually did cause Lunardi much distress.'

Of Lunardi's visit to Scotland, Hodgson tells us:

> Lunardi travelled to Edinburgh—where he arrived on September 12th—drawn thither partly by reason of the sincere friends he had found amongst Scotchmen in England, and partly by the 'inspiring thought' of being the first aeronaut in Scotland. His attractive personality appears to have made a ready appeal to men of distinction in the ancient capital, and within a fortnight Lunardi—with the help of Henry Erskine, who had been Lord Advocate of Scotland in 1783—had arranged for an ascent from George's Square on October 5th. His hopes, however, were destined to be disappointed, owing to the opposition of a lady resident, but after some little delay he obtained permission, thanks to the good offices of Sir William Forbes, to ascend from the gardens of Heriot's Hospital. But again Lunardi was plunged into despair owing to the slow transit (over ten days from Liverpool) of his hydrogen 'apparatus'—'the airy Vehicle', as he facetiously termed his balloon, having previously arrived— and eventually he was obliged to get a plumber to make two lead cisterns, 14 feet long, 4 deep, and 4 wide, for use in place of barrels.[2]

This use of lead is of great interest in the light of Dr. Roebuck's use of lead in the vitriol works at Prestonpans. Although it was known that lead was resistant to the corrosive effect of sulphuric acid, there are, as far as we know, no illustrations of a generating apparatus using lead. Barrels were apparently universally employed.

[1] Aimé Argand to Matthew Boulton, 16th September, 1784.

[2] Hodgson, *op. cit.*, p. 132.

On the 7th January, 1785, the Channel was crossed by J. P. Blanchard (1753–1809), a French aeronaut, and John Jeffries (1744–1819), an American physicist. The interest in balloons was enormous. Bowden[1] tells us that popular journals[2] introduced departments dealing with 'Aerostatic Experiments'. A play produced at Covent Garden in 1784 was entitled *Aerostation*, and the plot centred around 'the passion of a lady of fortune for balloons'. According to a rhymester of the time, 'Admirals forsake the swelling tide', and 'surgeons leave their patients to their fate', in order 'high on the wings of mighty winds to ride'. The 'celebrated aeronauts' of the time were 'sumptuously entertained'. The exploits of a balloonist in 1785 were said to have been witnessed by more than 40,000 persons. Not content with the amusement they afforded, those of prophetic inclinations foresaw the time when 'the inquisitive turn of mind which distinguishes the present era will improve . . . the art of ascending and exploring the upper regions . . . (so as to apply it) to many useful purposes of which at present we have no conception'.

Members of the Soho group in Birmingham were among those who became interested in balloons, particularly John Southern, F.R.S. (1758–1815).

In the Boulton and Watt Collection (Birmingham Reference Library) there is a copy of the *Journal de Paris* (23rd December, 1783) which had been sent to James Watt. It contains calculations of the powers of ascension of Charles's balloon, and this may have inspired the publication of Southern's pamphlet. He was also the designer of a lead-lined box for the generation of hydrogen for the inflation of balloons, but we have no evidence of its ever having been used. Then Aimé Argand, already mentioned, sent drawings of balloons to Boulton, and tells him of Montgolfier's generation of hydrogen from water:

> Mr. Montgolfier was the first to convert water into inflammable air at least in France, for I have an idea that Mr. James Watt found it a consequence of his system.[3]

Boulton also made enquiries at M. Pradeaux, who referred him to Faujas de St. Fond's book, and gave him details of the experiments of Pilâtre de Rozier at Versailles, etc.[4]

[1] W. Bowden, *Industrial England in the Eighteenth Century*, p. 17.

[2] *New London Magazine*, I, 1785, 93, 301; *European Magazine*, VI, 394; *Gazetteer*, 1785, April 2nd; *Gentleman's Magazine*, LV, 522, 1002; *Annual Register*, 1784–5, 223 (Chron.). 323 (Chron.).

[3] Aimé Argand to Matthew Boulton, 1st May, 1784, 4th July, 1784, A.O.L., Birmingham.

[4] M. Pradeaux to Matthew Boulton, 30th July, 1784, A.O.L., Birmingham.

On behalf of James Watt, Southern carried out experiments on latent heat, and as a result of his interest in balloons, published a tract on them at Birmingham in 1785.[1] T. Cavallo in his *History and Practice of Aerostation*, 1785, mentions a letter from James Watt describing an aerostatic experiment made by Boulton in 1784.[2] On this occasion, Watt and Boulton used a balloon filled with a mixture of hydrogen and oxygen to investigate the cause of the growling of thunder, i.e. to determine whether it was caused by repeated flashes of lightning, or by echoes. Here are the details, as communicated by Watt to J. Lind:

> The history of Mr. Boulton's experimental balloon, is as follows: He made a balloon of thin paper, and varnished it with oil varnish. The size was about five feet diameter. It was filled with a mixture of about one part common air, and two parts inflammable air from iron. In the neck of the balloon he tied a common squib, or serpent, to which was fastened a match about two feet long, which was made very quick at the end next the serpent. When the balloon was filled, the match was lighted and the balloon launched.[3]

Erasmus Darwin, too, frequently makes references to balloons. The experiments carried out by James Watt and Joseph Priestley on the composition of water enabled balloons to be filled with hydrogen at less cost than that originally involved. Priestley devised a furnace for making hydrogen in considerable quantities from water vapour and a reducing agent. He observed to Faujas de St. Fond that by increasing its size, balloons might be filled at small expense and without the trouble and cost involved in the use of sulphuric acid.[4]

Of Priestley, Hodgson says: 'Joseph Priestley—though not so directly interested in aerostation, must always be associated with the discovery of the balloon, if only because Montgolfier admitted his indebtedness to this great English chemist, for ideas derived from his *Experiments and Observations on Different Kinds of Air*, 1774. Though it is said that Priestley's experimental work on gases was inspired by the researches of Black and Cavendish, his own contributions include the independent discovery, in 1774, of what he termed 'dephlogisticated air', or oxygen, as it was afterwards called by Lavoisier.

[1] J. Southern, *A Treatise upon Aerostatic Machines, containing Rules for calculating their Powers of Ascension.* [2] T. Cavallo, *op. cit.*, p. 152.

[3] J. Watt to J. Lind, 26th December, 1784.

[4] *Travels of Faujas de St. Fond*, Vol. 2, p. 351; *Description des Expériences de la Machine de MM. de Montgolfier*, etc., 2 Vols., Chez Cuchet, 1783–4.

That Priestley was interested in assisting the earliest aeronauts in England in the matter of obtaining hydrogen, is evidenced in a letter from Richard Price—the political and economic writer—to his friend Benjamin Franklin, which relates that 'Priestley had discovered a method of filling the largest balloon with the lightest inflammable air in a very short time and at a very small expense'—two factors which gave the pioneers of ballooning considerable trouble, and which it was very desirable to overcome.

This process Priestley himself explained to Faujas de St. Fond—allowing him to take a drawing of his apparatus for the benefit of French chemists working at the subject—when the French scientist visited him near Birmingham in the autumn of 1784. The 'simple and ingenious' apparatus was described by Faujas (in his narrative of *A Journey through England*) as a tube, thick and long, made of copper and cast in one piece to avoid joints.

> The part exposed to the fire was thicker than the rest. Into this tube (Priestley) introduced filings or slips of iron, and instead of dropping in the water, he preferred making it enter as vapour. The furnace destined for this operation was heated with coke made from coal, the best of all fuels for the intensity and equality of its heat. By these means he obtained a considerable quantity of inflammable gas of great lightness and without any smell. He observed to me, that by increasing the apparatus and using iron or copper tubes of a larger calibre, aerostatic balloons might be filled at small expense and without the trouble and cost involved in the use of vitriolic acid.[1]

Aimé Argand wrote to Boulton telling him of his expectations from this method:

> You have heard, I suppose, of Blanchard's experiment, which was committed to my direction. I was very successful in filling the balloon in two hours, and with half the expense. But now we may hope to get inflammable air very cheap by obtaining it from *steam* passing through red hot iron, which in that case yields one more of very pure inflammable air than when dissolved in the vitriolic acid.[2]

From the foregoing it will be seen that the balloon was early used for scientific purposes, the first time being by the physicist John Jeffries to collect specimens of the upper air for Cavendish. The specimens were analysed by the latter chemist, and found not to differ in composition from air at sea-level.

[1] F. de St. Fond, *Journey through England in 1784*, p. 351.

[2] A. Argand to M. Boulton, 23rd October, 1784, Assay Office Library, Birmingham.

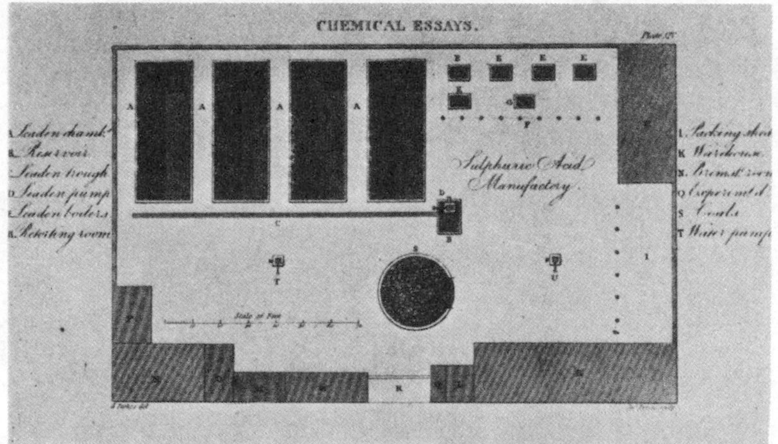

S. Parkes, *Chemical Essays*, 1815

28. Earliest known plan of a sulphuric acid manufactory.

29. The Lead Chamber Process *c.* 1840. Probably the earliest view of the plant used for the process which was, for many years, the focus of the whole heavy chemical industry.

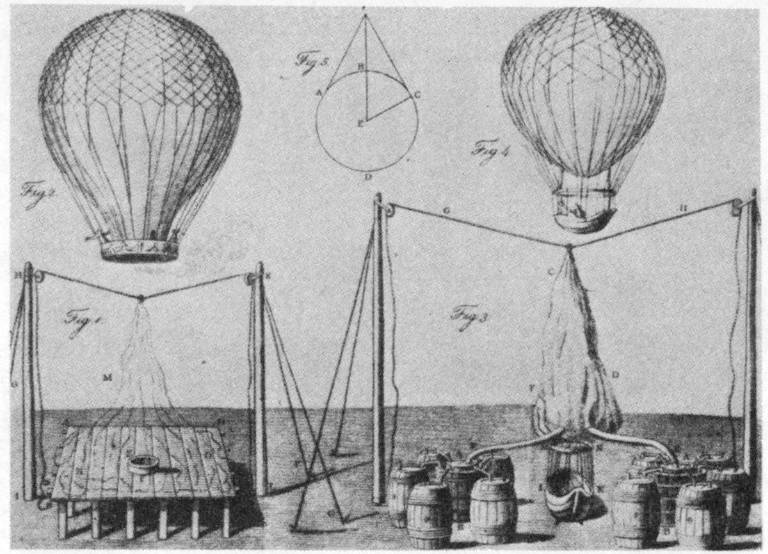

T. Cavallo, *History of Aerostation*, 1785

30. Physical and chemical balloons. Note the barrels used for generating hydrogen from sulphuric acid and zinc or iron.

31. Cartoons depicting the ascent and descent of the first balloon carrying living freight, 1783.

The large quantities of materials consumed in the inflation of a balloon were mentioned at the beginning of this chapter. Cavallo gives some more detailed figures:

> Suppose that the balloon is thirty feet in diameter, then its capacity is 14,137 cubic feet; and for the production of such a bulk of inflammable air, there are required about 3,900 pounds of iron turnings, 3,900 pounds of vitriolic acid and 19,500 pounds of water. As the balloon should not be above three-quarters filled, it is evident that the above-mentioned quantities are rather greater than required; but it is always proper to have more materials than what are just sufficient.[1]

The equation which represents the reaction between the iron and the sulphuric acid is:

$$Fe + H_2SO_4 = H_2 + FeSO_4$$

iron + sulphuric acid = hydrogen + ferrous sulphate

It will be seen that an equivalent quantity of copperas (ferrous sulphate, $FeSO_4$) is produced. This was evidently sold to dyers and printers, because Cavallo suggests that zinc might be used in place of iron, since the white vitriol (zinc sulphate, $ZnSO_4$) could be sold at a higher price than the copperas.

> White vitriol is said to be sold much dearer than the vitriol of iron. If this is true, it will be a saving to make the inflammable air by means of zinc and vitriolic acid, rather than from this acid and iron; because the sale of the white vitriol, arising from the former, will, in a great measure, compensate for the expense of the materials.[2]

Only five days after the first human ascent, and less than four months after the practical realization of the invention, André Giraud de Vilette, who accompanied Rozier on one of his early flights, suggested use of balloons in another direction, for reconnaissance in warfare. A pamphlet on the military value of the balloon, *The Air Balloon: Or a Treatise on the Aerostatic Globe* was published in November 1783. Benjamin Franklin, the American scientist, perceived the immense tactical value of an airborne army. Writing to John Ingenhausz on 16th January, 1784, he said:

> The invention of the balloon appears, as you observe, to be a discovery of great importance. Convincing sovereigns of the folly of wars may perhaps be one effect of it, since it will be impossible for the most potent of them to guard his dominions. Five thousand balloons, capable of raising two men each, could not cost more

[1] T. Cavallo, *op. cit.*, p. 303. [2] *Ibid.*, p. 325.

> than five ships of the line, and where is there a prince who could afford to cover his country with troops for its defence as that ten thousand men descending from the clouds might not in many places do an infinite amount of damage before a force could be brought together to repel them?[1]

It was not until the French Revolution that advantage was taken of the balloon as an aid to military operation. Early in 1793, Guyton de Morveau pointed out the usefulness of balloons to Lazare Carnot, and one of the Montgolfiers suggested their use for dropping explosives. Gaspard Monge (1746–1818), mathematician and professor at Mézières, pointed out that captive balloons might be used by the Republican armies, and on 25th October, 1793, an order was passed authorizing the construction of a balloon. The chemist, Jean Marie-Joseph Coutelle, together with other scientists and engineers, was put in charge of operations.

According to F. S. Haydon[2] they were not allowed to use sulphuric acid for generating the hydrogen as 'every available supply of this substance was sorely needed for the manufacture of munitions.' Consequently, Coutelle is supposed to have used the method of preparation suggested by Priestley, i.e. the decomposition of water, but an original watercolour of the ascent of the balloon at the battle of Fleurus, by Nicholas Conte, one of Coutelle's collaborators, clearly shows barrels used for the generation of hydrogen from acid, like those figured in Hodgson's *History of Aeronautics*.[3] The whole project in any case was received with indifference in the field, and more experiments followed. As a result a balloon company, the *1er Compagnie d'Aérostatiers*, was constituted by an Act of April 1794. A special knowledge of chemistry, sketching, carpentry, and masonry was the qualification for admission to the company. At the battle of Fleurus it went into action for the first time and played a decisive part in the course of the operations. So successful were the *Aérostatiers* during the following winter, that in the summer of 1794, an *Ecole Nationale Aérostatique* was established. Governmental interest in military aeronautics subsequently declined, however, and the school was closed before the end of the century.

Aeronautics of sorts continued, however, in other countries. The son of the Earl of Dundonald, then Lieutenant Lord Cochrane, towed kites from his brig *Pallas*, releasing pamphlets from

[1] *The Complete Works of Benjamin Franklin*, VIII, 432.

[2] *Aeronautics in the Union and Confederate Armies*, *passim*.

[3] Fig. 10. (*See also Coutelle, J.M-J.; Sur l'aérostat employée aux armées de Sambre-et-Meuse et du Rhin* (*Paris*, *n.d.*), pp. 3–4.)

them along the French coast.[1] In his *Aeronautics in the Union and Confederate Armies*, from which much of the above military information is taken, Stanbury Haydon tells us:

> In America, the Employment of military aeronautics remained practically unknown until the Civil War. The science of penetrating and navigating the upper air, however, had attracted attention very soon after the successful experiments of the Montgolfiers and Charles in France. Thomas Jefferson mentions experiments with fairly large balloons in Philadelphia as early as May 1784. The following year, an American, Dr. John Jeffries, made thc first aerial voyage across the English Channel in company with the French aeronaut, Jean Pierre Blanchard; and experiments with hydrogen balloons and envelopes inflated with coal gas were conducted at the College of William and Mary in 1786. The first successful ascension by a man in the United States did not take place until January 9, 1793, when Jeffries' protégé went aloft at Philadelphia. Though these examples indicate that interest in aeronautics had quickly spread to America, the application of the science of military tactics there was slow to develop.[2]

Returning again to Europe, in the 1830 campaign in Algeria, French military aeronautics re-enter the picture, and this time we know that sulphuric acid was involved in the generation of the hydrogen because:

> Jean Margat, a professional balloonist of some reputation, was employed by the Minister of War to serve with the African expedition. His service began rather inauspiciously. While his train was *en route* to Africa in the brig *Vittoria*, his carboys of sulphuric acid were accidentally broken by the rolling of the ship, and caused a fire that resulted in damage amounting to some 80,000 francs.[3]

The above statement that coal gas was experimented with as an inflating medium for balloons is open to suspicion because, although the idea of obtaining a 'spirit of coals' by destructively distilling coal, was known, there is no record of a large-scale preparation of coal-gas till Wm. Murdock's experiments at the Works of Boulton and Watt in Birmingham. When the commercialization of coal-gas lighting was effected, it made a new material available for the inflation of balloons.[4]

When John Clayton (1693–1773) carried out his experiments

[1] Thomas Cochrane, Tenth Earl of Dundonald, *The Autobiography of a Seaman*, Vol. 1, p. 201.

[2] Haydon, *op. cit.*, p. 23. [3] *Ibid.*, p. 17. [4] Ch. XIX.

on the destructive distillation of coal, and filled bladders with the gaseous products of the decomposition, he must have been very near to observing the buoyant effects of the less dense coal-gas. This, however, does not appear to have been done until M. Thysbaert filled a balloon with coal-gas at Louvain in 1784. (The lecture-hall of the University of Louvain was one of the first buildings to be illuminated with coal-gas lighting.) In the next year Cavallo recommended the use of coal-gas in aerostatics, but no use appears to have been made of it until Charles Green (1785–1870) adopted this new filling. According to Hodgson, 'it is said that Green's attention was first directed to aerostation as the outcome of experiments he tried with an apparatus for making gas, by means of which he proposed to light his own premises. He observed that the first distillation gave the best results for illumination, and that towards the end of the process the flame was scarcely visible. "Suspecting"—to quote from the obituary notice of Green in the *Fourth Annual Report of the Aeronautical Society*—"that this must be nearly pure hydrogen, he inflated some small balloons with the gas obtained at various stages of distillation".'

Green's first ascent was from the Green Park on the occasion of the coronation of George IV, with gas supplied from the Piccadilly main of the 'Gas-Light Company', the date 19th July, 1821. A contemporary advertisement throws an interesting light on the difficulties of handling sulphuric acid. Of a proposed ascent by Green it was said:

> As this Stupendous BALLOON will be inflated by a large Pipe from the Company's Main, all Accidents to Ladies' Dresses, &c., from the action of Acid will be avoided.

He later published further reasons for choosing coal-gas:

(1) Coal-gas is less injurious to the silk envelope.
(2) It renders the inflation of the balloon a quicker and more certain operation.
(3) It is not so penetrating, i.e. does not diffuse as quickly through the envelope.
(4) It is much cheaper than hydrogen. (Green computed that at this time the cost of 6 inflations with coal-gas was not more than 1 with hydrogen.)

Another of Green's innovations was the use of an 'elastic Indian rubber cord from the factory of Mr. Sievier' in place of the usual cable.[1]

[1] Hodgson, *op. cit.*, Ch. XI, *passim.*

Chapter VIII
COGNATES TO THE TEXTILE INDUSTRY

Science and industry form a power to which it is dangerous to present impediments.

LIEBIG.

THE following four chapters on bleaching, dyeing, calico printing, and mordant production, deal with technological development in industries dependent on the application of chemistry to textile finishing. This took place over the same period as the application of mechanical technology to them. In the specific textiles under discussion, namely linen and cotton, three overlapping phases of development are discernible. These are determined in large measure by the relative state of mechanical invention. As it is desirable to view the evolution of textile manufacture from within the framework of both mechanical and chemical technology, it is necessary to look at the content of the three phases in turn. To do so adequately would require space beyond our disposal, and for a detailed economic analysis the reader should refer to Henry Hamilton's *Industrial Revolution in Scotland*, or to the more specific treatises of Warden, Gill, Daniels, *et al.*

The first phase is coextensive with the rise of the Scottish linen industry, but naturally does not follow it throughout its whole history. It may be said to date[1] from the same period as the other eotechnic foundations in Scotland which were discussed in Chapter I. It passes generally into a second phase between 1760 and 1780.

In 1727 two important Acts became law: (*a*) an Act for the better Regulation of the Linen and Hempen Manufactures in Scotland, and (*b*) an Act for encouraging and promoting Fisheries and other Manufactures and Improvements in Scotland.

The latter brought into being the Board of Trustees for Manufactures (1727–1832), which made it its business, together with the Society of Improvers in Agriculture (1723), and the British Linen Company (1746), to foster the linen industry. This industry was thus fortunate in receiving State encouragement for the greater part of a century.

[1] Baron Nicolas Dupin made a contract between an English Linen Company and the Convention of Royal Burghs in 1696.

At no time does Scotland appear to have been self-supporting in flax, the raw material of the linen industry. Ever increasing quantities were imported through the eighteenth century, from Holland, Flanders, and the Baltic. At the same time, however, considerable quantities were grown in certain areas of Scotland, particularly in Perth, Forfar, Fife, and Lanarkshire. Flax, whether home-grown or imported, was prepared for the spinner in water-powered lint—mills, where operatives severally performed the various distinct operations of scutching, heckling, etc. Fibre so prepared was spun by women and children on the rock-reel till 1735, when in the Lowlands the spinning-wheel was introduced. Weaving, less a part-time occupation than spinning, was carried out in the crofter-weaver's home, or in small loom-shops belonging to a master-weaver. These shops contained perhaps six looms at most. The organization was that of a typical domestic craft, and the occupation was prosecuted in the main in districts associated with flax-growing. Forfar specialized in coarser fabrics in imitation of German osnaburghs and silesias; Renfrew and Lanark in fine lawns and cambrics.

Apart from the attention which the Board of Trustees gave to organizational problems, it subsidized the laying-down of many bleachfields, and collaborated with scientific personnel in the Universities. The result of this association was the substitution of sulphuric acid (H_2SO_4) for sour milk as a souring agent (1756), in consequence of researches by Dr. Francis Home of Edinburgh University, an advance of first-class significance in the art of bleaching.

> In the early stages of the industry when the country people grew flax and made cloth for domestic uses they bleached it themselves, but as the trade developed this primitive method of finishing cloth tended to die out. Certainly in the second half of the century it was a paying proposition to specialize in bleaching. As the market for cloth extended, the crude methods of the farmer-weaver were no longer sufficient. The increase in the volume of trade and the demand for finer cloths, such as cambric, lawns, and damask, made bleaching a highly specialized business, requiring much skill and capital. Moreover, considerable advances were being made in the art of bleaching in the eighteenth century, and this hastened the tendency already noticed. The country people had been in the habit of using buttermilk for 'souring' the linen, and this often took weeks. Then the cloth was washed in the fields with water baled from trenches by means of wooden scoops, and before the cloth was ready usually eight months elapsed. Such a tedious and slow process would have prevented

the development of the industry had it not been that great progress was made in the art of bleaching.[1]

As we shall see in Chapter IX, Perth became a great centre of bleaching, where work was carried out not only for Scottish merchants but for distant English firms as well. On occasion two million yards were bleached at Perth in the course of a season.

To co-ordinate the diverse and necessarily scattered parts of the linen industry, the merchant-capitalist was a necessity. Not infrequently bleachers, being 'substantial people with capital', acted in this capacity.

> In the Barony Parish of Glasgow there were 3,000 linen and cotton weavers in 1790, mostly employed by manufacturers and bleachers.[2]

Sandeman and Company, important Perth bleachers, had agents for handling yarn as far off as Tarbat in Cromarty. In this way bleachers secured work for themselves and manufacturers supplies of material for export to England and the Plantations.

As an index of expansion during this first phase, figures for linen stamped by the Board of Trustees over the period in question may be taken.

1730	.. ..	3·7 million yards.
1740	.. ..	4·6 " "
1750	.. ..	7·8 " "
1760	.. ..	11·7 " "
1770	.. ..	13·0 " "
1780	.. ..	13·4 " "
1790	.. ..	18·1 " "
1800	.. ..	24·2 " " [3]

Mechanical technology contributed practically nothing to this great efflorescence, but it was during this phase that fundamental inventions were devised which were to alter the Scottish textile trade so profoundly in the ensuing decades. The primary invention, based on Black's researches on heat, was James Watt's separate condenser steam-engine (1769). It freed industry from reliance on the power of water or human muscles. It was followed by an invention for multiplying output per man, the spinning-jenny of J. Hargreaves (1770). While effecting increased output, the product spun by the jenny was defective in that it was not sufficiently strong for warp, and so had to be used solely for weft.

[1] H. Hamilton, *Industrial Revolution in Scotland*, p. 103.
[2] *Ibid.*, p. 101. [3] A. J. Warden, *The Linen Trade*, p. 480.

This weakness was surmounted by the invention of roller-spinning with the water-frame, an invention credited to R. Arkwright and his Warrington clockmaker-amanuensis (1769 and 1780). Yet neither jenny nor water-frame produced a thread sufficiently fine to satisfy the demands of lawn and cambric weavers. To do so it was necessary to wait for the fusion of principles involved in both jenny and water-frame as combined by S. Crompton in his spinning-mule (1779).

The effect of these inventions taken together led to a second phase in the evolution of the textile industry, a phase which witnessed many changes, including the transition from linen to cotton as Scotland's premier textile product, and the rise of the factory system in the cotton-spinning branch.

The second phase of development includes the boom period of Scottish cotton (1785–95) and extends till approximately 1830. The basis of the boom was mechanical invention in the preceding phase, in particular mechanization of cotton-spinning. Before an all-cotton fabric became a practical proposition, small quantities of cotton were used as weft, along with a linen warp. Arkwright's water-frame made an all-cotton fabric possible. Crompton's mule spun a fine, strong thread, which enabled J. Monteith of Anderston to pioneer the muslin trade.

These inventions were readily absorbed into the Glasgow district, already conspicuous for the success which its weaving personnel had made of the production of fine linen lawns and cambrics.

The first spinning-mill in Scotland was at Rothesay (1779); the first really large undertaking, the founding of New Lanark Mills (1785) by David Dale, who became the father-in-law of Robert Owen. In 1785 Gordon Barron and Company of Woodside Works on the River Don, who had been linen bleachers and printers since 1775, began to spin cotton. Other spinning-mills followed at Deanston on the Teith in Perthshire (1785), and Catrine in Ayrshire (1787). Expansion was of almost incredible rapidity. There was one mill in 1779, nineteen in 1787, thirty-nine in 1796, and no less than a hundred-and-twenty in 1812. They were widely distributed, but concentration was always highest in Lanark and Renfrewshire. Ownership was largely in Glasgow hands.

At first, water-power determined the locus of the cotton-spinning mills, a reminder that it was not the steam-engine that brought about the Industrial Revolution; but the steam-engine soon contributed its share to the disintegration of the domestic system. In 1800 there were already eight Boulton and Watt engines in Scottish spinning-mills, and in 1831 Glasgow alone had a hundred-and-seven steam-powered mills, vast barracks with frosted glass windows.[1]

Just as the first phase was characterized by an important advance in bleaching, so also was the second. In 1774 the Swedish chemist, C. W. Scheele, discovered the new element chlorine (Cl), and in 1785 a French chemist, C. L. Berthollet, discovered that it had marked bleaching powers. Knowledge of this was obtained by James Watt in 1786, and by Professor Patrick Copland of Aberdeen University in 1787. Both contributed to the introduction of chemical bleaching into industry, Copland in 1787, and Watt a little later. Furthermore, at the end of the century, Charles Tennant of St. Rollox, Glasgow, marketed the same reagent in a more readily transportable condition, viz. bleaching powder ($CaOCl_2$).

> These vast improvements not only made possible the enormous expansion of the linen as well as of the cotton industry, they made bleaching a highly specialized and important trade.[2]

Another contemporary development dependent on chemical technology was the discovery, in this country, of an effective process for Turkey Red dyeing (1785). This was commercialized by George Macintosh, father of the inventor of waterproofed fabrics, and David Dale, assisted to a certain extent by imported technical skill. Together, Turkey Red dyeing and chlorine bleaching, used on the same fabric, increased the variety of effects possible in the finishing trade by giving discharge printing.

George Macintosh and his son Charles were responsible for many cognate developments, particularly those needing more advanced chemical knowledge. They included cudbear manufacture, the manufacture of textile ancillaries such as alum, copperas, lead and aluminium acetates, and, lastly, waterproofing.

During the second phase linen technology also advanced.

> Competition between the two textiles, cotton and linen, especially after 1780, directed attention to the new methods of production. In 1787 an important invention was made by Kendrew, an optician, and Porthouse, a clockmaker, both of Darlington, for spinning flax by machinery. . . . The first works (in Scotland) was erected by Sim and Thom on the haughs of Bervie in Kincardine for the spinning of linen yarn in 1787. They used the invention of Kendrew and Porthouse, and the machinery was

[1] J. Sinclair, *General Report*, App. 2. p. 308.
[2] H. Hamilton, *op. cit.*, p. 103.

> procured from England and erected under the supervision of the patentees. The mill was driven by water-power, though later a steam-engine was employed, and the yarn spun was used in the manufacture of thread. . . . The first mill to spin yarn for cloth was set up at Brigton in Forfarshire by James Ivory and Company, in 1788 or 1789, and they too used the Darlington patent.[1]

But progress was very slow till 1825, when Kay 'showed that by steeping the flax for six hours the fibres were made more slippery and so could be drawn by machinery into finer threads'. By 1836 there were a hundred-and-seventy linen-spinning mills in Scotland, concentrated for the most part in Forfar, Fife, and Perth. With the establishment of these a corresponding contraction in domestic production took place. The same districts, Forfar, Fife, and Perth, were the seats of the linen-weaving industry, still domestic, though now much more in the clutches of merchant-capitalists.

As is to be expected, the linen industry felt the effects of the advance of cotton in the west, as is seen by the following figures for linen stamped by the Board of Trustees.

	1788	*1822*
Lanark	1·4 million yards	23 thousand yards
Renfrew	1·8 ,, ,,	26 ,, ,,

During the second phase, while a factory system of spinning was being established for both linen and cotton, mechanical technology was devising further mechanisms which were to break up finally, in the following decades, the remaining remnants of the old order. Joseph Marie Jacquard exhibited the principles of his loom at a National Industrial Exposition in Paris in 1801. This loom, in the hands of Dunfermline weavers, produced the figured damask for which they became famous. About 1806 the loom which Edmund Cartwright invented in 1787 was perfected. These had a profound effect on textile production, but not for many years, and before they had, new factors were operating in the economic life of Scotland.

A third phase of economic development started about 1830, during which there was an expansion not only in textiles but also an even more striking one in metallurgical industries. The latter took pride of place in the end, especially after the collapse of the Scottish cotton industry as a result of the American Civil War (1861–66). Expansion in output of iron depended on David Mushet's discovery of blackband ironstone (1801) and J. B. Neilson's invention of the heated air blast-furnace (1829) which

[1] H. Hamilton, *op. cit.*, p. 106.

made the blackband workable. In textiles, expansion depended on power-loom weaving.

It was the highly skilled hand-loom weavers of the Glasgow area who, by transfer of skill from flax to cotton, made possible the rapid expansion when the latter industry was grafted on to the existing linen industry. In turn, hand-loom weavers gave place to power-operated looms, but their replacement, even in cotton, was slow. The power-loom was introduced at Catrine in 1807, and soon after at Deanston and Gordon Barron and Company, till in 1830 there were 10,000 in Scotland turning out millions of yards of calicos and fustians. Yet in 1838, 48,000 hand-looms were still devoted to cotton in the Lowlands alone. In the 1840s, 'the factory system in cotton-weaving was becoming typical of the industry', and factories were erected to engage solely in cotton-weaving.

The introduction of power-weaving in the linen industry was an even more protracted transition, as the following citation from Hamilton shows:

> The earliest instance of power-looms in the linen trade of Scotland is in the case of one Wilson of Brechin, who evidently used them with success from 1810 onwards. His reason for adopting them was that the labour of hand-loom weavers was scarce and expensive, probably owing to the great demands of the Government during the war for coarse cloths. Experiments were made with the power-loom in Dundee in 1821, but the pioneers met with little success. One large Dundee manufacturer, Messrs. W. Baxter and Company, in 1826 proposed setting up ninety power-looms, but they appear to have had misgivings about the practicability of machine-weaving, so they did not proceed any further with their scheme. . . . However, the firm of Baxter and Company built a power-loom factory in 1836, and this was the first of its kind in Dundee.[1]

It was not till late in the nineteenth century that linen-weavers felt the full impact of mechanical technology. As that is too late to be discussed in the present study, we must now pass to a more detailed examination of bleaching, dyeing, calico-printing, and the manufacture of textile ancillaries.

[1] H. Hamilton, *op. cit.*, p. 115.

Chapter IX

THE SCOTTISH BLEACHING INDUSTRY

> 'Raikes, Near Bolton, April 7th, 1789. As it has been industriously reported that I am quitting the business of a bleacher, I think it necessary to step forward and thus publicly to declare that so far from my quitting business, I have lately made an engagement for an establishment in the chemical mode of bleaching, which will enable me to return all goods in a short space of time and in condition and colour that shall do us credit.'
>
> Advertisement by JOHN HORRIDGE.

WHEN Robert Peel was asked by a Committee of the House of Commons in 1750 what the different stages in the manufacture of cotton were, he replied: 'They are four in number, spinning, weaving, bleaching, and printing'. The former pair have been discoursed on by every economic historian; the full significance of the latter is seldom appreciated, though as far as technological progress is concerned, the distinction between them is trivial.

> Radical changes in the bleaching of cottons were due to chemical discoveries more than to the invention of machines, but the distinction is unimportant: technical improvements of various kinds were manifestations of the same general spirit of progress; and the results of such improvements were similar.[1]

The capitalization of bleaching is every bit as important as the capitalization of the cognate spinning and weaving industries. In certain districts of Scotland hundreds of acres of field covered with webs and yarn in the process of being whitened is as much an indication of eotechnic capitalism as pollution of atmosphere and stream is a sign of the transition to palaeotechnics. So extensive was the area devoted to bleachfields that, with the introduction of chemical bleaching materials which rendered sunlight a non-essential, the release of land for agricultural purposes was heralded as one of the great benefits conferred by chemistry upon the community.

Bleaching, like all processes in textile industry, was originally carried on as a domestic craft employing the simplest equipment and materials. Although the organization of the trade altered, technique made no appreciable progress till palaeotechnic

[1] Witt Bowden, *Industrial Society in England towards the End of the Eighteenth Century*, p. 80.

bleaching agents (sulphuric acid and chlorine) became available. Writing in 1733, Patrick Lindsay stated:

> At present the cloth bleached by every private hand differs so much one piece from another, that scarce one pack of goods, of the same fineness is to be had of the same colour; but was all our cloth whitened in publick fields, all of the same staple would be also of the same colour.[1]

Since linen was Scotland's premier industry in the eighteenth century, it was specifically to the art of bleaching linen that attention was directed in Scotland. Moreover, bleaching linen was a much more lengthy process than bleaching wool.

> The linen yarn had to lye without as well in the night as in the day constantly for the space of one halfe yere to be whited before it can be made clothe.

While the finishing trades in England and Scotland were largely undeveloped at the beginning of the eighteenth century, considerable progress in the art had been made in Holland and Ireland. Cloth was sent from Scotland to Holland in the early spring, and only returned to Scotland in October, or even the next spring if the summer had been unpropitious for bleaching. *Hollands* made of Silesian thread of superior quality was highly esteemed, and sold throughout Europe. It was therefore by the importation of foreign skill from these countries that Scotland made up the lag in her industrial development. Dutch bleachers came to Cameron Field, Loch Lomond, and to Gorgie, near Edinburgh.[2]

The rapid progress achieved by Scottish bleachers during the eighteenth century can be ascribed to encouragement and assistance given by the *Board of Trustees for Manufactures, Fisheries, and Improvements in Scotland*, founded in 1727 to administer the *Equivalent* provided for in the Treaty of Union. For the first fifty years of its existence this body discharged a function which, if less ambitious than, is comparable with, that of the Royal Society in seventeenth-century London. Its first plan was to concentrate on the improvement of the linen trade. This was effected by sending Scots bleachers to Ireland to study Irish methods, by subsidizing the laying down of extensive bleachfields, and by awarding premiums to Scottish academic scientists for researches on the production of potashes and on various bleaching processes.

[1] *The Interest of Scotland Considered*, p. 165.
[2] G. Chalmers, *Caledonia*, Vol. 2, p. 742, *n.*; Vol. 3, p. 896, *n.*

On the commercial side the Board acted on behalf of the individual bleachers as an importer of foreign bleaching materials. The *British Linen Company*, incorporated in 1746, and afterwards to become the *British Linen Bank* on relinquishing its commercial activities, also imported potash and tallow. The first bleachfield of the British Linen Company was at Saltoun, Haddingtonshire, under the 'willing eye' of Fletcher, Lord Milton.[1]

In the year following its foundation the Board granted £2,000 to James Adair of Belfast to establish a bleachfield in Galloway. The next step was also of Irish origin. In 1732 Richard Holden had introduced bleaching with kelp into Scotland, and the Board helped him start a bleachfield at Pitkerro, near Dundee. Lindsay notices this as follows:

> Mr. Holden has lately undertaken to lay out a very large bleachfield in the neighbourhood of Dundee, for bleaching of much of their lowest staples as they are in use to export white, with kelp.[2]

This field of Holden's became the focus from which knowledge of the art of bleaching radiated. His work, too, was reckoned as a standard of quality. Samples of it were sent to the Board of Trustees by the Dundee Dean of Guild to demonstrate to them that whitening could be effected more easily, if less effectively, with kelp than with other materials. This development coincides with the beginning of the manufacture of kelp in the Firth of Forth area, and is probably not unconnected with it.[3]

> The purging and washing of yarn is now pretty well understood, since Mr. Holden's receits for that purpose have been dispersed over the country, and the people in several parts taught by him to practise it.
>
> We have by the help of publick encouragement introduced and brought to perfection the art of bleaching fine linens as practised in Holland. We miscarry now and then in several pieces: but we now know that this is not owing to the want of skill and pains in the bleacher, but to the badness of the flax whereof the cloath is made.

He was also the inventor of a machine for subjecting cloth to a 'severe and just trial which discovers at once the least fault in it'.[4]

Samuel Parkes states that it was also an Irishman who intro-

[1] G. Chalmers, *Caledonia*, Vol. 2, p. 495; *Board of Trustees for Manufactures, Fisheries, and Improvements in Scotland.* Minutes of Trustees' Meetings, 39 Vols., H.M. Register House, Edinburgh; S. H. Higgins, *A History of Bleaching*, p. 19.

[2] P. Lindsay, *op. cit.*, p. 178.

[3] *Annual Report of the Board of Trustees to the King, 1732.*

[4] P. Lindsay, *op. cit.*, pp. 165, 173 and 183.

BLEACHING INDUSTRY

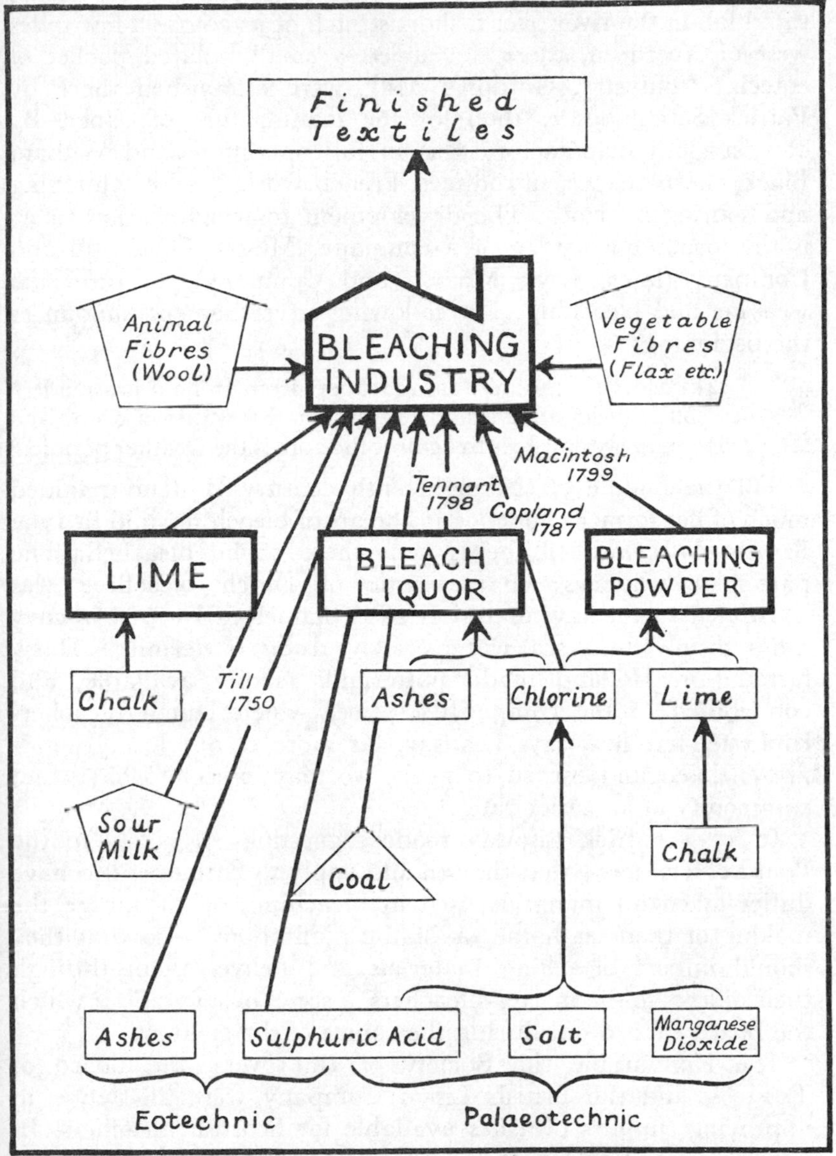

duced bleaching into the north of Scotland about the middle of the eighteenth century, but there is evidence that bleaching was in operation on the Don at an earlier date. Making use of the rapid fall in the river over a short stretch of its course a few miles west of Aberdeen, there developed a small isolated pocket of eotechnic industry. Gordon's Mills were established there by Patrick Sandilands *c.* 1696 for the manufacture of paper. By 1703 a cloth manufactory was in full operation, and William Black, the manager, introduced French workmen for whitening and souring his cloth. The development to which Parkes refers is the formation *c.* 1749 of a company, Messrs. Leys, Still and Company (later, Leys, Masson and Company), to prosecute weaving and bleaching. The following advertisement announces the beginning:

> 14th March, 1749: William Heaney, from Ireland has made a new bleachfield at Gordon's Mills, on the water of Don, and proposes to lay down cloth against the 20th if the weather permits.[1]

Till the middle of the eighteenth century Holland retained much of her former eminence in the art of bleaching, and in 1752 British cloth was still being sent there to be bleached. The particular whiteness characteristic of Dutch bleaching was attributed to the raw materials used, namely a lye of Muscovy ashes dissolved in sea water, followed by buttermilk.[2] Dairy farming in Holland made buttermilk readily available, and contributed to the competitive prices which Dutch bleachers could tender. But, says Lindsay, 'as more of our low grounds are enclosed and turned to grass, we may expect to have that commodity at an easier rate'.[3]

In 1733 Patrick Lindsay made suggestions of policy to the Board of Trustees—that they should apply to Parliament to have duties taken off materials used in bleaching, or encourage the making of them at home or in the plantations; also that they should import bleaching materials and deliver them through their officers to the master-bleachers, a series of suggestions which the Board carried out during the course of the century.

The Honourable the Society of Improvers, the Board of Trustees, and the British Linen Company were all active in improving supplies of ashes available for Scottish bleachers. In 1734 the Board of Trustees petitioned for an Act of Parliament

[1] S. Parkes, *Chemical Essays*, Vol. 4, p. 24; Patrick Morgan, *Annals of Woodside*, p. 71; R. Chambers, *Domestic Annals of Scotland*, Vol. 3, p. 156.
[2] Francis Home, *Experiments on Bleaching*, p. 20.
[3] P. Lindsay, *op. cit.*, p. 174.

32. Spinning and weaving as a domestic craft in the eighteenth century.

33. Hargreaves' spinning jenny (1770) which, while multiplying the output per operative, produced a yarn only sufficiently strong for the weft.

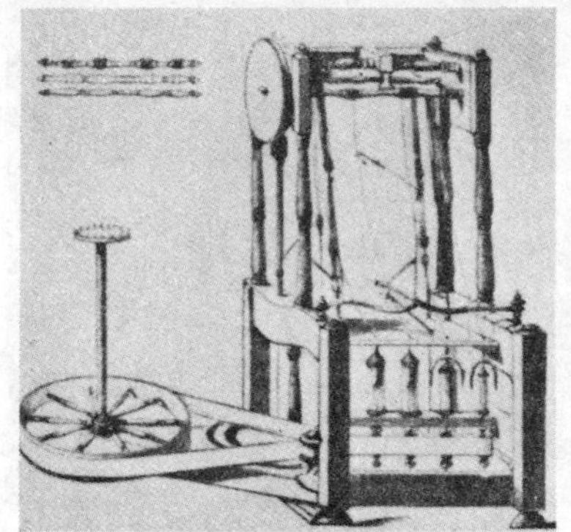

34. Specification drawings: *Left*, Arkwright's water-frame spinning machine (1769–80), which gave a stronger thread and made an all cotton fabric possible, though it was not sufficiently fine to suit the lawn and cambric weavers. *Right*, Diagram illustrating Kendrew and Porthouse's specification for a mill for spinning flax, etc. (1787). This invention had an important influence throughout Europe.

35. Specification drawing of Crompton's mule, which combined the principles of the jenny and the water-frame and gave a fine strong thread, thus furthering the development of the muslin trade.

exempting from duties the bleaching materials used at such bleachfields as should be licensed by the Trustees.[1] Two years later James Dunbar dedicated his *Smegmatalogia* to the Honourable the Society of Improvers. In it he discusses burning vegetables suitable for making potash for bleaching linen cloth and yarn. He also describes a method of rendering the potash lye caustic. This will be referred to later. A further indication of increasing interest in improvement is a premium of £150 to James Spalding, Bonymilns, for experiments from 24th February, 1744, to 18th March, 1747, on bleaching coarse linen.[2]

Francis Home also experimented on burning ferns, tobacco, peat, kelp, etc., and extracting potash from the residue. He found that kelp gave the richest yield, but that it also contained something which gave the linen a yellow colour. This restricted its use to coarse materials. Another of these early experimentalists was Dr. William Cullen, then of Glasgow. Cullen had several contacts with the Board of Trustees, and on 13th June, 1755, he was awarded a premium in the form of three suits of table linen to the value of £21 for his ingenious experiments and observations on the art of bleaching.[3] Later, on 17th March, 1763, a Memorial was received from Cullen reviewing results of experiments to find substitutes manufactured from local raw materials to supplant foreign potashes and alkaline salts imported for use in the bleaching industry.[4]

Yet another investigator of potash at this time was no other than Dr. Joseph Black. Black analysed samples sent in in consequence of premiums offered.[5] In 1782–4 Memorials on burning of vegetable plants to ashes for bleaching were submitted for the perusal of the Right Hon. the Lord Advocate by Ebenezer McCulloch.[6]

How the process of bleaching was carried out in Scotland at the middle of the eighteenth century can best be gathered from instructions in the art published by the *Honourable the Society of Improvers in the Knowledge of Agriculture*, in their *Select Transactions*, edited by Robert Maxwell.

> The yarn is soaked about nine hours in cold water. This is run off and fresh water added till the loose dirt is removed. It is then wrung dry and put to dry in the bleaching yard. The yarn is then bucked (i.e. heated) with lye of ashes and heated by a slow fire to

[1] A. Warden, *The Linen Trade*, p. 450.
[2] *Board of Trustees: Finance (Particular)*, Vol. 121. [3] *Ibid.*
[4] MS., *Records of Manufacturers in Scotland, Register of Theses and Memorials on Technical Subjects.* [5] *Board of Trustees, op. cit.*, April 1783.
[6] National Library of Scotland, MS. 640 ff. 1–10.

> the boil. Boil three hours. Rinse well and wring. Bleach the yarn three to four days watering during exposure. The yarn or cloth is then again soaked in warm lye. As the cloth becomes whiter use weaker lye. Again expose to weather seeing that yarn does not dry between operations.[1]

Yarn was bleached for all kinds of cloth except cambrics and lawns, where the material itself was bleached. The object of this process was to remove as much as possible of the colouring matter which gave the yarn a dingy appearance, and was thus in the modern sense only partially a bleaching operation (i.e. oxidation). This is even more true of the second process, the *souring*. Its object was to remove, by solution in acid, inorganic substances which would act as mordants in the course of use of the cloth.[2] These mordants if left in the fibres would form insoluble coloured matter with dirt which could not be removed by ordinary washing.

The souring was carried out initially in sour or buttermilk, in which the cloth had to soak for at least another forty-eight hours. As alternatives, fermented bran, the juice of crab-apples, and, in the East, lemon juice, were sometimes employed.[3] After soaking in milk, the cloth was again spread on the green, and watered continually so that it did not dry with milk on it. It was then rinsed and washed with soap, and the whole series of operations repeated six or seven times:

> That is to say, you must milk, then grass, then wash, then soap and buck by turns, until you attain your end.[4]

After the last milking, the webs were laid in an open cistern or tub where 'people with clean feet should tramp them well for half an hour or thereabouts'. When the bleaching process was complete, the cloth was *blued*, with azure in preference to indigo, and finally starched. According to Maxwell, the milking of cloth was not introduced into Scotland till well on in the eighteenth century, but, once adopted, experience showed that it contributed much to the production of a good colour on the cloth. Lindsay says that the Irish were not in the habit of using milk.[5]

[1] R. Maxwell, *op. cit.*, p. 343 *et seq.* [2] Ch. XII.
[3] S. H. Higgins, *op. cit.*, 71. [4] R. Maxwell, *op. cit.*, p. 348.
[5] J. Dunbar, *Smegmatolia*, p. 25; P. Lindsay, *op. cit.*, pp. 35 and 176; cf. Conrad Gill's *Rise of the Irish Linen Industry*, who attributed to the Irish use of buttermilk, but omits reference to sulphuric acid. If Gill is right, the bleaching of linen in Ireland could be accomplished in three months (p. 51), and of cotton in less than a week (p. 91).

The amount of handling to which the cloth was subjected makes it quite apparent that considerable economy of labour was achieved with properly laid out bleachfields supplied with the requisite water. It is said that four men and six women could tend the same amount of cloth on a bleachfield as sixty women bleaching on the banks of a stream.[1] The paramount importance of water supply is illustrated by the fact that Messrs. John and Thomas Barland cut an eighteen-foot canal three miles long to supply their Stormond bleachfield.[2]

Such was the operation of bleaching in Scotland in 1750. The only sign of the coming revolution in the art was a sporadic reference to the employment of sulphuric acid in place of sour milk.

The Irish seem to have been in advance of the Scots in the way of bleaching machinery, as the Board of Trustees sent Samuel Read of Leven bleachfield to Ireland to collect mechanical ideas, paying him in April 1763 £30 for expenses, with a further £15 two years later for improvements which he had carried out.[3]

From early prints of bleachworks, heavy wooden machinery was used. The works were surrounded with extensive areas of field for the grassing of the yarn and cloth. Little is known of the early bleaching works in Scotland, but R. Stephenson in his journal *On a Tour of Inspection* (1756) gives an interesting account of Irish equipment about that time:

> They had four main buildings: two buck houses, a large drying-house, and a water mill. One buck house contained two pans, two wooden kiers, and one brick kier for lees (*n.* to resist effect of alkali). In the mill house there was a large water-wheel, two pairs of washing stocks, two rubbing boards, a mill for grinding ashes, two kiers for souring, a calendar, and a beetling engine.[4]

The capital outlay for such a well-equipped bleachfield was about £3,000.[5]

In the bleachfields there was a possible market for the engines of Boulton and Watt. Writing to Boulton, David Melvill of Dublin remarked that there was a great scarcity of water at many of the bleachfields, and that if Boulton were to go

[1] P. Lindsay, *op. cit.*, p. 177. [2] *Statistical Account*, Vol. 18, p. 78.
[3] 25th January, 1765. *Board of Trustees: Finance (Particular)*, Vol. 121.
[4] R. Stephenson, *op. cit.*, p. 166, quoted by C. Gill, *op. cit.*, p. 88.
[5] C. Gill, *ibid.*, p. 246.

to Ireland with a model engine 'for reversing the water of a mill', he might reasonably expect a reward from the Irish Linen Board.[1]

Bleachers refer to Irish and Dutch methods of bleaching. As far as can be ascertained, the difference lay in the use of a rubbing-board in the Irish method.[2] The Dutch method, which, according to the *Statistical Account*, was first used by Messrs. Gray at their Old Monkland bleachfields, was only for fine goods, cloth costing less than 3*s.* per yard not being 'able to bear the expense of it'. The Irish method succeeded very well for 'middling' cloths 'at such rates as those kinds of cloth would bear'. The colour attained this way was good enough for shirtings at 1*s.* 6*d.* to 3*s.* per yard. Bleaching, incidentally, cost between 2*d.* and 3*d.* per yard in the eighteenth century. The cloth of between 1*s.* and 2*s.* a yard was then sent to London for printing, with further intolerable delays. It was to prevent this excessive waste of time that improvements were undertaken in Scotland and in Ireland, where there was also a *Board of Trustees* for the Linen Trade.[3]

The Scottish Board of Trustees continued to subsidize the laying-out of bleachfields till 1772, when there were about ninety fields throughout Scotland, distributed as follows:

> Aberdeenshire 5; Ayr 3; Banff 5; Berwick 2; Dumbarton 3; Dumfries 3; Edinburgh 14; Elgin 1; Fife 13; Haddington 6; Inverness 1; Kincardine 1; Kinross 1; Lanark 8; Linlithgow 1; Orkney 1; Perth 6; Renfrew 3; Ross 1; Roxburgh 1; Selkirk 1; Stirling 3; Forfar 9.[4]

The majority of these lay in areas associated with the linen trade and usually catered for local interests. But this was by no means universal. In 1775, Messrs. Richardson and Company established a bleachfield at Tibbermuir, Perth, where they operated mainly on cloth from Perth, Dundee, Dunfermline, Edinburgh, and Glasgow, but latterly they built up an extensive connection with Darlington manufacturers.[5]

Several bleachfields which were founded while the Board of

[1] D. Melvill to M. Boulton, 1st February, 1773, Assay Office Library, Birmingham.

[2] Wadsworth and Mann, *The Cotton Trade and Industrial Lancashire, 1600–1780*, p. 178.

[3] *Statistical Account*, Vol. 7, p. 385; P. Lindsay, *op. cit.*, p. 176; Conrad Gill, *op. cit.*, p. 245, *n.*

[4] *Board of Trustees: Finance (Particular)*, Vol. 121.

[5] *Statistical Account*, Vol. 17, p. 638.

Trustees was active continued their operations into the present century. Among them were:

1750: James Burt-Marshall, Luncarty Bleachfield, nr. Perth.
1769: The Bowfield Bleaching Company, Howwood, Renfrewshire.
1782: Blackwoods Ltd., Craigton Bleachworks, Milngavie.
1792: Stevenson McKeller & Co., Newlandsfield, Pollokshaws.
1817: Jas. McHaffie & Son, Kirktonfield, Neilston, Glasgow.
1825: John McNab & Co., Midtownfield, Howwood, Renfrewshire.[1]

The organization of the bleaching industry in the early stages of its evolution varies from the one-man concern in which the individual weaver bleached his own web, to the full-scale capitalistic enterprise. Bleaching by individual weavers continued in some districts right to the end of the eighteenth century, and many of the bleaching establishments remained quite small. When Pitt, in 1785, proposed the introduction of a bleacher's licence costing £2, it was then stated that most bleachers were men of small property, who did little work, worked hard, and yet were very poor, to whom £2 was a matter of great importance.[2] There was no uniformity of organization in Scotland, Ireland, or England. In Scotland the costliness of equipment kept the bleaching and finishing trades in capitalistic hands. Some of the varied modes are illustrated by the following examples.

At Perth, there were *public* bleachfields. In some districts proprietors encouraged and contributed to the establishment of bleachfields, or leased land for that purpose direct to manufacturers, as for example at Cullen.[3] At Logie, Forfar, the bleachfield belonged to a Montrose company which had the field in tack from the proprietor of Craigo. At Tyrie in Aberdeenshire, the tacksman paid £30 per annum for the field.

The first of two revolutions in bleaching arose out of the commercialization of sulphuric acid by Dr. John Roebuck at Prestonpans in 1749. According to Parkes[4] the use of sulphuric acid instead of sour milk was introduced about 1750 by Dr. Francis Home of Edinburgh University on account of the newly available low-priced acid. It is probable, however, that Roebuck knew of the bleaching action of sulphuric acid, a property which was first

[1] A. J. Sykes, *Concerning the Bleaching Industry*, *passim*.

[2] Henry Hamilton, *The Industrial Revolution in Scotland*, p. 102; *Statistical Account*, Vol. 4, p. 189; Wadsworth and Mann, *op. cit.*, p. 306.

[3] *Statistical Account*, Vol. 4, p. 365; Vol. 5, p. 175; Vol. 18, p. 516.

[4] S. Parkes, *Chemical Essays*, Vol. 4, p. 17.

noticed by the French chemist, C. L. Berthollet.[1] In Scotland an investigation of its bleaching properties was undertaken by Francis Home in conjunction with the Board of Trustees. By the end of 1755, a petition by sundry master-bleachers and others who had attended Dr. Home's lectures, had been presented to the Board praying that the experiments might be published.[2] Some doubts were expressed as to the advisability of doing this, as the information would thereby fall into the hands of all and sundry. Lord Belhaven and Henry Home, Lord Kames, were deputed to converse with Francis Home on the subject.[3] They decided in favour of publication, and *Experiments on Bleaching* appeared in 1756. On 15th April of the same year, Home was voted a premium of £100 'for the expenses and trouble in making experiments in consequence of former resolutions of the Board'.[4] In his publication, he reveals the revolution wrought by the substitution of sulphuric acid for milk. Here are his own words:

> The mineral acids are exceedingly cheap. I will fairly own that, at first, I had no great opinion of the success; from two reasons; through want of all fermentation, which I then looked on as necessary; and the extreme corrosiveness. But the experience of two different summers, in two different bleachfields (one of them John Chrystie of Perth) has convinced me, that they will answer all the purposes of the milk and bran sours; nay, in several respects, be much preferable to them. I have seen many pieces of fine cloth, which had no other sours, but these of vitriol, and were as white and strong as those bleached in the common way. I have cut several webs through the middle, and bleached one half with milk and the other with vitriol, gave both the same number of operations and the latter were as whole and strong as the former. . . . It is observed that this sour performs its task much sooner than those of milk or bran; so that Mr. John Chrystie, in making the trial, used to lay the milk some 24 hours before the vitriol. . . . It takes milk 5 days to perform its task; but the vitriol sours do it in as many hours; nay, perhaps in as many minutes.[5]

We have no specific reference to the first use of sulphuric acid in Scotland, but we know that within a decade J. and N. Philips and Company of Tean, Staffordshire, were using it, purchased

[1] *Annales de Chimie*, 1789, *2*, 159.

[2] *Board of Trustees*; Minutes of Trustees' Meetings, Vol. 13, 10th December, 1755.

[3] NOTE: Berthollet, in his *Elements of the Art of Dyeing* (translation by Andrew Ure, p. 29), confuses the two Homes: Henry Home, Lord Kames (1705–82), with Francis Home, M.D. (1719–1813), Professor of *Materia Medica*, Edinburgh University.

[4] *Board of Trustees: Finance* (*Particular*), Vol. 121.

[5] F. Home, *Experiments on Bleaching*, p. 83.

from Dr. Roebuck's first vitriol works at Steelhouse Lane, Birmingham. It is most unlikely that they were the first users even in England, as Wadsworth and Mann state.

In the autumn of 1754, the same J. and N. Philips of Tean bought from the Coalbrookdale Company a 'large iron furnace' of 140–150 gallons' capacity to boil yarn in for bleaching, at a cost of £8 14*s.* 10½*d.*[1]

We can follow the progressive changes in bleaching technique at the Wellhouse bleachfield of John Gray, which was founded *c.* 1730. To begin with, the raw materials used were a mixture of Hungarian pearl- and pot-ash, followed by liquid buttermilk.

> This system kept its ground till about 1762, when Mr. James Macgregor acquired and adopted the Irish method of beetling by water machinery, and booking with oil of vitriol diluted in water. This strong acid superseded and restored to man his cool and natural draught of butter-milk in the heat of summer.[2]

The only other materials used in bleaching, up to the time of the second revolution in the art, were hydrochloric acid and lime. When added to a solution of potash (potassium carbonate), lime (calcium hydroxide) renders the solution caustic.

$$K_2CO_3 + Ca(OH)_2 = 2KOH + CaCO_3$$
$$\text{potash} + \text{lime} = \text{caustic potash} + \text{chalk}$$

The chalk is insoluble, and can be filtered off, leaving a solution of caustic potash. This, in concentrated solution, has a much more powerful detergent, and also destructive, effect than potash.

It may have been this that caused the passing of a series of Acts, both in Scotland and Ireland, forbidding the use of lime in bleaching.[3] Under later Acts, officers of the Board of Trade were given powers of search and seizure, any bleacher found using lime being prohibited from bleaching for several years.[4] Regulations in Ireland were even more stringent than in Scotland.[5] Probably the cause of the trouble was simply the use of excess lime, as J. Dunbar, in his *Smegmatologia*, gives details for the use of lime-water, a saturated solution of calcium hydroxide, without any excess solid in suspension. He says:

> I know there are a great many who are enemies to bucking of Cloath with Lime (and so am I), because the Lime in the bucking penetrates and lurks in the Cloath, which burns it, and makes it

[1] Wadsworth and Mann, *op. cit.*, pp. 178 and 296.
[2] Andrew Brown, *History of Glasgow*, Vol. 2, p. 249.
[3] *Acts of the Parliament of Scotland*, Vol. 5, pp. 49 and 597; Vol. 9, p. 312.
[4] R. Maxwell, *op. cit.*, p. 399. [5] S. H. Higgins, *op. cit.*, p. 14.

a yellow colour. . . . But Lime-Water is of a different Quality, for it can be taken inwardly by men and women, and it will whiten Cloath sooner than the best Lee you can make.[1]

With the general improvement in bleaching, and most likely also with improvement in the quality of yarn bleached, lime was again experimented with in the second half of the century. About the time when renewed prohibitions were imposed in Ireland in 1764 (21 and 22 Geo. III, *c.* 35), Dr. James Ferguson of Belfast was again experimenting with lime, and was in correspondence with Dr. Black of Edinburgh on the subject. Here is the substance of Black's communication:

All the chemists agree that lime loses a great part of its weight on being burnt. That this loss in weight is occasioned by the dissipation of an aerial matter is clear from a consideration of Markgraaf's.' Black then refers to his own experiments made in 1752–4, in which he established the nature of lime and of 'fixed air'. He is doubtful as regards the use of quicklime in bleaching. After discussion the conversion of quicklime applied to cloth into chalk, by the fixed air (carbon dioxide) in the atmosphere, he dissuades Dr. Ferguson from using dilute vitriol for this purpose; the gypsum ($CaSO_4$) would be very difficult to remove from the fibre, owing to its sparing solubility in water; moreover, 'soaping the cloth would make it become greasy, the acid deserting the earth to unite with the salt of the soap, and that separating the oil'. He does not recommend 'buttermilk sous' (lactic acid) or vinegar, because their attack of chalk is so slow. 'The reason for this too is plain and easy. There is no substance that attacks and retains mephitic air so strongly as do the calcareous earths; they therefore attract it very near as strongly as they do these weak acids (by weak I do not mean dilute, but less active or powerful) and hence these acids expel the air slowly and as it were with difficulty.[2]

The upshot of this correspondence was that the Irish Linen Board prohibited the use of lime in bleaching, as mentioned above. This opinion only persisted for a few years, and by 1770 Dr. Ferguson had changed his opinions. His conclusion was that it was 'a safe, effective, and economical' bleaching agent. He received in 1770 a premium of £300 from the Irish Linen Board for the experiments he carried out. Although prohibited, lime, according to Hoyle, a Lancashire bleacher, was in constant use between 1760 and 1800, but was only used once in each series of operations.[3]

[1] Dunbar, *op. cit.*, p. 30.
[2] W. Ramsay, *Life and Letters of Joseph Black*, p. 52.
[3] S. H. Higgins, *op. cit.*, p. 16.

Apparently the later laws were not strictly observed, because after the unsuccessful prosecution of an Armagh bleacher in 1815, under the Act of 1782, no more is heard of lime being prohibited. Eleven local bleachers then openly declared that they believed lime to be the safest and best method, and the prohibitory Acts were finally repealed in 1825.[1]

When potash became very expensive, W. Higgins introduced sulphuret of lime (calcium sulphide, CaS) as a bleaching agent. It could be prepared cheaply from lime and sulphur. When dissolved in water it decomposes (hydrolyses) to give an alkaline solution like caustic alkali.

$$CaS + 2H_2O = Ca^{+2} + S^{-2} + 2H^{+1} + 2OH^{-1} = Ca^{+2} + 2OH^{-1} + H_2S$$

This means that we are left with virtually a solution of lime-water. Thus the advantage of the method was legal rather than technical: it allowed of the use of lime indirectly. At the same time, a solution of calcium sulphide would not become exhausted as quickly as a solution of lime, and in consequence smaller quantities of liquid would have to be used.[2]

The other bleaching material, hydrochloric acid (HCl), was probably rendered necessary in consequence of the adoption of lime in the bleaching process. Where souring with sulphuric acid followed treatment with lime, the lime left in the fabric was converted into calcium sulphate ($CaSO_4$), which is insoluble and which spoils the finish or feel of the cloth.

$$Ca(OH)_2 + H_2SO_4 = CaSO_4 + 2H_2O$$

If hydrochloric acid be used instead, a soluble calcium salt (calcium chloride, $CaCl_2$), is formed, and this can be washed out, leaving the fabric intact.

$$Ca(OH)_2 + 2HCl = CaCl_2 + 2H_2O$$

So sulphuric acid gave place to hydrochloric acid, but demand for the former was not diminished, as it was required in the preparation of the hydrochloric acid (or Spirit of Salt).

$$\begin{array}{c} 2NaCl + H_2SO_4 = 2HCl + Na_2SO_4 \\ \text{Common salt + vitriol = Spirit of salt + Glauber's salt} \end{array}$$

In the finer fabrics of the muslins, the common vitriol is retiring from the stage, and making way for its essence. A few years ago, chemistry lent the bleacher her acid. He now extracts the essence

[1] C. Gill, *op. cit.*, p. 291.

[2] W. Higgins, *The Theory and Practice of Bleaching*, p. 53.

in steam which is both safer and more expeditious than the former process.[1]

Dr. Eason says 'the vitriolic acid is that which has usually been employed, not because it is preferable to hydrochloric acid but because it was to be bought in large quantities and at a small expense'. Since hydrochloric acid had dropped to 3½*d.* per lb. in 1782 he recommended its substitution for sulphuric acid.[2]

The next advance in the art of bleaching was far-reaching in its effects, even when compared with the introduction of sulphuric acid, which had reduced the time of bleaching from so many months to as many days. The result was that, in the last quarter of the eighteenth century, no greater advance was made in the finishing sections of the textile trade than in the art of bleaching.

In 1774 a new gaseous element, chlorine, was discovered by the Swedish apothecary Carl Scheele. After the lapse of a decade Berthollet, who had already worked on sulphuric acid as an alternative to sour milk, suggested the use of chlorine as a bleaching agent.[3] The new bleaching agent was introduced into Scotland by two routes. In 1786 the acquisition of privileges and patents on the Continent led James Watt to Paris, where he met several of the great French scientists of his day, Lavoisier, Laplace, Monge, de Prony, and especially Berthollet, who demonstrated to him the bleaching action of chlorine. This was of twofold interest for Watt. Watt was no mean chemist, and as such the friend and collaborator of Dr. Black and Dr. Roebuck. Moreover, his father-in-law, James MacGregor, was a Glasgow bleacher. When he returned to his country some considerable time elapsed before he could personally supervise a demonstration of the use of chlorine.[4] Watt wrote to Thomas Henry on 25th February, 1788, saying that 1,500 yards were bleaching.[5] When Watt and MacGregor ultimately did get tests completed, it appears that they hesitated to adopt chlorine as a bleach on account of difficulties in handling it.[6]

But in 1787, Patrick Copland, Professor of Natural Philosophy at Marischal College, Aberdeen, was travelling on the Continent with the Duke of Gordon, and passed some weeks with Professor De Saussure at Geneva, De Saussure having been the Duke's tutor.

[1] A. Brown, *op. cit.*, Vol. 2, p. 249.

[2] *Observations on the Use of Acids in Bleaching of Linen*, *Mem. Manch. Lit. and Phil. Soc.*, 1785, *1*, 240.

[3] *Journal de Physique*, 1785, *26*, 325.

[4] Watt to Boulton. 25th February, 1787, quoted by P. Mantoux, *The Industrial Revolution in the Eighteenth Century*, p. 251; *Watt to Boulton*, 30th December, 1787, B.R.L.; S. Smiles, *Lives of Boulton and Watt*, p. 381.

[5] Rees, *Cyclopaedia*; Article, Oxymuriatic acid.

[6] H. W. Dickinson, *James Watt*, p. 151.

De Saussure demonstrated the use of chlorine bleaching to Copland.[1] On returning to Aberdeen, Copland communicated his information to Messrs. Gordon Barron and Company, who took up the process commercially in July 1787. Gordon Barron and Company first introduced cotton into Aberdeen in 1779. This was probably the only house in Scotland to import, spin, weave, bleach and print cotton.[2] One of the partners was a friend of Richard Arkwright (1732–92), who offered to teach some of their employees in his works in Derbyshire.[3] They made their chlorine, as did Scheele, from muriatic acid (hydrochloric acid) and manganese dioxide (MnO_2, often called, simply, manganese) in white wood vessels.

There is little doubt that Messrs. Gordon Barron and Company have six months' priority over the large-scale experiments of Watt and MacGregor, and were employing chlorine at Aberdeen in preparing goods for sale many months prior to any such application of it at Manchester, or any other place in Great Britain, MacGregor's works at Glasgow alone excepted. Watt, however, is not the only other claimant to the honour. Robert Hall of Basford, Nottingham, is one, and a controversy arose as a result of a letter in the *Annals of Philosophy* from William Henry of Manchester concerning his father's part in the development.[4] In the course of the argument, Samuel Parkes communicated with Copland and received the following reply, which we give at length, since Watt is almost invariably credited with introducing this second palæotechnic revolution in bleaching.

Marischal College,
Aberdeen.
April 27, 1814.

. . . . It was in the early part of 1787 I had the honour of accompanying the present Duke of Gordon on a tour to the Continent, during which we passed several weeks at Geneva, chiefly in company with Professor de Saussure, under whose direction His Grace had studied there, in the early part of his life. Among much valuable information I received from Saussure, he showed me the experiment of discharging vegetable colours by oxymuriatic acid, which though I had met accounts of it (I think in M. de la Metheric's Journal) I had never before seen tried. Impressed with its importance to our manufactures, and well acquainted with the

[1] Probably also to Dr. Richard Kirwan of Dublin.

[2] R. Wilson, *Delineation of Aberdeen*, p. 215.

[3] P. Morgan, *Annals of Woodside*, p. 20.

[4] *Annals of Philosophy*, 1815, *6*, 421.

chemical knowledge of the Mr. Milnes, I immediately on my return communicated it to them, and perfectly recollect our instantly trying it on a hank of yarn directly from the spinner, to which in less than an hour we gave a good white colour. To the best of my recollection this was about the end of July, 1787, and from that time I was frequently informed by Mr. Milne and his late brother that they always continued to use this new mode of bleaching in their manufactory, and particularly for finishing orders where they were limited as to time. I also think that they were enabled to extend its application to larger quantities, by using vessels of white wood in place of glass, as at first. Mr. Milne is, therefore, in my opinion, perfectly correct in stating that theirs was the first manufactory in Great Britain where the new method of bleaching was introduced and continued to be practised. As His Grace dines with me tomorrow, on his way to London, before sealing my letter I shall ask his opinion as to the dates, &c., and get him to direct it.

I am with great regard,
Sir, your obedient humble servant,
PAT. COPLAND.

To Samuel Parkes, Esq.,
Goswell Street, London.

28th.—P.S.—His Grace having read the above, perfectly recollects the experiment shown by Saussure, with the opinion we both entertained of its importance; and as it may add to the authenticity of your account, permits you to use his name also in your publication.[1]

Following on the introduction of the new knowledge by Copland and Watt, the next mention of chlorine is at the Dollar field of Messrs. William Haig. The Dollar bleachfield, established in 1787, covered, within a few years, twenty acres of ground. It was here that the celebrated diaper or table-linen of Dunfermline was bleached. Chlorine bleaching was introduced before 1790.

The chemical method of bleaching by *oxygenated muriatic acid* has been successfully employed here by Mr. Haig, the founder of this establishment; and he gained, in 1790, the premium for this method of bleaching from the Honourable Board of Trustees. Since that period he has made many valuable discoveries, both with regard to the preparation and application of that acid; and finds it very useful, particularly at the end of the season, when the influence of the sun has become diminished. He finds this method better adapted for cotton cloths than for linens.[2]

[1] *Annals of Philosophy*, 1816, 7, 98.
[2] *General View, Clackmannan*, p. 362; *Statistical Account*, Vol. 15, p. 164.

R. C. Clapham, in his *Account of the Commencement of the Soda Manufacture on the Tyne*,[1] summarizes the introduction of chlorine bleaching as follows:

> Berthollet was the first to suggest, in a paper to the French Academy, the use of chlorine in bleaching. This was in 1785. In 1787, Professor Copland and the Duke of Gordon commenced works in Aberdeen to make chlorine in large glass Woolf's apparatus. This was afterwards improved by the apparatus being made of hard wood.
>
> Bleaching by means of chlorine was introduced to the manufacturers of Glasgow in 1789 by Mr. Watt the engineer, and it soon found its way into Lancashire.

One of James Watt Junior's friends was Thomas Cooper, M.D., and it was probably through young Watt that Cooper got his knowledge of chlorine bleaching.[2] Be that as it may, in 1788 Cooper, in conjunction with Dr. Charles Taylor, also of Manchester, and Mr. Baker, bleached, printed and calendered a piece of cotton in three days, and so impressed was Cooper with the result that he established a bleach-works at Raikes, near Bolton. The reason for the choice of Bolton was that it was then easier to take cloth to Bolton than coal to Manchester.[3] Both Watt Junior and Cooper were members of the Constitutional Society in Manchester, and were sent as delegates to present an address to the Société des Amis de la Constitution. They were in Paris on 13th April, 1792.

Under Dr. Thomas Henry, the College of Arts and Sciences in Manchester took up the scientific development of the new agent, and as a direct result of Henry's work, the Ridgways of Horwich, near Bolton, built Wallsuches to operate the process.[4] Other famous Manchester scientists contributed to the development of the bleaching industry in Lancashire—John Dalton, for instance, carried out water analysis for Sykes of Edgeley.[5]

The use of a gaseous bleaching agent presented considerable difficulties in handling. Already in 1789 Antony Bourboulon de Bonnueil and Company, Liverpool, obtained a patent for an improved method with less escape of gas,[6] a patent which seems to have caused some disturbance among Manchester merchants. Opposition to Bourboulon came from Birmingham as well, even

[1] *Trans. Newcastle Chemical Society*, 1871, *1*, 43.
[2] H. W. Dickinson, *James Watt*, p. 151.
[3] A. J. Sykes, *op. cit.*, p. 12.
[4] S. H. Higgins, *op. cit.*, p. 80.
[5] *Ibid.*, p. 116.
[6] *Register of the Great Seal*, July 1789, Vol. 20, No. 375, G.R.H.E.

before the patent was granted. Matthew Boulton wrote to Comte Claude Louis Berthollet (1748–1822):

> Mr. Watt with your Permission hath made a successful *en grand* at his Father in Laws (Mr. MacGregor, Glasgow) in Scotland. But it now appears that one Monsr. Bourboulon de (Boninsil?) and Co. have applied to the British Parliament for a reward or an exclusive Priviledge for Whiteing Linen or Cotton by your invention. I thought it proper to acquaint you of this proceeding and to request you will inform me if it is with your consent or whether you wish me to oppose it. At all events it appears from Publications in the news papers that it will be opposed by the manufacturers and bleachers in Manchester.[1]

The gaseous chlorine still remained unruly. Pajot des Charmes, in *The Art of Bleaching*,[2] mentions that he had been forced to design a mask with a glass eye-piece to be worn by the bleaching operatives. The corrosiveness of chlorine on the metal parts of bleaching machinery was also a difficulty. Two French chemists, Valette and Tenant, employed by Ainsworth of Halliwell, tried to use paper instead of metal pipes to convey the gas, but without much success. Yet despite these difficulties improvements continued to be made in the handling of the gas, and the use of chlorine continued to expand. In 1792, Clement and George Taylor obtained a patent for bleaching rags for making paper with it.[3]

From Javelle, in France, certain manufacturers announced that they had perfected a liquid bleach which they called *eau de Javelle*. C. L. Berthollet analysed the product, and quickly realized that chlorine could be absorbed in a solution of potash to produce a bleaching liquid. As a commercial proposition, the manufacture of *eau de Javelle* was a failure, and its makers emigrated to Liverpool to endeavour to get exclusive rights to work the process in England for twenty-eight years, but without success. They therefore made the solution without protection of a patent, but the difficulties of transport were so great that they could not distribute their product, and failed. According to W. Higgins, speaking with reference to Ireland:

> Although the oxygenated muriate of potash (i.e. the *eau de Javelle*) has been used with great advantage by the paper-makers, it does not appear to have made any great progress in our bleach-greens.[4]

[1] M. Boulton to C. L. Berthollet, 1788 (March), A.O.L.B.
[2] Translated by Nicholson, London, 1799.
[3] *Register of the Great Seal*, Vol. 20, No. 442.
[4] W. Higgins, *op. cit.*, p. 49.

A Mr. Foy, probably an employee of the *eau de Javelle* manufacturers, then took up the idea of supplying equipment to enable manufacturers to make their own bleaching liquor. It was probably the same Hugh Foy who, in the course of a legal action, claimed that it was he who taught Charles Tennant to make bleaching liquor in 1796.[1] T. L. Rupp described an apparatus to use the gas simply dissolved in water, thus saving the cost of the alkali used in the *eau de Javelle*.[2]

The bleaching action of chlorine is quite different, chemically, from that of sulphuric acid. It will be remembered that the acid was used to remove insoluble substances in the fibres which, with the passage of time, would pick up dirt and fix it on the fibres. Chlorine acts by oxidation.

Moist chlorine, or chlorine gas in solution, is a powerful oxidizing agent. In sunlight this solution of chlorine decomposes into oxygen and hydrochloric acid.

$$2Cl_2 + 2H_2O = O_2 + 4HCl$$

When unbleached cloth is placed in the moist chlorine or in the chlorine water the colour is removed. The oxygen has formed colourless oxidation products which can be washed out.

The absorption of chlorine in potash leads to the formation of a hypochlorite (potassium hypochlorite, KOCl), along with an equivalent quantity of potassium chloride (KCl), which is inactive.

$$Cl_2 + 2KOH = KOCl + KCl + H_2O$$

In the process of bleaching with *eau de Javelle* the potassium hypochlorite decomposes to give oxygen which then reacts in the same way as with chlorine itself.

The next great advance in the art of bleaching was the production of a solid hypochlorite (bleaching powder) which, while as efficacious as gaseous chlorine, was much more easily handled and transported. This advance is associated with the name of Charles Tennant of St. Rollox Works, Glasgow.

Charles Tennant (1768–1838) was born at Ochiltree, Ayrshire. After being educated at home, and at the parish school, he was sent to Kilbarchan to learn the manufacture of silk, and from there to the bleachfields at Wellmeadow, where he studied bleaching methods. Ultimately he set up as a bleacher himself, in partnership with Cochrane of Paisley. At the height of the kelp boom, Tennant

[1] S. H. Higgins, *op. cit.*, p. 88; S. Parkes, *op. cit.*, Vol. 4, p. 63.
[2] *Mem. Manch. Lit. and Phil. Soc.*, 1798, 5, 298, with diagram.

took out a patent for making a bleach liquor from chlorine (Cl_2) and a sludge of lime ($Ca(OH)_2$), instead of the usual chlorine and potash (KOH). Tennant is supposed to have acknowledged his indebtedness to Hugh Foy for the idea, but he took out his patent unknown to Foy.[1] Transport difficulties still existed, so Tennant allowed manufacturers to work his patent at their own bleach-works, for £200 a time. W. Higgins describes an apparatus for the use of chlorine, where it was passed into 'eighty lb. of well slaked and sifted quick-lime in eight hundred gallons of water'.

> This short description will give a sufficient idea of the apparatus, and of the expense attending it. Those who use it, I understand, find it very convenient, but doubtless practice will improve it. The apparatus itself may be seen at work at the bleach-green of Charles Duffin, at Dungannon. A Mr. Tennant who works with him, and who it seems, is very expert at the process, may be employed at the different bleachers, until they get into the method of managing it themselves.[2]

It is little surprising that Mr. Tennant 'was very expert at the process', for it was in all probability Charles Tennant demonstrating his first method. Higgins says that there were thirty such plants in operation in the north of Ireland in 1799. That the process was widely used is very probable, because when Tennant raised an action for infringement of his patent in 1802 so much evidence was brought against him that the patent was declared bad. Among those upon whom he served subpoenas were many prominent bleachers in England, including the Slaters, Varleys, *et al.*

In the year following the first patent, 1799, Tennant took out a second patent (30th April), this time for the production of a dry *bleaching powder* which more or less completely solved the transport problems. It is quite definite, however, that Charles Macintosh, and not Tennant, devised the second or dry process.[3] It appears that the patent was only taken out in Tennant's name for ease of working.

At the time of both patents Tennant was in partnership with Charles Macintosh, James Knox of Hurlet, Alexander Dunlop, and Dr. William Couper.

> James Knox of Hurlet, Paisley, was a great friend of Charles Tennant and as the London House was known at first as Tennant,

[1] S. H. Higgins, *op. cit.*, p. 88; *Register of the Great Seal*, Vol. 20, No. 628, January 18, 1798.

[2] W. Higgins, *op. cit.*, p. 42.

[3] T. Thomson, *History of Chemistry*, p. 155; *New Statistical Account*, Article: *Glasgow*; George Macintosh, *Memoir of Charles Macintosh*, p. 38.

Jacob van Ruisdael, 1628–1682

36. Bleaching-ground near Haarlem, to which British cloth was sent to be bleached till at least the middle of the eighteenth century.

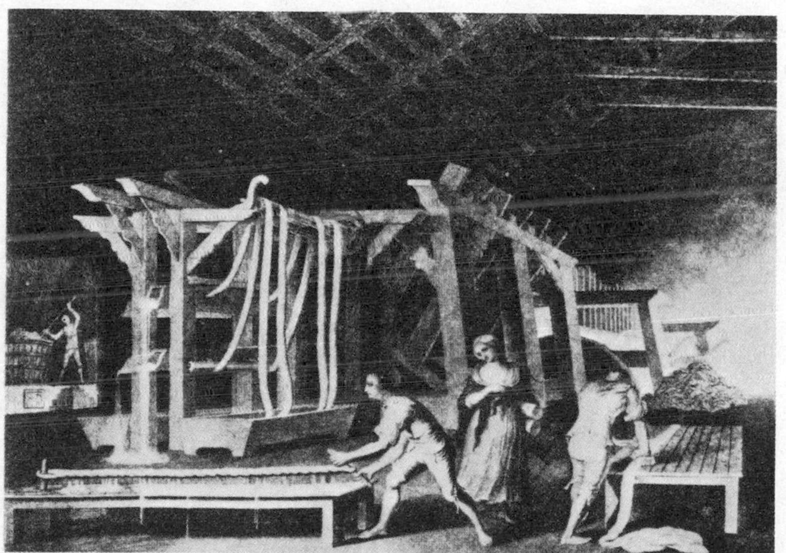

37. Interior of an Irish bleaching-works, showing the wash-mill, beetling engine, rubbing-boards, and on the left a view of the boiling-house.

38. Scottish bleachfield, probably near Perth, illustrating the competition between industry and agriculture for the use of land.

39. Gordon Barron Co.'s Woodside Works, near Aberdeen, where chlorine bleaching was introduced in July, 1787.

Knox and Co. it is probable that he took an active interest in the business.

Alexander Dunlop of Drums, was related to Tennant by marriage, as his brother James Dunlop, married Margaret, daughter of Charles Tennant. Their son, Charles Tennant Dunlop, became head chemist at St. Rollox.

Dr. William Couper was a W.S. and surgeon. He was also connected with Tennant through the marriage of his three sons to three daughters of Tennant's.

Charles Macintosh was the son of George Macintosh of Dunchattan.

It is also stated that Anderson, founder of the Andersonian College in Glasgow, was an original partner, but there is no existing proof of this.[1]

The firm moved from Darnley to St. Rollox about 1799 to establish works to prosecute the second patent, purchasing the land for £5,000 from John and Robert Tennent, of the famous, and still existing, brewery. The St. Rollox works expanded rapidly, soon becoming one of the largest chemical works in Europe. By 1830 they covered ten acres, and were producing, in addition to bleaching-powder, sulphuric acid, soda and soap. The increasing production of bleaching-powder and the fall in price is illustrated by the following figures:

						£	s.	
1799–1800	Production was	52	tons	bleaching-powder	@	140	0	ton
1805	,,	,, 147	,,	,,	,,	112	0	,,
1820	,,	,, 333	,,	,,	,,	60	0	,,
1825	,,	,, 910	,,	,,	,,	27	0	,,
1870	,,	,, 9,251	,,	,,	,,	8	10	,,

The process of manufacturing chloride of lime, and a discussion of its composition, were published by Dr. Andrew Ure.[2] James Knox went to Ireland to introduce it to the linen-bleachers, and the revolutionary nature of the innovation can be assessed by the computation that in one year from the introduction they saved £166,800 in Ireland alone.[3]

One of the by-products of bleaching-powder manufacture was sulphate of soda (Na_2SO_4), but since it was subject to an excise duty of £30 per ton till 1815, it was only after that that it could be used freely as a raw material in making alkali. One of the first

[1] E. W. D. Tennant, *One Hundred and Forty Years of the Tennant Companies 1797–1937*, p. 12.

[2] *Quarterly Journal*, 1822, *13*, 1.

[3] G. Stewart *Glasgow Curios*, p. 40.

to use it was Charles Tennant; considerable quantities were made and many patents taken out covering the process.

Losh, Wilson, and Bell were the first to make bleaching-powder on the Tyne in 1830. Their apparatus consisted of lead stills and wooden chambers covered with pitch.[1]

Some bleachers continued to make their own *chemical* or *chemic*; e.g. in 1817 a bleaching foreman was advertised for in Dundee, 'acquainted with all the new improvements made in "*making* and using" the new bleaching agent.'[2] The price of Tennant's bleaching-powder, however, was adjusted so that it was just cheaper for a bleacher to use it rather than make his own.[3]

Tennant's contribution to technology is summed up by Liebig as follows:

> By the combination of chlorine with lime it could be transported to distances without inconvenience. Therefore it was used for bleaching cotton and but for this new bleaching process it would scarcely have been possible for cotton manufacturers in Great Britain to have attained the enormous extent which it did during the nineteenth century, nor could it have competed in price with France and Germany. . . . Had not chlorine bleaching been introduced, finding capital to purchase land for the old methods of bleaching would have presented a considerable problem, especially when it is realised that by 1840 a single establishment near Glasgow was bleaching 1,400 pieces of cotton daily throughout the year.[4]

Nothing could illustrate better the lopsidedness of the emphasis that historians put on the purely mechanical aspect of industrial expansion in the 1750–1830 period.

One property of chlorine of much significance to modern society, its disinfectant property, was known to and used by the Scottish bleachers.[5]

We cannot here trace in detail the evolution of Tennant, Knox and Company, but its rapid expansion can be illustrated diagrammatically as on the opposite page.

The rapidity of these developments is all the more remarkable when we take into account the difficulties under which the bleachers worked. The duty on soap used by bleachers was removed in 1782, but that on ashes remained. It also remained on salt, which became, in a few years' time, the raw material not only

[1] R. C. Clapham, *Trans. Newcastle Chemical Society*, 1871, *1*, 44.
[2] A. Warden, *The Linen Trade*, p. 721.
[3] S. H. Higgins, *op. cit.*, p. 91.
[4] J. Liebig, *Familiar Letters*, p. 28.
[5] S. Parkes, *Chemical Essays*, Vol. 4, p. 163.

CHARLES TENNANT AND COMPANY, OF ST. ROLLOX, 1797[1]

CHEMICAL AND SOAP MANUFACTURERS.

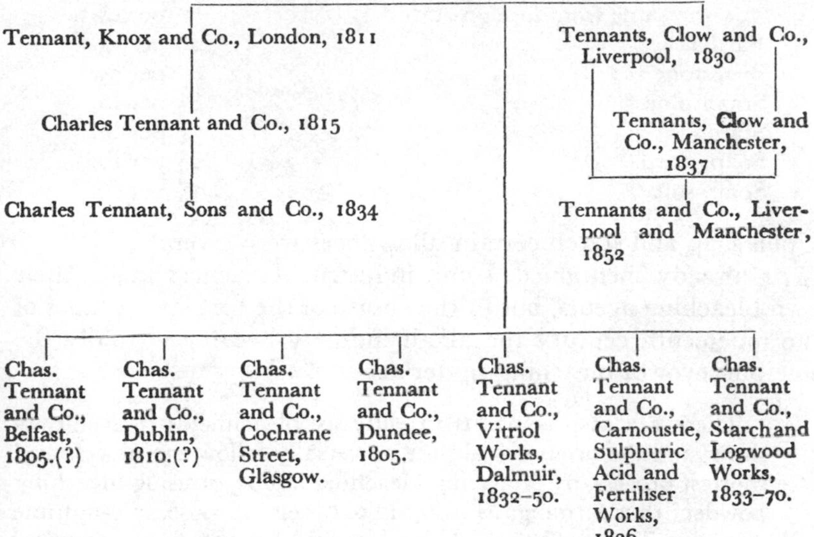

for the manufacture of soda but for chlorine as well.[2] In 1789 bleachers succeeded in getting a rebate of duty provided they took out a bleacher's licence.[3] This by no means removed all their difficulties. Due notice had to be given to a salt officer that salt was to be used, and he had to be present at the mixing. This had to begin within half-an-hour of his arrival. Moreover, the operation could not be carried out oftener than once a week.[4]

Impediments to progress varied from country to country. As late as 1810 Scottish bleachers were complaining of duties which they had to pay, and from which the Irish were exempt. They were all the more aggrieved when they remembered that the sixth Article of the Treaty of Union with Ireland expressly provided 'that the subjects of Great Britain and Ireland shall be entitled to the same privileges and be on the same footing as to encourage the bounties on the like articles being the growth, produce, or manufacture of either'. The various duties which they had to pay were:

[1] E. W. D. Tennant, *One Hundred and Forty Years of the Tennant Companies*, p. 4.

[2] A. Warden, *op. cit.*, p. 539.

[3] Such a licence, belonging to the Slaters of Dunscar, is illustrated in S. H. Higgins's *History of Bleaching*.

[4] Ed. Hughes, *Studies in Administration and Finance*, p. 432.

	s.	d.	
Pot and Pearl Ash, the produce of the British Colonies	1	4	per cwt.
,, the same from foreign states . .	4	8	per cwt.
Barilla	11	4	per cwt.
Brimstone	12	8	per cwt.
Small blue		8	per lb.
Starch		3¼	per lb.
Soap, hard		2¼	per lb.
Soap, soft		1¾	per lb.

Upon soap and starch certain allowances were given.[1]

As already mentioned, some industrial bleachers made their own bleaching agents, but in the course of the first few decades of the nineteenth century the alkali industry became virtually the sole purveyor of bleaching materials.

> There are at present (1812) fifty-six bleachfields in Renfrewshire; and the occupiers almost universally follow the newest and shortest process, by preparing bleaching liquor, or using bleaching powder; thus forming, as a liquid for their purpose, oxy-muriate of lime. Their art has derived much benefit from chemical discoveries, and they have wisely availed themselves of the aids which science affords. Their process is carried on within doors; and the whole operation is often finished without having recourse at all to the field; so that bleaching grounds are now not much wanted, and consequently portions of them have been thrown into cultivation.[2]

In the manufacture of soda from salt, which was widely commercialized in the third decade,[3] hydrochloric acid is produced to the extent of one-and-a-half times the amount of sulphuric acid used. At first the profit on the soda was so great that the hydrochloric acid was allowed to go to waste, but it was soon found that chlorine produced from the by-product acid was purer than that from any other source, and it consequently became the raw material from which the chlorine for bleaching-powder was obtained. Thus salt duties, which held back commercialization of the conversion of salt into soda, had a dual effect on bleaching. Indeed, it seems likely that by-product soda, made from the residues of the manufacture of chlorine as the primary product, was introduced by an Aberdeen bleacher, Alexander Drimmie, before

[1] *Memorial of Manufacturers and Bleachers in Edinburgh stating the Articles on which they were subject to duty and from which Irish were exempted.* N.L.S. MS. 640, f. 155 *et seq.*

[2] J. Wilson, *General View of the Agriculture of Renfrew*, p. 267. [3] Ch. IV.

synthetic soda was manufactured in Britain. The substitution of soda for potash was, according to A. Warden, an event of considerable importance in the bleaching trade.[1] With the repeal of the salt taxes the synthetic products completely replaced plant ashes.

William Higgins draws a parallel between mechanical and chemical technology:

> How gradually and yet how progressively the steam engine, from its first invention by the Marquis of Worcester, was brought to its present degree of perfection! Undoubtedly, it is just so with respect to alkalies, the substances now used by the bleachers, it must have taken a considerable time after their first application in bleaching, before they could be made the most of.[2]

By 1815 there were more than two hundred and fifty bleachers in Scotland, of which at least a hundred were in the neighbourhood of Perth, Glasgow, and Paisley. These required for their operations some 4,000 hands, of which women were to men in the ratio four to one.[3] According to Sinclair, wages in Scotland were 18*s.* for men, and 6*s.* for women, but these probably represent maximum rates. The writers in the *Statistical Account* give wages at the Forfar bleachfields as 3*s.* to 5*s* per week 'finding their own victuals', while at a small bleachfield in Renfrew men got 7*s.* 6*d.* to 12*s.* per week, women 2*s.* 6*d.* to 6*s.* per week.[4] In England, the wages paid by Ridgways of Wallsuches can be compared with those cited by Sinclair, for in a pay-sheet preserved by them, men are shown to receive 10*s.* 6*d.* to 16*s.* a week, and women 6*s.* 6*d.*[5] According to D. F. Macdonald, the bleaching and dyeing industries as they developed in central Scotland attracted labour, particularly female labour, from the north.

> They were of necessity of migratory habits, and were described as 'not having families nor residing in families, but in women-houses so-called, erected by the masters on purpose near almost every field (for bleaching), where they lodge only during the working season.'[6]

These Scottish bleach-works were of sufficient importance to attract foreign visitors, and when a Mr. Preckel of Westphalia wanted to see one, Thomas Thomson recommended him to go and see Thornliebank, it being 'as perfect as any where else'.[7]

[1] A. Warden, *op. cit.*, p. 721.
[2] W. Higgins, *op. cit.*, p. 65.
[3] Sir J. Sinclair, *General Report*, Appendix 2, p. 284.
[4] *Statistical Account*, Vol. 2, p. 168; Vol. 9, p. 142.
[5] A. J. Sykes, *Concerning the Bleaching Industry*, p. 24
[6] D. F. Macdonald, *Scotland's Shifting Population, 1770–1850*, p. 65.
[7] Letter from Thomas Thomson, July 26, 1825, N.L.S., MS. 640, f. 38.

The relationship between chemistry and bleaching may be fittingly summed up in the words of Thomas Henry, the Manchester chemist (1734–1816):

> Bleaching is a chemical operation. The end of it is to abstract the oily parts of the yarn or cloth, whereby it is rendered more fit for acquiring a greater degree of whiteness, and absorbing the particles of any colouring material to which it may be exposed.
>
> The materials for this process are also the creatures of chemistry, and some degree of chemical knowledge is required to operate and judge of their goodness. Quick-lime is prepared by a chemical process. Pot-ash is a product of the same art; to which vitriolic, and all the other acids owe their existence. The manufacture of soap is also a branch of this science. All the operations of the whitster; the steeping, washing and boiling in alkaline lixiviums; exposing to the sun's light, scouring, rubbing and blueing are chemical processes, or founded on chemical principles. The same may be said of the arts of dyeing and printing, by which those beautiful colours are impressed on cloths, which have contributed so largely to the extension of the manufactures of this place (Manchester). How few of the workmen, employed in them, possess the least knowledge of the science to which their profession owes its origin and support! If random chance has stumbled on so many improvements, what might industry and experience have effected, when guided by elementary knowledge? The misfortune is that few dyers are chemists and few chemists dyers. Practical knowledge should be united to theory, in order to produce the most beneficial discoveries. The chemist is often prevented from availing himself of the results of his experiments, by want of opportunities of repeating them at large; and the workman generally looks down with contempt on any proposals, the subject of which is new to him. Yet under all these disadvantages, I believe it will be confessed that the arts of dyeing and printing owe much of their recent progress to the improvements of men who have made chemistry their study. Much remains to be done; and perhaps in no respect are the manufactures of this country more defective than in the permanency of their permanent colours. Sensible as our manufacturers are of this defect, is it not strange, that so few of them should attempt to acquire a knowledge of those principles which would most probably supply them with the means of improving and fixing their dyes.[1]

[1] Thomas Henry, *On the Advantages of Literature and Philosophy, and especially on the consistency of Literary and Philosophical with Commercial Pursuits, Mem. Manch. Lit. and Phil. Soc.*, 1785, *1*, 7.

Chapter X
THE PHILOSOPHY OF COLOUR

At present there are a greater number of intelligent practical chemists in Scotland, in proportion to the population, than perhaps in any other country in the world.
SIR JOHN SINCLAIR in 1814.

TO students of economic history it is a commonplace that there was a profound change of English fashion towards the end of the seventeenth century. This was a demand for lighter and brighter clothes. English weavers and dyers found difficulty in supplying it. Consequently, foreign textiles had to be imported. The East India Company seized the opportunity to flood the English market with silks, calicos, muslin, and chintzes, of better quality and cheaper than the English were formerly familiar with. Poor people, anxious to ape the more fortunate, were the first to avail themselves of the eastern imports, but the well-to-do soon found in them a welcome addition to their wardrobes.

> Elegance was thus combined with cheapness; and little more is needed to make a commodity popular.

Vested interests in the wool and silk trade were disturbed by the increasing unpopularity of their merchandise, and petitioned Parliament to prevent the new importation, but, once established, the demand for more attractive garments and furnishings had to be met. The result was the establishment of textile printing in this country.[1]

This reaction was a direct parallel to what took place earlier in the supply of dye-stuffs. New dye-stuffs, such as logwood and cochineal, were introduced from America in the sixteenth century. The English woad-cultivators opposed their adoption and use. They petitioned Parliament to prohibit importation, were successful in 1580, and a ban remained operative till 1661. It was removed when the English were importing foreign skill and operatives.[2] The same prejudice was meted out by woad-cultivators when indigo began to be imported from the east.[3]

[1] P. J. Thomas, The Beginnings of Calico Printing in England, *English Historical Review*, 1924, *39*, 206.

[2] For the early history of cochineal see J. Beckmann, *History of Inventions and Discoveries*, Vol. 2, p. 171.

[3] For the indigo trade, see W. H. Morland, *From Akbar to Aurangzeb, passim.*

The following citation from E. Baines' *History of the Cotton Manufacture in Great Britain*, shows how the various Acts influenced the development of dyeing and textile printing in Great Britain.

> The development of calico printing in England depended on a Parliamentary prohibition of the importation of cheap and beautiful prints from India, Persia, and China, which was brought into operation in 1700. (11 and 12 William III, c. X.) This Act was introduced as the result of pressure from woollen and silk interests. It however, stimulated the embryonic printing trade, the English having become accustomed to printed calicos and chintzes. Unprinted Indian calicos were still admitted on payment of a duty. By 1712 it was worth while for Parliament to impose a duty of 3*d*. per square yard on calicos printed, stained, painted or dyed, and in 1714 the duty was raised to 6*d*. (10 Anne, c. 19; 12 Anne, sec. 2, c. 9.) At the same time a duty equal to half the above was laid on linens. These Acts had not the desired suppressive effects, so in 1720 a further Act was passed prohibiting the wearing of any printed or dyed calicos whatsoever, whether printed or dyed at home or abroad, and even of any printed cotton goods of which cotton formed a part; excepting only calicos dyed all blue, and one or two other exceptions. This diverted printers from cotton to linen. (7 George I, c. 7.)

This latter Act is referred to as the *Calico Act*, 'an act to preserve and encourage the woollen and silk manufactures and for the more satisfactory employment of the poor, by prohibiting the use and wear of all printed, painted, flowered or dyed calicos in apparel, household stuffs, furniture or otherwise'. The Act made it illegal to wear or use any printed calico under a penalty of £20, with a reward of £5 to the informer. Some of these prohibitions were repealed in 1736 and the printing of mixed fabrics was developed, that is, fabrics with a linen warp and cotton weft.[1] This change arose out of attempts to apply the *Calico Act* to home manufactures. The *Manchester Act*, as it is called, provided that the Act of 1720 should not be interpreted to prohibit the use of material made of cotton with the addition of linen or wool, whether printed or painted.

These prohibitions and regulations continued in operation till 1774, when the printing of calico was permitted again on payment of a duty of 3*d*. per yard. By this time the *Calico Act* was standing in the way of the expansion of the English cotton industry. In 1779 and 1782 three additions were made to this duty, each of 5 per cent.[2]

[1] 9 George II, c. 4. [2] E. Baines, *op. cit.*, p. 259.

The established silk and woollen manufacturers had resented the increasing use of printed cottons, and been successful in their appeal to Parliament to forbid importation. Popular demand, however, prevented a return to the old-fashioned drab materials, and instead, the printing of textiles was undertaken in England. This stimulated a new interest in the production and utilization of dyeing materials.

The earliest known publication on dyeing in English is a translation from the Dutch, printed in 1605. The first written in English is by Sir William Petty. It was presented to the Royal Society in 1662 and printed with Thomas Sprat's *History of the Royal Society of London* in 1667. Petty lists the following dyes then used in England:

> . . . for blacks, copperas and filings of steel or slippe; for reds, red-wood, brazil, madder, cochineal, saffron, kermes-berries, and sanders; for oranges, arnotto and young fustick; for yellows, weld, wood-wax, old fustick and turmerick; for blues, woad, indigo and logwood.

Nearly two years after Petty's publication, Robert Boyle presented a paper to the Royal Society on *Experiments and Considerations, touching Colour*, and in 1669 Robert Hooke produced a piece of calico printed by a method of his own. The Royal Society desired him to extend it to other colours, besides those used on the first piece, and next month he produced another specimen of printing with yellow, red, green, blue and purple colours, which, he claimed, would endure washing with warm water and soap. This claim indicates the difficulties experienced in getting fast colours.[1]

Indeed, there is no question that the quality of goods then dyed in England was poor, African negroes invariably preferring Indian cottons if they could get them.[2] Even into the eighteenth century, Lucock attributed poor results in the dyeing of wool to unscientific methods:

> But indeed what can we expect but faint, muddy, and uncertain colours, where wool is dyed, as is too much the custom in Yorkshire, without being scoured, in pans unwashed, and with materials mixed together upon a floor unswept, where a little before perhaps have been mixed ingredients calculated to produce a totally different tint.[3]

[1] T. Birch, *History of the Royal Society*, Vol. 2, p. 401.

[2] Wadsworth and Mann, *The Cotton Trade and Industrial Lancashire*, p. 158.

[3] *North Riding Session Records*, vii, p. 172; Thirske, 1709: quoted by H. Heaton, *Yorkshire Woollen and Worsted Industries from the Earliest Times up to the Industrial Revolution*, p. 332.

The principles of dyeing and calico printing formed part of eotechnic craft knowledge. Before we can appreciate the change which took place during the process of industrial evolution, it is necessary to consider for a moment the theory of dyeing as then understood. Colouring matters can be divided into two groups, *substantive* and *adjective*. A substantive dye colours material itself. Adjective dyes are incapable of giving permanent colours without being fixed by a *mordant*.[1] Andrew Ure in his *Dictionary of Chemistry* explains the fixing of an adjective dye by a mordant as follows. If a piece of cotton is dipped in ink made by mixing a solution of ferrous sulphate ($FeSO_4$) and a concoction of oak galls, it takes a black colour, but the result is neither good nor durable. If, however, the cotton is first dipped in the tannin solution extracted from the galls, then dried, and subsequently immersed in a solution of ferrous sulphate, the cotton takes a deeper, more permanent colour, because the ink (a colloidal iron tannate) is formed *in the fibres*, and does not merely coat them. The tannin solution extracted from the galls is called a *mordant*, and the solution of iron is the adjective dye. If the mordant is applied over the whole of the cloth before it is dipped into the dye-bath, the article is uniformly dyed. If the mordant is applied only to parts of the material, the latter alone become coloured on immersion in the dye-bath. This, in brief, is the process of printing textiles.[2]

The process of dyeing was thus as follows. Cloth to be dyed with an adjective dye was first immersed in a solution of the appropriate mordant, dried, and washed or *dunged* to remove the excess mordant. The amount of cow-dung used was so great that it became an article of great expense. Some calico-printers kept as many as twenty-eight to thirty cows for no other purpose than to supply dung.[3] Its importance is mentioned by Parkes, who says: 'the brightness of the colours, and the purity of the whites, always depend on the quantity of the dung employed'.[4] Palaeotechnics did away with the necessity for the biological material.

> Calico-printers used for a long time the solid excrement of the cow, in order to brighten and fasten colours on cotton goods; this material appeared quite indispensable, and its action was ascribed to a latent principle which it had obtained from the living organism. But since its action was known to depend on the phosphates

[1] Ch. X.

[2] Andrew Ure, *Dictionary of Chemistry*, 8th edition, 1821, Article, *Dyeing*; also Thos. Thomson's Article on Dyeing, *Encyclopaedia Britannica*, 7th edition.

[3] *Annals of Philosophy*, 1814, *4*, 235.

[4] S. Parkes, *Chemical Essays*, Vol. 4, p. 104 *n*.

contained in it, it has been completely replaced by a mixture of salts, in which the principal constituents are the phosphates of soda and lime.[1]

The theory of mordant action attracted attention at an early date. Regarding alum ($K_2SO_4.Al_2(SO_4)_3.24H_2O$), an important mordant, Petty himself remarked: 'the use of Allum is to be a *Vinculum* between the Cloth and the Colour'. Thomas Henry credits James Keir with being 'the first who suspected that (in dyeing) the earth of alum was precipitated, and in this form attached to the material prepared or dyed'. This explanation was adopted by P. J. Macquer, whose *Dictionnaire de Chimie* Keir translated, and published in 1778.

Macquer, the son of a Scottish emigrant painter, and Berthollet, whose connection with the art of bleaching was mentioned in the previous chapter, were two of the first chemists to interest themselves in scientific aspects of dyeing and calico-printing. A. Wolf, in his *History of Technology*, tells us:

> During the earlier part of the eighteenth century the traditional method of dyeing was still followed. Improvements were, however, introduced in the course of the century; and the practical improvements were intimately associated also with advances in the chemical theory of dyeing. The improvements were brought about in large measure by the special interest which the French Government took in the industry. Already, in the seventeenth century, under the enterprising Colbert,[2] steps had been taken to encourage progress in the art by making awards to those who helped to improve the technique of dyeing, and to increase the efficiency and skill of the craftsmen engaged in the industry. Eventually a special official was appointed to exercise general supervision over the whole industry. It is noteworthy that the two investigators who did most for the advancement of the practice and theory of dyeing were successively Directors of the Dyeing Industries in France. They were Macquer and Berthollet.
>
> The theory of dyeing held in the early decades of the eighteenth century, notably by Du Fay and Hellot, two of the earlier Directors of the Dyeing Industries of France, was purely mechanical. This was not unnatural in view of the predominance of mechanical theories during the seventeenth century. According to the mechanical theory of the phenomena of dyeing, particles of the dye used

[1] Justus Liebig, *Chemistry in its Application to Agriculture and Physiology*, p. 176. This mixture of salts is sold to calico-printers in large quantities under the name of *dung substitute*. It would be well worth experiment to try its effects as a manure upon land. Its cost is 3*d*. or 4*d*. per pound, and is not, therefore, dearer than nitrate of soda, which is now so extensively used. (Translator's note in above.)

[2] Founded Académie des Sciences in 1666.

entered the pores of the dyed material. The preparatory processes to dyeing were largely regarded as helping to open the pores of the fabric, either by heat or by some chemical action, so as to enable the particles of colouring matter to enter and fill them. The retention of the dye of the dyed fabric was similarly explained as due chiefly to the action of cold in contracting or closing the pores of the fabric, and thereby imprisoning the particles of colouring matter in the texture of the material. Difficulties in dyeing certain substances were likewise explained by reference to the alleged presence of salts which blocked up the pores of those substances and thereby prevented the entry of the particles of dyes. There was no chemical theory of the process of dyeing till Macquer and Berthollet took up the problem.[1]

The following citation illustrates Berthollet's advance beyond the mechanistic conception.

'The colouring particles possess chemical properties that distinguish them from all other substances, they have attractions peculiar to themselves, by means of which they unite with acids, alkalis, metallic oxyds, and some earths principally alumine. They frequently precipitate oxyds and alumine, from the acids which hold them in solution; at other times they unite with the salts and form supra-compounds, which combine with the wool, silk, cotton, or linen. And with these, their union is rendered much more close by means of alumine or a metallic oxyd, than it would be without their intervention.'[2]

Here we have most explicitly an attempt to explain the phenomena of dyeing in strictly chemical terms, in terms of chemical affinity. The solution of the particles of a dye depends on the affinity of the dye for the solvent. Particles of dye may combine with a fabric either directly or through the mediation of a mordant, like alum for instance. The original colour of a dye might be modified by the colour of the material with which it combines. And conditions such as air and light might also influence the colour eventually obtained.[3]

Individual behaviour of different fibres to the same dye made the composition of textile fibres a fit subject for scientific investiga-

[1] For Du Fay's and Hellot's view see *Mem. de l'Académie Royale des Sciences*, 1737, p. 253; and 1740, p. 126; and 1741, p. 38, respectively. A Wolf, *History of Science, Technology and Philosophy in the Eighteenth Century*, p. 512.

[2] *Elements de l'Art de la Teinture*, Paris, 1791, Vol. I, p. 20; English translation by W. Hamilton, *Elements of the Art of Dyeing*, London, 1791, Vol. I, p. 22.

[3] A. Wolf, *op. cit.*, p. 514. For an extended account of the contributions of Hellot, Macquer, Berthollet and Chaptal, and the close connexions between the dye industry in France, and the advance of theoretical chemistry, see *Great Masters of Dyeing in Eighteenth Century France.* (Ciba Review, 1939, 2, 618.)

tion. Woollen fibres are more easily dyed than cotton, and dyeing of woollen and silken goods was the first to attain any degree of excellence, while dyed cottons long remained of poor quality, because of the small attraction of the fibres for the colouring matter. Chemical analysis revealed that there were differences between the fibres. The following results are given by Andrew Ure in his translation of Berthollet's *Elements of the Art of Dyeing*:

	Per cent. Carbon	*Per cent. Hydrogen*	*Per cent. Oxygen*	*Per cent. Nitrogen*
Wool . .	53·70	2·80	31·20	12·30
Silk . .	50·69	3·94	34·04	11·33
Cotton . .	42·11	5·06	52·83	—
Flax . .	42·80	5·50	51·70	—

The distinction between animal and vegetable fibres will be readily discerned by the absence of nitrogen in the latter.

These differences determined the dyer's *modus operandi.* Wool, which more readily adsorbs (or absorbs) colouring matter, was given but little preparation. A wash in weak alkali to free it from natural greasiness was usually sufficient. Stale urine was the detergent generally used, because the alkali which it contains, ammonium hydroxide (NH_4OH), though strong enough to cleanse the wool, does not attack the fibres. When taken from the cone, silk is covered with a kind of varnish, making it harsh to touch, and easily tarnished. A solution of good soap is a satisfactory cleansing agent. The fibre loses a quarter of its weight in the process. To whiten it further before dyeing, and so to brighten the final colour, silk was bleached with fumes of burning sulphur (i.e. with sulphur dioxide, SO_2).[1]

Cotton and linen were much more difficult to prepare. Cotton contains some natural earthy matter which acts as a mordant. Being unequally distributed, this earthy matter gives rise to a patchy effect after dyeing. To remove it, the material was soured by soaking in weak oil of vitriol (H_2SO_4). As we have seen,[2] this process was one of the first which furnished a market for Roebuck's sulphuric acid.

Cottons were, in fact, so difficult to dye, that in desperation a woollen thread was sometimes woven into cotton cloth, to give a striped effect.[3] Till 1750, however, little pure cotton cloth was woven in England, though a fair amount of ready-woven cotton

[1] M. Baumé, *Bleaching Raw Silk*, Philosophic Journal, 1797, *1*, pp. 32 and 88; A. Ure, *Chemical Dictionary*, Article, *Dyeing*.

[2] Ch. VI.

[3] Wadsworth and Mann, *op. cit.*, pp. 159 and 178.

was imported from India and dyed. The earliest references to *cottons* usually mean woollen cloth, or sometimes fustian (i.e. linen warp and cotton weft).

The original dyes, as previously mentioned, were simple affairs, usually of vegetable origin. From the *Breton Laws* and the *Book of Rights*, published by the Celtic Society in 1847, Professor Eugene O'Curry has made out the complete process of the manufacture and dyeing of textiles by the ancient Celts.[1] In Scotland in 1684 the people gathered the 'excrescences which grew on craigs at Minnigaff' and formed it into balls for dyeing a purple colour which they called *corklit*. They also made an orange-brown or *philamort* colour from woodraw, a kind of *forg* with a broad leaf which grew at the foot of trees.[2] According to the author of the *General View of the Agriculture of Caithness*, about 1770, 'the *gudewife* was generally competent to dye the woollen yarn, either a blue, red, green, yellow, or black, colour as might be required'. With the coming of the industrial revolution this craft-skill was no longer required.

> That simplicity of life and industry are now gone (1810) and instead of these native fabricks, nothing will do but broadcloth from Leeds, and blankets and flannels from the southern markets.[3]

The method of using dyes was elementary, as can be seen by instructions given for 'setting of a blue vat', in the *Select Transactions of the Society of Improvers*.[4] Indigo was ground with water till it was 'as fine as oil', then heated, with subsequent additions of 'about a chopin of wheat bran, three pounds of best English potatoes and half a pound of bright madder'. When boiled, this mixture was ready for use. Its efficacy as a dye is uncertain. Indeed, the opinion of the Linen Committee of the Board of Trustees in 1755 was:

> . . . that it is of capital importance to the linen manufacture to get the Art of Dyeing a proper Blue as much improven, and the skill of it as much diffused over the Country as possible.

In order to do this they directed the Secretary to enquire of Provost Ayton of Glasgow and Provost White of Kirkcaldy about dyeing practice and prices. They remitted to Lord Strichen the task of finding what method of dyeing Mr. George Lind used and

[1] R. W. Cochran-Patrick, *Mediaeval Scotland*, p. 32; *see also* A Ross, *Highland Dyes*, Northern Notes and Queries, Vol. 1, p. 10.

[2] G. Chalmers, *Caledonia*, Vol. 3, p. 295, *n.*

[3] *General View: Caithness*, p. 246.

[4] R. Maxwell, *op. cit.*, p. 365.

to report upon it.[1] If this is the same Ayton as is mentioned in a *Session Paper*,[2] he was the co-proprietor of the Harlem Dye Company, the first dyeing factory to be set up in Glasgow. It was carried on by Ayton and Allan and Smellie, who were also the proprietors of the Dovehill Dyeing Manufactory. They were connected with the linen trade by 1733 and set up the dye works in 1740.

For wool, indigo was the most satisfactory blue dye. It was obtained by fermenting the indigo plant.[3] Woad, which could be grown in the British Isles, was originally used, but woad alone was gradually displaced by mixtures containing a quantity of indigo depending on its current price. The latter was imported by the East India Company, until a rival supply came from the West Indies and decreased the production in the east.[4] A valuable paper on the cultivation of woad in England was published by John Parrish, in the *Letters of the Bath Agricultural Society*.

Experiments in growing substitutes for indigo at home were undertaken as part of the general scheme of improvements in Scotland. In 1756 Dr. Francis Home reported to the Board of Trustees upon a petition which was presented by James Ross with regard to a proposal to make indigo from a plant growing in Scotland. Home's opinion was that the material produced from the plant was:

> . . . neither of the colour, nature, nor quality of the true Indigo, nor would it strike a blue colour on either Linen, Cotton or Woollen, and therefore, could be of no use in any of these manufactures nor in dyeing and probably not in any other way.[5]

Home also submitted a report on blue checked material sent in from Dunfermline, probably by Messrs. Hogg and Mosman. In this case he stated that it was dyed with logwood, and recommended that the method of dyeing should be discouraged.[6] Dr. Black, too, experimented with dyes. His letters, used by his biographer Sir Wm. Ramsay, contained samples of tape dyed various shades of blue and labelled 'the first experiments I made to compare indigos'.[7]

Attempts to improve colours were made in other directions as well. In 1704, Prussian Blue was discovered accidentally by

[1] *Board of Trustees Minutes*, Vol. 13, p. 83, 12th December, 1755.

[2] Signet Library, 55:4, January 1759.

[3] For early history of indigo, see J. Beckmann, *History of Inventions and Discoveries*, Vol. 4, p. 101.

[4] W. H. Morland, *From Akbar to Aurangzeb*, p. 113.

[5] *Board of Trustees Minutes*, Vol. 13, p. 114, 27th February, 1756.

[6] *Ibid.*, 15th January.

[7] W. Ramsay, *Life and Letters of Joseph Black*, p. 103.

Diesbach, a Berlin colour manufacturer.[1] His method of manufacture was kept secret till 1724, when Woodward published it in the *Philosophical Transactions*.[2] It was made by a complicated process involving the calcination of alum, green vitriol, alkali, with bullock's blood, flesh, or other animal matter. At first it was used as an artist's pigment. Shortly after this, however, Macquer was experimenting with pigments in the hope of finding one suitable for a blue dye, and announced that he had found in Prussian Blue a dye for both silk and wool.[3] Wolf tells us:

> About the middle of the eighteenth century, when Macquer took up the problems of dyeing, the only fast blue dyes known were indigo and pastel. Macquer introduced the use of Prussian blue. He boiled skeins of cotton, flax, silk, and a piece of cloth in a solution of alum and green vitriol. He then dyed them by immersing them in a solution of alkali calcined with organic matter. Next he dried them in the air, and then dipped them in very dilute but hot sulphuric acid. The result was a bright blue; and the colour was fast in the case of silk and the wool. He found also that the colour could be made deeper by repeating the immersion in the dye, so that a large range of shades of blue could be obtained by varying the number of dippings. Moreover, the Prussian blue seemed to penetrate the dyed material completely; whereas the indigo and pastel only coloured the surface, so that when the surface of the dyed material wore off the undyed texture under it showed through. Prussian blue consequently came to be used extensively.[4]

It was not, however, a satisfactory dye, for although it possessed a colour whose:

> . . . lustre greatly surpasses everything before seen in wool, and emulates even the *transparency* and brilliance of the finest sapphire to such an extent that the eye, which has once seen Prussian blue so communicated, disdains after to fix itself upon the common indigo blue. . . .

it did not give an even effect, and it was discharged, i.e. removed, by the alkali of soap.[5] Seven years later Hellot published ten methods of making it, and in 1770 it was being manufactured in England at Newcastle at £2 2*s*. per pound.[6]

Blood for the use of a firm of Prussian Blue manufacturers was

[1] L. J. M. Coleby, *Annals of Science*, 1939, *4*, 206; Ed. Bancroft, *The Philosophy of Permanent Colours*, Vol. 2, p. 60. [2] *Phil. Trans.*, 1724, *33*, 15.

[3] C. L. Berthollet, *Elements of Art of Dyeing*, p. 29.

[4] A. Wolf, *op. cit.*, p. 513. [5] Ed. Bancroft, *op. cit.*, Vol. 1, pp. 60–90.

[6] L. J. M. Coleby, *op. cit.*, *4*, 206.

40. The bowking keirs (which work on the principle of the modern coffee percolator) in which the materials were boiled in the process of bleaching. Dash-wheels can be seen in the background.

C. F. Kreisig, *Turkey Red Dyeing*, 1820

41. Furnace and benches set up for the preparation of various chemical products for bleachers and Turkey Red dyers; for example, hydrochloric acid, wet and dry bleaching compounds, and the tin mordant used in dyeing scarlet.

Encyclopédie Méthodique, 1782–1832

42. Dyehouses arranged for dyeing different colours towards the end of the eighteenth century.

collected in Edinburgh. Their factory was situated at Portobello, and was erected in 1785 by Davidson and Davenport of Newcastle. Some years later, when the feuars in the neighbourhood raised an action in the Court of Session on account of the offensive smell of the operations, it was owned by Edward Colston and his sons.[1]

In Scotland, Charles Macintosh started the manufacture of Prussian Blue at his Alum Works at Campsie, Stirlingshire. He calcined blood, hooves, hair, woollen rags, etc., with potash in iron vessels, and followed by dissolving the mass in sulphuric acid. On addition of iron sulphate, Prussian Blue precipitated. By his method, Macintosh also reduced the price of prussiate of potash ($K_4Fe(CN)_6$), which is formed before the addition of the iron sulphate, from 5*s*. or 6*s*. per ounce to about 5*s*. per pound in 1825. It subsequently fell to 2*s*. 6*d*. in 1835, and to 1*s*. 5*d*. in 1846.[2] Of the product it was said: 'His crystals of prussiate of potash are supposed to be unrivalled for their beauty and purity, and his Prussian Blue cannot be excelled',[3] yet calico-printers were slow to avail themselves of them. William Tate wrote Charles Macintosh:

> You will conclude that I have been inattentive to the introduction of your blue liquor, by my silence on the subject; but the truth is, I sent the liquor to Mr. Ainsworth, as you requested, and applied to him afterwards for his opinion. I also consulted him on the probability of any material consumption taking place in his neighbourhood; to none of these inquiries could I obtain anything decisive.[4]

Macintosh followed up Macquer's idea of using Prussian blue as a dye. So successful was the manufacture in this country that large quantities were exported to China and the East India Company.[5]

The manufacture of Prussian blue was but one of many chemical ventures of the Macintosh family. George, son of Lachlan Macintosh, tacksman of a farm at Auchinluich, Newmore, Ross-shire, was born in 1739. He was given a fair education, and started work as a junior clerk in the Glasgow Tan Work Company, whose partners included Glassford of Dougalston, Campbell of Clathic, Speirs of Elderslie, Boyle of Daldownie, Provost Bowman, *et al.*

[1] Session Papers, Signet Library, *207*, 19.
[2] G. Macintosh, *Biographical Memoir of Charles Macintosh, F.R.S.*, p. 49.
[3] J. M'Ure, *Glasghu Facies*, p. 1234.
[4] William Tate to Charles Macintosh, Manchester, 6th March, 1810: quoted, G. Macintosh, *op. cit.*, p. 50.
[5] S. Parkes, *Chemical Essays*, Vol. I, p. 91; J. R. McCullock, *Dictionary of Commerce*, Vol. 2, p. 956.

Young Macintosh soon realized that the most remunerative branch of the trade was the manufacture of boots and shoes. He started on his own, and by 1773 had 500 employees. He also had interests in glass manufacture and the East India trade, but abandoned these to take up the manufacture of Cudbear, a dye for silk and wool.[1]

The story of the discovery of cudbear is as follows. George Gordon, a London coppersmith belonging to Foddaletter, Banffshire, while repairing a copper boiler in a London dye-house, noticed the similarity between the method used there (Archella dyeing) and that used in the Highlands for making the dye known as crottel (*Lichen tartareus*).[2] His knowledge as a botanist and chemist was considerable. In 1758 he entered into partnership with his *brother*, Cuthbert (not his nephew as is commonly stated), and Messrs. William Alexander and Company of Edinburgh to manufacture this dye, calling it Cudbear after his mother, whose maiden name was Cuthbert. According to the *General View of Banffshire*, prior to the year 1755 George Gordon discovered that by a simple preparation of a species of moss, a purple dye could be obtained. His brother, Cuthbert Gordon, published in the *Scots Magazine* (1755) certificates from dyers in Leeds, Halifax, Edinburgh and Glasgow, that 'they frequently continued its use, and found it answer their purposes well for dyeing linen, cotton, silk and woollen goods, to their great satisfaction both as to the elegance of the colours produced and the indigo saved; and they think that Mr. Gordon ought to be encouraged by every lover of his country for such a curious and valuable improvement'.[3]

George Gordon died on 21st September, 1764, and two younger brothers James and William joined the firm. In 1775 Cuthbert Gordon took sole possession, but soon failed, even a loan of £300 in 1778 from a Glasgow dyer, Adam Grant, failing to save him from bankruptcy.[4]

In 1771 a similar Glasgow manufactory had been formed, and purchased ground at the foot of the Craigs. The partners were John Glassford, James Gordon, John Robertson, *et al.*, under the management of George Macintosh, who, on the failure of Cuthbert Gordon's venture in Leith, bought his business and transferred it to Dunchattan, near Glasgow. The Gordons continued partners.[5] Everything was done at these works to ensure secrecy.

[1] *Glasgow Curios*, pp. 66–68; G. Macintosh, *op. cit.*, p. 117; *Statistical Account*, Vol. 12, p. 113.
[2] *Statistical Account*, Vol. 12, pp. 113 and 442.
[3] *Scots Magazine*, 1755, p. 61.
[4] *Session Papers*, 194: 32, Signet Library.
[5] G. Macintosh, *op. cit.*, p. 118.

Walls ten feet high were built round everything, and Highlanders speaking only Gaelic were employed.[1] The initial partnership was followed by Messrs. George Macintosh and Company, one of the leading partners of which was Henry Glassford, son of the above John Glassford. Further ground for works was acquired in 1784, until some seventeen acres were included under the general name of Dunchattan.[2] By 1786 their output must have reached considerable dimensions, and George Macintosh sent his son Charles on a continental tour to introduce the new dye, and effect sales at the same time for the Prestonpans Vitriol Company, then owned by John Glassford.[3]

To prepare the dye a lichen, *lichen tartareus*, was scraped from rocks, macerated, and stirred in wooden troughs with ammonia. The latter was obtained by distilling human urine, 'that of graminiverous animals being deficient in the volatile alkali'. The patent registered under the Great Seal by George and Cuthbert Gordon[4] gives as ingredients: (*a*) *lichen*, (*b*) *Muscus Rupibus admiscens* or *Coroloides*, (*c*) *Muscus pexidatus*. The method of preparing the cudbear was to clean the lichen with water, dry it, and pound with spirit of soot (i.e. ammonia). Lime was added, and the mass digested for fourteen days. The purity of the ammonia was apparently important, and probably a good deal of Macintosh's success was due to the fact that he used a pure solution. Bancroft afterwards demonstrated by many experiments the importance of pure ammonia for getting good colours.[5] The demand for cudbear increased until 250 tons of lichen were used annually,[6] and when the Scottish supply became exhausted it was imported from Norway, Sweden, the Canary Islands, and Malta.[7] Sweden exported 130 tons annually from 1770, and in 1793 Charles Macintosh visited Gustavus Adolphus IV of Sweden to get an export duty removed.[8] The price of the lichen rose progressively from £3 to £25, and during the wars to £45 a ton. It is estimated that £306,000 in all must have been spent on it between 1778 and 1838. In 1823, *lichen pustulatus* was successfully substituted for *tartareus*.[9]

To supply ammonia about 2,000 gallons of 'the waste product of the town' were used daily, and men 'with a few horses collected the urine about the suburbs, and the collectors were furnished with

[1] *Glasgow Curios*, p. 70. [2] J. McUre, *Glasghu Facies*, p. 767.
[3] G. Macintosh, *op. cit.*, p. 20.
[4] Register of the Great Seal, Vol. 19, No. 198, General Register House, Edinburgh. [5] Ed. Bancroft, *op. cit.*, Vol. 1, p. 300.
[6] *Glasgow Curios*, p. 70. [7] J. Sinclair, *General Report*, App. 2, p. 311.
[8] Ed. Bancroft, *loc. cit.* [9] G. Macintosh, *op. cit.*, p. 119.

pocket hydrometers, to prevent them being imposed upon by a spurious and inferior article'. This hydrometer, which was made for Macintosh by an instrument-maker of the name of Twaddle, is now known as Twaddle's hydrometer. Sir William Ramsay tells us that the other hydrometer which has long been in general use was described in a note in Dr. Joseph Black's handwriting as 'Hydrometer or Peseliqueur employed by Pierre Jacques Papillon'.

> It is exactly like an ordinary hydrometer, with a glass stem, graduated, however, in arbitrary degrees, with a bulb full of air, and beneath the bulb a smaller bulb containing the requisite amount of mercury to make it float to the height desired.[1]

They also had casks in the loom-shops. This urine cost the firm £800 per annum.[2] It was distilled for the ammonia which can be got from it, and we find Charles Macintosh trying to get a market for the residue. Some of this '*caput mortuum*' had been sent to Dr. Black by George Macintosh, with requests for suggestions about mordanting cloth.[3]

Writing to Matthew Boulton, Charles Macintosh says:

> In our Cudbear Works here, we make use of a very large quantity of Urine, not less than 3,000 gallons a day, from this we separate only, the disengaged volatile alkali; nearly the whole of the microcosmic salt (sodium ammonium hydrogen phosphate) remaining in the residuum of our stills, and which might be applied to use; as from experiments I have made, I apprehend it would not be a very difficult matter to obtain the salt in question, in pretty large quantities, in case there would be any demand for it in your, or any other part of England, as a flux.
>
> Of this I am at present ignorant, and have to intreat you would be kind enough to give me your sentiments on the subject, and if it would not be too much trouble, would esteem it a most particular favour, your procuring any other information which might not perhaps come immediately under our knowledge as whether or not it might be substituted with effect for Borax, and if this salt is used in considerable quantities in your neighbourhood.[4]

Attempts to preserve secrecy regarding the cudbear manufacture met with failure. An employee, a Highlander named McBraine, went to London and disclosed the process to Mr. Grant and General Horace Churchill, who erected large cudbear works in Great Peter Street in 1793. These did not succeed, and were

[1] W. Ramsay, *op. cit.*, p. 131.

[2] *Statistical Account*, Vol. 12, p. 114; *Glasgow Curios*, p. 71; J. Sinclair, *General Report*, Appendix 2, p. 311; Andrew Brown, *History of Glasgow*, Vol. 2, p. 252.

[3] *Ibid.*, p. 71.

[4] C. Macintosh to M. Boulton, 17th June, 1794, A.O.L.B.

given up in 1803. The *London Gazette* for 1803,[1] mentions the bankruptcy of James Gordon of the Cudbear Company, Great Peter Street, and the *Gentleman's Magazine* for 1811 says, 'Mr. J. Gordon, late of the Cudbear Company, Great Peter Street, Westminster, died in James Street, Buckingham Gate, on 25th May, 1811, aged 75'.[2]

Although Macintosh is reputed to have bought up the Cuthbert Gordon concern, it must have been revived at some subsequent date, the *General View of Banff* says on the cessation of commercial intercourse between this country and Norway and Denmark. It was in operation in 1808–9, moss being obtained from the Cairngorms.[3]

Other works built in Liverpool and Ireland were unsuccessful. Indeed, the best protection Macintosh had was not his ten-feet walls and his Gaelic-speaking workmen, but his scientific method of production.[4]

The colours produced by cudbear varied from bright pinks and reds through purples to bright blues. It dyed silk and wool, but not cotton. While not as permanent as indigo, it was reputed to give a brilliance and lustre that indigo alone could not.[5] It was also used to improve madder reds, which were generally too yellow.[6] The end of the cudbear trade did not come because better dyes took its place, but because fashion favoured blacks and greys in the 1840s. The change was progressive. Elizabeth Grant (1797–1885) in her *Memoirs of a Highland Lady* tells us:

> We were inundated this whole winter (1816) with a deluge of a dull ugly colour called Waterloo Blue, copied from the dye used in Flanders for calico of which the peasantry made their smock-frocks or blouses. Everything new was 'Waterloo', not unreasonably, but to deluge us with that vile indigo, so unbecoming even to the fairest! It was really a punishment; none of us were sufficiently patriotic to deform ourselves by wearing it.[7]

How right is Mumford when he says:

> Iron and coal dominated the palaeotechnic period. Their colour spread everywhere, from grey to black: the black boots, the black stove-pipe hat, the black coach or carriage, the black iron frame of the hearth, the black cooking pots and pans and stoves. Was it

[1] *London Gazette*, p. 1059.

[2] *Gentleman's Magazine*, 1811, *II*, Part 1, p. 675, quoted in *Scottish Notes and Queries*, 1st series, Vol. 1, p. 97; 2nd series, Vol. 3, p. 92.

[3] D. Souter, *General View: Banff*, p. 61.

[4] G. Macintosh, *op. cit.*, p. 119.

[5] A. Ure, *Dictionary of Chemistry*: Article, Archil; A. Brown, *History of Glasgow*, Vol. 2, p. 254.

[6] Ed. Bancroft, *op. cit.*, Vol. 1, p. 354.

[7] Quoted J. G. Fyfe, *Scottish Diaries*, etc., p. 500.

> mourning? Was it protective coloration? Was it mere depression of the senses? No matter what the original colour of the palaeotechnic milieu might be, it was soon reduced by reason of the soot and cinders that accompanied its activities, to its characteristic tones, grey, dirty brown, black.[1]

The Macintosh cudbear works, however, continued till 1852,[2] producing some cudbear, as well as other dye-wares such as sugar of lead, cream of tartar, and Rochelle salt. The firm also tried making yeast from malt, but this venture was commercially unsuccessful owing to opposition from London brewers.[3]

After retiring from the original Cudbear Company in Leith, Cuthbert Gordon did not cease in his endeavours to produce new dyes. About 1790 he drew the attention of the Privy Council to the use of yellow *ladies' bedstraw*, as a substitute for madder and cochineal, its root giving a red, and its stem a yellow, dye. He claimed that the dye was almost equal to cochineal in colour and better in durability. He was rewarded by the Government with £200 for his experiments. Efforts were made by the Privy Council to promote the cultivation of the plant, but nothing more is heard of the venture.[4]

Success in one direction made progress possible in others. In Scotland the first to manufacture good coloured sewing thread was Mungo Dick of Dundee. He

> . . . so effectually established the character of his manufacture, that in all places of Great Britain, Dundee threads and good threads are phrases of the same import. Hence the retailers generally put over their doors, 'Dundee threads sold here'.[5]

For long, however, the production of one cotton colour, *Turkey Red*, defeated the dyers of this country. The difficulty of producing a bright red is mentioned in a letter dated April 1738 from Escrike, a Lancashire manufacturer, to Wilson and Harrop of London:

> I think those (fustians) you order for rose-colour and scarlet ingraine had much better been sent up in the gray cut . . . You dye them much better than we can in this country. We pay a guinea an end for dyeing scarlet and a very poor colour. They will not dye for washing red a good full colour for 7*s*. 6*d*. per end, but shall leave it to you.[6]

[1] L. Mumford, *Technics and Civilization*, p. 163. [2] *Glasgow Curios*, p. 71.
[3] J. Sinclair, *General Report*, App. 2, pp. 309–314.
[4] Ed. Bancroft, *op. cit.*, Vol. 1, p. 305.
[5] F. Douglas, *General Description of the East Coast of Scotland*, p. 41.
[6] Quoted by Wadsworth and Mann, *op. cit.*, p. 263.

This difficulty led Negroes to prefer Indian cottons. They found that British dyed cottons would not wash. After the Seven Years' War, African merchants refused Manchester goods, and the failure of Manchester industry to supply reliable goods led to a Memorial from Liverpool merchants to the Board of Trustees. In it they indicate that imitation goods were exported when Indian goods supplied by the East India Company were too expensive or not obtainable, and the merchants proposed importing Indian cottons from other European countries if the monopolist East India Company could not supply the demand. The powerful East India Company was able to get a Bill passed, however, forbidding this unless their supply failed.[1] It is evident that the export market in dyed cottons was an important one, and it was imperative that a suitable bright red should be found. In despair, manufacturers tried weaving in worsted stripes which, being wool, would take a red more easily.[2]

To obtain a satisfactory dye was an expensive business. It had to withstand bleaching so that it could be woven into stripes and checks with grey thread which was bleached after weaving. If this were impossible the yarn had to be bleached before weaving, at increased expense. Dyed yarn was sometimes brought from the East, and on occasion West Indian cotton was even sent to the Levant to be dyed with *Turkey Red*. This dye was produced from the madder (*Rubia tinctorum*).

All western European nations had the same difficulties in producing a fast red dye of quality equal to that produced in the Orient. In 1747 works were started at Darnetal near Rouen with operatives transferred from Adrianople to carry out the method used in the Levant. The Dutch next tried to establish the method at Leyden, but results were unsatisfactory. In 1751 the English Minister at Copenhagen forwarded a petition to the Secretary of State from a Dane who claimed to understand the method and wished either a grant or to be sent to Manchester to demonstrate his process. Neither request seems to have been granted. In 1756 the Society of Arts took up the quest and made several awards, but 'as the competitors did not know the nature of the problem, success was not achieved'. Several people[3] described the process as operated in Turkey, but the price of the materials used was too great to allow of its adoption in this country.

A dyer from Glasgow was experimenting in Manchester in

[1] 5 Geo. III, c. 30: quoted by Wadsworth and Mann, *ibid.*, p. 159.

[2] R. Dossie, *Memoirs of Agriculture and other Oeconomical Arts*, 1768, Vol. 1, p. 180.

[3] *e.g.* Prof. Pallas in the *Philosophical Magazine*, 1798, *1*, 4.

1767, but with unknown results.[1] This was probably the dyer referred to by James Watt in a letter to Dr. Small, which incidentally contains a reference to the waterproofing of textiles many decades before the patent of Charles Macintosh, whose dyeing ventures are dealt with in the present chapter.

> I have just now got a curious book, being an account of all the machines, furnaces, methods of working, profits, etc., of the mines of the Upper Hartz. It is unluckily in German, which I understand little of, but am improving it by the help of a truly chemical Swiss dyer, who is come here to dye standing red on linen or cotton, in which he is successful. He is, according to the custom of philosophers, *ennuye* to a great degree, but seems to be more modest than is usual with them; and, what is still more unusual, is attached only to his dyeing, though he has a tolerable knowledge of the rest of chemistry. He promises to make me a coat that will not wet, though boiled in water. I enclose you some of Chaillet's red yarn, boiled in soap, and some as it is dyed.[2]

John Wilson of Ainsworth, near Manchester, to whom the velvet industry owed much for improved dyeing and finishing, has given an account of his efforts and failures, illustrating the lengths to which an enterprising manufacturer was willing to go to solve the problem.

> This valuable colour (Turkey Red) cost me several hundred pounds. In the year 1753, I sent a young man to Turkey on purpose to learn to dye it. He had lived with a Mr. Richard Dobs, a merchant in Smyrna, some time before he had learned the language of those Greeks who dye it; and by Mr. Dobs' interest, got admittance into their dyehouses, and was instructed; and on his return brought the true method, and with him many bales of the best Madder Root, pronounced by them 'Choke Biaugh'; and a letter from Mr. Dobs to assure me he would buy for me everything I might want on a very moderate commission.[3]

Unfortunately, the type of trade in which Wilson was engaged was not the most suitable for the application of Turkey Red, and he goes on:

> He executed the business I sent him about, and I rewarded him for his trouble; but when I got it, to my great disappointment, it would not suit my purpose, that is, for cotton velvets, nor any

[1] *Manchester Mercury*, 17th March, 1767.

[2] Glasgow, 28th May, 1769, quoted in J. P. Muirhead, *Origin and Progress of Mechanical Inventions of James Watt*, Vol. 1, p. 63.

[3] J. Wilson, *An Essay on Light and Colours, and what Colouring Matters are that dye Cotton and Linen*, 1786, quoted by Wadsworth and Mann, *op. cit.*, p. 180.

> other sort of piece work I then made. The tediousness of so many operations and the exactness required every time, rendered it of no more value to me than the Madder Red . . . which is so easily dyed, whereas the Turkey Red requires 12 or 13 operations in repeated steepings, dryings, washings and dyeings.

In 1761, the Society of Arts premium was awarded to Wilson. In 1779, the Government awarded £5,000 to Mr. Berkenhout for experiments with cochineal as a cotton colour,[1] and in 1786 awarded a further £2,500 to Louis Borelle and his brother who had come to Manchester and offered a process to the Manchester Committee of Trade.[2] Another Frenchman, P. J. Papillon from Rouen, under the name of Cigale, who had been a dyer in Turkey at one time, also made overtures in Manchester, but ultimately had to go to Glasgow to find backers.[3]

The backers found by Papillon in Glasgow in 1785 were George Macintosh, David Dale, founder of New Lanark Mills, *et al.* Macintosh met Papillon in London and suggested that he should come to Glasgow, where extensive Turkey Red works were being built at Dalmarnock on the Clyde. According to the *Statistical Account*, this was earlier than in any part of Great Britain. About the same time, Charles Taylor set up works in Manchester for Turkey Red dyeing, probably using Borelle's method. One of the Borelle brothers obtained a premium from Parliament for Turkey Red dyeing, but when specimens were presented to a Committee of the House of Commons, 'manufactured pulicates of a superior quality were shown by George Macintosh, and only cotton yarn by Borelles'.[4] By 1787, Papillon and Macintosh had parted company on account of the former's unruly temper. Having left Macintosh, Papillon went to manage another dye-house, and in 1790 received a premium from the Board of Trustees for communicating to Dr. Black his method of dyeing Turkey Red, on condition that the method was kept secret for a number of years. There was little need for such secrecy, as in December 1786 Thomas Henry had published full details in the Memoirs of the Manchester Literary and Philosophical Society.[5] The description given by Papillon to Dr. Black was said to be 'so incongruous as to lead any scientific reader to suppose either that M. Papillon wished wilfully to mislead, or that he possessed no chemical

[1] *Journal of the House of Commons*, Vol. 37, pp. 392, 422.
[2] *Ibid.*, Vol. 41, pp. 289, 467 and 882.
[3] *Beauties of Scotland*, Vol. 3, p. 289; Cleland, *Abridged Annals of Glasgow*, p. 346.
[4] *Statistical Account*, Vol. 12, p. 114.
[5] *Mem. Manch. Lit. and Phil. Soc.*, 1790, *3* 380.

science whatever'.[1] So it is doubtful whether Macintosh learned anything from him.

The Dalmarnock Turkey Red Works prospered, though even Macintosh's grandson admitted that their products did not equal those produced on the Continent. This was so, he said, since Adam Smith, McCulloch, and McGregor of the *Board of Trustees* induced them to use inferior alkali, in order to reduce prices.

> Since 1785 this manufacture, though still inferior to that of Continental Europe, has prodigiously increased in Britain; and the beautiful modification of the process in the department of bandanas, and in the various details in which it is combined with the process of calico-printing, afford interesting illustrations of the application of chemistry to the arts, and of the advance of madder red dyeing, since its origin in the East and its progress through Armenia and the Levant to Europe.[2]

Andrew Brown tells us that in 1796 more than five thousand looms were employed in the neighbourhood of Glasgow making 'policates' for the Turkey Red dyers. These pullicates became known throughout Europe as *Monteiths*, after Henry Monteith, who acquired the Dalmarnock Works in 1805.[3]

The process of Turkey Red, or Adrianople, dyeing, long known in this country as Dale's Red, after Macintosh's partner, was complicated. A madder-bearing plant (*Rubia peregrina*) grew wild in a few places in England, but for use in dyeing was imported from Provence, Smyrna, and Cyprus.[4] The imported madder was got from *Rubia tinctorum*. It had been cultivated in the Levant for centuries, and was introduced into France by Jean Altken in 1766, an event which is commemorated by his statue at Avignon. From the end of the eighteenth century, madder-growing was considered to be one of the principal sources of wealth in France. Seeds imported from Cyprus were planted in Alsace, and the industry received a great stimulus from the decree of Louis Philippe (1830–48) that the soldiers of the French army should wear red trousers dyed with madder.[5]

At the first meeting of the *Society for the Encouragement of Arts, Manufactures, and Commerce*, held in a Covent Garden coffee-house in 1754, it was decided to raise a fund to provide prizes for the discovery of cobalt in England, and for the growth of madder. In 1789 the *Bath Agricultural Society* attempted to cultivate it near

[1] *Glasgow Curios*, p. 73. [2] G. Macintosh, *op. cit.*, p. 121.
[3] A. Brown, *op. cit.*, Vol. 2, p. 255; Thos. Henry, *op. cit.*, p. 362.
[4] Thos. Henry, *Mems. Manchester Lit. and Phil. Soc.*, 1780, *3*, 386.
[5] *Ciba Review*, 1937, *1*, 65.

Bristol, hoping that it would eventually be produced in large quantities and sold to English dyers cheaper than imported roots.[1]

For best results, roots were pounded and inferior parts removed. The *crop madders* thus produced gave brighter colours than *madder roots* in their native state. About 1813, mills were erected in England to give the roots this treatment.

The preparation of the fabric was more than usually complicated. The process consisted of repeated boiling and steeping in mixtures of mineral alkali, oil, and animal excrement.[2] According to Bancroft and Ure, Papillon's method consisted of the following operations:

(1) Cleansing of cloth by ashes and lime.
(2) Grey steep with barilla, sheep's dung, oil of vitriol, olive oil, etc. This steep was stated to contain three ingredients not previously used, but the use of oil of vitriol seems to indicate a want of chemical knowledge, as, by neutralizing the soda, the oil of vitriol was rendered useless.
(3) White steep, similar to (2) but without dung.
(4) Gall steep.
(5) and (6) Alum steeps with alum and barilla.
(7) Dyeing steep with madder and ox-blood.
(8) Fixing steep with equal parts grey and white steep.
(9) Brightening steep by boiling with soap and strong barilla.

According to M. Chaptal, cotton does not dye a permanent red with madder unless sufficiently impregnated with oil.[3] Boiling with soap was claimed to brighten the colour. In calico-printing madder reds were usually duller because they could not be boiled to remove yellow and brown tints.[4] Henry's method was similar to that of M. Chaptal, but involved fifteen separate operations. It is little wonder that Dr. W. Withering wrote to James Watt, Jr.:

> I shall some time or other hope to hear from you a rationale of the art of dyeing; a subject which I never could at all understand, the multifarious ingredients in every formula baffling every attempt I have made to develop the subject.[5]

The red dye produced by these processes proved satisfactory, and was so fast that it could be woven with brown cotton or linen yarn and subsequently bleached, the Turkey Red threads actually being improved in the process.[6]

[1] For some interesting details concerning madder, *see* J. Bechmann, *History of Inventions and Discoveries*, Vol. 3, p. 271.
[2] Thos. Henry, *op. cit.*, p. 362. [3] *Philosophical Magazine*, 1798, *1*, 274.
[4] A. Ure, *op. cit.*, Article, Madder; Ed. Bancroft, *op. cit.*, Vol. 2, p. 249.
[5] W. Withering to James Watt, Jr., 8th March, 1790, B.R.L.
[6] *Statistical Account*, Vol. 12, p. 115.

The extraordinary series of operations which constituted the Turkey Red process did not go unexamined by the chemists of the time. John Thomson of Glasgow tried to develop a theory of the process in a paper which was published in the *Annals of Philosophy*.[1] In it he examined the successive operations, and in collaboration with an eminent calico manufacturer concluded that the only essential mordants were the oil and alumina (aluminium hydroxide, $Al(OH)_3$), and that bright and fast reds could be obtained without employing dung, galls, blood, etc.

> To this day the precise nature of the Turkey Red pigment has not been established beyond doubt; from the results of recent research it may, however, be described as a lake for the formation of which aluminium and calcium must be present. The process of Turkey Red dyeing is made up of a number of distinct stages, the first three of which—oiling, sumaching, and mordanting—serve to prepare the fibre; the next stage is the actual dyeing, which in turn is followed by the clearing process. Whereas only an alum mordant was used in ordinary madder dyeing, the Turkey Red process is thus complicated by a number of additional treatments.[2]

The most characteristic change of textile production in the transition from eotechnic to palaeotechnic economy is the substitution of power, generated centrally, by the improved steam-engine, for the power of human muscles, and the consequent aggregation of spindle and loom in mill or factory. The centralized generation of power made possible the adoption of steam in other directions. Samuel Parkes remarks 'that steam is now employed in a vast variety of ways in the different manufactories of this kingdom'. Among the first to make use of a central supply of steam were the dyers. By this means Messrs. Gott and Company of Leeds boiled thirty large coppers from one boiler, doing away with much cumbersome individual heating of dye-baths.[3]

Benjamin Gott (1762–1840) was the first dyer to attempt scientific control of his industrial processes. Born in 1762, he built the first big woollen mill in England and experimented in new methods of manufacture and in finishing goods for the market. Utilizing steam for heating, instead of individual fires, he carried out a long and expensive series of experiments in the dye-house. Dyeing processes were, as has just been described, entirely empirical and frequently defective. Gott spent many thousands of pounds in research from which he evolved 'the most advantageous

[1] *Annals of Philosophy* 1824, *8*, 463. [2] *Ciba Review*, 1941, *3*, 1409.
[3] S. Parkes, *Chemical Catechism*, p. 88, *n*.

scheme of dyeing by steam instead of water'. W. B. Crump tells us in *The Leeds Woollen Industry*:

> Of Gott's many experiments his most famous were in the use of steam in his dyehouse and drying-sheds. The details are not known, but he seems to have heated the dye-liquids by injecting steam into them instead of heating them over a fire. Whatever his improved technique was, he spent thousands of pounds developing it, and won fame among his contemporaries for his persistence, resourcefulness, and willingness to spend money on the experiments. In all his work he was frequently appealing to Southern, of the Soho Foundry, for advice, suggestions, and criticisms of his own idea. Southern's replies are copied in the Foundry Letter Books, and they positively glow with friendly personal feeling at times. One letter ran over twelve sides of paper, full of detailed suggestions and criticism. The steam expert did more than supply steam engines.[1]

He sought physical aids to control his processes. Thermometers were installed to keep the temperature of the dye-baths level. By these means there emerged much improved products, especially those dyed blue, scarlet and black.[2] Important documentary material concerning Gott's activities has survived and is in the Library of the University of Leeds.

> One is a substantial foolscap book, two inches thick, that was presented by Mr. John Gott in 1928. It is a pattern book containing samples of dyed cloth and yarn with corresponding recipes or working instructions, and is clearly the work of Benjamin Gott's third son, for he has inscribed in it his name and the date—'Wm. Gott, Augt. 1815'. He was then eighteen years of age, and the book reveals him qualifying himself in the dye-house, no doubt under the instruction of the head dyer, and keeping a record of his work.[3]

Here, for example, is one of the recipes dated and titled

> Aug. 1815. Pelisse No. 79111. 10½ yds. Crimson.
>
> Purify the water with 1 pint of nitrate of tin, ½ dish of bran, then boil and put in ½ lb. of Alum, 1½ lb. of Tartar, 1 pint of Spirit, 8 oz. of Cochineal, put in the cloth and boil ¾ hour, wind up, beck.[4]
>
> Clean out the vessel, put in fresh water, boil 3 lb. Cudbear, after the water has boiled a few minutes turn off the steam and put in 1 gallon of wash,[5] let the cloth remain in ½ hour without steam, wind up, beck.[6]

[1] W. B. Crump, *The Leeds Woollen Industry, 1780–1820*, p. 188.
[2] H. Heaton, *Econ. Historical Review*, 1931, *3*, 45.
[3] W. B. Crump, *op. cit.*, p. 168. [4] Wash. [5] Stale urine.
[6] W. B. Crump, *op. cit.*, p. 308.

Another of the documents is the Bean Ing Mill Note-book, a record of mill practice, and in fact a veritable 'woollen manufacturer's *vade-mecum*'. There is no name on it, but Alexander Yewdall, of the Textile Department, Leeds University, is of the opinion that it was begun by one of Gott's sons, most likely the eldest, John. From the record contained in it we get 'not only a vivid picture of the Mill and its working, but also a collection of technical and economic data relating to the woollen industry of the period under review, that is undoubtedly unique'. It shows too the emphasis placed on science in the training of young Gott, for the note-book contains a copy of an experiment on the properties of steam taken from William Henry's *Elements of Experimental Chemistry*, with additional exercises devised by Joshua Dixon, under whose care John Gott was placed. Here in detail is the record of an experiment made by Joshua Dixon and William Pritchard in the Dye-House, Park Mill, on the 9th September, 1800:

> At 7 o'clock the Steam Gauge standing at 4 inches, I opened the cocks of the several dying vessels as under; their Contents were previously taken as accurately as possible, as well as the steam, and a mark was made on the side of the vessel at the surface of the water, that the increase of condensation might be known. The steam cocks were opened one after another, so that as soon as the vessel arrived at the temperature 180° or thereabout, the cock was partly shut and another opened. In this manner I proceeded till the whole arrived at the temperature below, which occupied exactly six hours. By this means the whole number of vessels were nearly boiling together, at the conclusion of the Experiment. The boiler was filled with water the night before so high that it wanted no fresh supply till 10 o'clock, so that as will appear from the quantity of water evaporated, it must have contained nearly 750 gallons (wine measure) more than when it is at its usual level. At 10 o'clock the boiler was supplied with 236 gals. at 98°, at ¼ past 11 with 266 gals, and at ¼ past 12 with 240 gals. at the same temperature which was found sufficient to furnish steam till the conclusion of the Experiment, when the water in the boiler was at the level as at the expiration of the first 3 hours.
>
> From the above it appears that the quantity of water reduced to the state of Steam from the boiler was nearly 250 gals. per hour. 3 Cwt. of coals raised the water from about 100° to the pressure first mentioned viz. 2 lb. per sq. inch, and the quantity consumed during the experiment was 1 Ton ½ Cwt. or nearly 3½ Cwt. per hour.[1]

Gott's interests did not stop at the application of chemistry to his processes. On 6th April, 1795, he wrote James Lawson of

[1] W. B. Crump, *op. cit.*, p. 305.

Soho, thanking him for sending 'Fourcroy's expln of ye new Nomenclr—it is one of ye clearest books I have read'. This was *The Philosophy of Chemistry, or Fundamental Truths of Modern Chemical Science, arranged in a new order*, by Antoine François Fourcroy (1755–1809). The importance of chemistry to dyers was early appreciated in France. The Gobelins studios had a chemist in charge of the dye department even in the eighteenth century, and in 1824 Eugène Chevreul (1786–1889), whose researches in connexion with soap are discussed in Chapter V, was appointed head of the dyeing department. In this field Chevreul made an equally important contribution in his invention of the chromatic circle and the law of colour contrasts. No similar development took place in Great Britain.[1]

None of the changes introduced by such men as Gott can be regarded as establishing the palaeotechnic phase in the dye industry. This came as late as 1856, when William Perkin made the first dyestuff from a coal-tar derivative.

[1] *Ciba Review*, 1938, *1*, 158.

Chapter XI

CALICO-PRINTING

You can go to Scotland for mere information if that appeals to you.

GIDEON OUSELEY in 'Tumbling in the Hay'.

WHILE separately mechanical and chemical invention brought about changes in textile production, together they formed the basis of advances in textile printing. '*Taste*, *chemistry*, and *mechanics*,' it was said, 'were the three legs of calico-printing'.

The first textile-printing factory in England was established in 1690 by Rene Grillet, a Frenchman, who took out a patent for painting and printing calicos, and started work at Richmond. A second factory followed at Bromley Hall, Essex,[1] but not till 1729 did such work begin in Scotland. In June of that year it was announced that linens were stamped and printed in all colours at the Gorgie Bleachfield, on the Water of Leith, near Edinburgh.[2] According to A. McLean, by 1738 calico-printing was being carried on in the neighbourhood of Glasgow as well, but he does not specify by whom.[3] Certainly one of the earliest works in Scotland was at Pollockshaws, but it is usually reckoned to have been founded a little later, *c.* 1742. The printfield was started by Archibald Ingram and some West Indian merchants. According to Andrew Brown, they started under every possible disadvantage. First they had to buy raw cotton which they gave out to be spun. They received it back as yarn for the wefts. For warps they bought linen yarn. The master-printer who was in charge of the printing was not well qualified for his post. His skill was 'annually acquired by stealth from the working printers in London, where the manager was supposed to resort in the winter, and return to work in the spring, as full of information in his art as a London dancing-master from Paris, with half-a-dozen new lessons at his heels'. Despite this inauspicious beginning, all difficulties were eventually surmounted.[4]

[1] P. J. Thomas, *The Introduction of Calico-Printing in England*, English Historical Review, 1924, *39*, 208.

[2] G. Chalmers, *Caledonia*, Vol. 2, p. 743, *n.*; A. Brown, *History of Glasgow*, Vol. 2, p. 212.

[3] A. McLean, *Industries of Glasgow*, p. 141.

[4] A. Brown, *op. cit.*, Vol. 2, p. 212.

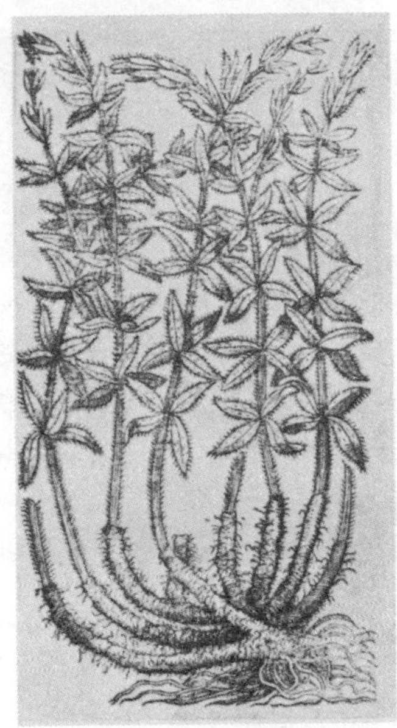

43. The madder plant (*Rubia tinctorum*), from whose roots the dye for the celebrated Turkey Red was extracted. *Left: from* John Gerarde's *Herball* (1636); *right: from* Revillius' *Historia generalis plantarum* (1587).

44. Indigo vats in the Blue Dyehouse of the printworks at Messrs. Lees, Manchester, *c.* 1840. The calico was stretched on frames and immersed for seven and a half minutes. It was then removed so as to oxidize the reduced indigo and the process repeated in increasingly strong vats, several of which can be seen in the figure.

Picture Post Library

45. Calico-printing by hand, showing men engaged in blocking, and, on the left, women grounders finishing off the printed textile.

Encyclopédie Méthodique, 1782–1832

46. Textile-printing: a move towards the factory system.

Calico-printing was next taken up by the Stirlings of Cordale Printfield and Dalquhurn Dye Works. William Stirling formed a co-partnery and erected works on the Kelvin at Dalsholm, a village in New Kilpatrick parish, Dumbartonshire, where successful printing of handkerchiefs was engaged in. In 1764 William Stirling and Company were advertising in the *Glasgow Journal* that they would print linen and cotton cloth for their customers, who could select their own patterns from pattern-books kept by the firm's agents in Edinburgh, Greenock, Ayr, Paisley, etc. They had extended their activities to cloth for garments and furniture. Since the price of labour was not favourable for expansion at Dalsholm, where the firm started about 1750, they removed in 1770 to Cordale to the vicinity of an uncle's bleachfield at Dalquhurn, on the banks of the Leven in Renton parish. On further expansion in 1791 they acquired this bleachfield as well. There they employed about six hundred people.[1] It was in the same year that calico-printing was introduced into Lancashire. It is attributed to Messrs. Clatron, of Bamber Bridge, near Preston. The first Sir Robert Peel, father of the Member of Parliament, developed the industry there.[2]

According to the *General View of the Agriculture of Renfrew*, textile-printing was undertaken at Corsemill, or Levern printfield, in the parish of Paisley. The greater part of the Levern valley was inhabited by a manufacturing population, with centres at the villages of Neilston, Barrhead, and Hurlet.[3] In 1773 a similar work was begun at Fereneze, on the borders of Abbey and Neilston parishes, Renfrew. Some time afterwards (1779), works of the same kind were established at Thornliebank, a manufacturing village in Eastwood parish, Renfrew. They belonged to Messrs. Crum, who engaged in cotton-spinning and weaving, calico-printing and bleaching. About 1810 a small establishment was started on the Lochar Water, in the parish of Kilbarchan, where a linen factory had been established in 1739. The works at Kilbarchan, Fereneze and Corsemill employed 230 persons, paid £7,500 in wages, and contributed £11,000 to the excise in the Paisley district. The goods printed at all these works were shawls and handkerchiefs: at Kilbarchan and Fereneze for the home and foreign markets, at Corsemill only for the former.[4]

By 1796 there were thirty large-scale printfields in the neigh-

[1] G. Chalmers, *op. cit.*, Vol. 3, p. 897, *n.*

[2] E. Baines, *History of the Cotton Manufacture in Great Britain*, p. 262.

[3] Hurlet was the scene of operation of the Hurlet and Campsie Alum Company, which supplied printers' ancillaries. See Ch. XII.

[4] J. Wilson, *General View, Renfrew*, p. 273.

bourhood of Glasgow. The main centres of the industry in Scotland were Dumbarton, Lanark, Perth, and Aberdeen.[1]

There is an MS. account of the *Chemistry of Calico-Printing from 1790–1835* in the Manchester Reference Library. Its author is John Graham, a brother of Thomas Graham (1805–69), one of the founders of colloid chemistry, who was at one time on the staff of the Andersonian Institution in Glasgow.

Methods of printing have been described by Parkes.[2] The cloth was first dressed to remove the dry nap (fluff) by passing it quickly over a red-hot roller. It was then steeped for twenty-four hours, and boiled with weak alkali to remove the grease which came from the reeds used in weaving. This process was called *ashing*. The cloth was next well washed and bleached in the usual manner. This finished the preparation, and the goods were then *calendered* (passed between hot rollers) to glaze the surface. The material to be printed was placed on stout tables covered with a woollen cloth. Mordants, made up with gum arabic or gum senegal, and flour paste, were applied by contact with engraved wooden blocks, originally made of holly or sycamore, five to ten inches square. Brass or copper faces were subsequently added, because the life of a wooden face was too short, twenty-eight yards of cotton taking about 500 applications of the block.[3] Printing from engraved copper plates was started at Pollockshaws about 1769, and at Carmile about 1771.[4] When the price of gum senegal rose from £150 to £400 per ton during the Napoleonic Wars, Lord Dundonald convened Scottish calico-printers and communicated to them his ideas for making gums suitable for calico-printing from lichens or tree-mosses, to get over the difficulties of having to rely on imported products.[5]

The mordant, with whatever paste they used, was spread on the surface of a sieve covered with a fine woollen mat. The pattern block was then applied to pick up paste, which was transferred to the cloth by gentle pressure. When the mordant was colourless, a little fugitive dye was added to it, e.g. Brazil wood, to make the pattern visible, an operation called *sight-seeing*. The mordanted cloth was dried in a warm room for twenty-four hours to fix the mordant.

The next process was *dunging*, that is, passing the cloth through mixtures of cow-dung and water by means of winches to remove

[1] J. Sinclair, *General Report*, Appendix 2, p. 280.
[2] S. Parkes, *Chemical Essays*, Vol. 4, p. 109 *et seq.*
[3] *New Statistical Account*, Vol. 6, p. 150.
[4] A. Brown, *op. cit.*, Vol. 2, p. 216.
[5] *Philosophical Journal*, 1st series, 1802, 5, 228 *n.*

the excess mordant and effect a subsequent brightening of the colours. Washing followed. Much of the dyeing was done with madder. Almost any colour could be obtained with this dye-stuff by the use of different mordants. The cloth, treated as described above, was placed in a copper of cold water along with the ground-up madder and brought to the boil, with constant stirring. After removal from the dye-bath, any madder which had inadvertently fixed to the white ground was removed by treating with a mash of wheat bran and water, and if this was insufficient to complete the clearing, ordinary bleaching was a last resort, i.e. exposure to the sun. Later, chlorine was used. S. Parkes tell us that about 1815 'a Scotch house of great consequence' had introduced an innovation by soaking their pieces in a weak solution of bleach. This effected in a few minutes what had formerly taken days with the result that 'it promises very soon to supersede crofting entirely'.

Such was the process of calico-printing during most of the eighteenth century. A revolutionary change, printing by means of cylinders instead of blocks, was introduced in 1785 by a Scot, Thomas Bell, *the elder* (copper plate printer), of Walton-in-the-Dale, near Preston, in the County Palatine of Lancaster.[1] This method was first successfully applied by Livesay, Hargreaves, Hall and Company at Mosney, near Preston. Patterns were engraved on copper rollers 18 to 42 inches long, 3½ to 5 inches in diameter, and to prevent their having a join in them they were bored from a solid casting like cannon. More than one colour could be printed at one time by the use of several rollers. From the *General View of the Agriculture of Renfrew* we learn that 'both copperplate and block-printing are carried on at all those (Renfrew) Works, and ably conducted. At Thornliebank, not only are great varieties of printed goods of those descriptions produced, but also fine chintz pieces, for gowns and furniture, chiefly for exportation'.[2] Although cylinder-printing enabled large quantities of cloth to be treated at small cost, often as little as a penny per yard, including dung, colour, paste, and printing, poor printing remained common, colours fugitive, mordants were often omitted, and a shower of rain was enough to remove the design.[3]

Another method of printing for small goods, such as handkerchiefs, was *resist-work*. Baines tells us that it was the invention of a person of the name of Grouse, traveller for a London house, 'possessing little practical and less scientific knowledge, fond of experiments and dabbling by the fire-side in the processes of

[1] *New Statistical Account*, Vol. 5, p. 150; E. Baines, *op. cit.*, p. 265.
[2] J. Wilson, *op. cit.*, p. 273. [3] S. Parkes, *op. cit.*, Vol. 4, p. 179.

printing. His process for resist work he sold for *five pounds*'.[1] To a mixture of copper sulphate, nitrate, chloride or acetate was added flour paste or pipe-clay and gum, and this was painted on to the cotton. When dry the cloth was immersed in a blue-vat. The result was a white pattern, often spots, on a blue background. By subsequent dyeing in madder, weld, or bark, patterns in red or yellow on a blue background were obtained. Silks were patterned in a similar manner. Heated tallow and rosin was sometimes used in place of paste to resist the bath. Such silk handkerchiefs were at one time very popular, and much wax was used in their manufacture. When only a single unit of a design was to be printed, it was put in by hand in indigo by the process known as *pencil-blue*.

From the end of the eighteenth century, patterns were obtained by *chemical-discharge*, that is, the cloth was dyed a uniform colour and a design bleached out by forcing bleaching liquor through a metal stencil. This method of printing Bandana handkerchiefs, a Scottish attempt to imitate by another method an Indian process, was introduced by Henry Monteith and Company in 1802. The handkerchiefs were dyed with Turkey Red, clamped between lead plates with holes in them, and treated with bleaching liquor. In 1813 a patent was obtained by an English calico-printer for imitating this design in another way by using the old method of printing from a block. He was James Thomson, a Fellow of the Royal Society, of Primrose Hill near Clitheroe, Lancashire. The details of his patent, and a second one which he obtained in 1815, are given in C. L. Berthollet's *Elements of the Art of Dyeing*.[2] According to Baines, Thomson combined in an eminent degree, scientific, with practical, knowledge.[3] He stamped his cloth with a vegetable acid and passed it through a solution of chloride of lime. Chlorine was liberated when the cloth was printed with acid, and so the pattern was bleached out.

A further modification of this process was devised in 1820. An ore of chromium was then found in abundance in North America, and 'a beautiful salt of it' made by Messrs. Turnbull and Ramsay. A lead salt and a vegetable acid were applied to the Turkey red dyed cloth, which was passed, first through a chloride of lime solution to bleach it, and then through a solution of potassium chromate which gave a yellow instead of a white pattern. Another modification gave green patterns.[4] Andrew Ure in his notes to Berthollet's *Elements of the Art of Dyeing* tells us that these modifications resulted in a brilliant style of calico-printing which was

[1] E. Baines, *op. cit.*, p. 276.
[2] C. L. Berthollet, *op. cit.*, Vol. 2, p. 332.
[3] E. Baines, *op. cit.*, p. 279.
[4] A. McUre, *Glasghu Facies*, p. 1232.

carried to high perfection at the establishment of Messrs. Monteith at Glasgow, but, judged by modern standards, it is probable that these modifications were relatively unsatisfactory.

The manufacture of chromium compounds used in these processes has been associated with Glasgow since early in the nineteenth century. The raw material for practically all the compounds is chrome iron ore, which is converted into soluble chromium compounds by calcination in a reverberatory furnace with lime and potash. For a long time the only manufacturers were J. and J. White of Shawfield, Rutherglen, who started in 1808, but they were joined later by Stevenson, Carlile and Company, Limited, at the Millburn Chemical Works, Glasgow, and by the Eglinton Chemical Company, Irvine.[1]

By the beginning of the nineteenth century, calico-printing had become an important industry. Although there was no essential change in the method of printing prior to Bell's substitution of cylinders for blocks, the organization of the calico-printing industry had early begun to answer to the characteristics of a factory system. There had been established a number of large printing enterprises which employed men and women not in their homes but in central plants. It was, by necessity, organized on a capitalist basis. It had no guild traditions, and it could not be carried out as a domestic industry, because the work necessitated a considerable degree of specialization. From the start it was, therefore, forced to become a factory industry. Originally all calico-printers were bleachers, but near London the trades started to separate. In North England and Scotland, however, most printers were bleachers as well, and in Ireland, especially, calico-printers were known for the whiteness of their goods.[2]

Internal specialization soon took place. Patterns were designed, and *cutters* reproduced them on blocks or cylinders; colours were applied and *grounders*, mainly women, put in the finishing touches; *tearers*, usually children, and *fieldmen*, bleached the cloth after it was printed. There were also job-printers who renewed and reprinted old calicos and linens. Their work was said to give 'great encouragement to servants to rob their masters and mistresses, for, by getting it (i.e. stolen cloth) printed over alters it so much as cannot be known'.[3]

On the artistic side, improvement was effected by the Society for the Encouragement of Arts, Manufactures, and Commerce, founded in 1754 by William Shipley, a drawing-master. In the

[1] A. McLean, *op. cit.*, p. 167.
[2] S. Parkes, *op. cit.*, Vol. 4, p. 106.
[3] P. J. Thomas, *op. cit.*, p. 211.

first list of premiums drawn up by the 'Committee of the Polite Arts', prizes were offered for designs for printed calico. Apparently, by 1778, the Society were pleased with the results of this activity, for it was claimed that:

> . . . the elegance of pattern adopted by them (weavers and calico-printers) may with justice be attributed in a very great degree to the rewards and attention bestowed upon them by the Society.

A number of Scottish print-works employed about two hundred people—men, women and children.[1] They were usually well-equipped, and located within easy reach of an abundant water supply.[2] At the Kincaid printfield at Campsie, Stirlingshire, Messrs. Henderson, Semple and Company dug a reservoir 120 by 70 yards to ensure a good water supply. In 1772 William Gillespie established a printfield at Anderston, near Glasgow, and employed three to four hundred workers. As typical of a printfield may be given the numbers employed and the wages paid at the Kincaid field, which was established in 1785.[3]

Block-printers	36	18–21*s.* per week
Copperplate printers	22	17–21*s.* ,, ,,
Pencillers	160	4–6*s.* ,, ,,
Tearing boys	34	2*s.* ,, ,,
Bleachers	26	8*s.* ,, ,,
Engravers	16	18–22*s.* ,, ,,
Miln Wrights	2	12*s.* ,, ,,
Labourers	6	7*s.* ,, ,,
Furnace men	8	7*s.* ,, ,,
Excise Officers	2	£50 ,, annum
Revenue to Government	–	£8,000 ,, ,,

Apprenticeships were for seven years, during which boys received 3*s.* a week for the first three years, and 4*s.* a week for the last four years.[4] They were later allowed to earn what they could at piece-work, being paid half a journeyman's wages.[5] Piece-work proved more satisfactory, at least to the masters. Many disputes arose through the masters trying to reduce wages and the operatives trying to keep them up.

> Among other manoeuvers, they appointed a committee of their number from the different printfields in the West of Scotland, to meet and to regulate the prices, when they were to oblige their

[1] *Statistical Account*, Vol. 9, p. 7; Vol. 14, p. 555.
[2] *Ibid.*, Vol. 15, p. 354. [3] *Ibid.*, Vol. 15, p. 356.
[4] *Ibid.*, Vol. 3, p. 445; Vol. 4, p. 356. [5] *Ibid.*, Vol. 15, p. 356.

masters to give for their different pieces and work. They were to allow no persons to be employed but such as came under certain regulations which they had framed . . . These measures obliged the masters to commence prosecutions, and to imprison some other hands last summer, and a kind of compromise has been made between the masters and servants for a time; but it will be easily foreseen, that one of the parties must be in complete subjection to the other, before the trade can be upon a proper and sure footing.[1]

According to the master, wages were high, but:

. . . it may be proper to notice, that, although the wages of calico-printers seem to be the highest of any in the country; no doubt, when the long apprenticeship is considered, along with the unwholesome nature of the work, the wages perhaps should be greater than of most other operative people; at the same time when it is considered that the highest wages do not always make the wealthiest tradesmen; perhaps, if some method could be fallen upon to reduce the prices, both the tradesmen and the public would be gainers.[2]

Or this plaint from Young, Ross, Richardson and Company, of Ruthven, who employed some six hundred persons:

For several years past, the petitioners, and others engaged in the printing trade, have experienced almost insuperable difficulties in the prosecution of their business, in consequence of a general combination which now exists among the operative calico-printers in all three kingdoms, the objects of which are, to raise the prices of work, and for that purpose, to limit the number of apprentices which the masters may employ; and in a word, by every means in their power, to render the masters in a great degree dependent upon the workmen in the prosecution of their business.

With this view, a well regulated and uniform system is organised, the country is divided into districts; secret meetings are held, at which appointed officers preside; regular correspondences are kept up, and a general fund is provided for prosecution the purposes of the combination, one of which purposes is, that when a journeyman, or number of journeymen, strike work, or are turned off by their employers in consequence of their following out the objects of the combination; they are entitled to, and do draw from this fund, a sum corresponding to their situation and circumstances, so long as they continued unemployed. Every journeyman and apprentice, too, must become a member of this combination, and pay certain dues of admission, and unless so entered, he is not able to procure work, in consequence of the

[1] *Statistical Account*, Vol. 3, p. 458. [2] *Ibid.*, Vol. 15, p. 354.

> determined opposition and persecution which he everywhere meets with from the combination.[1]

The evolution of calico-printing, like that of its counterpart, dyeing, was influenced by opposition from vested interests in the textile trade, and by the fiscal policy of the Government. When the woollen and silk manufacturers became alarmed at the popularity of cotton, they succeeded in getting Parliament to prohibit the use of printed cottons in 1720. This applied to the use of cotton both for clothing and furnishings. As an alternative, fustian printing was developed, and also the printing of linen. This, of course, was just as disadvantageous to the woollen trade as was the printing of cotton. A partial removal took place about 1736, but the prohibitions against pure cotton remained operative till 1774.

Bleachers, printers, and dyers had to pay for the American War. In 1784, Pitt imposed a tax on both printed and bleached goods and bleachers, printers and dyers were compelled to take out a licence costing £2 per annum. The same Act[2] laid a duty on all cottons if bleached or printed of 1*d.* or 2*d.* per yard, depending on value. There was also an additional 15 per cent. on the new duties. These levies created great dissatisfaction, and in the following year they were withdrawn.[3] After some further changes the duties stood, on cloth valued at between 1*s.* 8*d.* and 3*s.* per yard, at 5¾*d.*, and, on cloth over 3*s.*, at 1*s.* 5*d.*

According to Baines, these taxes which were in continuous operation for over a century would have killed the calico trade had it not been that modern science contributed a 'series of wonderful inventions and discoveries' which resulted in such an increase in the trade that no less than 28,621,797 yards were printed in 1796, and 128,340,004 yards in 1829. The repeal of the duties two years later was a renewed stimulus.

> To the consumer it is a great relief, especially to the poor, as a woman can now buy a useful and respectable printed dress for *half-a-crown*, which, before the repeal of the duty, would have cost nearly four shillings. Indeed a printed dress of good materials and a neat pattern, with fast colour, may be bought for two shillings.[4]

We may sum up the mechanico-chemical nexus as applied to the art of calico-printing in the words of E. Baines: 'Chemical

[1] *Session Papers*, Signet Library, 187: 22, Dec. 1808.
[2] 24 George III, c. 4.
[3] 24 George III, c. 24.
[4] Baines, *op. cit.*, p. 282.

science has done at least as much to facilitate and perfect these processes, as mechanical science to facilitate and perfect the operations of manufacturing.'[1] The managers or proprietors were scientific men who utilized their capital to obtain the best machinery and well-furnished laboratories.

> A printing establishment, like a cotton mill, is a wonderful triumph of modern science; and when the mechanical and chemical improvements of both are viewed together, they form a splendid and matchless exhibition of science applied to the arts, and easily account for the rapidity of growth and a vastness of extension in the manufacture, which has no parallel in the records of industry.[2]

[1] Baines, *op. cit.*, p. 246. [2] *Ibid.*, p. 285.

Chapter XII

MORDANTS AND THE MACINTOSHES

> I know nothing that can best be taught by lectures, except where experiments are shown. You can teach chymistry by lectures—you might teach making of shoes by lectures.
>
> BOSWELL.

JUST as sulphuric acid, in the latter half of the eighteenth century, became an indispensable adjunct to one branch of the finishing trades, so also the supplying of other ancillaries, in particular, *mordants*, was one important function of early chemical industry. According to C. L. Berthollet:

> Mordants merit the greatest attention. It is by them chiefly that we diversify the colours, give them their brilliance, fix them on stuffs, and render them more durable.[1]

The group of salts known as *alums* constitute the commonest class of mordant for vegetable dyes. The permanent dye in this case is an insoluble complex salt of the metal with the organic molecule of the dyestuff. The metal is frequently, but not necessarily, aluminium, derived from aluminous minerals. What are called *alums* are examples of the class of isomorphous salts of general formula $M_2SO_4.R_2(SO_4)_3.24H_2O$. R is usually aluminium (Al), but may be chromium (Cr), or iron (Fe). If the alum were made from soaper's waste, or potash, M would be sodium (Na) or potassium (K). On the other hand, if stale urine were included in the preparation, M represents the ammonium (NH_4) radicle. Hence Dan. Colwall's description, 'alum is made of a stone digged out of a mine, of a seaweed, and urine'.[2] The so-called ammonium alum was in common use in the period under discussion.

Alum was an important eotechnic mordant;[3] so also was copperas (iron or ferrous sulphate, $FeSO_4.7H_2O$), also referred to as green *vitriol*. With the emergence of palaeotechnics others were introduced. These, together with alum and copperas, will be discussed in the following pages.

Since mediaeval times alum has been one of the products handled by Mediterranean traders.

[1] C. L. Berthollet, *Elements of the Art of Dyeing*, p. 65.

[2] *Phil. Trans.*, 1678, *12*, 1052.

[3] Since this was written Dr. C. Singer has published an elaborate history of alum under the title '*The First Chemical Industry*'.

> The Italian merchants disposed of the alum which they brought from the Orient at the fairs of Champagne. The dyers and tanners of Flanders and other continental centres of industry frequented the fairs to buy the necessary stocks of this indispensable product. When the Italians did not bring their goods over the Alpine passes by means of pack-horses, they were shipped to Marseilles and reloaded for land transport. As early as 1229, the customs regulations of Marseilles distinguished between the following kinds of alum: 'cequerin' (sugar alum), 'de Castilla' (from Castile), 'blanc' (from Egypt), and 'Dalap' (from Aleppo); on other occasions the inferior variety known as 'de bolcan' (from the Liparian island of Volcano) is mentioned.[1]

According to J. U. Nef, Bristol dyers used 'Alym de Wyght' as early as 1364, but our first record of the starting of alum works in England comes during the reign of Queen Elizabeth.[2] The making of alum was then a papal monopoly, and Sir Thomas Chaloner brought workmen from the Pope's alum-works to Guisborough, near Middlesbrough. This he did, it is said, 'notwithstanding the bulls and anathemas which His Holiness used in abundance against him'. The manufactory was at a later date deemed *Mine Royal* by King James, and leased to Sir Paul Pindar at an annual rent of between £12,500 and £15,000, but even at this rent the mines proved a profitable speculation. Sir Paul soon employed 800 workmen, and sold his alum at £26 a ton.[3]

By 1678 a number of other alum-works had been formed. The writer of the *General View of the Agriculture of Renfrew* recounts the details of the spread of this industry in Europe:

> From Professor Beckmann's *History of Inventions and Discoveries*[4] it appears that, previous to the middle of the fifteenth century, Europe was furnished with alum from Turkey, and in particular from Constantinople and Smyrna. The manufacture, according to that author was introduced into the Pope's dominion in 1458, where it is still continued at Tolfa, near Civita Vecchia. It was next established in Spain, at works near Carthagena; in Germany, about the middle of the sixteenth century; at Andrarum in Sweden, in 1630; and in England, in the reign of Queen Elizabeth. For the introduction of this manufacture, England is indebted to Sir Thomas Chaloner of Gisborough in Yorkshire, who, observing 'the trees tinged with an unusual colour', naturally suspected that this was owing to some mineral in the neighbourhood, which,

[1] *Ciba Review*, 1941, *3*, 1430.

[2] J. U. Nef, *Rise of the British Coal Industry*, Vol. 1, p. 184.

[3] T. Pennant, *A Tour in Scotland*, Vol. 1, p. 21; *British Association Report*, 1863, p. 709.

[4] J. Beckmann, *op. cit.*, Vol. 1, p. 288.

being searched for and discovered, proved to be aluminous. It was near the end of the sixteenth century that he established his works, about twenty miles from Whitby, for the manufacture of alum. These were succeeded by others in 1615, near that town, and in a short period, there were in that district, no fewer than sixteen works for the manufacture of that salt.[1]

According to Thomas Thomson, writing in the Supplement to the IV–VI Editions of the *Encyclopaedia Britannica*:

> England possessed no alum works till the reign of Charles I. Thomas Chaloner, Esq., son of Dr. Chaloner, who had been tutor to Charles, while hunting on a common in Yorkshire, took notice of the soil and herbage, and tasted the water. He found them similar to what he had seen in Germany, where alum works were established. In consequence of this, he got a patent from Charles for an alum works. This manufactory was worth two thousand a year, or perhaps more. But some of the Courtiers thinking this was too much for him, prevailed the King, notwithstanding the patent, to grant a moiety of it to another person. This was the reason why Mr. Chaloner was such a partizan of the Parliament, and such an enemy of the King, that, at the end of the civil war, he was one of those who sat in judgment upon his Majesty and condemned him.[2]

The methods used by English alum-makers were described by Dan. Colwall in the *Philosophical Transactions*.[3] According to S. Parkes, the methods did not change much throughout the eighteenth century, and a similar description by Dr. Winter will be found in the *Philosophic Journal*.[4]

> The ancient Grecian and Italian process, which is probably still very general, consisted simply in collecting the aluminous mineral (already calcined by the great volcanic processes which take place in those countries), in lixivating and in boiling the lixivium to the requisite degree of concentration. Alum appears to have been procured very abundantly, by this simple method; for the important fact due to the modern researches of chemical science, of the necessity of an alkali, and of potash in particular, to the formation of this salt, seems to have escaped observation; in consequence of ammonia, or an alkaline substance, being in fact combined with the aluminous materials.[5]

Pennant, in his *Tour*, describes the Guisborough works. There he saw 'vast mountains of alum stone' from which the alum was

[1] J. Wilson, *General View: Renfrew*, p. 276.

[2] Vol. 1, p. 225. *See also* John Aubrey, *Letters written by eminent persons in the 17th and 18th. Centuries*, Vol. 2, p. 281.

[3] *Phil. Trans.*, *Loc. cit.*

[4] *Phil. Journ.*, 1810, *25*, 241.

[5] J. Wilson, *loc. cit.*

extracted as above. The stone was calcined (oxidized) in great heaps which continued to burn by their own heat if once lit by a coal fire. Coal, according to Beckmann, was substituted about 1678, before which either wood, furze, or even cinders of coal was the fuel used. The combustion continued for six, ten, or even fourteen months, depending on the size of the heap. The calcined matter was then thrown into pits and lixiviated with water to dissolve out the saline part. The resulting solution was boiled with kelp, or black ash, made from the waste lyes of the soap-boilers, or with stale urine.[1] It was then evaporated like brine in large metal pans, approximately 9 by 5 by 2½ feet, and run into casks to crystallize out. Eight pans were required to maintain an output of five tons of alum per week, and at least three tons of coal were required to evaporate sufficient solution to yield one ton of alum.[2] By the end of the seventeenth century, probably some 10,000 tons of coal were used annually in English alum works.[3]

By 1770 alum had become a commodity of such importance to emergent palaeotechnics that it was considered worth while by an adventurer, Sir George Colebrook, to try and create a world corner in it. He started his scheme in 1771 when the price of alum was unusually low and production localized. The best alum still came from the papal states, inferior qualities from Asia Minor, Sweden, Liège and England. The annual consumption was steady, but less, probably, than the 7,000 tons upon which Colebrook based his calculations.

Two things were necessary to effect his world corner: (*a*) control of production, (*b*) control of distribution. These he endeavoured to obtain by purchasing large quantities, by entering into contracts with large manufacturers, and by leasing or buying-up English companies. The only important producer who remained independent was Sir Lawrence Dundas. Prices rose, as Colebrook hoped, but with the rise manufacturers produced more, and Colebrook failed to absorb the increased production. The monopoly could not hold out. His scheme failed.[4]

There is an early mention of alum in Scotland in a sign manual dated June 1663, granting to 'Lieut. Collonoll Timothie Langley the gift of making Allome'.[5] About 1796 the attention of Charles Macintosh, son of Geo. Macintosh of The Cudbear and Turkey Red Works, turned to the manufacture of this product so necessary

[1] T. Pennant, *loc. cit.*; S. Parkes, *Chemical Catechism*, p. 143, *n.*
[2] Phil. Trans., *loc. cit.*
[3] J. U. Nef, *op. cit.*, Vol. 1, p. 209.
[4] L. Stuart Sutherland, *Economic History*, 1936, *3*, 236.
[5] National Library of Scotland, *Miscellaneous Law Tracts*, pp. 25–34.

to dyers and others. Charles Macintosh was a chemist of considerable ability, and realized that it would be possible to manufacture alum from aluminous shales found in exhausted coal wastes at Hurlet, near Paisley, Renfrewshire, where tentative efforts had already been made by Messrs. Nicholson and Lightbody between 1766–9.

The situation and nature of the raw materials is described as follows:

> The seam of schistus situated between the lime and coal at Hurlet is very variable in thickness, being sometimes only six inches, and in several instances 3½ feet. Its specific gravity is 2·404; and, though extremely hard when fresh dug, it decomposes by the action of the air. The extent of excavation or waste in these mines is about 1½ mile in length, and the greatest breadth about ¾ of a mile; the whole of the cavity is in general dry, and the temperature from 60° to 63°. In this situation the schistus gradually decomposes, and acquires a flaky or downy appearance. In some places, where the schistus may have lain for a very long period, the decomposition is complete. When thus completely decomposed, it is a beautiful vitriolic efflorescence resembling plume alum, but seems to be rather a sulphate of iron than of alumine. It appears in many instances to contain nearly equal quantities of each. The separation of the former, with a view to manufacture alum, was always found difficult. Many unsuccessful attempts to prepare alum from this material were made in 1768 and 1785: but from experiments made in 1795 and 1796 it appeared, that by proper application of the principles of chemistry, this separation might be effected.[1]

It appears that a large quantity of this shale had been laid down by the working of a coal-pit some two hundred years previously, and the action of the atmosphere had completely acidified the surface layer, though not actually converting it to alum, which requires the addition of an alkali.

> Dr. James Millar of Edinburgh, in the appendix to his edition of Williams' *History of the Mineral Kingdom*, has given some account of the mines at Quarreltoun and Hurlet, and of the productions of the latter. He assumes, that in consequence of gradual decomposition, the schistus is 'converted into a native alum', but this is erroneous. In tracing the circumstances which have produced the changes on the materials found in these mines, he observes: 'The pyrites of the coal is decomposed, and the sulphur being oxygenated, is converted into sulphuric acid, part of which combines with the oxide of iron, and thus forms sulphate of iron, or copperas;

[1] John Wilson, *op. cit.*, p. 26.

while another part of the sulphuric acid combines with the alumina of the schistus, and probably also with an alkali, thus yielding a *native alum.*' But no such combination with ammonia or with an alkali can take place, for none of these substances are to be found in the mines at Hurlet, which never afford a single specimen of native alum. Native copperas, however, is frequently found.[1]

Production of alum was started in 1797 by Macintosh, Knox and Company, and Hurlet soon became the largest alum works in the country. The partners included J. Findlay, John Wilson, James Knox, and later Charles Stirling. Their alum could be made at much less expense than in Yorkshire.

The method of preparation used by Macintosh, Knox and Company was very similar to that described by Pennant as taking place at Guisborough. The heaps of calcined alum were certainly, as Pennant said, almost the size of a small hill, being on occasion 200 feet square and 100 feet high; 130 tons of shale were required to produce a ton of alum.[2]

John Wilson in his *Agricultural Survey of Renfrew* gives a full account of the operations, and points out some peculiarities of the Hurlet deposits.

At Hurlet and Campsie the aluminous schistus rests on a *pyritous-coal,* which has been excavating for a long series of years, from mines which are extremely dry, and of which the temperature is seldom under 63° of Fahrenheit, and in places excluded from the current of air of the shafts, the temperature is above 70°. The sulphur of the ore may be conceived to be acidified (by the contact of atmospheric air aided by the gentle ventilation and heat) and to combine with its various other constituents; of consequence the most abundant products are the sulphates of iron and alumine, but those of magnesia and lime are also not infrequent: and thus the result of rapid calcination is attained more slowly and abundantly in the lapse of years. The mixed salts of iron and alumine thus formed, are conveyed from the mines to the works, then lixiviated, and the black insoluble residue is thrown aside to the hill, as that immense mass is technically termed, which has been thus formed in the fifteen years since the manufacture was begun. This hill, or mass of refuse, continues to undergo farther decomposition, still affording a product of some value.

The lixivium being next concentrated to the proper specific gravity, either copperas is first crystalised, and afterwards alum

[1] John Wilson, *op. cit.*

[2] Andrew Ure, *Dictionary of Chemistry*, Article, *Alum;* Winter, *Philosophic Journal*, 1810, *25*, 241.

by the addition of potash salts to what are termed the mother liquors; or, at once the mixed salts of alum and copperas, by a similar addition of potash materials, are deposited by cooling, and the latter is separated by continued solution and crystallisation. The final solution of alum for roaching, is brought to a high specific gravity, and to a high degree of heat; afterwards it is run into large vats where it continues to cool for fourteen days, and is then marketable alum.

The number of men employed in mining the alum-ore, carting materials, attending furnaces, and in the various manufacturing process, is at present forty-seven, and their wages from 2*s.* to 3*s.* 6*d.* per day: 3,000 tons of coals are consumed annually, and six horses are constantly employed. The works afford a ready market for above 300 tons of the potash-residuum of other chemical works, such as the sulphate of potash, muriate of potash, soapers' salts, etc., all of which are used in this manufacture. At the commencement of these works the price of alum in Glasgow was £28 per ton, but it immediately fell to £21, and has varied from that rate to £25 per ton.[1]

So successful was the Hurlet manufactory that, in 1808, similar works were started by the same company at Campsie, Stirlingshire, where a very pyritous alum shale occurred between the coal and limestone.[2] There was also a second attempt to start alum works at Campsie. From the *Session Papers*[3] we learn that Walter Logan, who acted as mercantile agent in Glasgow, and was associated with the Dalmuir Alkali Works, wished to 'speculate in practical chemistry', and for that purpose associated with one of the name of Laird. They attempted to establish the manufacture of alum at Campsie coal works, but were not successful.

The following citation from the *General View of the Agriculture of Stirling* gives an idea of the extending chemical activities of Macintosh and Knox:

In the immediate vicinity of Campsie, there are considerable chemical works carried on, where alum, copperas, soda, Prussian blue, &c. are manufactured on an extensive scale; and in which a very large capital appears to be embarked. The following account of these works, liberally communicated to the Reporter by a gentleman who is concerned in them, must be interesting to the reader.

'The company produces the alum and copperas from a decomposed aluminous schistus found in a considerable quantity in the

[1] J. Wilson, *op. cit.*, p. 279 *et seq.*
[2] G. Macintosh, *Memoir of Charles Macintosh*, p. 44.
[3] Signet Library, 241: 25.

47. Cylinder-printing of calico, an invention, introduced by Thomas Bell about 1785, which forged a further link between the mechanical and chemical revolutions.

48. Bandana press at Messrs. Henry Monteith's, Glasgow. A process was introduced about 1802 whereby bleach liquor was forced through perforated metal stencils which gripped uniformly dyed cloth, thus discharging patterns and giving, for example, the well-known Bandana handkerchief.

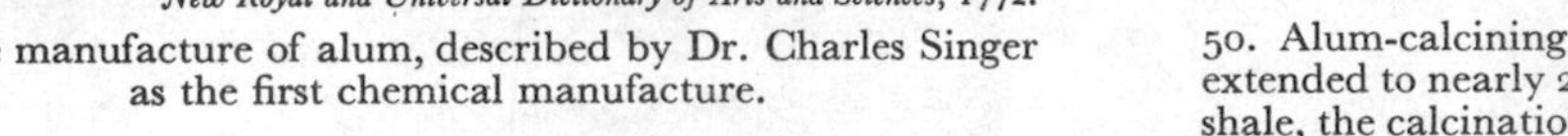
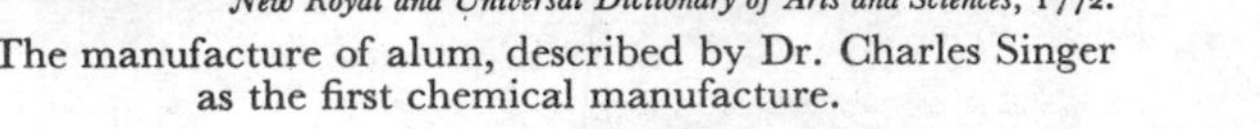

New Royal and Universal Dictionary of Arts and Sciences, 1772.

49. The manufacture of alum, described by Dr. Charles Singer as the first chemical manufacture.

50. Alum-calcining fields at Hurlet, near Paisley, Renfrew. The field extended to nearly 20 acres and contained some 20,000 tons of alum shale, the calcination of which lasted about 12 months. It was started by Macintosh, Knox and Co., in 1797, and was soon the largest in the country.

adjoining coal wastes. This schistus forms originally the covering or roof of the coal strata of the district, and is composed of silex, alumine, or clay, iron and sulphur; the two latter probably in a state of chemical union. Soon after the coal is wrought, this schistus, of various thickness, separates from a limestone stratum immediately above; thus falling down into the waste. In process of time, indeed, generally after the lapse of many years, owing to a constant circulation of air through these wastes, which, being level free, are always dry (an indispensable requisite to this operation of nature) the sulphur becomes oxygenated; and is converted into vitriolic or sulphuric acid; this, uniting with the iron, forms copperas, and with clay, sulphate of alumine, from which crystallised alum is afterwards made.

'The decomposed schistus, as taken out of the wastes, is lixiviated, and the lixivia evaporated. Upon cooling, pure sulphate of iron, or copperas, separates. The mother waters are then boiled with a solution of sulphate of potash, by which (the triple salt) crystallised alum is formed; this separates in its turn by cooling, and is purified by subsequent crystallisations.

'The making of Prussian blue, being a delicate and intricate process, although it is known that alum and copperas enter into its composition, the manipulation of this process is not divulged by the company; (the theory is no secret). Neither is that of the soda manufacture made public; for which it is presumable that the company has local facilities; amongst these, the abundance and moderate price of coal is no doubt to be reckoned.'[1]

By 1812 they were producing 1,000 tons of alum per annum, using 3,000 tons of coal, and 300 tons of potassic materials bought from other works, e.g. soapers' salt. By 1835 the annual production had increased to 2,000 tons, and the price had fallen to £12 a ton,[2] and by 1846 it was still further reduced to £9 10*s.* a ton.[3]

Alum was produced in Stirlingshire continuously till almost the end of the nineteenth century, when cheaper methods of manufacture in operation elsewhere made competition impossible. The development of these cheaper methods was due mainly to the removal of the salt duties in 1825, and, at the same time, to the production of cheap sulphuric acid, which enabled alum to be made from aluminous clay and acid. Amounts produced directly from shale thus progressively decreased, involving Macintosh in considerable losses.[4] However, the Hurlet and Campsie Alum Company, with works at Lennoxtown, was still in operation when the British Association met in Glasgow in 1901. Messrs. King were

[1] P. Graham, *General View: Stirling*, p. 344.
[2] *New Statistical Account*, Vol. 6, p. 166.
[3] G. Macintosh, *op. cit.*, p. 46.
[4] *Ibid.*, p. 181.

MORDANTS

Diagram illustrative of transition to products of advancing chemical industry

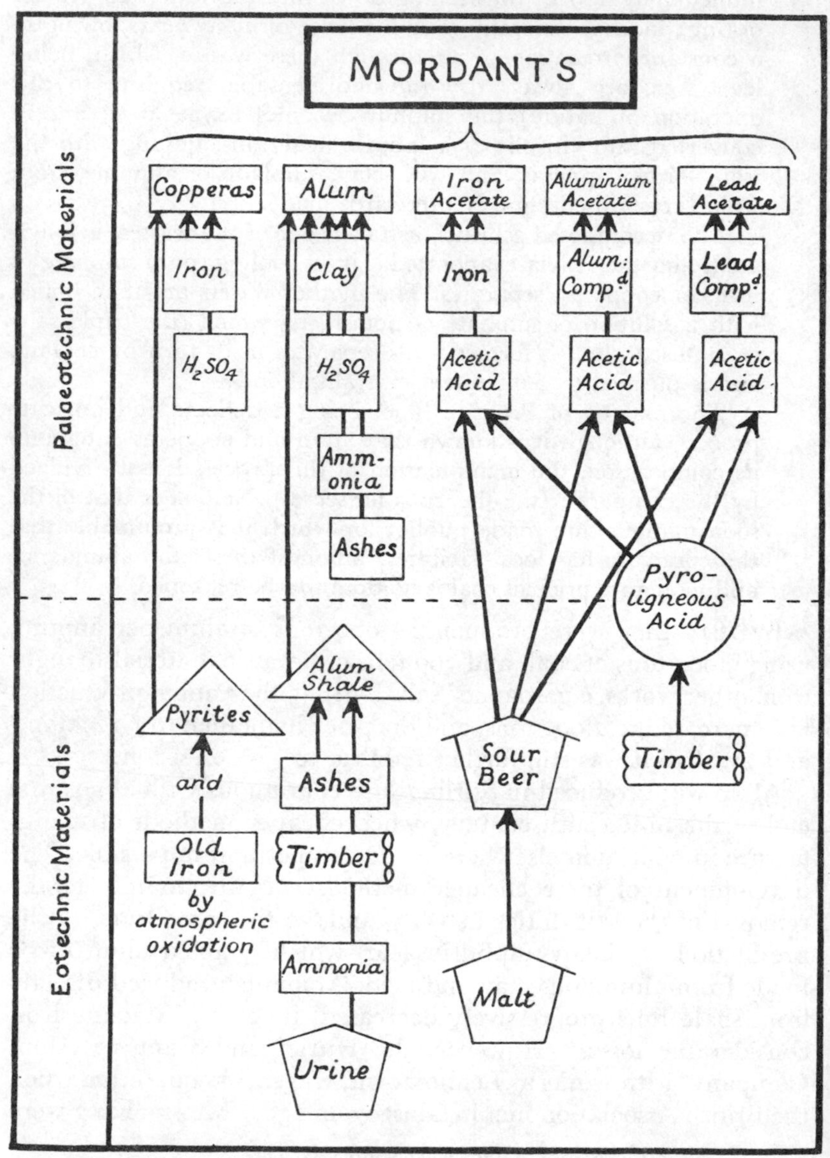

the sole partners, and the same company had similar works at Falkirk and Manchester. Alum was made from shale till about 1880, when the shale supply gave out, but the manufacture of potassium ferrocyanide, which had been carried out at Campsie for over eighty years, was continued.[1]

Alum was used for several things besides mordanting. It hardened tallow; printer's cushions and blocks were rubbed with burnt alum to remove greasiness which prevented the ink adhering properly; wood and paper were soaked with alum solution to make them less liable to catch fire, the paper then making a suitable container for gunpowder; paper soaked in alum was said to whiten silver and silver brass without heat; it cleared turbid water; it was used in tanning and dyeing for 'opening the pores'; crayons were treated with it; in medicine it was used as an astringent; it was also sometimes used for purifying sugar.[2]

For long after the foundation of the Hurlet and Campsie Alum Works, alum-making in Great Britain was carried on by another firm of Scottish origin, that of Peter Spence.

Peter Spence was born in Brechin, on 19th February, 1806, and spent his early years apprenticed to a grocer in Perth. His main interest, however, was in reading such scientific books as he might lay his hands on. The grocery business proved unsuccessful, and subsequent to his marriage, while Mrs. Spence ran a coffee-house, her husband obtained a situation in the Dundee Gas Works. Contacts with chemical science which he made in the prosecution of his duties in Dundee laid the foundation for his subsequent activities. He was one of the first industrialists to realize that waste products of gas manufacture were a source of valuable chemical raw materials. J. Fenwick Allen tells us:

> In the year 1834 he left Dundee and proceeded to London, where he established himself in a small way as a chemical manufacturer, and his earliest patent is dated July 27, 1836, his address being Henry Street, Commercial Road, in the County of Middlesex. His first essay in invention was to achieve, what Peter Spence constantly aimed at, the utilisation of waste products or refuse material; in the present instance he sought to manufacture Prussian Blue, and Plaster of Paris from the refuse lime and the refuse lime liquors of gas works.[3]

This venture was not a success, and he removed to Burgh, in Cumberland, where he interested himself in the manufacture of

[1] Angus McLean, *Local Industries of Glasgow*, p. 172.
[2] Andrew Ure, *Dictionary of Chemistry*, Article, *Alum*.
[3] J. Fenwick Allen, *Some Founders of the Chemical Industry*, p. 257.

alum and copperas. As the result of work done there, he took out a patent which completely revolutionized the alum trade and so cheapened alum that it became much more widely used. This was in 1845. Cumberland was not a suitable locus for a manufactory of this kind, and within a few years he moved to Pendleton, near Manchester, from which another patent concerning alum was taken out on 12th November, 1850.

> Spence's alum process was a great success, both chemically and commercially, and at Pendleton his business rapidly grew, till he was the chief alum manufacturer of the world. After a time he established two other works (at Birmingham and Goole), and entered into other branches of manufacture (copper smelting, copper precipitate, carbonate, muriate and sulphate of ammonia, sulphate of potash, aluminoferric, and sulphate of alumina).

The works which Spence founded in Manchester shortly after 1845 is still in operation.[1]

According to Nef,[2] the other eotechnic mordant, copperas (ferrous sulphate, $FeSO_4.7H_2O$), was certainly used in England as early as 1593, and it is probable that it was first made in England by Matthew Falconer, a Brabanter, in the Isle of Sheppey.[3] In his *Travels* Brereton described the Sheppey manufactory:

> A square plot of ground, about an acre, the earth hath been taken all away, and a kind of stone brought from Essex shore, which, falling into the sea, is tempered by the salt water; which stone, by rain and snow, is beaten and reduced to soil: the ground is clayed in the bottom, and is made of a fair bottom. Betwixt every bed is a trough made with three deane (deal) boards, bored full of holes; this is digged over summer, the bottom laid highest and the top lowest. This trough receives and conveys away all the liquor and moisture which doth flow from the soil in rainy and moist weather; old barrels, into which these troughs lead, are prepared and placed in the earth, in clay, which receives the moisture, out of which the liquor is conveyed into divers great cisterns of boards, two bayes of buildings, all which are laid and set in clay; out of which it runs into a great tub, placed in a more spacious house, and near unto a mighty cistern of lead, wherein this liquor is boiled. Under the cistern are five furnaces, a partition between them; these furnaces spend half a chaldron of coals per day, and this leaden cistern will last four years. There is half a barrel of old iron now boiled in this liquor, which is consumed to dust. When it is sufficiently boiled, it runs through a leaden pipe

[1] *See also* S. Miall, *History of the British Chemical Industry*, p. 23.
[2] *Op. cit.*, Vol. 1, p. 184.
[3] W. G. Armstrong, *The Industrial Resources of the Tyne, Wear & Tees*, p. 175.

> into cisterns of lead, six of them at least and it cools in these cisterns; the copperas matter thickens and adheres to birch twigs, or bushes, which they hang upon overcross poles into the cistern. It is worth £5 per ton; one boiling yields 3–4 tons.[1]

Nef states that the 'copperas stones' (i.e. pyrites, FeS_2) took five to six years to *ripen* in long wooden troughs filled with water before they were drained into boilers made to hold as much as twelve tons of liquid. *Ripening* in this context signifies oxidation leading to production of ferrous sulphate. The solution was heated with iron dust over a coal fire for about three weeks, fresh solution and iron being added from time to time. The works at Queenborough, Sheppey, used 300 tons of coal a year. After the Restoration, when inferior Newcastle pan-coal became available, this quantity was probably increased.[2]

In 1748 Thomas Delavel started the manufacture of copperas at Hartley, near Newcastle, and in 1789 Barnes and Foster opened a copperas works at Walker on the Tyne. This was still in operation a century later, and producing two thousand tons a year. Faujas de St. Fond describes the manufacture at Newcastle. The heaps of pyrites were sprinkled with water, the solution of ferrous sulphate ultimately produced by oxidation of the pyrites being collected on a clay floor and run into a reservoir prior to evaporation with old iron as described above. Much of the Newcastle production was exported to France, where the dyers of Rouen, Paris, Lyons and Marseilles, consumed 'astonishing' quantities.[3]

In Scotland the production of copperas is associated with the foci of alum manufacture, Hurlet district, and Campsie.

> Copperas is occasionally prepared at Hurlet alum work. The oldest establishment in this part of the kingdom for the manufacture of that salt, was at that place: it was begun in 1753, by a company from Liverpool; and, till 1807, when a similar manufacture commenced on the adjoining lands of Househill, was the only copperas work in Scotland.
>
> In the stratum of coal at Hurlet and Househill there are found considerable quantities of pyrites. These have since the year 1753 been carefully separated from the coal at Hurlet, and sold to a company established there, for manufacturing sulphate of iron or green vitriol. The price of these pyrites or copperas stones, by old contract, was 2½d. per hutch, of two hundredweight.
>
> The pyrites from which the copperas is obtained, are collected, for this manufacture from the neighbouring coal works, as already

[1] Sir W. Brereton, *Travels in Holland, etc.*, 1634, p. 2.
[2] Nef, *op. cit.*, Vol. 1, p. 210.
[3] B. Faujas de St. Fond, *Journey through England and Scotland, 1784*, pp. 142–5.

mentioned: they are exposed to the weather on beds contiguous to the works; and, after gradual decomposition and washing with rain water, afford liquor, which, upon being concentrated by boiling, with a small addition of iron, produces green copperas, or sulphate of iron. The process, as carried on at Deptford in 1666, is given by Mr. Colwall in the Philosophical Transactions for that year, and the same process is still employed with few alterations. The price at Hurlet varied about ten years ago from £7 to £9 per ton; in the year 1805–6, it rose to £10 or £11, and since that period it has fallen to £5 per ton. The number of men employed at the two copperas works at this place is seven, exclusive of persons who collect the ore, and cart materials to and from the works. The quantity of coals consumed is about 750 tons, and the quantity of copperas manufactured about 400 tons, yearly.[1]

In addition to the two works at Hurlet mentioned in these citations, copperas was made in Macintosh, Knox and Company's establishment at Campsie, and Sir J. Sinclair speaks of a fourth at Baldernoch, Stirling.[2] Pyrites was dug in small quantities at Longfauld, Dumbarton, and sold to the manufacturers of copperas at 15*s*. per ton.[3]

As mentioned elsewhere, Charles Macintosh, the first successful manufacturer of alum in Scotland, was the son of George Macintosh, manufacturer of Cudbear and Turkey Red. He was born in Glasgow in 1766, and as a boy devoted his spare time to the study of science, and especially to chemistry, attending the lectures of Dr. Wm. Irvine at Glasgow, and later, it is believed, those of Dr. Joseph Black at Edinburgh. He began his business life as a clerk with Glassford of Dougaldston, but gave that up when only twenty years of age to engage in the manufacture of sal-ammoniac (NH_4Cl) from soot and urine. A manufactory of this kind had been carried on near Charenton in France by M. Baumé (1728–1804), who, however, relinquished it as unprofitable in 1787. Sal-ammoniac was at this time extensively used by dyers, braziers, and tin-plate makers.[4] Macintosh carried on his business in partnership with his father and William Couper, an eminent Glasgow surgeon, who managed the concern. To begin with, all did not go well with the process, and we learn that Macintosh had consulted John Finlay, brother of Kirkman Finlay, and Secretary to the Duke of Richmond, about his difficulties.

You ask my opinion about the cause of the want of success in your last experiment, and with regard to the probability of glass

[1] J. Wilson, *op. cit.*, p. 281. [2] *General Report*, Appendix 2, p. 310.
[3] A. Whyte and D. Macfarlan, *General View: Dumbarton*, p. 10.
[4] S. Parkes, *op. cit.*, p. 153.

> vessels being made use of at the sal ammoniac manufactory at Edinburgh.[1] With regard to the first, it is probable that the failure might proceed from two causes, either separately or combined, viz. from too grêat heat being employed in drying it, by which part of the ammoniacal salt might have been dissipated, and from the want of a sufficient degree of heat to raise the sal ammoniac, being applied in the last process, as you seem to think; and the cause, which I believe has contributed not a little to the failure of your last experiment, has been the employing of too low vessels, by which the sal ammoniac raised could not be condensed, and escaped at the mouth of the vessel. From what Mr. Geddes[2] has mentioned to you, it appears certain that the Edinburgh people employ glass vessels; and the expense attending the employment of them appears to me to be the only objection to their use. Sal ammoniac, as you know, requires almost red heat to raise it; but vessels of green glass are very capable of sustaining that heat. I should imagine that the cracked vitriol bottles are luted on as heads to the vessels which contain the matter to be sublimed, and that when their inside is coated to a sufficient thickness with sal ammoniac, they are broken to get it out.[3]

Some, at least, of the difficulties must have been overcome, because the business was prosecuted till 1792, when it was given up as having become unprofitable. By this time Lord Dundonald was obtaining sal-ammoniac by the distillation of coal, a method almost certainly more economic than that employed by Macintosh.[4]

In the same year as he began making sal-ammoniac, 1786, Macintosh undertook a continental tour for the purpose of introducing the cudbear made by his father, and effecting sales of sulphuric acid for the Prestonpans Vitriol Company, of which Mr. Glassford had become principal partner.[5] When he returned he introduced into this country the Dutch method of preparing another mordant, sugar of lead or *saccharum saturni* (i.e. lead acetate, $Pb(OOC.CH_3)_2$).[6] He was soon exporting it to Rotterdam, from whence it had previously been imported. It was used mainly in calico-printing in the form called red colour liquor.[7] Prior to this, nearly all the sugar of lead used in Great Britain had

[1] The Edinburgh manufactory of sal-ammoniac, that of Dr. James Hutton, more celebrated as a geologist, in the study of which he was enabled to engage in the leisure afforded by the proceeds of the sale of sal-ammoniac. (Ch. XIX.)

[2] The Geddes referred to in the above citation was doubtless one of the family of Geddes connected with the Verreville Glass Works. (Ch. XIV.)

[3] J. Finlay to C. Macintosh, 18th February, 1786. Quoted in G. Macintosh, *op. cit.*, p. 18. [4] Ch. XVIII. [5] G. Macintosh, *op. cit.*, p. 21. [6] Ch. XVII.

[7] *New Statistical Account*, Vol. 6, p. 165; *Quarterly Journal*, 1819, *6*, 255.

been imported from Holland, where it had been produced more cheaply than at home, despite a duty of threepence a pound, and the fact that the lead used was of British origin. Iron acetate ($Fe(OOC.CH_3)_2$) was also in demand by printers. At one time it was made by digesting old iron hoops in sour beer (weak, impure acetic acid, CH_3COOH), but when pyroligneous acid became available it was substituted, 'the oleaginous impurities of which tend, in some cases, to improve the mordant'.[1]

Still in partnership with his father, Macintosh started preparing in 1793 'a newly discovered chymical preparation which answers as a real substitute in dyeing and printing' as an alternative to sugar of lead and alum. According to the *Statistical Account of Scotland*, a factory had actually been founded to manufacture sugar of lead, but in the course of experiments the improvement was discovered. The 'newly discovered chymical' was aluminium acetate ($Al(OOC.CH_3)_3$), which could be prepared more cheaply than lead acetate and proved even more satisfactory to the dyers.[2] It rose rapidly in importance, and many thousands of pounds' worth of malt and barley were used annually in the production of the acetic acid necessary for its manufacture.

The demand for aluminium acetate was so great that from about 1820 it had to be made from pyroligneous acid and lime, the calcium acetate being decomposed by alum, or aluminium sulphate.[3]

$$(a)\quad 2CH_3COOH + Ca(OH)_2 = Ca(COO.CH_3)_2 + 2H_2O$$
$$(b)\quad 3Ca(COO.CH_3)_2 + 2Al_2(SO_4)_3 = 2Al_2(COO.CH_3)_3 + 3CaSO_4$$

Pyroligneous acid is produced by destructively distilling wood in cast-iron retorts. 'The gas passing through a worm contained in a refrigatory, is condensed in the same manner as common spirits'. One ton of wood gave eighty–ninety gallons acid and ten gallons tar, charred wood or charcoal remaining in the retort.[4] Sprays of all trees, if not resinous, could be used for the production of pyroligneous acid. The operation was carried out as follows:

> The distillation is carried out in a cast or malleable iron boiler, which should be from five to seven feet long, three feet wide, and say four feet deep from the top of the arch, built with fire-brick. The wood is split or round, not more than three inches square in thickness, and of any length, so as to go into the boiler at the door.

[1] S. Parkes, *Chemical Essays*, Vol. 4, p. 137.
[2] J. Sinclair, *Statistical Account*, Vol. 12, p. 115.
[3] *New Statistical Account*, Vol. 6, p. 166.
[4] *General Report*, App. 2, p. 308.

When full, the boiler door is properly secured, to keep in the steam; then the fire is put to it in the furnace below, and the liquid comes off in the pipe above, which is condensed in a worm, in a stand filled with cold water, by a spout, and empties itself, first into the gutter below, and from that it is let into barrels, or any other vessel; and thus the liquid is prepared. One English ton weight of any wood, or refuse of oak, will make upwards of eighty gallons of the liquid.

The distillation apparatus is illustrated in Loudon's *Encyclopaedia of Agriculture*.[1] According to Monteith, one of the principal calico-printers of Glasgow, the acid sold there in 1819 at £1 2*s*. to £1 10*s*. per ton.

In the Government gunpowder manufactory at Waltham Abbey, where charcoal was made expressly for the manufacture of gunpowder, pyroligneous acid resulted as a by-product. John Finlay, secretary to the Duke of Richmond, sent Charles Macintosh a sample for investigation.

> I have got some of the acid of wood (pyroligneous acid), and shall send you a couple of bottles that you may see its strength. I shall at the same time send you a bottle of the same acid, which has been once distilled from the coarse kind, and shall beg of you to let me know what you would give per gallon for each kind, or what they would be sold for at Glasgow. I must endeavour to dispose of it for the Board of Ordnance to the best advantage I can.[2]

Macintosh eventually took up the manufacture of his own acid, producing it at a cost of about fourpence a gallon. By about 1815 there were seven such distilleries in Scotland—four in Lanarkshire (at Camlachie, Tradestown, Brownfield and Lanark); one near Torryburn in Fife; and two in Dumbarton (at Millburn and Cordale).

> The distillery of pyroligneous acid at Milburn, employs about seven hands, and consumes daily a ton of small timber, chiefly oak, from which the liquor, a kind of coarse vinegar, is extracted. The process is beautifully simple. A number of iron ovens, or retorts, are placed in a row, and filled with the timber cut into small pieces. A fire of coals or charcoal is kindled in a furnace attached to each, and by its heat, forces the acid to fly off in the form of vapour. This vapour is conducted by a small tube proceeding from each retort into a refrigeratory or long metal pipe,

[1] T. C. Loudon, *An Encyclopaedia of Agriculture*, p. 657.

[2] J. Finlay to C. Macintosh, 8th November, 1791: quoted by G. Macintosh, *op. cit.*, p. 32.

> on which a jet of cold water from above is continually falling. Here the acid is condensed, and runs from the end of the pipe in a considerable stream of reddish brown colour. Besides the liquor thus procured, which is employed in making colours for the calico printers, there is a considerable quantity of tar and charcoal produced during the process, the value of which is esteemed equal to the expence of fuel.[1]

According to A. McLean, the oldest-established firm of wood distillers in Scotland at the beginning of the present century was Turnbull and Company of Camlachie Chemical Works, Glasgow, whose first works was erected at Millburn in the Vale of Leven towards the end of the eighteenth century. They subsequently established works in other parts of the country, the principal being that at Camlachie, which was opened in 1808.[2]

Macintosh was apparently also interested in the new method of bleaching introduced by Copland and Watt, and corresponded with German industrialists concerning difficulties in managing it.

> Vous m'obligerez infiniment, si vous donnez la paine pour l'amour de moi, de me communiquer vos éclaircissements sur la nouvelle manière de blanchir le linge. J'ai fait plusieurs essais, et même dépense de l'argent pour traiter la chose au fond. J'ai commencé à un peu étudier la chymie pour mieux pouvoir juger de la chose, et pour la traiter pertinentment, mais toujours je trouve mes essais brulés. C'est là un point essential sur lequel je vous prie d'étendre un peu vos réflections. Quant a l'odeur que le preparation donne, j'en imagine qu'on peut facilement y remédier, du moins en partie, par un apparat apprechant celui de Wolfe, lequel vous connoissez sans doute.[3]

Shortly after his return from a second Continental tour made in 1789 he met Charles Tennant, with whom he subsequently became associated in business. Tennant, as has been described,[4] was a bleacher, and was experimenting with a liquid bleach easier to handle than gaseous chlorine. In 1797, Tennant took out a patent for manufacturing a liquid bleach. Difficulties arose over the patent, which was declared void. Two years later, a further patent was taken out, also in Tennant's name, this time for dry bleaching powder. There seems little doubt, however, that the credit for the solid bleach is due to Macintosh. The details are set forth in the *Memoir*.[5]

[1] A. Whyte and D. Macfarlan, *op. cit.*, p. 275.
[2] A. McLean, *op. cit.*, p. 176.
[3] Monsieur le Baron d'Aescher to C. Macintosh, 1789, quoted by G. Macintosh, *op. cit.*, p. 28. [4] Ch. IX. [5] G. Macintosh, *op. cit.*, pp. 36–44.

For many of the manufactures in which Macintosh was interested, ammonia was a raw material. So large was his consumption of ammonia that in 1819 he contracted with the recently established Glasgow Gas Works to take all their by-products for the sake of ammonia which they contained. To begin with, gas-tar and ammoniacal liquor were nothing but a nuisance to gas companies, and were dumped wherever convenient. The Glasgow authorities were thus, no doubt, pleased to accept Macintosh's offer. Having used the ammonia, he was faced with the economic utilization of the remaining products. By distillation he separated a fraction from the tar called naphtha. This, he realized, resembled turpentine, which he knew would dissolve rubber, and as such might open up new fields for exploration—one in particular, the use of the rubber solution to render fabrics waterproof.

Methods of waterproofing were known in the seventeenth century. In 1691, William Sutton and George Hager invented a 'new and extraordinary art of ordering all sorts of stuffs, silks, hats, and leather, so as to make them hold out water and also preventing them from damage by moths and mildew'.[1]

In the middle of the eighteenth century, waterproof garments were, however, still novelties. In a letter dated 28th May, 1769, James Watt mentions having been helped to translate from German a book on furnaces, machines, etc., by 'a truly chymical Swiss dyer of name Chaillet who came to Glasgow to dye red on linen and cotton'. Chaillet promised to make him a coat that would not wet though boiled in water—a promise which was fulfilled.[2] Umbrellas gave added protection as well.

> From the year 1783, when a Glasgow doctor displayed for the first time a yellow umbrella which he brought from Paris, there were seen everywhere the bulky rainproof implements of yellow and green glazed linen.[3]

No mention is made of the use of rubber in these early references to waterproofing, but rubber had begun to interest British industrialists. The following account, sent to Matthew Boulton by Sir John Hort, Consul-General, Lisbon, is full of interest.

> I sent to you by the last Packet Boat, by Mr. Glover of your town, a very little flask, made of an extraordinary substance, commonly sold here by the name of Borracha, which means no more than leathern bottle, but in Brazil from whence it is brought,

[1] W. R. Scott, *Joint Stock Companies*, Vol. 2, p. 120.
[2] H. W. Dickinson, *James Watt*, p. 58.
[3] H. G. Graham, *Social Life in Scotland in the Eighteenth Century*, p. 144.

it is called by the Indians, Caoutchouc. It is here applied singly to the use of leathern bottles, in which shape it is brought of all sizes, up to about an English Gallon; but its qualities are so remarkable, that your inventive genius will I am persuaded be capable of finding purposes for it, less sufficiently performed by any other material hitherto known. In its natural, fresh state, it is a gummy liquid, and is obtained by wounding the tree which produces it: the utensil wanted is modeled in clay, and placed under the incision, from whence it is directed along the surface till the whole is covered. For some time it is susceptible to any impression; and the flasks are usually covered with rude capricious circular and other lines by way of ornament, such as you will see on the specimen I sent you. But in a few hours it becomes inaccessible to any lasting mark. Its elasticity exceeds greatly that of any natural substance yet discovered—And the Chymists know but a preparation of Ether and of two other substances capable of dissolving it: The ether it is said leaves it in a state from which it may be restored to its former texture and qualities. It communicates no flavour or smell to any common substance lodged in it, wine, brandy, etc. The last circumstance I remember of it, its price; from the common shops the little vessel sent to you and containing I suppose half a pint, cost ready mounted less than six pence; and if you find means of making it a commercial object, I conclude it might soon be afforded for less than half that sum. Speaking of its actual applications, I have made no account of whimsical uses tried by a few persons, which are various. Among others it is worn as boots, made by patterns sent to Brazil, balls for playthings, squirts, figures of cows, horses, etc. It also effaces with singular readiness every mark of lead pencil; to which I apply it perpetually and for that purpose. Lately a workman sold it in London at three shillings for half a cubick inch: If the specimen I delivered to Mr. Glover arrives distorted, it will presently return to its former shape by laying before the fire or in hot water.[1]

In 1791 a patent was taken out by Samuel Peal for the waterproofing of leather, silk, paper, etc., by means of rubber dissolved in turpentine.[2] Here is the description of the patent:

Take Caoutchouc or (what is called in this Country) Elastic Gum or India Rubber. Dissolve the same by distillation or, by infusion in a small quantity of Spirits of Turpentine over a brisk fire. It may also be dissolved by infusion in other Spirits and in most kinds of Oils. Or the gum may be used with equal advantage in its native fluid state. The gum thus prepared or in its native state is to be applied to the hide, skin, leather, cloth, silk, or any

[1] Sir John Hort to Matthew Boulton, 7th July, 1776, A.O.L.B.

[2] *Register of the Great Seal*, Vol. 20, Nos. 427 and 432. G.R.H.E.

> other article or manufacture intended to be rendered waterproof by means of a brush or other utensil capable of giving a regular coating.

The solubility of rubber in ether ($(C_2H_5)_2O$) was also known, and made use of in making small articles. It is mentioned in Macquer's *Dictionnaire de Chimie*, and was supposed to have been discovered by a London apothecary, Winch, who incidentally described the process to Faujas de St. Fond. Winch made small articles by first forming them of clay, and dipping the clay model into a solution of rubber in very pure ether.[1] This method of making rubber tubes, etc., was adapted *c.* 1814 to the making of elastic catheters.[2] Rubber gloves and boots were also made by James Howison, who dipped wax moulds into latex. The resulting articles would not keep shape, and stretched. To get over this difficulty he suggested the substitution of a fabric base upon which to build the rubber article. As an outlet for his products, he indicated that they might be used for sea-boots, coach-hoods, etc. J. Syme, Professor of Surgery at Edinburgh, substituted naphtha for ether.[3]

These developments all took place prior to Macintosh's 1822 patent (No. 4804), which was for cementing together two thicknesses of cloth by a solution of rubber in naphtha, thus rendering the textile impervious to water. To develop his patent he established a manufactory for waterproof materials in partnership with Messrs. Birley[4] under the name of Charles Macintosh and Company, first in Glasgow, later in Manchester.[5] Charles Goodyear says:

> The Macintosh firm started in 1821 and made principally wearing apparal such as coats and capes, also inflatable goods, cushins, pillows, beds—life preservers. The dissolved gum elastic was spread between two layers of cloth and this in the humid atmosphere of England, and in other cold countries, notwithstanding their imperfection, these goods have been found extremely useful: and the inventor not only attained a high reputation, but was thereby enabled to accumulate a very handsome fortune.[6]

In addition to using the by-product naphtha from the Glasgow Gas Works liquor, Macintosh bought light oils and spirit from one of the early tar distilleries, that of Dr. Longstaff and Dalston, near

[1] B. Faujas de St. Fond, *op. cit.*, p. 28.

[2] C. B. Rose, *Annals of Philosophy*, 1814, *4*, 312. [3] *Ibid.*, 1818, *12*, 112.

[4] One of the partners was H. H. Birley, at this time a director of the Manchester Gas Works.

[5] *Mechanics Magazine*, XXIV, p. 529 *et seq.*; G. Macintosh, *op. cit.*, pp. 83, 97.

[6] C. Goodyear, *Gum Elastic and its Varieties*, p. 82.

Leith.[1] The patent was infringed, and resulted in 1836 in a case against Evrington, Ellis, *et al.*, of London. During the hearing of the action Lord Campbell, the Attorney-General, told the jury:

> The cloaks are now very generally known, and have obtained great celebrity, and are of the greatest utility; so much so, that the patent has become almost as well-known as the celebrated Mr. Watt's for steam engines.[2]

Yet, for a time, it was thought that the *Macintosh* would lose its popularity:

> A change, however, has come over this state of matters, coincident with the increase of the railway system. The number of travellers by gigs, the outside of coaches, and on horseback, have, since the introduction of railways, been prodigiously diminished; and as, in addition, the members of the medical faculty having lent their aid to run down the use of water-proof (apparently from having found it a decided enemy to their best friends, colds and catarrhs), the use of the article in the form of cloaks, etc. has of late (*c.* 1847) become comparatively extinct.[3]

This state of affairs was only transitory, and as a result of increasing use of rubber, wild supplies became insufficient, and the West Indian traders started plantations for the cultivation of the rubber tree, with the result that its locus of production was entirely altered.

Large residues of tar still remained after extraction of ammonia and naphtha. To dispose of them Macintosh invented in 1826 a furnace for burning tar under gas retorts. H. H. Birley of the Manchester Gas Works wrote him:

> You were good enough to promise me a sketch of your method of burning tar under retorts; and I am now writing in a committee of the gas works, for the purpose of requesting that you will be good enough to furnish me with the sketch at your earliest convenience. We are very much plagued in the sale of the tar.[4]

This furnace proved successful, and was soon adopted by most of the gas works in Great Britain. It saved millions of pounds in fuel, and was used till tar became one of the most important palaeotechnic chemical raw materials.

In 1825 he obtained a patent (No. 5173) for converting malleable iron into steel by exposing it at white heat to a gas such as coal gas.

[1] *Journal of the Society of Chemical Industry*, 1881, *1*, 7.
[2] G. Macintosh, *op. cit.*, p. 83. [3] *Ibid.*, p. 89.
[4] H. H. Birley to C. Macintosh, 12th May, 1826, quoted by G. Macintosh, *ibid.*, p. 97.

The process was completed in a few hours, but was never a practical proposition as it was impossible to keep the plant gas-tight.[1] He also co-operated with J. B. Neilson in the introduction of the hot-blast in iron-smelting.[2]

Macintosh had many other interests. In 1806 he had suggested purifying saltpetre in India instead of bringing the crude material to Britain, but the East India Company replied that they were interested in increasing, not decreasing, the quantity of imports. At the alum works he invented a surface-evaporating reverbatory furnace for fluids which was later adopted by the manufacturers of synthetic alkali. He modified and improved a hydrometer for measuring the specific gravity of liquids. In 1807, Macintosh also devised a new method for evaporating citric acid, and shortly after it was discovered, of preparing iodine.

Macintosh was a keen agriculturist. He farmed two hundred acres at Crossbasket, near Glasgow. One thing in particular seems to have interested him—the production of sugar from beet, which he saw being carried out in France in 1829. His interests extended alike to practical and to theoretical chemistry. Among his early (1787) theoretical preoccupations were the relation between animal and vegetable substances, the theory of dyeing, and the nature of vegetable alkali.

[1] *New Statistical Account,* Vol. 6, p. 167. [2] Ch. XVI.

Chapter XIII

SCOTTISH PAPER-MILLS

Heaven prosper the manufacturers of paper.
TRISTRAM SHANDY.

TILL the fifties of the nineteenth century, when the spread of education among the masses put fresh demands on the paper-makers, linen rags and hempen waste were almost the exclusive materials from which paper was made in Great Britain. After linen, in the form of underclothing, was worn out its utility was by no means at an end; it began a second cycle of usefulness as raw material for the paper industry.

> The incalculable advantages which the moderns have derived from the art of printing, would have been only imperfectly known, but for the invention of linen rag paper. A more plentiful and economical substance could not be conceived than the tattered remnants of our clothes, linen worn out, and otherwise incapable of being applied to the least use, and of which the quantity every day increases: nor could a more ready operation be imagined, than a few hours trituration in a mill.[1]

Experiments to find substitutes were not lacking, for there was, at times, acute competition between paper-makers and other users of rags—witness the appeal to the Privy Council of Scotland to prohibit candlemakers from using 'clouts and rags for the wicks of candles'—but esparto grass and wood pulp were late palaeotechnic introductions.

Little has been done to elucidate the early history of paper manufacture but this much is certain, the manufacture was introduced into Europe by the Moors, and from Toledo in Spain spread to Hainault in France, and subsequently throughout the rest of Europe. The earliest extant English and Scottish manuscripts are on paper of French manufacture, the first mill in England being set up towards the end of the fifteenth century by John Tate of Stevenage, Hertfordshire, son of the Lord Mayor of London.[2] Early mills in Scotland were discussed in Chapter I. Output from pioneers referred to there was of necessity small but

[1] A. Rees, *Cyclopaedia*, Article, *Paper*.

[2] Rhys Jenkins, *Early Attempts at Paper-making in England*, 1495–1788, Lib. Ass. Record, 1900, 2, 479.

51. The evaporating houses at Hurlet, which could evaporate 5,000 gallons of alum solution in twenty-four hours.

J. C. Loudon, *Encyclopedia of Agriculture*, 1831

52. A pyroligneous-acid still, in which chopped-up wood, usually oak, was destructively distilled for the production of an impure acid. Large quantities of the acid were used for the production of acetates for the calico-printers.

53. Grinding colours for dyers and calico-printers, *c.* 1840.

The Northern Looking-Glass, 1825

54. Macintosh's Waterproof Life Preserver.

probably increased rapidly. By an examination of watermarks, Ed. Heawood has been able to demonstrate that, by the second half of the eighteenth century, Scottish papers were being exported to London. In fact, Midlothian became the greatest centre of paper production in Scotland, paper occupying an unusually high place in Scottish manufactures.[1] As little has been written on Scottish paper-mills, some details of economic significance might be given at this stage.

In the central counties of Scotland, the North Esk and the Water of Leith are to this day the sources of water for many mills. One centre on the Esk, which for over two centuries has been continuously associated with paper manufacture, is Penicuik in Midlothian, only a few miles south of Edinburgh. The first mill there, the Valleyfield Mill, was founded in 1707 by Anderson of Edinburgh. In 1779 it was purchased by Charles Cowan, and is still being operated by a Company under the name of Alex. Cowan and Son, Limited. So successful were the Cowans throughout the eighteenth century that, in 1803, they acquired a neighbouring cotton mill and converted it to paper. This mill was known as the Bank Mill, from the fact that they engaged there in the making of bank-note paper for Scottish bank-notes. Further expansion took place in 1815 when they added Low Mill, formerly owned by Nimmo of Edinburgh.[2] Cowan, with the help of William Cunningham, an Edinburgh chemist, obtained a patent for the use of chlorine in paper-bleaching.[3]

According to the *Statistical Account*, in Edinburgh itself there were three mills in 1763, producing 6,400 reams of paper. By 1795 the mills had increased to twelve, and the output to 100,000 reams. Of the exact location of these mills we are still uncertain. They were perhaps Dalry, where a mill was founded in 1675 on the Water of Leith, Duddingston Mills, and Colinton. At a later date there was certainly a Colinton paper mill, operated by Balfour and Sons.[4]

There were other mills on the various rivers in the country round Edinburgh. The parish of Currie, through which the Water of Leith runs, has also a long history of paper-making. Nisbet, Macniven and Company operated at Balerno, where there are still mills operated by James Galloway Ltd.[5] According to the *New Statistical Account*, there were three extensive paper manufactories

[1] Trans. Bib. Soc., 1930, *11*, 263 and 467.

[2] *Statistical Account*, Vol. 10, p. 422; *New Statistical Account*, Vol. 1, p. 44; D. Bremner, *Industries of Scotland*, pp. 322 and 330.

[3] *Vide infra.*

[4] *Statistical Account*, Vol. 6, p. 595; H. Arnot, *History of Edinburgh*, p. 465.

[5] *Statistical Account*, Vol. 5, p. 323.

in the parish, which circulated much money in the villages of Currie and Balerno.[1] The Currie mills are now owned by Henry Bruce and Sons, Limited. Walker and Company of Kenbeth, were in the same locality.[2]

Somewhat farther afield, but still on the east coast, were three Berwickshire mills, Millbank at Ayton on the Eye Water, Broomhouse on the Whiteadder, and Allanbank on the Blackadder, a trio of mills employing some two hundred operatives. These mills, in addition to giving employment to the male population, made use of 'unemployed wives and infants of the workmen', in sorting and cutting rags.

Two factors control the economic location of the industry. It was therefore natural to find in Edinburgh and the circumjacent country, the principal centres of paper production in Scotland. We may define paper, according to G. T. Morgan, as a fabric made by the deposition of finely-divided fibres from suspension in water, followed by heating, rolling and drying.[3] One necessity is therefore obvious, an abundant supply of clean water, for no paper mill can exist unless near the banks of a stream or river. Hence the importance of the Water of Leith and the Esk in this respect. The second factor is an available supply of fibre.

The type of fibre used has varied considerably since the industrial revolution, but up to 1850, the only fibre of importance was flax. The flax was not used directly but, as mentioned above, in the form of cast-off underclothing. Thus, the raw materials being a by-product of human habiliment, it was almost always in the neighbourhood of a large town, provided always that water was available in sufficient quantity, that a paper mill could be worked most economically. It was perhaps the fact that Glasgow developed as an industrial city only late in the eighteenth century, and long after Edinburgh was of considerable importance, that accounts for the slow extension of the paper industry into the west of Scotland. In fact, paper-making has never developed on the west coast to an extent comparable with the east. There were, it might be noted in passing, however, three mills in Dumbartonshire and one in the New Kilpatrick parish which had three vats and twenty-five workers.[4] In north-west Renfrew there was a mill at Cathcart, near the White Cart Water.[5] There was also an isolated single mill at Dalbeattie, in Kirkcudbright, either on the Urr Water or Dal-

[1] *New Statistical Account*, Vol. 1, p. 552. [2] *Session Papers*, 207, 3.

[3] G. T. Morgan, *British Chemical Industry*, p. 194.

[4] *New Statistical Account*, Vol. 8/1, p. 58; Whyte and Macfarlan, *General View of the Agriculture of Dumbarton*, p. 274.

[5] J. Wilson, *General View of the Agriculture of Renfrew*, p. 275.

beattie Burn, managed by Alexander Copland, on whose estate it was.[1]

Even in the most favourable situations, as has been noted, rags were, on occasion, in short supply, and chemistry made a vital contribution to the art of paper-making when chlorine bleaching of rags was introduced into the paper trade, because it increased the range of rags which could be used for white or light shades of paper.

To augment the supply of rags, which may have dwindled as a result of changes in fashion, experiments to find alternatives were made long before the introduction of wood pulp and esparto grass. There are two volumes in the National Library of Scotland of *Experiments and Samples for making Papers without rags or a very small addition, if any*, which were published in 1785 at Ratisbon by James Christian Schaeffer.[2] All sorts of alternatives were tried with apparent success, viz. poplar down, wasp nests, sawdust, shavings, beechwood, willow, tree moss. In the words of its author, he appears to have been able to produce 'a paper perfectly smooth and slightly beyond what I myself had imagined'. Such was the demand for cheap paper that by 1857 there were upwards of 200 patents covering inventions of alternatives to rags.[3] Mechanical wood pulp dates from 1850, esparto grass from 1860.[4]

It was pressure arising from the increased importance of the printing press that made itself felt at the time of Schaeffer's experiments. The first *daily* appeared in Dublin in 1700, followed by the earliest of the English dailies, the *Daily Courant*, in 1702. The first Aberdeen newspaper, the *Aberdeen Journal*, or *North British Magazine*, started in 1748, two years before the founding of the plethora of mills around Aberdeen. By 1776 there were several London dailies. The *Morning Post* dated from 1772; *The Times* from 1788.

> *The Glasgow Journal*, which flourished for over a century, dates from 1729. *The Glasgow Herald* was begun in 1783, though it then, and for twenty years after, bore the name of *The Advertiser*; *The Glasgow Courier* in 1791. The establishment of *The Edinburgh Advertiser* in 1764 was due to the enterprise of Alexander Donaldson, and afterwards his son, James Donaldson, who made a large fortune as printer and publisher, which he bequeathed for the endowment of Donaldson's Hospital. *The Edinburgh Gazette* was

[1] *Statistical Account*, Vol. 11, p. 75; for successive owners, *see* D. Frew, *The Parish of Urr*, p. 118. [2] N.L.S., 23.1.5.

[3] *See also* Matthias Koops, *Historical Account of the Substances used to describe Events from the Earliest Date* (1800).

[4] D. Bremner, *Industries of Scotland*, p. 323.

begun in 1699, though the official publication under this name was not started till near the close of the eighteenth century. *The Evening Courant*, which survived till 1886, followed in 1718; *The Caledonian Mercury*, incorporated in 1867 with *The Scotsman*, two years later; *The Edinburgh Weekly Journal* in 1744; and *The Aberdeen Journal*, which still survives, in 1748. Other provincial papers established within the century were *The Kelso Mail* (1797), and *The Greenock Advertiser* (1799). *The Scots Magazine* first appeared in 1739, to be followed by *The Weekly Magazine*, started by Walter Ruddiman at Edinburgh in 1768.[1]

Next to Midlothian, another well-watered area, Aberdeenshire, is the most important locality in Scotland for the production of paper. Abundant supplies of clean water and access to the rags of the north of Scotland doubtless determined the locus of the mills, the first of which dates from 1696 when Patrick Sandilands established the manufacture of paper at Gordon's Mills on the Don. The oldest with a history of continuous operation dates from 1750, when Bartholomew Smith's Culter Mills were founded on the Culter Burn, almost at its confluence with the River Dee. The other mills in Aberdeenshire are on the River Don, in one of the most interesting pockets of industrialization in Scotland. The mills at Stoneywood, founded by John Boyle and Richard Hyde, a dyer, date from 1770, and are still in operation (now Alex. Pirie and Sons). Charles Davidson founded Mugiemoss Mills in 1821.

A mill at Craiglug, on the Dee, had a short life. It was set up in 1803 by Alexander Brown (Provost of Aberdeen in 1822–3), and his father-in-law, James Chalmers, printer and publisher of the *Aberdeen Journal*.[2] It only lasted a few years, however, the buildings being bought in 1807 by William Black, quondam partner in the Gilcomston Brewery, who established the Devanha Brewery, which still survives.[3]

Another population-centre associated with paper-making was Perth, where there were three mills producing 9–10,000 reams of writing paper and 7–8,000 reams of blue, or packing, paper. The proprietors of some at least of these mills were Morrison and Lindsay.[4]

At Almondbank, in Methven parish seven miles from Perth, there were two other mills also belonging to Morrison and Lindsay.[5] About 1830 the mill at Woodend was changed over to power-loom weaving by Turnbull of Huntingtower Bleachfield.[6]

[1] J. Mackinnon, *Social and Economic History of Scotland*, p. 16.
[2] Ch. XXII. [3] *New Statistical Account*, Vol. 12, pp. 72 and 111.
[4] *Statistical Account*, Vol. 18, p. 517. [5] *Ibid.*, Vol. 10, p. 617.
[6] *New Statistical Account*, Vol. 10, p. 154.

In Markinch parish there were paper-mills at Rothes, Balbirnie and Auchmuty. Tullis, Russell and Company Limited of Markinch was founded in 1807 by Robert Tullis of Cupar, printer to St. Andrews University.

There is, to our knowledge, only one instance of a public body or corporation encouraging the manufacture of paper in Scotland during the period of industrial evolution. This was at Crieff, the birthplace of Thomas Thomson, Professor of Chemistry at Glasgow, where James Taylor erected a paper-mill in 1763. The Trustees for the Forfeited Estates managed lands at Crieff from 1752–84, devoting their attention as usual to the establishment of bleachfields, but in this instance adding tanning and paper-making. A second paper-mill was established there in 1780 for the manufacture of coarse paper. According to the *Statistical Account*, it processed 1,700 stones of rags per annum.[1]

Having considered the factors contributing to the location of the paper industry in Scotland, we must discuss its technology, to which mechanics and chemistry made equally important contributions. Till the introduction of chlorine-bleaching made possible the use of rags, coloured or otherwise, garments from which they were obtained were sorted into five categories, taken to pieces and cut into strips by hand. Colours were arranged to suit the type of paper to be made. Sorting was a skilled and important operation. The rags had next to be reduced to a fine, even suspension which would settle evenly and so yield a uniformly thick sheet of paper. With eotechnic mechanisms this was not easily accomplished by mechanical power alone, so the rags were soaked in water, and a process of fermentation which disintegrated the rags allowed to take place.

> The rags, being sorted, are put into a large stone vat, sixteen feet long, ten broad, and three deep; water is poured upon them to the top during ten days, and eight or ten times every day without stirring them. They are afterwards left to rest for the same number of days, and sometimes more or less, without pouring water upon them. Then, being turned over, the centre is brought to the surface, to facilitate the fermentation: and after being turned again, they are still left fifteen or twenty days in fermentation, so that the rotting may last five or six weeks; the term is not fixed, but when the heat becomes so great, that the hand, thrust in, cannot endure it above some seconds, it is judged that it is time to stop it.

In smaller mills the process was somewhat different.

[1] *Statistical Account*, Vol. 9, p. 592; A. Porteous, *History of Crieff*, p. 177.

> In mills, which have but few rags to work upon, they are left to rot longer, because the heaps being smaller, heat less, and with more difficulty; so that nothing can be justly determined of the time proper for the rotting part. It also depends on the quality of the rags; the finest linen does not rot so soon as the coarse; and linen that has been worn with more difficulty than new, because the internal humidity, that disposes the fibres to fermentation, is more considerable in new or coarse, than in fine or worn linen. Such as are more or less strong, more or less worn, resist the action of rotting in different degrees, some being spoiled, when others have not gone through the first fermentation: so that the rags, which have been sorted with great care, ought to be left to rot together, to avoid the risk of altering the whole parcel, by the mixture of a portion of rags quite different from the rest. When champignons grow on the heaps of rags, it is reckoned to be a sign of their being well rotted.[1]

The fermented rags were next passed through a series of *stamp-mills*, and gradually reduced to pulp. A constant stream of clean water flowed through the mills removing dust, etc., loss of the fibres being prevented by a horsehair sieve. When the required fineness had been attained the triturated rags, consisting of 10 per cent. fibre and 90 per cent. water, were passed into a vat where the dilute dispersion was maintained uniform by constant stirring. Into the vat the paper-maker dipped a shallow rectangular box with a wire gauze bottom, called a deckle. By removing it from the vat full of suspension and keeping it horizontal, a well-felted deposit of fibres could be made to settle on the gauze. The sheet of pulp so formed was turned out on to a woollen felt, a process called *couching*, and when a pile of alternating sheets of pulp and felt, called a *post*, had been built up, it was transferred to the vat-press where most of the water was removed. Pressing the sheets in a screw-press without the felts followed, and they were then dried in the air on hair ropes, to prevent marking of the paper.

When writing-paper was being made it had to be sized to prevent the ink flowing when it was written on. To do this the sheet was passed through a solution of vegetable or animal gelatine and dried again. A. Rees says alum and oil were used.[1]

The whole of a paper-maker's impedimenta are aptly illustrated by the following advertisement of 1807.

> The whole machinery and utensils of that extensive Paper Manufactory lately erected at Ferry-hill in the vicinity of Aberdeen consisting of four paper engines, complete; six vats, six stuff chests,

[1] A. Rees, *Cyclopaedia*, Article, *Paper*.

three vat presses, eight dry presses, one size press, Treble's treble posts and hair line, etc., a complete steam apparatus with boiler, etc., a blue mill with shafts, spur wheel, etc., wire cloth duster, complete; rag tables, boxes and knives, bleaching apparatus, water cisterns, and moulds of all sorts necessary for carrying on the business of a six-vat mill.[1]

Cellulose pulp was used not only for making paper but also from an early date for making articles of *papier mâché*. These were introduced into the Birmingham district about 1750 by Baskerville, and were first described in the Rev. T. H. Croker's *Complete Dictionary of Arts and Sciences* (1765). R. B. Prosser has a ten-page essay on the articles which were made from paper pulp.[2] Macintosh's Alum Works was roofed with tarred paper.

The basis of this new species of roof is common sheathing paper, so called from its being employed in sheathing ships. It is first dipped in tar, and heated to the boiling point, that it may penetrate the paper more readily. After being exposed to dry for two days, the tar is found to be completely imbibed. The sheets are, a second time, dipped in tar at a lower temperature, and then nailed on the roof in the same manner as slates, over-lapping one another, so as to be triple at the joinings, and double in every other part. Above the whole is laid a coat of tar boiled to the consistency of pitch, on which smithy ashes are passed through a sieve, to diminish the combustibility, and to prevent the liquefaction of the tar.

This kind of roof was introduced into Stirlingshire in 1807. In the parish of Campsie, a large pile of buildings, in which an alum manufacture is carried on, together with a village containing 50 families, is roofed entirely in this manner. Mr. Speirs of Culcruich, an enterprising agriculturist of this county, has lately set the first example of covering a house purely rural with paper.

With regard to the durability of this species of roof, it may suffice to observe, that it was first introduced into Scotland for covering a public store-house in Greenock, twenty years ago; during that period, it has received no repairs; and, at the present day, continues in perfect preservation.[3]

The hand-processes of making single sheets of paper which have been described were naturally slow, and paper-making, like other crafts, felt increasingly the impact of mechanical technology during the course of the eighteenth century.

The earliest improved contrivances introduced into paper-

[1] Quoted A. A. Cormack, in *The Paper Market*.

[2] R. B. Prosser, *Birmingham Inventors and Inventions*, p. 38.

[3] P. Graham, *General View: Stirling*, p. 81.

making were alternatives to stamp-mills. An oval trough with a set of revolving steel knives working against a stationary bedplate replaced the cruder stamp-mills. From its country of origin, this machine or beating-engine became known as a *Hollander*. So vanished the slow fermenting process and pounding in the stamp-mill, and as a result the paper-makers were enabled to produce greatly increased quantities of pulp to meet the demand for paper. Hollanders are first mentioned in 1718.

The next attempt at mechanization was a remarkable mechanical man invented by one of the Montgolfier brothers (J. M.) of aeronautical fame.[1] His idea was to mechanize the process of dipping the deckle into the vat, an arduous operation, let it be admitted, requiring nine men for large sheets of paper.[2] For more economical working, and in keeping with the general evolution of mechanical technology, a continuous process was needed, and in 1816 he bought a patent registered by Cameron of Edinburgh.[3]

By 1792 Nicholas Louis Robert, clerk at the Essonne Paper Works in France, had constructed a machine which would produce a continuous narrow strip of paper without the mechanical assistance of vatman and coucheur. The principle of Robert's machine was the same as that which to-day is employed in great paper machines. Pulp was fed to an endless band of wire gauze, and the resulting ribbon of dehydrated pulp was partly dried and pressed between rollers.

Just as Nicolas Leblanc found revolutionary France an unpropitious environment for his embryonic invention, so did Robert. Didot, the owner of the paper-mill where Robert was employed, came to England in 1800 and asked his brother-in-law, John Gamble, to find capital to develop the machine. As a result Didot was put in touch with two wealthy London stationers, Henry Fourdrinier (1766–1854) and Sealey Fourdrinier (*d.* 1847), and a patent was procured. The first machine in England was built by Bryan Donkin (1768–1855) at Bermondsey, and erected at Two Waters Mill, Herts. Both Gamble and Donkin became interested in the tinning of food, and will be referred to again in Chapter XXIV. This later development may have been the cause of Gamble giving up his interest in the paper machine. Be that as it may, he sold his share in the patent to the Fourdriniers, whose first machine was put into operation in Kent in 1803. The patent which they had acquired expired in 1807, and, not being the original inventors, they were unable to have it extended. As a

[1] Ch. VII. [2] R. Herring, *Paper and Paper-Making*, p. 43.
[3] J. Kay, *Paper, its History*, p. 40.

result, after spending £60,000, they went bankrupt, and despite a Parliamentary reward, died almost penniless. In so doing they were not by any means alone. Donkin, however, went on building machines till the middle of the century, and was the only one to profit by Robert's genius.

Changes wrought in paper-making as a result of mechanization have few parallels in the rest of mechanical technology. The paper-making machine reduced the time of making a given amount of paper from three weeks to three minutes, and even before such mechanization:

> . . . it was observed by a French writer, that the dispatch of the processes of paper-making is so great, that five workmen in a mill may furnish sufficient paper for the continual labour of 3,000 transcribers.[1]

The migration of Didot to England was not the only connexion between paper and the revolutionary state of Europe at the end of the eighteenth century. At Haughton Mill, near Hexham, Northumberland, paper was made in large quantities at the instigation of the British Government for forging French *assignats*. These the British Government circulated on the Continent in an attempt to embarrass the financial operations of the Revolutionaries.[2]

If mechanical technology contributed conspicuously to increased production of paper during the period of rapid industrial evolution, it was only following what chemical technology, whose major contribution was the introduction of chlorine-bleached rags, had already accomplished. In paper technology it enabled the paper-maker to use printed linen for the production of the white writing-paper which formerly was made solely from unprinted linen. The history of the use of chlorine in textile-bleaching was recounted in Chapter IX.

In May 1792 a patent[3] was granted to Clement and George Taylor of Maidstone, Kent, for the exclusive privilege of bleaching rags with chlorine in the form of hypochlorite solution for paper-making. This occasioned annoyance among Midlothian paper-makers who, in a Court of Session case,[4] adverted that Chaptal had shown that the reagent could be used for bleaching both pulp and rags, and that Dr. Joseph Black had been demonstrating the

[1] A. Rees, *Cyclopaedia*, Article, *Paper*.

[2] W. G. Armstrong, *The Industrial Resources of the Three Northern Rivers, Tyne, Wear and Tees*, p. 222.

[3] Register of the Great Seal, Edinburgh, Vol. 20, Nos. 442 and 456.

[4] *Session Papers*, S.L., 207, 3 Feb., 1796.

bleaching action of chlorine in his lectures in the 1789–90 session. As a result of this the Edinburgh paper-makers were familiar with the reagent, and to spread knowledge of it further, Mr. Kerr's translation of Berthollet's work on bleaching was distributed by the Board of Trustees. William Simpson of Polton Mill near Lasswade had it in use by 1791.[1]

This was the first mill in Scotland to use chlorine in the bleaching of rags for paper-making. Since 1745, when the first mill was founded there, Lasswade had been an important Scottish paper-producing district. By 1795 there were five mills employing 260 operatives and paying out £3,000 per annum in wages. In the Court of Session case over chlorine bleaching, several Lasswade and Penicuik district paper-makers were involved; William Simpson of Polton Mill, a mill now owned by Annandale and Son, Limited; and William Cadell and Company, Auchindinny, on the Glencorse Burn, who also had paper-mills at Roslyn and Cramond.[2] Another was John Pitcairn of Melville, possibly also in the Lasswade area.[3]

In 1792 Hector Campbell got a patent for the use of the gaseous element covering England only.

> To bleach the rags by the oxymuriatic acid gas, they are first washed in hot water, by a fulling mill, such as is used for scouring cloth: this removes the dirt, and they are put into a receiver or chamber, made of wood, in a cubical form, and the joints air-tight. It is provided with several stone retorts which being filled with a mixture of manganese, with two-thirds its quantity of sea-salt, and a quantity of sulphuric acid equal to the salt, will, when moderately heated by a small sand-bath furnace, throw into the receiver a gas which quickly discharges any colour the rags may contain. This process, which is very similar to that used for bleaching cotton thread, was invented by Mr. Campbell, who had a patent for it in 1792.[4]

There is a further connexion between paper and chlorine. The high reactivity of chlorine set early bleachers a difficult problem in designing apparatus to use it. Ainsworth of Halliwell Bleach-works, who employed two French chemists (Valette and Tenant), was persuaded by them to try paper pipes instead of metal. The innovation, though ingenious, was not successful. Use to excess often led to damage to the paper.[5]

[1] *Statistical Account*, Vol. 10, p. 279.
[2] *New Statistical Account*, Vol. 1, pp. 354 and 600.
[3] *Statistical Account*, Vol. 10, p. 279; *Session Papers, loc. cit.; New Statistical Account*, Vol. 1, p. 334.
[4] A. Rees, *Cyclopaedia*, Article, *Paper*.
[5] S. Higgins, *History of Bleaching*, p. 78.

Another chemical contribution to paper technology was a method of cleaning cotton waste so that it could be used for paper manufacture. As a result the Manchester area became one of the principal seats of the industry.[1]

When paper manufacture was well established in Great Britain, the Government saw fit to make it a subject of taxation. Although the excise regulations were perhaps less fantastic than those appertaining to the manufacture of glass, they were nevertheless very irksome, with the inevitable surveillance of operations;[2] labels for packages only issuable by the excise, red for England, blue for Scotland, green for Ireland, etc.[3]

All paper was divided into two categories: (*a*) first-class paper, which included printing, writing, coloured, and wrapping papers, and even card and pasteboard. Upon this the duty was 3*d.* per lb. First-class paper included in fact all papers, unless they were made wholly of old rope which had not been treated to remove tar and pitch. The product of such materials was called (*b*) second-class paper. It was brown, and often smelt of the tar which clung to the fibres of the ropes. On second-class paper the duty was 1½*d.* per lb.[4] Out of its social context, the curious practice of inhabitants of the old town of Edinburgh burning brown paper to neutralize the odour of domestic refuse when it was being collected, is not easily understood. As the distillation of phenolic matter from tarry rope, it takes on a new and interesting significance.

A flat rate like this meant that the tax varied from 25–200 per cent. *ad valorem*, depending on the quality of the paper taxed. There was a drawback of duties on paper exported; and also for books printed at Oxford or Cambridge, in the Latin, Greek, Oriental or Northern languages, and also for bibles, testaments, psalm-books or books of common prayer, printed in either of those Universities.[5]

Expansion in the industry can be traced in the total revenue for the following years:[6]

1784	£46,868
1815	£315,802
1830	£619,824

According to Sir John Sinclair, the Scottish contribution for the year ending January 1813 was £42,787 16*s.* 5¼*d.*[7] Some individual

[1] J. R. McCulloch, *Commercial Dictionary*, Vol. 2, p. 876.
[2] 34 Geo. 3, c. 20. [3] 26 Geo. 3, c. 77. [4] 42 Geo. 3, c. 94.
[5] 43 Geo. 3, c. 69, Schedule C; A. Rees, *Cyclopaedia*, Article, *Paper*.
[6] Bremner, *Industries of Scotland*, p. 328. [7] *General Report*, App. 2, p. 285.

districts made large contributions, e.g. Aberdeen (1810), £6,178 12*s.* 1½*d.*; Berwick, £4,000; New Kilpatrick, £2,000; Lasswade, £1,800. The Penicuik Mills were reckoned to contribute more to the Exchequer than the whole Irish industry.[1]

After the Napoleonic Wars, when iron chains in place of hempen ropes were extensively introduced in shipping, the quantity of old cordage available to the paper-makers decreased and the price rose. At the same period increased importation of sugar and cotton made large quantities of jute bags available and, since jute could not be bleached for use in white paper, these bags would have been available to the industry for the manufacture of second-class paper, had the law allowed it. However, the lower rate of duty only applied to paper made solely from old cordage, with the result that the absurdity of the excise put a premium on both white and brown paper. About 1830, as paper technology advanced, and a considerable degree of working outwith the law was engaged in, the various categories of paper began to be less clearly defined. As a result the duties were equalized in 1836, the duty being 1½*d.* on all qualities.[2]

Yet there was still a demand for raw materials in excess of supply, and all sorts of alternative substances were tried out, e.g. straw, peat, hop vines, tarpaulins, oilcloth, door mats, etc. Esparto grass and wood pulp were only introduced after the middle of the nineteenth century. The paper tax was then (1857) yielding some £1,244,652 per annum to the Exchequer, of which £263,786, a high proportion, came from Scotland. Arguments began to be marshalled against a tax which was felt to be a serious imposition.

> This places a great obstacle in the way of the progress of knowledge, of useful and necessary arts, and of sober, industrious habits. Books carry the productions of the human mind over the whole world, and may be truly called the raw materials of every kind of science and art, and all social improvements.[3]

Literacy was spreading; England began to increase her quota of universities: compulsory education was on its way.

[1] *New Statistical Account*, Vol. 1, p. 44.

[2] W. G. Armstrong, *op. cit.*, p. 223; J. H. Clapham, *Economic History of Modern Britain*, Vol. 1, p. 325.

[3] H. Parnell, *Financial Reform*, p. 30, quoted J. R. McCulloch, *Commercial Dictionary*, Vol. 2, p. 877.

Chapter XIV

THE ART OF GLASS

'Glasses were scarce, if bottles and casks were numerous, and in many households the ale or wine was drunk from the same glass, which went the round of the table.'

CONTINUING to explore fields of economic activity rendered arid by the rigours of the timber famine and refertilized by the solution of the salt-to-soda problem, we come to what Samuel Parkes calls the most truly surprising of all human arts, considering the materials from which it is formed, the Art of Glass. Lewis Mumford gives pride of place to glass as a major development in advancing eotechnic economy, and goes on to say:

> Through glass new worlds were conceived and brought within reach and unveiled. Far more significant for civilization and culture than progress in the metallurgical arts up to the eighteenth century was the great advance in glass-making.[1]

Glass and learning radiated in phase from the Mediterranean into less hospitable regions of Europe. Glass windows created an atmosphere fit for contemplative thinking. It was from Italy, too, that the glass tubes used by Pascal in 1646 to demonstrate the weight of the atmosphere came, and it was Italian facility in glass-manipulation that led to Galileo's thermoscope, precursor of thermometer and barometer alike. The perfection of lens-making made possible the exploration of the very small and the very distant.

> Through glass some of the mysteries of nature themselves became transparent. Is it any wonder then that perhaps the most comprehensive philosopher of the seventeenth century, at home alike in ethics and politics and science and religion, was Benedict Spinoza: not merely a Hollander, but a polisher of lenses.[2]

The new worlds revealed by glass have concentrated attention on glass as the basis of microscopic and astronomical exploration, but the debt which chemistry owes to glass cannot be overemphasized.

Professor J. L. Myres, the classic archaeologist, has even suggested that the backwardness of the Greeks in chemistry was

[1] L. Mumford, *Technics and Civilization*, p. 124. [2] *Ibid.*, p. 131.

due to the lack of good glass. For glass has unique properties: not merely can it be made transparent, but it is, for most elements and chemical compounds, resistant to chemical change: it has the great advantage of remaining neutral to the experiment itself, while it permits the observer to see what is going on in the vessel. Easy to clean, easy to seal, easy to transform in shape, strong enough so that fairly thin globes can withstand the pressure of the atmosphere when exhausted, glass has a combination of properties that no wood or metal or clay container can rival. In addition it can be subjected to relatively high heats and—what became important during the nineteenth century—it is an insulator. The retort, the distilling flask, the test-tube, the barometer, the thermometer, the lenses and the slide of the microscope—all these are products of glass technics, and where would the sciences be without them? A methodical analysis of temperature and pressure and the physical constitution of matter all awaited the development of glass: the accomplishments of Boyle, Torricelli, Pascal, Galileo, were specifically eotechnic works. Even in medicine glass has its triumph: the first instrument of precision to be used in diagnosis was the modification of Galileo's thermometer that Sanctorius introduced.[1]

In order to understand the fundamental importance of glass to western civilization in general, and the development of the industry in Great Britain in particular, one must consider several sets of conditions which operated upon the developments. Briefly, they are as follows:

(*a*) glass-making as an eotechnic operation, and the influence. of the timber shortage on locus and methods of production;
(*b*) modification necessitated by transition to pit-coal;
(*c*) successful organization of raw material supplies and their general relation to eotechnic shortcomings;
(*d*) transition to palaeotechnic economy resulting from availability of sodium carbonate (Na_2CO_3), and later sodium sulphate (Na_2SO_4);
(*e*) superimposed upon these is the influence of an ill-conceived and crippling fiscal policy.

Till the sixteenth century, while the great cathedrals and monastic houses were filled with glass, ordinary houses knew neither glass window, mirror, nor table vessel. During that century apothecaries' bottles, test-tubes, and proletariatized lenses, in the form of spectacles, became common. Windows and mirrors were introduced. The changes brought about by glass were subjective as well as objective.

[1] L. Mumford, *Technics and Civilization*, p. 127.

> If the outward world was changed by glass, the inner world was likewise modified. Glass had a profound effect upon the development of the personality: indeed, it helped to alter the very concept of the self.[1]

Further, speaking of the glass receptacle:

> The standardised glass bottle, so useful for medicines and wines, was a late eotechnic achievement. Before the finer forms of glass, goblets, alembics, mirrors, and distilling flasks had been created. Without the use of glass for mirrors, microscopes, telescopes, windows, and containers, our modern world, as revealed by physics and chemistry, could scarcely have been conceived.

In Scotland, bottle-making had an important influence on the development of glass technology. Leith was an important entrepôt for continental wine. Bottles were required for its distribution. Rum was made in Glasgow from the by-products of sugar boiling. It was smuggled away in Glasgow bottles.

Early glass houses depended on timber for fuel, and in consequence we find them centring in districts where timber was still available, e.g. the Weald. According to E. W. Hulme, the only authenticated seat of the glass industry in England before 1585 was at Chiddingford in Surrey. In Elizabeth's reign it spread rapidly.[2]

In the Forest of Dean, glass houses are said to have moved about in search of adequate fuel supplies. Thus the glass industry suffered, in the same way as other timber consumers, from the timber famine. The date at which coal was substituted for timber is not known accurately, but it probably occurred shortly after James VI was called to the throne of England. James, as an expedient to raise money without the aid of Parliament, granted to Sir Robert Mansell (1573–1653) an exclusive patent for making glass, and allowed it to be given out that this was afforded him in consideration of, and as a reward for, his having substituted pit-coal for timber in its manufacture.[3] James possibly had an interest in the coal-mines of Scotland. He was certainly interested in the substitution of coal as a fuel, and is reckoned to have spent some £5,000 on preliminary experiments.[4]

Scottish coal, exported from Prestonpans, was the chief fuel used in London glass houses, except for a brief period during the reign

[1] L. Mumford, *Technics and Civilization*, p. 128.

[2] Antiquary, *30*, 210, 259; *31*, 68, 102, 134. *See also* E. Straker, *Wealden Glass: The Surrey-Sussex Glass Industry* (1226–1615).

[3] Samuel Parkes, *Chemical Essays*, Vol. 3, p. 407.

[4] J. U. Nef, *Rise of the British Coal Industry*, Vol. 1, p. 218.

of James. In fact, for a considerable time the Scots held the key to the profitable prosecution of glass-making. When in 1619 the English Parliament held that Scottish glass came under the heading *foreign*, in terms of an Act of 1615 forbidding the importation of glass into England, the Scottish Privy Council threatened to prohibit the export of Scottish coal. This had the immediate effect of altering the ruling of the English Parliament.[1]

As glass houses followed timber supplies, so, while coal-working remained primitive and cost of production increased sharply with depth of workings, Mansell moved his glass houses from one coal outcrop to another to maintain a ready supply of cheap fuel. Thus the link between glass and coal developed; manufacture was begun in Scotland at Leith, Wemyss, and possibly also in Ayr, and in a similar manner it came to be centred on English coal-fields, e.g. at Newcastle, Stourbridge, etc. It is supposed that when the Scots withheld their coal, Lady Mansell, who managed her husband's concerns in his absence, experimented with Newcastle coal and found it suitable for glass-making. Ultimately, Newcastle itself became an important glass-producing centre, based on the available supplies of coal.[2] The changes necessary for the transition to the next phase in the evolution of glass-making were thus effected in the era of Mansell's patent. Of these changes, J. Arnold Fleming says:

> Although the same ingredients have been used from time immemorial, their purity and thoroughness of mixing and blending have been vastly improved. When wood was forbidden as fuel and coal introduced, there was no material alteration in the batch, but fundamental changes in the fusing process took place. The wood was burned on top of pebbles without any grating or firebars; the flame passed around the pots, which were about two feet high. . . Coal firing did in a few hours what wood took days to accomplish, but the sulphurous fumes and sooty particles caused a chemical reaction to take place inside the pots through the reactive reagency of the heat while the metal was in an iridescent frothy condition of fusion. The glass became more darkly tinged in consequence. To overcome this defect, covered or hooded pots became necessary.[3]

The covering-in of the glass pots is one of the most important changes which took place in the history of glass-making, because it brought other changes in its train. To compensate for lower temperature in the pot, which resulted from its being closed, an

[1] J. U. Nef, *Rise of the British Coal Industry*, p. 182.
[2] H. J. Powell, *Glass-making in England.*
[3] J. A. Fleming, *Scottish and Jacobite Glass*, p. 42.

55. Bartholomew Smith buying rags on the Castlegate, Aberdeen, 1751–8.

Picture Post Library

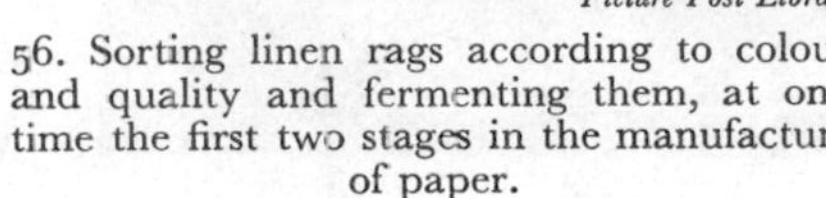

56. Sorting linen rags according to colour and quality and fermenting them, at one time the first two stages in the manufacture of paper.

Diderot and d'Alembert's *Encyclopédie*

57. The paper-maker at his vat (*left*); pressing the water from the made sheets (*right*).

Diderot and d'Alembert's *Encyclopédie*

58. Sizing and pressing paper.

increased quantity of lead flux was added. By the middle of the eighteenth century, the gradual replacement of lime by lead oxide brought about the evolution of a distinct type of glass, usually called *flint glass*, from the raw material used to make it as clear and sparkling as possible. Covered pots continued in use till the commercialization of coal distillation made clean gaseous fuel available. This reversion is, however, recent. At the time the substitution of coal for timber was regarded as a mixed blessing:

> The English glass-makers labour under an inconvenience which is not felt in France and some countries on the Continent of Europe. In consequence of their heating their furnaces with pit-coal, they are under the necessity of making their best flint glass in *covered* pots; and as the heat is much lower in the interior of a covered crucible than in an open one, we are obliged to make our glass of more fusible materials; and consequently by using a large portion of lead and less silex, we produce a softer glass than is prepared by those who have an abundance of wood fuel.[1]

Technical details of glass-making are too complicated to consider here, but it is important to study specifically the economics of the raw materials of which glass is made. Glass results from the fusion of sand or flints (both of which are silica or silicon dioxide (SiO_2)) with soda (Na_2CO_3), potash (K_2CO_3), or lime ($Ca(OH)_2$), during which operation chemical reaction takes place with the formation of silicates.

$$\text{Sand } (SiO_2) + \left\{\begin{array}{ll} \text{Soda} & (Na_2CO_3) \\ \text{Potash} & (K_2CO_3) \\ \text{Lime} & (Ca(OH)_2) \\ \text{or Chalk} & (CaCO_3) \end{array}\right. \rightarrow \left\{\begin{array}{ll} \text{Sodium Silicate} & (Na_4SiO_4) \\ \text{Potassium Silicate} & (K_4SiO_4) \\ \text{Calcium Silicate} & (Ca_2SiO_4) \end{array}\right.$$

Glass, then, is a complex mixture of silicates. When some lime is replaced by litharge (Pb_2O_3) or red lead (Pb_3O_4), highly refracting *flint glass* or *crystal* is produced. Glass-makers in the Forth area used Tranent sand, in England sand from Alum Bay, Isle of Wight, Maidstone, or King's Lynn. From Parkes we learn that:

> A century ago the best glass was always made from ground flints: hence it was that white glass acquired the name *flint* glass; but when it was discovered that good sand is nearly all silicious earth, this was substituted for flints.[2]

It is revealed by the above equation that the alkali used is capable of variation, but, while common names like soda, potash, and lime conceal the common factor in these compounds, formulæ

[1] S. Parkes, *op. cit.*, Vol. 3, p. 502 *n.*

[2] *Ibid.*, Vol. 3, p. 142.

show that three of them are carbonates, viz. sodium carbonate, potassium carbonate, or calcium carbonate. The latter systematic names were the outcome of the reforming spirit which was abroad at the time of the French Revolution.[1]

It was from revolutionary France, too, that the first comprehensive book on glass-making to be published since the eotechnic era came. According to S. Parkes:

> The best book on glass-making, which I have seen, is an octavo volume, by H. Blancourt, with plates, 1699. But a scientific work, published in Paris by Loysel in 1800, entitled *Essai sur l'Art de la Verrierie*, is, I believe, the best modern work on this subject. It gives a detailed account of the different processes in this manufacture, and treats of each chemically.[2]

While timber was still available, wood ash or potash was used as the alkali, but this could be replaced by ash of bracken, bramble, or briar. Fern ash was used in the production of *verre fougère* in France, and for the *waldglas* of the Flemings. As the tint of glass results from traces of impurity in the raw materials, the colour of the glass produced from them was beyond the control of the glass-maker. Much of the charm of ancient glass is fortuitous.[3] Parallel to the replacement of timber by pit-coal for firing, other forms of alkali were introduced as raw materials in the batch. These included barilla and, of particular importance to Scotland, kelp. The latter development took place during the second half of the eighteenth century.

In the eotechnic era the changes just described brought considerable prosperity to British glass-makers, and paved the way for the transition to the next phase. The beginning of the eighteenth century saw some ninety glass houses in operation in different parts of England, twenty-four in London and Southwark, seventeen in Stourbridge, eleven in Newcastle-on-Tyne, and nine in Bristol.[4] Summarizing the changes which took place, they were:

(*a*) the substitution of coal for wood fuel;
(*b*) introduction of covered pots to keep contamination from the glass;

[1] Expressions of need for reform were voiced about 1782, but the ground was hardly ready for them to take root. A few years later Lavoisier, Berthollet, and Fourcroy presented to the world of science their *Méthode de Nomenclature Chemique* (Paris, 1787, translated into English by J. St. John, 1788). This system forms the basis of the classification in use today. It was conceived so that a name reveals the chemical nature of a substance, rather than some accident of its generation.

[2] S. Parkes, *Chemical Catechism*, p. 106, *n.*

[3] H. J. Powell, *op. cit.*, p. 41.

[4] S. Parkes, *Chemical Essays*, Vol. 3, p. 408.

(*c*) increased use of lead oxide to compensate for the resulting lower temperature. The property which lead has of increasing fusibility was known from early times, but the discovery is often attributed to Ravenscroft. Its use was introduced at London about 1730, and soon after at Stourbridge, Bristol, and Newcastle.

There was still one difficulty with regard to raw material supplies—inadequate supplies of alkali. Pressure from glass-makers stimulated the development of the kelp industry in Scotland, as has been described.[1] The shortage of soda remained a severe restraint upon the development of the glass industry, especially while a heavy fiscal duty on salt still remained operative. Although synthetic soda did not come on to the market in large quantities until about 1825, for several years before this, glass-makers were making their own soda from salt, and had succeeded in getting salt used for this purpose duty-free.[2]

Mention has already been made of the methods whereby soap-makers augmented their supplies of soda before the commercialization of Leblanc's method by Muspratt. According to Parkes, the glass-makers obtained soda by mixing a solution of common salt with potashes:

$$2NaCl + K_2CO_3 \rightarrow 2KCl + Na_2CO_3$$

What is called *double decomposition* takes place, and soda or sodium carbonate is formed, along with potassium chloride. The latter is less soluble and crystallizes out, and the soda is obtained by evaporating the remaining liquid. A special Act of Parliament was required to permit glass-makers to dispose of the potassium chloride to alum-makers on paying an impost of 20*s*. per ton.[3] Messrs. Chance Brothers and Company of Smethwick was one firm which made alkali as well as glass, using first of all as their raw material, bought sodium sulphate. Since the yields of alkali which they were able to produce from the bought sodium sulphate were very low, they took advice from the alkali manufacturers, Clay and Muspratt, and Adkins and Company, and as a result resolved to build their own sulphuric acid and alkali plant. At the end of 1834, William Neale Clay was appointed manager of this section of the enterprise. One of Clay's ideas was to recover the lime and sulphur from the waste, a problem which was not solved for half a century.[4] Such small-scale processes came to an end when synthetic soda became a saleable commodity. When Gamble and

[1] Ch. III. [2] *See* 38 Geo. III, cap. 89, §116.
[3] S. Parkes, *op. cit.*, Vol. 3, p. 462.
[4] J. F. Chance, *A History of the Firm of Chance Brothers and Company*, p. 18.

GLASS

(Silicates of Sodium, Potassium, Calcium, etc.)

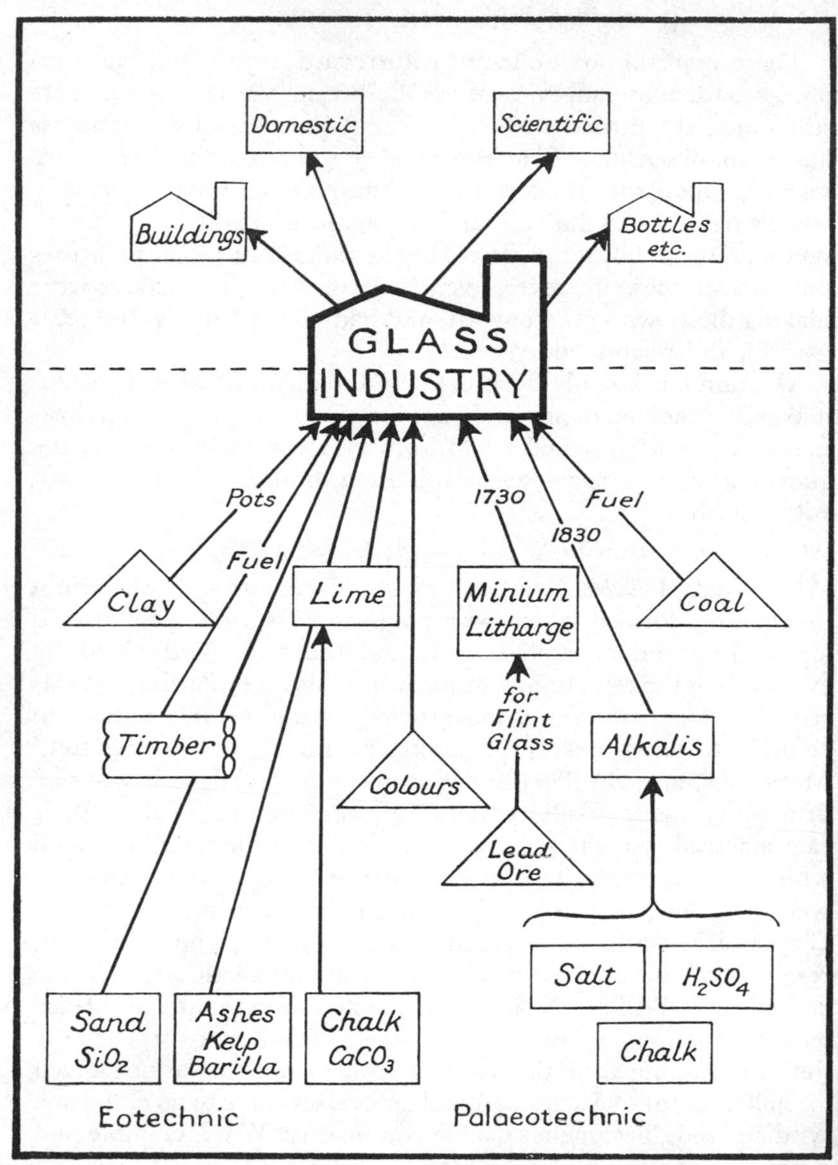

Muspratt brought their alkali works to St. Helens, the Plate, Crown, Flint, and Bottle Glass works there were all in operation, and it was not long before the Crown works began to use sodium sulphate supplied by the alkali makers.[1] In Scotland the names connected with this revolution are Charles Tennant and J. Stevenson, who adopted Leblanc's process.

The commercialization of sulphuric acid placed still another raw material at the disposal of the glass-makers, viz. sodium sulphate, or Glauber's Salt ($Na_2SO_4.10H_2O$). Experiments on this material, which could be produced sufficiently cheaply to be used in large quantities, were being carried out in the early years of the nineteenth century. Dr. Gehlen is associated with these. In 1810 he published his *Beytrage zur wissenschaftlichen Begründung der Glasmakerkunst.*[2] Pajot des Charmes also carried out experiments and found that while sodium sulphate alone would not do, with the addition of an equal quantity of lime a pale yellow glass resulted. De Serres later recommended the addition of a small quantity of carbon.[3] After these experiments were made in the first two decades of the nineteenth century, sodium sulphate became an important raw material to the glass industry, and it soon gained in favour over the carbonate. Notice of sodium sulphate (Na_2SO_4) being tried by Messrs. Chance of Smethwick is contained in a Board Minute of 20th November, 1832. Chances discarded carbonate in the Crown houses in 1841, and limited its use for sheet glass to 6 per cent., subsequently raised to twenty-five.[4]

The successful substitution of coal tended to concentrate glass houses on coal-fields, especially those at which supplies of other raw materials were available. Bristol, where the great Nailsea works were situated, is at first sight an exception. Here the demand was for bottles to hold Bristol and Bath waters, in addition to beer, cider, and perry. The industry there did decline, however, due to taxation, competition, and 'difficulties in obtaining adequate supplies of coal'. In Scotland the manufacture of glass centred in the Forth and Clyde areas; in England at Newcastle, and, after synthetic soda came on the market, on the verge of the Cheshire salt-fields. These changes are illustrated by the following north-of-England developments.

In 1772 there were sixteen important glass works in Newcastle, five making bottles, five broad or common, three crown, two flint,

[1] J. Fenwick Allen, *Some Founders of the Chemical Industry*, p. 53.
[2] *Attempt to Establish the Art of Glass-making on Scientific Basis.*
[3] Nicholson's *Philosophical Journal*, 2nd Series, 1812, *31*, 357.
[4] J. F. Chance, *op. cit.*, pp. 9 and 34.

and one plate, glass. By 1833 the number of glass houses had increased to thirty-eight, of which ten were owned by the Cookson dynasty of glass-makers. The following table outlines the development of their connexions:

1730. The name appears for the first time.
1746. Cookson, Jeffreys and Dixon.
1776. John Cookson and Dixon.
1795. John Cookson.
Cookson and Cuthberts, with two factories at South Shields.
1811. Isaac Cookson of The Close.
Cookson and Company of The Quay.

The family also had connections with the Glasgow glass-makers, and business relations with Charles Tennant and Company. Some idea of the magnitude of Isaac Cookson's transactions may be gained from the fact that he is reputed to have paid £60,000 per annum in excise dues. The contributions of some other firms are interesting—Ridley and Company of Newcastle with three houses, £40,000; Courthorpe and Company of Bristol with two houses, £39,000; Charles Atwood of Gateshead, also with two houses, £38,000. Actually, the second largest as a producer of Revenue was William Chance of Stourbridge, who paid £54,000 on a production of what is reckoned to have been seven hundred tons of glass.[1]

The material used by Newcastle bottle-houses was an extensive fluvial deposit at Jarrow Slake. This contained 'siliceous, calcareous, and argillaceous earths in excessive comminution, united with carbonaceous and saline matter'.[2] The strong position of Newcastle revealed by the above data was not maintained in the nineteenth century in consequence of the powerful attractions of Lancashire.

Before going on to consider how glass manufacture ministered to the social demands of the palaeotechnic era, it is necessary to look at the conditions in the industry arising from governmental control, because fiscal, not chemical, classification and regulation ordered the working of glass houses. In the first place, we must mention the state of stagnation into which the industry had sunk by the beginning of the nineteenth century, as instanced in the following quotation:

> The history of the glass trade in England illustrates among other things the tremendous effects upon home consumption

[1] J. H. Clapham, *Economic History of Modern Britain*, Vol. 1, p. 189; for a full list *see* H. J. Powell, *Glass Making in England*, p. 95.

[2] W. G. Armstrong, *The Industrial Resources of the Tyne, Wear and Tees*, p. 201.

which may be produced by taxing and so raising the price of any article not of absolute necessity. In 1801, with a population of sixteen millions, the quantity of glass used was 325,529 cwt., and in 1833, with a population of twenty-five millions, the quantity was no more than 363,468 cwt., an increase of less than one-eighth, while the population had increased in the proportion of one-half.[1]

Comparison may also be made between increase in glass production and other basic constructional materials. Between 1791 and 1821, while the production of bricks increased by 90 per cent., glass increased by only 2 per cent. Thus back-to-back houses are perhaps less a reflection of capitalist exploitation than of material necessity. How was it that these deplorable conditions arose?

(1) In addition to a window tax of 15*s.* 6*d.* on eight, and £3 17*s.* 6*d.* on sixteen windows, three payments had to be made by the glass houses: (*a*) an annual payment on each glass house for a licence to manufacture glass; (*b*) a payment per pound on all glass melted in the glass pots and ready for use; (*c*) a payment per pound on the excess in weight of manufactured glass over 40 per cent. (later 50 per cent.) of the calculated weight of molten glass. There were certain allowances for spoiled glass and glass broken during the annealing.

(2) The size of the glass melting pots was controlled by law. The inside measurements of each pot had to be registered. Parkes remarks that as the pots decrease in size on being heated and the excise officer gauges it on its original size, there is about ten per cent. extra duty to be paid every time.

(3) Annealing ovens had to be of peculiar shape, and only one opening was permitted per furnace. They were fitted with heavy doors and locks under the control of the excise officers.

(4) At least two excise men were quartered in the glass house throughout the twenty-four hours.

(5) The industry was partitioned into five fiscal, not scientific, categories, viz., crown, bottle, broad, plate, flint glass. Only one kind of glass was permitted to be made in any one glass house. Moreover, a common bottle-glass factory might not produce phials less in content than six ounces. Crown sheet had to be less than one-ninth inch, plate between one-eighth and five-eighths of an inch.[2]

[1] G. R. Porter, *The Progress of the Nation*, 1912 Edition, p. 283.

[2] The regulations are set out at length in J. R. McCulloch's *Commercial Dictionary*, Vol. 1, p. 604.

These restrictions produced certain obvious results. In general, the effect of excise regulations was to repress all inclination to experiment towards the improvement of glasses, since they forbade mixing of different types of glass. So the bottles of the time ranged from amber, through green, to almost black. When clear bottles were introduced, they seem to have been quite an innovation, which will be referred to later. Occasionally, excise did stimulate experiment. In order to evade the regulations that flint glass works might not produce window glass, either by blowing or rolling, the Whitefriars Company in 1844 introduced and patented a process for stamping small diamond-shaped panes of glass called quarries, with a mechanical press, for church windows. Duties also led to the founding of pirate works, which used up broken glass, remade it and sold it cheaply, but still at a considerable profit. It is reckoned that in London alone, sale of pirate glass brought in £65,000 per annum.[1]

Towards the end of the eighteenth century the glass houses were stirring for reform. Glasgow manufacturers addressed the Commissioners for the Glass Tax as follows:

> As we understand that Mr. Pitt has agreed to make Trial of the Plan suggested for levying the duties on Glass in the Annealling or Tempering ovens, in place of the present mode of charging it on the Materials, or Metal in the Pott . . . (they thought this would) prove the very Salvation of the Glass Manufactory in this Country.[2]

They were unanimously in favour of change, but it was not till 1823 that Commissioners were appointed to examine the situation. A further ten years elapsed before anything was done. The *Thirteenth Report of the Commissioners of Excise Enquiry on Glass* in 1835 contains the following message from a glass-maker:

> Our business and premises are placed under the arbitrary control of a class of men to whose will and caprice it is most irksome to have to submit and this under a system of regulation most ungratiously inquisitorial. We cannot enter into parts of our own premises without their permission: we can do no single act in the conduct of our business without having previously notified our intention to the officers placed over us. We have in the course of the week's operations to serve some sixty or seventy notices on these, our masters, and this under heavy penalty of from £200 to £500 for each separate neglect.[3]

[1] H. J. Powell, *op. cit.*, p. 154.

[2] *Memorial of the Glasswork Company at Leith, and the Glasswork Company at Dumbarton*, 1791. MS. 640, ff. 49 and 86, N.L.S.

[3] Quoted by H. J. Powell, p. 153.

The stultifying effect of taxation and regulation is seen by the great increase in flint glass production when conditions were eased in 1835. After much agitation, duties were removed on 24th April, 1845, though a licence duty of £20 per pot was retained for several years longer. Simultaneously, the import tariff was removed.

Sir Robert Peel, in his Budget speech on 14th February, 1845, referred to the condition of the glass industry:

> Your export of earthenware was last year double that of glass; it was to the value of £751,000; but the export of glass, subject, as I have said, to the duty, and the constant, vigilant and annoying interference with the manufacturer, in order that it may be collected, was only to the extent of £388,000. I am about to state another important fact in regard to glass: there is no excise duty in glass in France, Belgium, or Bohemia, and what are the consequences? That in Bohemia, in particular, the manufacture by the application of the chemical arts has been brought to a state of admirable perfection. There, glass under the application of the most beautiful chemical principles, is exposed at different stages to various degrees of heat, and thereby contracts a diversity of colours that produce the most beautiful effects. We have peculiar facilities for accomplishing the same ends; we command the alkali and the coal, and yet we cannot compete with the foreigners in the manufacture of glass. There is a great import of foreign glass into the bonded warehouses of this country, to be afterwards exported, and it is now beating our own manufacture, not only in foreign markets, but even in the markets of our own colonies. It you permit this article to be duty free, it is difficult to foresee, in the first place, to what perfection this beautiful fabric may not be brought; and, secondly, it is impossible to say to what new purposes glass, manufactured by our skill and capital, may not be applied. . . . It is to be borne in mind, that the cost of collecting the duty on flint glass is not less than 57 per cent. In order to prevent fraud, it is necessary that you should have a series of most minute and troublesome regulations as to the melting of glass; notice must be given to the excise officers respecting annealing and other parts of the process, which so encumber it as to make the application of additional skill and ingenuity almost impossible. My belief is, that with this change, if we do not supply almost the whole world with glass, we shall at least be able to enter into competition with other nations, who have hitherto had the benefit of that supply.[1]

The situation is summed up in the *Progress of the Nation*:

> The check on consumption was not wholly owing to the regulations whereby improvements were prevented, it was also

[1] Hansard's *Parliamentary Debates*, 1845, No. 3, p. 490.

occasioned, in part, by the excessive amount of duty imposed, as was proved by the demand which accompanied the various alterations in the rates of these duties. The pernicious effect of the glass duties was apparent from the contrast exhibited by this manufacture to others, not subjected to similar disadvantages. . .

There were, however, other reasons, distinct from these above which rendered the system under which the glass duties were levied particularly pernicious. The free progress of invention and improvement was by this means prevented, not only in the manufacture of glass, but also in many other arts and sciences to which glass is subsidiary. A manufacturer who by his skill had succeeded before 1845 in making great improvements in the quality of bottle glass, was stopped in his operations by the excise officers, on the plea that the articles which he produced were so good in quality as not to be readily distinguished from flint glass, to which description a higher rate was attached; the danger to the revenue being, that articles made of the less costly or less highly taxed ingredients would be used instead of flint glass.[1]

The result of the impeded development was that sixteenth-century houses were often closer to modern standards of light than average dwellings built in the late palaeotechnic era. Citing again from Lewis Mumford:

> In the fifteenth century glass, hitherto mainly used for public buildings, became more frequent; at first only in the upper part of the window. The glass would be heavy, irregular, feebly transparent; and the leads that held it would further reduce the amount of light. By the sixteenth century, however, glass had become cheap and fashionable. The popular saying in England about Hardwick House—'more glass than wall'—was equally true of the burgher houses. In North Germany and England a broad bank of window would extend across the whole house at each storey, front and rear, thus making up in effect for the tendency to deepen the house. The effects of governments in the eighteenth century to raise revenue by means of window taxes partly arrested this popular development: an atrocious stupidity.[2]

J. F. Chance tells us that Arthur Aiken (1773–1854), secretary to the Society of Arts, and a disciple of Priestley, showed how the degradation of painting and enamelling on glass in England was due to the illegality of necessary experiment, and how lenses for telescopes could not be made at all.

> I built a small furnace . . . for the purpose of investigating the action of some of the causes that affect the quality of optical

[1] G. R. Porter, *op. cit.*, p. 284.
[2] L. Mumford, *The Culture of Cities*, p. 38.

> glasses. On mentioning the circumstances to the late Mr. Carr, then secretary to the Excise, and with whom I was personally acquainted, I received such an answer as determined me to give up my intention.[1]

In order to make optical glass, firms were forced to make slabs in open defiance of excise regulations. At the end of the eighteenth century, Louis Guinard discovered the beneficial effect of stirring glass in the pot during the manufacture of glass for this purpose. His ideas were confirmed by Michael Faraday, but no British manufacturer was in a position to benefit by them. Consequently, while they were adopted in France and Switzerland, nothing was done to improve the manufacture of optical glass in this country till after 1848.[2]

Although glass bottles usually preceded windows in point of time, bioaesthetically, windows are more important. For example, some window glass was made at Monkwearmouth on the Wear and at Jarrow on the Tyne for the great monasteries. The first manufactory in Great Britain of window or crown glass was begun by Mansell at Newcastle in 1616. Little glass had been introduced into the windows of English farm-houses before then.[3] In 1661 in Scotland the ordinary country houses were without glazed windows, and in royal palaces only the upper halves of windows were glazed.

Window-glass was first made in Scotland in 1752. The *Scots Magazine* in that year wrote:

> The Glass-house Co. of Glasgow have arrived at so much perfection in the making of crown glass, that it meets a preference in the market to that made at Bristol, etc.

At this time in Scotland two materials were being substituted for the oiled paper then used in windows: (*a*) glass; but as often (*b*) by *stressan*, 'the envelope or covering of the quadruped *in utero*'.[4]

Window glass was made by fusing sand with barilla or kelp, the usual composition being sand to kelp in the ratio three to five by weight. Regarding the use of barilla or kelp in glass-making, Parkes tells us:

> Orkney kelp is preferred for window glass, because it is of a better colour than the Western Isle or the Scotch kelp. This is

[1] Letter from G. Dollond, March 17, 1835, quoted by J. F. Chance, *op. cit.*, p. 17. [2] H. J. Powell, *op. cit.*, p. 110.

[3] W. G. Armstrong, *op. cit.*, p. 198; J. R. McCulloch, *Commercial Dictionary*, Vol. 1, p. 603.

[4] A. Brown, *History of Glasgow*, Vol. 2, p. 265; J. M'Ure, *Glasgow Facies*, p. 1225.

> probably owing to the sand of the latter containing more iron. Mr. Bowles, the celebrated maker of crown glass at Cock-hill, Ratcliffe-highway, London, always used Spanish barilla instead of kelp, and his glass was ever preferred to that of the others.[1]

The raw materials were mixed and thrown into the fritting arch. Here the kelp was partially decomposed, and after four hours the resulting greyish mass was withdrawn and cut into brick-like lumps. These were stored for some time before melting, in the belief that the glass so made was of better quality than if made right away. If an imperfect glass resulted, common salt (sodium chloride, $NaCl$) was sometimes added to improve vitrification.[2] Sheets of this glass were obtained by blowing a large globule, one end of which was cut off, and the remaining bell spun out till it became a flat disc. Crown glass so produced had a brilliant surface, but its disk-like shape and 'bullion' at the centre meant that it could not be cut into large rectangular panes. Thickness was also limited to 18 oz. per square foot, the *usual* substance being 13 oz., and other, for special purposes, 9 oz. Its slight curvature also distorted objects seen through it. For many years, however, English glass-makers concentrated on improvement of crown glass, whereas Continental makers specialized in *broad* glass, which in time became known as *sheet* or *German* (Bohemian) glass.[3]

The only maker of crown glass in Scotland was the glass house at Dumbarton. Here there are connexions, on the one hand with glass production in England, and on the other with iron production in Scotland. The works was operated by John Dixon, who came from Sunderland, where Dixon, Austin and Company was famed. Another member of the Dixon family had Stourbridge connections. John Dixon first sank pits at Knightswood and Gartnavel, and subsequently purchased the glass works, his real object being to get an outlet for his coal. Sand he obtained from Sandspit at the estuary of the Leven, and he purchased kelp from the West Highlands. During 1776 the Dumbarton Glass Company was dissociated and operated as a separate concern with Jacob Dixon as manager.[4]

The glass turned out by this house is said to have equalled any made in Great Britain. By 1826 they had agents as far away as New York, much of their output going to America and the West Indies.[5] They employed over three hundred operatives, consumed

[1] S. Parkes, *Chemical Essays*, Vol. 3, p. 425 *n*.
[2] *Ibid.*, Vol. 3, p. 425. [3] H. J. Powell, *op. cit.*, p. 105.
[4] J. A. Fleming, *op. cit.*, p. 154 *et seq.*
[5] J. Sinclair, *General Report*, Appendix 2, p. 298.

15,000 tons of coal in the furnaces, and bought 1,200 tons of kelp. In addition they required 38,000 stones of hay and straw to pack their products, which included window, crown, and bottle glass. The sums paid in excise by Dumbarton may be compared with those of the Newcastle houses.[1] In 1789 they paid £119,000, and the average for several subsequent years was £115,000.[2]

In 1831 Jacob Dixon was succeeded by his second son, Joseph Dixon. The capital was £16,000, divided into fortieths, of which Andrew Houston held four; Joseph Dixon, Senr., four; William Dixon of Dumbarton, four; James Dunlop of Garnkirk, twenty-six; and Joseph Dixon, Jr., two.[3] The latter soon got into financial difficulties through, among other things, opposing Kirkman Finlay in a corrupt election. As a result he was forced to sever his connexions. James Christie, a partner of the Dixons of Calder Iron Works, purchased the plant for £12,000. The glass house continued to operate till 1853, when the site was sold and incorporated in Denny's shipbuilding yard.

Of the categories of glass mentioned above, broad, or inferior window glass was prepared from cheap raw materials, viz. kelp and sand. Glass could also be made from waste materials from other industries, for example, soap-boiler's waste. The waste consisted of lime used to render the alkali of the soap-boiler caustic:

$$\begin{array}{ccccc} Na_2CO_3 + Ca(OH)_2 & \rightarrow & 2NaOH & + & CaCO_3 \\ \text{soda} + \text{lime} & \rightarrow & \text{caustic soda} & + & \text{chalk} \end{array}$$

Sodium hydroxide (NaOH) is formed in place of sodium carbonate (Na_2CO_3), and is more caustic than it. In addition, soaper's waste included insoluble matter from kelp and barilla, along with some common salt, all in a pasty state. There was a trade in this waste between Leith soap-boilers and Newcastle glass-makers. The proportions in which it was used were two of waste, one of kelp, and one of sand. These were mixed, dried and fritted for twenty to thirty hours.[4] Evan Deer got a patent for using alum waste as well.[5]

On fusion the silica (sand) reacts with the constituents of the soaper's waste, producing a mixture of silicates, which constitutes the molten glass or *metal*, as it is called.

$$\begin{matrix} CaCO_3 \\ Na_2CO_3 \end{matrix} + SiO_2 \rightarrow \begin{matrix} \text{Calcium Silicate} \\ \text{Sodium Silicate} \end{matrix}$$

[1] P. 278.

[2] G. Chalmers, *Caledonia*, Vol. 3, p. 897; *see Diary of Joseph Farington*, 1792; the plans of the works c. 1792 are in the Dumbarton Public Library.

[3] *Session Papers*, 369: 4, S.L.

[4] S. Parkes, *op. cit.*, Vol. 3, p. 442.

[5] Register of the Great Seal, Vol. 19, No. 333. G.R.H.E.

This is blown into cylinders which are slit axially and flattened out on an iron plate covered with sand. Sheets up to about four feet square could be made this way, but they had obvious defects in consequence of being slit with shears and flattened out while hot. At a later date cylinders were blown to a larger size, and allowed to cool before slitting. The flattening then took place in a kiln. This gave larger, more perfect sheets, and caused broad glass to supersede crown glass.[1]

In 1830 a further development came with the making of sheet glass as the result of a mechanized process. Coloured sheet glass dates from 1835, patent plate from 1840, rolled plate from 1847. The firm of Chance Brothers and Company is, in the main, connected with these innovations. So great were the difficulties which had to be overcome, that, although they started experimental work in 1832, it was not properly established on an industrial scale till 1848.[2]

One might dwell for a moment on the glass and soap nexus. As glass became available for windows, more light was admitted, only to reveal the squalid interior. This created a demand for soap, and as a result still more raw material for glass-making became available.

It is difficult to decide whether manufacture of glass for windows or for glass containers had the greater effect on the trend of western civilization. Perhaps windows were the more revolutionary, because glass containers were preceded by opaque vessels made of clay or metal. Nevertheless, the mass production of cheap standardized bottles represents an important palaeotechnic development. Due to fiscal regulations, bottles had to be made of the coarsest glass.

> Government will not allow the makers of this species of glass to use any but the commonest river-sand, lest the glass should be too good, and the revenue be defrauded by its being applied to purposes for which the best glass is generally used, and which pays a higher duty. The greatest proportion of the sand which is used in bottle glass is obtained from the river Thames at Woolwich.

It was even forbidden to use up crown glass waste, although to do so improved the quality of bottle glass. When Frederick Fincham discovered how to make ordinary green bottle glass acid-resisting, and suitable for chemists' phials, his experiments were stopped by the Commissioners.

[1] H. J. Powell, *op. cit.*, p. 105. [2] W. G. Armstrong, *op. cit.*, p. 200.

> He was told that his work could not be allowed to continue, because he produced an article so good, that it could not be sufficiently distinguished from flint glass, the danger being that this article, which for a great variety of purposes was admitted to be in all respects as good as the comparatively highly taxed, and therefore high-priced, article of flint glass, would be substituted for that description of glass to the detriment of the revenue, however much the substitution might conduce to the convenience of the public.[1]

An old account of *The Art of Bottle-making*, quoted by H. J. Powell,[2] gives the raw materials as 'sharp sand, sharp lime, soap waste, red brick clay, and salt-dross waste from the salt pans'. The same materials were still in use when Andrew Ure was writing his *Dictionary of Chemistry*.[3]

The first bottle-house was built in Scotland on the Broomielaw Croft, Glasgow, in 1730. It was set up by a company of Glasgow merchants, who found that a three-pot furnace operated for four months of the year was sufficient to supply the demands of the west of Scotland and north of Ireland. In order to absorb labour when not required in Glasgow, Leith and Edinburgh, wine-merchants erected a bottle-house in Leith, to which Glasgow operatives resorted for the rest of the year. It was as a result of the need for pots in the bottle-house that manufacture of fire-brick was introduced into Scotland. An Englishman, John Clerk, who had worked in Holland, was brought to Glasgow in 1755, to teach the art to Glasgow bottle-makers.[4] From this conjunction a process of integration may have gone on. Glass and pottery-making were associated at Verreville (*vide infra*); Lancefield Pottery and Glass-works were almost one; and Hydepark Pottery and Glassworks were alongside one another.[5]

Bottle-making in Scotland has many cross-connexions with other industrial activities. For example, they form an interesting link between Bristol, Glasgow, and the West Indian trade. Molasses was imported from the West Indies into Bristol and there tran-shipped to Glasgow, where it was converted into rum. From Glasgow it was smuggled back to England in special bottles made in Glasgow. According to Arnold Fleming, the glass bottle proved too fragile for this hazardous journey, and leather bottles were substituted. The demand for these is given as the stimulus to the founding of the tanning industry in the south of Scotland. Irish

[1] J. F. Chance, *op. cit.*, p. 17.
[2] H. J. Powell, *op. cit.*, p. 55.
[3] S. Parkes, *op. cit.*, Vol. 3, p. 448.
[4] A. Brown, *op. cit.*, Vol. 1, p. 264.
[5] J. A. Fleming, *op. cit.*, p. 141.

distilleries, too, were supplied by the *Glasgow Bottle House Company*, founded in 1730, which had a long career of varied success. The *Session Papers* in the Signet Library contain an abstract balance sheet drawn up in 1798:

	£		£	£
Buildings, etc.	7,000	Borrowed money		10,300
Utensils	3,861	Accounts, etc.		8,262
Mfd. bottles	2,688	Allowances for debts,		607
		Stock, viz.,		
Flint glass	3,132	Hamilton, Brown, Wallace & Co., 8/17	4,458	
	£16,681	Glasgow Bottle Works 8/17	4,458	
Debts	11,961	J. Geddes, 1/17	557	
				9,473
	£28,642			£28,642

An apparent profit of £1,475 was shown, but this did not take into account bad debts and buildings, etc.[1]

This bottle-house was finally sold up in 1836. At the auction there was sold black ash, kelp, barilla, and soda-cake, an indication that synthetic alkalis were by that time in use in Glssgow.[2]

The introduction of clear bottles was an event of importance, because they found a ready market when, towards the end of the eighteenth century, Highland whisky was introduced into the Lowlands. A works specializing in these was founded by a glass-maker of the name of Jamieson at the foot of what is now King Street, Glasgow.

Throughout the greater part of the eighteenth century, beer was sold in barrels. It was first successfully bottled in 1736 to supply 'pale ale' for the India trade, but it was not till some time after 1800 that there was much bottling for the home trade.[3]

Other towns in the Clyde area, at Port Dundas, Port Glasgow, and Greenock, are associated with the making of bottles. The coal of Kilwinning attracted glass-making into Ayrshire, where a glass company continues to operate to this day at Irvine.[4]

The continental trade stimulated the making of bottles at Leith, where the *Bottle House Company* was formed in 1746. Their first bottle-house was at North Leith, but this was burned down in the year of its erection, and in the following year a new house was built on the sands at South Leith, and extensions added in 1764 to meet the increasing demand for bottles. By the end of the cen-

[1] *Session Papers*, 391: 22. S.L.
[2] J. A. Fleming, *Scottish and Jacobite Glass*, p. 146.
[3] J. C. Drummond, and A. Wilbraham, *The Englishman's Food*, p. 395.
[4] J. A. Fleming, *op. cit.*, p. 150.

59. Verreville Glass Works, Glasgow, *c.* 1836. It is possible that the *gabbert* on the river is carrying kelp to be used in the manufacture of the glass.

Diderot and d'Alembert's *Encyclopédie*

60. Glass-production, showing the blowing of various kinds of glass, and stocks of wood used for firing the furnaces.

61. Interior of glass house, *c.* 1840.

62. Blowing glass-bottles.

63. Making crown-glass. The sheet was obtained by blowing a large globule, one end of which was cut off. Crown-glass had a brilliant surface but did not yield convenient rectangular pieces.

64. Blowing cylinders for the production broad or sheet-glass. The large cylinde were slit and flattened out in a kiln. T large sheets which resulted enabled it supplant crown-glass in windows.

tury they were employing some eighty operatives in the making of nearly twenty thousand cwt. of green glass. The annual expense of carrying on both houses amounted to between £8,000 and £9,000 per annum.[1] An account of the expansion at Leith and Edinburgh is contained in the *Statistical Account*.[2]

Of increasing amenity from the middle of the eighteenth century, H. G. Graham tells us:

> In 1745 an adventurous tradesman began a business in painted paper for hanging walls in Edinburgh—the maker confining himself to two colours with designs of a rudimentary taste. The recess beds with plaiding curtains vanished from drawing-room and bedroom; the pewter plates and dishes went the way of their timber predecessors, and china and delft came in their stead, greatly to the encouragement of the struggling industry in Leith and Glasgow; the pewter stoups in which claret had been served, when bottles cost 4*d.* each, gave place to green glass bottles, which the glass-blowers in Leith were then making.[3]

Bottle-making is also associated with the art of brewing in Scotland, and on several occasions brewers helped to tide bottle-houses over times of financial stress. For example, when the *Edinburgh and Leith Glass Company* was overtaken by financial disaster, a group of brewers, with William Geddes as manager, took it over and restarted it. It is probable that the Youngers acted in a similar way at Alloa.[4]

The *Alloa Glass-house*, which still retains its name, was erected in 1750, utilizing fuel from a neighbouring colliery. About 1770 it became the property of the Stuarts of Tough, who extended it till it covered some six acres of ground. These extensions were more than the financial position of the Stuarts was able to carry, so in 1773 they advertised a half-share for sale.[5] Further difficulties followed, and in 1788 a George Younger advertised that the existing partnership was to be dissolved. This suggests a connexion with the famous brewery which was founded about 1750. A joint stock company was formed, and it may be from that date only that bottles were made.[6] In 1832 the business was sold again, and for the next seven years only crown and window glass were made. This apparently did not pay, and they went back to bottle-

[1] H. Arnot, *History of Edinburgh*, p. 456.
[2] *Statistical Account*, vol. 6, p. 595.
[3] H. G. Graham, *Social Life of Scotland in the Eighteenth Century*, p. 57.
[4] J. A. Fleming, *op. cit.*, p. 114.
[5] *Edinburgh Courant*, 22nd Nov., 1773.
[6] *Session Papers*, Signet Library, 197: 7, 274: 13, 286: 16.

making.[1] Throughout its career there have been many changes in ownership. At one time they had connexions with the *Nailsea Heath Glass-house*, Bristol, and interchanges of personnel took place.[2]

> John Donaldson arrived from Scotland to supervise the manufacture of the local glassworks in 1796.[3]

Bottles were also made in Dundee in a bottle-house with two cones erected in the Cowgate in 1789. They are mentioned in the *Newcastle Courant*, 25th August, 1798.[4]

It was mentioned when discussing the transition from eotechnics to palaeotechnics that to compensate for the covering-in of the pots, a new constituent had to be added to the glass, either litharge, or red lead, oxides of lead. As a result of this a new species of glass evolved, characterized by a great brilliance and clarity. According to E. W. Hulme, the earliest reference to flint glass is the specification of Oppenheim's patent of 1755. To keep it crystal-clear, flint or pure silicon dioxide was used, and so the glass became known as *flint glass*. It is 66 per cent. lead silicate. The raw materials were fused in the proportions, silica three cwt., litharge one cwt., and purified pearl ash one cwt. If a green tint developed, this was corrected by the addition of manganese dioxide, with sometimes, in addition, saltpetre and arsenic. To bring the mass to a proper state of fusion, thirty hours' heating were required.[5] Regarding the best flint glass, Parkes says:

> Rather more alkali is used than is necessary to flux the sand, and when the whole is in fusion the fire is continued so as to volatilise the superabundant quantity. If an excess of alkali be left in the glass, it will attract water from the atmosphere, and in a short time assume a liquid state.[6]

The glass was blown into shape, and facets were cut on the articles to increase their brilliance. This was a costly operation requiring skill and time. It was discovered that a semblance of cutting could be obtained by pressing glass into moulds while still plastic. This led to the use of glass by sections of the community which could not possibly pay the price of genuine cut glass.[7]

The making of this glass created a demand for large quantities of red lead, the making of which became a separate industry. Lead, too, serves to connect glass-making with ceramics, one of

[1] *Edinburgh Gazette*, 24th February, 1832.
[2] J. A. Fleming, *op. cit.*, p. 126.
[3] *Bristol Journal*, 5th June, 1802.
[4] J. A. Fleming, *op. cit.*, p. 129.
[5] A. Ure, *Dictionary of Chemistry*, Article, *Glass*.
[6] S. Parkes, *Chemical Catechism*, p. 170, *n.*
[7] W. G. Armstrong, *op. cit.*, p. 200.

the important methods of glazing clay objects depending on the use of a coating of lead silicate, i.e. virtually flint glass.[1]

In Glasgow there was an important works given over to production of flint glass, the *Verreville Works of Williams, Ritchie and Company, Flint Glass and Crystalware Manufacturers.* It was established in 1770 to make 'crystal table glass according to the finest manner of the Continent', and was the first of its kind in Scotland. The actual founder was Patrick Colquhoun, who was born in Dumbarton in 1745. After gaining practical experience in glass-making, he induced several Tyneside men to form a joint-stock company with him. They included Alexander and John Ritchie, of Glasgow, Charles Williams, Joseph Robinson, Isaac Cookson of Newcastle, and Evan Deer of North Shields, who was in addition connected with the Nailsea Heath Works at Bristol.[2]

Verreville was conveniently situated for raw materials, coal from Orr's Camlachie coal pits, sand locally, and kelp from the Islands, which was purified to make it suitable for this quality of glass. Glass-cutters and engravers were brought from Germany.[3]

The American war brought difficulties to Verreville, as it did to many other Glasgow enterprises. The result was that a new company, known as Hamilton, Brown, Wallace and Company, had to be formed. Ownership ultimately passed to John Geddes in 1806 by repurchase from the Dumbarton Glass Works, who had bought it from Gilbert Hamilton and Archibald Wallace, the principal partners. Geddes is said to have had an expert knowledge of physics and chemistry, and a practical knowledge of glass-making gained in the Edinburgh and Leith Glass Works which was owned by his elder brother William. He introduced steam for operating the cutting machinery at Verreville. The Geddes family had many connexions with glass-making, with Alloa through marriage, and with Lancefield Bottle Works (Geddes, Kitson and Company). On account of the crippling effect of the excise duties, Geddes withdrew from glass-making about 1840, and went over to pottery production.[4]

To produce a sheet which would not distort images and which was of equal thickness throughout, the glass had to be cast by throwing the molten metal on to an iron or copper table and then rolling it. This method was first adopted in Lancashire in 1771 though at an earlier date it had been tried on a smaller scale at

[1] J. Sinclair, *General Report*, Appendix 2, p. 309; for methods of production *see* R. Watson's *Chemical Essays*, Vol. 3, essays 8, 9 and 10.

[2] J. A. Fleming, *op. cit.*, p. 132.

[3] A. Brown, *op. cit.*, Vol. 2, p. 267.

[4] J. Sinclair, *op. cit.*, Appendix 2, p. 298; *see also* N.S.A., Vol. 6, p. 162.

South Shields. The materials used for this kind of glass were good carbonate of soda, produced by decomposing common salt with pearl ash, along with pure sand, pure quick-lime, saltpetre, and broken glass in the proportions:

Pure sand	43 per cent.
Soda	26·5 per cent.
Pure quick-lime	4 per cent.
Saltpetre	1·5 per cent.
Broken plate glass . . .	25 per cent.

> Lime is employed on account of the property which it gives to glass, of being a better conductor of heat, and consequently rendering it less liable to break by sudden change of temperature; and because such glass cuts more readily by a diamond than that which is purely silicious. It is important, however, to use quick-lime, and not the carbonate, as the latter occasions so much swelling of the materials as to endanger their flowing over the sides of the crucibles.[1]

Under an Act,[2] monopoly of manufacture of plate glass was given to the *Governor and Company of the Cast-plate Manufactory*, in 1771 or 1772. This company operated at Ravenhead, St. Helens, but failed in 1794. Isaac Cookson and William Cuthbert of Newcastle began to make plate as well, and South Shields became the main producing centre. Ravenhead passed to Pilkington, then wine and spirit merchants and rectifiers. They began making crown glass in 1827, and ultimately gave up the wine and spirit business and concentrated on glass-making.

After casting, plate glass was polished to a flat surface and principally used in making looking-glasses. This required power. At Ravenhead a steam engine was installed in 1789. The polishings, though considerable in quantity, are useless for working up into glass again, and were sold for scouring pewter.[3]

[1] S. Parkes, *Chemical Essays*, Vol. 3, p. 459. [2] 13 Geo. III, c. 38.
[3] J. H. Clapham, *op. cit.*, Vol. 1, p. 190, *n.;* J. R. McCulloch, *op. cit.*, Vol. 1, p. 604; H. J. Powell, *op cit.*, p. 124; *Victoria County Histories: Lancashire*, Vol. 2, p. 405.

Chapter XV
POTTERY

Science, for its progression, requires patronage.
HUMPHRY DAVY.

GLASS and ceramics are both furnace products. They frequently occupy complementary places in human economy. Shelter is built of brick and glass. Pewter and wood, eotechnic furnishing of the table, were replaced by the palaeotechnic products, glass and pottery, in the first instance, by imported Oriental porcelain, next by *delft*, a Dutch imitation of porcelain, and latterly by earthenware. Thomas Somerville, author of *My Own Life and Times, 1741–1814*, says that about 1750, wooden platters were in use in almost every house, and exclusively in those of the farmers and of many of the clergy. Fifty years later, a writer in the *Encyclopaedia Britannica* wrote:

> In many parts of Scotland, instead of ale and beer mugs, they use small hooped wooden vessels, of which the staves are feather-edged or dove-tailed into one another. This, as the staves are of different colours, increases the beauty of the vessel.[1]

While glass, in consequence of fiscal policy, evolved into several sharply defined categories, each more or less characterized by the raw materials used, pottery was made from a multitude of naturally-occurring argillaceous materials whose complexity makes it difficult to characterize ceramic products scientifically. Even to-day, in fact, it is difficult to find a satisfactory system by which to differentiate the varied products produced by the action of heat on clay. Scientific precision is lacking; recipes rather than formulae determine composition.

In his *Rise and Progress of the Staffordshire Potteries*,[2] Simeon Shaw says of the terms, *pottery* or *earthenware*, *china*, or *porcelain*:

> All are general terms for a kind of manufacture consisting of certain mineral productions, on one and the same principle

[1] *Encyclopaedia Britannica*, Supplement, Third Edition, Vol. 1, p. 454.

[2] Shaw's work cannot be considered as a reliable guide to the economic history of the pottery industry, although we have used it as a source-book for data concerning the scientific impetus in potting. John Ward's *History of Stoke-on-Trent* contains Shaw's local history.

properly combined and amalgamated, and ultimately subjected to most intense firing, or baking.[1]

Like so many technics already discussed, transition from wood to coal economy had a considerable influence on the pottery industry. J. U. Nef, quoting from Gerbier's *Council and Advice to all Builders,* says that of every 20,000 bricks burned with *seacoles,* no less than 5,000 were unfit for use. The difficulty, it appears, was to get coal sufficiently free from sulphur compounds. The cause of the localization of the Potteries in North Staffordshire was, according to John Thomas, the asset of a fuel supply. Next to labour, fuel was the principal charge in the making of pottery. 'The close connection of the development of the pottery industry with the coal-mining industry of North Staffordshire is symbolized by the Adam's Colliery, Tunstall, with its pithead engine within a stone's throw of the pottery ovens and kilns.'[2] Chemical reaction occurring between the combustion products of coal and the surface of articles being fired again became important when decorated articles came to be made. This difficulty led to the introduction of *saggars*—cases of fireproof clay enclosing pottery while being baked.[3]

Organization in early potteries was extremely simple. V. W. Bladen tells us that the average pottery consisted of a small thatched cottage with lean-to sheds covered with sods. The firing oven was of coped form, only six to eight feet high, surrounded by a hovel of sods for insulation. The clay, dug locally, was prepared in a tank or *sunpan* some twenty feet square and eighteen inches deep. One corner of it was partitioned off, and here clay was *blunged,* or beaten with water by a man using a sort of oar. Large stones were thus removed, and the purified clay was run through a sieve into the sunpan until it had attained a depth of about four inches. Here it was left to evaporate, after which a further quantity was added, and the process repeated till the pan contained a foot of purified clay. It was then removed from the sunpan and stacked in a corner. Before use it had to be *wedged,* i.e. cut and recut by a wire, pieces being slapped together to expel air bubbles.[4]

The clay was then kneaded, formed into vessels on the wheel, and set aside to dry. Additional parts like handles were added, and

[1] S. Shaw, *Rise and Progress of the Staffordshire Potteries,* p. 86.

[2] J. Thomas, *The Pottery Industry and the Industrial Revolution, Economic History* (Supp. to Econ. J.), 1934–7, *3,* 399.

[3] J. U. Nef, *Rise of the British Coal Industry,* Vol. 1, p. 216.

[4] V. W. Bladen, *The Potteries in the Industrial Revolution, Economic History* (Supp. to Econ. J.), 1926, *1,* 117.

possibly some decoration applied by running on a different liquid clay through a quill. By dusting the surface with beaten and sifted lead ore, or *smithum*, obtained from Lawton Park, near Burslem, a glass was formed on the surface when the ware was fired. Coarser articles were subjected to the naked flame; finer pieces were placed in saggars—*mettal'd pots of marle*—to prevent chemical reaction between sulphurous vapours (sulphur dioxide, SO_2) from the coal and the clay. One firing of the kiln per week was the usual arrangement. It began late in the week and the kiln was stoked for the last time early on Sunday morning, and then allowed to cool off slowly. When the oven was drawn, articles were sold directly to travelling cratemen, and so disposed of without transference to warehouses.[1] In such a pottery six to eight men and boys worked, with separate sheds for thrower and decorator. Each man, however, could carry out all the operations if required.

> Burslem was so much the centre of the pottery industry that there were few pot works elsewhere. Here then we have a small industry which, though it flourished on the Continent and was there organized as a craft guild, never seems to have reached that state here. The probable explanation of this is that wooden platters for the very poor, and metal utensils of baser or richer metal for the wealthier classes, sufficed in days when road transport was so bad as to endanger any frailer vessels.
>
> This suggestion is supported by the fact that though the pewterers, silversmiths, and glassmakers are mentioned in the Statute of Apprentices, the potters are not included. Nevertheless, the trade seems to have imitated the organization of the rest of industry as a matter of convenience, for it was considered the normal thing to serve an apprenticeship of seven years as in other trades. No large capital was as yet embarked in this trade, but it was free of all restrictions and was capable of easy expansion.[2]

In addition to requirements in clay, lead ore, salt, and coal, a pottery created a great demand for oat and wheat straw for packing, and for hazel rods and coppice wood for crate-making. Dr. Thos. Wedgwood, at the end of the seventeenth century, combined farming with pot-making, while his son was an innkeeper. Paper for both packing and printing later became an important adjunct and doubtless stimulated the Fourdriniers to set up their paper mill in Hanley.[3] The expenses for a week's working amounted to the following:

[1] V. W. Bladen, *op. cit.*, p.120. [2] J. Lord, *Capital and Steam Power*, p. 47.
[3] S. Shaw, *op. cit.*, pp. 10 and 43.

	£	s.	d.
3 men at 4s. per week and 3 at 6s. . . .	1	10	0
4 boys at 1s. 3d.		5	0
1 cwt. 2 qrs. lead ore at 8s.		12	0
Manganese		3	0
Clay, 2 cart loads at 2s.		4	0
Coals, 48 horse loads at 2d.		8	0
Carriage of, do. at 1½d..		6	0[1]

The influence of contemporary science on pottery at this stage of its development can only be brought to light by diligent searching, although the making of pottery is in itself a skilled scientific operation. In the words of J. G. Crowther:

> Pot-making provides the first example of the use of a chemical change for a constructive purpose, and it involves a series of difficult technical processes. Clay cannot be moulded satisfactorily unless it has the correct consistency. If it is too wet it disintegrates into mud, and if too dry it crumbles. If it contains no grit it will stick to the fingers in moulding, and if the bits are too large they will interfere with the moulding and weaken the material.
>
> If the damp, moulded clay is immediately fired, it will crack, so it must be dried first. Then it must be heated to 600° C. This produces the hardening, which is due to the expulsion of water chemically attached at lower temperatures to the aluminium silicate of which the clay is chiefly composed. The dried moulded vessel changes colour during the firing, and the resultant hue depends on the chemical composition. If the clay contains iron oxide and is exposed to air while heated, the oxide will be oxidised further into red ferric oxide, and produce a reddish hue. If the pot is heated in glowing charcoal, so that the air is excluded, the iron oxide in the clay will be reduced to the black ferroso-ferric oxide, which will make it grey.[2]

The fundamentally chemical nature of the operations was obviously apparent to Shaw, the nineteenth-century historian of the Staffordshire potteries, for of Enoch Wood's pottery he says:

> The very extensive manufactory of Enoch Wood and Sons, has such a judicious arrangement that it presents all the appearances of a most extensive Laboratory and the machinery of an Experimentalist.[3]

Shaw also refers to associated chemical operations in progress in the Potteries. At Clay Hills, near Tunstall, Smith Child had a

[1] Wedgwood MSS., quoted by V. W. Bladen.
[2] J. G. Crowther, *Social Relations of Science*, p. 19.
[3] S. Shaw, *op. cit.*, p. 30.

CERAMIC PRODUCTS
(Complex alumino-silicates)

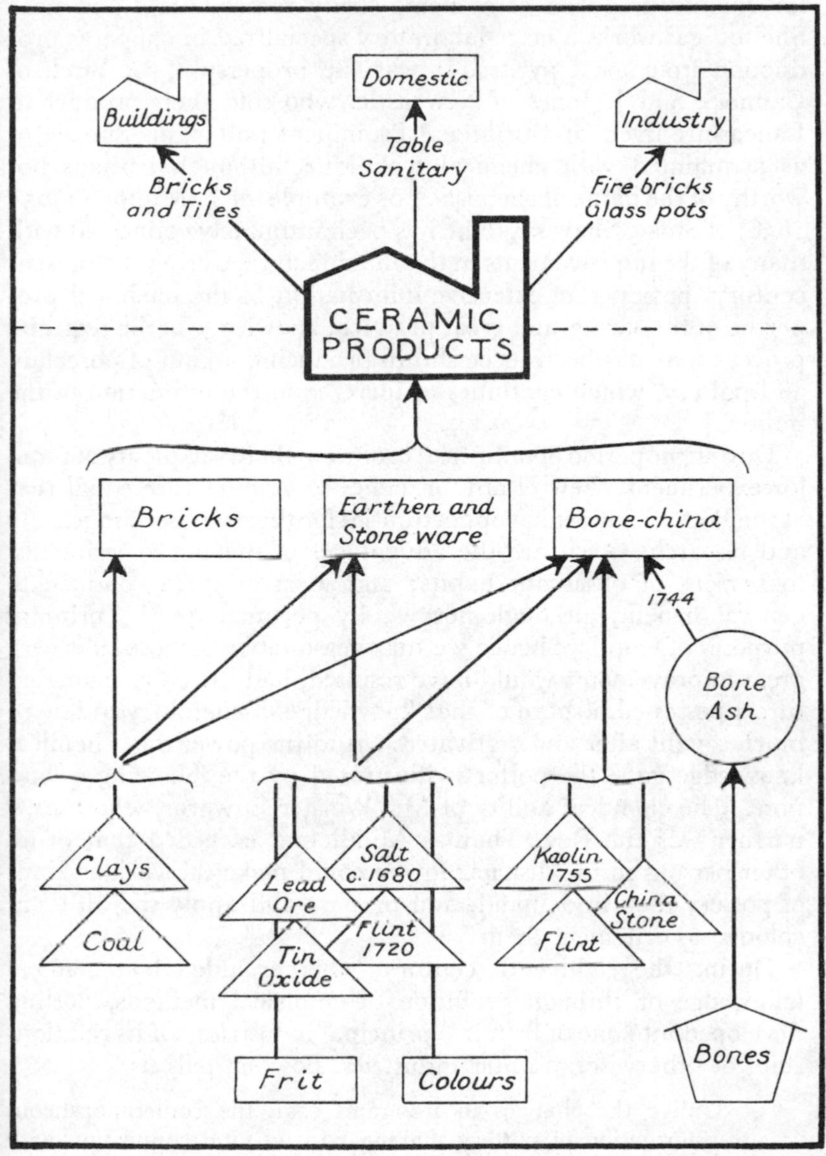

commodious manufactory of chemicals which, from its position, isolated from other manufactures, doubtless supplies ancillary materials to the potteries.[1] At Hanley, a Mr. Wilson had a shop for chemicals spoken of as being 'truly respectable',[2] and near Shelton gas works a large laboratory specialized in copperas production from local pyrites. It was the property of G. Birch of Cannock and J. Jones of Newcastle, who sold their product to Lancashire dyers and printers.[3] Prominent potters are spoken of as acquainted with chemical technique, although perhaps not worthy of the name of chemists. For example, of T. Minton (1765–1836) of Stoke, Shaw says: 'he has been intimately connected with many of the improvements in the manufacture, during the present century; possessed of extensive information of the chemical properties of the earths, and great practical knowledge of the requisite processes, he has been successful in producing a kind of porcelain and pottery, which continues to increase in the estimation of the public'.[4]

During the period of industrial evolution there was plenty of room for experiment. 'Few of any branches of manufacture equal that of the Potter in affording opportunities for the exercise of ingenuity and research. Great are the advantages of making experiments, to persons of observant habits; and most of them conduce to general benefit, although not wholly pertinent to the primary purposes of enquiry; hence we may reasonably suppose that very great improvements would have resulted, had the early manufacturers possessed a share of that knowledge of chemistry of late so much sought after and cultivated.'[5] And the power that chemical knowledge gave the potter is illustrated by the following quotation: 'The chemical ability of Mr. Warner Edwards, whose *secret* partner was the Rev. Thomas Middleton, exceeded that of all other persons in the district; for he could make the various kinds of pottery then in demand, and prepare and apply the different colours, to ornament them.'

During the eighteenth century, whether aided by chemical knowledge or through evolution of empirical methods, potting developed into one of Britain's principal industries. Of its relationship, or otherwise, to other industries, Bowden tells us:

> Unlike the changes in iron and coal, the contemporaneous transformation of pottery making had no vital connection with other industries. Its main distinctive contribution was in developing the artistic side of manufacturing; and its aesthetic influence

[1] S. Shaw, *op. cit.*, p. 18. [2] *Ibid.*, p. 40. [3] *Ibid.*, p. 49.
[4] *Ibid.*, p. 64. [5] *Ibid.*, p. 103.

> on other industries was negligible. But pottery-making was of great importance in itself: by means of improved processes and formulas it developed into a leading branch of English industry; and its rise involved a reorganization of economic society characteristic of modern large scale enterprises. In technical progress, in volume of business, in concentration of capital and labour, and in the newly acquired wealth and influence of its leading representatives, it became one of the three principal modern-type manufacturing industries which emerged immediately out of the era of invention. Though intrinsically less important than either of the other two—cotton, and the iron and ironwares industries—pottery-making, on account largely of the genius of Josiah Wedgwood, acquired at times an equal and even superior influence.[1]

The extent to which this is true is illustrated by the fact that as late as 1799, by which time the principal problems of pottery-making had been solved in England, the French National Institute was offering prizes for earthenware which would not break with sudden changes in temperature.[2]

It appears, however, that some of Bowden's generalizations are too sweeping: (*a*) pottery, though not occupying a basic position like coal and iron, was indeed connected with the coal-mining, salt-boiling, and lead-mining industries, and ultimately became completely dependent for its successful development upon coal. The industry too was one of the first to make use of platinum, and Wedgwood's jaspar ware necessitated the mining of barium compounds; (*b*) it is a common confusion to over emphasize the significance of *artistic* pottery. There were two quite distinct trends in the evolution of the industry; the first, the production of elaborate wares for the noble houses—the so-called artistic productions. Of much greater importance was the imitation of exclusive productions which placed on the market in large quantities wares once only available to the nobility. This democratization was achieved in two ways: (*a*) by continual experiment towards the fabrication of durable products, (*b*) by developments in the direction of mass production. The latter were mainly mechanical, and, while making the production of ceramics into the modern type industry, as referred to by Bowden, are foreign to the present study.

> As wealth increased, as refinement and education spread, as food became better in quality, the necessity for good and cheap crockery was absolute. Pewter dishes and porringers, wooden cups

[1] W. Bowden, *Industrial England in the Eighteenth Century*, p. 92.
[2] *Phil. Mag.*, 1798, *1*, 96.

> and platters, would do no longer for a generation whose industry was beginning to be backed by the greatest mechanical inventions. And when Wedgwood met this necessity, in improved fabrics, and higher-class forms, when he placed his excellent cream ware, his Egyptian tea-pots and cream-jugs, his prettily decorated tea-things, and his white jugs and basins upon homely tables, he did that for his age which in moral effects was far beyond price.[1]

Subsidiary to these lines of development there was the frank imitation of imported oriental wares. This led to the evolution of *English Bone China*, a definite subsidiary to the main line of evolution which will be discussed later. For the present we shall follow the scientific development which led to mass production of serviceable articles, the main line being the strivings towards durable white products, decoration of these being regarded as of secondary importance. The production of subscription pieces at £50 each does not concern us here.

Pottery industry in both England and Scotland starts with utilization of local clay dusted with lead ore to produce a glassy surface when fired. The operations were carried out in a pottery such as has been described already. The principal early products were jars, bottles, and butter pots. Variety in local clays suitable for potting was enormous. Dr. Martin Lister, who wrote in the second half of the seventeenth century, distinguished no less than twenty-two clays which he himself had examined.

About 1680, it is said as the result of an accidental observation, salt began to be used in addition to lead ore as a means of producing a glass film on the surface of fired articles. The ware resulting was the *common brown (or red) ware*.

This *brown ware* was but a poor substitute for white imported china, so Thomas Miles of Shelton mixed whitish Shelton clay with sand and thus produced a rude *white stone-ware*. A *brown stoneware* was also produced with can-marl from the coal-pits mixed with sand and fired. Variety in ceramics was thus increased with little searching for new raw materials. Glazing further added to the possible combinations. Some wares were glazed with salt, others with lead. The Staffordshire potteries got their lead ore from Derbyshire or from north Wales, beyond Chirk and Wrexham. Scottish potteries were supplied from Leadhills.[2]

Simulation of white oriental porcelain continued to stimulate potters in this country. By the utilization of Derby crouch clay they evolved *Crouch ware* about the end of the seventeenth century,

[1] E. Meteyard, *Life of Josiah Wedgwood*, Vol. 1, p. 169. [2] Ch. XVII.

at which time the industry was described by Dr. R. Plot in his *Natural History of Staffordshire* (1686). He recorded that:

> By far the greatest pottery they have in this country is carried on at Burslem, where for making their several sorts of pots they have as many sorts of clay, which they dig around the town, all within half-a-mile's distance, the best being found nearest the coal; and are distinguished by their colours as followeth:
> (1) Bottle clay, of a bright whitish streaked yellow colour.
> (2) Hard fire clay of a dull whitish colour, and fuller interspersed with a dark yellow, which they use for their black wares, being mixed with
> (3) Red blending clay, which is a dirty red colour.
> (4) White clay, so called, it seems, though a blewish colour and used for making a yellow coloured ware, because yellow is the lightest colour they make any ware of.

All these were subjected to a single firing, begun about Thursday and finished on Saturday, when salt was added to the kiln by a workman standing on a scaffold near the top of the oven. Salt glazing was in practice long before the introduction of white clay and flint, and so extensive was its use for glazing that it early led to industrial pollution.[1]

As far as Staffordshire is concerned, salt glazing is believed to have been introduced by John Philip Elers. His story and that of his brother David is recounted in detail in Llewellyn Jewitt's *The Wedgwoods*. They were of Saxony stock, and reputed to have come to this country with William of Orange, settling at Bradwell Wood, between Wolstanton and Burslem. In the last decade of the seventeenth century, the Elers were turning out work characterized by its high degree of perfection. By attending to the composition of clay for the body, J. P. Elers had a lasting effect on the subsequent productions of the Potteries. 'The careful levigation of the clays which he practised, the use of the lathe and of metal stamps, and the process of salt-glazing, were precious legacies to the district.'[2]

From 1710–40, *white dipped* or *slip ware* was in great demand. From crouch ware this was evolved by easy transition simply by washing over the insides of vessels with a cream made of pipe-clay. So was evolved *white dipped stone-ware*. Its invention is associated with the name of J. Astbury of Shelton. This was a further step in the direction of light-coloured earthenware as a substitute for porcelain or tin-enamelled faience used by the well-to-do. The impetus to devise such a dipped ware was economic. Pipe-clay

[1] S. Shaw, *op. cit.*, p. 109 *et seq.*
[2] A. H. Church, *English Earthenware*, p. 54.

had to be transported from Devon, no easy matter in those days, and so it had to be conserved.[1]

Pipe-clay was next mixed with fine grit and sand, and a *light-coloured and durable stone-ware* made for the first time. The heyday of this fine salt-glazed stoneware was 1690 to 1780, though its manufacture lingered on in Burslem for yet another half-century. It was almost a porcelain, being translucent in thin pieces, and so hard that it was only just scratched by quartz.

> The perfection of salt-glazing is sometimes reached in objects of common materials and common workmanship, even on drain-pipes, filters, and chemical apparatus. In the old Staffordshire ware it frequently combined perfect efficiency as a protective coating with that exquisite half-glass which, without interference from its own excessive brilliancy on the one hand, or course irregularity on the other, brings out both the form and the decoration of the body.[2]

It combined resistance to heat with resistance to chemical action. Of it the chemist, Macquer, said, 'The best common stoneware is the most perfect pottery that can be: for it has all the essential properties of the finest Japanese porcelain'. Vessels which would resist chemical action were of great importance to early chemists, who could not then buy ready-made glass-ware. Pott, author of *Lithogeognosie Pyrotechnique, ou, Examen Chymique des Pierres et les Terres* (Paris), 1753, wrote a work devoted to crucible-manufacture.

In the domestic sphere, this ware further brought the amenities of the rich within the grasp of the middle classes. The body of the ware changed progressively. Starting off with brick-earth and fine sand, it passed to can-marl and sand. Next grey clay from the coal measures was utilized, along with fine sand, and finally grey clay with ground flint. The next forward step is also due to John Astbury (1688?–1743). The date is about 1720. It is the addition to the body of the ware of ground flint. The traditional story of its introduction is as follows:

> A mere accident at this time (1720) caused another and important improvement. Mr. Astbury being on a journey to London, on horseback, had arrived at Dunstable,[3] when he was compelled to seek a remedy for the eyes of his horse, which seemed to be rapidly going blind. The hostler of the tavern at which he stayed, burned a flint stone till quite red, then he pulverized it very fine, and by blowing a little of the dust into each eye, occasioned both to discharge much matter and be greatly benefitted. Mr. Astbury,

[1] S. Shaw, *op. cit.*, p. 123 *et seq.*

[2] A. H. Church, *op. cit.*, p. 56.

[3] Meteyard says Banbury.

> having noticed the white colour of the calcined flint—the ease with which it was reduced to powder—and its clayey nature when discharged in the moisture from the horse's eyes, immediately conjectured that it might be usefully employed to render of a different colour the Pottery he made. On his return he availed himself of his observation; and soon obtained a preference for his ware, which produced considerable advantages. The specimens warrant the conclusion that he first employed the flint (after it had been calcined and pounded in a mortar) in a mixture with water to a thick pulp, as a *wash* or *dip*, which he applied to give a coating to the vessels, some time before he introduced it, along with the clay, into the body of his ware. For which method, a person, a few years afterwards, obtained a patent.[1]

Astbury occupies an important place in pottery technology as the precursor of Wedgwood. Both were men of fertile genius, eminently practical, who contributed to the advance of complementary aspects of their art. Wedgwood's contribution was to the scientific and artistic side, Astbury's to new adaptations of elementary principles.

> In the infancy of all arts, rude experiment solves the greatest problems; and some minds are so constituted as to arrive at positive conclusions more by rapid intuition than by laborious processes of thought; and this seems to have been the case with Astbury.[2] At that date chemical science, either in practice or theory, had made but little progress; and in a remote district like the Potteries, no means whatever existed for either self-instruction in the elements contingent to the potter's art, or for basing its advance upon scientific formulae. Thus every discovery was almost necessarily fortuitous.[3]

Further improvements in earthenware bodies were made by Thomas Astbury, son of the above John, who began potting at Shelton in 1723. It was he who first produced the 'cream colour' which, in the hands of Wedgwood, was transformed into one of the most universally used table wares.

Grinding flint introduced new industrial hazards into the potting industry.

> When it was first introduced, the potters put it to calcine in their ovens when fired; after which it was pulverised in large iron mortars, by men, and then passed through a fine hair sieve. These processes were, however, exceedingly laborious, and extremely deleterious; every possible precaution employed being ineffectual

[1] S. Shaw, *op. cit.*, p. 129.

[2] Samuel Astbury, the brother of John, married Elizabeth Wedgwood, Josiah Wedgwood's Aunt.

[3] E. Meteyard, *op. cit.*, Vol. 1, p. 148.

> to prevent great quantities of the finest particles of the silex floating in the air of the apartment, and being inhaled by the workmen, produce the most disastrous effects, by remaining in the lungs in spite of every expectorant, causing asthma, and often premature death.[1]

The difficulties caused by the dust were got over by grinding under water. Both Samuel Parkes and Richard Watson[2] credit James Brindley (1716–72), who had been employed by the Wedgwoods to construct flint mills before becoming connected with the Duke of Bridgewater in the capacity of canal-builder, with inventing the process, but it is more than likely that he merely contributed to its evolution, since he was only a boy of ten when Thomas Benson, a London painter, took out a patent[3] for grinding flints in water. This patent covered grinding with iron wheels, but the presence of iron in the ground flint was found to be injurious to the ware made with it, so a subsequent patent was added by Benson[4] covering grinding by means of stone balls. Detailed information is given by E. Meteyard, who says of the contributions of Astbury and Benson:

> To Astbury belongs the merit of having introduced two of the essential components of modern pottery, and this in their purest and best form; and from the date of Benson's improved method of flint-grinding may be traced the extra-ordinary development of the trade in all its branches. Men were lured from agricultural and other employments to an extent that, from about 1735 to 1763, it was difficult for the agriculturists of the surrounding country, and the tradesmen of the adjacent villages and towns, to meet with workmen, or boys for apprenticeship.

Yet Benson was no material gainer from his invention.

> .Thomas Benson shared the fate of too many patentees and inventors. The sums he borrowed to carry out his improvements, and secure his patents, led to embarrassments, and these to ruin; and it is said that he died in most reduced circumstances.[5]

Careless use of lead, which, on the improved bodies had replaced salt, was equally serious, but the difficulties were not entirely technical. A contemporary account of occupational disease in potters, and attempts to improve conditions, is given by Parkes in his *Chemical Catechism*.[6] Parkes tells us that French

[1] S. Shaw, *op. cit.*, p. 142. [2] *Chemical Essays*, Vol. 2, p. 262.
[3] 5th November, 1726, No. 437. [4] 14th January, 1732, No. 536.
[5] E. Meteyard, *op. cit.*, Vol. 1, p. 153.
[6] S. Parkes, *Chemical Catechism*, p. 111. Plumbism in Potters: Wm. Burton, *Mem. Manch. Lit. and Phil. Soc.*, 1900, *45*, 1.

65. Sunpan, in which the clay is *blunged*.

66. Slip-kiln, in which the purified clay is partially dried.

67. Old method of flint-crushing—a serious industrial hazard.

68. Improved method of flint-grinding under water, variously attributed to Thomas Benson and James Brindley. All the illustrations on this page are probably of the works of Minton or Copeland early in the 19th century.

Picture Post Library

69. Mill-room in which the various ingredients for pottery are mixed.

70. Saggars to hold the pottery during firing were introduced when the fuel was changed from wood to coal.

potters used common flint glass instead of lead, and found it safer and more economical. It was applied by being ground, mixed with clay and water, and goods dipped in it.[1] In England it is doubtful whether this could have been done, on account of the absurdly strict glass laws. Salt, as an alternative, was also restricted by fiscal duties.

> If the duties were removed from Salt it could be used as glaze for pottery and thus remove danger from men working with lead solutions who are liable to paralysis.[2]

Nieseman, a potter at Leipzig, proposed nitre, potash, salt and powdered glass for the same purpose.[3]

The Society for Promoting Arts, Manufactures and Commerce, which, from its inception, was interested in the suppression of dangerous trades, offered rewards for the production of leadless glazes.[4]

Job Neigh was awarded a gold medal of the Society for a glaze entirely free from the deleterious qualities of lead.

Reverting to the story of the evolution of white-bodied ware, we learn from Shaw of the difficulties of his time, i.e. first half of the eighteenth century:

> In the early processes of the *White Stone Pottery*, many obstacles required to be surmounted, and the prejudices of workmen presented various impediments. The manufacturers possessed little knowledge of the chemical properties of the various articles; neither had they any precedents for the kinds they severally attempted to make. But, as several persons at the same time were endeavouring to produce new and particular kinds, each experienced some degree of success.[5]

The difficulties can well be illustrated by those encountered by Thomas and John Wedgwood, sons of Aaron Wedgwood, who set up as potters at Burslem round about 1740. 'All round the sun-pans was thrown rubbish containing refuse salt, which mixed with the efflorescence from the salt glaze ware, was carried by the water from the falling showers into the sun-pans, and formed a saline liquid. This very important fact was forgotten by the Wedgwood brothers in using this almost saturated liquid to mix their flint and clay. As a result they sustained a loss by their pottery

[1] S. Parkes, *Chemical Essays*, Vol. 3, p. 324, *n.*
[2] R. Dacre, *Testimonies in Favour of Salt*, p. 206.
[3] *Philosophic Magazine*, 1799, *4*, 203.
[4] For details *see* A. M. Bailey's *One Hundred and Six Copperplates of Mechanical Machines, etc.*
[5] S. Shaw, *op. cit.*, p. 148.

fusing at a temperature much below that of other vitrescent kinds, even though glazed with salt, and prior to introducing the glaze.'[1]

The Wedgwoods, Thomas and John, carried out many experiments, and successfully produced durable articles without glazing at all. Such are referred to as *dry-bodies*. They subsequently experimented upon stone-ware treated by rubbing with manganese, cobalt, copper, iron, etc., before glazing. They made a considerable fortune, built the Big House, and were able to retire in 1763.

Lead ore glazing continued to be used in many potteries. When, however, lead ore, alone or mixed with a little flint, was applied as a glaze, the white clay, not being of the best quality, and the flint not being carefully prepared, the resulting pottery was yellowish in colour, and it was this pottery that was ultimately called *the cream-coloured*. (It must be remembered that this method of making cream-coloured was in operation in many potteries before the ware was so named.) Manufacturers of white stone-ware therefore experimented in various other directions, and the yellowing effect of lead glaze brought back salt glaze, formerly used for brown stone-ware, into prominence again, because lighter colours could be obtained with it. This constituted a further stimulus to improve salt glaze. Aaron Wedgwood and his brother-in-law, W. Littler, of Longton Hall, conceived the idea of dipping the clay article in a *slip* of fusible material when it was no more than partially dried. When fired in the salt-glaze oven, a shinier surface resulted than that obtained by salt-glazing alone.

White stone-ware was now in demand, and, with the increased demand, workmanship also improved up to a point. Many new potteries opened up in already established centres. So wrote, for example, Josiah Wedgwood (1730–95) of his early activities:

> White stone-ware was the principal article of our manufacture. But this had been made a long time, and the prices were reduced so low that the potters could not afford to bestow much expenditure upon it, or to make it so good in any respect as the ware would otherwise admit of. . . . And with Elegance of form, there was an object very little attended to.

By 1750 sixty factories were working for an extensive trade in this ware. They included Thomas and John Wedgwood, sons of Aaron Wedgwood, Burslem; R. and J. Baddeley, Shelton; Thomas and Joseph Johnston, Lane End; R. Bankes and John Turner, Stoke; John Barker and Robert Garner, Fenton; John Adams and John Price, Lane Delph. Over the period 1725 to 1755

[1] S. Shaw, *op. cit.*, p. 157.

all these potters and many others were making *crouch* ware and fine white stone-ware. In ten years, for reasons which we shall now examine, the trade had begun to decline and, according to Dr. Aiken, 'this ware began to be rejected from genteel tables'.

Gradual improvements in materials used in fabricating articles called for correlative changes, particularly more carefully controlled conditions of firing. This was first achieved by the introduction of two firings by Enoch Booth of Tunstall (*c.* 1750). He carried out a preliminary firing of the clay article, producing so-called *biscuit* ware, and followed it by *glost* firing, merely to glaze the already fired article. This operation was first put into common practice by the Warburtons of Hot Lane, Burslem, and the Baddeleys of Shelton.[1] Enoch Booth, in addition, united carefully prepared local clays with imports from Devon and Dorset, together with a proportion of flint as used by Astbury. This body he glazed, first with lead ore, later with lead ore in conjunction with some clay and flint, and finally with lead and flint according to Littler's method. The dipping was carried out at the biscuit stage. Shaw indicates that Booth's success is attributable principally to quality, and changes he brought about in the body of his ware, as analysed by the celebrated chemist Réaumur showed that different methods of glazing earthenware were already widely known. The changes which Booth introduced were, however, of great consequence, as they enabled Wedgwood to perfect *cream-coloured ware* under the name of *Queen's ware*.[2]

We may sum up the developments to date in the words of A. H. Church:

> Between 1710 and 1715 there were in Burslem six ovens turning out stoneware, in all probability glazed with salt. Between these dates and the middle of the century, much advance was made in the quality and quantity of this stoneware, and there can be no doubt that the introduction of salt-glazing into the Staffordshire potteries was the cause of the rapid expansion of the earthenware manufacture there during the first half of the eighteenth century. The commixture of clays, the addition of silica and especially ground flint, the repeated efforts to improve the colour, texture, hardness, and form of a ware which at the outset had much to recommend it, all stimulated further invention and progress. In the first quarter of the eighteenth century, indeed, this particular manufacture was not carried on with vigour and to the extent that one might have expected from our experience of to-day. But still the single kiln of each maker soon became insufficient to

[1] W. Burton, *Josiah Wedgwood and his Pottery*, p. 41.

[2] S. Shaw, *op. cit.*, p. 177.

supply the demand, and so the works were enlarged and more workmen were trained, until towards the middle of the century the manufacture was carried on in scores of potworks vying with one another in the perfection and variety of their products. In immediate succession to John Philip Elers we meet with the names of Astbury and Twyford as makers of salt-glazed stoneware. Astbury used Devon and Dorset clays as well as local materials for his body, adding afterwards a large proportion of ground flint. The patents of Thomas Billing in 1722 and of Ralph Shaw in 1732 were directed towards further improvements in the ware. Dr. Thomas Wedgwood, of Burslem, was another celebrated potter of that day.

In 1755 a discovery of great consequence to potters was made by a Plymouth chemist and druggist, William Cookworthy. It was the discovery in Cornwall of kaolin, or *china clay*. This clay is white in colour and results from decomposition of granitic felspar. Clay of this type had been in use for a decade before Cookworthy discovered it in England, but it had to be imported from Carolina and Virginia. Josiah Wedgwood was one of the importers. In connexion with importing it, Wedgwood asked his friend Lord Cathcart, High Commissioner for Scotland, to assist him, and was advised to apply to 'Garbett, a Scottish potter', who could no doubt give him the necessary information.[1] This was very likely Dr. Roebuck's partner at Prestonpans.

In 1768, Cookworthy obtained a patent for his discovery, but failed financially in 1774. He sold the patent to Richard Champion, Bristol, who attempted to get it extended to cover another fourteen years. Staffordshire potters opposed this. Wedgwood was particularly active. Their objections were sustained, and as a result supplies of china clay became generally available, except in that it could not be used in the manufacture of porcelain. We next find Wedgwood in correspondence with James Watt with reference to managers for mines which he and John Turner had leased at St. Austell. Using this material, Wedgwood made his light and dark fine earthenware, the colour being varied by the percentage of iron in the glaze.[2]

The demand for cream-coloured ware made by Enoch Booth's method with the fluid lead glaze increased, and its quality was further improved about 1765 by John Greatbach's glaze, and ultimately by addition of Cornish stone as well as china clay from about 1768. It was this improved cream ware that Josiah Wedg-

[1] J. A. Fleming, *Scottish Pottery*, p. 41.

[2] E. Meteyard, *op. cit.*, Vol. 2, pp. 67–68. *See also The Case of the Staffordshire Manufacturers.*

wood began making. The ware became under royal patronage *Queen's ware*. Its composition has remained unaltered to the present day: a body of ball-clay, china clay, ground flint, and china stone; glazed with glassy frit of fused borax, soda, and a little potash, with the addition of china clay, whiting, and flint.[1]

Many records of experiments by Josiah Wedgwood are extant, and were used by Samuel Smiles when he wrote his biography of the potter. His notebooks date from 15th February, 1759.[2]

As already mentioned, Wedgwood records that his principal manufacture was white stone-ware, glazed with salt. Its cheapness led him to experiment with coloured glazes, with composition of clays for bodies and on form for his products. As patents could not be relied on to give protection, he kept his notes in cipher. By continual experiment on clays, kaolin, barium carbonate, etc., he succeeded in producing a wide range of beautifully finished wares.

By 1763 he had evolved a species of stone-ware with a rich and brilliant glaze which, at the same time, had the added advantage of being cheap. It was moreover easy to manufacture, and resistant to thermal change. This became *Queen's ware*. Its composition then was roughly four measures of flint, twenty to twenty-four of whitest Dorset, Devon, or Cornish clay. It was fired twice, and covered with a glaze made from six pounds of good flint glass, twenty-four pounds flint, and one hundred and twelve pounds of white lead. The glaze was virtually flint glass.[3]

How great the influence of Queen's ware really was, even in Scotland, may be gathered from the following citation:

> But perhaps, among all the improvements made in the household furniture and utensilry, the greatest about this time was the introduction of a new species of dishes from England, instead of the old, clumsy, Dutch delft-ware, and the more ancient pewter plates. This was the elegant cream-coloured stoneware, invented in 1763, by Josias Wedgewood [*sic*], in Staffordshire, from whence it took its name. In the course of a very few years it spread over the whole country; and being fully as cheap as any of those kinds of table service mentioned, was very readily adopted in the farmers' families, as it displaced none of their own handy-works, and was highly agreeable to the females of the house.[4]

Experimenting was not confined to perfecting Queen's ware. About 1760, Wedgwood produced an unglazed black porcelain

[1] W. Burton, *op. cit.*, p. 41.
[2] S. Smiles, *Josiah Wedgwood, F.R.S., his Personal History, passim.*
[3] A. Ure, *Chemical Dictionary*, Article: *Pottery.*
[4] Quoted J. G. Fyfe, *Scottish Diaries and Memoirs, 1746–1843*, p. 276.

which he named *basaltes*. This was effected by the admixture of a considerable quantity of clay ironstone ($FeCO_3$—clay) and manganese ore, with the clay for the body.[1] To increase the range of his products, he assiduously collected varied raw materials from all over the country; cobalt from Mr. Tolcher in Plymouth, growan clay from Redruth, soap-stone from Lord Falmouth's property. He experimented with them, and analysed them in so far as he was able. Tests of their applicability to practical pottery then followed. At his death he left no less than seven thousand specimens, classified according to utility. Many of these experimental and trial pieces were given by Josiah Wedgwood II (1769–1843) to Elijah Mayer, and they now form part of the Mayer Collection in Liverpool Museum.

In the application of chemistry to clays and mineral substances, he foresaw possibilities of making elaborate ornamental goods. Writing in 1770 to his partner, Thomas Bentley (1731–80), he says:

> I am going on with my experiments upon various earths, clays, etc., for different bodies, and shall next go upon glazes.

He continued to apply himself with great industry to the study of chemistry, and ultimately produced a number of different, previously unknown, wares which brought him lasting fame. He systematically experimented with every likely mineral, many collected by himself or sent to him by scientific friends. About 1773, his notebooks reveal that he was experimenting with witherite or barium carbonate ($BaCO_3$), and heavy spar or barium sulphate ($BaSO_4$), known to Derbyshire miners as *cawk*. Devising a ceramic mixture containing barium sulphate was Wedgwood's greatest achievement, and a fitting reward for an adventurous journey to Stony Middleton in 1774 in search of it. The ware passed by the name of *jaspar ware*, and chemical analysis reveals its composition as:

Barium Sulphate . . .	59 per cent.
Clay	29 per cent.
Flint	10 per cent.
Barium Carbonate . . .	2 per cent.

A few years later he invented a *jaspar dip* to economize in the use of such expensive raw materials.[2]

[1] W. Burton, *op. cit.*, p. 63.
[2] For analysis, *see* A. H. Church, *Josiah Wedgwood; also* S. Parkes, *Chemical Essays*, Vol. 2, p. 244.

Down to the discovery of this jaspar ware, Wedgwood confined his attentions to *earthenware*. In this field he succeeded in giving to hard pottery a brilliant glaze which, until then, had been an exclusive property of porcelain. A. H. Church is not uncritical of some aspects of Wedgwood's work, however:

> His improvements in the potting or fashioning of his wares, and in their body or paste, were very great and perfectly legitimate. So much as this cannot be said of the artistic value of his work. Accepting, and even encouraging, the prevailing fashion of his day, Wedgwood adopted the rather shallow conceptions of classic art then in vogue. . . . In a word, Wedgwood was a great potter, but not a great artist. In the former capacity he influenced favourably the whole subsequent course of English ceramic industry; less happy in their results have been his fondness for the antique and his lack of originality. . . . The improvements which he effected in the ceramic industry of the country were too substantial to be seriously compromised by the want of spontaneity in the artistic character of much of his choicer ornamental ware. Indeed, the latter formed in reality but a small proportion of the array of different productions which emanated from the works of Josiah Wedgwood. His 'useful' and 'table' ware it was made his fortune and influenced the whole subsequent manufacture of pottery in England.[1]

Wedgwood's connexion with contemporary scientists and his own scientific labours led him to experiment in making crucibles, retorts, mortars, and pestles. The experimental work necessary for making these successfully covered a long period, and only succeeded after the adoption of Cornish clay in making them. Samuel Parkes speaks of his retorts as the best obtainable.[2] Among their users was Dr. Joseph Priestley, Wedgwood's personal friend, who was being supplied with them even before he left Leeds.[3] Of Wedgwood and Priestley, J. W. Mellor, of the Stoke-on-Trent Technical College, which, incidentally, contains one of the world's most important ceramic libraries, said:

> Josiah Wedgwood was not a trained chemist, neither was Priestley. However, both had the necessary instinct and natural aptitude of a chemist, and both developed their special gifts. . . . It is a pity that a great mass of work done by Wedgwood in what might be called the theory of pottery has not been published, since all will have to be repeated. If Wedgwood's work were

[1] A. H. Church, *English Earthenware*, p. 80.

[2] S. Parkes, *Chemical Catechism*, p. 166, *n.*

[3] J. Priestley to J. Wedgwood, 26th May, 31st May, 26th June, 1781, and many others. B.R.L.

available, the newer school could start where Wedgwood left off, instead of spending decades in arriving at the same stage as he had attained.[1]

In the education of his family, Wedgwood put a strong emphasis upon scientific training. We hear of him sending his son a long account of *The Natural History and Uses of Lead.* In choosing a tutor for his sons, he strengthened his connexions with Scotland. One tutor resident with the Wedgwood family was John Leslie, who became Professor of Natural Philosophy in Edinburgh, and a distinguished Scottish scientist. Leslie was interested in heat and light, and it is not unlikely that he influenced Wedgwood's son, Thomas, who became a pioneer in photography.[2] Later, the family studied chemistry under Mr. Walton, a great admirer of Priestley, who may have influenced Wedgwood in the production of his Priestley medallion.

In October 1785, Wedgwood sent his son John to Edinburgh University, where he met Dr. Joseph Black, Professor in Chemistry and friend of James Watt and Dr. Roebuck. John is said to have attended lectures by Drs. Rutherford, Duncan, Hutton, and Robison. It was William Playfair of Edinburgh who pointed out to Wedgwood the desirability of relating the calibrations of his pyrometer to the scale already established by Fahrenheit. Wedgwood followed his advice, and in his second and subsequent papers to the Royal Society, temperatures are given in degrees Fahrenheit.[3]

Although they were well received by men of note in Scotland, and associated with such men as Black, Dugald Stewart, and Leslie, the Wedgwoods did not like Edinburgh, and complained of the dirt, moors, mountains, wind, smoky chimneys, and criticized the plays, concerts, and assemblies. They were waited upon by Lord Dundonald, who presented them with specimens from a pottery he had established.[4]

Later, John went to the Continent to continue his studies. We hear of him writing his father to tell him that he has been dining with M. Lavoisier. A subsequent letter from Geneva is dated from the house of Professor Pictet, and mentions that his brother Thomas was following him at Edinburgh. While in Edinburgh, they stayed with Professor Blacklock. Even the Wedgwood family portrait modelled by Stubbs shows one of the boys making

[1] J. Wedgwood, *The Personal Life of Josiah Wedgwood.* Postcript by J. W. Mellor, quoted by J. Thomas, *op cit.*

[2] W. Burton, *op. cit.*, p. 145.

[3] S. Smiles, *op. cit.*, p. 245.

[4] E. Meteyard, *A Group of Englishmen* (1795–1815), p. 35.

carbon dioxide in a glass apparatus. Wedgwood himself describes it as:

> Jack standing at a table making fixable air with his glass apparat, and his two brothers accompanying him. Tom jumping up and clapping his hands in joy and surprise at seeing the stream of bubbles rise up, just as Jack has put a little chalk to the acid. Joss with the Chemical Dictionary before him in a thoughtful mood, which actions will be descriptive of the respective characters.[1]

Much of Wedgwood's success was due to his scientific outlook, in an industry where there was much need for scientific control. S. Parkes tells us:

> Notwithstanding the energy with which Mr. Wedgwood devoted his inventive talents and chemical knowledge, to the improvement of his favourite manufacture, during the course of a long and industrious life, I have no doubt but that chemistry might still confer many benefits on the trade, if the manufacturers in general would pay a proper attention to the cultivation of the science.

Yet Wedgwood was not by any means the only scientific potter. His favourite pupil, William Adams (1745–1805), was an indefatigable experimenter, and built a private laboratory in his house in Tunstall, where he continued to make trials and researches in ceramic chemistry to the end of his life. There is ample evidence that he, too, was successful in the application of science to pottery.[2]

We must now consider another facet of ceramic art. Demand for a white ware and attempts to copy imported oriental wares led to developments in another direction—the making of that characteristic ware, English bone china, or porcelain. Porcelain is without doubt of Chinese origin, but the date at which this ceramic species was invented is uncertain. It is also difficult to classify. A. H. Church gives the following information:

> It must be owned that it is impossible to formulate a perfect definition of porcelain, for the term embraces many species of ceramic wares, and not only what may be termed varieties. At one end of the series of porcelains may be placed the fine stonewares, hard and distinctly translucent in their thinner parts: at the other end will be found the soft, brittle, and absorbent silicious pastes (such as the Persian) held together by the coats of glaze with which they are covered. Somewhere between the two extremes will come the glassy porcelains, in which a fine solid and opaque substance is suspended, so to speak, in a translucent vitrious medium.[3]

[1] S. Smiles, *op. cit.*, p. 233 [2] W. Burton, *op. cit.*, p. 164.
[3] A. H. Church, *English Porcelain*, p. 8.

The introduction of porcelain into Europe is thought to have been effected through Egypt, the knowledge of it crossing the Mediterranean to Florence, where, about 1568, the first recorded European manufactory was founded. From Florence the art spread throughout Europe, becoming established in such famous centres as St. Cloud, Vincennes, Sèvres, etc. The wares produced in the early stages of these manufactories were, however, more or less of the nature of an *artificial* porcelain, and it was not till 1709 that Böttger produced the first true hard porcelain at Meissen (Dresden china). He was able to do so on account of his knowledge of a china-clay found in Aue, Schneeberg, Erzegebirge. Although he attempted to keep his process a secret, the manufacture of hard porcelain spread throughout Europe, reaching St. Petersburg by 1745, and Berlin by 1750. It must be appreciated that it was the discovery of the necessary raw materials in Europe that made the manufacture of even an artificial porcelain possible. There were two such raw materials: (*a*) china-clay (hydrated aluminium silicate); (*b*) china-stone, or growan-stone, both discovered in England by William Cookworthy. Much porcelain has been made, however, from related chemical materials.

In 1768, kaolin or china-clay was identified at St. Vrieix in France by Macquer, but the credit of first making hard porcelain from the kaolin of Alençon goes to Comte Brancas-Lauraguais (1733–1824). He may also have been familiar with the kaolin of Cornwall, the use of which in the making of porcelain was patented in 1768 by William Cookworthy. Cookworthy, as mentioned above, also discovered china-stone, a second important ingredient of porcelain, which, unlike kaolin, is not a definite mineral species, but a complex (felspathic, silicious, micaceous) mixture of altered minerals with some kaolin as well.

On the 1st April, 1766, Louis-Félicité de Brancas, Comte de Lauraguais, wrote to Matthew Boulton offering to sell him the patent he had been granted for the manufacture of porcelain. He enquired at the same time whether Boulton would join the Stratford-le-Bow Company in working the patent.[1]

About this time Lauraguais appears to have tried to assert influence on the potters of Staffordshire. Isaac Panchaud (*d.* 1789) of Paris wrote Boulton telling him that:

> Count Lauraguais called on me lately and wanted me to go down into Staffordshire with him to endeavour to persuade the manufacturers to make china instead of earthenware: you will suppose I desired to be excused, as a very improper person: and

[1] Lauraguais to M. Boulton, 1st April, 1766, A.O.L.B.

yet if his china is as good as the Old Japan and will bear heat, it would be worth some ingenious man's attention.[1]

Thus we must bear in mind that neither china-clay nor china-stone was known in England when the first *porcelain* manufactories were begun. These—Chelsea, Stratford-by-Bow, Derby, and Worcester—started some time about 1745, but the exact date is not known. From 1750, manufacture spread over England and Scotland. In Scotland one of the earliest records of china being made is a note in the *London Chronicle* of 1755 which announced that 'four potters well skilled in the workings of English china were engaged to go up to Scotland where a new porcelain factory is going to be established in the manner and process of that now carried on at Chelsea, Stratford-by-Bow'. We know nothing definite of the locality of this pottery, but it was probably at Prestonpans.[2]

The porcelain made at this time was not a true porcelain, but a factitious product which nevertheless had very desirable specific characteristics. These ceramic products contained as one of their principal constituents, bone-ash (tri-calcium phosphate, $Ca_3(PO_4)_2$), prepared by calcining bones in contact with air, after removal of organic matter, viz. gelatin and fat. Distinctive character marks this product. It arises from the use of bones in making it. The early factories made no secret of imitating imported products, even marking them with Chinese characters.

BONE-ASH IN ENGLISH BONE CHINA.

Table showing the ultimate adoption of bone-ash as an ingredient at all the English Porcelain Works.[3]

Glassy x x x x   Hard paste = = =
Soap-stone o o o o   Bone-ash – – –

	1745-50-55-60-65-70-75-80-85-90-95-1800-10-15-20
CHELSEA	x x x x x x – – – –
BOW	x x x – – – – – – – – –
DERBY	. . . . x x x x x x x x – – – – – – – – – – – – – – – –
LONGTON HALL	. . . . x x x x
BRISTOL and WORCESTER	o o o o o o o o o o o o o o o o o o o – – – – – – –
PLYMOUTH and BRISTOL	. . . . = = =
NEW HALL	. . . . = = = = = = = = = = – – – – –
SPODE	. . . . ————————

[1] Isaac Panchaud to M. Boulton, 1st April, 1766, A.O.L.B.

[2] *Vide infra*, p. 323.

[3] From W. B. Honey, *Old English Porcelain*, p. 278.

Of the four manufactories mentioned above, Stratford-by-Bow is the most noteworthy to us, since the consensus of scientific opinion is in favour of regarding it as the locus for the introduction of bone-ash, which innovation influenced the whole subsequent trend of English china production. With regard to the loci of china production in general, the following citation from A. H. Church is important:

> It should be noted that so long as wood was the only fuel used in the kilns, and so long as the productions of the works were intended to be looked upon as mainly decorative or ornamental, any place provided with facilities of water carriage was adopted for the site of a factory. It was, in fact, mainly by reason of the accident of their residence in Plymouth and Bristol respectively, that Cookworthy and Champion established their works in localities not far removed from the quarries of china-clay and china-stone which they employed. The Chelsea factory, it is supposed, originated in some glass works long before carried on in that place. The personal element comes in with reference to the establishments at Worcester and Derby, where the inventors of the ware happened to live. Not until the extended use of porcelain for daily requirements had stimulated the competition of cheap manufacture, was a single great ceramic centre established. Cheap coal and many varieties of clay, together with many facilities for the carriage of the raw materials and finished products, have at last concentrated china-making in the Staffordshire potteries. Bow, Chelsea, Plymouth, and Bristol have long ceased to make porcelain.[1]

Returning to the introduction of bone-ash at Stratford-by-Bow, the origin of this manufacture is obscure, but possibly it arose out of a patent for the production of porcelain taken out on 6th December, 1744, by Edward Heylyn and Thomas Frye. Frye was the first manager of the works. He took out a second patent on 17th November, 1748, for a new method of making 'a certain Ware which is not inferior in Beauty and Fineness, and is rather superior in Strength than the Earthen Ware that is brought from the East Indies and is commonly known by the Name of China, Japan or Porcelain Ware'. Frye speaks of using *virgin earth* in his patent, and this is identified by Church as none other than bone-ash. Here are Church's own words:

> I suggested more than a quarter of a century ago (1875) that there could be no difficulty in identifying the special kind of white or virgin earth really used by Frye with *bone-earth*, that is calcined bones, which consist mainly of phosphate of lime. Frye, indeed,

[1] A. H. Church, *English Porcelain*, p. 14.

> names animal matter in the first place, as a source of his earth, and there is, in fact, but one animal matter, bone, which does yield by calcination a white, abundant, and insoluble ash. . . . I am convinced that Frye did introduce bone-ash into a porcelain-body at least as early as the year 1748, no less than seven years before the first experiment in this direction by the French chemist, Macquer. The persistent attribution of this discovery to Spode has misled for a long time English and Continental writers on ceramics. . . . So the frequent attribution to Josiah Spode the younger (about the years 1797–1800) of the first introduction of bone-ash into the paste of English porcelain must be regarded as destitute of any basis of fact. To assume that he at least fixed the best proportion of bone-ash to be employed is negatived by the evidence afforded by the chemical analysis of specimens of English soft porcelain made at many famous factories long before Spode's time.[1]

We can thus safely conclude that bone-ash was in use some years before the discovery of china-clay and china-stone in England.

The result of adding bones was to make china very white, but less dense and more liable to fracture. Parkes attributes it to economy, and describes it as 'injurious to the texture of the ware', adding, 'its use cannot be justified on any account whatsoever'.[2] Parkes, however, is censured for some of his statements about pottery by Shaw in his *History of the Staffordshire Potteries*.[3]

The discovery of kaolin and china-stone opened up developments in two directions: (*a*) the production of hybrid chinas containing china-clay, china-stone, and bone-ash; (*b*) the manufacture of true porcelains, the necessary materials then being available. The former of the raw materials, the china-clay, was discovered, as has been mentioned, about 1755 by William Cookworthy (1705–80).

William Cookworthy was a chemist and druggist of Plymouth. His interest in porcelain manufacture dates from as early as 1745, and for many years after he carefully and scientifically experimented with Devon and Cornish clays. He was an experienced chemist, and the first in the country to make cobalt direct from ore. He took up manufacture of finely modelled pieces, fired solely with wood, coal not being available in Plymouth.

By 1768, he had satisfied himself that clay for manufacture of genuine porcelain existed in his own locality. In that year he took out a patent for manufacturing at Coxside, Plymouth, a kind of porcelain, then newly invented by himself, composed of the second raw material which he discovered, moor-stone, or growan. The

[1] A. H. Church, *English Porcelain*, p. 39.

[2] S. Parkes, *Chemical Essays*, Vol. 3, p. 317. [3] S. Shaw, *op. cit.*, pp. 111, 145, 178.

patent is dated 17th March, 1768. Shortage of fuel in Plymouth led him to remove to Bristol after three years. Messrs. Cookworthy and Company operated there from 1771–3. Of Cookworthy, Shaw says: 'Mr. Cookworthy was doubtless a person of considerable ability; but according to the information concerning him from relations and Mr. Champion, he was constantly so very eager in acquiring knowledge, that he could seldom find leisure to communicate to others his own stores of information. Hence all there is to commemorate him are a few letters and essays in the periodicals of that day; and his discovery of materials for making porcelain'.[1] In 1774, he was forced to make over the business and patent rights to Richard Champion of Bristol.[2] Champion's success, commercially at least, was not much greater than Cookworthy's, as we shall see.

When Champion applied to Parliament in February 1775 for an extension of the patent for a further fourteen years, Josiah Wedgwood, John Turner, and other Staffordshire earthenware manufacturers engaged in pamphleteering against Champion's monopoly. As a result, clauses were added in the House of Lords throwing the raw materials open for use by potters, except in the manufacture of porcelain. This was of great advantage to Staffordshire, as Cornish clay was utilized to improve their earthenwares. In 1777 Champion sold his patent to a Staffordshire company.[3]

> After the fight in 1775 with Richard Champion over the china clay patent, Wedgwood visited Cornwall in company with John Turner of Lane End. There they saw Newcomen engines in operation, and so impressed was Turner that he installed one on his return. When Spode took over Turner's pottery it is probable that the engine passed over to him. Wedgwood got James Watt to survey flint-mills at Trentham and Lane End, and in particular to visit Spode's works. In 1782, Wedgwood ordered his first engine from Boulton and Watt. This was fortunate for Watt, since after he had displaced many Newcomen engines in Cornwall, Stafford was ripe for mechanisation. It was only after the general adoption of mechanical power in the Potteries that the market for engines developed in the mills of Lancashire and Yorkshire.[4]

By the end of the eighteenth century a variety of different wares was thus available. The last twenty-five years of the century had seen salt-glazed white stone-ware give place to fluid-glazed cream-

[1] S. Shaw, *op. cit.*, p. 292.

[2] The history of porcelain manufacture in Bristol can be traced in Hugh Owen's monograph, *Two Centuries of Ceramic Art in Bristol.*

[3] S. Smiles, *op. cit.*, p. 176. [4] J. Thomas, *op. cit.*, p. 407.

coloured ware. While the stone-ware continued to be used on middle-class tables, it was rejected by the genteel in favour of something better, which had evolved through the labours of Cookworthy and others.

As a rough guide, one can say that to-day the difference between earthenware and porcelain is as follows:

	Earthenware	*Porcelain*
Flint	35 per cent.	—
Ball clay	25 per cent.	—
China clay	25 per cent.	25 per cent.
Stone	15 per cent.	25 per cent.
Bone-ash	—	50 per cent.

We must now discuss methods evolved for decorating various kinds of pottery formed with the various bodies which have been described.

Before pattern-printing on china was introduced, table services of home manufacture were either plain Queen's ware or were decorated with a coloured edge, not always characterized by the elegance of its execution. Minton, an engraver who became a master potter, did much to further this development, which led in turn to ceramic lithography by which, in one process, several colours can be applied to pottery at the same time. Thus paper-making and colour-grinding in North Staffordshire were ancillary to the pottery industry. After some improvement had been effected, Lardner wrote:

> This modern improvement added materially to the decent comforts of the middle classes in England and has more than any other circumstance contributed to the great extension of our trade in earthenware with the continent of Europe.[1]

Most printing, in early days, was carried out with cobalt. For blue printing, cobalt manufacture was developed at the hands of William Cookworthy and Roger Kinnaston, who about 1772 set up at Cobridge a furnace for preparing blue from zaffre. The first blue-printed table service made in England was completed, it is said, by John Turner of Caughley, near Broseley, Salop, for the Parliamentary representative of Bridgnorth.

In 1801, M. Brogniart observed that at the time his essay was written, no attempt had been made to apply principles of chemical science to glazing. Before this, however, Edward Hussey Delaval had published his *Experimental Investigation into the Causes of the Changes in Opake and Coloured Bodies* (London, 1777). A year or two

[1] *Cabinet Encyclopaedia*, Vol. 26, p. 96.

later, Alexander Brogniart of Sèvres, and Professor Proust of Madrid, published remarks on glazes.[1]

From the methods of carrying it out, salt-glazing was not subject to modification, but dipping, on the other hand, could easily be modified to give coloured glazes. Colour was achieved by mixing various oxides with glaze, thereby forming coloured glasses. There is a tradition that bright green and yellow glazes, much used in Staffordshire, were Wedgwood's first inventions, an advance on effects obtained simply by dusting the surface with iron (smithy scale), copper, or manganese oxide.[2] The oxides used were cobalt for blue, antimony and silver for yellow and orange, lead for a silver colour, gold for gilding and purple, copper for green, and iron for red, brown, and black.[3]

In Scotland, cobalt from silver mines at Alva, Stirlingshire, was used in the Prestonpans porcelain works. It was first imagined that this cobalt was a silver ore, but a chemist showed it to be otherwise.[4] Here is the story of the mines as related in the *General View of the Agriculture of Stirlingshire*:

> About the year 1759, Charles Erskine of Alva, Lord Justice Clerk, in company with some other enterprising gentlemen, renewed the search for silver ore in these hills with considerable industry and exertion. The course of the vein was pursued a great way beyond the old workings. A shaft was made to the depth of several fathoms, immediately below the waste, from which the rich mass of ore that has been mentioned was taken, and a drift carried on upon that level in the direction of that vein. None of these operations, however, was on that occasion accompanied with success. But in driving a level at a considerable distance, nearer the bottom of the hill, for the purpose of carrying off the water from the works that were situate above, a large mass of *cobalt* was discovered, a great part of which was employed in a manufactore of porcelain which had been established about that time at Prestonpans in East Lothian. When this cobalt is deprived of the arsenic with which it is strongly impregnated, and otherwise properly prepared, it produces a powder of a beautiful deep blue, with which a variety of useful and ornamental pieces of china and glass have been coloured. There is, indeed, reason to believe that the cobalt of the Hills of Alva is, in no respect, inferior to that which is procured from the mines of Saxony.[5]

Rathbone of Portobello produced a mulberry shade by mixing iron oxide and manganese dioxide. The latter he obtained from

[1] *See Philosophic Magazine*, *13*, 342; *14*, 17; 1805, *21*, 313.
[2] W. Burton, *op. cit.*, p. 27. [3] S. Parkes, *Chemical Essays*, Vol. 3, p. 330.
[4] *Statistical Account*, Vol. 18, p. 142.
[5] P. Graham, *General View: Stirling*, p. 57.

71. Wedgwood's Pottery at Etruria, Staffordshire, *c.* 1840.

72. Wedgwood portrait medallion of Dr. Joseph Priestley by Hackwood.

73. Wedgwood portrait medallion of Benjamin Franklin.

I. Clark, 1824

74. View of Carron Iron Works from Falkirk.

75. The Upper Works at Coalbrook Dale, Salop, 1758, showing on right-hand side, heaps of ore being calcined.

manufacturers of bleaching powder, probably the Prestonpans Vitriol Company. By the use of manganese, black could also be obtained.

Other metals were subsequently introduced in the decoration of ceramic products. Of special interest are the four metals, gold, silver, copper and platinum, which were used for giving a metallic lustre to pottery. Of these, the most successful was platinum. In 1810, Peter Warburton took out a patent for printing in platinum and gold. This was one of the earliest industrial uses of platinum. Applied to the surface of clay articles, it was used to simulate silver plate. Much *silvered* ware frankly imitated in form and decorative treatment the silver or plated ware it was intended to replace. 'The plainest pieces are most agreeable, but the fluted and gadrooned patterns are often commendable.'[1]

Platinum attracted considerable attention during the second half of the eighteenth century. It was first described as a compact metal by William Watson, who got specimens of it from the explorer Charles Wood.[2] A more detailed description by Scheffer appeared two years later in Memoirs of Stockholm Academy under the title *On White Gold, termed by the Spanish 'platina del Pinto'*. In pottery industry it was dissolved in *aqua regia*, precipitated, washed, and taken up in a solution of sulphur and Venice turpentine, i.e. ordinary turpentine to which oil of lavender was added. Wedgwood obtained information about it from Dr. John Fothergill, F.R.S., in the year 1776, and in 1806 one Henri Bertrand was writing to Matthew Boulton to tell him that he could work platinum, with the further suggestion that Boulton should 'enter more into the subject of it'. On account of declining health the latter expressed himself disinclined to engage in the new venture, but owned 'I consider the refining and perfecting of that most obstinate metal Platinum of vast importance to the arts'.[3]

The making of lustre wares was general in the Potteries before the close of the eighteenth century, and increased considerably in the first quarter of the nineteenth. It was developed there by John Hancock, and John Gardner, who was associated with Josiah Spode III.[4]

Supplying materials for glazing became a separate industry.

[1] A. H. Church, *English Earthenware*, p. 128.

[2] *Roy. Soc. Phil. Trans.*, 1750, *46*, 584.

[3] H. Bertrand to M. Boulton, 10th February, 1806; M. Boulton to H. Bertrand, 14th February, 1806, A.O.L.B.

[4] For further details of its use in pottery, *see* Klaproth, *Phil. Mag.*, 1799, 5, 135; and for an extended account of pottery colours, Lardner's *Cabinet Encyclopaedia*, Vol. 26, p. 74.

'The greatest consumption of cobalt is by the potters and porcelain manufacturers, some of whom make their own colour from foreign zaffre. But in Staffordshire there are several people who make an entire trade of preparing this colour for the earthenware manufacturers, and who conduct the process with great secrecy. The usual price is two guineas the pound.'[1] Parkes speaks of the vein of zaffre for cobalt in Cornwall soon being exhausted, and states that it was principally obtained from Saxony, at a cost of 40*s.* to 60*s.* per lb.

Many large firms had developed in England before potting became a conspicuous industry in Scotland. The principal English works were: Wedgwood's *Etruria*, Spode's *Stoke*, Wood's *Burslem*, Davenport's *Longport*, Minton's *Stoke*, Bourne's *Fenton*, Ridgway's *Shelton*, Dimmoch's *Hanley*, Hick's and Meigh's *Shelton*, Meigh's *Hanley*.[2]

Attracted by available coal and clay, the period of industrial revolution in Scotland was marked by the development of potting mainly at Bo'ness, Prestonpans and district, Stevenson, Cumnock, Edinburgh, and in Glasgow.[3] It will be noted that the Scottish pottery districts are more spread out than those of England. This makes it more difficult to bring out their importance.

The following is a summary of details concerning the principal Scottish potteries.[4]

Delftfield: Pride of place with regard to date of foundation is usually given to Delftfield, Glasgow, founded in 1748. We know that by February 1749, the first good kiln of ware had been drawn, its quality being considered to be as good as, if not superior to, that of Bristol and London. It appears that they were much impeded by lack of technical knowledge at the beginning, but rapidly adopted Queen's ware when this improved ware became available. This is the pottery with which James Watt became associated in 1763. There are legal papers concerning the pottery in the Faculty of Procurators' Library, Glasgow.

About the same time, Robert Dinwiddie, a London merchant, Laurence Dinwiddie and Patrick Nisbet, Glasgow merchants, and Robert Finlay, a Glasgow tanner, thought to take up the pottery line and engaged John Bird, a London potter, to experiment with Scottish clays. In this, however, he was unsuccessful, and it is doubtful if they were able to get a pottery into operation.[5]

[1] S. Parkes, *Chemical Catechism*, p. 330, *n.* [2] S. Shaw, *op. cit.*, p. 2.
[3] J. Sinclair, *General Report*, Appendix 2, p. 297; J. A. Fleming, *op. cit.*, p. 118.
[4] For full account *see* J. A. Fleming, *Scottish Pottery*, *passim.*
[5] *Session Papers*, Signet Library, 40: 7.

Calton: Next comes Calton, founded in 1778, which became one of the most important Scottish potteries. *Bagnal's, Mill Road* (1780–1852), followed in 1780, and then in 1786 *Old Cumnock Pottery*, Ayr.

The year 1789 saw *Verreville Glass Works*, Glasgow, go over to pottery production. One of the associates of this pottery, John Geddes, was a Scottish pioneer in technical education. Next year saw the start of the *Caledonian Pottery*, on the north bank of the Monkland Canal at St. Rollox, conveniently situated for obtaining supplies of coal and clay. It was managed by Josiah Rowley of Stafford, and was probably the first pottery in Scotland to produce decorated china like that of Worcester. One of the partners, Archibald Patterson, had business associations with David Dale.[1] Before the end of the century (1797), *Wellington Pottery*, Glasgow, was in operation, and *Holden's Pottery* near Gorbal's Kirk soon followed. The beginning of the century saw the *North British Pottery* (1810) started, and *Lancefield of Anderston* (1820), the latter an offshoot of Verreville, owned jointly by Kitson and Geddes. John Geddes was interested in chemistry and natural philosophy, and soon after taking over Verreville, introduced one of Watt's engines to drive machinery and flint mills. He induced workers to come to Glasgow from Holland, Flanders, and Lambeth, and opened a technical school in the factory to teach the *Art of Potting*.

In the meantime *Bell Brothers* had established a pottery at Port Dundas on the Forth and Clyde Canal, where they specialized in making salt-glazed water pipes for the conveyance of gravitation water.

Greenock also had two potteries, *Clyde Pottery* (1814) and *Greenock Pottery* (*c.* 1820). These produced crockery for the Newfoundland market, seal fishermen taking their produce in exchange for oil, etc. One of the sons of John Geddes had a share in the latter pottery.

A similar development can be traced at the other end of the Scottish industrial belt where at one time Prestonpans and district vied with Glasgow as an industrial centre. Potteries were carried on in practically every village on the Firth of Forth: Cuttle, Joppa, Newbigging, Musselburgh, Portobello. This district, associated with several manufactures discussed elsewhere, e.g. salt-boiling, vitriol manufacture, brewing, etc., was without doubt one of the most prosperous parts of eighteenth-century Scotland. At Prestonpans it is recorded that by 1754 over seventy potters were employed in two potteries, one *Watson's*, the other *Gordon's*. Both salt

[1] J. A. Fleming, *op. cit.*, p. 221.

and flint-glazed wares were made, the flint coming from Strichen Hill.[1]

A ledger and an inventory (taken on 20th September, 1801) belonging to Prestonpans Pottery Company have survived.[2] This was probably Gordon's pottery. It reveals that stock in hand consisted of 3,724 dozen cream-coloured articles, 355 dozen best-dipped ware, 899 dozen best painted, and 2,217 dozen second ware, together with clay, flint, white lead, grinding stones, crates and cordage, the whole valued at £711 15*s.* 6*d.* To this was added for tools, etc., £215 14*s.* ½*d.*; bricks and tiles £140 2*s.* 11*d.*; biscuit or green ware £108 10*s.* 4*d.*, a total of £1,176 2*s.* 9½*d.*

The company traded with most of the east of Scotland, sending wares as far north as Lerwick. Other towns in which they had customers included Edinburgh, Glasgow, Aberdeen, Dundee, Inverness, Forres, Crail, Prestonpans, Thurso, Leith, Perth, Kirkcaldy, Huntly, Banff, Meigle, St. Andrews, Brechin, Biggar, etc.

The monthly make of pottery and the raw materials used for the period covered by the ledger are given below:

Month	*Coal*	*Clay*	*Flint*	*Lead*	*Value* £	*s.*	*d.*
Oct. 1801 . .	171 carts	11 tons	5½ tons	15½ cwt.	189	15	2
Nov. ,, . .	238 ,,	15 ,,	7½ ,,	21 ,,	244	9	7
Dec. ,, . .	149 ,,	15 ,,	7½ ,,	22 ,,	260	0	9
Jan. 1802 . .	136 ,,	8 ,,	4 ,,	12 ,,	173	6	8
Feb. ,, . .	185 ,,	5 ,,	2½ ,,	22 ,,	113	17	8

Prices of materials were: coal 4*s.* 3*d.* per cart, clay 31*s.* per ton, flint 30*s.* per ton, and lead 40*s.* per cwt. Coal they bought from Dr. Roebuck's friends the Cadells of Cockenzie, and straw from Fowlers.

Economic conditions of labour can be gathered from the following account of wages for one of the above months, viz. November 1801. There was expended in the

	£	*s.*	*d.*
1st week on men . . .	20	13	11
,, ,, ,, boys . . .	3	4	9
2nd ,, ,, men . . .	22	10	4
,, ,, ,, boys . . .	3	3	6
3rd ,, ,, men . . .	22	2	8
,, ,, ,, boys . . .	2	19	9
4th ,, ,, men . . .	20	13	3
,, ,, ,, boys . . .	3	2	9

[1] *Statistical Account*, Vol. 17, p. 67. [2] Edinburgh Public Library.

During the sixties and seventies there were several off-shoots from Prestonpans. In 1764, *Portobello*, which gave the name to so-called ware; in 1770, *Waverley*, which first made bone-china in the district. There was also *Rosebank* and, in the next century, *Newbigging*, *Musselburgh*, and in Prestonpans, *Bellfields*, founded in 1832. Here white sanitary ware of superior quality was first made in Scotland.[1]

The Prestonpans brewing industry was a stimulus to production of stone-ware. Watson's Pottery made stone-ware bottles for holding Prestonpans ales. When in financial difficulties they sought assistance to tide them over their period of embarrassment from John Fowler and Company, Brewers. The Fowlers were related to the Cadells mentioned above, and one of them managed the pottery on behalf of the Fowlers. Later, one of the name of Thomson took over, but he also became involved financially, due to backing Laidlaw of the 'Saltpans and Sulphur Works'. This suggests that Roebuck's former vitriol company may have been in trouble.

In Scotland there has been no family of potters with a continuous reputation like the Wedgwoods in England, but pottery has served to link several famous names which we have mentioned in other sections. One of these is James Watt. In 1763 Watt extended his activities to an interest in Delftfield Pottery Company. By 1772 he had £475 invested. This pottery was situated on the land side of Anderston Walk, now known as Argyll Street, Glasgow, extending to the banks of the Clyde at Broomielaw. Delftfield Lane was renamed James Watt Street. Of their wares we know that by 1766 white stone-ware, harder in the *paste* than delft, had been introduced. Delftfield was indeed a Scottish pioneer, and without doubt Watt carried out experiments to effect improvements in their products, testing clays, advising on flint-grinding and furnace-construction.[2] Watt and Josiah Wedgwood were brought into contact, it is thought, through Dr. Small, an Edinburgh graduate with whom Watt had a life-long correspondence. This coterie brought Watt into contact with Dr. Erasmus Darwin, Wedgwood's medical adviser, and a keen experimental scientist. Darwin was interested in Watt's engine, and suggested that Wedgwood should instal one. In 1802 as a result we find Wedgwood paying Watt £2,165 13*s*. 7½*d*. for an engine and flint-mill. Even in 1789, Watt was being referred to on pottery problems. Marsiglio Landriani, in a letter to Matthew Boulton, wrote that 'if

[1] *Statistical Account*, Vol. 18, p. 366.

[2] H. W. Dickinson, *James Watt*, p. 29.

Mr. Watt can tell me of a method for making a fine black colour for porcelain he will oblige me'.[1]

It will be remembered that Dr. John Roebuck was associated with the original trials of Watt's engine. Roebuck himself was associated with pottery manufacture in the Forth area. A manufactory of stone-ware was begun at Bo'ness about 1766. It belonged to William Cadell, almost certainly connected with the Cadells who were partners with Roebuck in the Carron Iron Works.[2] In 1784, when he was sixty-six, Roebuck purchased this pottery, then making coarse pottery from local clays. He set about importing better clay and flint, and succeeded in producing a good-quality white and cream stone-ware.

To begin with, Roebuck availed himself of such chemical knowledge as was available by a query addressed to Dr. Joseph Black:

> In the reply, after instruction to use white lead instead of red lead, and to add borax as a flux, he goes on: 'I am doubtful if manganese or arsenic be necessary or useful in the composition of glazings, and the last is so very dangerous to the health of the poor workmen that it cannot be used tho' ever so cautiously without constant apprehensions of its doing mischief'. He then describes the preparation of gold for gilding: dissolving gold in *aqua regia*, evaporating the solution to dryness, and igniting; or precipitating with *sal martis* (ferrous sulphate). In either case the gold is mixed with borax and oil of lavender to give it the consistence of a paint. Black then goes on to describe the best method of preparing a brightly coloured 'Purple of Cassius'.[3]

Roebuck's shares, we are told, were bartered for rags, mainly in Edinburgh, which were brought back to the *secret works* of Robert W. Hughes. We do not know what works this refers to.

There were also Cowans, potters, in Bo'ness.[4]

It was said by a contemporary writer that no industry required more processes than the ceramic, and how right this statement is has been illustrated by the above long discussion. We must now pass to the consideration of a department of technics where the processes were few but the social effects immense.

[1] 8th September, 1789, A.O.L.B.
[2] T. S. Salmon, *Borrowstoneness and District, c.* 1550–1850, p. 153.
[3] W. Ramsay, *Life and Letters of Joseph Black*, p. 102.
[4] *Session Papers*, Signet Library, 368: 22.

Chapter XVI
IRON AND IRON-MASTERS

> The dependence of our knowledge on the skill of our hands causes the history of a science to be closely related to the history of practical technique.
>
> P. M. S. BLACKETT.

THE iron industry of Great Britain in general, and of Scotland in particular, has already been the subject of a number of detailed studies. Though some might regard this as a sufficient reason for excluding further discussion here, valid reasons can be adduced for reconsideration. One is that, although the iron industry forms a central theme in historical studies from the economic standpoint, emphasis is usually laid upon results, rather than methods whereby results were achieved. Another is that the development of the ferrous metals industry is intimately connected with correlative development of other industries already considered. In the words of Mantoux:

> The history of iron and steel is not that of a single industry, but can, from a certain point of view, be identified with that of 'Great Industry' itself.[1]

Indeed, when one studies the activities of the great iron-masters, one thing becomes evident. There was what we may call an 'interlocked inventorate' between iron industry and other industries which developed simultaneously, particularly in Scotland, where such men as Roebuck, Macintosh, Neilson, Dundonald, *et al.*, had connexions with several undertakings other than iron. Within the framework of private investment, marriage also helped to complicate the ramifications of the iron industry.

The date when iron ore was first reduced in Scotland is lost in antiquity, but there is sufficient evidence to demonstrate that although iron-working before the middle of the eighteenth century was not extensive, it was at least widespread. Moreover, certain districts early specialized in fabricating definite articles. For example, from 1669 till 1727, smiths in Culross enjoyed a monopoly of making *girdles* (iron plates for baking), said to be their invention.[2]

[1] P. Mantoux, *The Industrial Revolution in the Eighteenth Century*, p. 277.
[2] *Statistical Account*, Vol. 10, p. 141.

By the beginning of the eighteenth century the making of iron for domestic articles, and for such implements of agriculture and war as were made of iron, was carried out in two operations: (*a*) smelting of ore in an air- or later in a blast-furnace for production of pig-iron; (*b*) fining of iron so reduced, in a forge, for production of bar or malleable iron. The detailed evolution of these two processes cannot now be traced.

The ore of iron is usually ferric oxide (Fe_2O_3), which occurs as mineral haematite. The ore is introduced into the furnace mixed with coke (or formerly charcoal) and limestone (calcium carbonate, $CaCO_3$). The carbon of the coke (or charcoal) combines with oxygen of the air to produce carbon monoxide.

$$2C + O_2 = 2CO$$

This gas then reacts with ore and reduces it in stages to molten iron.

$$(1)\ Fe_2O_3 + CO = 2FeO + CO_2$$
$$(2)\ FeO + CO = Fe + CO_2$$

The iron so produced is not pure and has to be *fined*. At first this was done in the forge, next by *puddling*, which in turn was superseded by melting the pig-iron from the blast-furnace in a reverberatory furnace with a haematite lining.

Although the equipment required to produce iron was relatively simple, T. S. Ashton states that an iron-works required a volume of capital so great that few save landowners could command it. Furnace and forge were usually operated conjointly by a single owner. Where there was apparent division of labour, it usually indicated that the 'specialist' could not command sufficient capital to carry out both the smelting and forging.[1]

> A mine, one or two blast furnaces, and often an iron-works in the same hands, necessarily amounted to a capitalist undertaking.[2]

The operator of such an undertaking was often a landowner, who had at his command on his own estates the necessary raw materials, viz. ore, lime, charcoal derived from his timber, and later, coal from his mines. Indeed, iron-making was a method of 'improving' estates. It was this class of society which created simultaneously a demand for implements of war and agriculture. There is another point:

> In considering the eighteenth century changes in the iron industry, one should remember, too, that the accompanying reorganisation of economic groups was slighter than in the textile

[1] T. S. Ashton, *Iron and Steel in the Industrial Revolution*, p. 6.
[2] P. Mantoux, *op. cit.*, p. 283.

industries, for the relatively large iron-works had for a long time more nearly resembled the factory system than had the smaller and household reorganisation of textile manufacture.[1]

In both operations, smelting and forging, the fuel used till well into the eighteenth century was charcoal. Forest therefore determined the locus of the iron industry. Even in Scotland, where the industrial development lagged markedly behind that of the south, although the quantity of iron made before 1760 was small, the amount of timber consumed by the operation of smelters and forgers had been sufficient to necessitate the passing by the Parliament of Scotland of an Act prohibiting the making of 'yren from wode'. The districts of Scotland in which this Act was operative were most likely the Lowlands, that part of the country which has always supported the greatest concentration of population. As evidence one may cite the innumerable references in Chalmers' *Caledonia* to disafforestation in the Lowlands. In the Highlands, by contrast, forests remained for long untouched. Consequently when iron-masters in England could no longer get adequate supplies of charcoal in England, this Highland timber attained considerable economic importance, and furnaces for the production of charcoal iron made from imported ore were set up in Scotland. The details of the furnaces are as follows:[2]

1728. *Culnakyle*, near *Abernethy*, a short-lived venture of the York Buildings Company.

The York Buildings Company had purchased timber in the Abernethy forest in 1728 with the intention of using it for the masts of His Majesty's ships. It was, however, found to be unsuitable, and in all probability this turned them to the idea of converting it to charcoal. The following information is given in David Murray's *The York Buildings Company*:

> The Company set up iron furnaces in the neighbourhood under the charge of one Benjamin Lund, where were produced 'Glengarry' and 'Strathdown' pigs; and they also had four furnaces for making bar iron. For these charcoal was needed, and for this they cut down great quantities of timber. Besides supplying their own wants, they made large shipments of charcoal to England, to Holland, and other places. The iron venture was not more successful than the timber scheme, and at Christmas, 1732, there stood to the debit of 'Iron Works' account £6,935 6*s*. 11½*d*.

[1] Witt Bowden, *Industrial England*, p. 91.

[2] *See* W. Ivison Macadam, *The Ancient Iron Industry of Scotland*, Proc. Soc. of Antiquaries, 1887, *21*, 130.

The ore smelted at Abernethy was haematite carried by ponies from the Lecht mines, beyond Tomintoul, a distance of fourteen miles. For some other notices of the work see Sir Thomas Dick Lauder's account of *The Morayshire Floods*.

1729–36. *Invergarry*, Inverness-shire; on south side of River Garry, close to Loch Oich: founded by the Backbarrow Company of Furnace. The Backbarrow Company had connexions with Liverpool merchants, and received charter from John Macdonald of Invergarry.

1730. (1) *Bonawe* or *Lorne*, Taynuilt on Loch Etive, Argyllshire: founded by an Irish Company in 1730, which was
1753. (2) followed by Richard Ford and Company in 1753: worked till 1874.

1754–1813. *Goatfield*, eight miles from Inverary, Argyllshire (now called Furnace): founded by the Duddon Company of Lancashire: worked till 1813.

Reasons why these companies came to the Highlands are clearly indicated in the following extract:

> Some years ago (*c.* 1726) a company of Liverpool merchants contracted with the Chieftain of this tribe, at a great advantage to him, for the use of his woods and other conveniences for the smelting of iron, and soon after put their project in execution, by building furnâces, sending ore from Lancashire, etc. . . . By the way, I should tell you that those works were set up in this country merely for the sake of the woods because iron cannot be made from the ore with sea or pit coal to be malleable and fit for ordinary uses.[1]

When touring Scotland in 1784, the Bonawe furnace was visited by Faujas de St. Fond, who described it as follows:

> It stands in a charming situation, embellished around with woods, verdure, and well-cultivated land. . . . We learned that it belonged to an English company, which had been induced to erect works at this place in consequence of the abundance of wood and water, and its proximity to the sea. . . . The ore used in this foundry was brought in vessels from Cumberland. . . . This establishment appeared to be conducted with as much skill as economy, but the woods were beginning to be worked out. . . . It is therefore feared that this foundry cannot be carried on much longer.[2]

Both these writers clearly emphasize the importance of woods, water, and convenient transport to the expanding iron industry.

[1] Ed. Burt, *Letters from a Gentleman in the North of Scotland*, p. 264.
[2] Faujas de St. Fond, *Travels in England and Scotland*, p. 148.

Alteration of locus and migration to the Highlands of Scotland brought but a temporary alleviation of the difficulties caused by the timber shortage. Attention had to be directed to possibilities of finding in peat or coal an alternative to charcoal. This problem of finding a convenient substitute was one which was by no means confined to the iron industry, but was of equal importance to brick-makers, brewers, dyers, and brass-finishers. The solution of the problem had far-reaching consequences, as has repeatedly been demonstrated.

> The consumption of charcoal in fluxing iron from its ores, and in manufacturing it into bar iron is so very considerable, and the price of charcoal, from the great scarcity of underwood in this kingdom, is so great, that many attempts have been made to substitute in this business charred pit-coal in the room of charcoal. These attempts have in part succeeded, and iron is now very generally extracted from its ore by fires made with coak.

According to T. S. Ashton, there are several claimants to the honour of successful effecting of the above substitution in the iron industry. Dud Dudley (1599–1684) usually gets credit. Ashton says of his claims:

> That Dudley did produce some sort of iron with mineral fuel is probable enough, but that this was of sound mercantable quality is unlikely; and there is no valid reason why this Balliol undergraduate, rather than any one of a dozen other projectors of the seventeenth century, should have been singled out for fame.[1]

Dudley suffered many setbacks. Floods ruined his works. His patents were abolished by Parliament. The charcoal iron-masters drove him from Worcester County, and after he had removed himself to Staffordshire, a riot was organized and his apparatus destroyed. Tradition was heavily weighted against experiment.

An important technological advance did come, however, when Abraham Darby I (1677–1717), a Coalbrook Dale Quaker, first converted his mineral fuel into coke before using it in the blast-furnace. He also took out a patent for casting kitchen ware in sand, a process which he mastered on a trip to Holland in 1706.

Darby was by no means the first to produce coke. A process for making it had been patented as early as 1627, and success in its use was only achieved by him after many trials. Darby's contribution to technology was recognized by the offer of a Fellowship of the Royal Society. This he refused.[2]

[1] T. S. Ashton, *op. cit.*, p. 10. [2] *See Phil. Trans.*, 1746–7, *44*, 371.

The advantages of coke were summed up by Thomas Clark (1801–67), Professor of Chemistry, Marischal College, Aberdeen. He stated that they were 'the higher temperature of combustion in consequence of none of the resulting heat disappearing in latent form in the vapours arising from the coal'.[1] Larger charges could thus be handled in the furnace, and a longer contact between the fuel and ore resulted in a more fluid iron being produced. On the economic side, Faujas de St. Fond made the pertinent remark that 'this admirable means of preserving our great and valuable forests is a thousand times more efficacious than that crowd of laws, of regulations, and of employees, which tend only to destroy them'.[2] Darby's success also appears to have been connected with his having had a sufficiently strong blast to produce a properly fluid instead of a pasty iron. Abraham Darby II (1711–63) is credited with the adding of limestone (calcium carbonate, $CaCO_3$), to give a more easily-melted flux which removed impurities as a slag.

> The old charcoal furnaces, from twelve to eighteen feet high, or, where a good water supply existed, up to twenty-eight feet, gave place to coke furnaces of forty feet, fifty feet, and sixty feet. An attempt was made in South Wales to have one seventy feet, but in vain, and it was reduced to thirty feet.[3]

An immediate effect of producing this fluid metal was to give a fillip to the foundry side of iron-working, causing cast iron (i.e. pig-iron remelted in a founder's cupola) to be substituted for wood, copper, lead, brass, and particularly for bar or malleable iron, which was more expensive to produce than cast-iron.

Only certain coals are suitable for coking. Darby I was probably fortunate in his choice.

> Pitcoals, Kennelcoals and Scotch-coals which burn to a white-ash like wood and abound in a bitumen may be used for the first fluxing of the iron from the ore.[4]

This seems strange, as the high percentage of bitumen in Scotch coals made them expensive to coke, and it was probably freedom from sulphur that made the difference. Various explanations of 'hot and cold shortness' were forthcoming.

Throughout the eighteenth century coke continued to displace charcoal, though a few charcoal furnaces lingered on into the

[1] *Trans. Roy. Soc. Edin.*, 1836, *13*, 374.

[2] B. Faujas de St. Fond, *op. cit.*, p. 150.

[3] *See* Rees' *Cyclopaedia*, quoted H. Scrivenor, *A Comprehensive History of the Iron Trade*, p. 283.

[4] Cramer, *Elements of Assaying*, p. 347, *n.*: quoted by T. S. Ashton, *op. cit.*, p. 34.

nineteenth century. The Sussex iron industry came to an end in 1827, but production continued somewhat later at Ulverstone and in Scotland. Thus, these remnants apart, from following the forests the iron industry began to converge on the coal-fields. One of the four major British undertakings, namely Carron—the others were Coalbrook Dale, Bersham, Rotherham—settled on the carboniferous formations of Scotland.

Before the founding of Carron Iron Works any iron production in Scotland, apart from the few charcoal furnaces in the Highlands, was effected at widely scattered centres, too unimportant for information about them to have survived to the present day. But it is worth noting that, as Mantoux says:

> A description of the iron industry in the first half of the eighteenth century would be very inadequate if the part played in scores of market towns and villages by the tinker and the farrier was left out. In Scotland almost the whole metal industry was still in their hands.[1]

On account of its unique position as the first great iron-producing concern in Scotland, Carron Iron Works has been the subject of several detailed studies, in particular that of Henry Hamilton, from which much of our information has been drawn.[2] Hamilton's information was based upon manuscripts, etc., in the possession of H. M. Cadell of Grange, a descendant of one of the founders. The Carron Company, founded in 1759, was a pioneer in several ways. It was the first to use in large quantities iron-stone of the carboniferous formations of Scotland. It was also the first to introduce the reduction of the iron-stone with coke instead of charcoal. The existence of clayband iron-stone in Scotland had been known during the first half of the eighteenth century, and the idea of founding iron-works there must have been under consideration some time before the founding of Carron, since we know that samples of Scottish ore were sent to Birmingham for assay.[3] The personnel at Carron is itself important on account of its links with other Scottish enterprises. First there was William Cadell of Cockenzie (*d.* 1777), a Scottish shipowner and trader in timber and iron, who had already made several essays at iron-making.[4] The two others were Samuel Garbett (1717–1805) and John Roebuck, M.D. (1718–94), already associated in other chemical enterprises at Prestonpans, as discussed in an earlier chapter.[5]

[1] P. Mantoux, *op. cit.*, p. 281. [2] *Scot. Hist. Rev.*, 1928, *25*, 185.

[3] Cadell MSS., Garbett to Cadell, 13th January, 1759: quoted by H. Hamilton.

[4] S. Smiles, *Industrial Biography*, p. 135. [5] Ch. VI.

The deed of partnership was signed at the beginning of 1760, seven partners dividing the capital of £12,000 as follows: John Roebuck, a quarter; his brothers Benjamin, Thomas, and Ebenezer, each one-twelfth; Samuel Garbett, one quarter; William Cadell, Senior and Junior, each one-eighth.[1] Before settling on Carron as the site for their works, the partners, or perhaps Roebuck, considered various alternatives, including Craigford in Stirling and Monymusk in Aberdeenshire, where the presence of iron-ore had been known since before 1760. In his *Tours in Scotland* Pococke says, 'They find also Iron Ore on Kern William (Cairn William) which yields twelve and a half out of twenty',[2] and it seems that Archibald Grant was anxious to develop the deposit, for we find him writing some years later, probably to Matthew Boulton, on the 13th December, 1769, stating that 'I am now ready to treat for Iron Works on my Estate'.[3] The site ultimately chosen at Carron was one which combined the advantages of being near not only coal and ironstone mines, but wood supplies, and water transport.[4] The partners varied somewhat in their ideas of the proper scale on which to start. Cadell favoured small charcoal furnaces and sent his son to study the methods in operation at the recently erected Goatfield furnace.[5] Garbett, on the other hand, wanted to set up works which would rival those of the Darbys at Coalbrook Dale, both with regard to size and quality of products. This most likely was also Roebuck's conception, since from the beginning the Carron Works was planned to be like that of the Darbys, from which both labour and materials were imported to start it. The initial stages in this great palaeotechnic undertaking were arduous. Their potential source of labour, whether English or Scottish, was still, as in the case of the Potteries, from the land.

> In the early days, as already pointed out, a close connexion existed with iron-making and agriculture; and in the eighteenth century it was still from the ranks of men bred on the soil, rather than those of the town, that labour for the iron industry was recruited. Furnaces and forges were often stopped during the summer months in order that the workers might assist with the harvest; and many labourers themselves held small plots of land which they cultivated in their spare time and which saved them from idleness when shortage of water or other cause brought the iron-works to a stand.[6]

[1] H. Hamilton, *op. cit.*, p. 189. [2] Richard Pococke, *Tours in Scotland*, p. 200.
[3] A. Grant, 13th December, 1769, A.O.L.B. [4] H. Hamilton, *op. cit.*, p. 188.
[5] H. M. Cadell, *The Story of the Forth*, p. 147. [6] T. S. Ashton, *op. cit.*, p. 197.

D. F. Macdonald says:

> When the Carron works was established, the total lack of skilled labour represented a serious difficulty, and recourse was again had to English masons, bricklayers, millwrights, bellows-makers, furnace-men, and indeed, all the operatives required were brought thence, chiefly from Coalbrook Dale Ironworks in Shropshire. They were paid high wages, and were expected to teach the ignorant Scots the essentials of the work.[1]

The general plan of Carron can be gathered from Hamilton's extracts from the Memo. Book of W. Cadell, the first Manager. The initial idea was to set up:

4 blast-furnaces, and assisting air furnaces when required.
4 air-furnaces to be constantly employed.
1 boring mill.
1 double forge.
1 forge for drawing salt pans.
1 slit or rolling mill.

By March 1760 two at least of the air-furnaces were in production, products made from the iron produced being marketed in London by Thomas Roebuck and Company.[2] These air-furnaces were followed in turn by the erection of blast-furnaces and the forge. The quality of the iron turned out by the air-furnaces was not up to Coalbrook Dale standard, so Garbett pressed for the erection of the blast furnaces. With the use of coke, and a consequent higher temperature, they gave a product which compared better with what was turned out by the Darbys.

> Roebuck was indefatigable in improving the manufacture; from the beginning he employed pit-coal for smelting his iron, and to do this, he had need of Smeaton's advice, in order to increase his blowing apparatus. Moreover, in order to improve the transport to Glasgow, the Carron Company surveyed a line for the Forth and Clyde Canal, which, though at the time abandoned owing to the objections of the landlords, was later carried into execution on the lines of the original suggestion.[3]

In 1763 Dr. Black visited Carron. To his brother he wrote:

> I perceive by your last that you had got from brother Robert an account of his journey hither, and of the agreeable time we spent together when he was in this country. When we parted at Edinburgh, Jamie Burnet and I came west by Falkirk, saw the

[1] D. F. Macdonald, *Scotland's Shifting Population*, 1770–1850, p. 66.
[2] *Roebuck to Cadell*, 11th March, 1760; *Garbett to Cadell*, 22nd March; quoted Hamilton.
[3] J. Lord, *Capital and Steam Power*, p. 79.

IRON

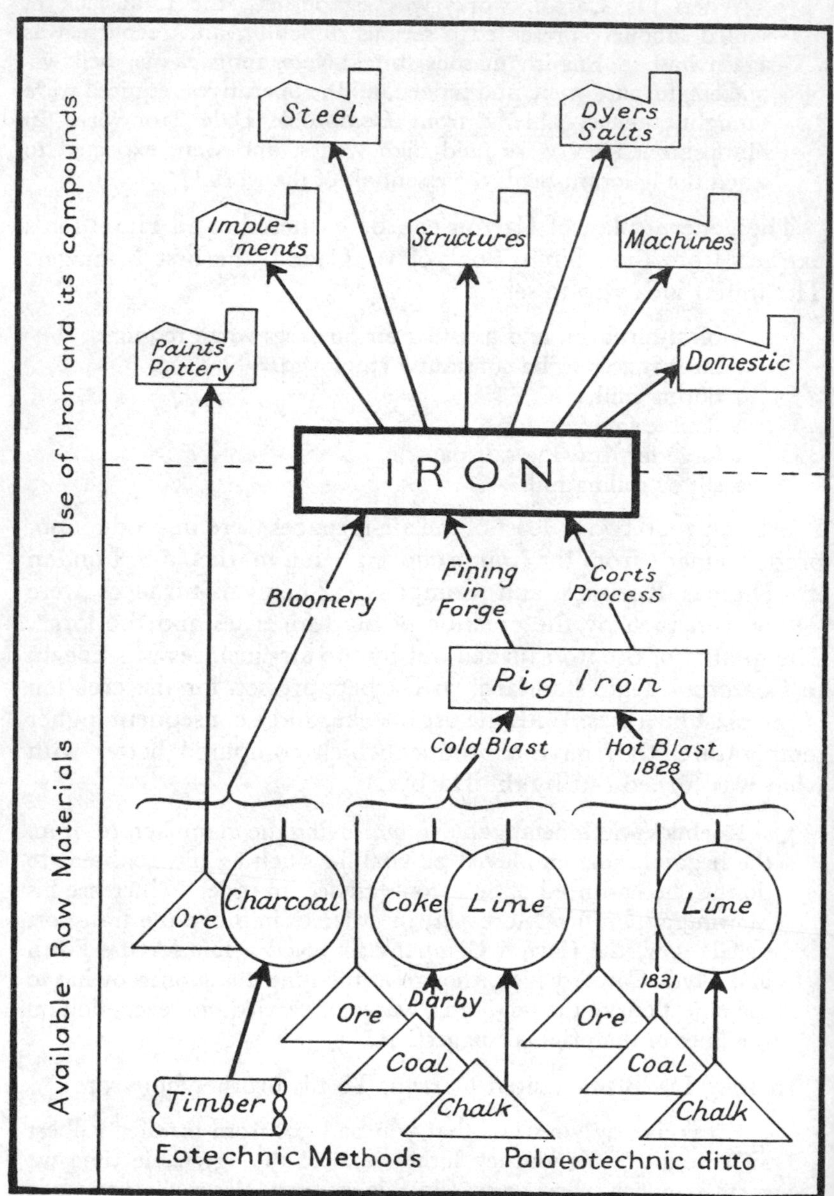
Use of Iron and its compounds
Available Raw Materials
Steel
Dyers Salts
Imple-ments
Structures
Machines
Paints Pottery
Domestic
IRON
Fining in Forge
Cort's Process
Bloomery
Pig Iron
Cold Blast
Hot Blast 1828
Ore
Charcoal
Coke
Lime
Lime
1831
Ore
Darby
Ore
Coal
Coal
Chalk
Chalk
Timber
Eotechnic Methods
Palaeotechnic ditto

works in the neighbourhood of that town, lately set on foot by a brother Doctor and chemist, for manufacturing iron from the ore in a very large and extensive way.[1]

The preparation of the coke and the charging of the furnaces were described by B. Faujas de St. Fond, who was conducted over the works by Dr. Swediaur of Prestonpans, who had already taken him to the Vitriol Works. Swediaur, it will be remembered, was at Prestonpans endeavouring to make soda from salt. Faujas de St. Fond has left us a description of the coking process:

> At Carron Foundry, this operation takes place in the open air, and is of the greatest simplicity. A quantity of coal is placed on the ground, in a round heap, of from 12 to 15 feet in diameter, and about 2 feet in height. As many as possible of the large pieces are set on end, to form passages for the air; above them are thrown the smaller pieces and coal-dust, and in the midst of this circular heap a vacancy is left about a foot wide, where a few faggots are placed to kindle it. Four or five apertures of this kind are formed round the ring, particularly on the side exposed to the wind. There is seldom, indeed, occasion to light it with wood; for these works being always in action, they generally use a few shovels of coal already burning which acts more rapidly than wood, and soon kindles the surrounding pile. As the fire spreads, the mass increases in bulk, swells up, becomes spongy and light, cakes into one body, and at length loses its bitumen, and emits no smoke. It then acquires a red uniform colour, inclining a little to white; in which state it begins to crack and split open, and to get distorted into the shape of a mushroom. At this moment, the heap must be quickly covered with ashes, of which there is always a sufficient provision around the numerous fires where the coal is prepared in this manner.

It must be remembered that Roebuck had had such formal training in chemistry as was possible at that time, so we find quantitative operation in progress. Regarding the scientific control of the furnaces, Faujas continued:

> The ore is methodically mixed, carefully weighed, and put into baskets of equal dimensions. The same attention is given to the coal. Every thing is placed in regular order, within reach of the foundrymen, under the sheds appropriated to that service. The baskets for each charge are always counted; a time-piece, which strikes the hour beside the large furnaces, determines the precise moment for putting in the charge. It is the same with respect to

[1] Joseph Black to John Black, Glasgow, 20th September, 1763, quoted by W. Ramsay, *Life and Letters of Joseph Black*.

> the outflow of the melted iron; the clock announces when they should proceed to that operation; and each workman then flies to his post.[1]

There is one point which is worthy of comment here. Carron seems to have been late in the introduction of coking in ovens. Ashton[2] attributes to David Mushet the statement that by 1763 ovens were in general use and, true, 1763 appears to be the earliest period in which coke ovens are mentioned. When Faujas de St. Fond was at Carron he described the operation still carried out in the open, but a letter from Samuel Garbett to Matthew Boulton tells that coking in ovens was carried out in Carron in 1773.[3] G. Jars in his *Voyages Métallurgique* (1774) gives a drawing of kilns at Newcastle used for removing sulphur and reducing coal to coke. The Register of the Great Seal in Edinburgh mentions the granting of a patent to Jean de Canolles for the coking of coal in 1784.[4] Their operation is described by Parkes in his *Chemical Catechism*, 1818.[5] According to David Mushet, the following ores were smelted at Carron in 1768: Brighton (Polmont), Bonnyhill (Falkirk), Kennaird (Larbert), Castleranky (Denny), Dysart, Pitfirrane (Dunfermline), Stonehaven, Orkney, and Cumberland.[6]

Only by slow degrees did Carron achieve success, but the increasing cost of Russian and Swedish iron was an incentive to continued effort. In 1767 it was showing a loss of £10,000, and by 1771 a necessary reorganization had taken place and new partners been brought in. Even by 1770, Dr. Roebuck's name had disappeared, a point which is worthy of notice, as it was not until 1773 that he was involved in bankruptcy. Carron stocks were very low. Glassford of Dougaldston even refused to take an interest in the Company when it was offered him. Garbett, too, was ruined by 1780.[7] Roebuck's departure must have been a considerable loss, as the comment of Faujas de St. Fond on the state of affairs after he left shows:

> We may well believe that it was only by repeated groping experiments, and expenses, often fruitless, that this establishment has at length reached its present high state of perfection, where everything is arranged and carried on, with exact precision, and nothing is left to mere routine or chance.[8]

[1] B. Faujas de St. Fond, *op. cit.*, pp. 182–188.
[2] T. S. Ashton, *op. cit.*, p. 37.
[3] 7th April, 1786, A.O.L.B.
[4] Vol. 20, No. 255, January, 1784.
[5] Additional Note 50.
[6] D. Mushet, *Papers on Iron and Steel*, p. 117.
[7] H. M. Cadell, *op. cit.*, p. 163.
[8] B. Faujas de St. Fond, *op. cit.*, p. 187.

Their main business was the casting of cannon, for which, in the second half of the eighteenth century, there was a considerable demand. Carron exemplifies Mumford's dictum, 'the partnership between the soldier, the miner, the technician, and the scientist is an ancient one'.[1] In addition, Carron produced 'cast boilers five feet in diameter, for the making of sugar in the West Indies . . . hoes of different sorts for cultivating the sugar cane, which were ground to a sharp edge on large grindstones'. They also made the first piece of cast-iron mill-gearing for, incidentally, the father of William Murdock, the gas-lighting pioneer. They made the first iron plough to be used in Scotland, for James Small (1763).

From the *Agricultural View of Stirling*, we learn of the effect of Carron in another direction—as a consumer of agricultural products:

> That these extensive works (i.e. Carron), in which so many hands are employed, have some influence on the agriculture of this district, cannot be doubted. They increase the demand, and give an additional spring to industry. In this view, they have led to the general improvement of the country; but the Reporter feels himself bound to add, in the words of an intelligent friend well acquainted with that neighbourhood, 'That they have contributed little to the improvement of the particular district in which they are situated'. They consume, he remarks, 'A great quantity of oats, but the oats of Hull or Aberdeen, being cheaper than those of Stirlingshire, they procure them from these places; so that the agriculture of these is more benefited by the Carron-works than that of Stirlingshire. If the barley of this county bears in general a better price, it arises from our vicinity to Glasgow, where pot barley is in great demand. I have no hesitation, therefore,' he adds, 'to say, that the Carron-works have only, in a very remote, and not in an immediate degree, affected the agriculture of this part of the country',[2]

As has been mentioned already, the possibility of iron-making in the hands of a landed proprietor was a potential means of increasing his income. The following letter from Sir John Dalrymple to Matthew Boulton shows the slender knowledge on which some proprietors were prepared to engage in this operation:

> I have on my Estate in the county of Edinburgh, Iron Stone, Lime and Coal, and the command of water for mills, which has

[1] L. Mumford, *Technics and Civilization*, p. 87.
[2] P. Graham, *General View: Stirling*, p. 349.

> made me think of establishing an Iron Works for pig iron and coarse goods. I can lay down the iron stone at the forge for 6*s.* 6*d.* a ton, the coal 4*s.* 9*d.*, and the Lime 1*s.* 6*d.* I am distant ten miles from the port of Musselburgh, and ten from Leith, on a turnpike road and down hill. I am six miles from woods, so I think of using only coal. I take the liberty of asking your advice. There is lately discovered in this country an art of extracting pitch and tar from coal. Mr. John Clark has brought it to such perfection that a ship has gone to sea payd with it, and the merchants and carpenters much pleased with it. The residues after extracting the pitch by distillation is charcoal for iron works, and more valuable than the coal that was thrown into the vessel. So that the pitch is all clear profit.[1]

The letter goes on to ask for advice on specific points on iron-making. Dalrymple went carefully into cost of production, and three years later published his *Addresses and Proposals on the Subject of the Coal, Tar, and Iron Branches of Trade* (*1784*).[2] He gives the cost at which raw materials could be put down on his estate, and cost of labour at Carron Foundry. These figures are of importance, as they give a detailed picture of the various operations required for iron-production.

Dalrymple's calculations were based on a probable output of seven hundred and fifty tons of pig-iron per annum, which, if retailed at £6 per ton, would bring in a total of £4,500. He reckoned that the cost per ton of producing this would be as follows:

	£	s.	d.
5 tons 'great coal' @ 4*s.* 2½*d.* per ton . .	1	1	0½
4 tons iron-stone @ 7*s.* per ton . . .	1	8	0
17½ cwts. lime stone @ 2*s.* per ton . .		1	9
8 cwts. 'small coal'[3] @ 3*s.* per ton . .		1	2½
	£2	12	0

Thus he calculated that the seven hundred and fifty tons of iron would cost £1,950 to produce, to which he added £150, being cost of carriage to seaport, bringing the total to £2,100.

Cost of labour worked out at £6 12*s.* per week, made up as follows:

[1] Sir J. Dalrymple to Matthew Boulton, 21st July, 1781, A.O.L.B.

[2] National Library of Scotland. [3] For calcining the iron-stone.

	£ s.
2 keepers @ 15*s*. each	1 10
2 fillers @ 8*s*. each	16
2 lime breakers	12
1 coaker	10
1 servant to do.	6
2 iron-stone burners	12
1 coak wheeler	6
1 coak filler	5
1 mine wheeler	8
1 cinder wheeler	6
1 smith	8
1 striker	8
1 carpenter	8
Total	£6 12[1]

Over the year's working this represented a cost of £343 4*s*., to which £70 had to be added for clerks, and £300 for interest, wear and tear on buildings, machinery, etc., at 15 per cent. on a capital sum of £2,000. This made the total annual running cost of the proposed iron-works £2,813 4*s*., leaving a profit of £1,686 16*s*. to accrue to the proprietor.

Carron became a focal point from which other iron-works developed, both in Scotland and farther afield. Charles Gascoigne, who had married Samuel Garbett's daughter, and involved his father-in-law in ruin, removed himself to Russia where he founded an iron-works.[2] The first extension from Carron in Scotland was the founding, by two brothers of the name of Wilson, of *Wilsontown Iron Works*. Their main production was pig-iron, and they enjoyed prosperity until 1810. From 1812 to 1821 the works was closed. When reopened it had passed under the ownership of William Dixon of Calder. It finally ceased operation in 1842.[3]

From Wilsontown the iron industry began to move west, particularly into the south-west corner of Old Monkland parish, Lanarkshire. Here the *Clyde Iron Works* was opened in 1786. In the same year the French chemists Monge, Vandermonde, and Berthollet published a memoir proving that the difference between the various kinds of iron and steel is mainly determined by the

[1] Note.—There is an arithmetical error here.

[2] *New Statistical Account*, Vol. 8, p. 355.

[3] Details of the coal, iron, and limestone worked by the Carron Company and by Wilsontown will be found in a paper by Charles Forsyth, *On the Mines, Minerals, and Geology of West Lothian*, in the *Transactions of the Highland and Agricultural Society of Scotland*, 1847, 2, 299 (3rd series).

amount of carbon they contain.[1] This was a subject which puzzled the illustrious Dr. Black.[2] The founders of Clyde were William Cadell, who had resigned from Carron in 1769, and Thomas Edington of Cramond Iron Works, a subsidiary works founded to fabricate articles with iron produced at Carron. Clyde Iron Works later passed under the control of Colin Dunlop (later J. Dunlop and Company, Tollcross). It has several claims to fame. It was here that the illustrious David Mushet, discoverer of black-band iron-stone, began his career, only to be dismissed by his conservative masters on account of his progressive outlook on the scientific aspects of iron-making. Yet a generation later, one of the most revolutionary changes which the iron industry experienced, the heating of the air before it was blown into the blast-furnace, was introduced for the first time at Clyde.[3] Their original blowing engine was erected by Watt. Of the raw materials which they obtained from the neighbouring county, the following note speaks:

> It is only within the last seventeen years that any iron-stone has been wrought in the county. On the banks of Cart, on the estate of Blackhall, considerable quantities have been dug and conveyed, by land carriage, to Clyde Iron Works in Lanarkshire, a distance of about eleven miles. The expense of working has commonly been from 4*s*. to 5*s*. per ton; the rate of carriage from 5*s*. to 7*s*. per ton, and the proprietor obtains a royalty or lordship of 10*d*. or 1*s*. per ton. The bands of ironstone at Blackhall are numerous, of considerable thickness, and afford iron of very good quality.[4]

In the next year, 1787, *Omoa Iron Works* started near Cleland. It was established by Col. William Dalrymple who, having distinguished himself at the capture of Omoa in the West Indies, bestowed that name on his new establishment. During a long career the works changed hands several times, passing into the hands of R. Stewart's trustees, and was finally abandoned in 1866.

The total number of furnaces in blast in Scotland at the founding of Omoa Iron Works were, according to Scrivenor:

Goatfield . . .	1	charcoal furnace producing	700 tons
Bonawe . . .	1	,, ,, ,,	700 ,,
Carron. . . .	4	Coke furnaces producing	4,000 ,,
Wilsontown (or Cleugh) .	2	,, ,, ,,	1,600 ,,

In the same year as the founding of Omoa, iron-stone was utilized locally in the north-east of Kyle district, Ayrshire, and

[1] *Journal de Physique*, 1793, *43*, 373.
[2] J. Black, *Chemical Lectures*, Vol. 1, p. 138
[3] *New Statistical Account*, Vol. 6, pp. 160 *et seq.*
[4] J. Wilson, *General View of Renfrew*, p. 24.

this led to the opening up of a new industrial area with the founding of *Muirkirk Iron Works* in 1790. It is now operated by the Eglinton Iron Company. The *General View of Ayrshire* gives the following account of its beginning:

> The manufacturing of iron was carried on for some time by Lord Cathcart and others, at Muirkirk, in the early part of last century. It was made from ore dug there, and sent to Bunawe, in the county of Argyle, to be formed into pig-iron, which was afterwards brought to Muirkirk, where it was made into bar-iron. In the last operation, charred peat was used; the art of coking coals not being then known. These operations were found, however, to be too expensive, and the work was abandoned.
>
> In the year 1787, some respectable gentlemen in Glasgow entered into a copartnery, and erected very extensive works in the parish, which are still carried on by another company, who lately purchased these works from those who established them. The works consist of three large blast-furnaces for the making of pig-iron, an extensive forge for making bar-iron, with a foundry for inferior work. Between three and four hundred workmen are constantly employed at these works, who, with their families, will exceed 1,000 souls.
>
> The pig-iron made at Muirkirk works is esteemed by the founders in Scotland and Ireland, soft, easily melted, and of the very best quality. The bar-iron is also superior to any other made in Britain, and is allowed by the best judges, to be no way inferior in quality to the Swedish iron. This superior quality of the Muirkirk iron proceeds partly, from that of the coals used in manufacturing it, which are less impregnated with sulphur, and partly from the mode of beating out the iron. The bar-iron at Muirkirk, is not, like that made at the generality of the other iron-works in Britain, drawn out and formed by rollers;[1] but is beaten by very heavy hammers, which give it much greater solidity, toughness, and durability in wearing, on cart wheels, horse shoes, etc., than any other British manufactured iron.[2]

Muirkirk was followed by *Devon Iron Works* in the Sauchie section of the Clackmannan parish of Clackmannan, two and a half miles from Alloa. The early furnaces there were interesting because hollowed out of a steep escarpment of rock. They were managed by Mr. John Roebuck, son of Dr. Roebuck. It was there that young Roebuck determined that there were definite conditions of volume and pressure of air supplied to a furnace for optimal working. The nature of the fuel used, it should be noted, also limits the size of the furnace for maximum production. If the

[1] *Vide infra*, Cort's patent. [2] W. Aiton, *General View of Ayrshire*, p. 603.

weight of the superincumbent charge is too great, the coke will crush and obstruct the free passage of the blast through the furnace. Devon was finally closed in the middle of the eighteenth century.

Out of the same rock a hollow chamber was carved to act as an equalizer of the air pressure. In it Roebuck carried out some daring experiments while the furnace was in operation. The conservative owners viewed his scientific investigations with misgivings, and after some disagreements he severed his connexion with them.[1]

Andrew Brown, in his *History of Glasgow*,[2] gives the following figures for the iron works in Scotland after the founding of Devon:

	Blast-Furnaces	*Air-Furnaces*	*Founder's Cupolas*	*Boring Mills*	*Bar-iron Forges*
Carron . .	5	8	4	3	1
Clyde . .	3	5	2	3	0
Muirkirk .	3	1	0	0	1
Cleland .	2	1	0	0	0
Wilsontown .	1	1	0	0	1
Devon . .	2	0	0	0	0
	16	16	6	6	3

He also gives figures for the profits of the operation, etc.

Chronologically the next iron works was one founded at *Shotts*, Lanarkshire, in 1802, later Shotts Iron Company. It was long managed by John Baird, who was connected with other iron-working activities in Scotland.

The next foundation, at Coatbridge in 1805, is the first of a long series of iron-works in the parish of Old Monkland, the first being that of *Calder*, later Wm. Dixon, Limited. This works is notable, since it was founded by David Mushet and partners. By contrast with the usual run of iron-masters, their outlook seems to have been progressive, e.g. they quickly realized the advantages of the hot-blast. At the time the *New Statistical Account* was being written, they were building extensive puddling furnaces to operate Henry Cort's method for making malleable iron, of which more later.

[1] *Trans. Roy. Soc.* Edin., 1805, *5*, 31.

[2] A. Brown, *History of Glasgow,* Vol. 2, p. 290.

The *Monkland* (or *Calderbank*) *Iron Works* followed at Airdrie in 1825, and *Gartsherrie*, Coatbridge, in 1828. The first furnace there was put into blast on the 4th May, 1830, so two years had been required for its construction. The founders were William Baird and Company, who in time became Scotland's principal iron-masters. The invention of one of the Bairds for heating the blast will be dealt with later.[1] Other Coatbridge works followed in quick succession: *Dundyvan*, founded by Colin Dunlop of Clyde and Wilson in 1833; *Summerlee*, by Wilson and Company in 1836; and *Carnbroe* (later Merry and Cunningham, Limited), by Alison and Company in 1840. Thus, in the words of Clapham:

> A real and serious concentration of industry developed in Scotland where civilization had always been concentrated. The four contiguous counties of Midlothian, Linlithgow, Lanark and Renfrew contained in 1831 between one-third and one-quarter of the total population. The industries round Edinburgh were those required for a capital city. The new industries were round Glasgow and Clydeside.[2]

The following table from *The Iron Trade*, by H. Scrivenor,[3] summarizes data about blast-furnaces in Scotland in the third decade of the nineteenth century.

BLAST-FURNACES IN SCOTLAND

Name	*1823 Total No.*	*Quantity in Tons*	*Erected in the years:*							*1830 Total No.*	*Quantity in Tons*
			1824	*'25*	*'26*	*'27*	*'28*	*'29*	*'30*		
Clyde .	3	2,500	–	–	–	–	1	–	–	4	8,000
Calder .	3	4,000	1	–	–	–	–	–	–	4	9,000
Monkland .	–	—	–	–	1	–	–	1	–	2	2,000
Muirkirk .	3	3,500	–	–	–	–	–	–	–	3	4,000
Gartsherrie .	–	—	–	–	–	–	–	1	–	1	—
Shotts .	1	2,000	–	–	–	–	–	–	–	1	2,000
Carron .	5	7,000	–	–	–	–	–	–	–	5	7,000
Devon .	3	3,000	–	–	–	–	–	–	–	3	3,500
Wilsontown .	2	—	–	–	–	–	–	–	–	2	2,000
Omoa .	2	2,500	–	–	–	–	–	–	–	2	—
Totals .	22	24,500	1	–	1	–	1	2	–	27	37,500

The above iron-works smelted ores from the following Scottish deposits: *Wilsontown*, Hogg Fence iron-stone mined in Mouse

[1] *New Statistical Account*, Vol. 6, p. 662.
[2] J. H. Clapham, *Economic History of Modern Britain*, Vol. 1, p. 51.
[3] H. Scrivenor, *op. cit.*, p. 134.

Water and Mossat Burn, later Curdly and Ginstone iron-stone from Levenseat; *Clyde*, clayband from Crossbasket (High, Blantyre), Calderwood (E. Kilbride), Glazert and Shiel Glen (Lennoxtown); *Muirkirk*, Pennel Burn haematite; *Devon*, clayband from Clackmannan coal-field, Tillicoultry and Dollar; *Calder*, ironstone from Woodhall; *Omoa*, clayband from roof of Clelandbrae coal; *Shotts*, a nodular layer in the local Furnace and Ball coal.[1]

The main product of the various iron-works enumerated above was pig-iron. By far the greater part of it was subsequently remelted and converted into cast-iron or foundry goods. These two kinds of iron are described by Ure in the ensuing passage:

> The iron which is obtained from the smelting furnaces is not pure, and may be distinguished into three states: white crude iron, which is brilliant in its fracture, and exhibits a crystallized texture, more brittle than the other kinds, not at all malleable, and so hard as perfectly to withstand the file; grey crude iron, which exhibits a granulated and dark texture when broken, this substance is not so hard and brittle as the former, and is used in the fabrication of artillery and other articles which require to be bored, tarred, or repaired; and black cast-iron, which is still rougher in its fracture, its parts adhere together less perfectly than those of grey crude iron.[2]

In re-melting pig-iron for casting, a purification of it takes place, and the quality of the resulting cast-iron is greatly improved. Remelting is carried out in a founder's cupola, which was the logical application of a blast of air to the founder's furnace working with natural draught. Ashton described its invention and advantages in the following terms:

> William Wilkinson devised the cupola, an obvious extension of the blast-furnace to the remelting of pig-iron. The higher temperature produced a more fluid metal which when cast was less liable to break, and particularly suited to the production of machine parts as well as hollow-ware.[3]

One of the most characteristic of modern culinary furnishings, enamelled hollow-ware, dates from the end of the eighteenth century. In 1799 the *Society of Emulation* in Paris suggested as the subject for one of their prizes:

> . . . to discover a composition fit for making kitchen utensils which should be free from the disadvantages attending copper, lead,

[1] *Memoirs of the Geological Survey of Scotland*—Special Reports on the Mineral Resources of Great Britain, Vol. 11: *The Iron Ores of Scotland*, p. 7.

[2] Andrew Ure, *Dictionary of Chemistry*: Article, *Iron*.

[3] T. S. Ashton, *Iron and Steel in the Industrial Revolution*, p. 102.

> tinned vessels, glazed earthenware, etc., which should be as strong as possible, less costly than the vessels used at present, and which should be able to bear the highest degree of kitchen fire, and the most sudden changes from heat to cold.

Whether stimulated by the offer of this prize or not, such an *enamel* was patented by Samuel S. Hickling in 1799, the patent being described as for 'the improving and beautifying of vessels of cast-iron or other metals for culinary, chemical, and other purposes'. Within a very short time, Sven Rinman, of the Royal Academy of Stockholm, submitted to the Academy a series of experiments showing that successful enamels could be made from a mixture of flint glass, red lead, potash, nitre, cobalt oxide, and tin oxide. Of this enamel it was reported that:

> As a vessel when coated with this enamel bears, without any injury, sudden changes of heat and cold, and also to have any greasy mixtures baked or boiled in it, it may be applied to vessels of various kinds, among others to tea-cups; particularly as it is neither brittle nor subject to crack, provided it is not exposed to violent blows.

So originated the ubiquitous enamel mug.[1]

Another process, better than Hickling's in that it did not depend on the use of lead, was patented by Thomas and Charles Clark of Wolverhampton in 1839.[2]

One outstanding problem remained. It was still difficult to convert cast-iron into malleable iron by the use of coal, and the purchases of timber (to convert into charcoal) which were made by the Carron Company were doubtless to effect this conversion by means of the latter fuel. John Walker in his *Economical History of the Hebrides* described the purchase:

> The Carron Company, many years ago, purchased a wood on the estate of Glenmoriston, on the north side of Loch Ness, for the purpose of charcoal. For this wood they paid nine hundred pounds, though it was distant eight computed miles of very bad road from water carriage on Loch Ness, and was to be transported from thence to Carron.[3]

Dealing with the difficulties experienced in substituting coke, Richard Watson, Bishop of Llandaff, wrote in 1782:

> Iron thus procured (i.e. by coke) is very fragile and coarse and cannot, without great difficulty, be rendered as malleable as that

[1] *Encyclopaedia Britannica*, Third Edition, Vol. 1, p. 626.
[2] R. B. Prosser, *The Birmingham Inventors*, p. 152.
[3] T. Walker, *Economical History of the Hebrides*, p. 209.

> which is fluxed with charcoal, it may answer nearly as well for casting, but it cannot be made into good bar-iron by any art hitherto known.[1]

Premiums were offered for a process in 1762 and 1766, but were never claimed.

The time-lag in the successful application of mineral fuel to the second, or fining, process created in the iron industry a situation parallel to that in the textile industry after the invention of the spinning-jenny when the weavers could no longer absorb the yarn mechanically spun.

Although the premiums mentioned above were apparently not claimed, iron-masters were actively engaged in attempting to invent a process whereby coke cast-iron could be converted into malleable or bar-iron. Various inventions contributed to make possible the process which was ultimately operated successfully. For example, the mechanical part of the operation was patented by John Purnell in 1728. In Scotland the co-founder of Carron, Dr. John Roebuck, patented a process for converting pig into malleable iron. According to Mantoux,[2] it gave encouraging results. It is also probable that Roebuck's partner, Samuel Garbett, paid a visit to Sweden in 1763 to see if he could collect any ideas on the subject.[3] In 1766, two Coalbrook Dale workmen, Thomas and George Cranage, obtained a patent for the furnace in which the operation was carried out.[4] The chemical reaction whereby the iron is purified was patented by Peter Onions of Merthyr Tydfil on 7th May, 1783. The following is a contemporary description of Peter Onions' patent for a reverberatory furnace and puddling process:

> There are two furnaces used in this operation or invention, to wit, a common furnace, in which the iron ore or metal is put and there smelted or melted, and another furnace which is made of stone and brick and other materials, as usual, and fit to resist the force of fire, and bound with iron work and well annealed, and into which the fluid iron or metal is received from the common furnace or smelting blast in its hot liquid state, and when so received is worked or refined as follows: a quantity or stream of cold water must then be run or be put into the cistern or trough under the ash grate of the refining furnace, and the doors thereof closed and luted with sand or lome, and the fire place filled with

[1] R. Watson, *Chemical Essays*, Vol. II, p. 344

[2] P. Mantoux, *The Industrial Revolution in the 17th Century*, p. 299.

[3] *See* Josiah Wedgwood's papers in the British Museum; Addit. MSS. 28311, p. 9; and Calendar of Home Office Papers 1760–65, No. 1359, for documents re Garbett.

[4] R.G.S., Vol. 20, No. 20, 1766. (No. 851.)

fuel of pit-coal, coaks, or wood charcoal, from time to time as occasion requires, and then the common bellows, cylinder or usual machine for blowing or pumping air into the space below the ash grate through the tubes, is begun to be worked, and the fire excited by the air until the cavity is sufficiently heated, and then the hot liquid iron metal is taken and carried in iron ladles from the above common furnace and poured into the refining furnace through an iron door or apperature raised by a lever; then the said apperature is stopped, and the blast of air and the fire used until the metal becomes less fluid and thickens into a kind of paste, which the workman, by opening the door, turns and stirs with a bar or other instrument or tool, and then closes the apperature again, and must apply the blast of air and fire until there is a ferment in the metal: and if no ferment ensues, then he must turn or convey the blast of cold air through the tube upon the matter, which will excite a kind of ferment or scoriafication in the matter or metal; and as the workman stirs or turns the metal it will discharge or separate a portion of scoria or cinder from it, and then the particles of iron will adhere and separate from the scoria, which particles the workman must collect or gather into a mass or lump, and then shut the door and head the mass until the same become of a white colour, and then take or convey out of the furnace, with a bar of iron or tongs, the said mass or lump to the forge hammer, and there by repeated blows, squeeze or beat out the remaining scoria or cinder, when a mass of malleable iron will be formed into an octagonal or other bar called a loop, which bar may be then or at any time heated in a fire, and worked by the workman and the forge hammer into rods and bars of iron, for various purposes.[1]

The reaction is called *puddling.*

The various germinal ideas contained in these patents were combined by Henry Cort into two patents which he obtained on 13th February, 1784.[2] The first patent covered the chemical reactions taking place when pig-iron is heated in a reverberatory furnace heated with mineral fuel. Thus it became possible to carry out the second operation in malleable iron-making with coal. Cort's second patent covered the forging and rolling of the iron so produced.

Cort consulted Dr. Black on the subject of iron-making in February 1787. The rolling of the billets, which was a characteris-

[1] *British Patents,* Vol. 14, No. 1370: Quoted by A. P. Usher, *The Industrial History of England,* p. 330.

[2] Nos. 1351 and 1420. *See Memoir of Henry Cort* by Thom. Webster, *Mech. Mag.,* 1859, 2, 53 (2nd series); Register of the Great Seal, Edinburgh, Vol. 20 No. 263, 1784.

tic part of Cort's process was effected by rollers actuated by a Boulton and Watt engine, and Black's biographer, Ramsay, infers that it was for this reason that Cort applied to Black.[1]

The Cort processes were speedily taken up, especially in Wales, where they were modified and improved, more especially by the Homfreys of Penydaren and Robert Gardner. So general was the Welsh adoption from about 1790 that the process was often called the *Welsh Process*. Its introduction in Scotland dates from 1830–6, when technical skill was imported into Lanarkshire from England and Wales. Thus we learn that at the time the *New Statistical Account* was being written (*c.* 1836), at Calder, Dixon was building near his blast-furnaces some forty-two puddling furnaces designed to have an output of four hundred tons of bar-iron per week. The Monkland Company, too, was procuring mills and forges capable of producing 220–230 tons per week, and at Dundyvan, similar arrangements were being made.[2] Once again malleable iron could be made at a rate sufficient to absorb the production of the blast-furnaces.

Bremner, in his *Industries of Scotland*, gives a graphic description of puddling.[3]

In the following letter Dr. James Hutton makes some comments on the effects of Cort's invention and of the parallels between the iron and textile industries.

> You (Matthew Boulton) were to enquire after the Lancashire iron ore, when I had the pleasure of seeing you, it then appeared to us to be a most valuable speculation. I now think their things are considerably altered with regard to that subject, on account of Mr. Cort's new method of making bar iron. By his perfect finery the worst pig-iron is made bar of the most valuable kind, and fit for every purpose, at least for most purposes. We have seen him take the Carron cold short pig, and of this make iron that works when hot in the highest perfection, and when cold it breaks most tough and fibrous. You will understand the value of this discovery; the land of coal and ironstone will be the land of bar iron. What a field for buying power at your Shop. I think you say that the slip mill gave the origin to the cotton mill; now the cotton mill may give a lesson to its parent, and learn Britain to spin iron for all the World. You may throw in a piece of ironstone into the furnace and see it come out a nail, string, or hoop without ever losing its heat.
>
> The Carron people and Mr. Cort have not agreed about their terms. I have advised Mr. Cort to give to the iron masters an

[1] W. Ramsay, *op. cit.*, p. 101.

[2] *New Statistical Account*, Vol. 66, pp. 659 and 697.

[3] D. Bremner, *op. cit.*, p. 51.

obligation that he will lower his tax in general whenever his total income shall exceed a certain sum; this will make all his dealers interested in his success. Mr. Cort appears to relish the proposal very well, and has promised to adopt it. I think it will be interesting to have the pig-iron from the different species of mines tryed by refining in this method, and converted into steel. What most amazed us was the curing of the cold short; for, as to the process for making malleable iron, nothing can be more rational. I imagine that the siderum is evaporated from the pig iron in the perfect calcination that it undergoes. It is a beautifull sight to see the iron at the end of the operation all come into a blaze like the brightest phosphorus when newly illuminated with the sun.[1]

Although the production of pig-iron in Scotland increased rapidly during the second half of the eighteenth century as the result of the founding of the iron-works described above, the pace in the north did not quite keep up with that in England. In fact the percentage of British iron made in Scotland tended to fall off at the beginning of the nineteenth century, as will be seen from the following tabulation given by T. S. Ashton:

OUTPUT OF PIG-IRON IN TONS IN GREAT BRITAIN, WITH PERCENTAGE PRODUCED IN SCOTLAND

	1788			*1796*			*1806*		
	Furnaces	*Output*	%	*Furnaces*	*Output*	%	*Furnaces*	*Output*	%
Total .	85	68,000	100	121	125,400	100	221	250,400	100
Scotland .	8	7,000	10·2	17	16,100	12·8	27	23,200	9·2

Indubitably this decline would have been progressive, and Scottish iron industry might have suffered permanent eclipse, had not two important new developments taken place in the first thirty years of the nineteenth century. The effect of these revived the Scottish industry, gave it new life and unprecedented vigour. The first development was the discovery of a new ore of iron, on the River Calder above Craighill, at Old Mill, namely blackband iron-stone; the second was a technical innovation in effecting its reduction to pig-iron. The ore itself was discovered in 1800 by David Mushet, who ultimately became one of the most important personalities Scottish iron industry has produced. Further discoveries at Airdrie and Calderbraes or Kennetburn, added to its economic importance.

[1] *J. Hutton to M. Boulton*, 24th May, 1784, A.O.L.B.

Mushet makes the following observations on Scottish iron ores just before he himself discovered the famous blackband iron-stone:

> In several places in Scotland iron ores have been discovered resembling in point of appearance those of Cumberland and Lancashire.
>
> A vein which indicated a large field of supply, was some years ago traced in the neighbourhood of Muirkirk iron-works, and its course followed for several miles. The kidney pieces of this ore are fully as rich and ponderous as those of the Cumberland vein.
>
> Salisbury Crags, in the neighbourhood of Edinburgh, afford also some very good specimens of an irregular vein, which still remains untraced; apparently the same ore is found on an estate belonging to the family of Dundonald, called La Mancha, ten miles south of Edinburgh; the lands of Cranston also afford a similar ore: the quality of the ore in both of these places may be reckoned upon the average to yield 42 per cent. of iron.
>
> The finest Scotch ore I have seen, and which possesses mixtures congenial to the existence of carbon in the blast-furnace, is found in the Ochil-hills, not above two miles from the Devon iron-works. This ore is soft, loosely striated, of a reddish colour, not very ponderous, but possessing a superior quality of iron.[1]

David Mushet (1772–1847) entered the Clyde Iron Company as an accountant in 1791–2, but was more attracted by the scientific side of iron-making, with the result that he began to make experiments on his own. The possible utility of his researches was not understood by his employers and he was discharged by the Clyde Iron Company in 1800. The shortsightedness of such behaviour is revealed when one considers that, in his time, Mushet was the most important single investigator in Great Britain into the technology of iron and steel-making, and contributed no less than thirty papers on the subject to the *Philosophical Magazine*. He did not let his discharge from Clyde sever his connexion with iron industry. In partnership with others he proceeded to build *Calder Ironworks*, but was involved in heavy financial liability and forced to sever his connexion with them about 1805. Following the footsteps of James Watt he found his way to England, where he settled in the Forest of Dean. Among his most notable work was that done on crucible steel which led directly to Heath's Process. Through his son, Robert Forester Mushet (1811–91), he is the precursor of those who brought about vast extensions in steel-making about 1860.[2]

[1] D. Mushet, *On Primary Ores of Iron, Phil. Mag.*, 1799, *3*, 350.

[2] Ernest F. Lange, *Mem. Manch. Lit. and Phil. Soc.*, 1913, *57*, No. 17.

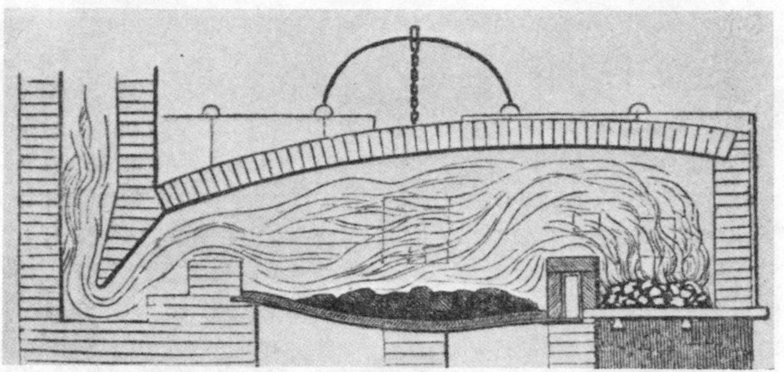

76. Puddling furnace for the production of malleable iron by Cort's Process (patented 1784); (*below*) view of exterior; (*above*) section of furnace.

Puddling was one of the most arduous palaeotechnic operations. When the metal began to melt, the puddlers had to keep it stirred with a special tool. The mass gradually thickened and was formed into lumps called 'puddler's balls'. About six charges were worked off in twelve hours. The 'balls' were removed and treated as illustrated in Fig. 76, above.

77. The Forge at the Coalbrook Dale Co.'s Works. Outside are the puddling furnaces and on the extreme right the shingling hammer which with twenty to thirty blows renders the 'bloom' fit for the next process, *viz.* rolling in puddle-rolls. This completes the operation carried out in the forge.

78. The world's first cast-iron bridge, made at Coalbrook Dale by Abraham Darby III and erected near Broseley, Salop, 1777–79. It has a span of 100 feet and a total weight of 378 tons.

The adoption of the blackband iron-stone, which Mushet discovered, was slow. Although it contained, in addition to iron, a high proportion of carbon, sufficient in fact for its reduction, it was at first regarded with contempt by the conservative iron-masters, and a quarter of a century elapsed before Monkland Company, then just formed, used it without admixture with other ore. Once adopted, a rapid expansion in Scottish iron production took place, particularly in the Old Monkland parish of Lanarkshire. It led to a veritable new era in ironworks foundation: *Monkland*, as mentioned, in 1825; *Gartsherrie*, by William Baird and Company in 1828; *Dundyvan*, by Dunlop and Company in 1834; followed by *Summerlee*, *Carnbroe*, *Chapel Hall*, etc. The dominating influence of blackband iron-stone in the Old Monkland parish will be appreciated if it is recalled that when the *New Statistical Account* was written, sixty-five of Scotland's eighty-eight furnaces were in this single parish.[1]

The expansion of iron production at this date is associated with a technical advance of first-class importance. For many generations it had been known that iron-making was less satisfactory during the summer months. This was erroneously attributed by the iron-masters to the variation in temperature between the seasons. As opposed to this, James Beaumont Neilson, manager of the Glasgow Gas Works, thought it might be caused by the increase in vapour pressure, there being more water in the air in summer when the temperature is high. He suggested that improvement might be effected by drying the air with quick-lime (calcium oxide, CaO) before passing it into the furnace.

$$\begin{array}{c} CaO + H_2O \rightarrow Ca(OH)_2 \\ \text{quicklime} + \text{water} \rightarrow \text{slaked lime} \end{array}$$

James Beaumont Neilson (1792–1865) was the son of Walter Neilson, once works engineer to Dr. John Roebuck at the latter's Borrowstoneness colliery. His mother was Marion Smith, the daughter of a Renfrewshire bleacher. When appointed to the Gas Company in 1817, Neilson can have had little scientific knowledge other than that which he might have picked up from his parents, but to acquire the chemical knowledge necessary for the efficient managing of the Gas Company he attended classes in physics, chemistry, and mathematics, at the Andersonian Institution, Glasgow.[2]

[1] *New Statistical Account*, Vol. 6, p. 658.

[2] T. B. Mackenzie, *Life of J. B. Neilson, F.R.S.*, *passim*, from which much of the following information is taken.

About 1824 his attention was drawn to problems of smelting by James Ewing, of Muirkirk Iron Works, who, in an attempt to explain the seasonal variation, suspected that sulphur was the cause of the poor yield from blast-furnaces in summer, and enquired of Neilson if it would be possible to purify air supplied to furnaces in a manner similar to that used in purifying coal gas before it was supplied to consumers. Neilson considered the problem and formed a hypothesis suggesting alternative causes, either (*a*) insufficient air or rather oxygen, or (*b*) excess water vapour in the blast fed to the furnace, which was picked up from water vaults that were added to the blowing equipment to equalize the pressure of the blast.

Time has shown that both were of importance, but unfortunately Neilson only followed up the first. Otherwise his name would be associated with a second important discovery.[1] It is probable that Neilson's attention was deflected from considering the influence of water vapour on the operation of blast-furnaces by the specific nature of James Ewing's problem. This was the unsatisfactory working of a furnace situated about half a mile from the blowing engine at Muirkirk. The idea of heating the blast to increase its volume occurred to Neilson. In the social context of the early aeronauts, Jacques Alexandre Charles (1746–1823) about 1787 had shown that air, oxygen, nitrogen, etc., expand equally when heated between 0° and 80° C., and in 1802 a young engineering student at the Ecole Nationale des Ponts et Chaussées, in Paris, Joseph Louis Gay-Lussac (1778–1850), determined the amount of the expansion. Neilson tested out his idea experimentally at the Glasgow Gas Works, first studying the effect of heated air on the combustion of coal gas and then of supplying heated blast to a common smith's forge.

Convinced that his experiments pointed the way to improvement, Neilson endeavoured to get his ideas applied to a full-scale plant, but found this very difficult owing to the strong prejudice among iron-masters against what they described as 'meddling with the furnaces'. The forces of reaction may be gauged by the fact that they were apparently unmoved by many Scottish furnaces standing idle because there was not a sufficient blast to smelt the ore, an operation which was not profitable unless a price of £6 per ton could be obtained for the iron. Colin Dunlop and John Wilson of Clyde Iron Works and Charles Macintosh first gave him opportunities for a large-scale trial, though even with them he had to proceed slowly. Indeed, it took him between two

[1] *Glasgow Mech. Mag.*, 1838, *3*, 41.

and three years to persuade them to permit a bend being put in the blast-pipe.[1] Once in operation on a large scale, the soundness of Neilson's suggestions became obvious. Neilson the scientist succeeded where the practical iron-masters failed, because he was willing to put his hypothesis to test.

In 1828 the process involving what has ever since been known as the 'hot blast' was patented. The specification was drawn up by Lord Brougham, whose offices were also used on occasion by Boulton and Watt. The partners who shared in the patent were J. B. Neilson with a three-fifths share, Charles Macintosh with three-tenths, and J. Wilson of Dundyvan with one-tenth. Colin Dunlop withdrew. The royalty for the use of the patent was one shilling per ton of hot-blast iron made.[2]

When the full significance of Neilson's work had been clearly demonstrated, the process was adopted more quickly in Scotland than in England. Calder, Wilsontown, and Gartsherrie were among the first to put it into operation. At first the temperature of the blast was only raised to 250° F. This nevertheless effected a saving of some three-sevenths of the fuel formerly used. To Baird of Gartsherrie is due an improvement in methods of heating blast so that the temperature could be raised to 600° F. and still more fuel saved.[3] The greatly increased temperatures of working (*c.* 1,400° F.) brought practical difficulties in their train. The solution of many of these came from the Scottish iron-works. Thus the nozzles or *tuyères* through which the blast was blown into the furnace gave trouble as a result of the greatly increased temperature, so John Condie of Blair Iron Works invented a water-cooled tuyère, often called the Scottish tuyère, which has continued in use to the present day practically unchanged. The first to introduce the hot blast into England were Messrs. Lloyd, Fisher and Company of Wednesbury in 1834. They also made an independent contribution to the success of the hot blast, by utilizing waste heat from furnaces to heat the blast. Using waste heat dates from 1811, when a French iron-master, M. Auberton, first put the idea into operation. Then two years before Lloyd and Fisher started to utilize waste heat, a patent for the process was obtained in England by James Palmer Budd.[4] This important technical advance culminated in the application, in 1860, of Siemens' regenerative principle to hot-blast stoves by E. A. Cowper of Middlesbrough.

[1] *New Statistical Account*, Vol. 6, p. 160.
[2] G. Macintosh, *Memoir of Charles Macintosh*, p. 106.
[3] *New Statistical Account*, Vol. 6, pp. 160 and 662.
[4] H. Cadell, *Trans. Roy. Scot. Soc. of Arts*, 1883, *10*, 506.

The hot blast proved to be the salvation of the Scottish iron industry which, as already shown, was developing more slowly at the beginning of the nineteenth century than was its English counterpart. With the coming of railways and steamships, it is doubtful if Scottish industry could have competed successfully with the English industry which was more advantageously situated. The influence which the hot blast had upon the production of iron in Scotland is illustrated by the following figures. Previous to its introduction, Clyde Iron Works was making 6,000 tons of iron per annum, each ton requiring for its production eight tons of coal and fifteen cwt. of limestone. In 1833, by which time they had adopted the hot blast, their production had risen to 12,500 tons, each ton now requiring only three tons of coal and eight cwt. of limestone.[1] Similar results were obtained at Calder, where Dixon, in 1831, cut down the consumption of coal by two-thirds while the yield of iron per furnace increased from forty to eighty tons per week. Pig-iron could be produced in some Scottish works for as little as £2 per ton. This led to economic depression in North Wales, and even the high quality of the iron produced in South Wales did not enable the Welsh iron-masters to withstand Scottish competition. Their disadvantages were offset to a certain extent by the rapid development of bar-iron production in Wales.[2]

It is little surprising that Andrew Ure speaks of Neilson's hot blast as one of the greatest discoveries ever made in the technology of iron-smelting and founding. Like so many other inventions, Neilson's process was really the culmination of the work of several earlier experimenters. A chemist to the Admiralty, of the name of James Sadler, referred to in the Chapter on balloons, had observed the effect of supplying heated air to a furnace and described his work in the *Philosophical Journal* in 1798, while a Kilmarnock minister called Stirling obtained a patent in 1816 for lessening the consumption of fuel by heating the air in a manner similar to Neilson.[3]

The improvements resulting from the application, in the hands of Neilson, of these ideas may be summarized as follows: as the temperature of the furnace was increased, fuel consumption fell and the yield of iron per furnace increased by more than one-third; the amount of limestone as flux could be reduced, and the same steam engine could be used to blow at least four furnaces instead of three. Finally, coal could be substituted for coke, i.e. a

[1] *New Statistical Account*, Vol. 6, p. 162. [2] H. Scrivenor, *op. cit.*, p. 299.
[3] *New Statistical Account*, Vol. 6, p. 660.

reversion of the previous important advance, a point to be discussed later. Here is a historical review of the two great discoveries, blackband iron-stone, and the hot blast, written by David Mushet:

> For several years after its discovery, the use of this iron-stone was confined to the Calder iron-works, erected by me in the years 1800, 1801 and 1802, where it was employed in mixture, with other iron-stones of the argillaceous class. It was afterwards used in mixture at the Clyde iron-works, and, I believe, no where else; there existed on the part of the iron trade a strong feeling of prejudice against it. About the year 1825, the Monkland Company was the first to use it alone, and without any other mixture than the necessary quantity of limestone for a flux. The success of the company soon gave rise to the Gartsherrie and Dundyvan furnaces, in the midst of which progress, came the use of raw pit coal and hot blast—the latter, one of the greatest discoveries in metallurgy of the present age, and above every other process, admirably adapted for smelting the blackband iron-stone. The greatest produce in iron per furnace, with the blackband and cold blast, never exceeded 60 tons a week. The produce per furnace now averages 90 tons per week.[1]

At first, great difficulties were experienced in giving a scientific explanation to these changes. According to Thomas Clark, who had been associated with Neilson in Glasgow, before going to fill the Chair of Chemistry, Marischal College, Aberdeen, the cause of the improvement could be explained in the following terms:

> The air is intended, no doubt, and answers to support the combustion; but its beneficial effect is, in the case of the cold blast, counteracted by the cooling power of six tons of air per hour, or two cwts. per minute, which when forced in at the ordinary temperature of the air, cannot be conceived otherwise than as a prodigious refrigatory passing through the hottest part of the furnace and repressing the temperature.[2]

This explanation, according to Thomas B. Mackenzie,[3] anticipates by half a century the explanation of the hot blast put forward by Johnson in his book, *Principles, Operation, and Products of the Blast Furnace*. Mackenzie gives extracts from Clark and Johnson for comparison:

[1] David Mushet, *Papers on Iron and Steel*, p. 128.
[2] T. Clark, *Trans. Roy. Soc. Edin.*, 1836, *13*, 380.
[3] T. B. Mackenzie, *Life of J. B. Neilson*, p. 28.

Thos. Clark in 1835.	*Johnson in 1904.*
In the manufacture of pig-iron experience has taught us that a certain temperature is required to work the Furnace favourably, and all fuel consumed to produce a lower temperature is fuel consumed in vain.	In 1889 it began to appear to me that there might be a critical temperature in the Blast Furnace above which some of the heat had to be supplied.

These are mainly physical considerations. Of chemical principles, Mackenzie goes on:

> The chemical reactions take place in the bosh and shaft of the Blast Furnace where the ore is being reduced by the highly heated ascending current of Carbon Monoxide gas and while the gangue is being fluxed by the lime-stone. From this point of view also the higher temperature, rendered possible by Mr. Neilson's invention of the hot blast, was a great boon to the ironmasters, as it is well known that the reactions in question proceed more vigorously and at a higher speed, as the temperature is raised. These two points of view, the physical and the chemical, are quite independent of each other, and their effect, in increasing output and in reducing cost, is therefore cumulative.[1]

Like many other inventors, Neilson was forced to defend his patent in the Law Courts, because an Association of Scottish Ironmasters was formed to evade or set aside his patent. The signatories to the Association were: (*a*) William Dixon; (*b*) William Baird and Company; (*c*) William Dixon and Company; (*d*) J. Macdonald, Junr.; (*e*) Alison, Merry and Cunningham; (*f*) The Househill Coal and Iron Company. In 1843 there was a ten-day Court of Session case against the Bairds of Gartsherrie over the latter's refusal to continue paying royalties. A verdict in favour of Neilson was returned. The cost of the action is reputed to have been £40,000. One hundred and two witnesses were examined. In evidence it was revealed that the Bairds made £400,000 profit out of the process between 1839 and 1842.[2]

Both in point of time and by his genius, J. B. Neilson occupies a central position in Scottish iron industry. He was connected through his father with the days of Dr. Roebuck and the founding of Carron. His brother, John Neilson, built the first iron steamer that sailed on the Clyde, and ultimately became co-founder of *Summerlee Iron Works*, near Coatbridge. His son, Walter Montgomerie Neilson, was the founder of the *Hydepark Locomotive Works*, later absorbed by the North British Company.

[1] T. B. Mackenzie, *Life of J. B. Neilson*, p. 30.
[2] For details *see* G. Macintosh, *Memoir of Charles Macintosh*, p. 108.

A subsidiary but important consequence of the introduction of the hot blast was the substitution of raw coal for coke in the blast furnace, thus circumventing the coking operation which had been of such prime importance in the eighteenth century. That this was possible was the discovery, in 1831, of Dixon of Calder Iron Works.[1] The change was of great consequence to Scotland, as few Scottish coals made good metallurgical coke. The remarkable fall in the consumption of fuel is illustrated by the following figures. To produce one ton of iron it required in:

1829 using coke and cold air . .	8 tons,	1 cwt.,	1 qr.
1830 ,, ,, ,, hot ,, . .	5 ,,	13 ,,	1 ,,
1833 ,, coal ,, ,, ,, . .	2 ,,	5 ,,	1 ,,

One significant point in the reduced fuel consumption was that the Scottish coal from which the coke was made lost no less than 55 per cent. in the coking operation.[2]

Angus McLean gives the following details of the coal used in Scotland:

> The coal used is of the variety called splint coal. It is often hard and stony, and may be brown or black in colour. On heating it gives off a very large quantity of volatile matter, and therefore burns with a long, bright flame, and leaves a residue of from 50 to 60 per cent. of coke, which is soft and friable and shows little sign of having softened or become semi-fused. The coal therefore does not swell up on heating—which is an essential character for blast-furnace use—the fragments of coke retain the form of the original pieces of coal, and are friable and easily crushed. It must be remembered that only the coke or fixed carbon is of any use in the blast-furnace, the volatile constituents being expelled at the upper part of the furnace, where there is no air for the combustion of them, and they pass away with the waste gases. The amount of fuel used is therefore much larger in the case of coal than of coke.[3]

The hot blast also made it possible to use the blackband iron-stone discovered by David Mushet at the beginning of the century. This iron-stone was virtually useless till the introduction of the hot blast. One proprietor whose land contained blackband iron-stone was not obtaining a penny for it in 1829, but in 1839 he was receiving royalties amounting to no less than £16,500 per annum. The advantages to be derived from heating the blast were, to begin with, confined to Scotland. A man of the name of Crane built a furnace on the borders of Brecknockshire, and got a patent for the

[1] *New Statistical Account*. Vol. 6, p. 661. [2] H. Scrivenor, *op. cit.*, p. 295.
[3] A. McLean, *Local Industries of Glasgow and the West of Scotland*, p. 18.

smelting of iron with anthracite. The process was, however, not successful, and the furnace only continued in operation for a few months.[1] In England the hot blast ultimately made anthracite available for smelters there. By 1840 the majority of Scottish iron-works were using raw coal. At first it was doubted if raw-coal pig-iron could be successfully converted into malleable iron, but, by a process of boiling instead of puddling, Messrs. Beecroft, Butler and Company, at Kirkstall, near Leeds, successfully converted Scottish pig-iron into malleable iron of good quality.[2]

APPENDIX TO CHAPTER 16

SCOTTISH PIG-IRON PRODUCTION[3]

Year	*Tons per Annum*	*Percentage of G.B. production*	*No. of furnaces in blast*	*Comments*
1760–88	less than 1,500		8	Founding of *Carron*
1788	7,000	10·25		
1796	*c.*17,000	14·9	17	Value of *Scot. Linen* 8 times as great. Carron using as much coal as Edinburgh (1793).
1806	23,000	9·3	20	
1814	32,760		21	
1829	37,500	5·5	22	Hot blast
1836	75,000	10·7	29	Period of boom
1839	195,600	17	54	
1843	300,000	25·6	62	
1845	475,000	26 (1848)	88	
1853	685,700		114	
1867	1,031,000		128	

Thus we may conclude with the words of Paul Mantoux: 'The evolution of the textile industry is due to mechanical inventions, that of the metal-working industry to chemical inventions'.[4]

[1] H. Scrivenor, *op. cit.*, p. 301. [2] *New Statistical Account*, Vol. 6, p. 663.
[3] Dr. Cleland, *Statistics of Glasgow and Lanarkshire*, p. 140.
H. Hamilton, *Scot. Hist. Rev.*, 1928, *25*, 193.
J. Sinclair, *General Report*, Vol. I, p. 69.
D. Bremner, *Industries of Scotland*, p. 32.
Macgregor, Lee and Wilson, *Iron Ores of Scotland*, p. 11.
[4] P. Mantoux, *op. cit.*, p. 304.

Chapter XVII

SCOTLAND'S TREASURE-HOUSE

EXCLUDING coinage metals, which had a special significance as standards of value, lead was the important metal in eotechnic economy. It followed thatch as a roofing material, and had a long life when used for certain special structural features, for example, rain-pipes, gutters, and for covering domes. As protection against ravages of marine life it was tried as a sheathing for ships before coal tar became available. The alloy which it forms with tin, namely pewter, was used to furnish the eotechnic table. The progressive decline in the consumption of pewter as it gave way to ceramic products may well have contributed to the fall in the production of lead which characterized certain phases of social evolution. Solder and type metal are also lead alloys, and for these there is a constant demand. Depending upon the type of ore used, considerable quantities of lead were on occasion produced as a by-product, when an ore containing both lead and silver was mined for the latter metal. This was often done although the percentage of silver was low.[1]

Some Scottish lead mines were of great antiquity. For example, those in the parish of Glenorchy and Inishail were declared *mines royal* as early as 1424.[2] The long eotechnic history of the principal Scottish lead mines, those at Leadhills, can be followed in G. V. Irvine and A. Murray's *Upper Ward of Lanarkshire*,[3] a district known for its mineral wealth as *God's Treasure-House in Scotland*.[4]

Lead-mining was confined, as far as really extensive operation was concerned, to a single area in Scotland, a wild stretch of country on the borders of Lanark and Dumfries-shire, and it is with operation and organization there that we shall be mainly concerned in this chapter. Small-scale operation was, however, widespread, but lack of both technical and business enterprise made many of these smaller mines transitory ventures. We see this, for example, in the account of lead-mines in Forfar-

[1] J. U. Nef, *The Rise of the British Coal Industry*, Vol. 1, p. 167; J. Sinclair, *General Report*, Appendix 2, p. 288.

[2] *Statistical Account*, Vol. 8, p. 351.

[3] Vol. 1, *passim*.

[4] *See also* R. W. Cochran-Patrick's *Early Records Relating to Mining in Scotland*, and *Session Papers*, Signet Library, 234: 6.

shire in James Headrick's *General View of the Agriculture of Angus*:

> A lead-mine was wrought at Gilfianan, in the upper part of the parish of Lochlee, during the forfeiture of these estates, after 1715; and another was wrought at Ardoch, near Mill-den, on the Esk. It appears from Edward's Account of Forfarshire, that the mine at Gilfianan, above the old castle of Invermark, had been discovered and wrought at a much earlier period: for he states, that in his time (A.D. 1678), 'eighteen miners were employed in this mine which seemed to be inexhaustible; and that the lead, when extracted and properly refined, yielded 1/64th part of silver'. These mines were abandoned after they got under water.[1]

The mines were in the news again in 1728, when a band of miners took a *bargain* to explore the possibilities of mining lead. They were no more successful than their predecessors. It was said that they lacked a *developer*.[2] One of the troubles was that palaeotechnics had not developed sufficiently to overcome technical difficulties. Here is another instance. At Kemback in Fife, rich lead-ore was discovered in 1722. John Bethune of Blebo and some of his friends formed a company to work it, but the veins were found to pass through rock so hard that it could only be wrought with the aid of gunpowder. This was considered to be too expensive, and the company was forced to disband. A further trial was made in 1748. Any lead produced was exported to Holland.[3] From the Ochils above Dollar, a company exported both lead and copper to Holland, but 'disagreement among the adventurers led to them giving up'.[4]

On occasion, technical skill was imported into Scotland. Derbyshire miners were contracted with to work lead found on a farm belonging to the Duke of Buccleuch at Westwater, Langholm.[5]

The discovery of lead-ore on his land was profitable to a landowner, and we find the Earl of Stair actively prospecting for lead on his land at New Luce, Wigtown.[6] Sir John Sinclair gives a full account of proposals he made to a Lead Company at 9 Martin's Lane, Cannon Street, to work ore discovered on his estate. We quote them *in extenso*:

> Proposals transmitted by Sir John Sinclair to the Lead Company, for letting a Lease of the Mine of Skinnet, in the County of Caithness, North Britain, anno 1790.
>
> ---
>
> The Hill of Skinnet is about four English miles from the town and harbour of Thurso, in the county of Caithness. It is the

[1] J. Headrick, *Gen. View of Angus*, p. 15.
[2] *Statistical Account*, Vol. 5, p. 367. [3] *Ibid.*, Vol. 1 4, p. 305.
[4] *Ibid.*, Vol. 15, p. 161. [5] *Ibid.*, Vol. 13, p. 584. [6] *Ibid.*, Vol. 13, p. 584.

property of Sir John Sinclair of Ulbster, and an idea being very prevalent, that some mines existed in that place, Sir John was led, in the course of the year 1787, to endeavour to discover the nature of the ore which might be found there; and in tracing the course of what is called in that country a burn (or rivulet), he accidentally hit upon a small vein of yellow mundick, of about 3 inches in breath; and upon digging a little deeper, he met with a great mass of white mundick, several cart-loads of which were dug up without the smallest difficulty.

Upon shewing specimens of these articles to persons skilled in mineralogy, particularly to some Cornish miners, they told him, that the mundicks he had found, however brilliant, were in themselves of no real value; but they informed him, in their technical language, 'that mundick, in such quantities, was a good sign of more valuable veins: that the white mundick, in particular, was a good horseman, and always rode on a good land'. And, in short, pressed him to make further trials and inquiries.

Mr. Raspe, a German mineralogist, having come into the county of Caithness, last autumn (anno 1789), was employed by Sir John Sinclair to make trials in the same place; and not far from the mundick, he discovered a regular vein of heavy spar, mixed with lead and crystals, three feet in breadth, and very near the spot where the mundick was found. No further progress was made, than merely to ascertain the size of the vein, and the nature of the metal which it contained.

Sir John does not propose to work this mine himself, and is very well disposed to give every reasonable encouragement to any respectable Company that would undertake it.

The Company would have several advantages in carrying on this mine. The miners and other workmen, and any tools that might be necessary, might be transported by sea to the town of Thurso, within four miles of the vein.

Any additional workmen that might be required, might be got at an easy rate, labour not being very dear in the county.

The road, at present, from Skinnet to Thurso is not very good, but it might be completed at a small expense.

Peats or turf, for fuel, are scarce, but of water there is abundance and coals may be transported by sea to Thurso.

The value of the mine might be tried at a small expense, as it lies on the side of a hill, gently sloping about half-a-mile higher than the river Thurso; but with such a descent to the river, that no engine would be necessary for clearing off the water.

Were this mine not to answer, there are many other appearances on the estate of Sir John Sinclair, and on the property of other gentlemen in the neighbourhood, which might be well worthy the attention of the Company.[1]

[1] *Statistical Account*, Vol. 20, p. 537.

The ore from the Hill of Skinnet was assayed by the illustrious chemist, Sir Humphry Davy, who wrote to Sinclair:

> I have completed the analysis of the lead ore from your estate. It gives by a common assay, sixty-one in the hundred parts of lead. Accurate analysis, however, proves that it consists of seventy-one of lead, and twenty-nine of sulphur. I congratulate you on being possessor of so valuable a substance: it is one of the richest species of potter's lead, and the price of lead is now excessively high.

Professor Jameson of Edinburgh also analysed the same ore, and found 69 per cent. lead, 14 per cent. sulphur, with traces of silver and earthy matter.[1]

Lead ore was sometimes discovered in consequence of operations carried out for other purposes, as at the lime quarry of the Earl of Dumfries.[2] In 1763 lead was discovered on the property of Mr. Heron of Heron, at Minnigaff, Kirkcudbright, while making a military road from London to Dublin [*sic*]. Heron developed the deposit and at one time was producing 400 tons of ore per annum. This had fallen to 30 in 1793. Lead ore was also mined to the extent of 400–500 tons per annum on the neighbouring property of Mr. Pat Dunbar of Machermore. The market price of the ore at Chester, where much of it was sent, was £8, and of the lead £18 per ton. Three tons of the ore produced two of lead.[3]

Similar quantities were produced at Sir James Riddel's Strontian lead mine at Sunart in Argyllshire (opened 1722), where two hundred men were employed in 1813. Riddel got one-eighth of the production free of charge. This mine is of more than passing interest, since it was in a mineral from this mine that Thomas Charles Hope (1766–1844) discovered a hitherto unknown element which was called *strontium* (Sr).

The principal lead-mines in Scotland were situated at Leadhills and Wanlockhead on the border of Lanark and Dumfries-shire in the parishes of Crawford and Sanquhar. Over the period which we have under discussion, those at Leadhills were the property of the Earl of Hopetoun, those at Wanlockhead of the Duke of Queensberry. We get some idea of the lead-producing district from the following contemporary description:

> The little village of Leadhills consists of numbers of mean houses, inhabited by about fifteen hundred souls, supported by the mines; for five hundred are employed in the rich *sous terrains*

[1] J. Henderson, *General View: Caithness*, p. 338.
[2] *Statistical Account*, Vol. 6, p. 408.
[3] *Ibid.*, Vol. 7, p. 54; *Session Papers*, Signet Library, 144: 1, 445: 41.

of this tract. . . . The space that has yielded ore is little more than a square mile, and is a flat or pass among the mountains; the veins of lead run north and south, vary, as in other places, in their depth, and are from two to four feet thick; some have been found filled with ore within two fathoms of the surface, others sink to the depth of ninety fathoms.

Archibald Stirling of Garden, the mathematician, who was an agent for one of the two companies at Leadhills in the 1790s, relates that, from a register then in the possession of the company, the mines were discovered by one Martin Templeton in 1513. They were first wrought by George Douglas of Parkhead, probably about the beginning of the seventeenth century. About the same time they passed into the hands of Thomas Foulis. Anne, daughter of Foulis, married Sir James Hope, King's Advocate, and his heirs became proprietors.[1] Detailed accounts of the lead-working were published by J. M. Porteous, in his account of the district, from which we take the following citations:

Two distinct companies obtained the mining-field on lease in 1747. The *Scots Mining Company* held the north-western portion. Its shareholders were chiefly gentlemen in London, whose descendants originated the Sun Fire Office. Although embracing but a fourth of the whole, the capital subscribed amounted to £10,000 in £100 shares. Mr. Marchbank and Company held the south-western portion. In addition to these, the Hopetoun family reserved and worked a portion east of Glengonar Burn. Mr. Marchbank and Company having resigned for want of success, the *Scots Company* obtained the south-western portion in 1772, and carried on the works with vigour. At that same time Mr. Popham, a Master in Chancery, and others obtained that part reserved by the Hopetoun family, and they termed themselves the *Leadhills Mining Company*. Having but poor success, they ceased operations in 1805. Then the *Scots Company* obtained the north-eastern portion; and the south-eastern, after a rest of three years, was leased by Mr. Horner, from Darlington, and others, who carried on the works with vigour till 1817, when Mr. Horner died. The family of that gentleman being large, and the property divided, want of capital and mining knowledge led to the abandonment of the mines by his sons in 1828.

The produce of the mines (Leadhills) has been known to vary from 10,000 to 18,000 bars of from 112 to 120 lbs. each. The sale also has been various. After the commencement of the French Revolution, the demand failed, so that £40,000 worth remained at Biggar, half-way, and an equal quantity at Leith, for a time. Soon after the demand raised the price to double that amount.

[1] *Statistical Account*, Vol. 21, p. 97.

In the zenith of the lead trade at Leadhills in 1810, about 1,400 tons annually were produced, which, according to the current price, valued about £45,000. Price and quantity decreased to 700 or 800 annually. Workmen were also reduced till only eight men were employed by the *Scots Mining Company*, whose Manager was Mr. Borron. The rent of the Earl of Hopetoun at that time was every sixth bar produced.

The Leadhills Mining Company purchased the lease from Mr. Thomas Horner. They were, however, unable to commence work, as they could not obtain water for their engines and wheels, the use of a watercourse having been interdicted them by the *Scots Company*. This resulted in a lawsuit between the *Leadhills Mining Company* and the Earl of Hopetoun on the one hand, and the *Scots Mining Company* on the other. This lawsuit, which lasted upwards of twenty years, and cost some £25,000, led to no satisfactory result. At length—after the mines had been thoroughly inspected, reports and suggestions presented by Captain Vivian and others—a compromise was entered into in 1861, by which the *Scots Mining Company* relinquished their lease, and the *Leadhills Mining Company* obtained possession of the entire mining-field.[1]

The *Statistical Account* gives, in addition, the following data:

There are nearly two hundred men employed by the *Scotch Mining Company*. They are divided into peckmen, smelters, washers, and labourers, besides carpenters and smiths. The payment of the peckmen depends on the quantity of lead found. Five or six of them join, and take what is called a bargain and according as it proves more or less abundant in lead, their wages are more or less. But at an average I am told every man receives from £18–£20 a year. A smelter receives fully as much.[2]

The operation of the smelting company was in the hands of a President and twelve directors, together with a secretary and clerk. Control was exercised from London. At Leadhills they kept an agent, two overseers, and two or three clerks.[3] The scale of these operations is comparable with the great copper mines of Anglesey, where, about 1830, there were in employment 500–1,000 men.

Details of output are given in the Agricultural Account of Dumfries:

The lead-mines occupy very barren grounds, remarkably bleak and elevated; but they are a great fund of industry and riches, and they furnish a part of the country with an excellent market for the surplus grain produced in that part. Lead-hills, with the

[1] J. N. Porteous, *God's Treasure House in Scotland*, p. 69 *et seq.*
[2] *Statistical Account*, Vol. 4, p. 511. [3] *Ibid.*, Vol. 21, p. 97.

mines, are in the county of Lanark, and belong to the Earl of Hopetoun, who draws about £7,000 a year from these mines. Wanlockhead is in Dumfriesshire, belongs to the Duke of Queensberry, and returns to the proprietor, from the lead-mines, near £5,000 a year. The produce of Lead-hills for the year 1809 was,

in bars, ...	25,200
and that of Wanlockhead was, in bars . .	15,552
Of course the united number of bars .	40,752

at 8 stones and 4 lbs. Amsterdam, or 9 stones avoirdupois each bar, were the produce of the mines in both places; and the price having been £32 per ton, the gross produce must have exceeded £80,000, in a year so favourable as that year was. It is unfortunate that, in 1811, the price has fallen to £24; and the lead cannot be sold with any advantage at that rate, as it is computed that at Wanlockhead the lead of crop 1810 has cost, in works and charges, £25 per ton.[1]

We must now follow the development of the neighbouring mines at Wanlockhead, and to do so, quote again from Porteous's very full account, which incorporates details given by T. Montgomery:

The subsequent history of these works, which include a circle of about two and a half miles, and join at the boundary with those of Leadhills, was graphically told by the Rev. Thomas Montgomery in 1835:—

The lead mines at Wanlockhead were opened by Sir James Stampfield about the year 1680, and were wrought by him with some success, but not to any great extent, till the Revolution in 1691. He was succeeded by Matthew Wilson, who procured a lease for nineteen years, and wrought the vein called Margaret's in the Dodhill. He carried his workings quite through that hill, from Whitecleuch to Wanlock stream, and was very successful in his discoveries. He was again succeeded in 1710 by a company for smelting lead ore with pit coal. They had a lease for thirty-one years, and wrought to a considerable extent in the veins of Old Glencrieff and Belton-grain, but were not very successful in their operations. At length, after much discouragement, they had the good fortune to find out the veins of New Glencrieff, where in a very short time they raised a great quantity of lead ore. In 1721 a numerous company was formed of persons residing in different parts of the kingdom, under the name of the *Friendly Mining Society*. They entered into partnership with the smelting company. . . . The two companies thus united carried on their operations in all the four principal veins then known, to a considerable extent, till 1727. They then separated from each other, and prose-

[1] Dr. Singer, *General View: Dumfries*, p. 27.

cuted their works in different grounds. The smelting company entered on the east side on the Wanlock stream, and vigorously continued their operations till 1734, when, having suffered great loss, though they had raised much lead, they resigned their lease. An individual partner in the company, however, Mr. Wightman, retained liberty to work in the southern part of their boundary. He confined his operations to the south end of Margaret's vein. But they were very unprofitable, and terminated with his death in 1747. The mining liberty which had been possessed by him was unoccupied till 1755, when it was entered on by Messrs. Ronald Crawford, Meason and Co. . . . They continued their operations in Margaret's vein forty-three years. For carrying off the water they erected three steam-engines. The first was in 1778, and supposed to be the second erected by Mr. Watt in Scotland. The *Friendly Mining Society*, or *Quaker Company*, having resigned their lease in 1734, were succeeded by Alexander and William Telfer. . . . They succeeded in raising great quantities of ore, which sold at a high price. At the expiration of their lease in 1755, Messrs. Ronald Crawford, Meason and Co. were also their successors, who now possessed the whole of the mining liberties at Wanlockhead. This enterprising and eminently successful company continued the works which had been left by Messrs. Telfer till 1775. About this period they discovered good ore in Belton-grain vein above water-level, and continued working there till 1800. Then finding the ore above water-level to be mostly wrought out, they were under the necessity of erecting one steam-engine, a second in 1812, and a third in 1817. The works were profitable till the free-trade system was introduced, and foreign lead was allowed to be imported to Britain, without being subject to the payment of duty. From that period the company must have suffered great loss. The veins also presented great poverty. . . . During the period which elapsed from 1823 to 1827, the company sank forty fathoms under level, and erected two steam-engines underground, but the quantity of ore found did not answer their expectations. . . . The five engines last mentioned possessed collectively 268-horse power. Previous to the erection of these and of those on Margaret's vein, the water was raised from the mines by hand-pumps and water-wheels. The steam-engines have now all been removed, and a water-pressure engine has been erected, which is succeeding remarkably well. It carries away all the water which was formerly removed by the two steam-engines underground; it works with little attention, requiring merely that the water be kept regularly upon it, and thus greatly lessens the expense which was formerly incurred. The company at their commencement in 1755 had a lease only of nineteen years; but an Act of Parliament was obtained afterwards authorising the extension of it till 1812, which was subsequently extended thirty years farther. The lessees in 1835 were the

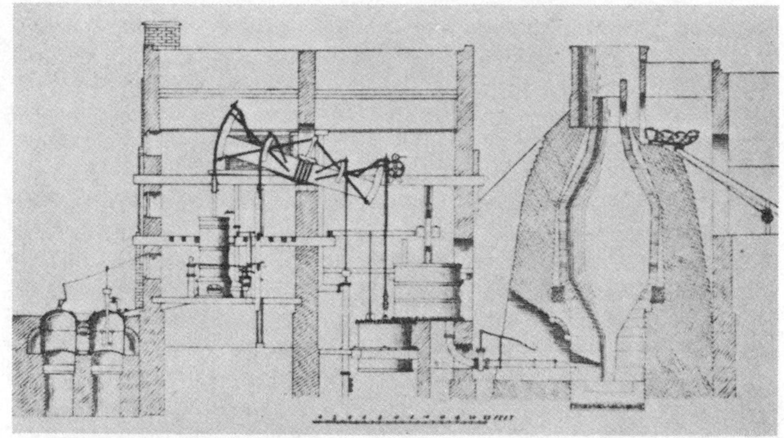

A drawing of 1788

79. Original Newcomen blast engine and furnace at Clyde Iron Works.

Sketch by W. D. Scott-Moncrieff

80. Summerlee Iron Works at night. Founded by Wilson & Co. in 1836, it is representative of the great concentration of blast furnaces in the Old Monkland parish of Lanarkshire, appropriately known as 'The Land of Fire.'

Diderot and d'Alembert's *Encyclopédie*

81. Hearth for smelting lead,

82. Preparation of lead compounds. (*left*) Operative placing strips of metallic lead over pots of acetic acid in the preparation of white lead by the Dutch, Vinegar, or Stack Process. The 'stack' consisted of alternate layers of pots and lead strips as illustrated, with dung or, later, tan bark which in decomposing gave off carbon dioxide. (*right*) Preparation of red lead, the main users of which were the flint-glass-makers. A description of the process by the Bishop of Llandaff will be found on p. 381.

Marquis of Bute, three shares; and Mr. M'Leod, one—in all, four shares. The company during fifty years have expended at Wanlockhead the sum of £500,000. By the terms of their lease, they delivered *a sixth part* of the lead raised to the proprietor as rent or lordship. But by a new agreement . . . they delivered a much less quantity. During fifty years, 47,420 tons of lead were raised. The success of the company was various. In the year 1809 were raised nearly 1,037 tons of lead, sold at £32 per ton. In 1811 its price was £24 per ton. In 1829 and 1830 respectively were raised 596 and 461 tons, when its price was only £13 per ton. The number of persons employed in the works in 1835 were 4 overseers and clerks, 154 miners, 12 washers, 8 smelters, 10 smiths, carpenters, and engineers, 20 boys who assist in washing—amounting in all to 208. The work was, and still is, let by bargains, generally for three months—that is, the workmen receive a certain stipulated sum for the quantity of ore per ton which they raise, or for the fathom of dead work which they perform in that time. They relieved each other by courses every six hours (now every eight), and in twenty-four hours the same course does not go to work more than once. Each miner, on an average, was supposed to earn about £20 during the year. Agreeably to the contract of lease, some spare pickmen are always at work for making new discoveries of lead.

The structure of the veins is very various. The ore frequently lies in a regular form, but sometimes it is irregular, and mixed with what are called vein-stones, as lamellar heavy spar, calcspar, rockcork, etc. The ores are lead glance, blende, manganese ochre, copper pyrites, green lead ore, white lead ore, lead vitriol, and brown hematite, all in small quantities except the lead ore. The contents of a bearing vein are often found as follows: On the under or lying side is lead glance or galena, then a layer of ochre of manganese several inches thick, above it a layer of quartz interspersed with iron pyrites, then another layer of manganese mixed with quartz, pieces of lead glance and carbonate of lime, followed by greywacke, which constitutes the walls of the mine. Besides lead glance or common galena, the following minerals are also found, viz., sulphate, phosphate, carbonate, and arseniate of lead. The vanadiate of lead has been found in the refuse of the old workings, where it was for a time taken for arseniate of lead. These different specimens are now and then found occupying the same drusy cavity, and when seen before being injured or removed from their relative position exhibit a fine lustre and beautiful crystallisation. The druses, or laeugh holes, as they are termed by the miners, are also frequently studded with quartz, carbonate of zinc, etc. Sulphuret of zinc or zinc-blende is found in considerable quantities in some of the veins, particularly in Margaret's. Specimens of iron also occur, as ochry-red iron ore, but iron is always reckoned by

> the miners as unfavourable to their prospects in procuring lead. The lead glance at Wanlockhead was found by analysis to contain a small quantity of arsenic, antimony, and silver. The last mentioned was from eight to ten ounces in the ton of lead.[1]

The reference to the steam-engine in the above citation is of considerable significance, since it relates the needs of palaeotechnics in English tin mines with similar needs in Scottish lead-mines. John Lord in his *Capital and Steam Power* gives a list of the Watt engines erected prior to the end of 1782 from a memorandum compiled by Boulton. The position of one of these engines is given as 'Wanlockhead, Edinburgh', and its function said to be pumping at a colliery. It was stated to be working in 1782 and bringing in £100 per annum to Boulton and Watt.[2] An agreement dated August 1777 shows that the 'Wanlockhead colliery' was owned by Ronald Crawford and Company, of which the general manager was Gilbert Meason, Esq., of Edinburgh. Other partners included the Countess of Dumfries and Sir Peter Crawford.[3] Since the Wanlockhead Lead Mines were leased in 1755 to a company with Ronald Crawford at the head, it appears unlikely that there was such a thing as a Wanlockhead colliery. In the *General View of Dumfries* there is no mention of a colliery at Wanlockhead, trial borings in the county only dating from 1792–5, and reference to the Boulton and Watt Papers in the Reference Library, Birmingham, leaves no doubt that the Watt engine in question was installed at the Wanlockhead Lead Mines. There are more than fifty-three letters from Boulton and Watt to Gilbert Meason dated from 13th September, 1776, till 3rd February, 1797, and a copy of a letter of agreement dated 4th July, 1777.

There is no doubt that this was one of the first Watt engines in Scotland, perhaps second to that erected at a colliery at Torryburn. Other enquiries about engines, which, however, came to naught, were addressed to Boulton and Watt from Samuel Garbett of Carron (16th February, 1776), Sir A. Hope (30th August, 1776), and G. Meason (24th April, 1777). In 1784, Dr. Black examined a sample of water for James Watt. This water was thought to be the cause of corrosion of a piston rod belonging to George Meason. However, according to Sir Wm. Ramsay, the corrosion was due to acidity of oil or tallow used as lubricant, a fault which was rectified by the addition of a little alkaline soap.[4]

[1] J. M. Porteous, *op. cit.*, p. 78 *et seq.*
[2] J. Lord, *Capital and Steam Power*, p. 158. [3] *Ibid.*, p. 191.
[4] W. Ramsay, *Life and Letters of Joseph Black*, p. 99.

Some of Dr. Black's connexions with Leadhills are illustrated by the following citations from Sir Wm. Ramsay's *Life*:

> In 1770 Black received from Lord Hopetoun samples of the soil from the Leadhills, a district in the upper part of Lanarkshire, for examination. He found it to contain 179 grains of gold per ton, besides lead ore (galena), iron pyrites and magnetic ore, as well as small garnets. Another batch contained spar with veins of 'black-jack' or zinc-blend; there was also kupfer-nickel, and from this Black prepared some 'niccolum', which he sent to Lord Hopetoun.
>
> In 1772, at the wish of Lord Hopetoun, Black visited the Leadhills, and made a report on the gold in the streams of that neighbourhood. He found gold in the river sand, and recommended a trial on the rock of the country, by the aid of a stamping-mill. In the sand he found ninety-seven grains per ton. This was obtained by washing and then cupelling the galena left in the dish; to remove the silver the operation of parting was resorted to. Each particle of galena had adhering to it a particle of a 'compound of arsenic, iron, and some cobalt, from which it appears that there is a vein of cobalt somewhere in these hills'. He gave Lord Hopetoun designs of a mechanical washer which he had invented, and which, he says, would cost three to four pounds.[1]

In contrast to its early introduction at Wanlockhead, it appears that steam power was not introduced in the Leadhills mines till 1842. Yet there is a description, by Dorothy Wordsworth, in her *Recollections of a Tour made in Scotland, A.D. 1803*, which reads very like that of a steam-engine:

> Our road turned to the right, and we saw, at the distance of less than a mile, a tall upright building of grey stone, with several men standing upon the roof, as if they were looking out over battlements. It stood beyond the village, upon higher ground, as if presiding over it—a kind of enchanter's castle, which it might have been, a place where Don Quixote would have gloried in. When we drew nearer we saw, coming out of the side of the building, a large machine or lever, in appearance like a great forge-hammer, as we supposed for raising water out of the mines. It heaved upwards once in half-a-minute with a slow motion, and seemed to rest to take a breath at the bottom, its motion being accompanied with a sound between a groan and 'jike'.[2]

Irvine and Murray make no mention of steam power in 1864 so, if introduced in 1842, it must have been abandoned twenty

[1] W. Ramsay, *op. cit.*, pp. 55 and 56.

[2] D. Wordsworth, *Recollections of a Tour made in Scotland, A.D. 1803*, p. 17.

years later. It is more than likely that fuel supply proved a difficult proposition, and that it was more economical to utilize available water. When J. R. S. Hunter visited Leadhills in 1884, he speaks of having seen a gallery at forty-four fathoms 'where formerly an engine was placed to draw water and ore and other materials to and from the surface'.[1]

The following, in brief, are the processes through which lead ore was put after being mined. (*a*) It was first taken to the *coups* or departments into which the produce of each partnership was deposited. (*b*) The ore was classified. (*c*) Ore, mixed with quartz (SiO_2) was crushed with water-power-operated crushers. (*d*) Crushed ore was washed by several processes, and then (1) trucked, (2) sieved, and (3) buddled. (*e*) Now dressed, it was weighed, and (*f*) roasted; (*g*) smelted and cast into ingots. Where desilverization was practised, bars were remelted and (*h*) a fractional crystallization carried out so that one part of the melt contained the most of the silver, while the rest became a purified lead. The argentiferous fraction was (*i*) cupelled, i.e. the lead was oxidized to litharge (lead oxide, PbO), and pure metallic silver left. Wanlockhead was the only lead-works in Scotland at which desilverization was carried out.[2] Coal was first used in Scotland in lead-smelting in 1727.

Pococke visited the Scottish lead mines and in his *Tours in Scotland* speaks of having seen the lead smelted on 'common hearths' with coal, lime and turf. Both coal and lime were transported from Douglas, some eight miles distant.[3] Of the coal situation in Dumfries-shire in general we learn from the Agricultural Survey:

> The upper part of Annandale is provided from the coal works of Douglas, in Lanarkshire, at the distance of thirty miles, conveyed overland, and in Moffat selling at 1*s.* 6*d.* per cwt. Occasionally a supply of what is called the cannel, or candle-coal, is brought into the same district for the purpose of lighting cottages and fires, from Lesmahago, a still greater distance. In the head of Nithsdale, blind coal (being mineral and native charcoal) is brought from Ayrshire, and used for drying oats on kilns, to be ground into meal, this coal answering very well, having neither smoke nor flame.[4]

There does not seem to be a contemporary account of smelting

[1] *Trans. Geol. Soc., Glasgow*, 1883–5, 7, 373. Full details of the lead mines of Scotland will be found in Wilson and Flett's *Memoir of the Geological Survey* dealing with the *Lead Ores of Scotland.*

[2] J. M. Porteous, *op. cit.*, p. 84.

[3] R. Pococke, *Tours in Scotland in 1747, 1750, and 1760*, p. 41.

[4] Dr. Singer, *Agricultural Survey of Dumfries*, p. 31.

plant in Scotland, so we will utilize this description dating from about the middle of last century:

> From far below we had seen smoke hanging about an opening before us. This was from the smelting-houses, the driver informed us; and the village lay about a mile and a half further on. The road crosses the valley near the smelting-houses; and they lay before us on the right—the turbid little stream oozing away from the works, and men and boys with hoes, spades, and scapers, washing the soil, on stage below stage, so that what escaped from one set of channels might run into the one below. It seemed a piece of unnecessary toil to place the square tower of the smelting-house—the tower whence the smoke belched forth—so high up the steep and stony breast of the hill. It afterwards appeared that nobody had occasion to go up there. The smoke was driven, by the blast of the furnace, through the interior of the hill, to issue forth from a chimney, which looks like a tower from below.
>
> Descending from the successive platforms where the bruised ore is washed, till it is almost pure dust of lead, we put our heads into the noisy vault, where the great water-wheel was revolving and letting fall a drip which filled the place with the sound of mighty splashings. The blast of the furnaces roared under our feet, and all around about us, every light substance, such as coal dust and shreds of peat, was blown about like chaff. At the furnace were men, enduring the blaze of the red heat on this sultry day. . . . They were piling up the glowing coals upon the bruised and washed ore in its receptacle in the furnace; and from under the front of the fire we saw the molten lead running down its little channels into its own reservoir, leaving behind the less heavy dross, which was afterwards to be cast out in a heap in the yard. We waited till there was lead enough in the reservoir to make a pig. One man ladled out the molten metal into the mould, while another skimmed off the ashes with two pieces of wood.[1]

Shortly after the date of the above account the mines were being worked by 'a company of Scotch gentlemen of which the principal partner is William Muir, Esq., Leith'. The company then had four hydraulic engines for pumping, one hydraulic engine and four water-wheels for drawing work, one water-wheel for crushing and dressing the ores, and one for driving the blasts at the smelting works, an aggregate of five hundred and fifty horse-power.

At Leadhills there were two roasting furnaces, four ore-hearths, one slag hearth, and a reverberatory furnace. These produced in all some fifty tons of lead per week.

[1] G. V. Irving and A. Murray, *The Upper Ward of Lanarkshire*, Vol. 3, p. 39 *et seq.*

LEAD

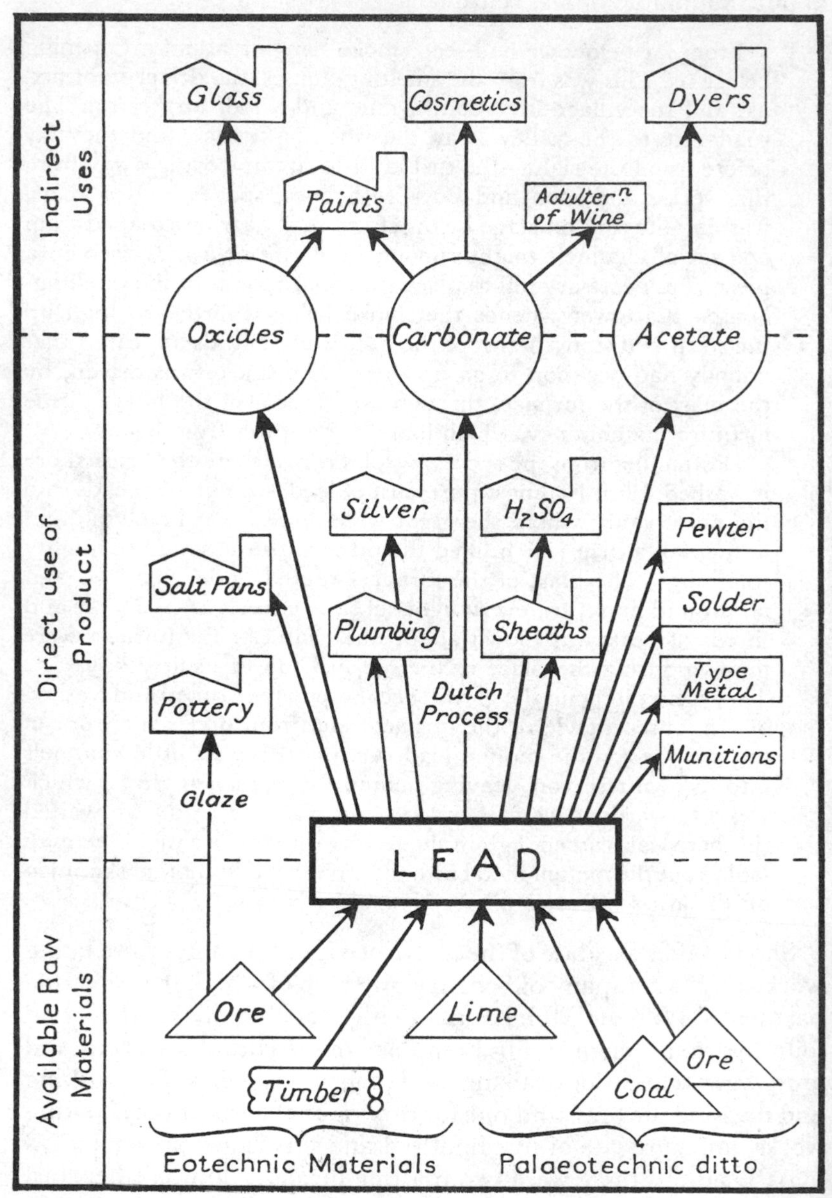

Indirect Uses
Direct use of Product
Available Raw Materials
Glass
Cosmetics
Dyers
Paints
Adulterⁿ of Wine
Oxides
Carbonate
Acetate
Silver
H_2SO_4
Pewter
Salt Pans
Plumbing
Sheaths
Solder
Type Metal
Pottery
Dutch Process
Munitions
Glaze
LEAD
Ore
Lime
Timber
Ore
Coal
Eotechnic Materials
Palaeotechnic ditto

The ore yields in general about seventy pounds of lead from a hundred and twenty of ore, but affords very little silver; the varieties are the common-plated ore, vulgarly called potters'; the small or steel-grained ore, and the curious white ores, lamellated and fibrous, so much searched after for the cabinets of the curious. The last yields from fifty-eight to sixty-eight pounds from the hundred, but the working of this species is much more pernicious to the health of the workmen than the common. The ores are smelted in hearths, blown by a great bellows, and fluxed with lime. The lead is sent to Leith in small carts, that carry about seven hundred-weight, and exported free from duty.

In many metallurgical processes a preliminary roasting of ore was effected to get it into a condition suitable for subsequent treatment. We shall see subsequently that roasting was an essential first step in the production of lead with a reverberatory furnace, and that ore even before going to the ore-hearth, may also have been so treated.

In this furnace ore and charcoal, or ore and what they call white coal, which is wood dried but not charred, being placed in alternate layers, upon a hearth properly constructed, the fire is raised by the blast of a bellows, moved by a water-wheel; the ore is soon smelted by the violence of the fire, and the lead as it is produced trickles down a proper channel, into a place contrived for its reception.[1]

Lead-ore is usually sulphide (PbS), and when smelted in contact with carbonaceous matter is first oxidized by contact with excess air, and then reduced by the glowing carbon.

$$
\begin{aligned}
&(a)\quad 2PbS + 3O_2 &= 2PbO + 2SO_2\\
&(b)\quad PbO + C &= Pb + CO\\
&\quad\quad 2PbO + C &= 2Pb + CO_2
\end{aligned}
$$

Richard Watson tells us that the *hearth furnace* was the only one in use in Derbyshire in 1730, but that by 1780 it had completely disappeared and given way to the *cupola* or *reverberatory furnace* in which the fuel used was pit coal. This furnace is said to have been the invention of a physician called Wright, who devised it *c.* 1698. It was introduced into Derbyshire *c.* 1747, and had moved north to Alston, Northumberland, by 1810. Of its introduction into Scotland we are not certain. It became known as the *English* Furnace.[2]

[1] R. Watson, *Chemical Essays*, Vol. 3, p. 272.
[2] M. Hellot, *Essais des Mines*, Vol. 2: quoted by R. Watson, *op. cit.*, Vol. 3, p. 274.

Ore in the reverberatory furnace is roasted at a low temperature. This converts the ore (lead sulphide, PbS) into a mixture of the oxide (PbO), sulphate ($PbSO_4$), and sulphide. As the operation proceeds, the air is shut off, and with the rise in temperature these substances react together with the formation of metallic lead and gaseous sulphur dioxide.

$$\begin{array}{c} 2PbS + PbSO_4 + 2PbO = 5Pb + 3SO_2 \\ \text{sulphide} + \text{sulphate} + \text{oxide} = \text{metal} + \text{sulphur dioxide.} \end{array}$$

Here is Watson's description:

> This furnace is so contrived, that the ore is melted, not by coming into immediate contact with the fuel, but by the reverberation of the flame upon it. The bottom of the furnace on which the lead ore is placed is somewhat concave, shelved from the sides towards the middle; its roof is low and arched, resembling the roof of a baker's oven; the fire is placed at one end of the furnace, upon an iron grate, to the bottom of which the air has free access; at the other end, opposite to the fire place, is a high perpendicular chimney; the direction of the flame, when all the apertures at the sides of the furnace are closed, is necessarily determined by the stream of air which enters at the grate, towards the chimney, and in tending thither it strikes upon the roof of the furnace, and being reverberated from thence upon the ore it soon melts it.[1]

This type of furnace does not have a bellows to force the draught. Consequently, small particles of lead ore were not driven about as was the case with the ore hearth, and this was an advantage, says Watson, 'superior to all the rest—the preservation of the workmen's lives. The noxious particles of lead are carried up the chimney in a cupola, whilst they are driven in the face of the hearth-smelter at every blast of the bellows'. The charge of the furnace amounted to about a ton and reduction was completed in six hours. A slag which floats on the top of the molten lead is formed simultaneously, and to free the lead from the slag, lime is thrown in. This causes the previously fluid slag to solidify, whereupon it is drawn to the side and the metallic lead run off at the bottom. Slag contains a considerable amount of residual lead and is worked up in a *slag hearth*, which is the same as the ore hearth, but fired with coke.

Until coextensive technology had sufficiently advanced to utilize at home lead produced in Scotland, it was carried to Leith and shipped to Holland and Russia, where it was desilvered and

[1] R. Watson, *op. cit.*, Vol. 3, p. 274.

converted into white and red lead. At the principal mines in Scotland the quantity of silver in lead-ore was given as 5 oz. per ton. This was considered too low for economic desilvering in Scotland till comparatively late. The ore from some deposits, however, had a very much higher silver content. From Hoy in Orkney, Dr. Joseph Black got a sample which, on assay, showed that it contained 46 oz. silver per ton of ore, but we do not hear of the deposit being worked, possibly on account of lack of fuel.[1] A high percentage was also found in ore from the Ochils.

The following table gives the production of the Scottish lead mines at various dates, with certain comments on the cause of rise and fall in price and production. A *bar* of lead weighed between 1 cwt. and 1 cwt. 1 qr. 2 lb.[2] The usual arrangement was for the mining and smelting company to allow the proprietor every sixth bar of lead. 'The men work, as in Cornwall, on tribute—sharing the success or failure of their enterprise with the proprietor'.[3] Hugh Stewart of Tonderghie, at Whithorn, was offered this but in addition he was to 'be compensated for the damage done to his land', a rather unusual arrangement in palaeotechnic economy.[4]

LEAD PRODUCTION[5]

(in bars)

Date	*Mines at*		*Remarks*
	(a) Leadhills	*(b) Wanlockhead*	
1786	10,080	—	
1790	18,000	—	Start of boom period with rise in price following American War.
1809	25,200	15,552	Price up to £32/ton.
1810	22,400	—	Value £45,000.
1811	—	—	Price fall to £24/ton.
1815	35,000	25,000	
1829	—	9,530	} Price £13/ton.
1830	—	7,380	

[1] *Statistical Account*, Vol. 20, p. 248.
[2] *Ibid.*, Vol. 3, p. 301.
[3] *Ibid.*, Vol. 8, p. 351.
[4] *Ibid.*, Vol. 16, p. 284.
[5] *Statistical Account*, Vol. 4, p. 512; *General Report*, Vol. 1, p. 69.

Although lead was still a British export in the early thirties of the nineteenth century, the industry was declining on account of competition from the Spanish industry in the Sierra Morena.[1]

In the preceding pages, several instances of the occupational disease associated with lead-working have been mentioned. Pennant speaks of the disease as follows:

> The miners and smelters are subject here, as in other places, to the lead distemper, or *mill-reek*, as it is called here, which brings on palsies, and sometimes madness, terminating in death in about ten days.[2]

More than that, water below the smelting kilns was dangerous, due to its containing arsenic (As), zinc (Zn), and sulphur (S),[3] yet, on the other hand, the author of the *General View: Dumfries*, comments: 'It is remarkable that though below the smelting-mills the waters are poisonous, they are pure and salubrious as they flow from the veins'.[4]

About the middle of the eighteenth century, most smelters died mad or idiots. One reason given for this is curious. It was said to be caused by the rising cost of spirits which prevented miners from drinking at work.[5] Animals were also affected. 'Fowls of any kind will not live many days at Leadhills. They pick up arsenical particals with their food, which soon kills them. Horses, cows, dogs and cats are liable to the *lead brast*. A cat, when seized with that distemper, springs like lightening through every corner of the house, falls into convulsions, and dies. A dog falls into strong convulsions, but sometimes recovers. A cow grows perfectly mad in an instant and must be immediately killed.'[6]

In spite of this, several travellers who visited Leadhills remarked on the large number of cattle kept there. Here is one instance:

> There are no less than eighty or ninety cows in this village—a very large proportion for the number of people. It is explained by the fact that the customary diet of the population is that which we saw the two quarry-men enjoying by the roadside—oatcake and milk.

It is interesting to link this old observation with what we now know of plumbism, it having become a statutory obligation to supply free milk to operatives in factories handling lead. Had it been discovered empirically that milk would offset the deleterious effects

[1] J. H. Clapham, *Economic History of Modern Britain*, Vol. 1, p. 241.
[2] T. Pennant, *Tour in Scotland in 1769*, Vol. 2, p. 129.
[3] *Statistical Account*, Vol. 21, p. 97.
[4] Dr. Singer, *op. cit.*, p. 30.
[5] *Session Papers*, Signet Library, 76: 4.
[6] *Statistical Account, loc. cit.*

of lead? That was not the contemporary explanation, but the high consumption of milk is certainly suggestive.

Sir H. Davy makes the following comment on the food of *Derbyshire* lead miners:

> I have been informed by Sir Joseph Banks, that the Derbyshire miners in winter, prefer oatcakes to wheaten bread; finding that this kind of nourishment enables them to support their strength and perform their labour better.[1]

Whatever the explanation, the inhabitants of the lead-mining districts of Scotland are said to have looked 'remarkably healthy, notwithstanding the presence of the fumes of the smelting'. When John Ramsay of Ochtertyre (1736–1814), who visited Leadhills in 1790, remarked on the miners being well clothed, he was told by Stirling of Garden, the manager, that a set of *trockhers*[2] came annually from Ireland and bartered Irish linen for the miners' old clothes.[3]

The village of Leadhills was characterized by the amount of cultivation round it. Land was granted free to the miners, who had reclaimed (*c.* 1850) some three hundred acres, and brought it to a high state of cultivation. This is not surprising, as the men appear to have worked five to six-hour shifts for only five days in the week.[4] Miners are traditionally part-time workers.

Briefly, economic conditions of labour in the Scottish lead-mines were as follows: from the *Session Papers* we learn that the *Scots Mining Company* worked a rich, bluish ore till 1740. Workmen were paid 10*d.* for dressing a bing (72 stones) of this ore, and 5*s.* 6*d.* for smelting a *tun* of lead, one *tun* of which was produced from three bings; that is to say, 8*s.* was the price of smelting and dressing one *tun.*[5] After 1740, a poorer ore was found in the Portobello vein which increased the cost of dressing and smelting to 18*s.* In the following century, in 1852, the only date at which we have figures, average income in money was nine shillings a week, 'which leaves so many with only six'. Writing some years later, when wages were said to be double what they were in 1857, *J.N.* (under whom the mines were conducted) said that the privilege of getting rent-free land, together with houses built by themselves, and wages at 15*s.* per week, 'made the mining population of Leadhills much more comfortable than the generality of workmen in other districts'. Although living in an isolated quarter, they are universally spoken

[1] Sir H. Davy, *Lectures*, p. 132. [2] Fr. *troquer* = barter.
[3] J. G. Fyfe, *Scottish Diaries and Memoirs, 1746–1843*, p. 189.
[4] G. V. Irving and A. Murray, *op. cit.*, p. 40 *et seq.*
[5] *Session Papers*, Signet Library, 76: 4.

of as being knowledgeable in the affairs of the world, and in possession of one of the first circulating libraries in Scotland, of which they appear to have been extremely proud. It was founded in 1741 and reputed to contain, at the middle of last century, 'a choice selection of books on biography, travels, chemistry, etc.

> Mr. Stirling, who had been professor of mathematics in Venice and was afterwards agent at Lead-hills, founded a library there, which consists at present of 1,200 volumes. There is another of 700 volumes at Wanlockhead. These libraries are accessible to the workmen for a small sum; and no mines in the kingdom give employment to workmen in general so well-informed, orderly, and respectable. The hours of labour are six out of the twenty-four.[1]

Some uses of metallic lead were listed at the beginning of the chapter. These, however, were not the principal uses to which Scottish lead was put. According to the author of the *General View of the Agriculture of Dumfries*:

> The great sources of demand for the lead are the works for the manufacture of red and white lead for paints. The plumbers also make use of very considerable quantities; and some of the lead is exported by the India ships, and finds its way as far as China.[2]

We have not yet details of the proportions in which the various lead compounds were used at the boom period of Scottish lead-mining, but the following figures for production in the Newcastle area at a somewhat later date can be used as a guide.[3]

White lead and paint	7,500	tons
Red lead (Pb_3O_4)	4,500	,,
Litharge (PbO)	800	,,
Sheet lead	4,500	,,
Lead pipes	1,500	,,
Shot	750	,,
Total	19,550	,,

Red lead or lead oxide (Pb_3O_4) was prepared either from the lower oxide, Litharge (PbO), or from lead (Pb). The main users of it were flint glass makers, and it was their opinion that red lead from litharge did not flux so well as that made from lead by direct oxidation. The process as carried out in Derbyshire, in 1782, in one of the nine furnaces for the purpose there, was described by Richard Watson, Bishop of Llandaff.

[1] Irving and Murray, *op. cit.*, pp. 41, 45, and 196. [2] Dr. Singer, *op. cit.*, p. 30.
[3] W. G. Armstrong, *The Industrial Resources of the Tyne, Wear, and Tees*, p. 141.

> The furnace was very like a baker's oven, its vaulted roof is not at a great distance from the bottom or floor, on each side of the furnace there are two party walls, rising from the floor of the furnace, but not reaching to the roof; into the intervals between these walls and the sides of the furnace the pit coal is put, the flame of which being drawn over the party walls and striking upon the roof, is from thence reflected down upon the lead, which is placed in a cavity at the bottom, by which means the lead is soon melted. The surface of the melted lead, when exposed to the open air, instantly becomes covered with a dusty pellicle; and this pellicle being removed another is formed, and thus by removing the pellicle, as fast as it forms, the greatest part of the lead is changed into a yellowish green powder. This yellowish powder is then ground very fine in a mill, and being washed, in order to separate it from such parts of the lead as are still in their metallic state, it becomes of a uniform yellow colour, and, when it is dried to a proper consistency, it is thrown back into the furnace, and being constantly stirred, so that all its parts may be exposed to the action of the flame of the pit coal, in about 48 hours it becomes *red lead*, and is taken out for use.[1]

In the process there is an increase in weight, twenty cwt. of lead giving twenty-two of red lead. Others give the ratio as 100:120. Concerning the increase in weight, Bishop Watson says:

> There have been great disputes among philosophers, to what principle this increase of weight should be ascribed; some have attributed it to what they call the principle of fire; others are upon good grounds convinced, that it is owing to the absorption of the air itself, or of some of the principles of which the air consists. This hypothesis concerning the fixation of air during the calcination of metals, is said to have been advanced by *Jean Rey*, French physician, in 1630; Dr. Hales was partly of the same opinion; and Dr. Pemberton very expressly affirms, that calcined metals receive their increase of weight from the air. . . . The ingenious labours of Dr. Priestley and M. Lavoisier have confirmed the conjectures and experiments of former philosophers, for they have clearly proved two points—first, that a large portion of the air may be *separated* from red lead, by *reducing* it to the state of a metal; and secondly, that a large portion of the air is *absorbed* by *lead* during calcination, by which it is reduced to the state of red lead.[2]

These observations show that chemical operations in industry were laying the foundations of data for the new theories of combustion and oxidation which were introduced at the end of the eighteenth century. In fact there is in the above passage a distinct

[1] R. Watson, *Chemical Essays*, Vol. 3, p. 339. [2] *Ibid.*, Vol. 3, p. 346.

feeling that industry was in a more receptive state for new ideas than was purely philosophical chemistry.

In addition to red lead, another compound, white lead, was of considerable importance as an absorber of the produce of the lead mines. From the time of Queen Elizabeth, lead was converted into white lead at Newcastle. The process used was that operated for generations, namely, the *Dutch*, *Vinegar*, or *Stack Process*, in which sheets of metallic lead are exposed to the action of acetic acid vapour (CH_3COOH), and carbon dioxide (CO_2) made by the fermentation of animal or vegetable matter. Tanner's bark, from which the tanning principles had been extracted, was substituted for dung by Richard Fishwick of Newcastle in 1787,[1] and the use of bark became general early in the nineteenth century. Here is the Bishop of Llandaff's description of the operation:

> Thin plates of lead are rolled up in spiral form, and placed in earthen pots containing vinegar; these pots being ranged on proper stages, and their mouths being covered in such a manner, as to permit the vapour of the vinegar to escape, and at the same time to prevent any impurity from falling into them, a quantity of horse dung is thrown in amongst them; by the heat of which, as it grows putrid, the vinegar is raised in vapour, and this vapour attaches itself to both sides of the lead, which is so placed as not to touch the vinegar, it corrodes the lead into white scales, which being beaten off from the plates, washed and ground in a mill, constitute the white lead of the shops, excepting that this is generally, even before it gets into the hands of the painter, adulterated with chalk.[2]

The reactions occurring in the process are assumed to be representable as follows. During fermentation of biological material, heat is generated and this vaporizes the acetic acid which attacks lead, forming a basic lead acetate.

$$(A = CH_3COO.)2HA + 2Pb + O_2 = PbA_2.Pb(OH)_2$$

But carbon dioxide is formed simultaneously and reacts with basic acetate, forming a mixture of normal lead acetate and basic lead carbonate.

$$3PbA_2.Pb(OH)_2 + 2CO_2 = 3PbA_2 + 2PbCO_3.Pb(OH)_2 + 2H_2O$$

The lead acetate so produced reacts with more lead to form basic acetate, and so the cycle of operations continues, a small quantity of acetate sufficing, in theory at least, for the conversion of a large quantity of lead into white lead.

[1] *Register of the Great Seal, Edinburgh*, Vol. 20, No. 343, 1788.
[2] R. Watson, *op. cit.*, Vol. 3, p. 362.

The stack process was handicapped by being slow. Three to four months were required to effect the complete conversion of the lead. Other processes consequently were tried: for example, Lord Dundonald patented a process in which lead oxide was treated with 'oxygenated muriate of potash' and subjected to the action of carbon dioxide.[1] But the high quality which characterized white lead made by the stack process retained it in operation for generations. When Sir John Sinclair was writing the *General Report*, it was being so manufactured at Portobello, and one firm, Alexander Ferguson and Company Limited, Glasgow Lead and Colour Works, the only one left, was still using the process when the British Association met in Glasgow in 1901.[2]

Another important lead compound, sugar of lead, *Saccharum saturni*, or lead acetate ($Pb(OOC.CH_3)_2$), was made at Rutherglen Bridge, Airdrie, Linlithgow, and at the Cudbear Works of the Macintoshes, where in fact it was first made in Great Britain. In a long letter, C. Macintosh recounts the antecedents of his introduction of it. The letter merits citation *in extenso*:

Glasgow, 20th January, 1800.

> When in Holland I was admitted to see a sugar of lead work, and was struck with the circumstance of both the lead and coal, and frequently the malt used in making the vinegar employed in it, being imported into Holland from Britain; and that the manufactured articles, when sent back to us, should become loaded, upon its arrival, with a duty of 3*d.* per lb. On my return to Glasgow, I attempted to make sugar of lead, and was successful in making a salt equal in quality to the Dutch. I established a manufactory of it, in the year 1786, which as you know, has been going on pretty successfully ever since. *Sacharum Saturni*, (now acetate of lead), which latter is, however, in my opinion, a defective name, according to the rules of the new nomenclature; for I am confident it is a *carbono-acetate* of lead,[3] is chiefly used by the

[1] *Register of the Great Seal, Edinburgh*, Vol. 20, Nos. 595 and 604, 8th June, 1796.

[2] J. Sinclair, *General Report*, App. 2, p. 309; A. McLean, *Local Industries of Glasgow and the West of Scotland*, p. 184.

[3] Note by Editor of *Memoir*.—This conjecture, considering the state in which chemistry was at the time when Mr. Macintosh wrote, marks the astuteness of his perception, and talents for observation as a chemist. When the acetate of protoxide of lead, or *Saccharum Saturni*, is dissolved in water, it is converted into a carbonate of oxide of lead; and the same effect is produced upon this salt by exposure to the air of the atmosphere. In neither case, however, does the carbonic acid exist originally as a constituent of the salt; but, in the one case, is derived from the carbonic acid contained in the water of solution; and in the other, from the carbonic acid contained in the air of the atmosphere. The very error into which Mr. Macintosh, in this instance, falls, proves him to have been in advance of other chemists.

calico-printers in making a mordant for preparing their cloth for the reception, and fixing (where it has been left unprinted) of the yellow, chocolate, and olive dyes, etc. For doing this, in respect to the greater number of these colours, alum is added in solution, to the solution of sugar of lead, when a double elective attraction takes place, the sulphuric acid of the alum combining with the oxide of lead, and falling down with it in an insoluble state; whilst the acetic acid forms a new union with the earthy part of the alum, and remaining dissolved, constitutes the acetate of alumine, or red colour liquor, as it is called. The mordant above alluded to (acetate of alumine), cannot be obtained by direct mixture of vinegar, even in its most concentrated state, with pure clay, a substance nearly corresponding to the earth of alum, properly so called; but by double affinity alone can this compound be formed.

It afterwards occurred to me, that a salt, possessing similar properties with the acetate of lead, in so far as related to the decomposing of alum, might be obtained by dissolving lime in vinegar. Upon putting this theory to the test of experiment, I found it to answer completely; when alum and this acetate of lime were presented to each other, a double decomposition being instantly effected, the sulphuric acid combining with the lime forming the solution of a salt to all intents and purposes the same as was obtained by using sugar of lead and alum. Now, as this was a much more economical process than the other, I immediately adopted it, and commenced making the above mordant for sale, which you know we have continued doing these ten years, with similar success to the sugar of lead. One reason for our continuing to make both articles is, in some measure, to humour the caprice, or perhaps to flatter the vanity, of our customers, many of whom use the sugar of lead in preference, notwithstanding the trouble it gives them, from the notion of their being thus possessed of valuable arcana, which enable them to eclipse all others of their profession in the composition of their colours. Lime has long been a favourite nostrum of mine, having first used it with success many years ago, when engaged with Couper in the manufacture of sal ammoniac, and for decomposing the ammoniacal salts contained in soot and urine, which it does with surprising facility, and which you may easily prove by mixing a little soot and hot water in a glass; and upon adding lime the strong pungent smell of hartshorn is immediately evolved.[1]

The principal market for Macintosh's lead acetate was the textile trade, but there were other less creditable uses to which it was put, constituting a serious menace to public health.

Although the industrial hazards of lead-smelting were well

[1] G. Macintosh, *Memoir of Charles Macintosh*, p. 24.

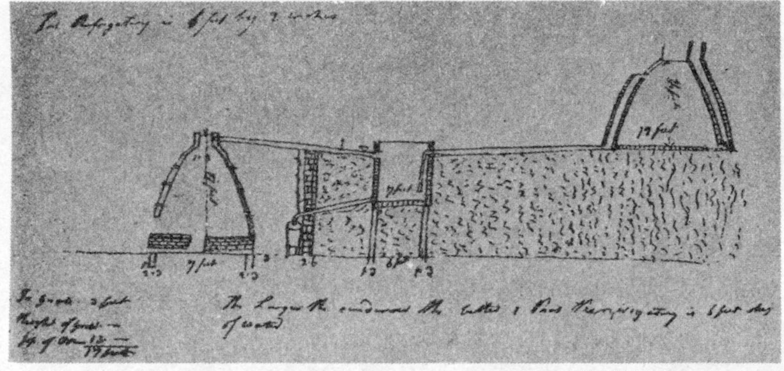

Boulton and Watt Collection, Birmingham

83. Lord Dundonald's Tar Ovens.

G. Jars, *Voyages Métallurgiques*

84. Kiln for the removal of sulphur from coal and its conversion into coke at Newcastle, 1774.

85. Spark-mill, one of the forms of lighting used in coal-mines prior to the introduction of the safety lamp. Rotation of the handle on the left produced a shower of sparks from the piece of flint on the right.

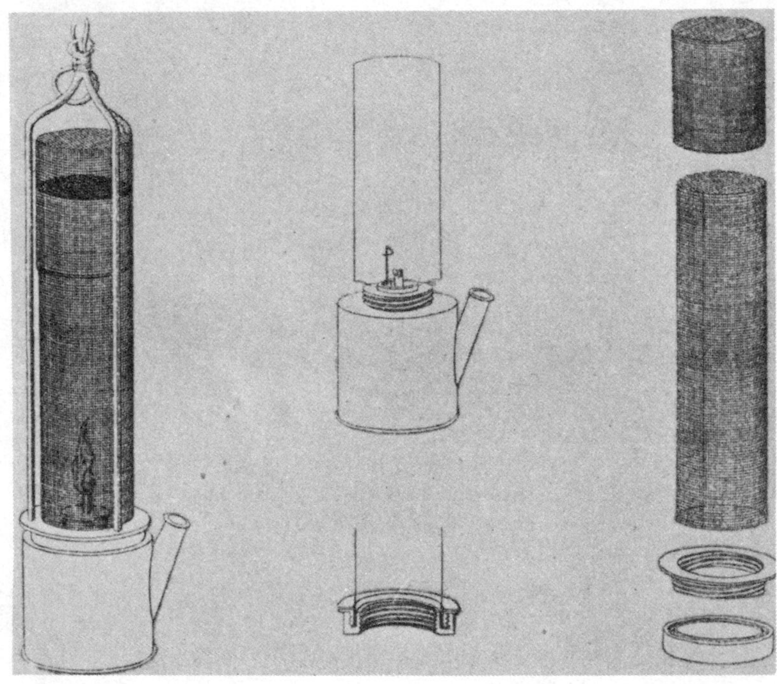

86. Davy's safety lamp, the vitally important outcome of the activities of the *Society for the Prevention of Accidents in Coal Mines*, which was founded in 1813.

known, it must be added that lead compounds were at one time extensively used as cosmetics and in the *doctoring* of wines. Of the former, the Bishop of Llandaff makes some pertinent remarks:

> Without presuming to explore the ARCANA of a lady's toilet, or to reveal the arts by which my fair country-women endeavour to improve charms, naturally irresistible, I would add to the admonition of St. Jerome (who had inveighed very forcibly against the use of rouge, etc., by Roman ladies) a caution more likely in these degenerate times to be attended to—the certain ruin of the complection, to say nothing of more serious maladies, which must ever attend the constant use of this drug. . . . But if, as is most probable, they will neglect this caution, I warn them, to forbear the use of such washes at Harrowgate, Moffat, and other places of the same kind lest they should be in the state of the unlucky fair one whose neck, face, and arms, were suddenly despoiled of all their beauty, and changed quite black by the sulphureous waters.[1]

The oxychloride and oxynitrate of bismuth (BiOCl and $BiONO_3$) sold as Spanish White, were also used as cosmetics. They stood 'in such repute in London, that the chemists can hardly prepare it fast enough to supply the demand'.[2] Maria Gunning, Countess of Coventry, whom Margaret Calderwood (1715–74) in her *Journal* describes as 'a pert, stinking-like hussy', died, it is said, from the effects of the over-application of cosmetics.

Much more serious was the deliberate addition of lead compounds to food. When litharge (lead oxide, PbO) is added to vinegar or an acid wine, it dissolves, with the production of lead acetate ($Pb(OOC.CH_3)_2$). This substance, also known as sugar of lead, or *Saccharum saturni*, was one of the sweetest substances known. At the same time, it is a poison. Nevertheless its saccharine properties were made use of by unscrupulous wine-dealers, who, 'respecting their own profits more than the lives of their customers', added large quantities of litharge or red lead to wine that had soured. Pennant, in his *Tour of London*, describes one such establishment operated by Mark Beaufroy near Cuper's Gardens, 'a recreation place of ill-fame on the south side of the Thames'.[3] J. C. Drummond tells us that the methods for preparing these crude mixtures were frankly described in a book, *The Retail Compounder and Publicans' Friend*, by John Hardy (*prob.* 1794). Colour was provided by burnt sugar, sassafras, logwood, cochineal, turn-

[1] R. Watson, *op. cit.*, Vol. 3, p. 365. [2] *Ibid.*, p. 364.
[3] Quoted J. C. Drummond and Wilbraham, *The Englishman's Food*, p. 240.

sole, blackberry juice, etc.; flavouring by orange peel and *terra japonica*; sweetness by sugar, and astringency by hops or oak-chips. Anything from a Malmsey to a hock could be compounded on the basis of a thin solution of spirit of wine. A good deal of the latter was prepared by fermenting molasses, imported from the West Indies. Crude wines were 'modified' by adding alum, and rendered less sour by plaster of Paris and chalk.

The reaction of different national states to this criminal practice varied considerably. In some parts of Germany it was a capital offence, but in France about 1750, no less than 30,000 hogshead of *vin gâté* were being treated in Paris and sold as genuine wine. The nefarious practice was also engaged in in England.[1] Frederick Accum, in his *Treatise on the Adulteration of Food*, tells of some of the methods:

> There is in this city (London), a certain fraternity of chemical operators, who work underground in holes, caverns, and dark retirements, to conceal their mysteries from the eyes and observations of mankind. These subterraneous philosophers are daily employed in the transmutation of liquors, and by the power of magical drugs and incantations, raising under the streets of London the choicest products of the hills and valleys of France. They can squeeze Bourdeaux out of the sloe, and draw Champagne from the apple.[2]

Fortunately, the presence of lead in wine was easily detected by the addition of calcium polysulphide (CaS_x), which in the presence of lead adulteration yields a black precipitate of lead sulphide (PbS).

> It is lamentable that the extensive application of chemistry to the useful purposes of life, should have been perverted into an auxiliary to this nefarious traffic. But, happily for the science, it may, without difficulty, be converted into a means of detecting the abuse; to effect which, very little chemical skill is required.[3]

Contamination with lead was also liable to occur in domestic economy. Accum refers to the lead oxide glaze on the common cream-coloured earthenware as a potential danger.[4] He condemns the use of earthenware jars and pots for storing jellies, marmalade, and pickles, since the glaze was soluble in the acids contained, either in the fruits, or in the vinegar used in making the pickles (acetic acid, CH_3COOH). He says that even the baking of fruit

[1] R. Watson, *op. cit.*, Vol. 3, p. 370.
[2] F. Accum, *Treatise on the Adulteration of Food and Culinary Poisons*, p. 95.
[3] *Ibid.* [4] Ch. XV.

pies in earthenware pie-dishes might lead to poisoning. Lead poisoning was also liable to occur in the dairy. Long after the poisonous nature of lead was recognized, reactionary rustics stuck to their traditional vessels. For example in Lancashire, where the milk vessels were of lead, it was believed that cream separated better in such vessels.[1] There are also cases recorded of cheese dyed with annotto being found to contain red lead. The annotto had been adulterated with vermilion, the vermilion with red lead.

Copper, too, was a common poison in pickles. Copper salts give a bright green, and therefore attractive, colour to vegetables. Though poisonous, it was recommended in several culinary guides published prior to 1820. One of these, E. Raffeld's *The English Housekeeper*, ran into no less than eighteen editions! The vinegar used in making the pickles was also often adulterated with sulphuric acid. Other instances of adulteration are quoted in Accum's *Treatise*.[2]

The publication of Frederick Accum's *Treatise on the Adulteration of Food and Culinary Poisons* (1820) at a time when the adulteration of food was most flagrant, caused a first-class sensation which continued off and on for upwards of thirty years, and culminated in the appointment of an Analytical and Sanitary Commission. Small thanks, however, did Accum get from the British public for exposing the appalling condition of its food supply. The following brief account of Accum is taken from J. C. Drummond's *The Englishman's Food*.[3]

Accum, whose family name was Marcus, was a German Jew, born in Bückeburg in 1769, who came to this country in 1793 and entered Brande's Pharmacy in Arlington Street. Through the good offices of William Nicholson, he was able to establish himself as a consultant in 11 Old Compton Street in 1800. From there he became 'chemical operator' at the Royal Institution, and subsequently lecturer in chemistry at the Surrey Institution, Blackfriars Bridge Road.

He wrote one of the first books on gas-lights,[4] but food analysis was his principal occupation. The *Treatise* referred to above brought him both fame and disaster. In it he included a list of those fined for selling adulterated beer. This caused a great outcry among brewers, and Accum was subjected to a campaign of violent abuse by those whose nefarious activities he so courageously exposed. They succeeded in persuading the managers of the Royal

[1] S. Parkes, *Chemical Essays*, Vol. 5, p. 193. [2] F. Accum, *op. cit.*, p. 295.
[3] J. C. Drummond, and A. Wilbraham, *op. cit.*, p. 341 *et seq.*
[4] Ch. XIX.

Institution to bring against him, on the flimsiest evidence, a charge of mutilating books in the Royal Institution Library. As a result, Accum fled the country, and returned to Germany embittered and disillusioned.

For a time the food scare blew over, but it was followed by others. John Mitchell, in his *Treatise on the Falsification of Food and the Chemical Means of Detecting them*, claimed that the same adulterants were still in use at the middle of the nineteenth century; alum and ammonium carbonate in bread, lead in wine and cider, kaolin and plaster of Paris in flour, chalk in milk. As a result, the Commission referred to above was appointed to report on the quality of the food 'consumed by all classes of the public'.

Chapter XVIII
THE BRITISH TAR COMPANY

Eighteenth Century Commercialization of Coal Distillation by Lord Dundonald

AMONG the modern chemical industries which have roots in the eighteenth century, one is coal gas. Just into the nineteenth century Boulton and Watt illuminated the front of their factory at Soho with coal gas to celebrate the Peace of Amiens. Thereafter gas became foremost in the commercial products of the destructive distillation of coal. So much is commonplace to students of the history of industry and of chemistry. What is less readily recognized is that the destructive distillation of coal was not a process first undertaken with that end in view. In this chapter we shall see that one of the foremost industrial developments during the second half of the eighteenth century was the manufacture from coal of a saleable product which had a pivotal place in the preceding wood economy. Indeed, the disposal of other by-products of a new type antedated the manufacture of coal gas itself. Speaking of these, John Maiben says:

> There are other two as much its inseparable results as the gas itself. The one is the coke, as the carbonaceous substance is called, which is left in the retorts after the gas is expelled. This is so far from being reduced in consequence of the process to a useless mass, that it is still more valuable as a fuel than the coal from which it is produced, in so much that, in many places, immense quantities are reduced to this state solely for the purpose of rendering them a better fuel than they are in the natural state; the expense and the profit of the process to be found in the greater price they bring in the market in their new form.
>
> The other valuable product is the tar which is deposited in the purification of the gas, which, when rectified by a slight distillation becomes an article of commerce, enabling those engaged in making it, to be at the expense of the fuel consumed in their furnaces, while they permit the gas to waste, and notwithstanding make a profit sufficient to remunerate them for their trouble.[1]

[1] John Maiben, *A Statement of the Advantages to be derived from the Introduction of Coal Gas into Factories and Dwelling Houses*, etc., p. 5.

In various departments of technics, including smelting, glass-making, ceramics, soap-making, dyeing, brewing and distilling, details of the transition from what Patrick Geddes called the wood-and-water eotechnic economy to the coal-and-iron palaeotechnic economy have been set out. Other fields of technics remain where the effects of the same transformation have yet to be explored. One of these is the utilization of coal as a chemical raw material, contemporaneous with James Watt's steam-engine, and parallel to its adoption as a source of heat and power.

At first sight the opinion expressed by a contemporary writer, that the patent granted in 1781 to Archibald Cochrane, ninth Earl of Dundonald, for producing tar was of greater national importance than Watt's engine patent, may seem extravagant. But the primary position now occupied by coal tar derivatives, with their wholesale displacement of natural by *made* products, signalizes visionary rather than imprudent thinking.

In studying the antecedents of gas lighting, developed by William Murdock and commercialized by Matthew Boulton and James Watt, our attention focused on the position occupied by Dundonald in this development, and the causes which put him in a position to make initial observations in this field. The Historiographer Royal for Scotland, Dr. Henry Meikle, then informed us that hitherto-unpublished documents were in the National Library of Scotland dealing with Dundonald's activities in that direction. On examination these documents were found to contain material out of which has been constructed a full account of this unexplored aspect of technics, the distillation of coal to produce a variety of products, including ammonia and tar, the latter as an alternative for wood-tar and pitch, products formerly imported from the forests of Europe.

Throughout the later eotechnic era, timber protected with tar or pitch was the exclusive raw material of shipbuilding, till iron was introduced for that purpose in the nineteenth century. The main producers of tar and pitch were States in the north of Europe. These two commodities were essential to a maritime power, such as Great Britain, then entering on the threshold of a turbulent century. In 1703, the countries in possession of the monopoly raised prices of tar and pitch so high that England was forced to look to her North American colonies to provide an alternative supply of marine stores. To this end she offered considerable premiums, and

> . . . dispatched to America men of sufficient ability to convince the Inhabitants how necessary it was for them to assist the view

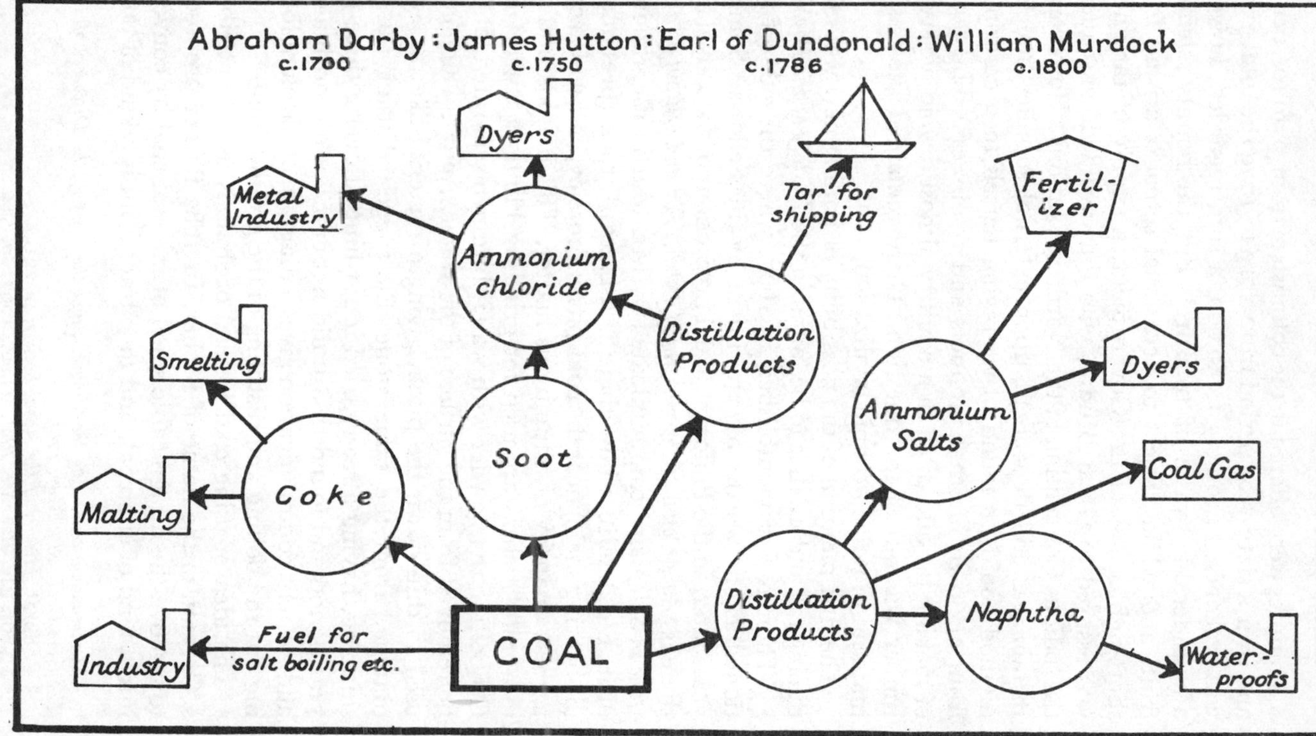
PROGRESSIVE UTILIZATION OF
COAL DISTILLATION PRODUCTS
Abraham Darby : James Hutton : Earl of Dundonald : William Murdock
c.1700
c.1750
c.1786
c.1800
Dyers
Metal Industry
Tar for shipping
Fertil-izer
Ammonium chloride
Distillation Products
Smelting
Dyers
Ammonium Salts
Soot
Malting
Coke
Coal Gas
Distillation Products
Naphtha
Industry
Fuel for salt boiling etc.
COAL
Water-proofs

of the Mother Country, and of sufficient experience to direct their attempts towards such objects.

Bounties were offered to the colonists of from 5*s*. to 10*s*. per barrel, and over a sixty-year period from 1719 to 1779 the total expended was reckoned at some £1,250,000, a sum regarded, however, as a considerable 'waste of treasure'. But the bounty scheme bore fruit, and while subject to occasional setbacks arising from the short-sightedness of the Government, a supply of tar and pitch was assured to Great Britain, who thus avoided consequences of the 'odious Convention of the Northern Nations'. However, Great Britain was still dependent upon external supplies 'unless some method could be found of supplying herself to a certain extent from her own internal products and resources'.[1] The possibility of making tar and pitch in a country almost devoid of serviceable timber was out of the question. There remained the basic raw-material of palaeotechnic civilization, coal.

The first mention of the possibility of finding new ways for producing tar and pitch is to be found in the *History of the Royal Society* by Thomas Sprat, published in 1667, where the subject is mentioned as one worthy of experiment by the Fellows.[2] Some thirty years later, in the Philosophical Transactions of the Royal Society, there is '*An Account of the making of Pitch, Tar and Oil out of a blackish stone in Shropshire*' communicated by Mr. Martin Ele, the inventor of it. Ele describes how there is a blackish rock (presumably an oil-bearing shale) near the coal-pits of Broseley, etc., which contains a great quantity of bituminous matter. This he took and ground in a mill 'such as are used for grinding Flints to make Glass of'. On boiling the powder with water, bitumen separated, and could be removed from the surface of the water. An oil, which could be used to thin down the pitch, giving a material like tar, was also obtained from the same stone. Ele suggested using this tar for shipping, having already tried it out himself over a three-to-four-year period and found that it did not crack as did ordinary pitch and tar. His communication gives a diagram of his plant for carrying out the boiling and distilling of the ground stone.[3]

The idea of using coal instead of shale is usually attributed to a German of the name of Becher. In 1682 Becher proposed both to remove the disagreeable smell of mineral coal by converting it into a kind of charcoal, and to extract from it a kind of tar which

[1] *Memorandum concerning the Progress, and the Uses of the Discovery of Extracting Tar from Coal, 28th May, 1785.*

[2] T. Sprat, *History of the Royal Society*, p. 221.

[3] Martin Ele, *Roy. Soc. Phil. Trans.*, 1697, *20*, 544.

he regarded as superior to that of Sweden.[1] Then followed a long series of attempts to make pitch and tar from coal at various centres in Great Britain and on the Continent, viz. at Coalbrook Dale, Rochester, Newcastle, Edinburgh, Liège, and different places in Germany. Among the adventurers who attempted to commercialize the process in Great Britain was the Marquis of Rockingham. Near Sheffield, the latter:

> . . . showed himself to be a real Patriot on nothing more strongly than in encouraging useful undertakings, patronised several unsuccessful attempts which were made in the neighbourhood of Sheffield for the Extraction of Tar from Coal.[2]

Similar undertakings were initiated by several persons at Coalbrook Dale. Among other experimenters were a Mr. Champion, John Clark (brother of Sir James), and lastly, Dr. John Roebuck, the founding of whose Vitriol Works at Prestonpans, near Edinburgh, in 1749, may be said to herald the palaeotechnic era in Scotland.[3] One can gather from the following extract from the diary of Miss Henrietta Zipporah Roebuck that Dr. Roebuck and Lord Rockingham were co-experimenters:

> All these three glasses were bordered with a pattern by means of sulphuric acid, the process invented by Dr. Roebuck. Unfortunately the large glass with the crest on it was broken on our move from Hampshire. *With these relics came a pocket-book given to the Doctor by Lord Rockingham in memory of many experiments made at Wentworth.*[4]

In 1782 Richard Watson, Bishop of Llandaff, stated that a patent had been granted to someone in Bristol for extracting tar from pit-coal, and that anyone wishing to work the patent had to pay the patentee a shilling a barrel.[5] All failed:

> . . . in one very material article, that the expense of the process was considerably greater than would allow the Tar made to be sold at the price of the Foreign Tar, which therefore precluded the further prosecution of the business.[6]

Successful commercialization, which, however, brought him no material gain, was due to Archibald Cochrane, ninth Earl of Dundonald (1749–1831).

[1] B. Faujas de St. Fond, *Travels*, p. 148; *also Essai sur le Goudron de Charbon de terre, etc.*

[2] *Memorandum concerning the Progress.*

[3] Letter from Sir John Dalrymple to Mr. Andrew Stuart, Edin., 12th December, 1782.

[4] Private communication from Francis E. Hyde. (Our italics.)

[5] R. Watson, *Chem. Essays*, Vol. II, p. 347.—For details of the processes, *see Patent Specifications.*

[6] *Memorandum concerning the Progress.*

When Archibald Cochrane succeeded to the ancient earldom of Dundonald in 1778, he acquired little more than the title and the entailed estate of Culross Abbey, Perthshire. The family had lost nearly all its once extensive possessions in Scotland. Indeed, it is said that all there remained as a memorial of the seventh Earl was the right of presentation of some bursars in the University of Glasgow. Archibald Cochrane's father, the eighth Earl:

> . . . taking it for granted the Estate of Culross would be secured as the Seat of his family expended considerable Sums in making very large plantations on the Muirs of the Estate . . .

the ravages of eotechnics having necessitated reafforestation in central Scotland as elsewhere. He hoped that if they throve they would bring, in time, by reason of their nearness to the sea, a large sum to the family.[1] At Culross there was also good coal, and the family hoped that:

> . . . by working these coals which are on the Sea side properly, and carrying on a Salt Work, very great Benefit in all probability would arise.[2]

Archibald Cochrane began his career in the Army and Navy, but in neither of these did he find much to interest him. They had, however, an important influence on the scientific activities of his later life. His son Thomas Cochrane said of him:

> While serving on the west coast of Africa, my father remarked the destructive ravages made on ships' bottoms by worms, and, from his chemical knowledge, it occurred to him that an extract from pit coal in the form of tar, might be employed as a preventative of the evil.[3]

In the manufactures depending on advancing chemical knowledge which mark the transition from eotechnic to palaeotechnic civilization, Dundonald saw 'sanguine expectations of retrieving the family estates by his discoveries'. This 'led him to engage in a host of manufacturing pursuits', in particular the extraction of tar, pitch, etc., from coal.[4] Dundonald appears to have inherited some of his scientific capabilities from his mother, Jean Stuart, who was more than usually knowledgeable in the technics of her time, as can be gathered from her letters. For example, writing to

[1] *Memorial from Lord Auchinleck about Earl of Dundonald (Eighth) and Lord Cochrane, etc., 1771.*

[2] Archibald Cochrane, *Description of Culross and its Minerals.*

[3] Thomas Cochrane, Tenth Earl of Dundonald, *Autobiography of a Seaman*, p. 36.

[4] *Ibid.*, p. 37.

her brother Andrew Stuart, she relates of a visit to Lancashire:

> We stay'd some days at Ulverston and other parts of Lancashire. Saw many wonders there. Visit'd the Pitts of Iron Ore the very same sort as our ore here, got some information in ev'ry point, took all down in writing, brought Samples of Ore from all Workings—this part of my jaunt has been of great service, and I hope will appear so to you in a short time. . . . The night we left Hamilton we reached the Cleugh, the great Iron Work, where I saw all their operations. When at Dougaldston, Mr. Glassford sent for his friend who is very knowing in the ironstone and ore; I shew'd him samples of the Lancashire, and of our ore, both Mr. Glassford and that Man think them of the same sort. He had then commission for two thousand ton of Lancashire ore, to be sent to some Iron Work in Argyllshire. He told me the long sea and land carriage make it come so high, that if our ore could be furnished cheap and as good the Iron Works in Scotland would certainly prefer getting supplies from us. . . . Mr. Glassford was exceedingly kind, told me he had so good an opinion of the ore, that he would probably write to you his sentiments on the subject. . . . He hinted at a price far above my most sanguine expectations.[1]

When Matthew Boulton, commercializer of James Watt's condensing engine, made his tour of Scotland and Ireland in 1783, he stayed for a month with Dr. John Roebuck, who by this time, in conjunction with William Cadell and Samuel Garbett, had initiated Carron Iron Works.[2] He dined with Lady Dundonald, who appears to have discussed manufacturing projects on equal terms with 'the famous man at Birmingham', as she calls him.

> He din'd with me here, saw the iron ore, has order'd 20 tons of it to be sent to him to Hull, and told me the discovery of the iron ore will turn to great profit, now that the important discovery is made how to make bar iron with cinders of pitt coal. Mr. Boulton told a gentleman that having iron ore, coal, limestone and fireclay in our property at Lamancha, would undoubtedly turn out to great advantage and would bring about many offers to us, so soon as the proprietors of the foundaries had carried some particular object which they mean soon to apply for to Parliament. Mr. Walker of Rotherham (one of four brother proprietors of the great Iron Works at that place, and lessees of the Duke of Portland and Lord Fitzwilliam's coals in Yorkshire) told me if this take place, the bar-iron and tar will exceed all profits we enjoyed by trade with America.[3]

[1] Jean Dundonald to Andrew Stuart, Lamancha, 25th October, 1783.
[2] H. W. Dickinson, *Matthew Boulton*, p. 119.
[3] Jean Dundonald to Andrew Stuart, Belleville, 23rd December, 1782.

Dundonald wrote asking him to visit Culross Abbey and see the works there:

> If your stay in this country will permit I shall be happy to see you at Culross Abbey when you will have an opportunity of seeing my method of extracting Tar and several other articles from coal.
>
> It is a matter of so great moment to the Proprietors of coal and foundries in the rest [west?] of England that I shall be happy of an opportunity of conveying them authentick information thro' so respectable a channell.[1]

That the discovery of the iron ore was due to Lady Dundonald herself is revealed in a further letter. In this she speaks of:

> . . . my discovery of as fine iron ore as is to be found in Lancashire, Sweden or Denmark or even Biscay, where the finest iron in the world is made.[2]

The iron ore at Lamancha was in such a form that it could also be ground and used as an ingredient of paint. In 1797 or thereabouts this venture was in the hands of the Hon. Capt. Cochrane.[3]

By 1781 the Earl of Dundonald had made his preliminary investigations into the manufacture of coal tar, and in February of that year was granted Letters Patent under the Great Seal of Scotland granting him:

> . . . the sole Power, Privilege, and Authority, of making, using, exercising, and vending, within His Majesty's Kingdom of Great Britain and Dominions, his Invention of a Method of Extracting or making Tar, Pitch, Essential Oils, Volatile Alkali, Mineral Acids, Salts, and Cinders, from Pit Coal, for the Term of Fourteen Years.[4]

Of this, Thomas Cochrane says:

> In the prosecution of his coal-tar patent, my father went to reside at the family estate of Culross Abbey, the better to superintend the works on his own collieries, as well as others on the adjacent estates of Valleyfield and Kincardine. In addition to these works an experimental kiln was erected near the Abbey.[5]

During 1782 the tar works at Culross were put into operation, but Dundonald's affairs had become more than ever involved.

[1] Lord Dundonald to Matthew Boulton, 6th October, 1783.

[2] Jean Dundonald to Andrew Stuart, 27th December, 1783.

[3] *Statistical Account*, Vol. 1, p. 149; C. Findlater, *General View: Peebles*, p. 25. *See also Statistical Account*, Vol. 21, p. 52, for other minerals at Lamancha.

[4] *Commissions, Patents and Inventories; Register of Great Seal, 1781*, Vol. 20, No. 185, G.R.H., Edin.

[5] Thomas Cochrane, *op. cit.*, p. 39.

Threats of foreclosure by one of the most insistent of his creditors, by name Cuthbert, made it a matter of urgency that he should obtain some additional backing for the tar project. Thus was formed the *British Tar Company*. In addition to Dundonald, the partners were Mr. Liddle of Newcastle, a Cumberland gentleman of fortune who was reputed to be worth £20,000 sterling, Mr. Crauford and Mr. Chapman, probably also of Newcastle, the former of these worth £10,000, and 'a very cautious man'.[1] Of this Dundonald says:

> A want of Funds to carry on the Undertaking on my own Bottom made me assume as partners to the extent of 3/5 of the Undertaking, a Company of Gentlemen from Newcastle who are taken bound to lay out the whole Money necessary and to account to me for 2/5 of the neat Gains.

The success of the venture exceeded even the sanguine hopes of Dundonald, and £900 having been expended on 'partial erections at Culross', sales were opened in September 1782.[2]

Dundonald made no claim to the invention of extracting tar from coal in closed vessels, but he was the first to put it on a commercial footing. In modern gas and coke oven industries tar is one of the products of the destructive distillation of coal in closed vessels, i.e. subjected to external application of heat. The readily condensed fraction, called the tar liquor, also contains ammonia and the aromatic hydrocarbons essential to coal-tar dye and synthetic chemical production. His description of the process as set out by himself in terms of the Letters Patent is as follows:

> The Method I have invented for the extracting of Tar, Pitch, Essential Oils, Volatile Alkalies, Mineral Acids, and Salts, and the making of Cinders from Pit Coal, consists in admitting the external air to have a Passage or Passages through the Vessels or Buildings in which the Coal, from which any of the above Substances are to be distilled, is put, whether by itself, or along with Lime, Stone, Flints, Iron Ore, Bricks, or any other Substance, by which Means the said Coals, after being kindled, are enabled, by their own Heat, and without the Assistance of any other Fire, to throw off, in Distillation or Vapour, the Tar, Oil, Alkalies, Acids, and Salts they contain, into Receivers or condensing Vessels, communicating with the Vessels or Buildings containing the Coals, and at the same Time of roasting, calcining, or burning any

[1] Lord Dundonald to his Mother, *Jean Dundonald*, Culross, 19th November, 1782: Sir John Dalrymple to Andrew Stuart, Edinburgh, 12th December, 1782.

[2] Lord Dundonald to Andrew Stuart, Edinburgh, 11th December, 1782.

Substances that may be mixed with them; it appears to me necessary, lest others encroach on my Patent, to describe, as above, the Principle upon which I act, in as few Words as possible, and in such a Manner as will admit of no Ambiguity; Therefore, according to what is above set forth and declared, Persons who shall extract Tar, et cetera, from Pit Coal in Vessels or Buildings (it matters not their Shape or Size), whereby the Coals are made to burn or ignite, without flaming, by a regulated Admission of the external Air through different Apertures in the Buildings, so as by their own Heat to throw off the Tar, Oils, etc., that they may contain; Persons who do so, without my Permission, are deemed to encroach upon my Patent; as the only Method used or known until my new Discovery, was a Distillation of Coal in close Vessels, where the Admission of the external Air was prevented; and where other Fuel or Coals were required besides the Coals contained in the close Vessel to produce the Heat necessary to pervade the same, and to cause the Coals contained therein to throw off the Tar, Oils, et cetera, that they contained. I do not think it any ways of Moment to subjoin any Drawings of the Buildings or Kilns that may be used according to my new Invention for the making of Tar, et cetera, because these Buildings may be made either square, circular, or oval, as Fancy may direct, the Art depending upon the Management of the Air admitted into the Kilns, which can only be acquired by Experience; and as it is by no Means meant to keep the Manufacture hid or concealed, those who want to see the Practical Part will have an Opportunity so to do at the different Places where the Manufacture is to be carried on. Exclusive of the above Invention, for which only the Patent has been obtained, I promote the Condensation of the less coercible Part of the Vapour that comes off in Distillation, by commixing it with the Steam of boiling Water, and complete the Condensation by the Means of cold Water, either in Contact with the Vapour, or applied externally to the Vessels through which it passes; and by an Admission of the external Air into the condensing Vessels when needful, I also cause the Vapour to pass through more condensing Vessels than one, to separate by that Means the different Oils and Substances, according to the different Degrees of Cold and Moisture requisite to condense them; or occasionally I follow the usual and common Modes in Practice for condensing the Vapours thrown off from any Substances by the Action of Heat.[1]

This patent was obviously drawn up to cover all possible contingencies, but fortunately there has survived a clear account of the actual works erected at Upper Cranston, near Dalkeith.[2]

[1] *A Bill for vesting in Archibald, Earl of Dundonald, etc.*

[2] *Statistical Account of Scotland*, Vol. X, p. 412.

At Upper Cranston, works have been lately erected for extracting tar from pitcoal; an ingenious process, which reflects much honour on the Earl of Dundonald, who made the discovery, and enjoys by patent an exclusive privilege for the manufacture.

The coals are put into ovens, and after being kindled, are slowly decomposed; while the volatile parts fly off into separate reservoirs, where they are condensed. Each condenser has two ovens appropriated to it, and between the ovens and condensers is placed a long leaden cistern filled with water, in order to hasten the process. The condensed fluid is then conveyed from the condensers, by a cock into wooden pipes, which lead it into a pit; from whence it is pumped into the still.

By the act of distillation, or boiling, the steam flies off into another large wooden vessel, where it is condensed into an oil and volatile spirit. The distillation continues for the space of 4½ days, when the residuum in the still makes excellent tar. The oil and spirit are then drawn off into a smaller vessel, and as the oil swims on the top, a separation is easily made, by drawing off the spirit. If the tar is boiled in the still for 5½ days, the stuff remaining in the still is then called half pitch; and should the process be continued a day longer, it would become as brittle as glass. The volatile spirit is distilled a second time before it is exposed for sale, and afterwards made into hartshorn by the chymist.

In a very short time it looked as if the tar works would retrieve the fortunes of the Dundonald Family. Writing to her brother Andrew Stuart, Dundonald's mother said:

Sir John Dalrymple and I went to Culross Friday and return'd Saturday. Sir John went to view the Tar and Pitch Works, and was greatly surprized to see the perfection it is now brought to with the almost next to certainty of it turning out to be a very profitable work, with a ready sale for all the tar that the four present kilns can produce, and for the produce of sixteen more immediately to be erected by the same people, who have the lease of the coal and the patent for the tar-work.

Unfortunately, Dundonald's affairs were still in a hopeless muddle. We soon hear that:

. . . the instant and peremptory demand of Cuthbert is about to frustrate all the happy prospect of relief from the tar works. Mr. Wanchop his agent threatens an immediate sequestration and sale of the estate, unless he gets his money paid up, of which there remains of Principal and Interest about £11,000.[1]

Alone of the creditors, Cuthbert insisted on settlement.

[1] Jean Dundonald to Andrew Stuart, Belleville, 7th December, 1782.

> The other creditors seeing the favourable prospect of great gain by the Tar and Pitch Works are willing to give Lord Dundonald some proper respite, the more so as he refus'd accepting an allowance of £200 a year for the support of his family and has resign'd the very 2/5th of the share belonging to him on the Tar Works. . . If Cuthbert was paid off, I think sunshine might yet break in upon his affairs.[1]

Dundonald envisaged an expansion far beyond the twenty kilns at Culross.

> We are encouraged to proceed in establishing the manufacture upon a very large scale in different parts of Great Britain and as the cinders or coke after the tar &c., has been extracted, have been found to answer for the iron founderys, it is at these places that we mean principally to extend our works, but a capital of thirty to forty thousand pounds will in the course of a few years need to be expended.[2]

This sum was difficult to find even with twelve years of a patent still to run. So two courses were open to the inventor: (*a*) to apply to Parliament for a prolongation of the patent for a further fourteen-year period, or (*b*) to endeavour to get the grant of 'a reward adequate to the importance of the discovery, by much the greatest in the chymical, commercial and naval line that has been made this century' in exchange for opening the patent to the public at large. To this end he approached his uncle, Andrew Stuart, indicating that one of his partners, Mr. Liddle, was intimate with the Earl of Surrey and would get him to bring the Bill into the House, 'this having the best possible effect and taking away the idea of its being what the English are too ready to call a Scots Job'. He played upon Uncle Andrew, saying:

> From your extensive knowledge and acquaintance of the Members of both Houses of Parliament you have it in your power to render me very essential service. . . . May I therefore sollicit your interest towards bringing about an end so material to the future wellfare of my family.[3]

Sir John Dalrymple also endeavoured to win over Andrew Stuart, telling him:

> About a fortnight ago Lady Dundonald sent me a very pressing letter from her son Lord Dundonald to herself entreating that I

[1] Jean Dundonald to Andrew Stuart, Belleville, 7th December, 1782.
[2] Lord Dundonald to Andrew Stuart, Edinburgh, 11th December, 1782.
[3] *Ibid.*

From the original in the Assay Office, Birmingham, by permission of the Assay Master

87. Scheme for the illumination of the façade of Boulton & Watt's Soho Works in celebration of the Peace of Amiens, in 1802.

88. Gas-lighting at the Andersonian Institution, Glasgow. It was installed there by Dr. Andrew Ure in 1805.

89. Gas apparatus designed by Samuel Clegg. At one time Clegg was employed by Boulton & Watt at Birmingham.

F. Accum, *Treatise on Gaslight*

90. Gas-light plant installed at the Royal Mint by Frederick Accum.

should come with her to Culross to witness some transactions between him and his Partners in the Patent for turning coal into tar and pitch.

Dalrymple went unwillingly, but acknowledged after his visit:

> I was surprized to find that Lord Dundonald has made his process compleat and that there was a considerable work going, and that it was very beneficial.

To form a true estimate of the probable profits of Dundonald's activities, Dalrymple on his own 'took accounts of things from his partners, clerks and servants of the work separately', and on returning to Edinburgh went to consult Dr. Joseph Black (1728–1799), Professor of Chemistry there, who had already made the acquaintance of Dundonald and who had been twice at Culross and expressed himself as 'perfectly satisfied from the convincing proofs he saw that the Tar Scheme was a solid and profitable one'.[1] As Dalrymple's own and Dr. Black's estimates did not quite tally, Dundonald was asked 'to bring over the Company's Books for Doctor Black, our friend Adam Smith and me to look at'. This re-examination showed that the production of coal-tar was 'one of the most profitable trades in the world'. Dalrymple then went on to outline his ideas of the various courses open to Dundonald, whether to endeavour to obtain (*a*) a prolongation of his patent, (*b*) a Parliamentary reward, (*c*) Naval Contract, etc. Of the possibilities in connexion with the extension of the patent he says:

> I was Lawyer in the House of Commons for Watt's Patent for the Steam Engine, an invention of national importance a trifle to this one, and I remember how easily we got a prolongation; Adam's Cement was still more unimportant, and yet he got it. I consider very well what I say, when I say that from a Naval Nation Lord Dundonald deserves a Statue of Gold.[2]

In view of the neglect of the chemical aspect of the so-called industrial revolution, it is pertinent to point out that Dalrymple's comparison is not so much an exaggeration as we might be tempted to assume at first sight. While, to quote Lewis Mumford:

> Popular historians usually date the great transformation in modern industry from Watt's supposed invention of the steam engine; and in the conventional economics text-book the application of automatic machinery to spinning and weaving is often treated as an equally critical turning-point . . .

[1] Jean Dundonald to Andrew Stuart, Belleville, 7th December, 1782.
[2] Sir John Dalrymple to Andrew Stuart, Edinburgh, 12th December, 1782.

it must be considered that coal-tar is now the raw material of vast industries: dyeing, explosives, drugs, and chemicals, in almost endless variety. We shall see later that the utilization of by-products of refining the coal-tar liquor had already become a business proposition by the latter half of the eighteenth century.[1]

As yet, Cuthbert had not been disposed of. He was still bent on sequestrating the estate. This, as Lord Dundonald pointed out, would carry off all the coal and wood, both necessary for making tar and barrel staves at his works. Dalrymple suggested consulting Mr. Glassford, a prominent Glasgow entrepreneur and a successor to Dr. Roebuck as partner of the Prestonpans Vitriol Works, a suggestion with which Andrew Stuart concurred, replying to Sir John at great length. The following is only part of the letter, but it reveals Andrew Stuart's sentiments.

10th January, 1783.

Dear Sir John,

Mr. Glassford and I observe from the perusal of your letter, that you are strongly prepossessed in favour of the solidity, utility, and profitableness of the works Lord Dundonald has been carrying on, for extracting Tar and Pitch, etc., from Coal, and you refer to Dr. Black for his opinion of the matter, but you have not sent us any Paper containing Dr. Black's sentiments, nor the Estimate drawn up by him, of which mention is made in one part of your letter. This estimate from the Authority of such a Man as Dr. Black, must be of great weight, as his distinguished knowledge of subjects of this nature, as well as his established character for sound judgment and integrity, would produce a stronger impression in favour of a new discovery or project, than could be acquired by any other means.[2] . . . Mr. Glassford therefore suspends his opinion until he sees that Estimate, and if you'll take the trouble of fowarding it to me, I shall have an opportunity of shewing it to Mr. Glassford before he leaves this place. I wish to have the opportunity of talking it over with him while on the spot here, for I am quite ignorant of these matters myself, and have really a great reluctance to enter upon subjects of this sort, not only on Account of my ignorance, but because my observation of the numbers of persons and familys ruined by projects, has led me to be extremely diffident of all projects, and to have a strong degree of prejudice against the character of a Sanguine Projector.[3]

[1] Lewis Mumford, *Technics and Civilization*, p. 3.

[2] There was also the possibility that prices would vary greatly between times of peace and war.

[3] In addition to this, Andrew Stuart had been opposed to his nephew terminating his career in the service of the State.

Lord Dundonald has long had occasion to know my sentiments on these subjects. It was my earnest wish that he should have abandoned all his schemes above or below ground, which have been attended with so much Distress to himself and family, and that he should have adhered to an honorable Profession, for which he was extremely well qualified.

At the same time I gave him warning, that if he changed his mind he must not expect it of me that I should ever thereafter give myself any trouble about his affairs, and therefore that it would be quite unnecessary for him ever to write to me about them, or about his plans or projects of any sort.

My reason for withdrawing myself from any further interference in Lord Dundonald's affairs, does not proceed from ill humour, or exceptions taken to him personally on account of his not having followed my advice on repeated occasions, for as the only object of these advices was his own welfare, so the neglect of them, did not produce peevish resentment, but regret. But having during several years of my life bestowed much time and trouble for extricating the affairs of the Dundonald family, and for preventing the distresses in which the present Lord has unhappily plunged himself, I perceived, that these efforts were in vain, and that he had now put it totally out of my power to be of any material service to him.

My sister Lady Dundonald, from her anxiety about her son's affairs, and from the goodness of her own heart, harrasses me with long letters from time to time, about different plans of extricating these affairs; I mean to write her very soon, and to repeat to her, what I formerly wrote and told her, that it is impossible for me to enter upon these matters of business with her; If there is any plan of extricating them, the proper channel through which that should come to me, should be from Lord Dundonald's men of business; with whom a Correspondence can take place in language of business, as it is impossible that Lady Dundonald, or any woman can be qualified to understand properly, and to write upon these matters. I shall be most ready and willing to contribute my best aid with others in promoting the success of such an application, but before I embark in it I must be clearly convinced and satisfied that We proceed upon ground perfectly solid, and where there is no chance of the business being justly liable either to the imputation of its being a Scots Job, or a Projectors Job, either of which circumstances you know carry along with them a great weight of unfavourable prejudice in the House of Commons.

The grounds of the Application must, therefore, be very thoroughly weighed, for if during the progress of it there, any unexpected facts should occur to shew that the application was illfounded and improper, I should have no hesitation to rise up in the House and declare that I had been mistaken or misled and therefore renounced all further concern in the business.

I shall make no apology to you for the length of this Letter as you have brought it all upon yourself.

Adieu

My Dear Sir,

Yours most affectionately,

ANDW. STUART.

London,
10th Janry., 1783.
Sir John Dalrymple Bart.

To this outspoken epistle Sir John Dalrymple sent a very quiet reply:

> I am as great an enemy of visionary projects as you can be, because my family has suffered more by them. But I would not on that account be a drone or a brute if in a definite case, I saw the chance of a great gain; and I will be the last person in the world to advise Lady Dundonald to venture fifty pounds on *any project whatever*, without the approbation of Mr. Glassford, of whom I think more highly in point of prudence and heart than I ever did of any man, my Uncle Mr. Crauford excepted, and one other person whom I would name to any other person than yourself.[1]

With Sir John's letter an estimate from Dr. Black was enclosed, and a full statement followed a few days later.

25th January, 1783.

Dear Sir,

A conversation which I had lately with Sir John Dalrymple, and a Letter which I have received from my Lord Dundonald, have occasioned my giving you this trouble. They have both expressed their desire that I would give you my opinion with respect to his Lordship's Manufacture of Tar. Without further apology therefore, I shall communicate all I know upon the Subject.

His Lordship did me the honour to call upon me several times, since he first contrived his method for Extracting Tar from Coal; and he consulted with me on the terms of the Specification of his Patent. I must also add, that so far as I know he is undoubtedly the first Inventor of this method of extracting Tar from Coal, and the first person who has reduced it to Practice.

In some of his Lordship's visits to me, he invited me to go to Culross and see his operations; and towards the end of last Summer he gave me a particular and pressing Invitation—As I had formerly promised this visit, I complied and went to Culross; tho' to be plain with you, I had little inclination to go.

[1] *Sir John Dalrymple to Andrew Stuart*, Edinburgh, 16th January, 1783.

In my preceding conversations with his Lordship, he had appeared to me very ingenious, and acute; but at the same time, so overflowing with Projects, and so extremely sanguine, and I may say, extravagant in his estimation of them, that I neither expected to find any thing solid or satisfactory, nor had I the least hope that my advice or opinion would be attended to.

But at my arrival at Culross I found reason to admire the Ingenuity with which the Furnaces were contrived; and in my conversation with his Lordship, I became satisfied that he understood the nature of this Operation, and the Principles upon which it might be rendered successful better than any other person with whom I had talked upon the Subject. As it was his desire that I should form an opinion of the utility of his Inventions, I employed all the time I was there in collecting facts, and can now give you the following account of Profits as made up with all the attention that is in my power. It is made up on the Supposition that Twenty Furnaces are to be employed.

Expence Weekly

	£	s.	d.
20 Furnaces with all apurtenances cost £1,200, the weekly Interest of which is	1	3	1
Repairs of Do. (which appears to me highly stated) .	3	15	0
140 Tons of Chow Coal at 3/6	24	10	0
To workmen for charging and Drawing the Furnaces at 8*d*. p. Ton of Coal	4	13	4
56 Tar Barrels at 3/–	8	8	0
6¼ cwts. of Rosin for Varnish at 24/– p. cwt. .	7	10	0
	£49	19	5

Weekly Produce in time of War.

	£	s.	d.
56 Barrels Boiled Tar at 21/–	58	16	0
140 Measures of Cinders at 1/6	10	10	0
134 Gallons Varnish made of the Volatile Oils and the Rosin at 3/–	20	2	0
	[1] £88	8	0

In time of Peace the price of Tar will fall as I am informed to about 14/– p. Barrel—the 56 Barrels therefore, will bring only £39 4*s*., and the whole weekly produce of the Tar, Cinders and

[1] There is an error in Black's summation: the total should be £89 8*s*.

Varnish will be £69 16*s*., but the profit will still be very considerable—thus:

	£	*s.*	*d.*
Weekly Profit in time of War	38	9	6
Annual Do., at 48 Weeks p. Ann. . . .	1,846	16	0
Weekly Profit in time of Peace . . .	19	16	6
Annual Do., at 48 Weeks p. Annum . .	951	12	0

The year is stated at 48 weeks on account of unavoidable Interruptions.

The Tar is in great repute at present, and is bought for the bottoms of Ships at the above price of 21/–, as fast as he can make it. The Varnish is made of the Rosin and the Volatile Oil which is distilled from the Tar in boiling it, and Serves for the Upper Works and for the Masts and Yards etc., of Ships; it is also sold very fast at the above Price.

I found my Lord until he could dispose of his Cinders was employing them in the Boiling of Salt. But my Lord, in his Letter to me dated the 20th Currt., acquaints me that he has been selling them for some time past to the Salt Pans at 1/10, and that he has now opened a Sale for them to Leith and other places at 3/– for Drying Malt, etc., which if it continues will make a considerable addition to the Profit.

But there is one Article omitted above, which is a quantity of Volatile Alkali (one of the component parts of Sal-ammoniac) which is formed in the Water condensed along with the Tar from the Vapour of the Coals.

This Volatile Alkali, after being somewhat refined or purified from the oil and water, can be employed in the manufacture of Sal ammoniac; and my Lord, having sent a specimen of his prepared or purified Alkali to Messrs. Hutton and Dairy[1] who manufacture Sal ammoniac here, I know that they found it good for the purpose and made his Lordship an offer for a certain quantity annually, the consumption of Sal ammoniac being limited. But I do not know what offer they have made, or what profit this article is capable of yielding. They have hitherto extracted their Volatile Alkali from Soot, which is also produced by the condensation of Vapours from Coals.

I endeavoured to dissuade him from the pursuit of these for the present, and advised him to attend to the branches of his manufacture which had already succeeded and were bringing in money. He writes me accordingly that he has kept his four Furnaces going, and that the produce continues as much as is above stated.

These four Furnaces contain altogether 14 Tons of Coal and can be wrought off easily twice a week.

[1] Cf. Kay's *Portraits*, I, p. 55.

This is all that occurs to me at present on this subject. If there is any part of it which you desire to have more fully explained, I shall with pleasure give you all further Information that is in my power.

I am
Dear Sir,
Your faithful humble Servt.,
JOSEPH BLACK.

Edinburgh,
25th January, 1783.
Andrew Stuart, Esqr.

The details given by Dr. Black in this letter still left some doubt, principally concerning variation of the price of tar between times of peace and war, in the minds of Stuart and Glassford. Stuart replied:

It was agreed that Mr. Glassford was to get accurate information of the price of Tarr and Pitch in time of Peace, which was the only part of the calculations mentioned in Your Letter that seemed overrated.

This Glassford did, and prices of tar and pitch 'taken from real sales' were forwarded to Dr. Black and Andrew Stuart, who, while now wondering if the estimates drawn up on the assumption that twenty furnaces would be built, would be the same for only four, admitted that he now thought that:

. . . upon the whole the work would be profitable and carried along with it this additional advantage, that the outlay of money for producing that profit was not very considerable.[1]

Dalrymple conveyed details of Dundonald's approach to Andrew Stuart to Matthew Boulton, and discloses that (C?) Gascoigne was for some reason opposed to Dundonald.

I could not be free last night, before Mr. Gascoigne, who considers every man as a foe who has coal to turn into tar or stone into iron.[2]

It still remained for Lord Dundonald and his partners to decide which course, i.e. extension of the patent or request for premium, would be to their greatest advantage. To this end James Cochrane, Dundonald's brother, after consultation with Adam Smith and Dr. Webster, prepared a long document,[3] setting out the pros

[1] Andrew Stuart to Dr. Black, London, 7th March 1783, with enclosure.
[2] Sir J. Dalrymple to Matthew Boulton (1783?).
[3] James Cochrane, *Facts and Hints concerning Lord Dundonald's Extract of Tar from Coal*, 1783.

and cons of the various courses, and arriving at the conclusion:

> In such a discovery as Lord Dundonald's extract of tar from coal, it is more to the advantage of the publick,—that Government should grant him an immediate pecuniary compensation in lieu of his patent being thrown open to all proprietors of coal. Such sum should amount to £40,000 to be paid in three years; which is not an unreasonable demand, when upon the one hand the extract of tar from coal may save upon a medium during the time of peace and war, at least £60,000 from being sent abroad each year, and anually £100,000 to the different Foundries and proprietors of coal works at the places already mentioned.

An item which weighed heavily in arriving at this decision was that it would throw the patent open to the great foundries at Sheffield, Birmingham, etc., and thus gain their support.

> At present in all the Foundries of Great Britain it is necessary to *char* Coal before it is fit for the various purposes it is used in working up and tempering Metal. The process of *charing* such Coal at present throws off in Airs what by Lord Dundonald's Invention is converted into Tar and charred equally well at the same Time.
>
> Upon a moderate Computation in the three Capital Foundries of both Countries, viz. Coalbrookedale, Carron and Rotherham they charr above 150,000 Tons of Coal every year. Likewise in the inferior Foundries and the Works carried on at Birmingham, Sheffield, &c. &c. at a very moderate Computation they charr above 200,000 Tons more. According to these Computations, therefore, the Tar that might be extracted from 350,000 Tons of Coal is now expended in *Airs* which allowing at the Rate of 3 Tons of Coal to the Extract of one Barrel of *Tar* will produce 116,666 Barrels of Tar which is more by 16,000 Barrels than will be consumed of such Tar in the whole *Naval* and Merchant Service of Great Britain.
>
> At the Rate of one Guinea pr. Barrel which is the price that Dundonald now receives for His, such Waste in money now amounts to 116,666 Guineas.

Dr. Black throws an interesting light on the probability of Dundonald selling the coke from Culross Tar Works to Carron Iron Works in his letter to Andrew Stuart:

> My Lord at first valued these much more highly, by supposing that the Carron or other such Companys would take them for melting Iron Ore. But that operation requires Cinders of very particular Qualities and goodness, and it would not be easy to persuade the Carron Company to venture on the use of his Lordship's Cinders, or to please them as Customers suppose they were inclined to make use of them.

The inter-relationship between tar and iron trades was dealt with by Sir John Dalrymple.[1] His opinions are based upon the work of Dr. Black.

> The best judge, perhaps in Europe, of the merit of such inventions, I mean Dr. Black, professor of chymistry in Edinburgh, who has examined Lord Dundonald's and Mr. Cort's processes[2] gives me leave to say that, so far as he can judge from the opportunities he had of examining them, they appear to him compleat, and easy and cheap in execution, and that Lord Dundonald produces excellent tar for the bottoms of ships.
>
> The advantages of the above three discoveries combined, give the command of the iron trade of the world to Britain, and take it for ever, or, at least as long as the industry and liberty of Britons remains, from the northern kingdoms, and from America; because Britain is the only country hitherto known, in which seams of coal, iron-stone, or iron-ore, and lime-stone (the three component parts, or raw materials, of which the iron is made) are frequently found in the same field, and in the neighbourhood of the sea, or of short watter carriage to the sea.

Sir John Dalrymple was himself sufficiently impressed with the soundness of Dundonald's schemes to erect tar kilns on his own estate. He wrote James Watt about his activities:

> I am busy erecting Tar Works on my estate under Lord Dundonald's patent. I am to lay out £900 for the kilns, in return for which, Captain Cochrane, who has his brother's patent, gives me ten per cent. yearly, charrs for me 7,500 tons of coal yearly for nothing, and gives me 6*d*. a ton. He takes the tar to himself. Six tons of cinders, in point of measure, burns as much lime as 10 tons of sea coal. Capt. Cochrane will give you the premium of 6*d*. a ton, and coke the raw coal gratis for you. Half the carriage is saved for a ton of coal when reduced to cinders weighs only 1,100 weight, yet keeps it bulk, and so is counted a ton of cinders. I supported Lord Dundonald's family in their distresses, and they will listen readily to any person whom I recommend. If you or Mr. Boulton like the idea, I will propose it to Capt. Cochrane. The only persons I have recommended for this project are Mr. Wedgwood and my friend Sir Grey Cooper.[3]

James Cochrane further suggested that a pamphlet should be published setting forth the advantages of Dundonald's tar, and

[1] F. Dalrymple, *Addresses and Proposals on the Subject of the Coal, Tar and Iron Branches of Trade*, Edin., 1784.

[2] The two patents of Henry Cort for the manufacture of malleable iron from coke-smelted ore, by means of pit coal.

[3] Sir John Dalrymple to James Watt, 26th December, 1785.

sent to the principal coal proprietors and the proprietors of the foundries around Sheffield and Birmingham, the pamphlet to contain as an appendix the certificates ascertaining the quality of the coal tar, signed by Dr. Black, Professor of Chemistry in the University of Edinburgh, and the captains of ships who had made practical trials of the coal tar.[1] The opinion of the Attorney-General of England and the Lord Advocate of Scotland was to be sought, whether:

> Will it be of advantage to Lord Dundonald to get the proprietors of the different foundries and coal works to sign his Petition to Parliament either for the continuance of such bounty as was formerly granted upon the importation of tar from America, or for a temporary grant of such bounty by way of indemnification for his throwing open to the publick his patent of the extract of tar from coal?[2]

In 1785, as a result of these suggestions, Dundonald published *An Account of the Qualities and Uses of Coal Tar and Coal Varnish, with Certificates from Ship-masters and Others.* After giving an account of the success of Dundonald's method, the pamphlet goes on to enumerate the other by-products of coal distillation and suggests uses for them. The by-products include coke, lamp black, ammonia, sal ammoniac, Glauber's salt, and fossil alkali. Many uses for tar were mentioned, including preservation of wood in the form of ship-bottoms, buoys, jetties, cart and waggon wheels, shingle roofs (especially in the West Indies), and in preventing corrosion of iron objects such as ship's bolts and anchors, guns, shot and shells, iron rails in streets, and fire-engine cylinders.

The greater part of the pamphlet is taken up with letters or 'certificates' in favour of the tar, like the following:

> Sir,
>
> I am favoured with your letter, respecting the coal tar bought of you in October last. You will be pleased to acquaint the Company that, from the experience I have had of it, it certainly claims the preference to Norway, Swedish, Russian, or Plantation tar for a ship's bottom.
>
> I am etc.,
>
> JOHN HALL.[3]
>
> Swan Tender,
> Leith Roads, Jan. 23rd, 1783.

[1] James Cochrane, *op. cit.*

[2] *Questions for Attorney-General and Lord Advocate.*

[3] Letter to Edward Park, Agent to the British Tar Company at Culross, from Mr. John Hall, Captain of the Swan Tender.

Many similar letters were received, all speaking of the high resistance of coal tar, and revealing that vessels treated with it had made voyages to St. Petersburg, Gothenburg, Riga, Memel, Spain, Portugal, Madeira, and the West Indies. A quantity was also exported to Newfoundland for trial there. Some idea of the scale on which the trials were carried out can be gathered from a letter from James Gray, Shipbuilder, Kincardine, who reveals that he and a neighbouring shipbuilder had between them treated some fifty to sixty ships with tar from Culross. B. Faujas de St. Fond, when travelling north of Edinburgh and crossing the Forth, says:

> I saw several vessels belonging to Glasgow and Leith, which were coated with bitumen or tar, extracted from pit-coal at the manufactory of Lord Dundonald, who has introduced the making and using of this tar on a great scale in England (i.e. *Scotland*). The vessels covered with it appear of a fine shining black which distinguishes them from the others.[1]

A considerable part of Faujas de St. Fond's *Essai sur le Goudron de Charbon de Terre, etc.*, consists of a translation of Dundonald's *Account of the Qualities and Uses of Coal Tar and Coal Varnish.* Here is his description of Culross, or *Cukross*, as he calls it.

> Il existe à dix milles d'*Edinburgh*, et presque en face de la fameuse fonderie de *Caron*, sur la rive du bras de mer qui se termine à *Stirling*, un lieu nommé *Cukross* environné de toutes parts de mines abondantes de Charbon: c'est-là que le Lord Dundonald, propriétaire de cette terre, et aimant les Sciences et les Arts, méditoit dans le silence sur la multitude d'avantages que le Charbon de terre offroit aux usages de la vie, et sur le parti que l'industrie pourroit en retirer encore.
>
> Le Charbon de *Cukross*, semblable à celui des environs d'*Edinburgh*, de *Prestonpans*, de *Caron*, etc., est supérieur et préférable à tous les Charbons de terre de l'Angleterre, pour le chauffage domestique, par la propriété qu'il a de s'allumer promptement, de jeter une flamme brillante, vive et alongée, et de ne point se coller en brûlant, ce que donnant accès à l'air, fait que le feu est toujours animé sans qu'on y touche; son odeur n'est point desagreable, et peut-être celui de brûler un feu trop promptement: sa cendre est d'ailleurs aussi blanche, plus légère et moins terreuse que celle du meilleur bois.
>
> Le Lord Dundonald, travaillant sur un Charbon toujours en gros morceaux, et ainsi favorable aux recherches dont il s'occupoit, trouva de ce côté-là un avantage qui a contributé sans doute à le faire jouir plus promptement des succès où devoient le conduire

[1] B. Faujas de St. Fond, *op. cit.*, vol. 2, p. 221.

ses connoissances en Chemie, soustenues par un génie plein d'activité.

Aussi ne tarda-t-il, pas à obtenir par des appareils simples, commodes, et de son inventure, et dans lesquels les opérations se sont très en grande, du goudron et de l'alkali volatil.

Chaque fourneau contient au moins quatorze à quinze mille livres pesant de Charbon en morceaux, et non en poussière; le feu monte graduellement dans cette masse de matière combustible, et à mesure que les premières couches s'embrasent, elles s'épurent, et la chaleur vive dégage le goudron et l'alkali volatil des couches supérieures, qui s'épurant à leur tour, font le même office à l'égrade des derniers rangs: de manière que rien ne se perd, qu'il n'y a pas un morceau de Charbon consumé, et que l'opération achevée, on le retire converti en *Coaks* de la parfaite qualité.

La découverte du Lord Dundonald sur cette nouvelle manière de traiter le Charbon, est très-importante puisqu'elle donne un produit que personne n'avoit cherché à obtenir avant lui, l'alkali volatil si utile pour fabrication du sel ammoniac: or, comme les procédés mis en pratique à *Cukross* sont traités en grand, le Lord Dundonald pourroit, en multipliant les apperails, obtenir assez de goudron pour en fournir à toute la marine angloise, et de l'alkali volatil en assez grande abondance pour ne plus tirer du sel ammoniac d'Asie, et pouvoir même en vendre aux Etrangers.

The use of tar products was not confined exclusively to marine works. A japanner, John Stobie, of Mill's Close, Canongate, Edinburgh, stated that the 'hard black' which the Company had sent him 'answered his japan exceedingly well'.

Lord Dundonald then presented a petition to Parliament, and a Committee was appointed to examine the matter. Among those examined by the Committee were Dr. Bryan Higgings, Mr. George Dyer, a London merchant deputed to express the opinion of Mr. William Menish, 'principal Chymist in London', and the above Captain John Hall.[1] As a result, Dundonald's patent was extended to run for a further twenty-one years from 1785, and the British Tar Company continued to spend considerable sums of money on their undertakings. By 1788 they had expended £22,400 on Tar Works in England and Scotland, and these were reckoned to yield '£5,000 of clear annual profit'.[2]

The British Tar Company's method of working was to erect tar kilns in the neighbourhood of various iron works. Reference has been made to the kilns at Upper Cranston.[3] The first makers of

[1] Report of Committee on Petition of Archibald Earl of Dundonald, etc.
[2] Lord Dundonald to Andrew Stuart, 10th May, 1788.
[3] *Statistical Account*, Vol. X, p. 412.

tar in Scotland were Smith of Drongan, and Cunningham of Enterkine, Ayrshire, neither of whom, however, made it a financial success. Matthew Boulton visited Dundonald at the Ayrshire Tar Works when returning to England from his visit to Scotland in the autumn of 1783, and wrote to Gilbert, the Duke of Bridgwater's manager at Worsley:

> ... the tar is better for the bottoms of vessels than the vegetable tar; and the coal-oil hath many uses.

He then suggests that a tar work might be a useful addition to the Duke's collieries and canal.[1]

A company was tried in England, and there, as in Scotland, they associated with a foundry to absorb the coke they produced.[2]

The enterprise begun at Enterkine moved nearer to Muirkirk Iron Works, where forty-five kilns were put into operation, each charged some 90 to 100 times per annum, and yielding in the course of a year about 360 tons of tar.[3]

> Coal-tar was begun, about 1787, to be manufactured at Muirkirk, on the plan discovered by the Earl of Dundonald, and for several years, considerable quantities were made there. After the demand for the tar had decreased, it was burnt into lamp-black. Of late an oil has been extracted from it, by distilling the tar, which was found to make a good and cheap varnish, or paint, on wood and iron. It has also been found to be a good remedy for maggot on sheep.[4]

These Muirkirk kilns had an interesting history. In a letter from Georgina Keith Macadam, quoted by Roy Devereux in his *Biography of J. Loudon Macadam*—famous for his theory of highway construction—it is stated that Macadam, who was Dundonald's second cousin, co-operated with his kinsman in connexion with his experiments.[5] The tar works at Muirkirk were established in 1789 on the lands of Kaims adjacent to the Muirkirk Iron Works. Macadam was tacksman of the Kaims Colliery, and the British Tar Company was to obtain enough coal from the colliery for their kilns, and Muirkirk Iron Works, founded the previous year, was to take the coke. Matters did not always run

[1] S. Smiles, *Lives of Boulton and Watt*, p. 329.

[2] *Session Papers*, Signet Library, 220: 20, 443: 61.

[3] Sir J. Sinclair, *General Report*, Appendix II, p. 312; *Statistical Account*, VII, p. 606.

[4] William Aiton, *General View of the Agriculture of Ayr*, p. 605.

[5] 'Tar-macadam' was a later development with which Macadam was not connected. Indeed it is antithetic to Macadam's conception of road structure.

smoothly between the two enterprises, and the differences eventually led to litigation.[1]

The tar kilns appear to have proved profitable for a number of years. By the beginning of the nineteenth century, however, the British Tar Company, at least at Muirkirk, was in financial trouble and Macadam was suing Dundonald for money.[2] Probably as a result of this, Macadam became the possessor of the Muirkirk Tar Works. By 1829 the Muirkirk Company had given up its lease and abandoned the kilns, the ruins of which were still to be seen in 1936, when, from the stones of which they were built, a cairn was erected to the memory of J. L. Macadam.[3]

The stumbling-block to the advancement that Dundonald had hoped for was the refusal of the Admiralty 'to make use of his preservative'. However, Thomas Cochrane tells us that:

> It was at once adopted by the Dutch and elsewhere in the North, and in the case of small coasting vessels is to this day used in our own country as less expensive than coppering. Had not the coppering of vessels become common shortly afterwards, the discovery must have proved of incalculable value.[4]

An interesting sidelight on the reasons why *coal varnish* was not more frequently employed in 'paying' bottoms of ships is the answer which Richard Kirwan obtained to his enquiries: 'I have been informed the principal reason is, that it succeeds too well; the ships not requiring such frequent repair'.[5]

What Dundonald lost by the adoption of copper sheathing by the Admiralty was gained by another Scot, William Forbes of Aberdeen. Forbes was brought up as a tinsmith in Aberdeen, but migrated to London as a young man. He obtained some information concerning the proposed use of copper by the Admiralty, and was in a position to buy up all the available copper, so that when tenders were advertised for, he was in control of all existing stocks. Iron nails were used initially to fix the sheathing, in consequence of which it was a failure, so Forbes bought back his copper for next to nothing. He then demonstrated that copper nails would lead to success, and followed this up with a resale of the copper, thereby netting a second handsome profit.[6]

[1] *Session Papers, loc. cit.*

[2] *Session Papers*, Signet Library, Edinburgh, 209, No. 6, 18th January, 1800.

[3] The production of tar according to Lord Dundonald's method was also in operation at Calcutts, Shropshire, before 1805. (*Nicholson's Journal*, vii, 1805, p. 74.)

[4] Thomas Cochrane, *op. cit.*, p. 38.

[5] Richard Kirwan, *Experiments on the Comp. and Prop. of Carbon in Bitumens, etc.*, *Phil. Jour.*, 1797, *1*, 487.

[6] David Murray, *The York Buildings Company*, p. 101.

It was only in 1822 that the British Admiralty saw fit to enquire into the utility of tar as a preservative, and requested the Royal Society to appoint a Committee to report.[1] This the Society did, appointing to consider the question Sir H. Davy and other eminent men of science, who reported favourably, but too late to retrieve Dundonald's fortune.[2] His son says:

> His motive was excellent, but his pecuniary means being incommensurate with the magnitude of his transactions, its object was frustrated and our remaining patrimony melted like flux in his crucibles; his scientific knowledge, as often happens, being unaccompanied by the self-knowledge which would have taught him that he was not, either by habit or inclination, a 'man of business'. Many who were so, knew how to profit by his inventions without the trouble of discovery, whilst their originator was occupied in developing new practical facts to be turned to their advantage and his consequent loss.[3]

This was exactly Dr. Black's estimate of Dundonald.

> When I was with his Lordship at Culross his mind was much engaged with the prospect of further profits of this manufacture, which he thought might be made from other articles of produce beside the above. Some of these schemes were without foundation, and I hope I have dissuaded him from thinking any more of them, or if he still hopes any thing from them, he will be undeceived when he makes the experiment, which he may do at little expense.[4]

Dundonald was, however, by no means the only 'projector' in Scotland, as we learn from the following letter from Dr. John Hope (1725–85), Professor of Botany, Edinburgh, to Matthew Boulton:

> Last night we had a full meeting at the Oyster Cellar, Mr. Cort, Lord Dundonald, Hutton, Black, MacGowan, etc. Dr. Hutton whispered to me, what a number of projectors, and Black said I was a fool of one myself. We had as usual a great deal of pleasantry, and every now and then some useful and interesting information.[5]

Dispersion of his energy was undoubtedly one of Dundonald's faults. While busily engaged in the development of the British Tar Company, he found time to write a pamphlet on *The Present State*

[1] C. R. Weld, *History of the Royal Society*, Vol. II, 362.
[2] *Council Minutes*, Vol. X, p. 4; quoted by Weld.
[3] Thomas Cochrane, *op. cit.*, p. 37.
[4] Dr. Black to Andrew Stuart, 25th January, 1783.
[5] J. Hope to M. Boulton, 22nd May, 1784, A.O.L.B.

of the Manufacture of Salt.[1] Andrew Stuart relented so far as to help him with it. Dundonald thanked him in a letter:

> . . . for the troubles he has been so good as take in arranging and in making some valuable additions to the publication on the Salt Manufacture which after reading with attention meet with his entire approbation.[2]

Thomas Cochrane tells us that his father carried on extensively and simultaneously a wide range of manufacturing transactions resulting from applications of chemical science to technics. Among these were the conversion of sea salt into soda as a substitute for barilla, till then the principal alkali used in soap and glass-making.[3] The focus of another group of his manufactures was the production of ancillaries for dyers and printers. These include a manufactory to carry out an improved method of making alum as a mordant for silk and calico printers, and an establishment for preparing *British Gum* as a substitute for *Gum Senegal*, a French monopoly produced exclusively in Senagambia, the price of which advanced from £150 to £400 per ton during the French wars.

> It does not appear from such trials as Lord Dundonald has hitherto made that there is any very great difficulty in the production of gum from Lichen collected from different trees or shrubs; all of them answer equally well for yielding a gum fit for calico printing.[4]

Early in 1801 he convened the calico printers of Glasgow and gave them an account of his discoveries. At the meeting several resolutions were passed so that they could be used to the best advantage and benefit to the inventor.[5] From the residue of the lichen it was proposed to make a kind of soap, and it is possible that Dundonald may have had some connexion with the patent for making soap from fish which in 1798 was granted to Sir John Dalrymple in the name of John Crooks.[6]

Dundonald's interest in paints and pigments extended beyond the production of pitch and coal varnish. He introduced the production of ceruse or white lead by a process then unknown in this country.[7] None of these enterprises brought him pecuniary

[1] Among the Dundonald Papers in the National Library of Scotland there is a fragment of an amended copy of *The Present State*—probably that referred to above. [2] Lord Dundonald to Andrew Stuart, 20th April, 1783.

[3] Thos. Cochrane, *op. cit.*, p. 37.

[4] *Directions by Lord Dundonald for Extracting Gum from the Lichen, or Tree Moss, etc.*, Glasgow, 1801. [5] *Phil. Journal*, 1st series, 1802, *5*, 228.

[6] *Ibid.*, 1st series, 1800, *3*, 108.

[7] *Comm. Inv. and Patents, Reg. of Great Seal*, 1796, Vol. 20, Nos. 598 and 604, G.R.H.E.

advancement. In 1799 he presented to his friends and to men of science a single quarto sheet entitled '*Prospectus or Index to Lord Dundonald's intended Publication*'. This publication was to give a 'detailed account of his discoveries, processes and patents, and of the circumstances which have hitherto prevented their becoming beneficial either to himself or the public'. In this prospectus he complains of '*the most cruel and oppressive usage from individuals and neglect on the part of the government*'. Enumerated in it, in addition to those already discussed, are improvements to manufacture of iron, analysis of wheat, production of sugar, treatment of starch-maker's liquor, etc.[1]

Indeed, Dundonald came near making several important contributions to technics which were left to others to develop. One such is described by his son:

> An experimental kiln was erected near the Abbey (Culross), and here coal-gas became accidentally employed in illumination. Having noticed the inflammable nature of a vapour arising during the distillation of tar, the Earl, by way of experiment, fitted a gun-barrel to the eduction pipe leading from the condensor. On applying fire to the muzzle, a vivid light blazed forth across the waters of the Firth, becoming, as was afterwards ascertained, distinctly visible on the opposite shore.

This was, however, not utilized, and some years later William Murdock, an employee of Boulton and Watt at Birmingham, independently developed and put into practice the utilization of gas made from coal. Murdock, himself a Scot, maintained that he was ignorant of Dundonald's activities,[2] but this much is certain. On one of his visits to England (*c.* 1782) Dundonald visited James Watt at Handsworth, near Birmingham. There, we know, that in the presence of Thomas Cochrane, the utilization of gas-lighting was discussed,[3] so it seems rather more than a coincidence that Murdock was entirely ignorant of even the germinal idea of this development. Needless to say, this does not detract from the credit due to Murdock for his part in stimulating the large-scale application of gas for lighting.

When Murdock suggested patenting his invention, he was advised not to do it by James Watt, Junr., and in 1804, Frederic A. Winsor obtained a patent. His patent, excepting the additional reference to collection of gas, is almost the same as Dundonald's tar patent. According to Arthur Elton,[4] Winsor specified:

[1] *Phil. Journal*, 1st series, 1800, *3*, 474.
[2] *Utility and Advantages of Gas-lights, Phil. Trans.*, 1808, 124.
[3] Thos. Cochrane, *op. cit.*, p. 40. [4] Private communication.

> An improved oven, stove, or apparatus, for the purpose of extracting inflammable air, oil, pitch, tar, and acids from and reducing into coke and charcoal all kinds of fuel.

After describing how coke, tar, and other by-products can be obtained by distillation of coal, Winsor goes on to say:

> The inflammable gas or air being purified from that of carbon so pernicious to respiration and dwellings may be led and conducted in a cold state through tubes of silk, paper, earth, wood, or metal to any distance in houses, rooms, gardens, places, parks and streets, to produce light and heat.

Winsor lists the uses to which the various products can be put: charcoal for gunpowder; pitch and tar for preserving wood; wood or vegetable acid for making alum, vitriol and copperas, for dyeing and tanning; ammonia for many purposes. Winsor was primarily interested in coke, and proposed building ovens in the neighbourhood of coal-fields, after the manner of the British Tar Company. To begin with, he only mentions lighting streets with gas as a subsidiary activity.[1]

Considering how near all this was to the activities of the British Tar Company, it is not surprising that the latter intimated their intention to oppose the Bill which Winsor attempted to get through the House for the formation of the Chartered Gas Light and Coke Company. The Bill was, however, lost at the third reading.

From manufacturing pursuits Dundonald finally turned to the application of chemical science to agriculture. His publications in that field are in the direct line of the advance of scientific agriculture initiated by Francis Home, Professor of Materia Medica in Edinburgh, and developed in the later works of Davy and Liebig, the former of whom had the advantage of an official position which was denied to Dundonald.

Even in this field he kept the production of tar and coke in view. Accum tells us:

> The Earl of Dundonald has shown that, in the application of burning lime, a quantity of coke uniformly burned a given portion of lime-stone in one third part of the time that the quantity of coal from which the coke had been made could do.
>
> This effect is to be accounted for from having previously freed the coal, or rather its coke, from the moisture and the tar, which it sends out during combustion, and which condenses on the middle and upper strata of stratified limestone and coal in the

[1] F. A. Winsor, *Pamphlet*, London, 1804.

> lime-kiln, and impedes the whole mass of materials from coming into a rapid and complete ignition; because the greater quantity of materials, and the sooner the whole is ignited, the better and more economically the lime is burned, both as to coals and time; the saving of which last is a material object, especially at lime-kilns where there is in summer time a great demand for lime, the coke occasioning the kilns to hold a *third more lime* at the *same time*.[1]

In other processes as well, coke became a desirable substitute for coal, particularly the baking of bricks and the drying of malt. An account of these, and also the making of plaster of paris, was published by Davis in the *Philosophical Magazine*.[2] At Cambridge large quantities of coal were converted into coke and sold in the town and neighbouring villages for the drying of malt. Here the coking was done in ovens in 1792, although many decades passed before the vastly greater quantities of *metallurgical* coke were produced in ovens and the by-products collected.[3]

From coal as a raw material, yet another product was obtained by Dundonald, sal-ammoniac or ammonium chloride (NH_4Cl), the main users of which were metal workers in tinning cast-iron for culinary utensils and also for tinning of brass and copper. A certain amount was also absorbed by pharmacists and calico-printers. Beckmann, in his *History of Inventions and Discoveries*, says:

> If I am not mistaken, the first real manufactories of sal ammoniac were established in Scotland; and the oldest of these, perhaps, was that erected by Dovin and Hutton at Edinburgh in 1756, and which, like many in England, manufactures this salt on a large scale.[4]

This refers to a co-partnery formed some time before 1749 by Dr. James Hutton and a Mr. Davie in Edinburgh, according to a plan conceived when they were Edinburgh students.[5] James Hutton was born in Edinburgh, and studied medicine there and later at Paris and Leyden, where he graduated M.D. in 1749. He did not practise. In 1752 he went to a Norfolk farm to study agricultural methods, returning to Berwickshire two years later. He remained there until he settled in Edinburgh, in the intellectual fellowship of Joseph Black, John Playfair, *et al.* His occupations included the study of chemistry, particularly sal-ammoniac, from which he derived sufficient income to pursue the geological

[1] Fred. Accum, *Treatise on Gas Light*, p. 161. [2] *Phil. Mag.*, *33*, 433.
[3] R. Watson, *Chemical Essays*, Vol. III, p. 339.
[4] J. Beckmann, *History of Inventions and Discoveries*, Vol. IV, p. 383.
[5] Kay's *Portraits*, Vol. I, p. 55.

studies which brought him lasting fame. It was when he was in Norfolk that Hutton was struck by the difference between the geological formation of East Anglia and his native heath. This led to his taking up the study of geology, and becoming the 'father of modern geology'.

Soot was then the material used as a source of sal-ammoniac, and the process was carried out as follows:

> Globular glass vessels, about a foot in diameter, are filled to within a few inches of their mouth with it (soot), and are then arranged in an oblong furnace, where they are exposed to a heat gradually increased. The upper part of the glass balloon stands out of the furnace, and is kept relatively cool by the air. On cooling, the upper parts of the globes are found to be lined with sal ammoniac in hemispherical lumps, about 2·5 inches thick. 26 lbs. of soot yield 6 of sal ammoniac.

For a number of years, Davie took all the soot collected by the Edinburgh *tronmen* (members of a chimney-sweeps' society formed in 1738).[1]

References to the manufacture of sal-ammoniac in Edinburgh and the green glass vessels is made in a letter from John Findlay, secretary to the Duke of Richmond, to George Macintosh (1786), and it is suggested that the vessels were probably old vitriol bottles from Dr. Roebuck's works at Prestonpans.[2] Messrs. Hutton and Davie were still in operation in 1783. In that year they began purchasing sal-ammoniac from Dundonald's tar works.[3] Dundonald's sal-ammoniac was, as already mentioned, one of the by-products of tar manufacture. For making it he obtained either hydrochloric or sulphuric acid from Glassford of Dougaldston, as one can infer from the following note in a letter from Dundonald to Andrew Stuart:

> My partners and I have it in our Power to render him (Glassford) very essential service by purchasing from him annually some Thousand Pounds worth of a commodity he manufactures concerning which I shall explain myself more fully when I have the pleasure of seeing him in Scotland.[4]

At this time Glassford had become a principal partner in the Prestonpans Vitriol Company, Roebuck having been forced to sever connexion with it on account of financial difficulties.

In 1784 Dundonald sent William Kinnaird, chemist and

[1] *Session Papers*, Signet Library, 282: 14.
[2] George Macintosh, *Memoir of Charles Macintosh*, p. 18.
[3] Dr. Joseph Black to Andrew Stuart, Edinburgh, 25th January, 1783.
[4] Lord Dundonald to Andrew Stuart, Edinburgh, 11th December, 1782.

apothecary, to London to learn the best way of making sal-ammoniac. Dundonald and his two brothers, John and Alexander, who were already in the British Tar Company, were to join Kinnaird in the manufacture, but difficulties arose over his salary, and it is said that he did everything he could to spoil the manufactory as he wished to start on his own.[1] The method of manufacture employed by Dundonald was probably that invented by M. Leblanc of St. Denis, near Paris, which consisted of using a brick kiln to vaporize the hydrochloric acid formed by the action of sulphuric acid upon sea salt, which he could obtain from his own pans. Dundonald's ammonia was produced by the action of heat upon nitrogenous matter in coal, in contradistinction to Leblanc who got it by destructive distillation of organic matter.[2] None the less, soot continued to be used as a raw material by Hutton and Davie. It was also used in a new manufactory which was founded by Charles Macintosh at Glasgow in 1786, in partnership with his father George Macintosh, founder of the Cudbear and Turkey Red Works, and Mr. William Couper, a surgeon, who became managing partner. In addition to coal, urine was utilized as a raw material on account of the ammonia which could be derived from it. According to Samuel Parkes, British manufacture of sal-ammoniac to replace what was formerly imported from Egypt, developed particularly in Scotland 'where it was formed by a peculiar process from soot; and also from a variety of refuse animal matter'.[3] The Macintosh factory continued for a number of years, but it is probable that more scientific methods of making sal-ammoniac led to its closing down.[4] At any rate, according to Sir John Sinclair, by 1814 there was only one sal-ammoniac works of any size in Scotland, and that was at Bo'ness.[5] Here Joseph Astley operated a patent process (1807) similar to that described in the *Annales de Chimie*,[6] which used the *bittern* from the salt works as a raw material, i.e. the liquid left after the removal of salt from sea-water concentrated by evaporation.[7] From this manufactory the sal-ammoniac was retailed at £9 10*s.* 10*d.* per cwt. Thomas Thomson gives the following details about Astley.

> One process more deserves to be mentioned, on account of its ingenuity and simplicity. It is the invention of Mr. Astley, who has secured the exclusive privilege by a patent, and has a

[1] *Session Papers*, Signet Library, 380: 37, 7th September, 1797.

[2] Andrew Ure, *Chem. Dict.*, Article, *Ammonia*; *Annales de Chimie*, XIX, p. 61; Samuel Parkes, *Chem. Essays*, IV, p. 339–395.

[3] Samuel Parkes, *Chem. Catechism*, 8th Edition, p. 153.

[4] Geo. Macintosh, *op. cit.*

[5] Sir J. Sinclair, *General Report*, Appendix II, p. 310.

[6] *Annales de Chimie*, Vol. 20, p. 186.

[7] Andrew Ure, *loc. cit.*

> manufactory at Borrowstowness, on the Firth of Forth, and another at Portobello, near Edinburgh. He mixes together animal matter (chiefly woollen rags), and what in Scotland is called *spirit of salt*. It is the mother ley that remains after all the crystals of common salt that can be got have been separated from sea-water. It consists chiefly of muriate of magnesia. This mixture is burned in furnaces, and the produce received in small chambers placed over the furnaces. This produce contains abundance of sal ammoniac, which is obtained pure by sublimation. We conceive the theory of this process to be, that the carbonate of ammonia is formed by the combustion of the animal matter. This carbonate immediately decomposes the muriate of magnesia, and sal ammoniac sublimes. In principle, therefore, it does not differ from Baumé's original process, though, in point of economy, it is probably greatly superior to it. We have little doubt that Baumé's method yields a greater return from the same quantity of materials; but this is probably much more than counterbalanced by the much greater expense attending his process. Nothing can demonstrate this more clearly than the circumstance that his method was abandoned in France as too expensive, though labour be much cheaper in that country than in this, while Mr. Astley manufactures his sal ammoniac with profit in the neighbourhood of Edinburgh.[1]

One Astley, 'who is an extensive chemical manufacturer in Portobello', was involved in a legal action in the Court of Session, through one of his clerks inveigling two or three members of the Edinburgh Tronmen's Society (chimney-sweeps) into a public-house, and persuading them to sign an agreement on behalf of the Society to supply him with all the soot they collected and which, since the death of Davie, had been going to Walker of Beverley, Yorks. This looks very like the above Astley.[2]

The new information which has been set forth concerning the commercialization of coal distillation, and the utilization of the products so produced, throws new light on the economic background of the gas industry, which developed in the first two decades of the next century. It also gives us a new perspective of the growth of experimental chemistry during the latter half of the eighteenth. It is worth bearing in mind that ammonia was only discovered by Joseph Priestley in 1774, and its composition determined by Berthollet in 1785. Thus when Priestley introduced the important technique of collecting gases over mercury, he was

[1] *Encyclopaedia Britannica*, Supplement to IV–VI Editions, Vol. I, p. 243.

[2] *Session Papers*, Signet Library, 282: 14. For protracted case before the Second Division, Court of Session, between Astbury and John Taylor, soap-boiler, Bo'ness, *see* the *Session Papers*, Signet Library, 307: 1, 1813.

exploring the properties of a substance which *already had established its commercial utility as a saleable by-product of tar production.* This is the light in which one must interpret Mantoux's statement that:

> . . . it is a well-known fact that chemistry hardly existed before Priestley and Lavoisier.[1]

Dundonald spent a lifetime endeavouring to retrieve the fortunes of his family by the application of chemistry to technics. Two circumstances contributed to his failure. He was admittedly unsystematic, which led to his missing chances which others would have seized, and also, in many ways his outlook was too far in advance of the emergence of palaeotechnic civilization. He was one of the first coal-owners to cease using female labour for carrying coal,[2] and had in his genius, according to Sir John Dalrymple, a natural lively character and a very good heart.

A leader in *The Times* on 17th June, 1847, referred to the despicable neglect which Dundonald suffered at the hands of his contemporaries. Thus he joins the long list of pioneers, led by Dr. John Roebuck, whose activities brought great economic power to industrial Britain without fortune to themselves. Britain, rather than Roebuck and Dundonald themselves, reaped the rewards of the abstinence of these men and of their relatives.

[1] P. Mantoux, *The Industrial Revolution in the Eighteenth Century*, p. 300.
[2] A. Cochrane, *Description of the Estate of Culross, etc.*, p. 66.

Chapter XIX

LIGHT AND LABOUR

MANCHESTER COTTON MILL FINES, 1823

	s.	d.
Any spinner putting his gas out too soon	*1*	*0*
Any spinner slipping with his gas lighted	*2*	*0*
Any spinner spinning with his gas too long in the morning ..	*2*	*0*
Any spinner having his lights too large: for each light ..	*1*	*0*

from Hammond: *The Town Labourer.*

BEFORE Lord Dundonald observed the production of light by burning flares of coal gas at the Culross Tar Works, methods to extend available time for labour by lighting the hours of darkness had changed but little during the course of the preceding thousand years. Indeed, there were technical problems of light production to overcome before eotechnic lighting appliances could be supplanted by coal gas as an illuminant.

Eotechnic artificial light was a flame rising from a rush or twig dipping into olive, or other vegetable, oil in an open clay or metal vessel. Solid animal fat dispensed with the necessity for using a vessel, and the familiar candle became a competitor with the open lamp. Eighteenth-century science was preoccupied with problems arising in connexion with both types of artificial lighting, as the leaders of eighteenth-century industry were preoccupied with the hours and factory conditions.

Wooden wicks for candles were investigated by Count Rumford,[1] and the shape and composition of the combustible material also came under review.[2] When in 1825 plaited wicks, which got over the difficulties of snuffing, were introduced, a parson of the day described the invention as the greatest of his era. A certain G. Morris suggested to Matthew Boulton that his lamps might be improved by having wicks of asbestos.[3]

The series of brilliant researches into the nature of fats and waxes published in the *Annales de Chimie* by M. Chevreul (1786–1889) had an important effect upon the candle industry, because it enabled the candlemaker to understand the scientific principles underlying the operations handed down by craft knowledge. When Chevreul was carrying out his researches, Samuel Parkes wrote:

[1] Rumford, *Phil. Mag.*, 1799, *4*, 79.
[2] *Phil. Journ.*, 1802, *5*, 187.
[3] 30th December, 1795, A.O.L.B.

> The manufacture of candles may yet derive advantage from chemistry which would repay the study. Foreign tallows, which frequently contain a large proportion of acid rendering them inferior to English tallow, may be purified at the most insignificant expense by chemical means; and by the proper application of chemical agents, other brown tallows may be rendered beautifully white, and fit for the best purposes.[1]

Crude fat from which tallow candles were made was contaminated with glycerine and low-melting oleic acid, and when the chemical composition of fats became known, these undesirable impurities could be removed.

Early wax candles were made of beeswax or spermaceti from the heads of sperm whales. Chambers in his *Cyclopædia* (1751) mentions them as a recent introduction. Subsequently, though much later, *paraffin wax*, obtained from Scottish shale distilleries, was added to candles. For the production of tallow candles and soap, large quantities of tallow were imported from Russia. The quantities were sufficient to tax as a source of national revenue, as the imports from Russia alone were stated to be valued at one million pounds sterling.[2]

Lamps went through a parallel series of improvements, both with regard to fuel consumed and to the construction of the lamp itself. For example, it was shown that keeping qualities of lamp oil could be improved if it were treated with acid.[3] A scientifically constructed lamp, with a cylindrical wick between two concentric tubes, was devised by Aimé Argand, a Geneva physician, who came to England in 1783. Argand was put in touch with Matthew Boulton and James Watt by a Mr. Parker. A patent for this lamp was granted in 1784, but the validity of the patent was subsequently disputed, and in 1787 Argand died in poverty at Versoix, near Geneva, so another pioneer bequeathed to posterity the rewards of his industry. That the patent was disputed is not surprising, since a lamp of the same construction was in use in the chemistry department of Glasgow University as early as 1766.[4] Argand's lamp burned a cylindrical wick, confined between two concentric tubes, which deflected a current of air on the inner surface of the circular flame.[5] Although this invention brought no

[1] S. Parkes, *Chemical Catechism*, p. 11.

[2] *Chas. Wallace to Henry Dundas*, November, 1786, Scottish National Library, MS. 640, *ff.*13–18; *Phil. Journal*, 1800, *3*, 112, *n.;* Leo Field, *Solid and Liquid Illuminating Agents* (Cantor Lecture, 1883), *J. Soc. Arts*, 1883, *31*, 834.

[3] *Phil. Journal*, 1799, *2*, 46.

[4] *Encyclopaedia Britannica*, Supplement to Third Edition, Vol. 2, p. 64.

[5] For a detailed description of the lamp, *see* H. W. Dickinson, *Matthew Boulton*, p. 128.

reward to its inventor, it became a standard by which other lamps were calibrated, and it also supplied the basic idea for the gas-consuming burner used by Boulton and Watt. Other lamps followed the pioneer researches of Argand, each with its own special object in view, some designed to burn tallow, others to give a steady light without attention.[1] The great Lavoisier (1743–94) was attracted to science through an offer of a prize by the French Government for an improved method of lighting the streets of Paris. He investigated the problem, and was awarded a gold medal for his researches.

The closing years of the eighteenth century witnessed a new source of artificial light. The manufacture of coal gas laid the foundations of an industry destined to have far-reaching results on account of its by-products, which eventually furnished raw materials of several other important chemical industries.

There was no essential novelty in Dundonald's observation. References to the production of inflammable gas from coal date back to the early days of the Royal Society in the late seventeenth century. Between then and the commercialization of gas lighting by Boulton and Watt an ever-increasing interest in chemical analysis of Newcastle coal led numerous observers, e.g. Thos. Shirley, in 1667, Dr. Stephen Hales, and Richard Watson, Bishop of Llandaff, to remark the production of inflammable vapours when coal is heated in a closed vessel.[2] The occurrence of natural gas in association with coal measures attracted chemists to the idea of producing it artificially. One of the earliest was Rev. Dr. John Clayton who, *c.* 1688, having been attracted by gas issuing from a fissure in the earth near a coal pit at Wigan, found that he could produce a similar gas by heating coal out of contact with air. He filled bladders with his gas for the amusement of his friends, and must have come very near to inventing the ærostatic balloon.[3]

No idea of commercialization appears to have been entertained at this date. This is not surprising. The industrial revolution had not advanced sufficiently far to create a demand for longer hours of labour. The emergent palaeotechnic iron industry, too, would have found it still very difficult to supply the necessary equipment. Moreover, in all probability there was an instinctive prejudice against the gas made from coal which had hitherto made its presence known principally in disastrous pit explosions. One has only to recollect the facetious remarks made about gas lighting by

[1] *Phil. Journal*, 1800, *3*, 363, 467 and 547.
[2] R. Watson, *Chemical Essays*, Vol. 3, p 7.
[3] *Experiments concerning the Spirit of Coals*, *Phil. Trans.*, 1807, *16*, 83 and 170.

Humphry Davy and Walter Scott, and the incredulity of the House of Commons in finding that the pipes which conveyed the gas were not hot, to understand why it was that gas lighting developed in the nineteenth, instead of in the eighteenth century.

Not before 1780 do we hear of attempts to use gas as an illuminant. One of the first was the lighting of a lecture-room in the University of Louvain in 1784, by Professor Jean Pierre Minckeler. In Great Britain, the development of gas lighting is associated with the name of William Murdoch (1754–1839), and its commercialization with Boulton and Watt.[1] William Murdoch was born at Bellow Mill in the parish of Auchinleck, Ayrshire. He was the son of the local miller. Though his father has no claim to fame as an inventor, he was reputed to be progressive in his own profession, and is believed to have had cast for him the first piece of iron-toothed gearing ever used in mill work. This was made in 1760 at Carron Iron Works. During his youth William Murdoch displayed a strongly mechanical bent. In 1777 we find him seeking employment with Messrs. Boulton and Watt. After an interview with Boulton he was engaged, and ultimately became their principal manager in Cornwall. There he remained until 1799.[2]

Murdoch combined with his mechanical skill a genius for chemical experimentation common to James Watt and other members of the Lunar Society in Birmingham. His inventions cover a wide range of activities, first among them the one with which we are here concerned. Details of experiments which Murdoch carried out to this end when in Cornwall are given in a letter from M. S. Pearce to Samuel Smiles.[3] It tells how Murdoch and Dr. Boaze, a physician in Redruth, heated a kettle filled with coal and burned the gas which was evolved in 'a large metal case, such as is used for blasting purposes'. The flame was next improved by substitution of a perforated thimble in place of the metal case. By 1792 the work had gone so far that he was able to light his own house in Redruth, conducting the gas some seventy feet through tinned iron and copper tubes.

Two years later Murdoch suggested to James Watt, Junior, that they should take out a patent. At the time young Watt was involved in litigation arising out of his father's patents. While

[1] We have to acknowledge here the great help we have derived from our being permitted to read an unpublished history of the gas industry by Arthur Elton. Throughout this chapter we have retained the spelling Murdoch, which in time was changed to Murdock, on account of the inability of the English to pronounce his name.

[2] R. B. Prosser, *Birmingham Inventors and Inventions*, p. 23.

[3] S. Smiles, *Men of Invention* 8th edit., p. 136.

Murdoch maintained that he was unaware of work which had been previously carried out in the same field by other experimenters, Smiles states that young Watt knew of Lord Dundonald's observations, and those of Bishop Watson. Smiles also says that, owing to difficulties in which James Watt had been involved over steam-engine patents, young Watt considered that it would not be advisable for the firm to seek a patent for the production of coal gas. This tallies with a reference Thomas Cochrane makes to a visit with his father, the Earl of Dundonald, to James Watt. The possibility of using gas as an illuminant was discussed at this meeting.

Murdoch went forward with his invention, although the process remained unpatented, devising various methods of washing and purifying the gas. He also designed several different types of mechanical apparatus for collecting the gas. Retorts were originally vertical cast-iron pots, but about 1802 he tried horizontal retorts instead: To begin with they were circular in section, some twenty inches in diameter, and three to seven feet long. The circular design was unsatisfactory, as the retort soon became encrusted with a layer of gas-carbon, and this prevented free transfer of heat to the contents. To get over this, Murdoch substituted retorts of elliptical cross-section. A great improvement in the yield of gas resulted.[1]

It is now a matter of speculation what it was that attracted Murdoch to the idea of lighting with gas produced by the destructive distillation of coal. There is, however, a letter dated 15th June, 1790, from a John Champion of Bristol to Boulton, which may have drawn Murdoch's attention to the subject of gas lighting, and there is always Dundonald's conversation with James Watt, though Murdoch denied knowledge of it. Champion tells Boulton:

> I have from some inflammable matter in Pitt Coal made a light for Light Houses—Tryal of it was made by divers of the Trinity Corporation and at its expense. It appears to make a much better more Steady and much cheaper light than the oyle light now used, and by the Calculations I made it will be Two Thirds Cheaper and more. . . . I am now in my 86th year and am desirous of someone that will undertake the Conducting this Concern and thinking you a very proper person I offer you one half of the Concern for Three Hundred Pounds. A Patent is going Forwards.[2]

This probably led to Murdoch's experiments. They were brought before the public in 1802.

[1] *Encyclopaedia Britannica*, Supplement to IV-VI Editions, Article, *Gas Lights*, pp. 448-462.

[2] John Champion to Matthew Boulton, 15th June, 1790. B.R.L.

What had actually happened at the first display of gas light is uncertain. It is said that the counting-house and factory at Soho were lit by gas in 1798. What is certain is that there was a public display to celebrate the Peace of Amiens in 1802, with the use of flaming open burners or 'Bengal lights', but to an extent which is a matter of conjecture. Most probably only the two end burners were of this type. Designs for the illuminations are still extant, and are reproduced in this volume.

The motive which led Boulton and Watt to make this public display of lighting by gas may have been information to the effect that already in 1801, in Paris, M. Le Bon (1767–1804) had lit his garden and the Hotel Seignelay. At any rate, Gregory Watt wrote from Paris pointing out that it was high time to take advantage of Murdoch's invention if they were to derive any benefit from it.

Thereupon Boulton and Watt took up the manufacture of gas-lighting apparatus. They laid out some £4–5,000 on the new venture, and by the middle of 1806 they had in stock 1,300 burners and 10,000 feet of gas tubing. At the beginning of the next year this had increased threefold. They had also on order another 4,000 feet of tubing, 12 retorts, 100 patent lamps, iron work for two retorts, gearing for a retort crane, and 5,000 burners.[1] They were faced with many difficulties, such as the conversion of Argand's burner to burn gas. They also tried to purify the gas chemically, a task well within the scientific capabilities of the personnel of the Soho firm.

The practice adopted by Boulton and Watt was to supply apparatus for the illumination of factories, theatres, and other large buildings. In 1805 they lit the cotton mills of Phillips and Lee of Manchester by nearly 1,000 separate burners. Phillips and Lee was one of the first two large-scale industrial lighting schemes in Great Britain. Writing to Boulton and Watt in the middle of 1805, Messrs. Phillips and Lee said:

> We tried last night a room . . . lighted by Patent Lamps. . . We are now preparing our Gazometer Pits and have determined to appropriate 6 instead of 4 to Gazometers. . . . We shall be obliged if you will order for us the Best Fire-bricks of a proper form for seating the Centre Bottom of the Retorts and constructing the fire place as early as convenient. As the Season is now advancing we shall be glad to know what and when we may expect a further Supply of apparatus for the Interior of the Mill.

This pioneer plant was used for large-scale experimental purposes not only by its erectors, Boulton and Watt, but by Dr. Wm.

[1] *Gas Light Papers*, 21st January, 1807: *Memorandum*, 4th June, 1806. B.R.L.

Henry of Manchester, for experiments upon the preparation and purification of gas.[1] Figures obtained with it were used in the paper read before the Royal Society by their President, Sir Joseph Banks (1743–1820), in the name of Murdoch.[2] From letters in the Boulton and Watt Collection, it appears that James Watt Senr. and his son were jointly responsible for writing the paper. The experiments upon which it was based were carried out by Henry Creighton at the factory of Phillips and Lee. The original manuscript is in the Boulton and Watt Collection, with several added corrections. None of it is in the handwriting of Murdoch, who was at Portsmouth attending to an engine when the paper was being prepared.

We see then that it was not in the midst of the *first* Industrial Revolution, but with the evolution of palaeotechnics, that gas lighting became commercialized.

> Twenty-four hour operations, which characterized the mine and the blast-furnace, now came into other industries which had heretofore respected the limitations of day and night. Moved by a desire to earn every possible sum on their investments, the textile manufacturers lengthened the working day: and whereas in England in the fifteenth century it had been fourteen or fifteen hours long in mid-summer with from two-and-a-half to three hours allowed for recreation and meals, in the new mill-towns it was frequently sixteen hours long all the year round, with a single hour off for dinner. Operated by the steam engine, lighted by gas, the new mills could work for twenty-four hours. . . . Time took on the character of an enclosed space: it could be divided, it could be filled up, it could even be expanded by the invention of labour-saving instruments. . . . .
>
> Andrew Ure, the great British apologist for Victorian capitalism, was aghast at the excellent physician who testified before Sadler's Factory Investigating Commission on the basis of experiments made by Dr. Edwards in Paris with tadpoles, that sunlight was essential to the growth of children: a belief which he backed up—a century before the effect of sunlight in preventing rickets was established—by pointing to the absence of deformities in growth, such as were common in milltowns, among the Mexicans and Peruvians, regularly exposed to sunlight. In response to this Ure proudly exhibited the illustration of a factory room without windows as an example of the excellent gas lighting which served as a substitute for the sun.[3]

[1] Andrew Ure, *Chemical Dictionary*, 1821; Art. *Coal-gas*.

[2] *Phil. Trans*, 1808, 98, 124.

[3] Lewis Mumford, *Technics and Civilization*, pp. 17 and 161.

In the year when Murdoch lit the factory of Messrs. Phillips and Lee, the same Dr. Andrew Ure installed gas lighting in the lecture-threatre of the Anderson Institution in Glasgow.

Gas lighting was then attracting considerable attention in Glasgow, as the following letter in the *Glasgow Courier* shows:

> From the very numerous trials which have been made, there seems now to be every reason to believe, that the light, afforded by the combustion of gas, may be employed very advantageously in a great number of situations; such as the lighting of streets, manufactories, theatres, etc. In the prosecution of this idea the necessary apparatus has been fitted up in several shops in this city, and manufactories in the neighbourhood, and, for some months past, the public of Glasgow have been gratified, almost every evening, with seeing a number of different places beautifully illuminated in this manner. On the evenings of Saturday and Monday last, the front of the Exchange was lighted on this principle, with a brilliancy greatly superior to that of the common lamps.[1]

The shops referred to in this letter were: Lumsden, a bookseller, John and Robert Hart, bakers, and one of the manufactories, the Auldton Spinning Mills. All of these were gas-lit before 1805.[2]

During the course of the next twelve months, Boulton and Watt received inquiries from upwards of a dozen manufacturers, etc., respecting gas lighting, but few of the proposed schemes were put into operation. For example, they received an ambitious inquiry from the civil engineer, Robert Stephenson, asking whether they thought it would be possible to light with gas the proposed tunnel under the Firth of Forth.[3]

The equipment of Messrs. Lodge's factories at Halifax, soon followed;[4] Messrs. Burleigh and Kennedy of Manchester during 1808; then Messrs. Gott of Leeds. At Gott's a gasometer house was built. For a short time, about 1820–30, it was thought that gasometers should be constructed so that they could be protected from severe weather. Hence the gasometer house at Gott's, but very few of these were in fact constructed. In addition to that at Gott's, two octagonal houses were built at Warwick, and the only other similar houses were at Inverness and Banff. The Inverness structure was a plain rectangular building to contain two circular holders, the engineer's drawings for which are dated 1826.[5]

[1] *Glasgow Courier*, 11th October, 1805.
[2] A. M(urdoch), *Light without a Wick, A Century of Gas Lighting*, p. 45.
[3] *Robert Stephenson to Boulton and Watt*, Edinburgh, 20th April, 1807.
[4] Clegg, Jr., claims that this was working before Phillips and Lee.
[5] W. B. Crump, *The Leeds Woollen Industry*, p. 266, *n*.

The first works in Scotland to be lit was a mill in Fife in 1810, and by 1812 several cotton mills and a print-works in Glasgow were gas-lit.

> The cotton-spinning, and the looms moved by a water-wheel at Thornly-bank, and by a steam engine at Pollockshaws, have been already mentioned. But it may be here farther observed as to the latter; that all the buildings for cotton-spinning and for weaving, at that extensive establishment, are lighted by the combustion of coal-gas; that 420 lights, each of them of as great degree of intenseness as three of the candles commonly used at the mills, are disposed through the different large apartments, affording a brilliant and pleasing light to the numerous workers in that large factory;—that though sometimes an offensive smell arises from the escape of unconsumed gas; this defect may certainly be remedied;—and, that the beneficial effects of security against fire, as already noticed, are fully experienced.[1]

Messrs. Gordon Barron and Company, Woodside Works, Aberdeen, added to their already very complete equipment by making their own gas.[2]

Among other large buildings which early adopted gas lighting were theatres. The Lyceum was lit in 1806, and the lighting of the Glasgow theatre was described as a social event of great brilliance.[3] The laboratory of the Royal Institution had its own plant, which consumed paper, wood, coal, etc.

The paper in which Murdoch presented details of his discoveries to the Royal Society contains the relative cost of illuminating Messrs. Phillips and Lee's cotton manufactory with candles and gas. To light the factory required 2,500 candles, and candles to the value of £2,000 were used annually, but after allowing for wear and tear on the generating apparatus and deducting from cost value of the coke sold, an equivalent of gas light could be provided for £600. In Scotland, where owing to the nature of the coal a greater quantity of gas could be obtained from a given weight, Sir John Sinclair calculated that candle-power was then eleven times as costly as gas light. And according to Justus Liebig:

> The price of the materials from which gas is manufactured in England bears a direct proportion to the price of corn; there the cost of tallow and oil is twice as great as in Germany, but iron and coal are two-thirds cheaper.[4]

[1] J. Wilson, *General View: Renfrew* 1812, p. 272.

[2] S. Smiles, *Men of Invention*, p. 141; Sir J. Sinclair, *General Report*, Appendix II, p. 302.

[3] *See* J. McUre, *Glasghu Facies* for details. [4] J. Liebig, *Familiar Letters*, p. 45.

Sir J. Sinclair's description of the apparatus used in the production of gas in the factories is worth recording:

> The coals are put in a cast-iron retort holding two to three hundred-weights. This is placed in a furnace. The gazometer, 16′×10′×9′, is of thinly rolled sheets of iron rivetted together, with a bottom but no top. It is covered with iron or cloth and made air-tight. This is suspended, bottom uppermost, from the ceiling of the house by chains similar to a chandelier. Below it there is a pit lined with bricks. This is filled with water. The gazometer is immersed, and the gas led into it from the heated coal.[1]

A small unit is figured by Samuel Parkes in *The Chemical Catechism* which corresponds almost exactly with this description.[2]

The soft, unvarying light which was given by gas brought it into great favour with the workers. It was regarded also as lessening fire hazards, as a result of which insurance premiums were reduced.[3]

It was several years after Boulton and Watt had started making gas lighting apparatus that the idea of a public gas supply was mooted. Several famous scientists of the day regarded it as impracticable, and general impressions were not improved by a prospectus for a public gas company sent out by Frederic Albert Winsor, a Moravian promoter. Winsor, who had written several fantastic pamphlets, lectured on gas lighting, and lit the Lyceum, applied for, and obtained, a patent for gas lighting, in order to undertake public gas lighting. A description of his lectures is given in Rees' *Cyclopaedia*.[4] By 1808 Boulton and Watt began to fear that Winsor would encroach upon their activities. They therefore obtained the services of Lord Henry Brougham (1778–1868) as counsel to oppose a Bill which Winsor was attempting to put through the House in order to get the charter for a gas-lighting company. Among those whom Brougham examined was Lee of Manchester. Lee related that in his own house he had replaced thirty pairs of candles by gas lights. Sir Humphry Davy was also a witness for Boulton and Watt.[5] Apparently Murdoch himself played very little part in the proceedings, other than publishing '*A Letter to a Member of Parliament*', which deals point by point with Winsor's pamphlets. Besides Murdoch, other members of the Soho

[1] J. Sinclair, *General Report*, Appendix II, p. 303.

[2] Additional Note LXVII, p. 504.

[3] W. T. Brande, *Observations on the Application of Coal Gas to the Purposes of Illumination, Quarterly Journal*, 1817, *1*, 71; F. Accum, *On the Method of Illuminating Streets by Coal Gas, Annals of Philosophy*, 1815, *6*, 16.

[4] Rees' *Cyclopaedia:* Article, *Gas.*

[5] *Quarterly Journal*, 1817, *1*, 74.

firm engaged in pamphleteering. When Boulton and Watt decided to oppose Winsor they got James Pillans (1778–1864) of Edinburgh University to comment on the Winsor publications. This he did in an article published anonymously in the *Edinburgh Review* of January 1809.

According to the diary of Dr. Thomas Chalmers, on the 20th May, 1807, he:

> . . . attended a lecture and exhibition of gas lights at Pall Mall in the evening. The Lecturer, Mr. Winsor, is a mere empiric, not a particle of science, and even dull and uninteresting in his popular explanations. The Londoners listened with delight, and I pronounce the metropolis the best mart of impudence and folly. It is not worth attending, tho' it may be rendered so by a better lecturer.

Three years later, however, he had the new Manse at Kilmeny fitted with gas pipes and made his own gas in the kitchen.[1]

Arthur Elton sums up Winsor's lecturing, writing, and behaviour by saying:

> Absurd though these pamphlets are, they contain one very important idea. From the beginning Winsor seems to have thought of a public utility company, publicly financed, and supplying gas from a central source. In this he was in advance of his time, and far more imaginative than, for example, Boulton and Watt, who thought only in terms of isolated units for each house and factory.[2]

Although Winsor's Bill was defeated he continued his activities and obtained a patent in 1809. In the next year he applied to Parliament for a charter to found a company with a capital reduced to a fifth of that of his former grandiose scheme, to be called the Gas Light and Coke Company. This time, though only after some delays, a charter of incorporation was granted in 1812, and the Company started under the direction of Frederick Accum (1769–1838), quondam lecturer at the Surrey Institution, with Samuel Clegg as chief engineer.

Samuel Clegg, a native of Manchester, was born in 1781, and associated with Boulton and Watt in their early development of gas lighting. By the date when Murdoch was lighting the factory of Phillips and Lee, Clegg had left the parent firm and was already constituting himself a serious rival. Thus on 23rd December, 1805, Murdoch wrote to Boulton and Watt:

> If materials cannot be forwarded in a more expeditious way than they have hitherto been done, it is of no use to think of

[1] A. M(urdoch), *op. cit.*, p. 42. [2] Private Communications.

> taking orders here for your old servant Clegg is manufacturing them in a more speedy manner than it appears can be done at Soho.

In 1813 Clegg joined the Gas Light and Coke Company and became one of the most prominent early gas engineers. He contributed many improvements to gas industry, contemporary descriptions of which were given by Brande. They include a gas governor to maintain supply at constant pressure, the first gas meter, and a mechanical retort designed to obviate some of the difficulties encountered by Murdoch. These retorts were put into operation at Chester, Birmingham, Bristol, and at the Royal Mint. We shall have occasion to refer to Clegg again when discussing the purification of gas.[1]

Despite obstructive behaviour on the part of pavement committees, the subsequent expansion of the industry was rapid. According to Accum, by 1815 there were twenty-six miles of main in London supplying 4,000 Argand burners, and by 1819 the figures had increased to 288 miles, and 51,000 burners.[2] This may be an over-statement. Clapham says that there were just over 200 miles of main in 1822.[3] In 1831, twenty-three Londoners described themselves as gas-fitters.[4] The rate of progress was sufficient, indeed, to call for contemporary notice:

> The progress which the illumination from coal gas is making not only in the metropolis, but in various provincial towns, and the perfection to which the apparatus is now brought, cannot be considered among the least of the improvements of the present day.[5]

Development was particularly rapid in Scotland, where *cannel* (a very bituminous) coal was commoner than in England. The Glasgow Gas Company was founded in 1817, and both Edinburgh and Glasgow had gas lighting before Manchester, Liverpool, or Dublin.

The Glasgow Company was incorporated by Act of Parliament in 1817, with a capital of £40,000. Within the next thirty years, this was increased to £150,000. Street lighting was there put into operation on the 5th September, 1818.[6] The first chairman of the company was Henry Monteith of Carstairs; the first superintendent

[1] *Quarterly Journal*, 1817, *1*, 278; *2*, 132; 1819, *6*, 71.

[2] F. Accum, *A Practical Treatise on Gas-Lights; Description of the Manufacture of Coal Gas.*

[3] J. H. Clapham, *Economic History of Modern Britain*, Vol. 1, p. 191.

[4] *Ibid.*, p. 68.

[5] *Quarterly Journal*, 1818, *5*, xvii.

[6] *New Statistical Account*, Vol. *6*, p. 163.

and foreman, James Beaumont Neilson, who is better known for his improvements of the art of smelting iron.[1] At the time of his appointment, Neilson is supposed only to have seen gas lighting on one occasion—the illumination of the Govan Colliery offices after the Peace of Amiens, a display given with the simplest of apparatus, without either condenser, purifier, or gas-holder.[2] He was nevertheless selected from some twenty applicants, and given a contract for five years, with a salary of £90 per annum. In order to improve his knowledge of chemistry and physics, he attended classes at the Andersonian Institution, and before his five-year contract was up he had been appointed manager and engineer at a salary advanced first to £200, and ultimately to £400.[3]

The production of gas from coal prompted experiments on 'cracking' natural (*whale* and *fish*) oils at an early date. Though this involves no intrinsic economy of material, it offers the advantage of centralized distribution, and hence economy of labour.

Gas from oil was the invention of John and Philip Taylor, who claimed for it considerable advantages over coal gas, particularly its purity. Had it not been that methods of purifying coal gas rapidly improved, oil gas might have become of considerable importance. One reason why oil gas may have found favour in Scottish ports was the availability of fish oil at that time.[4] It is certain that oil companies were prepared to foster the extension of oil gas at the expense of coal. Gas made from oil was not, however, a technical success. The oil was run into a red-hot retort and there decomposed into gas, but after a comparatively short time the retort became inactive for this purpose, and had to be replaced —a very early instance of negative catalysis.[5]

The Edinburgh Gas Light Company was formed about the same time as Glasgow, supplying first (most probably with oil gas), the North and South Bridges and then the theatres. Sir Walter Scott was Chairman of the Company. In contrast to the success of the Glasgow company, those both in Edinburgh and Leith appear to have operated unsuccessfully for a number of years. The original works at Tanfield supplied oil gas, but when they failed and a proposal to go over to coal gas was mooted, the pro-

[1] Ch. XVI. [2] We have no further information about this display.

[3] For Neilson's efforts to interest his workmen in chemistry and physics, *see* an article by one Ballantyne in *Glasgow Mechanics' Magazine*, 1825, *53*, 1st January, and 1825, *67*, 9th April. [4] *Quarterly Journal*, 1819, *6*, 108; *7*, 312.

[5] For a detailed description of Taylor and Martineau's oil gas plant, *see* *Quarterly Journal*, 1820, *8*, 120.

moters encountered opposition from a variety of directions. The opponents of coal gas were able to enlist the support of Dr. Fyfe, who expressed the opinion that coal-gas works at Tanfield would be a nuisance, as he did not know of any method whereby the smell could be completely prevented, and that, while the tar could be used as a fuel instead of being carted through the streets, its burning would give rise to an offensive smell. Opposition also came from the Caledonian Horticultural Society.[1] In 1827 however the Edinburgh Oil Gas Company applied for an Act to allow them to go over to coal, and a new company, the Edinburgh and Leith Gas Light Company, was formed in 1839.

A supply of gas was initiated in Aberdeen in 1824, also using oil gas. For the first four years it was made and distributed at a price which to-day seems fabulous, viz. £2 to £2 10*s*. per thousand cubic feet. On going over to coal gas, the price was progressively reduced from 15*s*. to 10*s*. per thousand cubic feet. The coal used was the best *parrot*, at 13*s*. per ton.[2] Before the founding of Aberdeen public supply, several private attempts at the production of gas were tried both in Aberdeen and in Dundee, but it was not until a year after the Aberdeen company was founded that a similar public undertaking was put into operation in Dundee.[3]

The next Scottish town to be supplied with gas was Greenock, where a joint-stock company was formed in 1827. By 1848 the company was employing some twenty men at a wage of 13*s*. per week. After the main Scottish towns had been supplied, the production of gas began to be pushed into the less populous areas. Banff followed in 1830, Peterhead in 1833, Turriff in 1838.[4] At Hamilton in the summer of 1831, the gas works was erected 'on a very elegant plan', and besides supplying private lights, there were one hundred and thirty lamps in the streets which 'were illuminated throughout the town for nine months in the year, from sunset to sunrise, with the exception of five nights at each full moon'.[5] Perth had a very capably conducted gas works.

Early gas companies were in no sense public utilities. They were capitalistic enterprises whose function was to make as large a return as possible to owners of capital put into them. In consequence, dealings between companies and consumers were at times strained. From a pamphlet published as the result of one such

[1] *Considerations relative to Nuisance in Coal Gas Works* . . . Edinburgh, 1828, App. II, Signet Library.
[2] *New Statistical Account*, Vol. *11*, p. 78.
[3] *Ibid.*, Vol. *11*, p. 9; Paton and Millar, *B.A. Handbook*, Dundee, 1912, p. 170.
[4] *N.S.A.*, Vol. 12, pp. 367, 1013; Vol. 13, p. 45.
[5] *Ibid.*, Vol. 6, p. 284.

dispute, we get details of the economics of gas production in the second decade of the nineteenth century.

Using the 'old method of distillation', presumably Murdoch's, it was possible to obtain 12,000 cubic feet of gas from one London chaldron (thirty-six bushels) of coal at a cost of £3 13*s.* 6*d.* Clegg, at the Peter Street gas works, Westminster, could get 18,000 cubic feet from the same coal for £2 12*s.* 6*d.*

For a town the size of Newcastle, it was estimated that capital required to establish gas lighting would amount to £13,000. Such works would supply Argand lamps, which, lighted till eight o'clock, burned six hundred hours per annum. The consumption of gas per hour in an Argand burner was four cubic feet, so that the annual consumption per burner was 2,400 cubic feet per annum. This gas sold at £3 3*s.*, or £1 1*s.* per 800 cubic feet.

So we can draw up the following account of expenses of production:

	£	*s.*	*d.*
Cost of one Newcastle chaldron coal (68 bushels)		8	0
Allow 20 per cent. of this coal for firing .		1	7¼
Four firemen tending fire eight hours each		10	0
Wear and tear on ten grate bars . .		2	6
Ditto and ditto on 40 retorts @ 4½*d.* ditto		15	0
Luting		1	0
Men's time drawing and charging retorts .		2	6
Repairs to pokers, shovels, etc. . .		1	0
	£ 2	1	7¼

From this chaldron of coal there was obtained:

	£	*s.*	*d.*
22,000 cubic feet gas at 21*s.*/800 cub. ft. .	28	17	6
1½ chaldrons coke (Newcastle measure) .	1	3	6
43 gallons tar at 8*d.*/gallon . . .	1	8	8
60 gallons ammonia at 3½*d.*/gallon . .		17	6
	£32	7	2
Less cost of production	2	1	7¼
Profit on distilling one chaldron . .	£30	5	6¾

It was reckoned that the supply would be sold as follows:

	£
600 shops lighted till 8 o'clock @ £5 5s. each .	3,150
190 inns and public houses @ £10 10s. each .	1,995
350 manufactories and workshops @ £6 6s. each .	2,205
220 offices, etc. @ £5 5s. each	1,155
500 dwelling houses @ £3 each	1,500
Public lamps, Institutions, etc.	500
	£10,505
Deduct for errors in calculation for those who may not take gas lights, and for those who may make gas for their own consumption . . .	2,505
State Expence of Gas Works	600
	£3,105

So that the probable net income of such a gas company would be £7,400, and on the capital sunk this represents a return of no less than 57 per cent.![1]

After 1812 the demand for gas and water pipes led to many improvements in foundry practice. Drawn tubes were common by 1782, but were confined to the making of telescopes, and copper tube for distillers' worms. In the patent which was granted to Joseph Bramah (1748–1814) in 1797 for his beer engine, there is included the making of pipe by the extrusion of molten metal chilled at the die. Very soon the demand for this pipe increased on account of the increasing use of gas. The first mention of iron tube for gas is in the patent of James Russell of Wednesbury, dated 19th January, 1824.[2] The accumulative effect of these advances was almost as great as from the more spectacular developments which had taken place in forge and rolling-mill.[3]

There were other correlative advances. Even before the introduction of gas for lighting streets, attempts had been made in the way of improving the light by adding reflectors, etc., to the lamps. One such reflector was patented by Lord Cochrane. Like so many others this patent was disputed and declared bad. The reflectors were made of tinplate and as such would have had a short life, as the reflecting surface would soon tarnish. A substitute more

[1] *Brief Remarks on the Existing Differences between the Gas Company and the Inhabitants of Newcastle-upon-Tyne*, Newcastle, 1818.

[2] R. B. Prosser, *op. cit.*, pp. 85, 87, 99.

[3] T. S. Ashton, *Iron and Steel in The Industrial Revolution*, p. 103.

efficient than tinplate was introduced by John Millington in the form of white porcelain reflectors made in the Staffordshire Potteries. By 1820 these had been adopted for street lighting in Bath.[1] It is strange that the Potteries supplied adjuncts to gas lighting before the glass industry, but it was a long time before gas had any effect on the latter. Powell says:

> It was not until 1879 that it was realised that shades for oil-lamps and gas-jets might be varied from the globular form and might be coloured. In that year the gas-burners of the Haymarket Theatre were screened with tinted glass shades, not globular in form.[2]

Development of gas lighting and the shocking increase in the number of explosions in coal-mines due to the demands which palaeotechnics put upon the miners, gave a new impetus to scientific investigation of flame and combustion. This was attacked by Sir Humphry Davy in a series of papers published in the *Philosophical Transactions of the Royal Society*. Out of these investigations there were developed 'principles applicable to the purposes of common life'. One in particular, the safety lamp for mines.

> Urged on by the heart-rending cry of suffering humanity, Science turned aside from her speculations, and after an examination of the nature of the enemy with which she had to contend, traced with laborious and often dangerous perseverance, its recondite principles, and at length presented to the astonished and grateful miner the ignited elements of explosion fluttering harmless in a wire cage. Never was there an invention better calculated to prove to the ignorant the connection between Science and the arts of life.[3]

So menacing was the death-roll from mine accidents that on 1st October, 1813, a '*Society for Preventing Accidents in Coal Mines*' was formed, the outcome of whose activities was the introduction of the miner's safety lamp. The names of the members who formed the original Society are given in the *Annals of Philosophy*.[4] Davy was by no means the only experimenter. The attempts by various scientists to find a satisfactory lamp are summarized by J. R. Morgan in the *Annals of Science*.[5]

An interesting drawing of Davy's lamp by Andrew Ure (1778–1857) in a contemporary letter in which he sent a description of the lamp to France, is reproduced in *Manchester Memoirs*.[6]

[1] *Quarterly Journal*, 1818, *5*, 177; 1821, *10*, 170.
[2] H. J. Powell, *Glass Making in England*, p. 133.
[3] *Quarterly Journal*, 1817, *3*, xii.
[4] *Annals of Philosophy*, 1814, *4*, 315.
[5] *Annals of Science*, 1936, *1*, 302.
[6] *Mem. Manch. Lit. and Phil. Soc.*, 1913, *52*, No. 19.

The principle of the miner's lamp also had other applications. In the *Edinburgh New Philosophical Journal* for 1840 there is a description of a burner, the forerunner of Bunsen's burner, in which the flame is prevented from striking back by having a layer of wire gauze placed over the upper end of the burner tube. Sir John Robison, who described the burner, used it in his own house, and all modern heating appliances work upon a somewhat modified extension of the same principle.

The use of gas for purposes other than illumination developed collaterally. Wm. Thos. Brande (1788–1866), chemist at the Royal Institution, suggested its adoption for heating. He carried out a series of experiments with raw materials other than coal, such as wood or paper. The gas from wood was found to burn with a hot, non-luminous flame which, though not then suitable as a source of light—the gas mantle not having been invented—could be used as an easily controlled source of heat. This further development did not take place till the monopolistic position achieved by the gas industry as a purveyor of artificial light was threatened by the emergent neotechnic source of light, heat, and power—electricity.[1] While individual gas plants were still supplying private consumers, John Maiben invented a gas fire in which gas flames played upon pieces of cast metal,[2] and in 1839 Alex. Cruickshanks suggested the substitution of '*asbestos*' balls. The same inventor proposed the production of incandescent light, making the gas flame heat a mantle of platinum gauze covered with lime, but as mentioned above it was not until electricity began to give indications of becoming a serious competitor that the gas industry stirred itself and heating and cooking by gas, as well as improved lighting, were introduced into the Victorian home.

The slow transition from eotechnic to palaeotechnic lighting, apart from heating, is well illustrated by the fact that the Great Exhibition of 1851 was not lit by gas, and it was still some time after that, before the new source of light which palaeotechnics made available became widely adopted.

Perhaps Jeremiahs were not without their influence. One, writing in the *New Times* of 20th May, 1819, adduced no less than seven reasons why the lighting of the streets was objectionable, varying through the theological, fiscal, medical, philosophical and moral, to the civic, economic, and patriotic. He sees the populace satiated with the perpetual circus of illuminated streets, and concludes: 'Let us be careful to preserve the empire of darkness'!

[1] *Quarterly Journal*, 1817, 7, 78.

[2] John Maiben, *A Statement of the Advantages to be Derived from Coal Gas.*

This is the more surprising considering the comments which were current in other directions regarding the social effects of the introduction of gas lighting. The *Westminster Review* of 1829 said:

> Old Murdoch alone, has suppressed more vice than the Suppression Society; and has been a greater police officer into the bargain than Old Colquhoun and Sir Richard Birnie united. It is not only that men are afraid to be wicked when its light is looking at them, but they are ashamed also: the reformation is applied to the right place.... Why was not this new light preached to them long ago: twenty bushels of it would have been of more value than as many chauldrons of sermons.[1]

Arthur Elton sums up: 'Of all single influences which have changed the home of the eighteenth century into the home of the nineteenth, and the home of the twentieth, gas has perhaps been the most important, yet Murdoch, Boulton and Watt, with all their inventive imagination and commercial sense, did not grasp the significance of the fuel which they did so much to perfect. The development of the gas industry as we know it to-day is not so much due to them as to such men as Winsor, in spite of his craziness, and to the elder Clegg.'

By 1819 two books exclusively devoted to manufacture and use of coal gas had been published, F. Accum's *Description of the Process of Manufacturing Coal Gas*, and T. S. Peckston's *The Theory and Practice of Gas Lighting*. These focused the attention of scientists on the new medium. Again quoting Elton:

> Accum's book is a scientific treatise outlining the chemistry of the coal gas process, describing gas manufacturing appliances in detail, and discussing experimental methods of determining the best shape of retorts and optimum carbonising periods and temperatures. Peckston is more interested in how best to maintain and run a gas works.[2]

For the first time, gases, other than air, had to be handled in large quantities.

One of the problems which faced the pioneer gas engineers was removal of extraneous products which on combustion of the gas produced undesirable by-products. At the Lyceum these difficulties proved insuperable, and the use of gas was given up.[3] The smell was very bad, and condensed water made the ladies' furs damp.[4] The same difficulties must have been experienced at the

[1] *Westminster Review*, 1829, *11*, 290.
[2] Private communication.
[3] *Annals of Philosophy*, 1816, *7*, 237.
[4] *Monthly Magazine*, London, 1805, *19*, 427.

Soho Works, for Murdoch wrote from Phillips and Lee on the 1st January, 1806, saying:

> There is no Soho stink has yet offended them. Mrs. and the Miss Lees have visited this night, and their delicate noses have not been offended.

Young Watt replied that he hoped Murdoch would continue to pursue the good opinion of the 'females'.

The first mention of a method of purifying gas is contained in Frederick Accum's *System of Theoretical and Practical Chemistry*, published in 1803. This suggests the use of lime water for the removal of carbon dioxide from gas obtained from the distillation of wood. Carbon dioxide is, however, only of minor importance as an impurity in coal gas; more objectionable are ammonia, compounds of sulphur, particularly hydrogen sulphide, and tar. In addition to removing carbon dioxide, lime will remove the compounds of sulphur.

Who introduced lime into industrial gas practice is a matter of doubt. Murdoch makes no specific claim to have done so, but it is probable that he effected by this means the purification of gas at Phillips and Lee. Initially the lime was simply added to the water in the gas holder. In an apparatus erected by Clegg for Harris of Coventry, a paddle was introduced at the bottom of the gas holder to agitate the lime. Clegg claimed to have introduced a separate liming apparatus at Stoneyhurst College in 1811. The purifier, which is illustrated by Accum in his *Treatise on Gas-Light*,[1] contains not only a compartment for a lime slurry, but one for another alkaline solution as well. This idea was not, however, operated for long.

In addition to Clegg, Winsor and one of his quondam employees claimed credit for the introduction of lime. Their idea was to heat lime along with the coal in the retort. Dr. William Henry also dealt with the use of lime in a paper communicated to the Royal Society by Sir Humphry Davy.[2]

Although subject to several modifications and improvements, for many years liming continued to be an essential process in the manufacture of gas. The first modification was the substitution of lime in a semi-solid state. This was followed in 1817 by a patent of Reuben Phillips for a process using dry lime. Ultimately hydrated iron oxide replaced lime because it removed sulphur in a form which functioned as the raw material in industries which worked up the by-products of gas manufacture.

[1] F. Accum, *Treatise on Gas Light*, London, 4th Edition, 1818, p. 95.
[2] *Phil. Trans.*, 1808, *98*, 282.

Murdoch himself is said to have devised a process of purification which completely removed objectionable products from gas, but unfortunately at the same time it removed the hydrocarbon constituents which were responsible for its illuminating power as burned in the simple burners of his time. Had it then been possible to develop incandescent illumination, Murdoch's method could have been adapted with success. Almost half a century passed, however, before the researches of Auer von Welsbach made incandescent lighting a practical proposition.[1]

While, as mentioned above, the water in the gas holder was used as a purifying fluid, it served a dual purpose. The gas first of all was conducted through a series of pipes immersed in the water of the holder, thus removing the readily condensing tar. As gas plants increased in size, this became impossible on account of the distance between the retorts and the holders. To effect a separation of tarry products, Samuel Clegg introduced at Greenway's Mile, Manchester, the '*hydraulic main*'. When further air condensors were added between the retorts and the purifying apparatus, gas plant began to take on the shape which persisted throughout most of the nineteenth century.

Many of the improved methods of gas manufacture were first put into operation in Scotland. Right from its inception, Glasgow Gas Works has occupied an important place in relation to the application of scientific control to manufacturing processes, and on account of this has formed the focus from which several important developments radiated. The influence of scientific control in the manufacture of gas is evidenced by the following statement made by Andrew Ure, Professor of Chemistry in the Anderson Institution:

> In the Glasgow coal gas establishment, which is conducted by engineers skilled in the principles of chemistry and mechanics, fully 4 cubic feet of gas are extracted from every pound of coal of the splend kind in 4 hour charges, from retorts containing each 120 lbs.; which is about two-thirds of their capacity. The decomposing heat is much the same as that used in London, but the retorts are compressed cylinders, a little concave below. Hence in 8 hours, fully double the London quantity of gas, is obtained from a retort in Glasgow.

The Company had 152 of these retorts, with storage capacity capable of taking 6,000 tons of coal, the whole establishment covering an area of some 15,000 square feet.[2] With regard to the quality of the gas which was produced, Ure continued:

[1] A. M(urdoch), *op. cit.*, p. 27. [2] *New Statistical Account*, Vol. 6, p. 163.

> An ingenious pupil of mine [probably Neilson], lately employed to visit the principal factories of gas in England, made a series of accurate experiments on its illuminating quality in the different towns. . . He found that the average illuminating power of the gas in the English establishments, was to that of the Glasgow Company, as four to five; the worst being so low as three to five. . . If we therefore multiply this ratio, into Glasgow gas-work, we shall have the proportion of light generated here, and in London, from an equal-sized retort, in an equal time, as 100 to 40.

Purity was also maintained at a high level. 'As at other gas establishments, the gas is purified with lime; but in addition to this process, it is made to pass through a solution of sulphate of iron, by which it is very much improved in quality.'[1] Accum mentions this reagent, and also lead acetate, but does not indicate whether they had been tried on a large scale.[2]

Among the improvements introduced at Glasgow by J. B. Neilson, mention is made of the purification of the gas from traces of tar and oil by passing it over charcoal (adsorption), the addition of iron sulphate as a purifying agent, and, of particular importance, the introduction of fireclay retorts. His burning of tar in the furnaces will be referred to later.

It was not long before the gas industry reacted upon the iron industry in the west of Scotland. In 1835 Andrew Liddle introduced the manufacture of butt-welded iron tubes, for gas and water, at the Globe Foundry.

A short history of the Glasgow Gas Company was published by the directors in May 1835, in which are given details of the charges made by the Company. In 1818, the period at which the lighting of the city commenced, the charge for a single jet to eight o'clock was 12*s*. per annum, a charge which was reduced by almost half in twenty years. Higher charges were made for jets which were required to burn longer, and the officials of the Gas Company were liable to make surprise visits to ascertain whether subscribers were adhering to their contracts. According to the *New Statistical Account* the undertaking, from the beginning, supplied the city and suburbs of Glasgow with gas at prices below what was charged in any other city in the Empire.[3]

The disposal of by-products incidental to the manufacture of coal was a perplexing problem to the early gas companies. To begin with, tar and ammonia were cast into the Thames at London, the Forth at Edinburgh, and into disused coal-pits at

[1] *New Statistical Account*, Vol. 6, p. 164. [2] Accum, *op. cit.*, p. 140.
[3] *New Statistical Account*, Vol. 6, p. 165.

Glasgow. The Glasgow Company, however, were soon fortunate in finding a purchaser for their tar and ammonia. With a view to substituting ammonia for urine previously used in the manufacture of cudbear, Charles Macintosh in 1819 entered into a contract with the proprietors of the Glasgow Gas Works to purchase their tar and ammonia. While he had an immediate use for the ammonia and pitch derived from the tar, he was left with a residue of low-boiling naphtha, till the idea of using it as a solvent for caoutchouc occurred to him. His idea proved practicable, and he was granted a patent for using the resultant varnish in the manufacture of waterproofed textiles. The patent was initially put into operation in Glasgow, but subsequent developments took place in Manchester, where in partnership with Birley (an H. H. Birley was a director of the Manchester Gas Co.) Macintosh formed a Rubber Company under the name of Charles Macintosh and Company. Another of Macintosh's patents was for a furnace designed to burn by-product tar from the works under the retorts. This was adopted in practically every gas works in Great Britain, and saved large sums of money which otherwise would have been spent on fuel. As palaeotechnics developed, it was realized that coal tar was too valuable as chemical raw material to burn, and the practice fell into disuse.[1]

The Glasgow company were also fortunate in their personnel, one of whom was the celebrated James B. Neilson. 'To the scientific attainments of this distinguished manager,' says the *New Statistical Account*, 'the Company are chiefly indebted for their uncommon success, and for the most perfect and beautiful establishment of the kind in the kingdom.'[2] Neilson improved upon William Murdoch's gas burners by devising the '*batswing*' burner, so, till the introduction of the gas mantle many decades later, the world was indebted to two Scotsmen, Murdoch and Neilson, for the advance which had been made in artificial lighting.

Neilson's principal discovery is associated with the smelting of iron; but even that arose out of his first-hand knowledge of gas-works' practice. In 1824 an iron-master enquired of him whether he thought it would be possible to purify air used in blast furnaces in a manner similar to that used for coal gas. This directed his attention to iron-making and with the advice and assistance of Charles Macintosh he invented and patented in 1829 the '*hot-blast*' process, an improvement of the utmost economic importance to Scotland.

While Murdoch was experimenting with gas for lighting, and

[1] G. Macintosh, *Memoir*, p. 86. [2] *New Statistical Account*, Vol. 6, p. 165.

Dundonald was extracting tar for the preservation of wood, other experimenters were exploring the possibilities of using the explosion of 'spirits of tar', or gas, for doing mechanical work. Of these efforts J. W. Young says that in 1791:

> John Barber, an Englishman, obtained a patent No. 1833, for 'An Engine for using Inflammable Air for the purpose of procuring motion'. Barber calls for special mention as his patent covers the method being used to-day by almost all of the experimenters working on a problem of the internal combustion turbine. His proposal was briefly as follows: coal, wood, oil, or other combustible substance, was gasified in a retort, the gas generated being cooled in a water jacketted receiver, from which it was drawn by means of a pump and forced into the combustion chamber, termed by Barber, an 'exploder'.[1]

The next development was a patent of Robert Street in 1794, for an engine to operate by the explosion of 'spirits of tar or turpentine'. His design, however, seems to have been very primitive, an operator being required to fire the charge by hand at each stroke. The first successful engine for using 'the vapour of alcohol or other ardent spirits, oil of turpentine, the essential oils, naphtha and all other articles capable of producing inflammable steam or vapour' was patented by Samuel Brown between 1823 and 1826. Several engines built to the design of this inventor were put upon the market.

According to the above paper by Young, 'The *Mechanics' Magazine* (1824–8) contains a number of articles on Brown's engine, including an account of the trial of a marine engine of his make which successfully propelled a boat on the Thames, and also of another fitted to a road carriage.' The impediments to the adoption of an engine of this kind were not unreliability, but economic, as the cost of the fuel was high when compared with cheap coal used in Watt's steam engine. Nevertheless Brown's work was of fundamental importance as foreshadowing the emergence, at a much later date, of a neotechnic economy which might render the steam engine obsolete. According to Young, it is not without justice that Brown has been designated 'The Newcomen of the Gas Engine'.

The gas industry is the direct antecedent of a large section of the chemical industry whose products are now essential to civilization. The initial discovery of some of these is described by Stephen Miall:

> The gas industry was started here about the year 1812, and in 1819 the English chemist, Alexander Garden, discovered the solid

[1] J. M. Young, *Trans. Newcomen Soc.*, 1936, *17*, 109.

substance naphthalene in some of the products of the distillation of coal. In 1825, Faraday discovered benzene in gas prepared from whale oil by the Portable Gas Co. He called this new substance bicarburet of hydrogen; it was subsequently named benzine by Mitscherlich in 1833, benzol by Liebig in 1835, and phène by Laurent in 1834, and from it have been prepared many thousands of compounds, some of which are of the greatest value. In 1826 the chemist Unverdorben obtained the liquid aniline by the destructive distillation of indigo, and shortly afterwards it was found that aniline could readily be prepared from nitrobenzene. In the year 1832, Dumas and Laurent discovered anthracene in the residues from the distillation of coal, and coal-tars were closely investigated by Runge in 1834 and by Leigh in 1842, and by Hofmann in Germany in 1843 and 1844.

It is said that Leigh first discovered benzene in coal-tar in 1842, but it was Hofmann who first proved its occurrence conclusively and made public this fact in 1845. According to Lunge, 'The process for obtaining it in any quantity from coal-tar was worked out in his laboratory by one of his pupils, Charles Mansfield, who carried out the process on the large scale and minutely described the principle of dephlegmation for separating the various hydrocarbons; he distinctly pointed out that the apparatus employed in rectifying spirit might be employed for this purpose, even with greater advantage than for spirit of wine itself—a suggestion usually, but erroneously, ascribed to E. Kopp, who first mentions it in 1860'.[1]

[1] S. Miall, *History of British Chemical Industry*, p. 67 *et seq.*

91. John Walker's house, Stockton-on-Tees, where non-phosphoric 'friction lights' were invented about 1827.

92. Apparatus in which glue, phosphorus, etc., were mixed for the preparation of match-heads.

93. Examples of early matches: Jones' *Lucifers* (1829) and *Congreves* (*c.* 1830). *Lucifers* were an extension of Walker's idea with the addition of sulphur. *Congreves* were a phosphorus-containing match developed by Dr. Charles Sauria.

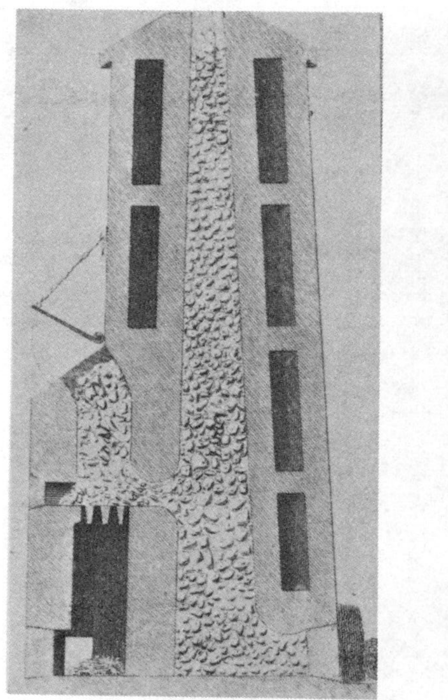

94. Count Rumford's lime-kiln, one of the many designed during the era of rapid agricultural improvement.

T. H. Croker, *Dictionary of Arts and Sciences*, 1764–6

95. Boiler of early steam engine, illustrating the similarity with the brewer's copper from which it was derived (*cf.* Fig. 103).

Chapter XX
INSTANTANEOUS LIGHTS

The progress of civilization, with its necessary increase of human wants, compelled man to invent means for their gratification.

LYON PLAYFAIR.

IT is significant to observe that the discovery of coal gas by Dr. Clayton and the discovery of phosphorus, a substance which could be used to provide an instantaneous light for it, were both made in the seventeenth century. In the social context of seventeenth-century England, neither discovery took root. Gas lighting had to wait for William Murdock and the rise of the cotton-mills, and it is perhaps not unconnected that the correlative need for a simple means of producing fire only became pressing during the same period.

> It is related that when No. 1 'Locomotion' was first placed on the rails at Aycliffe Level for a trial trip, her boiler filled with water, and wood and coal made ready for lighting, it was discovered that no one had a light. George Stephenson was just on the point of dispatching a man to Aycliffe for a lighted lantern, when a navvy stepped forward and presented a burning glass, said he often lit his pipe by its aid, and perhaps it might fire the engine. The glass was tried and found successful, and Robert Metcalf it was, who, with a common sun-glass and a piece of tarred oakum, actually lit the fire of 'Locomotion'.[1]

In 1673, Brand of Hamburg discovered the element phosphorus by distilling urine with sand. This discovery revealed the existence of a substance which numbered among its socially useful properties that of extremely easy ignition. That such a substance might supplement flint and steel for the production of fire is immediately apparent, but more than a century of development among technics already discussed had to take place before the problem of instantaneous fire-production was successfully solved.

In the first place, a source of readily available phosphorus was necessary. Now, in the middle of the eighteenth century, a phosphorus compound, viz. bone-ash, or calcium phosphate ($Ca_3(PO_4)_2$), had become an industrial raw material, and was

[1] M. Heavisides, *History of the Invention of the Lucifer Match*, p. 10.

LIGHTING DEVICES

(Matches, etc.)

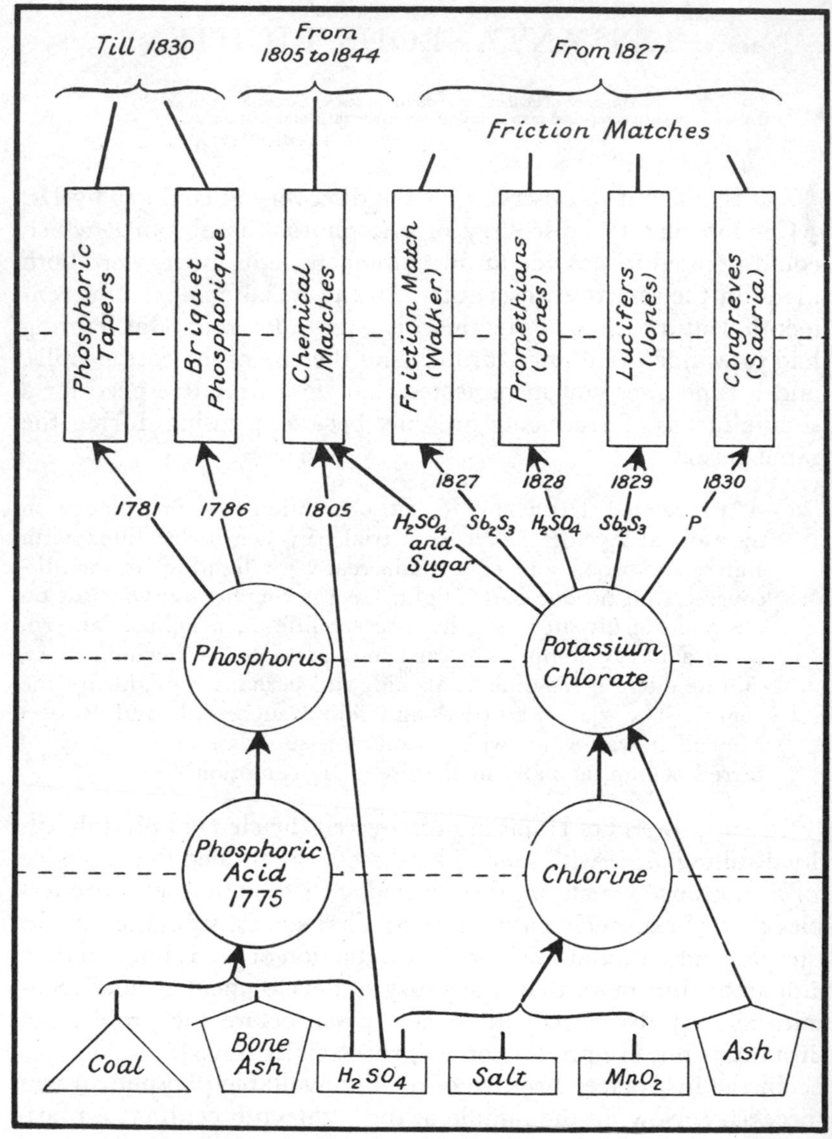

being used as an ingredient in English bone china at all English porcelain works by 1795.

In 1775, C. W. Scheele (1742–86) prepared phosphorus from bone-ash for the first time through the agency of another substance which by that date was available in commercial quantities, viz. sulphuric acid (H_2SO_4).

$$Ca_3(PO_4)_2 + 3H_2SO_4 = 2H_3PO_4 + 3CaSO_4$$

bone-ash + vitriol = ortho-phosphoric acid + calcium sulphate

When orthophosphoric acid so produced is heated with charcoal, metaphosphoric acid is first produced. This in turn is reduced to phosphorus according to the following equations:

$$H_3PO_4 = HPO_3 + H_2O$$

ortho-phosphoric = meta-phosphoric + water

$$4HPO_3 + 12C = P_4 + 2H_2 + 12CO$$

Shortly after these laboratory advances were made, experiments with phosphorus as a means of instantaneous combustion were started. A *phosphoric taper* appeared in 1781, and in 1786 in Paris a *briquet phosphorique* consisting of sulphur-tipped wood *spunks* and a bottle coated internally with phosphorus. The coating of the bottle is described in Nicholson's *Dictionary of Chemistry*. As kindling devices these were neither satisfactory nor safe. Robert Pearse Gillies (1788–1858) in his *Memoirs of a Literary Veteran*, says:

> Our most ingenious artizans and chemists had laboured to discover convenient means for causing instantaneous light, but in vain! I had once a round morooco leather case, bearing date 1788 and price £1 18*s*. 6*d*., containing a phosphorous [*sic*] bottle, a piece of wax candle, and other adjuncts; also a printed parchment of directions, on the back of which the disappointed purchaser has scrawled: 'A damned catchpenny! Tried fifty times, but never once could light the candle'.[1]

By the beginning of the nineteenth century, however, devices based on phosphorus had increased considerably in number, and they continued to be the principal instantaneous light devices till *c*. 1830. A box containing phosphorus and sticks dipped in sulphur is mentioned by Victor Hugo in *Les Misérables*. We have the record of a 'phosphoric light' being used by M. Testu of Paris, in the account of a remarkable aeronautical ascent which he made on the 18th June, 1786:

> In this region (3,000′), the voyager sailed till half past nine o'clock, at which time he observed from his 'watch tower in the

[1] Quoted by J. G. Fyfe, *Scottish Diaries and Memoirs 1746–1843*, p. 426.

> sky' the final setting of the sun. He was now quickly involved in darkness, and enveloped in the thickest mass of thunder-clouds. The lightenings flashed on all sides, and the loud claps were incessant. The thermometer, seen by the help of a phosphorus light which he struck, pointed at 21°, and snow and sleet fell copiously around him. In this most tremendous situation the intrepid adventurer remained the space of three hours, the time during which the storm lasted.

It need only be remarked that his balloon was of his own construction, made of glazed tiffany and filled with hydrogen![1]

The next step towards the solution of the problem centres round Berthollet's observation of the oxidizing power of potassium chlorate ($KClO_3$) on organic material (e.g. sugar) in the presence of an acid (H_2SO_4).

This reaction enabled Chancel of Paris to prepare in Thénard's laboratory in 1805, *oxymuriate or chemical matches*.[2] The head of these matches consisted of a mixture of six parts of potassium chlorate, two of sugar, and one of gum as a binding agent. The match was ignited by contact with sulphuric acid absorbed on asbestos in a small bottle. By 1812 there was a considerable sale to the well-to-do of these devices, which continued in use till 1844. Many of the boxes were stamped 'Berry's Patent'. An example may be seen in the Royal Scottish Museum, Edinburgh. The device, like its predecessor, had many shortcomings, e.g. the sulphuric acid was hygroscopic, but over the same period a *friction match* was being evolved.

Friction matches containing phosphorus are said to have been made in Paris from 1805–9. Derepas of Paris obtained a patent for a match composition rendered more manageable than phosphorus alone by the addition of magnesia (MgO). These were not particularly successful, and moreover, in view of the toxicity of white phosphorus, represent a socially retrograde step.

Although there are several claimants to the honour, there is little doubt that successful non-phosphoric 'friction-lights' are due to John Walker (*d.* 1859) of Stockton-on-Tees. Walker was an apothecary who was much interested in chemical experiment. At some date prior to 1826 he made a *percussion powder* consisting of a mixture of potassium chlorate ($KClO_3$) and antimony sulphide (Sb_2S_3). It then occurred to him to mould this powder into match-heads. These matches were sold locally in tin boxes at 1*s*. per 100

[1] *Encyclopaedia Britannica*, Supplement IV–VI Editions, Vol. I, p. 79.

[2] Chapoy, *L'Invention des Allumettes Chimiques;* W. Maigne, *Briquets, Allumettes*, etc.

plus 2*d*. for the box, and were struck by being drawn across glass paper. Walker's matches are often referred to as *lucifer* matches, but these were due to Samuel Jones, and come later. This misreference, and one to phosphorus which is also incorrect, occurs in the following citation:

> The inventor of the lucifer match certainly deserves well of his country, and his country would probably before now have recognized its indebtedness in this matter of striking a light if there had not been more than one claimant to the honour of being the first matchmaker. It has at last been proved by documentary evidence that the first maker of matches was John Walker, a chemist of Stockton-on-Tees. The matches soon became popular, and the people came from long distances to buy them. The poor of the town were employed to split the wood for these early matches, which were invariably dipped in the phosphorus compound [*sic*] by the inventor himself. This worthy man was pressed to form a company to work the invention and to patent it; but he refused, on the score that he had enough for his simple wants, and that he would put no obstacles in the way of a thing which promised to be a boon to the public.[1]

Walker's matches were, of course, non-phosphoric, and as such the lineal ancestors of twentieth-century matches. His day-book is preserved at Stockton, and records perhaps his first sale, on 7th April, 1827.

The idea of igniting a chlorate-containing match-head by reaction with sulphuric acid was not, however, immediately abandoned. In 1828 Samuel Jones patented his *prometheans*, which had a head of potassium chlorate, lycopodium, and sulphur, containing a small capillary filled with sulphuric acid coloured with indigo. To ignite the match, the capillary was crushed. In 1829 the same Jones was also selling *lucifers*, an adaptation and extension of Walker's principle with added sulphur. The name lucifer was never applied by Walker to his friction-lights, nor were his matches sulphur-coated. Other early retailers of lucifers were Richard Bell and Company of Wandsworth, and G. F. Watts of the Strand, who retailed Watts' Chlorate Matches.

The reign of the lucifer match was short since the solution of the friction match by using potassium chlorate again stimulated experiment with phosphorus. Replacement of antimony sulphide by phosphorus in friction matches took place at several places in the years 1830–4, and the resulting phosphorus matches came on the market as *Congreves*, although they were not invented by Sir

[1] *Chambers's Journal*, 1894, 5th Series, Vol. 11, p. 827.

William Congreve (1772–1828) of war-rocket fame, as is not infrequently stated. Their invention is usually credited to Dr. Charles Sauria of St. Lothair (*d.* 1895) who, in 1830, while a student at Collège d'Arc, Dole (Jura), made serviceable phosphorus friction matches. The claims of the other inventors need not be examined here.[1]

Improvement effected by introduction of phosphorus was offset by its poisonous nature, which leads to necrosis of the mandible, popularly known as *phossy jaw*. This distressing condition was first described in 1845 by Lorinser of Vienna. In the same year, Anton v. Schrötter, also of Vienna, discovered the non-toxic allotrope—red phosphorus.

Five years later non-poisonous allotropic red phosphorus was introduced into match manufacture, but complete prohibition of the use of white phosphorus was not effected till after the lapse of more than half a century, since matches made with it were of the strike-anywhere variety.

Safety-matches were simultaneously introduced, since red phosphorus could only be incorporated in the rubbing-surface on the box and so oxidizing agent was automatically separated from the oxidizable phosphorus. A British patent was taken out for their manufacture by Messrs. Albright and Wilson, then of Birmingham.[2] Phosphorus sequi-sulphide gives a non-toxic strike-anywhere match. It was introduced by Messrs. Sevene and Cahen in 1898.

Knowledge of the composition of bone-ash led to correlative developments in quite a different field, the foundation of the modern view on the cause of rickets which had greatly increased in incidence with the rise in prices of protective foods during the industrial revolution. According to Sir William Fordyce, writing in 1773:

> . . . there must be very near twenty thousand children in London and Westminster, and their suburbs, ill at the moment of the Hectic fever, attended with tun-bellies, swelled wrists and ancles, or crooked limbs.[3]

When it became known that the greater part of bone mineral was calcium phosphate and that this could be dissolved in acid, *Citoyen* Bonhomme presented a paper to the *Société Nationale de Médecine* connecting the softening of bones with development of

[1] J. R. Allen, *Proc. Roy. Soc. Antiq. Scotland*, 1879–80, 2, 253.

[2] C. S. Miall, *History of British Chemical Industry*, p. 248.

[3] William Fordyce, *A New Inquiry into the Causes of Fevers*, cited in J. C. Drummond, and A. Wilbraham, *The Englishman's Food*, p. 323.

'acescence' arising from improper feeding.[1] Although his hypothesis was wrong, he came near the mark by suggesting that a phosphate deficiency might be the cause of rickets, this conception being based on the experimental feeding of chickens on a diet with added calcium phosphate. Other of his studies were concerned with the calcium need during lactation. These remarkably advanced researches were introduced to the British public by William Nicholson in the first number of the *Journal of Natural Philosophy, Chemistry and the Arts*,[2] but unhappily it was many generations before the discovery of the vitamins led to the suppression of the disease that had earned for itself the unenviable title of 'the English ill'.[3]

[1] Bonhomme, *Annales de Chimie*, 1793, *18*, 113.
[2] *Journ. of Nat. Phil., etc.*, 1794, *1*, 174.
[3] J. C. Drummond and A. Wilbraham, *op. cit.*, p. 324.

Chapter XXI

FARM FACTORIES AND THE ECONOMY OF VEGETATION

PART ONE

> It has been said, and undoubtedly with great truth, that a philosophical chemist would most probably make a very unprofitable business of farming; and this certainly would be the case, if he were a mere philosophical chemist.
>
> H. DAVY.

THE four chapters which follow are concerned with agriculture and agricultural products. While chemistry undoubtedly contributed to the progress of the industrial revolution, there almost always having been a direct chemical parallel to related mechanical innovation, as far as the contemporary agrarian revolution is concerned, on cursory examination the contribution made by chemistry may seem open to question. If by the agrarian revolution we mean changing conditions in the production of food which arose through enclosure, drainage, improvement in leases and transport, coupled with such empirical practice, say, as marling, liming, fallowing, rotation, and the introduction of new crops, in fact all the attributes of capitalist farming with its various implications, then there is indeed considerable difficulty in sustaining the thesis that chemistry made any major contribution to these changes. If, on the other hand, we contrast the condition of agriculture in 1850 with that say a century earlier, then in perspective one sees at once that a radical advance was made during those years in the understanding of the more complicated problem of plant physiology, and that the basis had been laid of what we may at least call a quasi-scientific agriculture. The gross nutritional requirements of vegetation had been investigated, and the results so obtained established the rationale of fertilization with substances made available through advancing chemical technology. The same is true of initially empirical procedures such as the rotation of crops.

It was our primary intention in this part to deal with each of these as a specific study in itself, but, in the words of R. E. Prothero (Lord Ernle):

> . . . to trace in full detail one single point in which science has helped farmers would be the work of a separate volume.[1]

This having become amply evident, and in view of the slow change in agrarian background which took place in different localities at different times, we have reluctantly been compelled to content ourselves with an investigation of the gradual penetration of agriculture by scientific observation and experiment.

The need for a sound utilization of land was very great, e.g. the seventeenth century in Scotland ended with a series of bad seasons. The weather destroyed crops and stock alike. Then two years after the Union of 1707, there was famine throughout the land. Still another dearth followed in 1740. Just how great was the need for improved husbandry will be appreciated from the following brief summary of food supply and price conditions during the subsequent years of the eighteenth century.

1750	Prices rising.
1762	Grave dearth.
1772	Suspension of the duty on imported corn.
1776	Quartern loaf reaches unprecedented high price.
1778–9	Good harvests offset by the fall in value of money due to the discoveries of silver in America.
1793	Poor harvest and acute financial crisis.

These events naturally had an effect on agriculture. And so it continued. Here is what Mackinnon says in his *Social and Industrial History of Scotland*:

> A great impulse was given to Scottish agriculture by the French Revolution war. In 1795 the price of wheat rose from 50*s*. to 81*s*. 6*d*. per quarter, and in the following year to 96*s*. In 1812 it attained the record of 126*s*. 6*d*. A large amount of waste land passed under cultivation, and the rapid progress between 1795 and 1815 is apparent from the fact that the rent derived from agricultural land in Scotland rose from 2 millions to 5¼ millions sterling during these twenty years.[2]

But the country was never very far from famine. That was so in 1812. While the next three years saw bumper crops, and 1815 the introduction of Corn Laws to placate the protesting farmers, short harvests once again became the order of the day and from 1817–19 there was dearth, distress and rioting. Society was in a state of flux, and any convenient scapegoat was sought for and blamed as the source of the distress. The prevailing uncertainty

[1] R. E. Prothero, *English Farming, Past and Present*, p. 364.
[2] J. Mackinnon, *Social and Industrial History of Scotland*, p. 59.

is revealed in the following citation dating from the beginning of the nineteenth century.

> Much controversy has taken place among political economists, concerning the effect of manufactures upon agriculture. One sect maintains, that were the labour employed in manufactures exerted wholly in the cultivation of the soil, a much greater, and a more permanent amount of national wealth would be produced, than what arises from manufactures. They insist, that manufactures raise the wages of labour to an exorbitant pitch, and thus put it out of the farmer's power to improve his land, be he ever so much inclined.[1]

But consideration of such controversial topics must be deferred till we have assessed the contribution that chemistry made to practical agriculture during the period in which a manufacturing economy was established in Britain.

Agriculture, according to Justus Liebig, is both a science and an art. Knowledge of the life of vegetables, the origin of their elements, and the sources of their nourishment forms its scientific basis. The significant part which science plays in agriculture was early foreseen; for example, it was evident to the founders of the Royal Society. In 1664 they appointed a georgical committee of thirty-two members. The committee drew up a list of possible subjects for investigation which, under the title of *Heads of Enquiries*, was published in the *Philosophical Transactions*.[2]

As R. Lennard puts it:

> That the Royal Society made the attempt to collect descriptions of agricultural practices from all parts of the British Isles is in itself a significant fact both in agricultural history and in the history of English Science. It shows us how interested the scientific researchers of those days were in matters of practical utility, and that a brave attempt was made to link up book learning and scientific research with the experience of practical farmers. The scheme reveals an appreciation of the value of the comparative method which anticipates by more than a century the triumphant vindication of the method accomplished by Arthur Young and William Marshall, and in the Agricultural Surveys of 1793–95 and their successors.[3]

While the Royal Society scheme unfortunately did not materialize to the extent initially hoped, it did accomplish some useful work. It reveals, for example, that in the late seventeenth century

[1] J. Headrick, *General View of the Agriculture of Angus*, p. 547.
[2] *Phil. Trans.*, Vol. 1, pp. 91–4.
[3] R. Lennard, *Econ. Hist. Rev.* 1932–4, *4*, 23.

manures already included dung, lime, and marl, besides 'cole-ashes', 'soape-ashes', woollen rags, and sea sand. At that date seaweed was seldom used, despite its importance later on. Another point is significant. Already there were indications of doubts about the rationale of manuring practice—the start of controversy which lasted till the middle of the nineteenth century, only to be resolved by a proper understanding of plant physiology.[1]

In the eotechnic age it is not surprising that controversy should arise regarding manure, because the science of plant nutrition was still in the embryonic stage of its development. In the *Advancement of Learning*, Francis Bacon (1561–1626) stated his belief that water constituted the principal nourishment of green plants. This was not, however, based in any way on experimental observation. One of the first to experiment on plant growth was J. B. van Helmont (1577–1644), and unfortunately for the progress of agriculture, his results appeared to confirm the conjectures of Bacon. Similarly, they were acceded to by Robert Boyle in *The Sceptical Chymist* (1661). In these limited researches, two factors of prime importance were completely overlooked by the pioneer investigators: (*a*) the part contributed by the atmosphere to plant growth, and (*b*) the absorption of inorganic constituents (salts) from soil during the development of vegetation.

About Bacon's time a continental chemist, J. R. Glauber, did indeed discover in the course of other researches an important principle of vegetation, namely saltpetre (potassium nitrate, KNO_3). Russell tells us that:

> Having obtained saltpetre from the earth cleared out from cattle sheds, he argued that it must come from the urine or droppings of the animals and must, therefore, be contained in the animals' food, i.e. in plants. He also found that additions of saltpetre to the soil produced enormous increases in crop. He connected these two observations and supposed that saltpetre is the essential principle of vegetation.[2]

On the social context of Glauber's discovery, Lancelot Hogben makes the observation:

> This discovery which, like so many others in the history of science, was partly prompted by the search for new ways of destroying human life, led him to another which proved to be the means of conserving it. . . . Then Mayow (1674) in England seized on this discovery in the spirit of the new programme of

[1] Thomas Birch, *History of the Royal Society*, Vol. 1, p. 407.
[2] E. J. Russell, *Soil Conditions and Plant Growth*, p. 3.

> 'Heads of Enquiries', estimated the nitre content of the soil at different seasons by such methods as were then known, and showed that it is highest in spring when the crops begin to grow.[1]

About the same time, Nehemiah Grew published his first essay, *The Anatomy of Vegetables Begun* (1672), a further step towards the foundation of scientific agricultural chemistry. He examined the ash of plants by microscopical and chemical methods, and was able to demonstrate that salt solutions actually entered the soil and were subsequently absorbed by the plant rootlets.[2]

Before any further advance was made in the knowledge of plant nutrients, the first works on husbandry had appeared. The earliest Scottish work on agriculture was James Donaldson's *Husbandry Anatomised* (1697). It did not contribute any new ideas to husbandry, which is not surprising. *The Country Man's Rudiments or Advice to the Farmers of East Lothian* followed in 1699. It describes the backward condition of Scottish agriculture, with the infield and outfield system still in operation under which a return of three times the seed sown was considered a satisfactory yield. Farmers of Scotland were advised to conserve farmyard manure, to lime, and to fallow, and we know that summer fallowing was taken up in Scotland in the early 1700s by John Walker, of Beanston, in East Lothian. No mention is made of the beneficial effects of clover.

Contemporary with this publication, the part played by the mineral constituents of the soil was taken into account for the first time by John Woodward, who thus further advanced the work initiated by van Helmont and Boyle. Woodward's experiments are of classic importance because they extended the quantitative method of science to cover more completely the range of possible influential factors in the study of plant growth.

The importance of agricultural science was in fact already apparent to champions of improved husbandry, e.g. an attempt to publish a scientific agricultural paper for farmers was made by John Houghton, F.R.S. (1640–1705), when he issued his *Collections of Letters for the Improvement of Husbandry and Trade*. They appeared in two series, the first at approximately monthly intervals from 1681 to 1683, the second at weekly intervals for about ten years from March 1692.[3]

One other event of importance took place before the centre of

[1] L. Hogben, *Science for the Citizen*, p. 892.

[2] For biographical notice, *see Nature*, 26th May, 1941.

[3] R. E. Prothero, *op. cit.*, p. 133; G.E. Russell, *John Houghton, F.R.S.*, Notes and Queries, 1925, May 16.

gravity of agricultural research moved to Scotland. This was the publication in 1727 by Stephen Hales (1671–1761) of his *Vegetable Staticks*, which earned for its author the claim to be the founder of vegetable physiology.

> With the new air pump of Hooke and Boyle, the English physician Hales showed that air as well as salts and water is necessary for plant growth. In the same social context one of the predominant foci of enquiry was the nature of fluid pressure and flow. These (his experiments) showed that there is a continuous flow of water from the roots to the leaves. By fixing a pressure gauge to the cut end of the stem of a plant in moist soil or water, Hales measured the actual 'root pressure' of the ascending sap.[1]

Husbandry at the beginning of the eighteenth century was based upon relatively few empirical observations.

> It was one of the earliest discoveries in farming that one piece of land would grow better crops than another. It was also found that if the crops were grown upon, and removed from, a particular field, the productivity of the soil declined and the yields grew progressively worse. . . . A piece of land was cultivated until it was exhausted, when the farmer left it to revert to nature and started afresh elsewhere. . . . This was the system followed upon the outfields of Scottish farms till far on in the eighteenth century.[2]

As operated in Scotland it has been studied in detail by Henry Hamilton.[3]

Land was divided into two parts, *infield* and *outfield*. The infield was further subdivided, but only one subdivision received an annual dressing of manure. This single manuring provided the sole nourishment for three successive crops, probably oats, pease, and bear. Parts of the outfield, which made up perhaps four-fifths of the total arable land, were cropped continuously till the yield of grain no longer met the cost of seed and culture. When this happened, that part of the outfield was abandoned and cultivation shifted to another part, which was in turn subjected to the same continuous cropping till it also failed. Occasionally parts of the outfield were manured by confining cattle on it within turf walls, but on the other hand it was not unknown for as many as eighteen crops to be taken off a piece of land without its ever receiving a single load of manure. Henry Home, Lord Kames (1698–1782), mentions a field of nine or ten acres 'still most extraordinary' on

[1] L. Hogben, *op. cit.*, p. 893. [2] Watson and Hobbs, *Great Farmers*, p. 73.
[3] H. Hamilton, *The Industrial Revolution in Scotland*, p. 13. *See also* Lord Dundonald's *Treatise on Agriculture, etc.*, p. 172.

the river Carron in Stirlingshire. He saw on it a good crop of almost ripe oats, and was informed that it was the hundred and third crop of oats without intermission, and without manure, as far as was known. This was doubtless exceptional.[1]

A second feature of Scottish agricultural practice was the cultivation of arable land in rigs and ridges, a primitive way of effecting a certain amount of draining. Ridges were up to sixty feet wide and three feet higher at the centre than at the side. Exception need not be taken to this, but a complicated system of land tenure resulted, because neighbouring ridges were farmed by different cultivators, and were not held two years in succession. The appearance of the land under this type of cultivation was described by Dr. George Skene Keith in his *General View of the Agriculture of Aberdeenshire*.

> The old croft land, near the farm-steds, which has been long in culture, and is provincially called *infield*, from the quantity of animal manure, or rotted vegetables, has acquired a black colour, and may be called a sandy loam, in which the sand is more or less apparent, according as the land has been well or ill manured, or has been gently or severely cropped. The *outfields*, as they are called, on which black cattle were folded, and of which the soil was occasionally manured by the dung of these animals, have not acquired the black colour to the eye, nor the unctuous touch, which distinguish the kind that has been long under culture. Therefore a greater proportion of sand appears in them. And ever since the introduction of lime, their colour is not so black as that of the old croft land.[2]

To modern eyes such an arrangement seems curious, but Lord Dundonald explained the economic and social reasons for the development.

> The former system of management, however much it may now be disapproved of, and made the subject of reproach to our ancestors, was nevertheless the only one suited to their means. At a remote period, there were neither a sufficient number of men, horses, nor extent of capital, in Scotland, to admit the adoption of a different system; and probably at a still more distant period, causes equally efficient operated in England against improving the whole of the surface. Our forefathers, therefore, acted more wisely, by the application of the whole of the dung, to bring part of the land to a high state of cultivation, than if they had divided or applied it over the whole extent of the farms they cultivated,

[1] Henry Home, Lord Kames, *The Gentleman Farmer*, p. 325.
[2] G. S. Keith, *General View of the Agriculture of Aberdeenshire*, p. 47.

whence little benefit could have accrued to themselves or to posterity; whereas, by bringing certain portions only into a high state of fertility, a stock of materials has been accumulated, and left for their descendants to work upon, capable of repaying to the exhausted outfield lands, with abundant interest, the vegetable matters originally borrowed.[1]

The agrarian revolution saw an end of division into infield and outfield, and of the run-ridge system. Intermingled tenancy disappeared, though not necessarily the physical ridges. Shortage of labour continued, however, till the beginning of the nineteenth century. Robert Kerr, in his *General View of the Agriculture of Berwick*, says:

Perhaps one of the greatest inconveniences which agriculture experiences at present in Berwickshire, is owing to the want of sufficient population, for carrying on its operations, under the improved system of husbandry. This evil will naturally cure itself in time, by the encouragement which ample wages afford to population and marriage.[2]

Such was the system that was swept away by the agrarian revolution. Accompanying the changes, the whole face of the country altered. While at the beginning of the nineteenth century William Cobbett might speak of 'steam-engine farms', a description of the general aspect of the countryside at the beginning of the previous century, as given by Sir Archibald Grant of Monymusk, shows the long road that had to be traversed during the century of improvement to attain the status of farm factories:

I could not, in chariot, get my wife from Aberdeen to Monymusk. Col. Middleton (was) the first who used carts or wagons there; and he and I (were) the first benorth the Tay who had hay, except very little at Gordon Castle.[3] Mr. Lockhart of Carnwarth, author of Memoirs (was) the first that attempted raising or feeding cattle to size. By the indulgence of a very worthy father I was allowed (in) 1716, though then very young, to begin to enclose and plant, and provide and prepare nurseries. At that time there was not one acre upon the whole estate enclosed, nor any timber upon it but a few elm, sycamore, and ash, about a small

[1] A. Cochrane, Lord Dundonald, *Treatise on the Connection of Agriculture with Chemistry*, p. 177. [2] R. Kerr, *General View of the Agriculture of Berwick*, p. 457.

[3] Probably the first person to introduce agricultural improvements into Scotland was an English lady, Elizabeth Mordaunt, daughter of the Earl of Peterborough, who married the son of the Duke of Gordon. She introduced hay, fallowing, etc., and through her influence some land-owners introduced the sowing of French grasses. (*See An Essay on Means of Enclosing and Fallowing Scotland*, 1729.)

> kitchen garden adjoining to the house, and some straggling trees at some of the farmyards, with a small copse wood, not enclosed and dwarfish, and browsed with sheep and cattle. All the farms (were) ill-disposed and mixed, different persons having different ridges; not one wheel carriage on the estate, nor indeed any road that would allow it; and the rent was £600 sterling per annum, (when) grain and services (were) converted into money. What land was in culture belonged to the farms, by which their cattle and dung were always at the door. The whole land was raised and uneven and full of stones, many of them very large, of a hard iron quality, and all the ridges crooked in the shape of an S, and very high and full of noxious weeds, and poor, being worn out by culture, without proper manure of tillage. The farm-houses, and even cornmills, and manse, and school (were) all poor, dirty huts (occasionally) pulled in pieces for manure, or (which) fell by themselves almost every alternate year.[1]

In Scotland one of the first bodies to organize agricultural improvements was the *Honourable the Society of Improvers in the Knowledge of Agriculture in Scotland* (1723). Its Secretary and the moving spirit behind its activities was Robert Maxwell of Arkland, who by his own researches so depleted his exchequer that his wife was forced to keep a small shop in Edinburgh. The Society of Improvers did not survive the upheaval of the 'Forty-five. Maxwell, however, published selections from its correspondence, and continued to deliver courses of lectures. In 1757 he published his *Practical Husbandman.*

The cardinal position occupied by agricultural science was fully appreciated by Maxwell, who wrote, in somewhat visionary strain:

> I believe you are all satisfied, that Agriculture is not only a Science, but the Life and Support of all Arts and Sciences. . . . The Crown names Professors, and gives them salaries: which of them can be more useful to the Publick, than a Professor of Agriculture might be? Or a General Inspector of Improvements, who should be obliged to report annually the Husbandry of each County that errors might be rectified?[2]

It is not surprising in the circumstances that it was in Edinburgh that the first Chair of Agriculture in any country was founded. The pity is that it was not founded till the end of the century.

The period during which the *Improvers* were active in Scotland was not characterized by any outstanding scientific work. It was

[1] Quoted in R. Chalmers, *Domestic Annals*, Vol. 3, p. 418. *See also Spalding Club Miscellany*, Vol. 2, p. 97.

[2] R. Maxwell, *Select Transactions*, p. *x*.

T. H. Croker, *Dictionary of Arts and Sciences*, 1764–6

96. Cane-mill for extracting the juice from the sugar-cane.

97. Concentrating sugar-cane juice in open pans.

98. Howard's crystallizing pan—the application of reduced pressure to sugar-refining—patented in 1812. It carries a pressure gauge and safety valve like a steam engine.

99. Charcoal filters (1830). Carrying filter cloth and with a perforated double bottom, they were filled with charcoal granulated like gunpowder. The illustration is probably of one of Fairrie's sugar-houses.

a period of anticipation and consolidation. Maxwell's compilation shows, however, that a considerable advance in husbandry had already taken place. For example, he lays it down as bad to take two grain crops successively off the same ground. Such good advice was long neglected, since the practice in Scotland seems to have been to succeed a fallow with wheat, the wheat with peas, followed by successive crops of barley and oats. The oats were followed again by a fallow.

In spite of reaction, advances were nevertheless made in agricultural practice. One of the most important was the introduction of the drill and horse-hoe by Jethro Tull (1674–1741), who began farming in 1699 and became the protagonist of one of the rival schools of agricultural thought which developed during the eighteenth century. It is clear, however, that it was only after evolving his husbandry that he attempted to develop an ostensibly scientific theory to support it. It was only then in fact that he began to study the contemporary literature of his subject, since in the introduction to the *New Horse Hoeing Husbandry*, he confesses:

> . . . books of Agriculture coming to my hands—I never having read any of them before.

When he did turn to letters he achieved considerable success.

Tull maintained that the way to fertility was 'division of the parts of the earth . . . which division may be mathematically infinite'. His ideas were based on the erroneous belief that finely divided soil was the food of plants, and his theories ultimately led him to repudiate all fertilizers.

In Scotland, information about his methods was disseminated through Maxwell's *Select Transactions*. Number twenty-six of the *Select Transactions* is *A Letter concerning Mr. Tull's Method of Improvement, to a Person of Distinction in Scotland, and by him communicated to the Society*, and the next, *An Answer having been wrote to this Letter, desiring a Solution of several Doubts and Difficulties, Mr. Tull himself wrote as follows:*

> The only way to enrich the Earth, is to divide it into many Parts, by Manure, or by Tillage, or by both. This is called Pulveration. The Salts of the Dung divide or pulverise the Soil by Fermentation. Tillage by the Attrition or contusion of Instruments, of which the Plough is chief. The Superfices, or Surfaces, of these divided parts of earth, is the artificial pasture of plants, and affords the vegetabile pabulum to such roots as come into contact with it. 'Division is infinite'—so the more divided the soil, the more parts the roots come in contact with. . . . I found that 2*s*. bestowed on hoeing would much outdo 40*s*. in manure.

From Scotland the Tullian husbandry made its way south slowly. For example when the drill husbandry for turnips was introduced into Northumberland, it came from Scotland where farmers had the merit of first introducing Tull's management in the culture of this root.[1] Tull's popularity in England lasted for about thirty years, but, although exercising a considerable influence upon practical farmers, it was not long before his ideas were repudiated by agricultural scientists. Of Tull, Professor Francis Home (1719–1813) of Edinburgh, in his *Principles of Agriculture and Vegetation* (1757), wrote:

> Had Tull been a chymist, he would have known, that mere earth makes but a small part of all plants. . . . That the communition of the earth, by the mechanical action of the plough, is not the chief means of increasing the vegetable matter, as Tull asserts, appears plainly from these two facts: that even the lightest soil is the better of fallowing; and that when fallow ground is laid up in ridges, more benefit is received than when it is left quite flat.[2]

Tull, however, was not without his influence in scientific circles. Humphry Davy tells us of him:

> This ingenious author of the new system of agriculture having observed the excellent effects produced in farming by a minute division of the soil, and the pulverisation of it by exposure to dew and air, was misled by carrying his principles too far. Duhamel, in a work printed in 1754, adopted the opinion of Tull, and stated that by finely dividing the soil, any number of crops might be raised in succession from the same land. He attempted also to prove, by direct experiment, that vegetables of every kind were capable of being raised without manure. This celebrated horticulturist lived, however, sufficiently long to alter his opinion.[3]

The torch of science was passed on, but it never came into the hands of Tull.[4]

Yet another work dealing with mechanical means of improvement appeared before a truly scientific work on agriculture was written. It was Alexander Blackwell's *A New Method of Improving cold, wet, and barren Land* (1741). Blackwell was a native of Aberdeenshire, and a student of Boerhaave at Leyden, where he received his doctorate in medicine. He practised in Aberdeen and London, but without success. Going to Sweden, he proposed to

[1] *Agric. Survey, Northumberland*, p. 100.

[2] F. Home, *Principles of Agriculture and Vegetation*, pp. 36 and 107.

[3] H. Davy, *Elements of Agricultural Chemistry*, p. 13.

[4] T. H. Marshall, *Jethro Tull and his New Husbandry of the Eighteenth Century*, *Econ. Hist. Rev.*, 1929, 2, 40.

drain marshes and fens, but, being suspected of implication in a plot with Count Tessin, he was beheaded in 1748.

The next contribution to agricultural science also came from Scotland. It was a great advance on anything preceding it, and a milestone in scientific investigation. It was the publication referred to above of *The Principles of Agriculture and Vegetation*, by Francis Home, Professor of *Materia Medica* in the University of Edinburgh. This book was one of the first fruits of the activities of the *Edinburgh Society for the Promotion of Arts, Manufactures and Commerce*, and the Society awarded a gold medal to Home for his work.

> The date of Home's researches was opportune since with the accession of George III the tide of prosperity and comparative comfort began to ebb. The rapid increase in the population was to make enormous demands on agriculture by the end of the eighteenth century in order to feed the people. Moreover the need for improved husbandry was the more pressing in view of the series of bad harvests and wet seasons which contrast so vividly with the absence of bad harvests, wet summers and droughts during the first half of the century.[1]

At the date of his publication few scientific improvements in agriculture had been introduced into Scotland, and Home attributed the general lag in development to its leaders having had insufficient contact with science, to their being untutored in the art of observation, and unable to arrive at conclusions from any observations they might make. Pointing this out, he says in his introduction:

> The principles of all external arts must be deduced from mechanics and chymistry, or both together. Agriculture is in the last class: and although it depends very much on the powers of machinery, yet I'll venture to affirm, that it has a greater dependence on chymistry. Without a knowledge in the latter science, its principles can never be settled. As this science is but of late invention, and has not yet been cultivated with the regard to utility, and the improvement of trades and manufactures, as it ought and might, agriculture is hardly sensible of its dependency on it. The design of the following sheets is to make this appear; and to try how far chymistry will go in settling the principles of agriculture.[2]

Home's pioneer position is illustrated by the concluding sentence of quotation. He succeeded, too, in quickly getting at the crux of

[1] J. C. Drummond and A. Wilbraham, *The Englishman's Food*, p. 205.
[2] F. Home, *op. cit.*, Introduction.

his problem, the need for plant analysis and pot experiments. Indeed he was the first to think of systematically trying effects of salts on plant growth. He continues:

> As plants belong to the class of organized bodies, they thrive in proportion to the quantity of nourishment they receive at their roots. Hence arises a simple, but very comprehensive, view of husbandry. The whole of the art seems to centre in this point, *viz.* nourishing the plants. If we can discover what is the natural food of vegetables we shall the easier discover wherein consists their artificial food, or that given by art, and how it operates.

Experimental difficulties and the slow rate of progress attendant on such an investigation were clear to him.

> What a number of different observations of heat and cold, dry and wet, difference of soils, grains, seasons, etc. must be exactly made, before one can be certain of the general success of an experiment. How seldom can these experiments be repeated, which take a whole year before they can be brought to a conclusion.[1]

Some such pot experiments were initiated by Home himself.

That this research was limited and inadequate was realized by Home, but with such data as was available he endeavoured to draw certain inferences and check existing observations. Thus, the fertility of virgin earth was demonstrated. Also, the already observed manurial value of nitre or saltpetre (KNO_3) was confirmed. Hard water, i.e. water containing calcareous salts in solution, appeared to be beneficial. Olive oil, while initially promoting a quick vegetation, was not lasting. His results with nitre, nitre and olive oil, and with nitric acid, were not what he had been led to expect, so he adopted a non-committal attitude to the results of his first experiments. 'Its (i.e. KNO_3) great reputation for fertility would make one cautious in doubting that effect, without a sufficient number of experiments to support that opinion.' He found potassium sulphate (K_2SO_4) to be 'one of the greatest promoters of vegetation'. Common salt (NaCl) was 'an enemy of vegetation'. Attention to this observation on the part of his successors might have saved more than half a century of futile experimentation in the field, for it was only in the nineteenth century that it was finally concluded that salt, although cheap, was not of general utility as a manure.

From the results of his first research programme, Home was led to another in which the effects of natural fertilizers were more

[1] F. Home, *op. cit.*, p. 3.

accurately reproduced. 'In a natural way the fructifying principles must be bestowed on it by gentle degrees, and in very small quantities.'[1]

From his second series of experiments Francis Home arrived at certain wider conclusions, in particular the generalization that agriculture was a more certain and scientific art than was commonly thought, and that it could, like other arts, be reduced to fixed, unalterable principles.

> Agriculture is a branch of natural philosophy, and can only be improved from the knowledge of facts, as they happen in nature. It is by attending to these facts that the other branches of natural philosophy have been so much advanced during these two last ages. Chymistry is now reduced to a regular system, by the means of experiments made either by chance or design.[2]

The lack of facilities for making such agricultural experiments in this country was very apparent to him.

> When I look round for such, I can find few or none. I have read three volumes of experiments published by *Du Hamel*[3] on *Tull's* system of agriculture. They are distinct, exact, conclusive so far as they have gone, and stand a model for experiments in agriculture. What a shame for *Great Britain*, where agriculture is so much cultivated, and where that system took its rise, to leave its exact value to be determined by foreigners! Books in that art we are not deficient in; but the book which we want is a book of experiments.[4]

In order that available information might be correlated and distributed, he suggested the formation of a small committee of the *Edinburgh Society*, whose function it would be to collect experimental results and 'publish them to the world at stated times, like a public paper'. But the Edinburgh Society became more interested in the speaking of English, and so lost the opportunity of founding a useful agricultural periodical.

A note might be added here on oil as a promoter of vegetation. Undoubtedly it seems to have been an observation that treating some seeds with oil had a beneficial effect on subsequent growth, with the result that it was concluded that oil was a plant food. George Winter demonstrated that the effect was not a fertilizing action, i.e. the supplying of a plant nutrient.

> As oil so particularly promoted the growth of turnips and proved injurious to barley, wheat, peas and beans; curiosity induced me, first, to examine through a microscope, the structure

[1] F. Home, *op. cit.*, p. 92. [2] *Ibid.*, p. 201.
[3] *De la Culture des Terres.* [4] Home, *op. cit.*, p. 202.

> of turnip seed, which appeared spongy; from whence I concluded, that in consequence of turnip seed being generally sown in the hottest and driest season of the year, the heat and drying winds penetrate the pores, consequently greatly injure the vegetative properties of the seed. Whereas oil enters and fills the pores, protects the germinating properties of the seed from the scorching heat and drying winds; thus by being preserved, a little moisture makes the seed vegetate with such luxuriancy, as very speedily to grow into rough leaf, when it generally becomes too powerful to be materially injured by the fly.[1]

Although it is not possible to judge, in the light of modern knowledge, the individual merit of Home's experiments, at the time they enabled him to postulate conclusions of considerable utility. 'The food of vegetables is not all of the same kind.' He enumerates six kinds. Different plants, according to their natures, require the essential plant foods in varied proportions; thus different plants will remove specific constituents from the soil. This formed the scientific basis of an important agricultural practice. 'A succession of different kinds of grain has been found preferable to the sowing of one kind constantly on the same field.' This principle, the rotation of crops, was soon reinstated in practical husbandry, where it had fallen into disrepute, the reason being that, while Dr. Woodward pointed out advantages resulting from a change in the kind of crops grown on any piece of land, the influence of Tull and his pseudo-science led husbandmen to think it unnecessary. Woodward was one of Jethro Tull's contemporaries with whom he saw fit to enter into controversy to maintain support for his theories.

During the lifetime of Francis Home, industrial and agricultural changes were taking place side by side. This enabled him to make some observations on mineral (inorganic) manures of industrial origin.

> At the alum works near Scarborough, the farmer pays 2*s*. a cart-load for the refuse of the earth of these ashes, after almost all the salt is extracted from them. The refuse of the soap-manufacturers, and of the bleachfield, are rich manures. The ashes of peat, which are most used, afford salts equal only to the thirty-second part of the whole, and are the weakest of all those I know. . . . There is another manure which takes its rise from fire, and properly belongs to this class; that is, soot. This is found, by chymical experiments, to be a composition of volatile alkaline salt (NH_4Cl), oil, and a little earth. It is remarked, that the effects of this compost are very sudden, they being observable after the first rains.[2]

[1] G. Winter, *System of Husbandry*, p. 248. [2] F. Home, *op. cit.*, pp. 75–76.

Invention in industrial technique had, however, a clear-cut precision lacking in agrarian experiment. The reason for this is not far to seek. Conditions of the laboratory can more easily be reproduced in factory than field, where nature presents the agrarian experimenter with complex soil and weather conditions. In consequence of such complexities, opposed and often antagonistic schools of thought developed, basing their theories on the available half-truths. For the practical farmer this was unfortunate because he was driven by doubt and controversy to fall back on traditional methods. To counter some of this confusion, the *Museum Rusticum et Commerciale* was founded (1763–6). It selected papers on agriculture, commerce, arts and manufactures, had them revised by members of the *Society for the Encouragement of Arts, Manufacture and Commerce*, and published them. This Society encouraged agriculture in England before the founding of the *Agricultural Society* (1838), and chemistry and physics before there were specialist societies in these sciences. The *Royal Society* and the *Society of Antiquaries* alone antedate it. Dr. Stephen Hales was one of its founders.

Had more of the leaders of agriculture been endowed with the breadth of outlook of Francis Home, much of the unnecessary disputation might have been avoided. How true were his words:

> It is the common fate, in all disputed points, that each attaches himself to one side, without allowing the other any share of truth. I have found by experience, that each side has generally some truth in it; that mankind err by extending that particular truth to a general one; and that the real truth is generally made up of somewhat taken from each opinion.

Within a few years of the publication of Home's own treatise on agriculture, the book of experiments by a practical farmer that he wished for was published. Its author was the Rev. Adam Dickson, the stimulus to publication the activities of the *Edinburgh Society*. His *Treatise of Agriculture* (1762) was credited with being one of the best ever written on tillage. While Home realized that the crux of his problem was to investigate plant foods, he also paid considerable attention to soils and their chemical examination, but this called for criticism from Dickson. 'Dr. Home, in his treatise, begins with soils', says Dickson, 'and any person that reads with attention what he has wrote, will be convinced of the impropriety of beginning with this part of the subject'. Dickson relegates the study of soils to the end of his work, beginning with such factors as promote vegetation. Some of his conclusions are worth recording here, as they had a directive effect upon subsequent theory. The pre-chemical-revolution nomenclature should be noted in passing.

> Chymists assure us that plants contain all the ingredients mentioned, and no others (earth, air, fire, water, oil, salt). Now, if a person divests of prejudice, and attachment to any particular opinion, it would naturally occur, one would think, that the food of plants is compounded of the same things of which plants themselves are compounded. For it is not possible to conceive how any thing can be in the plant, unless something of the same kind is in the composition of its food.[1]

This treatise of Dickson's was the first to give due importance to the subject of rotations. Further important examples of profitable rotations are contained in the *Gentleman Farmer* of Lord Kames. Although it has been suggested that both Dickson and Kames were led to the principle of rotation by observations of Norfolk husbandry, it seems doubtful if Dickson, in his critical readings of the *Principles of Vegetation*, could have missed the implication contained in Francis Home's conclusions.

If one examines the chemical steps taken to initiate the agrarian revolution in Scotland, application of lime and marl figures largely in early stages. It was said that larger quantities of lime per acre were used in Scotland than in any other European country, though for generations the way in which it operated was not understood.

> The elder writers on agriculture had no correct notions of the nature of lime, limestone and marle, or of their effects; and this was the necessary consequence of the imperfection of the chemistry of the age. It is to Dr. Black of Edinburgh that our first distinct rudiments of knowledge on the subject are owing. About the year 1755, this celebrated professor proved, by the most decisive experiments, that limestone and all its modifications, marbles, chalks, and marles, consist principally of a peculiar earth (*calcium*) united to an aerial acid (*carbonic acid or carbon dioxide*).[2]

Black's work was of importance in other directions as well. For example, speaking of the effect of Dr. Black's researches on the development of the geological theory, John Playfair, in his *Illustrations of the Huttonian Theory of the Earth*, says:

> The discoveries of Dr. Black mark an era before which men were not qualified to judge of the nature of the powers that had

[1] A. Dickson, *Treatise of Agriculture*, p. 18. [2] H. Davy, *op. cit.*

> acted in the consolidation of mineral substances. Those discoveries were, indeed, destined to produce a memorable change in chemistry, and in all branches of knowledge allied to it; and have been the foundation of that brilliant progress, by which a collection of practical rules, and of insulated facts, has in a few years risen to the rank of a very perfect science. But even before they had explained the nature of carbonic, and its affinity to calcareous earth, I am not sure that Dr. Hutton's theory was, at least, partly formed, though it must certainly have remained, even in his own opinion, exposed to great difficulties. His active and penetrating genius soon perceived, in the experiments of his friend, the solution of those difficulties, and formed that happy combination of principles, which has enabled him to explain the most enigmatical appearances in the natural history of the earth.[1]

And we must not forget that it was agriculture that focused Hutton's attention on the problem of earth history.

Marl, according to George Winter, is composed of lime (i.e. calcium carbonate or hydroxide), clay, and sand; it is increased in value according to the quantity of pure lime it contains, and is best adapted for sands and light soils. In point of time it preceded lime as a fertilizer. The quantity applied was in proportion to the lightness of the land.[2] As mentioned in Chapter I, we have evidence that even in the latter half of the seventeenth century, application of lime was the usual way of 'gooding' land in East Lothian. So East Lothian possesses the honour of having led the way in Scotland to improved husbandry. In other parts of the south-eastern arable district, introduction of lime came later. In Berwickshire, the true era of efficient and lasting improvers is said to date from 1730, when Swinton of Swinton began to drain, marl, and enclose his land. He was immediately followed by Home of Eccles. Marling continued till about 1760, when it was gradually, superseded by the 'more active energizer', lime. About 1780 simple liming was in turn superseded to some extent by the more elaborate turnip husbandry.[3]

In Midlothian, where Thomas Hope of Rankeiler was one of the promoters of the Society of Improvers, although the benefits of manuring had previously been demonstrated, it was left to the Society to emphasize the beneficial effect of marl. It was not, however, widely adopted, partly on account of the shortness of leases, a frequent impediment to improvement, but also on account of expense.[4] Liming was in operation in Roxburgh by

[1] J. Playfair, *Illustrations of the Huttonian Theory of the Earth*, p. 523.
[2] G. Winter, *op. cit.*, p. 52.
[3] G. Chalmers, *Caledonia*, Vol. 2, p. 312. [4] *Ibid.*, p. 735.

about 1755, the date at which Sir Gilbert Elliot and Mr. Dawson introduced marl. In 1772 it was first offered for public sale.[1]

It was later in the century before the benefits of liming were realized in the southern pastoral counties; in fact, Wigtownshire was fifty years behind Ireland, just across the Channel.

In Kirkcudbright, real improvement dates from 1740, when shell marl was discovered, or at leased used, as a manure. To obtain adequate supplies, Mr. Gordon of Greenlaw encouraged the draining of Carlinwark Loch to get marl, and at his expense a canal was dug for its conveyance.[2]

In 1770, limestone was discovered in Dumfries on the lands of Sir James Kirkpatrick of Closeburn. He established large lime-kilns, but had considerable difficulty in overcoming 'prejudices of ignorance against the application of this potent fertilizer of a wretched soil'.[3] In the same county, lime from Sanquhar was supplied free to Buccleuch tenants.

In the south-west lowlands, the epoch of efficient improvements dates from 1757. In that year Margaret, Countess of Loudoun, took up residence at Sorn Castle, and started improving the land by liming, etc.[4] Her father, the First Earl of Stair was the first to raise turnips in open fields.[5]

Pennant in his *Tour* described agricultural conditions in the central counties.

> In respect to agriculture, there are difficulties to struggle with, for the country is without either coal or lime-stone; so that the lime is brought from the estate of the Earl of Elgin, near Dunfermline who, I was told, drew a considerable revenue from the kilns.[6]

Very large quantities of fuel were used in lime-burning, but it was not impossible to get over the want of coal. Pennant 'saw a stamping-mill calculated to reduce limestone to a fine powder in order to save the expense of burning for manure.[7]

In Renfrew no marl was found, but improvement was effected by application of lime. Difficulties in getting supplies of lime-stone and fuel were even greater farther north in Aberdeenshire, where there are very few deposits of limestone. Yet Aberdeen did not go without lime. Pennant spoke of a cave at Slains once filled with 'curious stalactial incrustations' which had been destroyed, as they supplied lime when burned.[8] The deficit was also made good by importation from the Earl of Elgin's works on the Firth of Forth,

[1] G. Chalmers, *Caledonia*, Vol. 2, p. 143. [2] *Ibid.*, Vol. 2, p. 286.
[3] *Ibid.*, Vol. 3, p. 126. [4] *Ibid.*, Vol. 2, p. 475. [5] *Ibid.*, Vol. 3, p. 484.
[6] T. Pennant, *Tour in Scotland*, p. 68. [7] *Ibid.*, p. 95. [8] *Ibid.*, p. 119.

and from Yorkshire. Walker, in his *Oeconomical History*, made some interesting comments on transport of lime.

> It is remarkable that *quicklime* is generally applied as a manure, with more industry, at a great distance from the lime kiln than in the immediate neighbourhood, where the lime may be had at the easiest rate. Lime is transported from Yorkshire to Aberdeenshire, at a high price, for improving the soil, while the lands of Yorkshire, well adapted for this manure, never receive the advantage of it. Draw kilns may be seen in Mid Lothian, surrounded by wild lands which might be highly improved by lime, but which has never received any, while from these very kilns, the lime is carried with great advantage to Berwickshire, by an expensive carriage of sixteen or eighteen miles. Though it may seem unaccountable, it is certainly true, that lime is most neglected as a manure in those parts of Scotland where it is in greatest abundance, and to be procured at the lowest price. The county of Fife is perhaps the chief exception.
>
> For these purposes, the want of coal is, no doubt, a great disadvantage. Even the abolition of the duty on coals will not render them sufficiently cheap for the burning of lime. But there are few places in the Highlands, where limestone and marble are found, where there is not also abundance of peat. . . . In a kiln, properly constructed, good limestone may be burned and calcined sufficiently with peat. This has been long an established practice, in several parts of Scotland, where there is no access to coal. But in the remote Highlands and Islands it is either not known or not practised. . . . It would perhaps be right to employ a skillful lime-burner to travel through the country for the purpose. . . . Such a man might be most beneficially employed for two or three years in this way. The expense could not be any great sum; and was it defrayed by the Highland Society, would be amply repaid in promoting their views.
>
> Some years ago, a Danish gentleman travelled in this country, at the King of Denmark's expense, and at the rate of some hundreds a year, for the purpose of importing into Denmark the useful oeconomical practices of other countries. He observed in Scotland, and with great attention, the custom of burning lime-stone with peats. In the countries of Jutland, Sleswick, and Holstein, they have plenty of limestone and plenty of peat; but no lime, except what is made with coal, brought from afar, or with wood. He judged that the communication of this simple practice, of calcining limestone with peat, would, to the inhabitants of these countries, repay more than tenfold all the expenses of his travels.[1]

At Deskford in Banffshire, lime was burned with peat. The consumption of peat must have been very considerable, since the

[1] J. Walker, *Oeconomical History of the Hebrides*, pp. 161–164.

author of the *Statistical Account* for that district attributes a decrease in the population which had taken place as due to the cessation of lime-burning in order to conserve the moss, i.e. peat, for domestic fuel.[1] Concerning the introduction of lime in Banffshire we learn from the *General View*:

> When lime was first introduced in this county as a manure, its stimulating effects were so powerful in producing heavy crops, that the farmer thought he could not over-do; and hence the system of oats after oats, for a succession of from 12 to 19 years, was so much persisted in, that the soil was reduced to so complete a state of sterility that it produced little else than thistles and other weeds. These effects, the natural consequence of such a management, brought the liming system so much into disrepute, that it was believed a second liming was hurtful rather than beneficial, and from this absurd opinion, it was nearly abandoned, until the late Lord Findlater, by his example as well as precept, introduced a more rational mode of management of lime on fallow and green crop, and afterwards laying out with grass seeds.[2]

As in most phases of eighteenth-century ecomony, coal played its part in agrarian development. According to Nef:

> Apart from the proof of the direct influence of coal upon the industrial growth and upon the size of the capitals employed in industry, there are reasons for believing that coal helped to bring about the agrarian and constitutional changes which appear to have been favourable to industrial capitalism. Coal provided a basis for improvements in methods of fertilization and therefore helped to increase the yield from the soil.[3]

At Closeburn, Dumfriesshire, lime was burned with coal.[4] General Dirom of Mount Annan, Dumfriesshire, improved the construction of the kilns, which were later perfected by Booker of Dublin.

In Inverchaolain, Argyllshire, 'the oppressive duty upon coal, and the distance of every other fuel make the expense of lime-burning so enormous, that the people cannot afford to lay it on their land'.[5] The duty on coal was an impediment, its removal a blessing.

> Now that the duty is happily taken off coals carried coastwise, would all farmers and proprietors who could conveniently get them, allow themselves to make more use of them for fuel, and

[1] *Statistical Account*, Vol. 4, p. 364. [2] D. Souter, *General View*, *Banff*, p. 233.
[3] J. U. Nef, *Rise of the British Coal Industry*, Vol. 1, p. 359
[4] Diagrams and details of the kilns employed are given in *An Encyclopaedia of Agriculture*, by J. C. Loudon, Second Edition, (London), 1831.
[5] *Statistical Account*, Vol. 5, p. 466.

employ the time which used to be consumed on peats, in draining and making compost manures, this would keep in their pockets those larger sums which now go for meal, and enable them to rear more cattle, and of a better kind. These things will come round in time. A few will be wise enough to set the example; and many, it is hoped, will follow it.[1]

Looking ahead to the end of the first phase of agricultural change, it becomes apparent that Scotland no longer suffered from a shortage of lime. Sinclair tells us in his *System of Husbandry* that many Scottish farmers spent ten shillings per acre per annum on lime alone, a sum not less than the average rent of land in many English counties.[2]

LIME PRODUCTION

		Bolls
1.	Lord Elgin's at Charlestown on Forth	400,000
2.	Closeburn in Dumfries	100,000
3.	Stob in Roxburgh	10,000
4.	Craigie etc. in Ayr, Loudon	150,000
5.	New Kilpatrick etc. in Dumbarton	60,000
6.	Campsie in Lennox etc. in Stirling	100,000
7.	Torphichin, Bathgate, etc. in West Lothian	80,000
8.	Gilmerton, the Raw, Crichton etc. in M. L.	140,000
9.	Salton, Pencaitland, etc. in East Lothian	150,000
10.	Forthan, the Bunyan etc. in Fife	50,000
11.	Hedderwick, Hill of Peat, etc. in Forfar	100,000
12.	Mathers etc. in Kincardine	30,000
13.	Ardonald etc. in Aberdeen	100,000
14.	Banff, Moray, Nairn, etc.	100,000
15.	Lanark and other Southern Counties	230,000
16.	Perth, Kinross, Clackmannan and Midlands	100,000
17.	Galloway and other Southern Counties	100,000
	Total	2,000,000

Such was the demand for lime in Scotland that considerable quantities were annually brought by sea from Sunderland and from Ireland, and by inland carriage from Northumberland and Cumberland.[3] The progressive development of coal-mining in Scotland during the eighteenth century was reflected by the increased availability of lime burned with it. This in turn extended the uses to which lime could be applied, so that in addition to manuring it was more widely used for building, and

[1] John Smith, *General View: Argyle*, p. 214.
[2] J. Sinclair, *System of Husbandry*, Vol. 1, p. 200.
[3] J. Sinclair, *General Report*, Vol. 1, p. 67.

in industry as a flux in iron-making, in sugar-refining, bleaching, and various minor processes. Although some districts of Scotland imported lime, others, curiously enough, exported it, mainly to the West Indies.[1]

The principal lime-works of Scotland, with their production at the end of the century, are given in the table on page 477.

To the total shown Sir John Sinclair added another million bolls for production of small, unidentified kilns, making a grand total of three million bolls.[2]

No matter how obvious the improvements in the husbandry of Scotland must have been, there are very few contemporary comparisons of the old and new husbandries. We therefore cite the following:

> The increase of produce in Islay, and the improved Hebrides in consequence of liming, is rather better than as three to one; i.e., land which formerly, and under the old system, produced in three years twelve pounds of value per acre, now produces thirty-six pounds value, or twelve pounds an acre per annum. But what is of still greater consequence is the fact, that the land is as good after the three crops have been raised upon it subsequent to liming, as it was before; while under the old management, it was exhausted, and, as it was termed, *run out*, for five or six years, and yielded nothing but a miserable species of herbage not worth 3*s*. per acre. To make the contrast more distinct, and the comparison more intelligible, we may state the two methods, during a complete rotation of eight years, as follows, mentioning merely the clear returns and the rent:

OLD MANAGEMENT

Year		£	*s.*	*d.*
1st	An acre Barley, 6 bolls at L.1 1*s*. . .	6	6	0
2nd	,, Oats, 4 b. poor quality, L.1 1*s*. .	4	4	0
3rd	,, Oats, 1½ b. bad quality, L.1 .	1	10	0
4th	An acre of bad grass		3	0
5th	,, rather better than last		5	0
6th	,, still improving		10	0
7th	,, ,, ...		18	0
8th	,, ,, ...	1	0	0
	Produce in eight years	£14	16	0
	Rent, 10*s*. per annum, deducted . . .	4	0	0
	Profit	£10	16	0

[1] J. Sinclair, *General Report*, Vol. 1, App. 2, p. 305. [2] *Ibid.*, App. 1, p. 155.

NEW MANAGEMENT

Year		£	*s.*	*d.*
1st	Barley limed, 8 bolls	8	8	0
2nd	Hay and pasture	6	6	0
3rd	Pasture, good	4	0	0
4th	Oats, good, 7 b.	7	7	0
5th	Turnips	8	0	0
6th	Barley, good, 8 b.	8	8	0
7th	Hay and pasture	6	6	0
8th	Pasture	4	0	0
	Produce in eight years	£52	15	0
	Rent, 10*s.* per annum	4	0	0
	Profit	£48	15	0

The above calculation is made from what was seen and learned in Islay in summer 1808, fixing the prices at a moderate average, not indeed of Islay (where barley is always immensely dear) but of all Scotland for ten years past. Were the calculation founded upon the Islay prices, the contrast, and the advantage of using lime and a proper rotation of crops, would be still more striking.[1]

As in all things, opposite views may be taken. The progress of agriculture throughout the course of the century made such vast strides that the traditional values of land were entirely altered. Many areas, from being wild uninhabited moors, were changed with the passage of time into important agricultural producing districts. With these changes, the relative values of rent and wages also altered. H. G. Graham gives us the reverse side of the above picture of Islay:

Troubles always attend economic measures which promote social progress, for hardship is the inevitable accompaniment of every process of development; and in the struggle for existence the weak must suffer, that the fittest may survive. The surest signs of progress in any industry are, however, to be found in the increase alike of gains of the master and wages of the servant, and here the evidence is striking, even startling. In a few years, land which had for more than a century been let at the same rent of 1*s.* 6*d.* to 3*s.* an acre, rose to 21*s.* in Berwickshire; land in Perthshire, which had brought at its highest 5*s.*, in nine years advanced to 17*s.* an acre, and in 1784 had bounded up to 45*s.*; and in Ayrshire, ground which had of old been let for 5 lbs. of butter per acre

[1] J. Macdonald, *General View, Hebrides*, p. 397.

easily let for 25*s.* after being drained and limed. In the Carse of Gowrie land which had let at the supposed high rent of 6*s.* 8*d.* an acre, had risen in twenty years (in 1783) to £6.[1]

To return to our main line of development; the accumulated knowledge of plant nutrition was still slight, but the men who spent their professional lives in the laboratory were also engaging in practical experiment on the farm as well.

> Scientific men were busy improving, like Dr. James Hutton, the geologist, who brought ploughmen from Dorset to initiate his workmen at Duns, and is said to have been the first to use the two-horse plough. Lawyers and judges joined the ranks of the farmers, such as Lord Kames, who at Blair Drummond began to cast the moss from the marshes in the swampy district of Kincardine, to drain the swampy soil, to encourage the tardy use of lime and marl, and at the venerable age of eighty published his shrewd, if whimsical, *Gentleman Farmer* to enlighten his countrymen; while a brother judge, Lord Stonefield, was striving to bribe tenants in Dumbartonshire to sow turnips.[2]

When, in 1776, Henry Home, Lord Kames, published *The Gentleman Farmer*, speculation about the manurial value of lime was rife.

> Much controversy has taken place, both among agriculturists and chemists, concerning the mode by which lime operates as a manure. The general opinion at present seems to be, that it only operates as a stimulus, in bringing into action all the dormant animal and vegetable matter already contained in the soil; but that it does not directly furnish any part of the food of plants. But as those who use the word stimulus, have not chosen to explain how lime produces the effect ascribed to it, the use of this word leaves us as much in the dark as we were before.[3]

The resulting confusion led Lord Kames to remark:

> In no brand of philosophy are imagination and conjecture more freely indulged, than in what concerns the food of plants. Every writer erects a system: if he can give it a plausible appearance, he inquires no further. It never enters into his thoughts, that his system ought to be subjected to the rigid touchstone of facts and experiments.[4]

We have an excellent example of this in the following citation:

> The Rev. Mr. Shepherd of Muirkirk, gravely states, in the Statistical Account of that parish, that lime is injurious to sheep,

[1] H. G. Graham, *Social Life in Scotland*, p. 211. [2] *Ibid.*, p. 204.
[3] J. Headrich, *General View: Angus*, p. 408.
[4] Kames, *Gentleman Farmer*, p. 18.

and unfavourable to the soil of Muirkirk!!!—The Rev. Gentleman knew nothing of agriculture himself, and only stated that opinion, for the information of the indolent store-farmer, of which his Kirk Session was composed.[1]

Lord Kames acknowledged his indebtedness to Dr. Joseph Black for guiding his writing. 'I am not, however, afraid', he said, 'of any gross error. An imprimatur from one of the ablest chymists of the present age (Dr. Black), has given me some confidence of being on the right track'. And, speaking of the relation between chemistry and agriculture: 'To be an expert farmer, it is not necessary that a gentleman be a profound chymist. There are, however, certain chymical principles relative to agriculture, that no farmer of education ought to be ignorant of.' For a time he and Black corresponded on the subject of solution.

Lord Kames imagined that an emulsion of clay in water differs only in degree from the solution of salt in water; Black tries to point out differences, but fails to convince Lord Kames. The last letter to Black, which finishes the correspondence, begins: 'Sir—, This is Sunday morning, and I figure myself applying to you as Abraham did to the Lord, "Behold, now I have taken upon me to speak upon the Lord, which am but dust and ashes". Your response has not satisfied me more than the response of the Lord did Abraham.' Black retorts that Lord Kames does not read his letters with care and attention; and the correspondence then ceases.[2]

That plants synthesize substances in their tissues was evident to Kames. He no longer regarded the finding of oil in vegetable tissue as an indication that the oil came from the earth, as had been suggested by Francis Home. Of relevant experimental results obtained by Dr. Stephen Hales, he says:

These facts are confirmed by many experiments; all of them convincing, that however different the juices may be that are imbibed by a plant, yet that the sap into which these juices are converted is the same, or nearly the same, even in plants of different species. If so, every plant must be endowed with proper powers; first, to imbibe juices; next, to convert into sap the juices imbibed; and last, to convert that sap into its own substance. With respect to the first two powers, all plants appear to be similar. The difference of species is carried on by the last power only, that of converting sap into the substance of a plant. Hence a peculiar texture, colour, smell, taste, in each species.[3]

[1] W. Aiton, *General View: Ayr*, p. 382.
[2] W. Ramsay, *Life and Letters of Joseph Black*, p. 57.
[3] Kames, *op. cit.*, p. 313.

Kames attacks supporters of the theory that, since oil and salt (i.e. mineral matter) are found in vegetables, these are the food of plants, or alternatively that salts and oil can be attracted from the air. His arguments were founded on experimental work carried out by Dr. Ainslie.[1]

He organized experiments to refute the hypothesis that each species of plant requires a peculiar nourishment. The experimental work in this case, was done for him by a 'student of physic in the college of Edinburgh, an ingenious young gentleman.' Kames applied to Dr. William Cullen for help, but Cullen was unable to undertake work additional to that already in hand, so put him in touch with Dr. James Bell (1755–1785) of Dumfries, then in his twentieth year. The experiments are described in *The Gentleman Farmer*.[2] Bell's researches show an advance on those of Home by his use of controls planted in pure sand and watered with pure water only. Water impregnated with putrescent animal and vegetable substances was the best manure, urine was also good, the more putrid the better. Excess lime was shown to be harmful, etc. Acquaintance between Kames and Bell thus begun matured into intimate friendship, and Bell spent considerable parts of several long vacations in the south of Scotland at the country seat of Lord Kames. Bell ultimately settled in Manchester.[3]

Kames also continued the criticism of Jethro Tull initiated in Scotland by Francis Home, handling him even more severely than his predecessor.

> Tull is one of the boldest theorists that have come under my inspection. He pronounces without hesitation, that all plants live on the same food or pabulum, which he says is pulverised earth; and upon that foundation, he pretends to raise perpetual crops of wheat in the same field, by the plough alone, without manure. This indeed would restore the golden age of ease and indolence; as there is no soil so barren, but what may readily be pulverised. Tull was a man of genius, but miserably defective in principles.[4]

As Sir John Sinclair almost always gets the credit for having conceived the idea of a Board of Agriculture, we feel that it is desirable to give the following long citation concerning Kames' prior claims.

> In justice to the late Lord Kames it may be proper to observe here, that, in the appendix to his valuable work, *The Gentleman Farmer*, Art. 2nd, which was first published in 1776, he proposed

[1] *Edinburgh Physical Essays*, Vol. 3, Art. 1.
[2] Kames, *op. cit.*, p. 320.
[3] *Mem. Manch. Lit. and Phil. Soc.*, 1785, 2, 384.
[4] Kames, *op. cit.*, p. 18.

the first idea of a Board of Agriculture, and suggested the propriety of county agricultural reports. These suggestions, although on a very different scale from the recently established institution, sufficiently refute the pretensions of a modern agricultural writer, who claims the merit of having been the *first* who recommended these great measures to public attention. In full proof of this, the following extracts from observations by Lord Kames on this subject, are here inserted:

> 'We have a board for manufactures and fisheries, a wise institution which had done much good. Why not also a board for agriculture? Is agriculture a less useful art than those mentioned, or does it less require instruction? Hartlib, in his legacy, laments that no public director of husbandry has ever been established in England', etc. etc.
>
> 'The plan I have in view is simple. Let the board consist of nine members, the most noted for skill in husbandry, and for patriotism. As I propose no reward to these gentlemen, but the honour of serving their country, the choice will not be difficult. To ease the board from the laborious branch of their business, they ought to be provided with an able Secretary', etc. etc.
>
> 'A regular meeting once a month may be sufficient. The things necessary to be undertaken by this board at the commencement of their operations, will require much labour and sagacity. The first is to make out a state of the husbandry practised in the different counties, in which notice must be taken of the climate, of the soil, of the mode of cropping, and of the instruments of husbandry; noting the prices of all the particulars that enter into farming', etc. etc.

Though Lord Kames could not accomplish the establishment of a board of agriculture, yet he got some county reports executed according to his idea of them, by a very respectable farmer in East Lothian, the late Mr. Andrew Wight of Ormiston. These were published in six volumes, and contain a very good account of the state of agriculture at that time, about 26 to 30 years ago. How different that was from the present state, would require more discussion than can be here admitted to explain; yet these reports certainly did material service to Scots husbandry. In regard to the establishment of the board itself, after it had been recommended in vain from other quarters, it is the more creditable to the person by whose exertions every difficulty was at length surmounted, and who has carried it on with so much zeal and success.[1]

[1] R. Kerr, *General View of the Agriculture of Berwick*, p. 89.

PART TWO

'All Britain is not Norfolk'
G. S. KEITH.

The next forward step with which we must deal is the development of plant physiology resulting from collateral advances in chemical technique both in laboratory and industry. In Scotland the important effect which advancing chemical knowledge had upon the development of industry was commented upon by John Naismith in his Agricultural View of Clydesdale:

> It ought here to be observed, that a number of the inhabitants of Glasgow have been remarkably zealous in making science an auxiliary to the arts. Chemistry, and other branches of experimental philosophy, have been studied with great assiduity, and applied with success to the advancement of different manufactures; and in every department much professional knowledge, much spirited enterprise and persevering industry, has always been displayed.[1]

Improvement in the manipulation of chemical substances made chemists realize that 'fumes and smells' were substances in the third, gaseous, state of matter. Some of these gases play important parts in the life-processes of plants, and in putrefaction several also play the leading roles.

> Black's work (1754) on carbon dioxide, and that of Cavendish (1766) on hydrogen, had been followed by a plethora of discoveries. Compounds which had been dismissed as fumes and smells were systematically investigated. Besides those of the four elementary gases (oxygen, nitrogen, hydrogen, and chlorine), the characteristics of about a dozen other gaseous substances had been studied by the end of the 'eighties. Priestley described the properties of sulphur dioxide (1770), hydrochloric acid gas, nitrous and nitric oxide (1772), ammonia (1774), and sulphuretted hydrogen (1777). Scheele and Priestley discovered silicon tetrafluoride in 1776. Scheele, who had first made chlorine in 1774, also made sulphur trioxide. Carbon monoxide was first studied by de Lassone in 1776. Methane and ethylene (two other constituents of the mixture that is called coal gas) were respectively investigated by Berthollet (1785) and by a group of Dutch chemists.[2]

Priestley's discoveries were published in his *Experiments and Observations on Different Kinds of Airs* (1775). From the point of view of plant physiology, his most important observation was that

[1] J. Naismith, *General View of the Agriculture of Clydesdale*, p. 195.
[2] L. Hogben, *Science for the Citizen*, p. 439.

plants possess the power of reviving or sweetening air which has been vitiated by animals breathing it. At the initial date of this observation, Priestley had not discovered the important gas, oxygen, with the result that he was not yet in a strong position to refute the contrary assertion made by Scheele, that plants as well as animals vitiate air. It is worth citing the substance of these important researches, as they appeared to a contemporary agriculturalist.

> In the year 1775, the ingenious and indefatigable Dr. Priestley, presented to the Royal Society, his experiments on the different kinds of air, which clearly proves that putrid air arising from dunghills, and the perspiration of animals, are not only absorbed by vegetables, but also add to their increase. As those experiments are of too interesting a nature to be passed over unnoticed, for the benefit of those who have not had an opportunity of seeing the original, I shall transcribe part, referring the curious to the learned Doctor's publication.
>
> 'When air has been freshly and strongly tainted with putrefaction, so as to smell through water, sprigs of mint have presently died, upon being put into it, their leaves turning black; but if they do not die presently, they thrive in a most surprising manner. In no other circumstances have I ever seen vegetation so vigorous as in this kind of air, which is immediately fatal to animal life. Though these plants have been crowded in jars filled with air, every leaf has been full of life; fresh shoots have branched out in various directions, and have grown much faster than other similar plants, growing in the same exposure as common air. This observation led me to conclude that plants, instead of affecting the air in the same manner with animal respiration, reverse the effects of breathing and tend to keep the atmosphere sweet and wholesome, when it is become noxious, in consequence of animals either living and breathing, or dying and putrefying in it.'[1]

In the same year as Priestley made this contribution to plant physiology, James Anderson published *Essays Relating to Agricultural and Rural Affairs*. The essays owe their origin to the persuasive efforts of John Gregory (1724–73), professor of medicine at Edinburgh, where they were published. According to Anderson, their aim was 'rather to instruct the ignorant by a simple enumeration of a few, well-known established facts, than to amuse the speculative by the idle display of useless ingenuity.' Anderson was born in 1730 at Hermiston, near Edinburgh, of parents who had been connected with the land for several generations. In

[1] G. Winter, *System of Husbandry*, p. 93.

Edinburgh he studied chemistry under William Cullen (1710–1790), and began practical farming on his own account by the improvement of some 1,300 acres in Aberdeenshire. Profits accruing from his improved husbandry ultimately enabled him to remove to London, where he busied himself designing a patent hothouse. He died in 1808.

The paradox of the opposed hypotheses of Priestley and Scheele was resolved by Lazaro Spallanzani (1729–1799), who observed that aquatic plants give off bubbles of oxygen when in sunlight, but do not do so in the dark. This observation was confirmed, and its consequences extended, by John Ingenhousz[1] (1730–1799) and by Jean Senebier of Geneva (1742–1809).

In 1782 Senebier concluded that the increasing weight of a plant during growth, which van Helmont had erroneously attributed to the plant living solely on water, was in fact due to absorption of carbon dioxide from air. Thus it became clear that the gaseous exchange between plants and air is of two kinds; one, a process of respiration comparable with that of animals, the other, only taking place in sunlight and peculiar to vegetation, viz., *photosynthesis*.

> In speaking of the ordinary operations of nature in promoting vegetation, it may be observed that the air which we breathe, and which is also necessary to the growth of plants, is composed of certain subtile substances called *gases* by chemists. It may be stated shortly, that oxygen gas, or vital air, constitutes nearly 25 parts in the 100 of the atmosphere; nearly 75 parts are composed of azote, which is unfriendly to animal life; but it enters copiously into the food of plants.
>
> It may be added, from the observation of facts, that there appears to be some inherent quality in solar light, which has a powerful effect in promoting and in perfecting vegetable life. Chemical science does not seem, as yet, to have advanced so far as to analyze this influence, or to account for it. Its effects, however, are remarked by the most unenlightened. Cultivate any plant, the potato, for example, in the dark, it will vegetate; but it will not possess either the colour or the flavour of its kind. Solar light seems to be indispensably necessary to communicate their proper qualities to vegetables.[2]

Chemical science had at last begun to make useful contributions to the theoretical aspects of plant growth. On the severely practical side also it was beginning to be applied to the scientific study of

[1] For biographical notice of Dr. Ingenhousz, *see Ann. of Phil.*, (1817, *10*, 161), which deals with his relations with Dr. Priestley, Sir J. Sinclair, *et al.*

[2] P. Graham, *General View of the Agriculture of Stirling*, p. 136 *et seq.*

rooting media. According to S. Parkes, the best contemporary account of this subject was contained in the *Essay on Soils* by Dr. John Alderson (1757–1829) of Hull. One of the earliest attempts at a scientific investigation of soil constitution was that of George Winter. 'The operations of manures upon different soils', said Winter, 'are to be accounted for from their known properties, ascertained by repeated experiments. Hence, every agriculturalist ought to study and know the qualities of different soils, as well as manures, that they may be properly applied to each other. This science I deem to be one of the most useful of all human arts; it is similar to a physician's knowing the causes of diseases, the properties of medicines, and their most effectual applications.'[1] In 1787, Winter published his *New and Compendious System of Husbandry*, in which he gave for the first time a systematic scheme of soil analysis. The detail, of course, does not now interest us. He experimented with dung, lime, coal ashes, etc., as manures, and came to the conclusion that soaper's waste ashes[2] had the most pronounced effect on vegetation, producing great quantities of white clover and sweet herbage. In his own words:

> . . . soaper's ashes are an excellent manure, superior to most, but inferior to none, for clay, still cold soils, and loams. Soaper's ashes are a composition of lime and wood ashes, and of equal service for arable land and pasture lands; the quantity from one hundred and sixty, to two hundred bushels per acre, is the most proper.[3]

In the Bath Society Transactions, it was recorded that soaper's ashes doubled the quantity of hay. 'The farmers here will readily give from twelve shillings to a guinea for a waggon-load and fetch it five or six miles, and find their account in so doing'.[4]

For many years after the publication of Francis Home's *Principles*, little was added to the information contained in his treaties. In the 'eighties of the century, however, leading French chemists effected a reformation of chemical nomenclature, necessitated by the rapid advance of the science, and the next agricultural treatise to be published, that of Archibald Cochrane, Ninth Earl of Dundonald, adopted the new French mode of expression. It may justly be regarded as the first modern treatise. In this field Dundonald has also another claim to distinction; he added to the nutritive salts known to be required by plants, alkaline phosphates. These salts, we shall see in the sequel, are of great economic consequence.

[1] G. Winter, *op. cit.*, p. 337.
[2] Ch. V.
[3] G. Winter, *op. cit.*, p. 49.
[4] *Bath Soc. Trans.*, p. 130.

Agricultural practice, however, still lagged considerably behind industrial developments. Dundonald attributed this slow progress to want of appreciation on the part of agricultural writers of the inter-connexion between agriculture and chemistry. Rapid advances in chemistry having been made by the 1790s, Dundonald considered the time opportune for publishing an agricultural chemistry. In the introduction to his *Treatise: shewing the Intimate Connexion which subsists between Agriculture and Chemistry*, he pointed out that:

> This discussion will come forward with peculiar advantage at a time when provisions bear so high a price, and when individuals, awakening from the golden dreams of manufactures and commerce, begin to see, and experimentally to feel, that the prosperity of a nation cannot be permanent, nor its inhabitants quiet and contented, in their respective situations, where Agriculture is neglected, and an unwise preference given to manufactures and commerce; occupations that produce very different effects on the bodies and minds of men, from those that are attendant on the sober and healthful employment of husbandry.[1]

Dundonald's own life as a pioneer chemical manufacturer was one of disillusionment. In later life, he found city life and its concomitants oppressive. For a by no means individual malaise he concluded that there was but one remedy, 'that the establishment of such branches of manufacture, as it may be wise to establish, be promoted only in scattered villages, resembling the townships in America. By this plan the diseases of the body and the mind would be rendered less contagious'.[2] Unfortunately, no one heeded him.

Dundonald's treatise was directly linked to the noteworthy advances in chemistry which preceded its publication. He acknowledged the assistance that he had derived from the researches of Priestley and Cavendish, particularly on the properties of airs or gases and on the composition of water. On this subject, James Watt,[3] as well as Cavendish, was one of the principal workers.

By the time that Dundonald came to write, the composition of the atmosphere was more fully known, particularly the fact that it contained carbon dioxide (CO_2), a gas involved in processes of

[1] Dundonald, *Treatise on Agriculture*, p. 2. [2] *Ibid.*, p. 5.

[3] Concerning Watt's part in discovering the composition of water, the following may be consulted: J. P. Muirhead's *Correspondence of the late James Watt on his discovery of the composition of water* (London), 1846; G. Wilson, *Life of Cavendish* (London), 1851.

plant life. The consequences of Senebier's researches thus clear to him were succinctly stated.

> Atmospheric air is equally necessary to the vegetation and growth of plants, as to the life of animals. By a most beautiful arrangement in the œconomy of nature, the different processes of animal and vegetable respiration are made mutually to assist each other. The particular gas or air thrown off by the respiration of the one contributes to the existence of the other.[1]

Dundonald advocated the accurate analysis of vegetable substances by what in modern terminology would be called *combustion*, an operation which, in the hands of Justus Liebig, became of great importance in the development of analytical chemistry in the early part of the nineteenth century. Combustion converts vegetable matter into a mixture of simple gases which is amenable to analysis, and leaves behind the mineral content of the plant as ash.

> By vegetables being thus reduced to their simple or elementary principles, they are found to be composed of gases, with a small proportion of calcareous matter. Although this discovery may appear of small moment to the practical farmer, yet it is well deserving of his attention and notice as it throws great light on the nature and food of vegetables, and proves that a large proportion of vegetables consists of the aeryform fluids and gases.[2]

The ease with which mineral constituents, or, as Dundonald calls them, calcareous matter, could be analysed after combustion led to the adoption of mineral fertilizers in the belief that it would be beneficial to add them to the soil in the ratio in which they were found in the plant ash.

Dundonald's main contribution to agricultural chemistry was his realization of the significant part played by phosphates in vegetable economy. It had been known for a long time that certain materials, for example bones, promoted vegetation, but it fell to Dundonald to associate this with the presence in bones, etc., of phosphate of lime ($Ca_3(PO_4)_2$). He showed that the sodium and potassium salts could be prepared from the calcium compound, and that 'these alkaline phosphates will be found to promote vegetation in a very high degree'.[3]

The fertilizing effect of sulphate of potash (K_2SO_4) was also dealt with by Dundonald. It had been shown by Dr. Francis Home to 'promote vegetation in an extraordinary manner', but the developing connexion between agriculture and industry was

[1] Dundonald, *Treatise on Agriculture*, p. 17. [2] *Ibid.*, p. 27. [3] *Ibid.*, p. 75.

now significant. According to Dundonald, 'this substance is a refuse article in some branches of manufacture; but the quantity produced is a mere trifle, in comparison to the quantity that might advantageously be applied to the purposes of Agriculture'.[1] He also spoke of several other substances that might have been used as artificial fertilizers had they been obtainable in sufficient quantity. These included sulphate of ammonia ($(NH_4)_2SO_4$), and nitrate of soda ($NaNO_3$), the former of which is now of the greatest importance. Nitrate of potash (KNO_3) might also have been used, but competition on the part of armament manufacturers increased the price so much that it was beyond the means of agriculturalists to purchase it.

Dundonald concluded his treatise with this peroration: 'More might have been said on the practical side of husbandry; but, unluckily for science, too much has already been written on that subject, and absurd theories have been too often blended with practice.'

Dundonald's treatise was followed by two considerable contributions to agricultural science: (*a*) Richard Kirwan's *On the Manures most advantageously applicable to various Sorts of Soils, and the Causes of their beneficial Influence* (London, 1796), and (*b*) Erasmus Darwin's *Phytologia, or the Philosophy of Agriculture and Gardening* (London, 1800).

It was Sir John Sinclair who persuaded Darwin to write this work, which Darwin in turn dedicated to him. We thus have some slender evidence of contacts being established between Sinclair and the Lunar Society of Birmingham. Maria Lovell Edgeworth, the daughter of another member of the Society, consulted Sinclair on culture of fruit trees. John Wilkinson, iron-master, who had many business relations with the Soho group, consulted Sinclair with regard to the improvement of his mossy land. In his correspondence, Sir John spoke of having gone to Castlehead, Lancashire, to observe at first hand the condition of the land on which Wilkinson desired advice.[2]

Samuel Parkes credited Erasmus Darwin with the discovery of the utility of ammonia to vegetation, but this is by no means clear to us in Darwin's own writing. True, of nitrogen and ammonia Darwin wrote:

> We come to the other ingredient, which constitutes a much greater part of the atmosphere than the oxygen, and this is the azote, or nitrogen; which also seems much to contribute to the

[1] Dundonald, *Treatise on Agriculture*, p. 82.
[2] J. Sinclair, *Correspondence*, Vol. 1, p. 481.

> food or sustenance of vegetables; for the azote, or nitrogen, enters into animal bodies in much greater quantities perhaps than into vegetables, so as to constitute according to some chemical philosophers the principal difference between these two great classes of organized nature; yet it enters also into the vegetable system, and is given out by their putrefaction; and also when lime is applied to moist vegetables it disengages from them both hydrogen and azote, forming volatile alkali (NH_3), as asserted in the ingenious work of Lord Dundonald on the Connexion of Agriculture with Chemistry.[1]

This mention of lime ($Ca(OH)_2$) and ammonia (NH_3) is important, since it was a point of controversy whether lime should be applied at the same time as dung. For example, Winter, in the nomenclature of his day, said:

> The objection I make to immediately mixing unslaked lime with dung, is, that as unslaked lime contains very caustic and absorbing properties, it destroys the oleaginous and vegetative particles of the dung, similar to that of fire in burning coal, which is reduced from its original. By the mode I have pointed out of applying the lime and dung separately, the soil is corrected by the lime, and enriched by the dung.[2]

Dundonald puts this in modern terminology:

> The effects produced on organic bodies by lime, clearly point out, that lime should never be mixed with dung, or with any substances which of themselves, or by the application of saline matters, would easily become putrid and rotten. Lime not only puts a stop to the putrefactive process, but disengages and throws off, in the gaseous state, a certain portion of the component parts of such substances; whilst, with those which remain, it forms insoluble compounds that are incapable of promoting vegetation, until they are again decomposed and brought into action by other substances.[3]

His objections could thus be supported by scientific evidence. About 1800, ammonia itself was tried as a manure, and it was not long before waste ammonia liquor from the gas establishments, the invention of William Murdock, another member of the Soho Group, was tried as a manure. Thus yet another synthetic material derived from coal which was initially unavailable as a manure in Dundonald's day became obtainable as a further advance in the application of chemistry to industry.[4]

[1] E. Darwin, *Phytologia*, p. 195. [2] G. Winter, *op. cit.*, p. 35.
[3] Dundonald, *Agriculture and Chemistry*, p. 120.
[4] S. Parkes, *Chemical Catechism*, p. 151 *n.*

Before discussing further chemical advances, we must digress to consider two events of signal importance which took place during the last decade of the eighteenth century. The first was the founding of a Chair of Agriculture in the University of Edinburgh, the oldest in the world. The second was the founding of the Board of Agriculture by Pitt, largely as the result of the stimulus of Sir John Sinclair.

First we must trace the evolution of agricultural education. About 1790, the Highland Society recommended its members to attend lectures on agriculture and other agrestic subjects to be given by Dr. James Walker in the University of Edinburgh. Walker was Professor of Natural History, so agriculture was not entirely foreign to his province. Shortly after Walker established these lectures, the first steps towards founding a Chair of Agriculture were taken, Lord Kames and the Highland Society supporting the proposal. As a result, an endowment of £50 towards the Chair was provided by Sir William Pultney. Walker, however, was not elected to the Agriculture Chair. It went instead to Dr. Andrew Coventry. Coventry was an experienced agriculturist who put his profession to practice in Kinross-shire. There he possessed an estate where he is said to have set an example of knowledge, industry, and good management to all his neighbours. A preamble to Coventry's lectures[1] was published in 1808, and may be consulted by those who wish to get an idea of the course in Agriculture provided by the pioneer University.

The realization in Edinburgh that agrarian economy was not beneath the notice of the academician did not pass without comment in other long-established quarters, but in 1830, speaking of Professors of Agriculture, it could still be said, of chairs outside Scotland, 'there is one in Dublin, supported by the Dublin Society; one in Cork; and one is destined to be established at some future period in Oxford, agreeable to the will and donation of Dr. Sibthorpe, professor of botany there'. By that time in Scotland, so august a person as the Professor of Logic and Belles Lettres in the University of St. Andrews, William Barron, had published essays on the mechanical principles of ploughs!

Following their success in promoting the Chair of Agriculture, the Highland Society suggested to the Edinburgh University authorities the founding of a Chair of Comparative Anatomy, to embrace the study of veterinary physics and surgery. The University refused, however, because they considered that such subjects

[1] *Discourses Explanatory of the Object and Plan of the Course of Lectures on Agriculture and Rural Economy.*

'implied adjuncts scarcely compatible with university life'. The result of this nice-mindedness was the evolution of the Royal (Dick) Veterinary College as an autonomous institution.

The endowment provided by Pultney for the Edinburgh Chair was inadequate, so for the greater part of the nineteenth century the Highland (and Agricultural) Society, as it became, contributed generously towards its upkeep. In Edinburgh, Professor Coventry was followed by David Low, who is remembered as a noted contributor to the literature of agriculture. He in turn was succeeded by James Wilson. Together these men brought some system to the subject they professed, and we can trace their influence in the innovations of the period. For example, the Edinburgh district, as is not unnatural, witnessed one of the first investigations which might properly be called a 'soil survey'. It was carried out in East Lothian at the beginning of the nineteenth century, and was incorporated in the re-edition of the Report on Agriculture. It shows that the fundamental importance of a knowledge of soil conditions was beginning to be appreciated. Here is a citation in evidence:

> Upon a subject so important as the soil of an extensive and valuable district, something more than a bare enumeration of the different kinds, and the quality of each, appears necessary; indeed, if it is ever expected that agriculture shall become a perfect science, and have its operations directed by fixed laws or principles, a correct knowledge of the nature and properties of different soils must appear indispensable, as without that, it will be found impossible either to remove the faults that occasion sterility, or employ such articles as have a tendency to render it fertile, with any degree of certainty or advantage.
>
> That point gained, the cultivator will no longer grope in the dark with regard to his operations; knowing the principles contained in his soil, he will readily understand the nature of the compounds formed between these principles and the manures he employs in that way, the application of fertilizing substances, will be brought to a degree of certainty greatly beyond what is at present known.[1]

Typical soil samples were collected and subjected to a simple scheme of analysis based on such knowledge as was available. While the results were expressed rather in the manner of Francis Home than in that of the later nineteenth-century chemists, they clearly demonstrated to the practical agriculturist of the time the analytical distinctions between the varied soils which he was called upon to cultivate.

[1] R. Somerville, *General View of East Lothian*, p. 279.

Although the significance of soil science was undoubtedly appreciated, the accurate analysis of soils did not develop with anything like the rapidity which was desirable. Even Humphry Davy was not spoken of too kindly as a soil analyst by the famous Justus von Liebig. Speaking of Davy's analysis of a very fertile sandy soil from the vicinity of Tonbridge, Kent, Liebig remarked:

> The great Davy, who was convinced of the importance of the inorganic constituents of soils, has omitted to detect the phosphoric acid, potash, soda, and manganese. All these must have been present in the soil, for we are informed that it produced good hops, for which these ingredients are indispensable. . . . Davy made several analyses of various fertile soils, and since his time numerous other analyses have been published; but they are all so superficial, and in most cases so inaccurate, that we possess no means of ascertaining the composition or nature of English arable land.[1]

Be that as it may, Davy was instrumental in emphasizing the importance of chemistry to the class of land-owner who was in a position to make use of it, through lectures he delivered before the Board of Agriculture, an institution about which a little must now be said.

It was Sir John Sinclair who was largely responsible if not for the conception, at least for bringing into being, the Board of Agriculture. Sinclair relates that initially Sir Joseph Banks, President of the Royal Society, which at its foundation gave a show of being interested in agriculture, was hostile to the former's idea of forming a Board of Agriculture, on the assumption, perhaps, that the new institution would detract from the comprehensive status of the Royal Society. Sinclair succeeded in winning him over by recommending that the President of the Royal Society should *ex officio* be a member, and ultimately Banks became a useful addition to the Board. He later induced Sinclair to write his *System of Husbandry of Scotland*.[2]

The first secretary of the Board was Arthur Young (1741–1820). Young was the son of the Rev. Arthur Young, and on his father's death took over an eighty-acre farm, the property of his mother. To obtain books to read he became an author, exchanging his own writings for books. Of his contributions to scientific agriculture, Richard Kirwan spoke highly. 'To the labours of Mr. Arthur

[1] Justus Liebig, *Chemistry in its Application to Agriculture and Physiology*, pp. 238 and 240.

[2] J. Sinclair, *Correspondence*, Vol. 1, p. 404. Part VII of Sinclair's *Correspondence* is devoted to Agriculture, and may be consulted for further details of his connexion with contemporary agriculture.

Young the world is more indebted for the diffusion of agricultural knowledge, than to any writer that has yet appeared.'[1] He corresponded with Joseph Priestley, and is said to have acquired a taste for pneumatic chemistry from him. An interesting account of his own experiments on manures will be found in the *Retrospect of Philosophical and Chemical Discoveries*. His wife's grandfather was reputed to be the first to use marl in Norfolk.[2]

One of the first activities of the Board of Agriculture was to arrange for the Assistant Professor of Chemistry at the Royal Institution to deliver a series of lectures emphasizing the connexion between Chemistry and Vegetable Physiology. The professor, although only twenty-three years of age, was already established as a brilliant lecturer. He was Humphry Davy (1778–1829). Although Sinclair gave the impression that it was the Board of Agriculture that brought the subject to the attention of Davy, he early evinced an interest in the importance of chemistry in its application to this field. This is evidenced by his letters and his introductory lecture in which he said:

> Agriculture, to which we owe our means of subsistence, is an art intimately connected with chemical science; for although the common soil of the earth will produce vegetable food, yet it can only be made to produce its greatest quantity, and of the best quality, by methods of cultivation dependent upon scientific principles.
>
> The knowledge of the composition of soils, of the food of vegetables, of the modes in which their products must be treated, so as to be fit for the nourishment of animals, is essential to the cultivator of land; and his exertions are profitable and useful to society, in proportion as he is more of a chemical professor.[3]

Before he went to London, Davy was fortunate in making contact with several scientists and industrialists who visited or had connexions in the south of England. One of the first of these was Davies Giddy or Gilbert, a Royal Society President with whom he formed a lasting friendship. The residence of Gregory Watt with Davy's mother brought him into direct contact with members of the Lunar Society. When Wedgwood and his brother spent a winter in Penzance, friendship developed between them and Davy, and it has been suggested that it was of substantial benefit to the latter. Gregory Watt, backed by Davies Gilbert, proposed Davy to Dr. Beddoes as an assistant at the Pneumatic

[1] R. Kirwan, *Trans. Roy. Irish Acad.*, Vol. 5.
[2] J. A. Paris, *Quart. Jour.*, 1820, *9*, 280.
[3] H. Davy, *Elements of Agricultural Chemistry*, p. 180.

Institution of Bristol. James Watt designed and made apparatus for handling the gases there. His friendship with Watt's son, Gregory, led to several visits to Birmingham, where 'his ambition was constantly excited by intercourse with congenial minds'. One of them was James Keir, to whom we have frequently referred. Writing to Gilbert on February 22, 1799, Davy said:

> I spent nine or ten days there (Birmingham), chiefly with Mr. Keir and Mr. Watt: I had a great deal of chemical conversation with them. Mr. Keir is one of the best-informed men I have ever met with, and extremely agreeable. Both he and Mr. Watt are still phlogistians: but Mr. Keir altogether disbelieved the doctrine of *calorique*.

While at Bristol, Davy was further influenced by the Birmingham group. According to Maria Lovell Edgeworth, her father's influence over Davy when he was still an unknown youth at Clifton was considerable. Edgeworth 'early distinguished and warmly admired his talents, and gave him much council, which sank deep into his mind'.

On 16th February, 1801, Davy was appointed Assistant Lecturer in Chemistry, Director of the Laboratory, and Assistant Editor of the Journals at the Royal Institution. In July the Managers requested that he should give some lectures, to begin in November, on the Chemical Principles of the Art of Tanning. He was given three months' leave of absence to make himself acquainted with the art, being supplied with necessary raw materials by Sir Joseph Banks. Upon the economics of this art he subsequently made a succinct remark which bears quotation: 'The great secret, on which the profit of the trade depends, is to give the hides the greatest increase of weight in the least possible time.'[1]

The first we hear from Davy of the agricultural lectures is a letter to Davies Gilbert written from the Royal Institution on 26th October, 1802:

> I am beginning to think of my Course of Lectures for the winter. In addition to the common course of the Institution, I have to deliver a few lectures on Vegetable Substances, and on the connexion of Chemistry with Vegetable Physiology, before the Board of Agriculture. I have begun some experiments on the powers of soils to absorb moisture, as connected with their fertility. I have, for this purpose, made a small collection of those of the *calcareous* and *secondary* countries, and I wish very much for some specimens from the *granitic* and *schistose* hills of Cornwall.[2]

[1] J. A. Paris, *Life of Humphry Davy*, Vol. 1, p. 171. [2] *Ibid.*, Vol. 1, p. 158.

Watt's son Gregory supplied him with many of the samples on which to work.[1]

The first lecture was delivered on Tuesday, 10th May, at twelve noon.

Davy's biographer, J. A. Paris, found the following allusion to the first lectures in Arthur Young's Memoranda: 'Mem. Two lectures by Mr. Davy have taken place, and have been very well attended; they intend retaining him by a salary of a hundred pounds a year—a very good plan. May 15th, 1803.' Five more lectures followed on succeeding Tuesdays and Fridays. The arrangements between Davy and the Board of Agriculture which preceded their delivery were given by Sir John Sinclair.

> In the year 1802 . . . the Board resolved to direct the attention of the celebrated lecturer, Mr. Davy, to agricultural subjects; and in the following year . . . he first delivered to the members of this Institution, a course of lectures on the *Chemistry of Agriculture*. The plan has succeeded to the extent which might have been expected from the abilities of the gentleman engaged to carry it into effect.[2]

Davy's success was followed by his appointment as Professor of Chemistry to the Board of Agriculture, an appointment which carried with it the obligation to lecture to members of the Board. For ten successive years he delivered a course of lectures, gradually modified to meet the changing outlook of chemistry. In 1813 they were published as *Elements of Agricultural Chemistry*, one chapter at least of which may still be read with interest, that on *Soils and their Analysis*.

Sinclair attributed the international reputation which Davy attained to his connexion with the Board of Agriculture, and claimed not a little of the glory for himself as Davy's patron. Be that as it may, he did, however, render Davy conspicuous service in getting a Bill through the House for improvements to the Royal Institution, where Davy carried out his experimental work. Here are Sinclair's own comments:

> No circumstance could be more gratifying to me, than to have a share in promoting the success of so eminent a philosopher as Sir Humphry Davy, whose talents have thrown such a lustre on his age and country. To his more early friends, Sir Joseph Banks and Mr. Davies Gilbert, he was greatly indebted; but no circumstance contributed more to his success in life, than his con-

[1] H. Davy to J. Watt, Jr., 6th February, 1803, Birmingham Reference Library.

[2] J. Sinclair, *Address to the Board of Agriculture*, April, 1806.

nexion with the Board of Agriculture, as he derived considerable emolument for his services to the Board, and there became acquainted with a number of the most distinguished characters of the kingdom.[1]

The 'considerable emolument' was one hundred pounds per annum; one of the 'distinguished characters' Sir Thomas Bernard (1750–1818), a zealous philanthropist. Shortly after arriving in London, Davy became acquainted with Sir Thomas, and as his interest in agriculture developed, Bernard put at his disposal for experimental purposes a considerable piece of land at Roehampton. Many successful experiments were conducted there, and were later used to illustrate the lectures on agricultural chemistry.[2]

The scientist Davy was not alone in engaging in practical experimentation. In fact, other scientists went even further than he did. His continental *vis-à-vis*, Lavoisier, cultivated two hundred acres on chemical principles in order to set a practical example to neighbouring farmers. His scientific mode of culture was attended with such success that he obtained at first a yield one-half greater than was procured by the traditional methods. After nine years cultivation, his annual yield of produce had doubled. Like Lavoisier, Priestley too was interested in agricultural improvement. From America, where he took refuge after the Birmingham riots, he communicated to Sir John Sinclair an account of agricultural discoveries made in America. In the postscript to a letter dated Philadelphia, 29th April, 1797, he spoke of directing experiments on the use of gypsum (calcium sulphate, $CaSO_4$) as a manure. His son became an American farmer.

Davy, in the preface to his lectures, made particular reference to Scottish contributions to agriculture, thus acknowledging his indebtedness to his predecessors. Here is what he said:

> I shall mention the Earl of Dundonald's Treatise on the connexion of Chemistry with Agriculture; Mr. Rennie's Dissertations on Peat; and the General Report of the Agriculture of Scotland. This last work did not come into my hands till the concluding sheets of these lectures were printing. Had it been in circulation before, I should have profited by many statements given in it, particularly those of the enlightened Professor of Agriculture in the University of Edinburgh; and I should have dwelt with satisfaction on the importance given to some chemical doctrines by his experience.[3]

[1] J. Sinclair, *Correspondence*, Vol. 1, p. 432.

[2] J. A. Paris, *op. cit.*, Vol. 1, p. 188.

[3] H. Davy, *op. cit.*, p. vi.

While Davy's lectures did not add to the principles already known to his predecessors he did sift and clarify much of the confused material which was handed down to him from them. As a brilliant lecturer, the association of his name made agricultural chemistry fashionable. On the other hand, when, as did happen occasionally, he supported hypotheses later shown to be untenable, they were given longer credence than would otherwise have been the case. As an example of this one might instance his failure to appreciate the significance of the work of Theodore de Saussure, Cruickshank, and Scheele, who showed that plants synthesize their carbohydrates from atmospheric carbon dioxide.

Davy's researches in the physics of soils was probably his most original and valuable contribution. In this field he undoubtedly extended the work of Richard Kirwan (1733–1812) and George Fordyce (1736–1802), the distinguished Aberdeen-born physician and teacher of medicine, whose nephew, George French (1783–1833), was the first Professor of Chemistry in the University of Aberdeen. After covering the whole gamut of agricultural operations, fallowing, rotations, manuring, etc., Davy concluded:

> I trust that this enquiry will be pursued by others; and that in proportion as the chemical philosphy advances towards perfection, it will afford new aids to agriculture: there are sufficient motives connected with both pleasure and profit, to encourage ingenious men to pursue this new path of investigation. Science cannot long be despised by any persons as the mere speculation of theorists, but must soon be considered by all ranks of men in its true point of view, as the refinement of common sense guided by experience, gradually substituting sound and rational principles for vague popular prejudices.
>
> The soil offers inexhaustible resources which, when properly appreciated and employed, must increase our wealth, our population, and our physical strength.
>
> We possess advantages in the use of machinery, and the division of labour, belonging to no other nation. Nothing is impossible to labour, aided by ingenuity. The true objects of the agriculturist are likewise those of the patriot.[1]

Humphry Davy had various connexions with Scotland. During his first years in London, much of his social success was achieved under the patronage of the Duchess of Gordon, who was herself keenly interested in agricultural improvement, to which end she founded a farming society in Badenoch which was reputed to have excited a great spirit of improvement in that part of the kingdom.

[1] H. Davy, *op. cit.*

According to her own statement she was inspired to do this by the example of Sir John Sinclair, to whom she wrote, 'I wish to make the inhabitants of the most beautiful country in the world, happy, rich, and independent'.[1]

In April 1812, Davy married the daughter of Charles Kerr, of Kelso, and widow of Shuckburgh Ashby Apreecee. After the marriage, Davy and his wife toured Scotland, where Lady Davy was already well known, particularly in Edinburgh, where she came in 1807 on the death of Apreecee. In Edinburgh she had opened her house to the *élite* of Edinburgh society, and her friendship with Madame de Staël made her a conquest with Edinburgh hostesses. Although it was reported that she and Davy were 'received everywhere with the most flattering marks of attention', little is known of the contacts they made. They were in Edinburgh in July, and at Dunrobin Castle, near Golspie, in August, where they resided with the family of Lord and Lady Stafford. As Elizabeth, Countess of Sutherland (1765–1839), Lady Stafford succeeded her father in 1766. In 1785 she married George Granville Leveson-Gower, who became second Marquess of Stafford on the death of his father in 1803. Golspie was the limit of their journey north. They returned to Edinburgh in October, having visited Gordon Castle and Dunkeld *en route*.

From Edinburgh, Davy wrote that 'there is little doing here', but spoke of meeting Playfair and Sir James Hall.[2] While there he took part in one of the important experiments undertaken to elucidate the chlorine controversy. Davy maintained that the substance chlorine, which had been used as a bleaching agent for a quarter of a century, was an element. In this he was supported by his brother, Dr. John Davy. Dr. John Murray, however, was the 'avowed partisan of the theory of Berthollet' (who incidentally had discovered the bleaching properties of chlorine), that the substance was an oxygen-containing compound. One of Murray's experiments was suspect by the Davys. While in Edinburgh Sir Humphry desired to repeat it with Dr. Thomas Charles Hope (1766–1844). This was effected in the College Laboratory, in presence of Sir George Mackenzie, Playfair, and several other gentlemen. The results of the experiment disproving Murray's contention were published by John Davy in *Nicholson's Journal*.[3]

Although Davy is often regarded as the first true agricultural

[1] J. A. Paris, *op. cit.*, Vol. 1, p. 137; J. Sinclair, *Correspondence*, Vol. 1, p. 160.

[2] J. A. Paris, *op. cit.*, Vol. 1, p. 350.

[3] *Nicholson's Journal*, 1812, *30*, 28 and subsequent letters; J. A. Paris, *op cit.*, Vol. 1, p. 335.

scientist, he 'still belonged to the old school who were largely groping in the dark', as we clearly recognize when we realize that some of Davy's beliefs were that 'oils were good manures because of the carbon and hydrogen that they contained', or that soot was good because its carbon was 'in a state in which it is capable of being rendered soluble by the action of oxygen and water'. Lime, he thought, was useful because 'it dissolved hard vegetable matter'.[1] It was only after 1815 that the effects of the application of new chemical methods of analysis began to be apparent; for example, we see it in the Report which Sinclair made to the Board of Agriculture on *The Agricultural State of the Netherlands compared with that of Great Britain.*

Besides encouraging Davy, another activity of the Board of Agriculture was the publication of a series of county *Reports* or *General Views of the State of Agriculture of the various counties*. To what extent these *General Views* give a true picture of conditions may be questioned on the strength of the following somewhat revealing comments by J. C. Loudon, on difficulties concerning the publication of the Peebles Report.

> The agricultural survey of this county is by the Rev. Charles Findlater, and it abounds in more valuable matter on political agriculture, or leases, prices, restrictions, markets, etc., than any survey that has been published, without a single exception. In fact, it was found to take such a masterly view of the moral incitements to agriculture; to expose the system of tithes, entails, lawyer's leases, etc., that it was rejected by the Board, as likely to offend the English clergy and the higher classes, and the author was reduced to publish it himself.[2]

Their publication certainly engendered friction in some quarters, as also did Sinclair's suggestion to undertake the production of a *Statistical Account for England* which had to be abandoned as 'the Archbishop of Canterbury refused to sanction it, lest it should interfere with the tithes of the clergy'.

Loudon did not spare himself in his censure of the Board.

> The Board of Agriculture was founded, under the authority of government, in 1793. Much was expected from this Board; but, except the publication of the county reports, and the general attention which it called to agriculture, it may well be asked what advantages arose from it. Their *Communications*, in several quarto volumes, contain fewer valuable papers, in proportion to their total number, than the publication of either the *London Society of*

[1] J. C. S. Watson and M. E. Hobbs, *Great Farmers*, p. 74.
[2] J. C. Loudon, *Encyclopaedia of Agriculture*, p. 1183.

> *Arts* or the *Bath Society*. In short it has been ably shown in *The Farmer's Magazine* and the article *Agriculture* in the supplement to the *Encyclopaedia Britannica* that the Board never directed its efforts in a manner suitable to its powers and consequence; and that, instead of discussing modes of culture, its attention ought to have been directed to the removal of political obstacles to agriculture, and to the eliciting of agricultural talent by honorary rewards, etc.

In attempts to curtail national expenditure following the Napoleonic War, the Government, despite the impulse given to agriculture by higher prices during the war, withdrew its support of the Board, and, according to Loudon, 'there being no longer funds for a handsome salary for a secretary, it soon fell to pieces, and is now only remembered at least by us for its lofty pretentions and its worse than inutility'.[1]

If Sinclair's Board failed in its mission, he himself was not slow to recognize some of the needs of contemporary agriculture, for example, the need for experimental farms conducted either by agricultural societies or by the State. 'The establishment of even one farm of that description, on a proper scale,' said Sinclair, 'would be more valuable to this country than the conquest of many distant possessions'. This, it might be mentioned, was voiced in 1814. He continued:

> Without experimental farms, carried on either by the great associations or at the public expense, theory on the one hand, and prejudice on the other, will long prevent agriculture from being reduced to a science. Experimental farms, properly conducted, would remove every doubt on the subject; and every material improvement, in the management of arable and grass land, in machinery, or in regard to livestock, would be speedily adopted. There are a great many desiderata, particularly with regard to livestock, that will not be fully determined without such farms, for many years to come, and respecting which conjecture, or at most probability, must in the meantime supply the place of established facts.[2]

Transfer of the laboratory pot technique initiated by Francis Home to large-scale field experiment was first systematically effected by J. B. Boussingault in Alsace. This marked a new forward step in agricultural science. The quantitative laboratory methods devised by de Saussure were operated on a large scale in such a way that a profit and loss account of manures and resulting crops could be struck. Thus Boussingault ascertained the gain

[1] J. C. Loudon, *Encyclopaedia of Agriculture*, p. 1224.
[2] J. Sinclair, *An Account of the Systems of Husbandry of Scotland*, Vol. 2, p. 140.

from the atmosphere, rain, the soil, etc. He examined the effect of varied rotation and argued that the best was the one that yielded the greatest amount of organic matter beyond that added as manure. As an example we may take the way in which soil nitrogen can be increased by rotation. This is illustrated by the following table. The increase in the case of the leguminous crop, lucerne, is significant, but consideration of it will be deferred till later.

BOUSSINGAULT'S NITROGEN BALANCE SHEET FOR DIFFERENT ROTATIONS—1841[1]

Rotation	*Nitrogen in Kilos per Hectare*			
	in Manure	*in Crop*	*Excess in Crop*	
			per Rotation	*per Annum*
A. (1) Potatoes, (2) Wheat, (3) Clover, (4) Wheat and Turnips (catch crop), (5) Oats . .	203·2	250·7	47·5	9·5
B. (1) Beet, (2) Wheat, (3) Clover, (4) Wheat and Turnips (catch crop), (5) Oats . .	203·2	254·2	51·0	10·2
C. (1) Potatoes, (2) Wheat, (3) Clover, (4) Wheat and Turnips, (5) Peas, (6) Rye, Jerusalem Artichokes (two years) .	243·8	353·6	109·8	18·3
D. (1) Dunged fallow, (2) Wheat, (3) Wheat .	82·8	87·4	4·6	1·5
E. Lucerne, five years .	224·0	1,078·0	854·0	170·8

It is quite impossible for us to go into the detail of rotation here, such is the bewildering variety of conditions which determine

[1] From E. J. Russell, *Soil Conditions and Plant Growth*, p. 14.

it. Moreover, some of the determining factors were outside the confines of agriculture altogether. Here is an excellent example of the unaccountably complicated effect of fiscal policy on technology, in this instance agriculture.

> The depriving the farmers of Aberdeenshire of the power of disposing of their bear[1] to distillers, by abolishing the privileges of the intermediate district, occasioned so low a price of that grain, and also so dull a sale, that the farmers gave up sowing bear on the second crop after breaking up out of grass; and the excessively huge malt-tax, joined with the imposing an unfair proportion of that tax on malt made from bigg, compared with that from barley, has now induced farmers to lay down their grass seeds very frequently with early oats.

Indeed, we may say with the writer of the first *General View of the Agriculture of Aberdeenshire*, 'There is no general system, nor plan of rotation of crops there adopted, every person varying his crops in the manner he thinks will turn out to best advantage for himself'.[2] Conditions varied throughout the country. After giving an instance of the failure of summer fallowing, Keith wrote:

> This circumstance is mentioned merely to show the absurdity of laying down an invariable system of agriculture, whether flatly condemning, or indiscriminately extolling, summer fallow for example. Great Britain is not all Norfolk, nor is it all the soil of East Lothian, any more than of Aberdeenshire.[3]

The large-scale experimentation initiated by Boussingault was unfortunately not continued, even in his own country. Boussingault, in collaboration with Dumas, published an important essay on his work in 1841. As far as Great Britain was concerned, it was some years before Sinclair's vision of experimental farms was realized, but once established, the research carried on by the experimental institutes was of great utility in spreading scientific knowledge.

Another new phase in agricultural science began with a publication in 1840 by Justus, Baron von Liebig. Liebig was born in 1803 and became Professor of Chemistry at Giessen in 1826. In 1838 he began investigating relations of chemistry to animal and vegetable life, and two years later presented to the British Association the now famous report on the state of organic chemistry, later published as *Chemistry in its Application to Agriculture and Physiology* (1840). Scottish professors helped to spread Liebig's

[1] Species of barley, also called *bigg*.

[2] G. S. Keith, *General View: Aberdeenshire*, p. 241.

[3] *Ibid.*, p. 230.

doctrines through their translations of this and others of his works. Lyon Playfair of Edinburgh edited *Chemistry in its Applications to Agriculture*, and William Gregory, Professor of Chemistry in King's College, Aberdeen, his *Animal Chemistry*.

Although Scotland lagged far behind in cultivation till the middle of the eighteenth century, by the 'forties of the nineteenth, it had outstripped England, especially in arable husbandry. This is evidenced by the fact that improvements introduced into England on arable land consisted, with few exceptions, of the practices and implements from Scotland.[1] In 1843, Midlothian farmers formed an Agricultural Chemistry Association,[2] whose chief promoter was John Finnie of Swanston. Numerous papers by their chemist, Professor Johnston of Durham, appeared in the *Transactions* of the Highland Society.

> The south-eastern arable district of Scotland was a land of big farms. East Lothian farms averaged from 200–300 for the smaller to 500–600 for the largest. In the pastoral counties of the south—Peebles, Selkirk, Dumfries, Kirkcudbright, and Wigtown, acreages were less. In the south-western lowlands—Ayr, Renfrew, Lanark, Dumbarton, farms were relatively small, but much bigger than the run-rig farms of 1760. The north-eastern lowlands from Kincardine to Nairn had mastered the turnip husbandry and the growing of artificial grasses. Aberdeenshire farms from 1790 were neat and tidy. In the central counties, Fife, Kinross, Clackmannan, Forfar, Stirling, and Perth conditions were exceedingly varied, and the Highlands and Islands had still a system of their own.[3]

From success in industrial capitalism, the new captains of industry often turned to agricultural improvement as a pastime in their declining years, a social phenomenon not by any means confined to Scotland. Land in the neighbourhood of Glasgow was improved by the iron-master, Colin Dunlop. At Tollcross, three miles out of Glasgow, the land was so poor that it would not support even rabbits in a warren. Dunlop formed this land into a profitable estate. His example was followed by circumjacent proprietors, with the result that the land in the vicinity of Clyde Iron Works was converted into some of the best in Scotland.[4]

[1] C. J. Loudon, *op. cit.*, p. 1178.

[2] When, in 1848, the Association came to an end, its work was taken over by the Highland Society, and a consultant chemist, Dr. Thomas Anderson (1819–1874), was appointed. He was followed by Mr. James Dewar, and he again by Dr. A. P. Aitken, a notable pioneer in field experimental methods. (Watson and Hobbs, *Great Farmers*, p. 230.)

[3] J. H. Clapham, *Economic History of Modern Britain*, Vol. 1, p. 108.

[4] D. Murray, *York Buildings Company*, p. 103.

In England, Matthew Boulton too was interested in agriculture. He was elected an honorary member of the Odiham Agricultural Society[1], and we learn of the Secretary of the Society sending him at Priestley's request *Queries relating to Agriculture.*[2]

We gather some details of the improving activities of the industrialist, Charles Macintosh, from a letter to Joseph Gordon, Esq., of Jamaica, dated 26th September, 1829.[3]

Macintosh carried out improvements on the two-hundred acre Crossbasket Farm, in the north-east corner of East Kilbride parish, Lanarkshire. They consisted of draining and summer-fallowing to extirpate weeds. He experimented with bones ground to the size of peas as manure applied at the rate of one ton per acre. *Greaves* (i.e. tallow chandler's refuse which we recalled *cracklings* in Scotland) were tried, at the rate of 12–15 cwt. per acre, along with a little lime. Macintosh found kelp a powerful manure when applied at the rate of half a ton per acie. He regarded the simultaneous application of lime as essential. The prices which he paid for various manures were:

Crushed bones	£5 per ton
Cracklings. . . .	£4 6*s*. per ton
Kelp . . .	£4 per ton
Woollen rags . . .	£2 10*s*. to £3 per ton
Horn chips . . .	£7 to £8 per ton

James Smith, of Deanston Cotton Mill, farmed two hundred acres of Perthshire. In 1831 he published *Remarks on Thorough Draining and Deep Ploughing.*[4]

By this time the full significance of Dr. Joseph Black's isolation of carbon dioxide had been realized. From the carbon dioxide of the atmosphere, and water, plants synthesized sugar, cellulose, starch, oil, etc. The age-old belief that 'corruption is the mother of vegetation' was scotched, since most of, or perhaps all, the carbonaceous food of plants comes from the atmosphere. The small mineral requirements (potash, phosphate, etc.) discoverable by the analysis of plant-remains, could be supplied by manuring. With regard to nitrogen, Liebig, like Davy, believed that the nitrogen, absolutely essential for plant growth, came from the atmosphere, and was absorbed as ammonia. The function of humus was to evolve carbon dioxide; phosphates were necessary for seed formation; potassium silicate for general development.

[1] J. Huntingford to M. Boulton, 21st June, 1786, Assay Office Library, Birmingham. [2] J. Huntingford to M. Boulton, 21st September, 1785.

[3] G. Macintosh, *Memoir of Charles Macintosh*, p. 100.

[4] R. Chambers, *Domestic Annals*, Vol. 1, pp. 18 and 253.

Writing in the *Farmers' Magazine*, Liebig said: 'If the soil is suitable, if it contains a sufficient quantity of alkalis, phosphates and sulphates, nothing will be wanting. The plants will derive their ammonia from the atmosphere, as they do carbonic acid'.[1] Of this, Sir John Russell says:

> So much did Liebig insist, and quite rightly, on the necessity for alkalis and phosphates, and so impressed was he by the gain of nitrogen in meadow land supplied with alkalis and phosphates alone, and by the continued fertility of some of the fields of Virginia and Hungary and the meadows of Holland, that he began more and more to regard the atmosphere as the source of nitrogen for plants. Some of the passages in the first and second editions urging the necessity of ammoniacal manures were deleted from the third and later editions.

That nitrogen is indeed absorbed as ammonia, which as we shall see may come either from the atmosphere, from added manure, or from the soil, was confirmed by Lawes, Gilbert, and Pugh, but at the time Liebig's beliefs lacked foundation. In the compounding of his manures, too, Liebig, by pushing his theories on too precipitously, made some unfortunate mistakes, but his ideas enabled a link-up between industry and agriculture to be effected. While he was at fault in his theory of mineral fertilizers, his writings helped to revolutionize the attitude that agriculturists had maintained towards chemistry, and in the hands of subsequent popularizers like Johnston of Durham and Augustus Voelcker of Cirencester, his discoveries had a great effect, particularly upon the gentleman-farmers of the country. It was largely through their efforts that the *Chemical Society* was founded in 1845. By that time an extensive chemical industry was in operation as the result, in the first instance, of the commercialization of sulphuric acid production, and the use of the acid in bringing about the conversion of common salt into soda. It is to personnel in the alkali industry that Liebig turned in his attempted commercialization of a 'patent mineral manure'. According to Fenwick Allen, from whom we cite at length:

> When Liebig was making his researches and working out his theories in Agricultural Chemistry, and when he thought he had discovered the secret of the refertilization of the soil, the principal thing being to restore to the soil, as manure, the inorganic constituents which it was found by the analysis of the ashes of the vegetation, had been taken out of the soil by the plant, he got

[1] *Farmer's Magazine*, 1847, *16*, 511.

James Muspratt to carry out his ideas by manufacturing certain manures. This manufacture was carried on at Newton, about the years 1843–44. In this venture, Muspratt was joined by Sir Joshua Walmsley, a gentleman of much ability and enterprise, who had only a few years before been Mayor of Liverpool; he was a member of Parliament, and a friend of Richard Cobden and George Stephenson. There is a letter of his in which he refers to this manufacture of Liebig's manures: 'Having read Liebig's work on agricultural chemistry, and being impressed with the force of the reasoning; in conjunction with Mr. James Muspratt of Liverpool, a man well versed in chemistry, I entered into an arrangement with Professor Liebig, to manufacture an article that would give back to land all that cropping had taken out of it. The ingredients were found too expensive for the returns, and after a fair trial on which was spent some thousands of pounds, the undertaking was relinquished. Yet I always felt pride in the thought that Mr. Muspratt and I had been the first in England to endeavour to put into practice Professor Liebig's evident theory. Liebig, in a note in his *Letters on Chemistry*, says: 'I do not conceal from myself that the discredit into which the employment of the constituents of the ashes of plants, as manure, may have fallen in England, arises in part from the failure of the so-called *Patent Mineral Manure*'.[1]

Liebig's failure made it abundantly evident that a series of planned field experiments, as envisaged by Sinclair, was needed to test out and confirm theories coming from chemical laboratories before passing them on to the practical husbandman. This idea had just been started in England by John Bennet Lawes of Rothamsted. Lawes studied chemistry under Dr. A. T. Thomson of University College, London, and on entering into possession of his paternal estate in 1834, began experimental work on plant-feeding, first in pots, later in fields. In 1835 Lawes was joined by Joseph Henry Gilbert, a London fellow-student who had continued his studies under Liebig. Lawes and Gilbert between them, in little more than a decade, worked out manuring principles. According to Fenwick Allen, Lawes studied de Saussure's researches on vegetation, and applied his own scientific knowledge to questions affecting practical agriculture. Thus culminated a century's research, initiated by Francis Home, carried on by Archibald Cochrane, elaborated by Humphry Davy, and systemized by Justus Liebig.

The name of John Bennet Lawes will be inseparably connected with the history of scientific agriculture in the past century. The

[1] Fenwick Allen, *Some Founders of the Chemical Industry*, p. 94.

> experiments which were carried on by him for so many years at Rothamsted have long since been accepted as the most complete and reliable in the world, and are as well known on the continent, and throughout America, as in this country.[1]

Lord Dacre, his neighbour, particularly called Lawes' attention, among other subjects, to the use of bones, which on some lands proved a valuable manure, but were ineffectual on others. In consequence hundreds of experiments were set agoing, some upon field crops, others with pot plants, in which constituents found in the ash of plants were supplied as fertilizers. Of all the Rothamsted experiments, those based upon the suggestion originally thrown out by Liebig, that neutral phosphate of lime (in bone, bone-ash, or the mineral apatite) be rendered soluble by sulphuric acid and the mixture applied to root crops, gave the most striking results. Results first obtained on a small scale in the years 1837–8–9 led to more extensive trials in 1840 and 1841.

The observation of the remarkable manurial effects of mineral phosphates (rendered soluble by treatment with sulphuric acid) upon turnips was thus one of the earliest rewards of their labours.

Lawes was quick to realize the implication, and in 1842 took out a patent for the manufacture of *superphosphate*. This became an important industry, and is to-day one of the chief absorbers of sulphuric acid. In 1843, he established a factory near London for the production of superphosphate of lime. In the next year, 1844, manufacture of superphosphate was begun by a Dr. Richardson at Blaydon, in the north of England. Among the raw materials used were bones, bone-ash from South Africa and America, as well as the refuse animal charcoal from the sugar refineries. At the end of thirty years, the quantity manufactured annually at Lawes' works alone was more than 40,000 tons.[2] A west of Scotland firm, Alexander Cross and Sons, founded in 1830, also owed their continued development to the production of fertilizer. They removed to Port Dundas in 1872.[3]

Throughout the nineteenth century Scotland continued to keep up a close connexion with agricultural research. One of the principal Rothamsted disciples was Dr. Andrew Aitken. He was born in Edinburgh, and educated as a chemist in Germany. In 1877 he became consulting chemist to the Highland Society, and contributed reports of his experiments to their *Transactions*. The summary of his reports in the *Transactions* for 1886 may be regarded

[1] S. Miall, *History of the British Chemical Industry*, p. 27.
[2] W. G. Armstrong, *Industrial Resources of Tyne, Wear and Tees*, p. 175.
[3] A. McLean, *Industries of Glasgow*, p. 190.

in reality as a comprehensive treatise on manuring. Another follower of Lawes, Augustus Voelcker, was born in Germany, but came to Johnston of Edinburgh for his chemical education. In 1849 he was appointed Professor of Agricultural Chemistry at the Royal Agricultural College, Cirencester. He was chemist first to the Society at Bath, and finally to the *Royal Agricultural Society*.

We cannot leave the subject of agriculture without reviewing briefly the new fertilizers which became available during the course of the nineteenth century, and alluding to their connexion with advancing chemical industry. Of subsequent developments in the use of manures, Prothero tells us:

> With increased knowledge of the wants of plant or animal life came the supply of new means to meet the requirements. Artificial manures may be roughly distinguished from dung as purchased manures. Of these fertilizing agencies, farmers in 1837 already knew soot, bones, salt, saltpetre, hoofs and horns, shoddy, and such substances as marl, clay, lime and chalk. But they knew little or nothing of nitrate of soda, of Peruvian guano, of superphosphates, kainit, muriate of potash, rape-dust, sulphate of ammonia, or basic slag. Though nitrate of soda was introduced in 1835, and experimentally employed in small quantities, it was in 1850 still a novelty.[1]

Guano samples had been examined by the Board of Agriculture in 1805—its utility had been pointed out by Davy, but the first large consignment was landed in England in 1835. In the space of a few years, 80,000 to 100,000 tons were being imported annually. In 1844 some Ayrshire members of the *Highland Society* purchased a vessel which they dispatched to Peru to bring back a cargo for their own use.[2] By 1847, imports reached 220,000 tons, and fears of shortage pushed the price up to £10 per ton. Its economic value was assessed by Liebig in the following terms:

> We believe that the importation of one hundred-weight of guano is equivalent to the importation of eight hundredweight of wheat . . . The same estimate is applicable in the valuation of bones. If it were possible to restore to the soil of England and Scotland the phosphates which during the last fifty years have been carried to the sea by the Thames and the Clyde, it would be equivalent to manuring with millions of hundredweights of bones, and the produce of the land would increase one-third, or perhaps double itself, in five to ten years. We cannot doubt that the same result would follow if the price of the guano admitted the application of

[1] Prothero (Lord Ernle), *English Farming Past and Present*, p. 369.
[2] Watson and Hobbs, *op. cit.*, p. 95.

a quantity to the surface of the fields, containing as much of the phosphates as have been withdrawn from them in the same period.[1]

Bones also came into use in the early nineteenth century. The value of imports rose from £15,395 in 1823 to £254,600 in 1837. At first they were simply broken small, in which state they reacted but slowly. Their rate of absorption could be increased somewhat by grinding to a coarse meal. In 1840, however, Liebig suggested treatment with sulphuric acid to render them more soluble, and in 1843 Lawes began the manufacture of superphosphates as recounted above. At a somewhat later date, further phosphatic manures were introduced. The merits of one of these, a by-product of metallurgy, was first demonstrated at Cockle Park, Northumberland. Watson and Hobbs tell us:

> For years basic slag, known of course to be rich in phosphate, had been cumbering the ground round the steel works because it was thought that its nutrient was too insoluble to be of use to the plant. When at last it was discovered that, with no other process than very fine grinding, it could be turned into the best of all fertilizers for poor clay pasture. On these it made grow not only the proverbial two blades of grass, but three of clover besides, and its use returned almost fantastic profits in the form of more and better sheep.[2]

This internal supply of phosphatic manure was of considerable economic importance to Great Britain.

Before concluding, it would be well to review briefly the changes which had taken place during the century under consideration. The facts about plant foods which had been established by 1855 are summed up by Sir John Russell as follows:

(*a*) Although Liebig was wrong in assuming that the composition of plant ash was a guide to its manurial needs, he was right in his belief that plants need alkaline salts and phosphates.

(*b*) To the manures thought necessary by Liebig must be added, except to legumes, nitrogenous compounds, either ammonium salts or nitrates.

To the mineral fertilizers enumerated by Liebig vegetable physiologists added several other elements necessary for plant life. By means of water cultures Knop showed that potassium, magnesium, calcium, iron, phosphorus, with the addition of sulphur, carbon, nitrogen, hydrogen and oxygen, are necessary for plant

[1] J. Liebig, *Familiar Letters on Chemistry*, p. 178.

[2] Watson and Hobbs, *op. cit.*, p. 81.

life. The only difference between this list and Liebig's is that it includes iron, but does not include silica. Even silica is now regarded as an advantageous addition in the culture of cereals.

(*c*) Chemical manures would maintain fertility at least for several years.

(*d*) Fallowing had been explained as a practice which enabled natural nitrogen compounds to accumulate in the idle ground.[1]

The great advances which resulted from the work of Liebig and the field demonstrations at Rothamsted depended upon 'the extended choice of fertilizing substances, in the scientific analysis of their composition and values, in their concentration and portability, and in the greater range of time at which they could be profitably applied'.[2] It was not, however, till some time after the appearance of Liebig's publications that an agricultural text-book freed from the erroneous ideas of earlier chemists was produced. According to J. A. Scott Watson, the nearest approach to a perfect agricultural text-book was written by R. Warington under the title of *Chemistry of the Farm*. This and six other hand-books were editorial products of J. C. Morton, a former student of Low at Edinburgh.[3]

At the conclusion of Liebig's researches, one line of investigation remained to be completed, namely, a full exploration of the source from which plants derive their nitrogen. The problem was complicated by the fact that, as became apparent later, the source of nitrogen for different plants is not always the same. Throughout the eighteenth century, observations were made which did in fact indicate this, though they were interpreted otherwise. So we may account for the inclusion of clover and other leguminosae in empirically developed rotations. For example, clover was introduced by the sixth Earl of Haddington into East Lothian about 1720, and the beneficial effects of another legume, Sainfoin, were known to the Improvers. Many attempts were made to explain the effect which the inclusion of leguminous plants in a rotation had upon subsequent crops. Francis Home, observing the effect, attempted to account for it in the context of his own scientific knowledge:

> The leguminous plants, by covering the soil, keep it moist, hinder the sun to consolidate, and destroy the weeds which help so much to bind it. Hence the reason why a change of species

[1] E. J. Russell, *Soil Conditions and Plant Growth*, p. 19.
[2] Prothero, *op. cit.*, p. 368. [3] Watson and Hobbs *op. cit.*, p. 252.

> meliorates the soil so much. When the ground is often sowed with white grain, it turns stiff. A crop of pease, beans, or clover pulverizes it again.[1]

Here is a later attempt, this time by George Winter, who remarked: 'Curious researches into the works of Providence, will investigate the cause why some crops improve lands more than others; let this enquiry be our present attempt'.

> Beans are a good fallow crop. Their tap roots will grow from twelve to eighteen inches or more, perpendicularly deep in a loamy ground. Their numerous fibres or lateral roots will extend themselves horizontally in a well-loosened soil, as long as their roots, and their size will be in proportion. Consequently the increase of production will also be in proportion to the growth of the roots and fibres. But when the stratum of loam is not more than six or eight inches deep on a clay, their roots cannot penetrate into, nor procure such store of nourishment, as when planted in a deep loam; and of course the crops cannot be so fruitful.[2]

The above quotation shows that Winter too had observed the beneficial effect of leguminous plants, but attributed it to still another cause.[3]

In more scientific hands the first important relevant observation was that of Priestley, who recorded that, when kept in a glass vessel, *Epilobium hirsutum* absorbed seven-eighths of the air present. Since air contains only one-fifth of its volume of oxygen, Priestley's experiment indicated that certain plants at least could absorb nitrogen directly from the air. This was denied by his contemporary, de Saussure. The varied behaviour of different plants, however, was brought out unmistakably by pot experiments undertaken by J. B. Boussingault in 1838. These revealed that, while peas and clover could absorb elemental atmospheric nitrogen, wheat on the other hand could not. In the field, experimental rotations confirmed the difference. The implications were fully realized by Dumas and Boussingault in 1841.[4]

It was Liebig's contention that the general run of plants could absorb a nitrous compound, ammonia, but not nitrogen itself. Under rigorous experimental conditions, this was confirmed at Rothamsted. Plants were kept under glass shades, away from all possible sources of ammonia, supplied with conditioned air and pure water. They all died. For all but leguminous plants, this was true in the field as well. They, while requiring no nitrogenous

[1] Francis Home, *Principles of Agriculture*, p. 137.
[2] G. Winter, *Systems of Husbandry*, p. 153. [3] *Ibid.*, pp. 155 and 159.
[4] *The Chemical and Physiological Balance of Organic Nature.*

fertilizer, flourished, contained nitrogen in their tissues, and even augmented the nitrogen content of soil in which they were grown. Only in 1881 did W. O. Atwater show that peas absorb large quantities of elemental nitrogen from the atmosphere aided by symbiotic nitrogen-fixing organisms. At last, the basic pattern concerning the economy of vegetation was complete.

Chapter XXII
SACCHAROPOLIS

Sugar: a sweet granulated substance, too well known to require any particular description.

McCULLOCH'S *Dictionary of Commerce.*

IN the preceding chapter the production of food in the raw state was dealt with at some length. It was seen that, not only was the quantity of food which could be produced from a given acreage increased by the application of science during the eighteenth century, but that new foods were introduced. Food processing was also developed on a commercial scale during that century. In the chapters that follow several industries will be discussed, all of which in a sense are food processing industries, though of varied origin and depending to quite different extents on the application of advancing science. There are, for example, crafts of brewing and distilling of ancient eotechnic origin, the rise of the sugar-boiling industry, and from the early years of the nineteenth century the preservation of food by canning.

The first of these chapters deals with sugar. It is fitting that sugar should be discussed first, since a parallel can be brought out between sugar, representing organic or biological materials, and salt, which was discussed in Ch. II, as an inorganic or mineral material. Within their own spheres of social utility they represent, in those two classes, the first examples of substances that were produced in large quantities in a state of comparative purity. They were, so to speak, though not usually so regarded, the first two chemicals to be isolated on an industrial scale. The parallel may be extended. Just as salt was of vital importance to eighteenth-century Firth of Forth economy while salt-boiling remained of comparative insignificance on the Clyde, so alternatively sugar figured prominently in the commerce of Glasgow, while on the east coast, although admittedly there were attempts to establish sugar-boiling, e.g. at Leith, it never attained anything like the importance which was attached to sugar-boiling in, say, Greenock.

This analogy between salt and sugar must not be pushed too far, however. It is true that along with soap-boiling, salt and sugar-boiling in many ways represent the basis from which the techniques of the chemical industry sprung, but there is an important distinction between the two commodities. Common salt (sodium chloride) belongs to the class of chemical compounds called *salts*, small quantities of which are essential for the maintenance of the balance of life at all stages of its evolution. Sugar, on the other hand, is a compound of carbon, oxygen and hydrogen, united to produce what is called a carbohydrate. There are many carbohydrates, our starchy food consists of them in the main, and among them cane sugar (sucrose) is not alone in being sweet to taste. Honey, which contains other carbohydrates (glucose and fructose) comes to mind immediately. One result of this difference between salt and sugar is that while the acquisition of salt by human communities recedes into prehistory, we can date roughly, and indeed follow up, the rise in sugar consumption in western Europe. It was not in the first instance as a food that sugar attracted attention, but as one of the *materia medica* under the name *saccharum*, introduced to render nauseating medicines more palatable. It was also used for external application. More will be said later of the reaction of the medical profession to increased ingestion of sugar.

Much as nostrums were resorted to they did not make great demands for sugar, and it is to changing social habits that we must look for the cause of the rise in sugar consumption during the eighteenth century. Drummond and Wilbraham tell us that:

> The consumption of sugar rose as the habit of tea-drinking spread. It had ceased to be a luxury as the trade with the West Indies developed and ordinary white sugar candy cost 6½*d.* per pound in the first half of the century, refined varieties being 2*d.* a pound dearer.[1]

John Galt in the *Annals of the Parish*, which documents the coming of the industrial revolution to south-west Scotland, spoke of the spread of tea-drinking about 1762, and ventured an opinion on the social changes which the habit brought in its train.

> What I thought most of it for, was, that it did no harm to the head of the drinkers, which was not always the case with possets[2] that were in fashion before. There is no meeting now in the summer evenings, as I remember often happened in my younger days, with decent ladies coming home with red faces, tosy[3] and cosh[4] from posset masking.[5]

[1] J. C. Drummond and A. Wilbraham, *The Englishman's Food*, p. 244.
[2] *Posset* = hot drink of milk and wine. [3] *Tosy* = confused.
[4] *Cosh* = comfortable. [5] J. Galt, *Annals of the Parish*, p. 23.

In another twenty-five years, yet other saccharine luxuries, as they were called at the time, were added to the diet of the common people of Scotland, viz. jams and jellies. Richard Pococke, in his *Tour*, spoke of having had *orange peel jam* for breakfast, but Galt gave a clearer picture of the changes:[1]

> By the opening of new roads, and the traffic thereon with carts and carriers, and by our young men going to the Clyde, and sailing to Jamaica and the West Indies, heaps of sugar and coffee-beans were brought home, while many, among the kail-stocks and cabbages in their yards, had planted grozet[2] and berry bushes; which two things coming together, the fashion to make jams and jellies, which hitherto had been only known in the kitchens and confectioneries of the gentry, came to be introduced into the clachan.[3]

Marmalade dates from 1797 and it is perhaps not unconnected that H. Arnot speaks of the increasing cultivation of rhubarb in his *History of Edinburgh*, and that some twenty pages are devoted to details of its culture in the *Letters of the Society for encouraging Agriculture at Bath*.[4] Among the letters is one from Dr. John Hope, Professor of Botany at Edinburgh.

Although the date given by Galt for the spread of the use of sugar shows that it only followed the establishment of large-scale sugar-boiling in Greenock, there had been in Scotland for more than a century the nucleus of the industry. To begin with, the raw cane sugar must have been obtained from Portuguese possessions in Brazil since it was only in 1643 that the first Barbados sugar reached this country.[5] The spread of English domination in the West Indies was encouraged by Portugal's unwillingness to sell sugar. And when the Plantations were opened up to Scottish traders subsequent to the Union of 1707, Scotland's commercial focus was diverted from Holland and Flanders to the islands of the west. Glasgow became Britain's greatest entrepôt for tobacco and in the wake of tobacco came sugar and cotton. The trade was by no means unilateral. Lime, possibly as ballast, was exported to the West Indies, and from Carron Iron works went hoes and spades required for the culture of the sugar cane, and iron pans for boiling down crude juice. One of the first products of Glasgow metallurgical industry was sugar machinery. Firms still or until recently existing, go back to the early days of sugar-boiling in

[1] J. Galt, *Annals of the Parish*, p. 146.
[2] *Grozet* = gooseberry. [3] *Clachan* = village. [4] P. 185, *et seq.*
[5] R. P. Stearns, *Production of Sugar in Barbados c. 1667*, *Annals of Science*, 1936, *I*, 173.

Glasgow. One such is M'Onie, Harvey and Company, Limited.[1] The international reputation of Glasgow sugar machinery is dealt with by C. A. Oakley.[2] Thus in the eighteenth-century social technology of sugar we must first look at the West Indies, the source of the raw material till the introduction of beet sugar in the nineteenth century.

The rise of the British sugar trade dates from shortly after the English settlements in St. Christopher (1625) and Barbados (1627). Sugar cane (*saccharum officinarum*) was introduced into Barbados in 1641 and after two years cultivation yielded the first sugar produced in a British possession. Jamaican sugar followed in 1664. Over the same period the French were colonizing the West Indies (St. Christopher 1625, Guadaloupe 1635, Martinique 1635, Grenada 1650), and the export of sugar which started from Guadaloupe in 1648 was a source of imperial rivalry, since the importation of sugar was an advantage to the contending governments as a source of duties, commissions, etc.[3] The production of sugar in the West Indies, particularly Barbados, Antigua and Jamaica, expanded rapidly, thus creating a demand, not only for shipping and commodities like sugar-pans, but also for labour.

> The introduction of sugar cane into the Barbados Islands in 1641 gave the first impulse to the slave traffic in the British West Indies. In 1662 the 'Company of Royal Adventurers Trading to Africa' was chartered and obligated itself to deliver three thousand slaves annually to the British West Indies. The West Indian sugar trade with the continental colonies of North America began to assume importance about the middle of the seventeenth century. Inseparably associated with it was the exportation of rum, molasses, and tobacco, and the importation of lumber, horses, and fish. Many cargoes of sugar, rum, and tobacco were reshipped from the mainland to England. Large consignments of rum were shipped to New England, where it was sold to the Indians for furs, which constituted one of the most important colonial exports from the northern colonies. Rum was at the same time the West Indian export most in demand in the African trade where it was exchanged for slaves, which were demanded by the sugar planters for increasing sugar production. The English, French, and Portuguese were actively engaged in the African slave trade. The West Indian Islands constituted the centre of the intercolonial

[1] A. McLean, *Industries of Glasgow*, p. 64.

[2] C. A. Oakley, *Scottish Industries To-day*, p. 271.

[3] B. Moseley, *A Treatise on Sugar*, p. 28; *The Importance of the Sugar Colonies to Great Britain stated, and some objections against the Sugar Colonies Bill Answered* (*1731*).

trade during the eighteenth century. All the economic and physical conditions contributed to make it so. Sugar and rum were the largest exports, with tobacco second in importance.[1]

Initially, the raw sugar imported into Britain was refined in London, where about 1544 the Venetian system of making sugar loaves had been introduced. The contribution made by continental personnel to the technology of sugar production throughout the whole of the period under discussion is a notable feature of the industry. Jealous of their craft, they left their mark on the refineries in the shape of 'secret rooms' in which they operated. Although Scotland did not participate in the colonial trade of England during the seventeenth century, we hear of sugar-boiling in Scotland from 1628. Details of the early sugar houses were given in Ch. I. Such houses as there were appear to have been well established and in possession of special privileges till after the Union.

> With regard to sugar-houses, although the colonies were not laid open to the Scotch until the Union, it appears that there were sugar-houses in Scotland long before that period; for in an action which the Crown brought against the sugar bakers in Glasgow and Leith, it was urged that they had not only enjoyed the exemption from the duties and customs on the import of materials for a great number of years, but also the duties of excise upon spirits and other commodities manufactured by them.[2]

A Court of Exchequer decision in 1719 found the sugar-houses liable for £40,000 of duty, and as they could not meet this unexpected cessation of their privileges they were forced to accept renunciation of their rights to the various exemptions in exchange for the demand being waived. This further extends the parallel between sugar and salt. It will be remembered that the effect of fiscal policy on salt production in the Forth area was analysed in Ch. I.[3] The analogy applies also to the preparation of raw sugar in the sugar islands. Here is a description of the process by G. Fairrie. The object of the operation is to neutralize the natural acidity of the juice which would decompose the sugar, and secondly to remove albumen. The process goes by the name of defecation.

> In the West Indies the juice from the cane mills used to be clarified in open pans, the heat being applied by a fire beneath. A

[1] G. Surface, *The Story of Sugar*, p. 26.
[2] *New Statistical Account*, Vol. 6, p. 135.
[3] For further details concerning the rights of the proprietors of sugar-houses in Scotland, *see* 1 Geo. I, St. 2, C. 19; 8 Geo. I, C. 4, S. 6.

> small quantity of lime[1] was added to the heated juice and, when the scum had risen to the surface, the fire was extinguished and after a short time the liquor was drawn off without disturbing the scum. The partly clarified juice was then removed to the largest of a series of open evaporating pans placed over a stove heated by a fire at one end; the juice was allowed to boil and the scum carefully removed from the surface. When the juice was sufficiently reduced in bulk, it was transferred to the second or middle-sized pan, and boiling and skimming were repeated until the juice could be contained in the third or smallest pan called the *teache* or *taich*. A group of these open evaporating pans was known as a *Jamaican train*.[2]

The resulting syrup was cooled in wooden boxes, when sugar crystals separated from the molasses, giving a magma which was transferred to *potting casks* in the *curing house*. The potting casks were crude filters consisting of hogsheads with small holes in the bottom plugged with a piece of cane through which the uncrystallized molasses could drain. After draining for some three weeks the crude sugar was ready for export. It appeared on the British market as *muscovado* sugar.[3] It might be thought strange that the process of preparing the sugar for the market was not completed in the sugar islands, but there were reasons for not doing so. Chemically and mechanically it was quite possible, but a factor of importance was water supply. Large quantities of pure water are required in sugar refining, and this was not generally available in the West Indies. Second in importance was the labour problem. The superior quality of labour required for a refinery was not available in the cane-growing countries. Moreover, ancillary materials could be more easily purchased in countries where sugar was consumed.[4] At the same time one cannot discount the fact that British West Indian interest did not wish sugar-refining to develop in the Plantations.[5]

Few changes were effected in sugar-boiling during the eighteenth century, but early in the nineteenth the correlative advance in mechanical technology made possible the introduction of radically different methods of refining. New methods of decolorising and clarifying the syrup were also derived from physical science. These innovations had an important effect on the set-up of the sugar industry in Great Britain. While eighteenth-century methods managed to hold their own for the first one or two decades

[1] Wood ash was used till *c.* 1750.
[2] G. Fairrie, *Sugar*, p. 152.
[3] A. Rees, *Cyclopaedia*, Article, *Sugar*.
[4] G. Fairrie, *op. cit.*, p. 56.
[5] J. M. Hutcheson, *Notes on the Sugar Trade*, p. 11.

of the nineteenth century, they were unable to continue for long in the face of the more scientific and modern processes which began to spread throughout the British industry. As a result, small and obsolescent sugar-houses were unable to compete with their better-equipped rivals, and either failed or were absorbed by the larger units. Thus, while there were 120 houses in England and Scotland in 1753, despite expanding output there were only 90 in 1863. How small some of the refineries were may be gauged by statistics given by J. M. Hutcheson in his *Notes on the Sugar Industry of the United Kingdom.* The average consumption of sugar in Great Britain over the last decade of the eighteenth century was 78,894 tons per annum, or 1,500 tons per week. The Scottish consumption was about a tenth. This was sufficient to keep about 150 refineries working, a good indication of their size. Consumption per head of population on the basis of these figures was 15–16 lb. per annum.[1] Some idea of the quantities imported into the Clyde will be gathered from the following figures for the year 1771. There were imported from Jamaica, St. Kitts, Granada, Antigua, St. Vincent, and Nevis, in that order, a total of 2,368 tons of brown or Muscovado sugar. In the same year, 1,581 tons of raw sugar were exported to Ireland, Maryland, and Sweden, while 107 of refined sugar were returned to Virginia, North Carolina, St. Kitts, St. Vincent, and six other West Indian Islands.[2] Yet it was on the basis of these small quantities that the industry developed, and it would be well, at this point, to look at the methods employed and at the histories of some of the principal houses founded before the introduction of new methods.

The following description of sugar refining is cited from G. Fairrie's *Sugar:*

> The raw sugar arrived at the refinery in hogsheads each weighing from 4 to 18 cwt. The raw sugar was mixed with lime-water, put in open pan over a fire and a quantity of *bullock's blood* called *spice* or *finings* was stirred in. The serum or watery part of the blood, consisting chiefly of albumen, of which white of egg is a familiar example, was curdled by the heat, and by entangling the impurities floating in the liquor brought them to the surface in the form of a thick scum which was carefully removed. This process was repeated several times and the partly clarified liquor was filtered through a basket lined with woollen cloth, and afterwards boiled in an open copper pan heated by a fire until sufficiently concentrated to crystallize. The mixture of crystals and

[1] B. Moseley, *op. cit.*, p. 162.

[2] Figures from Gibson's *History of Glasgow*, pp. 219-220.

syrup from these coppers was formed into conical *loaves*. The loaves were then broken up and redissolved in water and clarified with white of egg, the scum being carefully skimmed off. The liquor was again evaporated in an open pan and formed into loaves.[1]

Thomas Pennant on his *Tour* remarked on the large quantities of eggs collected round Berwick for sugar-refining there. He estimated the value of the trade at £14,000 per annum.[2]

The working off of a complete batch of sugar lasted the better part of three weeks, products becoming less pure towards the end of the operation. The way in which the various kinds of sugar were produced was described by Rees:

> The order of refining is uniformly this: to begin the first day with the finest sugar intended to be wrought, and to proceed daily with sugar of a lower quality, and of course to begin with small loaf-moulds, and to use larger moulds progressively; so that the brownish sugar will be put into large lump-moulds; for this sugar works best in large masses, and it is likewise more in demand in England than the finer kind. . . . The order of the first twelve days is usually as follows: first day, double loaves; second and third days, powder loaves; fourth, fifth and sixth, single loaves; seventh, Prussian lumps; eighth, Canary or pattern lumps; ninth, tenth, eleventh and twelfth days, large lumps. To these twelve days add four or five more, in a part of the process called *bastard-boiling;* and these sixteen or seventeen days constitute a complete series, denominated a complement of refining.[3]

Such were the operations as carried out before the impact of science on the sugar industry. It is opportune to look at some of the Scottish houses in which they were carried out.

Reference was made in Ch. I to early sugar-houses in Glasgow, Leith and Edinburgh. Several interesting manuscripts have survived concerning sugar-boiling in the Forth area. According to H. L. Roth's *Guide to the Literature of Sugar*, Articles of a Sugar Copartnery in Edinburgh (n.d.) are in the Manchester Free Public Library, and Hutcheson speaks of having seen several of the books kept by the *Edinburgh Sugar House Company*, viz., an Invoice book, Refining book, Day book, Letter book, and Minute book, so they were in existence at the beginning of the century. We have been able to trace what may be two of these. They were purchased in Harrogate some years ago and are now in the Edinburgh Public Library, where they are catalogued as being the Minute Book

[1] G. Fairrie, *op. cit.*, p. 154. [2] T. Pennant, *Tour in Scotland in 1769*, p. 39.
[3] A. Rees, *op. cit.*, Article, *Sugar*.

(27th April, 1752—26th March, 1763 (MS. 514, pp. P. 42614)) and Account Book (18th February, 1758—22nd November, 1759 (MS. 369, pp. P. 42613)). This is almost certainly the *Edinburgh Sugar House Company* founded in 1751 and still being operated by its founders at the time H. Arnot was writing his *History of Edinburgh* (1818).

There was also another concern, or so it is supposed, called the *New Edinburgh Sugar House Company*. It appears that in 1771 Robert Selby, a plumber, and Henry Gutzmer and Jack Somervile, sugar-boilers, started a sugar-boiling business in the house called St. Thomas's Chapel, at the foot of the Canongate. One of the name Livingston was a director. The day-book of the concern was entitled *Day-book for the New Sugar-House Co.*, and it begins with a contract of copartnership between Messrs. H. M. Gutzmer, Robert Selby, and Jack Somervile.[1] Selby withdrew in the year after the foundation, and the whole concern was bankrupt in 1773. This led to the company being sued by Charles Gordon of Abergeldie, Wm. Nairn *et al.*, yet they continued in operation till about 1778, when they finally closed down.[2] Hutcheson notes that the *Edinburgh Sugar Refinery*, then lying idle, was acquired by the Macfie concern in 1829 and worked for a time. Its address was 160 Canongate.

There are several mentions of sugar-boiling in Leith from 1757 onwards. There was a *Leith Sugar Refining Company*, Coburg Street, from 1800 to 1860, and William Macfie and Company built a refinery in Elbe Street in 1804 with material obtained from a dismantled refinery in Aberdeen, the last to be built and use the old method of sugar refining. Various later refineries were set up in Leith and Edinburgh, but they all had chequered careers and there were long periods during which sugar was being refined neither in Leith nor Edinburgh.

In two other Scottish ports, Dundee and Aberdeen, spasmodic attempts were made to establish sugar-boiling. The *Dundee Refining Company* was established in 1751 and continued to operate till 1834. William Wiedemann, one of the first boilers, was the great-grandfather of Robert Browning. It is rather surprising that Dundee did not maintain its position as a refinery town longer, in view of the establishment of industries there which used large quantities of sugar as raw material for marmalade and jam.

At Aberdeen two attempts were made in the second half of the

[1] *Session Papers*, Signet Library, Campbell's Collection, 25: 38.

[2] *Faculty Collection of Decisions*, January, 1775, Vol. 7, p. 1; Morison's *Dictionary of Decisions*, 14,551 (Edin. Univ. Lib.)

eighteenth century to introduce sugar-boiling. The first promoter was James Moir (*d.* 1784) of Stoneywood. Moir was forced to flee to the Continent on the collapse of the Rising of the 'forty-five. When he returned he started a sugar-house at Bellfield Terrace, Stoneywood, in 1762.

The other Aberdeen sugar-house was started in the neighbourhood of the harbour, and had a better chance of succeeding than Moir's. It was founded by a copartnery of Aberdeen merchants.[1] They imported raw sugar into Aberdeen, and began boiling in 1776, having obtained a charter from the Aberdeen Town Council on 25th April of that year. This sugar-house only continued in operation till 1798, and it was presumably plant from one or other of these houses that William Macfie and Company obtained for the Leith refinery mentioned above. The building of the second sugar-house still stands. For many years it formed part of the provision-preserving works of James Moir and Sons, who are not, however, connected with the James Moir referred to above.[2]

Edinburgh, Glasgow, Aberdeen and Dundee all had sugar-houses in the eighteenth century, but the town which became the *saccharopolis* of Scotland was Greenock. We shall follow the rise of the Greenock industry in greater detail. Data are taken from J. M. Hutcheson's *Notes on the Sugar Industry of the United Kingdom* (*passim*); R. Niccol's *Essay on Sugar*; and Dan. Weir's *History of Greenock*. Two facts should be noted in the following brief histories, increasing and detrimental influence of Liverpool, and absorption of smaller refineries by larger concerns. Other minor points of interest are the presence of foreign personnel, and the incidence of disastrous fires at many refineries.

The industry started in Greenock in 1765 with the founding of the *Greenock Sugar House*.

> The first sugar refinery built in Greenock stands at the bottom of Sugarhouse Lane, formerly part of the premises belonging to Messrs. Hoyle, Martin and Co. It was erected in 1765, and, although small in size, was then considered a rather extensive undertaking. The quantity of refined sugar at that time could not have been very large, as the gross customs revenue from all sources at Greenock did not much exceed, if it did exceed, £5,000!

It was built in 1765 by Mr. Mark Kuhll for a copartnery of gentlemen in Glasgow. It was enlarged in 1798, and by 1829 was owned by Alderston of Liverpool. In 1832 it changed to the ownership of Alex. M'Callum. A blank in its history occurs till

[1] James Burnett of Countesswells, John Burnett, Thomas Bannerman, Alex. Garioch, William Young and Alex. Young.

[2] Ch. XXIV.

1846 when, as it was idle, it was rented by James Fairrie and Co., and subsequently to Hoyle, Martin and Co., who became owners in 1851.

Before another house was founded, the sugar industry went through a period of depression following the loss of Grenada, St. Vincent and Dominica.[1] For the second Greenock sugar-house, a dwelling-house in High Sugarhouse Lane was converted in 1788 by Hunter Macalpine and Company. Nicholas Witt was master-boiler. This sugar-house changed hands frequently. It was in the hands of Harm Blancken and Company in 1845, but like the previous concern was absorbed by Hoyle, Martin and Company in 1847. These were the two pioneer houses in Greenock. After that there is a gap till the early nineteenth century, in the first decade of which two other houses started up. First came Robert Macfie and Sons of Bogle Street, who started in a small way with two pans in 1802. They continued to operate for half a century, when they transferred to Liverpool. The next house is one of considerable importance from the viewpoint of sugar technology in Scotland. It was built in 1809 by James Fairrie and Company of Cartsdyke. In 1819 the sugar-house was altered to make use of Howard's vacuum pans, being the first in Scotland to do so, and the third in Britain. The refinery was burned out in 1846 and not rebuilt, Fairrie and Company, like other refiners, concentrating on the expansion of sugar refining in Liverpool.

In 1812 Thorne and Sons Limited founded the *Glebe Sugar House*. It is of historical interest, since its name at foundation was later transferred to a different refinery which continued to operate till the present century. The name was derived from the fact that it was built on land which constituted part of the church glebe, purchased from the Rev. Robert Steele, of the Old Parish Church. The name was later transferred to Thomas Young and Company's refinery founded in 1831, which became known as the *Glebe Sugar Refining Company*.

No new refinery was established in Greenock between the founding of the *Glebe Sugar House* in 1812, and a whole series of foundations from 1826 onwards. During the interval, fundamental changes took place in the technique of sugar production, due to discovery in mechanical and chemical science. If we want to bring out an analogy with other phases of industrial change, then it can truly be said that the inventions necessary for the development of palaeotechnics, as far as sugar was concerned, were patented in the first decades of the nineteenth century. They may be

[1] *Report of Petition of Sugar Refiners, Journal of the House of Commons, 1781.*

classified under three heads: (*a*) sugar-boiling at lower temperatures and higher evaporation rates by working with closed pans at reduced pressure; (*b*) substitution of charcoal as a decolorizing agent in place of bullock's blood and white of eggs; (*c*) improved filtration and drying of separated sugar crystals in centrifugals in place of draining through earthen moulds.

The principle of sugar-refining is the process of recrystallization whereby pure sugar crystals are separated from a less pure syrup or *mother liquor*. Although water at room temperature dissolves twice its weight of sugar, large quantities of solution have to be handled, so that any invention which expedites concentration of the sugar solutions is of great value. By the beginning of the nineteenth century it was realized that evaporation could be accelerated by reducing pressure, indeed it had been suggested that reduced pressure stills should be introduced in spirit production.[1] The principle was applied to refining by E. C. Howard.[2] Instead of evaporating syrup in an open pan, Howard covered in the pan and reduced the pressure by pumping off the vapour. In effect he turned the sugar pan into a vacuum still, except in that it was the residue in the pan, and not the distillate, that interested him. His intention was to evaporate water from the sugar solution till a sufficient concentration was attained and then transfer the concentrated syrup to a granulating vessel where crystallization would take place on cooling. When constructed, however, the apparatus did not work quite as he intended. Crystallization took place in the pans without cooling, with the result that the early pans became known as *crystallizing* pans. For the few years between his invention and his death, Howard is reputed to have benefited from his invention to the extent of some £40,000 per annum.

The logical sequel to single vacuum pans was multiple-effect evaporation, which utilizes vapour from the first pan to heat the second, each successive pan working at a lower pressure and consequently boiling at a lower temperature. In 1802 the idea of using steam from one vessel to heat the contents of the next was patented by Charles Wyatt (No. 2639), but credit for applying the principle to sugar-refining usually goes to William Cleland. Although used on the refinery side, multiple-effect evaporation was of even more consequence in the raw-sugar factory where large quantities of dilute raw juice have to be handled. Cleland also patented an improved filter.

It will be recollected that the eotechnic method of purifying both salt and sugar was to add bullock's blood, or sometimes egg

[1] Ch. XXII. [2] Patent Nos. 3607 of 1812 and 1754 of 1813.

white. Albumen on heating coagulates, and, with the scum formed, impurities in salt and sugar are separated. In sugar refining the method continued ostensibly till 1850, when a patent was granted to Charles Derosne for decolorization of solutions by means of a new principle, adsorption on animal charcoal. Very little research shows, however, that Derosne's patent was, as so often, the culmination of a long line of discoveries, and there is ample evidence to show that his method of purification was in use before the great expansion in sugar refining at Greenock in the third decade of the nineteenth century. The chain of evidence goes back at least to 1785, for in that year Lowitz rediscovered the decolorizing properties of wood charcoal in the preparation of tartaric acid. It is conjectured by Lippmann that wood charcoal was in use in English refineries in 1794, and without doubt Guillon was using it in 1805. In 1812, Constant patented the use of charcoal in this country. His patent refers to *garret liquor*, from the place of treating the saccharine liquor during purification. Although the use of charcoal depended on adsorption of impurities on to the surface of carbon by a physical process, it was not so used to begin with. Charcoal was not used as a filter but was added to the syrup in the melting-pan, along with white of egg. The first charcoal was wood charcoal, and S. Parkes in his *Chemical Essays* spoke of large plantations of dogwood and alder being laid down in Kent, Sussex and Surrey, for making charcoal 'to remove the disagreeable taste from molasses'. In 1812, Figuier of Montpellier discovered that animal charcoal was a much more efficient adsorbant than wood charcoal, so animal charcoal was next applied to sugar-refining.

> A patent, dated May 8th, 1815, was granted to Peter Martineau the younger and John Martineau for a new method of refining or clarifying certain vegetable substances. For refining or clarifying sugar the patentees mention among other things, purified animal charcoal. This was added before the usual fining of eggs, blood and other albuminous matter and is separated when the albumen is coagulated and rises to the top from which it is skimmed. Sir Humphry Davy examined the scummy material and finding that it consisted largely of the bullock's blood, recommended its use as a manure, a suggestion that was followed to a considerable extent in some localities.[1]

At first the refiners did not appreciate that the decolorizing power of charcoal can be restored by heating, and it was thrown away, but in 1817 Joseph de Cavaillon patented revivification by

[1] A. Rees, *op. cit.*, Article, *Sugar*.

incineration in earthenware vessels placed in a kiln. Granulated animal charcoal used purely as a filter was suggested by Dumont in 1828 and patented in Britain by Charles Derosne in 1830. G. Fairrie describes the method of operation:

> The charcoal was placed upon cloth in layers until it reached a height of 15 to 16 inches. On the top of the charcoal was placed another perforated diaphragm covered by a linen cloth. The liquor to be refined was poured upon the top cloth to a depth of 4 to 8 inches. The charcoal filters used in the old refineries were rectangular wooden tanks. These tanks had open tops and, like the modern enclosed filters, were fitted with a perforated, false bottom covered with cloth. The liquor was run on to the top of a layer of charcoal a few feet [*sic*] deep and percolated downwards and out at the bottom. Sometimes the liquor was passed through several successive filters.[1]

Chemical bleaching as an alternative to adsorption was suggested in 1810, but did not come into use till about 1850.

Vacuum pans and a regenerable physical adsorbant were necessities for expansion in the sugar industry. In Scotland it followed almost immediately. In each of the years 1826, 1829 three times in 1830, in 1831, twice in 1832, and in 1833, new refineries were started either in Greenock or Glasgow.[2] According to the writer of the *New Statistical Account*, in 1845 there were eleven sugar-houses employing 350 persons in Greenock or thereby. In every house Howard's patent was in use, and combined annual production amounted to 14,000 tons. So the pace continued, till in the 1870s no less than twenty-four sugar houses operated in Greenock alone. It is not without justification that Greenock and the Clyde claim to be the birthplace of the palaeotechnic sugar industry in Great Britain. To the above refineries must be added the older houses in Glasgow and also those of Port Glasgow.

> At each of these towns (Greenock and Port Glasgow) the boiling or refining of sugars is carried on very extensively. At Port Glasgow alone there are four sugar houses belonging to three different

[1] G. Fairrie, *op. cit.*, p. 159.

[2] 1826. Angus, Balderston and Company, Greenock.
1829. Tasker, Young and Company, Greenock
1830. John Kirkland and Company, Glasgow.
1830. Reid and Pearson, Glasgow.
1830. Thomas Young and Company, Greenock.
1831. Alex. and Thos. Anderson, Greenock.
1832. M'Ewan, Son and Company, Glasgow.
1832. Hugh Hutton and Company, Greenock.
1833. Speirs and Wrede, Port Glasgow.

companies, which are frequently very fully employed: and when this is the case, they boil upwards of 3,000 hogsheads of sugar yearly, which at £45 per hogshead, the average price for the last five years, is £135,000 for the raw material alone.[1]

A further advance in sugar technology preceded another expansion in the number of Greenock refineries. Although this development, the drying of refined sugar in centrifugals in place of draining by gravity, is rather late for the present discussion, its importance ranks with Howard's invention of the vacuum pan, so it is desirable to give a brief account of its evolution. The first use of centrifugal force to remove moisture was its application in 1837 to wool-drying. Its introduction into sugar technology is recounted in the following citation from Fairrie's *Sugar*:

> According to Dr. H. C. Prinsen Geerlings, a centrifugal machine was first used for the separation of syrup and sugar crystals in 1844 by Schöttler in a sugar factory at Sudenburg, and he states that their general use on the Continent dates from 1850, when they were installed at a sugar factory at Lembeek in Belgium. The first British patent for the application of the principle of centrifugal force for the separation or dispersion of molasses from sugar was taken out by Laurence Hardman in 1843. The probability is that a centrifugal machine was first utilised in England on sugar very shortly after Hardman took out his patent, which also covered the liquoring of the crystals in the machine to remove adhering syrup.[2]

The Fairrie concern was the first to utilize the invention in England; Angus and Company (formerly Angus Balderston and Company), in Scotland.

After an interval of fifteen years (1832–47) during which no new refinery was built in Greenock, expansion began again. Three refineries were started in 1847, one in 1848, and one in each of the years 1851, 1852, and 1853. The figures on the next page reflect the effect of the foregoing technological advances.[3]

So far reference has been made exclusively to the sugar obtained from sugar cane (*saccharum officinarum*), but in addition sugar can be obtained commercially from sugar beet (*beta vulgaris*), a plant adaptable to a much more varied habitat than sugar cane, particularly the countries of Europe. Modern examples of sugar beet contain 16–18 per cent. sugar, plus 75–77 per cent. water, 5·5 per cent. fibre, and 2·5 per cent. salts. Throughout the later

[1] John Wilson, *General View of the Agriculture of Renfrew*, p. 236.
[2] G. Fairrie, *op. cit.*, p. 168.
[3] From D. Bremner, *Industries of Scotland*, p. 457.

Year	*Quantity used, in Tons*	*Lb./Head*	*Av. Duty cwt.*	*Av. Price incl. Duty cwt.*	*Total Duty*
			s. d.	*s. d.*	£
1700	10,000	3	—	—	—
1754	53,270	12	—	—	—
1801	159,916	22	20 0	79 5	3,066,163
1821	170,612	18	27 0	60 2	4,188,997
1841	202,899	17	25 2	64 1	5,114,390
1851	328,581	26	12 0	37 6	3,979,141
1861	446,865	34	13 10	40 4	6,104,325

decades of the nineteenth century, sugar beet had an important effect on the economics of sugar production.

The sweetness of beet juice was first observed by Olivier de Serres in 1590. In the middle of the eighteenth century Andreas Sigismund Marggraf (1709–82) experimented on different beets to find which would give the greatest yield of saccharine material. He also examined the possibilities of increasing the yield by selective breeding. In the laboratory he isolated sugar crystals from beet juice, but did not carry this experiment far enough to enter commercial production. Marggraf's work was continued by one of his pupils, Franz Karl Achard (1753–1821), a Berlin chemist of Huguenot descent, who announced about 1797 that he could produce beet sugar on a commercial scale. He was granted £8,000 by Frederick the Great, and Frederick William III placed him in charge of a sugar beet factory which was opened at Cunern, near Steinau in Silesia, in 1800. Some sixty acres of beet were cultivated. In 1811 Achard published an account of processes he adopted in beet sugar manufacture while he was at Cunern. At this time the most important technical development was the diffusion process devised by Göttling, which is still in use at the present day.[1]

When, *c.* 1810, Napoleon placed an embargo on the importation of English sugar, beet sugar entered the sphere of international economics. Information had already reached France of the recently founded Prussian sugar industry, but it was not till the blockade of the French ports following the Berlin and Milan decrees that serious attempts were made to imitate the Prussian example. Defeats at Jena and Auerstadt in 1806, and the occupation of Prussia by Napoleon brought about a temporary cessation

[1] *Phil. Journal*, *3*, 94, 237, 243, 474; *4*, 28, 335; *Phil. Magazine*, *3*, 291; *4*, 334.

in the embryonic Prussian industry. The consequent high price and scarcity of sugar led Napoleon to establish beet sugar schools, and demand from French chemists a substitute for cane sugar.

Much research was set on foot, particularly at the instance of the Paris sugar refiner and chemist Derosne. Between 1811 and 1816, some 213 small factories were opened. Derosne published an abridged account of Achard's treatise in 1812 under the title, *Traité Complet sur le Sucre Européen de Betteraves*. This work has an introduction by Derosne, and contains a detailed description of beet sugar manufacture in Prussia with extensive references to the use of sulphuric acid in the manufacturing processes, to which reference will be made later.

With the termination of the Napoleonic wars, economic conditions which were once favourable for the development of the new industry in France no longer existed, and it was unable to hold its own against West Indian sugar, which was readmitted. Attempts were made to keep the industry alive in metropolitan France by the imposition in 1822–5 of high taxes against colonial produce, but further less favourable legislation was introduced in 1836. To a certain extent these taxes were offset by the fact that five per cent. instead of two per cent. of sugar could be extracted from beet. McCulloch reckoned that French production was then in the neighbourhood of 8,000 tons,[1] but in his *Familiar Letters on Chemistry*, Liebig hazarded the opinion that the manufacture 'would ere long be in most places entirely discontinued'. Liebig based his argument on the rise in price of agricultural products between 1825 and 1843, in contrast with the price of beet sugar, which had remained constant, and with colonial sugar, which had fallen.

> Food and fuel have risen and wages to a less degree, while the price of colonial sugar has fallen. . . . at present it is more advantageous to cultivate wheat and to purchase sugar.[2]

This is how Charles Macintosh described what he saw in France in a letter to Joseph Gordon, Jamaica (26th September, 1829):

> It (the beet root) is taken up about the end of September, well washed, and cleaned from adhering earth. The leaves are separated as food for cattle, and the roots are rasped down by machinery into a moist, pulpy powder, or mass. This pulp is put into small bags, deposited in a chest perforated with holes, and exposed to

[1] McCulloch, *Commercial Dictionary*, Vol. 2, p. 1096.
[2] J. Liebig, *Familiar Letters on Chemistry*, p. 42.

the action of a powerful screw, or hydraulic press. By this means the whole juice is expressed, (the remaining mass is used for feeding cattle), which juice is ladled out into copper pans, brought nearly to the boil, and thoroughly skimmed; after which, a certain portion of quick lime is added, and the skimming continued. The clarified liquid is then transferred to *teches* where it is rapidly boiled down to a granulating point, or in some cases, contrary to all previous theory of sugar-making, this juice is placed in shallow pans, four inches deep, by three long and broad, which are placed upon racks made on purpose, heated up to about 200° (Fahrenheit), when the sugar crystallizes in strong grains.[1]

In Prussia legislation was more consistent than in France, and the industry continued to develop, particularly after 1830.

English reaction can perhaps be best summed up by citation from McCulloch.

We understand that a few small parcels of beet roots have recently been produced in this country; and with the present enormous duty on colonial sugar, we are not sure that the manufacture may not succeed. But, as the preservation of the revenue from sugar is of infinitely more importance than the introduction of this spurious business, the foundations of which must entirely rest on the miserable machinery of Custom-house regulations, sound policy would seem to dictate that the precedent established in the case of tobacco should be followed in this instance, and that the beet root sugar manufacture should be abolished. Inasmuch, too, as it is better to check an evil at the outset, than to grapple with it afterwards, as we trust that no time may be lost in taking vigorous measures, should there be any appearance of the business extending.[2]

Britain has on more than one occasion regretted the *sound policy* which McCulloch advocated. As Charles Macintosh aptly remarked:

If this beet sugar goes on successfully in France, you may depend upon it the manufacture will be attempted in England, when a conflict between the colonial and agricultural interests, of a keen and angry description, will follow.[3]

Beet sugar production in England to which McCulloch refers was doubtless the founding in 1832 by Marriage, Reed and Marriage, of the first English factory at Ulting, near Maldon Essex. It did not live long, due, it is sometimes said, to apathy on

[1] G. Macintosh, *Memoir of Charles Macintosh*, p. 100. *See* also article on *French Commercial System*, *Edin. Rev.*, No. 99.
[2] J. R. McCulloch, *loc. cit.* [3] G. Macintosh, *loc. cit.*

the part of farmers. Perhaps rural conservatism was a contributory factor, but it was by no means the only one operating against the early establishment of a British beet sugar industry, the subsequent history of which does not concern us here, and which may be followed in G. Fairrie's *Sugar*.[1]

Since the introduction of vacuum pans, charcoal, and centrifugals placed the sugar-boiler in the debt of the mechanical and chemical technologist in the early nineteenth century, it has become customary to subject sugar to processes of carbonation (Rousseau, 1849) and sulphitation. Any mineral acid will reduce the colour of juice, and this practice has its roots in the use by Achard of sulphuric acid to remove excess lime in the purification of beet sugar. The following details are taken from the *Histoire Centennale du Sucre de Betterave* published in 1912.[2]

> La défécation par l'acide sulphurique, suivie d'une saturation par la craie et d'une addition de chaux, a été inaginsé par Achard, et employée de 1802 à 1806 dans son usine de Kunern et dans celle de Krain (Silésie) dirigée par de Koppi, comme le prouvement les rapports de Baudet et du docteur de Neubeck et le traité d'Achard. Le jus était extrait à froid au moyen de la râpe et de la presse, et additioné, pour chaque litre, de 1 cc. 15 d'acide sulphurique concentré, étendu de quatre fois son volume d'eau, puis abandonné à lui-même sans être remué, dans un lieu frais pendant douze à seize heures, versé alors dans une bassine où on mettait de la craie (6 gr. par litre) et additioné de chaux en lait (1 gr. 80 CaO par litre); on chaufait, et quand le liquide était tiède, on ajoutait, par litre, 60 cc. de 'lait écrémé et prêt à aigrir'. On cuisait pendant une demi-heure et on filtrait sur 'une toile de drap serré'. Le procédé d'Achard fut critiqué par Deyeux dans son mémoire de 1811, mais adopté par Derosne en 1811–1812, et par Crespel, qui l'employa de 1811–1812 à 1818–1819 dans sa fabrique d'Arras. Dubrun faut proposa plus tard d'introduire l'acide sur la pulpe râpée.

Coextensive with the use of sulphuric acid, experiments were being tried with the reagent which is still in use, viz., sulphurous acid (H_2SO_3), or gaseous sulphur dioxide (SO_2), the function of which is twofold, to neutralize excess alkalinity, and bleach the juice.

> Drapiez fit en 1809–1810, intervenir pour la première fois l'acides sulphureux en sucrerie; il le produisait en attaquant des matières organique par l'acide sulphurique, dans un appareil

[1] G. Fairrie, *op. cit.*, p. 33 *et seq.*

[2] Supplied by John L. Mackie, author of thesis *British Sugar Taxation since the Napoleonic Wars* (London University, 1939)

dont nous possédons le dessin, et le faisait agir sur le jus, préalablement bouilli avec de la craie et du charbon et partiellement évaporé. Les pains de sucre qu'il présenta à la Société d'encouragement provenaient de jus semblablement traités.

L'acide sulphureux fut quelque temps délaissé par nos premiers fabricants. Nous le trouvons cependant employé, en 1838, par Stollé, en 1861 par Périeret Essoz, puis en 1870 par Seyfert.

Stollé was granted a British patent for the use of sulphurous acid and lime in 1838. In 1843 the use of carbon dioxide was patented in Germany, and in 1849 Cowper obtained an English patent. In 1852 Henry Bessemer, inventor of the steel process which bears his name, patented the use of gaseous sulphur dioxide. Bessemer, having made a large fortune in other branches of technics, thought to engage in sugar manufacture in 1852. In this, however, he was singularly unsuccessful. Bessemer's process used sulphur dioxide alone. The use of both lime and sulphur dioxide was patented by Sievier in 1857.

At the beginning of this chapter it was mentioned that sugar was in the first instance introduced by the medical profession, and we cannot do better than end with one or two citations of contemporary medical opinions on the effect of increased ingestion of sugar. Drummond and Wilbraham tell us that:

. . . the medical profession was expected to pronounce on the food value of sugar during the eighteenth century, since it became a cheap and common household commodity. They were, as usual, divided.[1]

Dr. Gregory maintained that it was hurtful to teeth, while other physicians recommended its addition to cow's milk when the latter was substituted for human milk in infant feeding, as human milk contains a higher percentage of sugar (lactose) than does cow's milk. Dr. William Cullen ranged himself against those who arraigned the new commodity, and Lord Dundonald supported him.

He that undertakes to argue against *sweets* in general, takes upon him a very difficult task, for nature seems to have recommended this taste to all sorts of creatures; the birds of the air, the beasts of the field, many reptiles and flies, seem to be pleased and delighted with the specific relish of all sweets, and to distaste its contrary. Now the sugar-cane, or sugar, I hold for the top and highest standard of vegetable sweets. Sugar is obtainable in some degree from most vegetables, and Dr. Cullen is of the opinion,

[1] J. C. Drummond and A. Wilbraham, *op. cit.*, p. 281.

> that sugar is *directly* nutritious. There is also good reason to suppose, that the general use of sugar in Europe has had the effect of extinguishing scurvy, the plague, and many other diseases formerly epidemical.[1]

But the prejudice against sugar continued for a long time, and Moseley had some pertinent remarks to make on the economy of the prejudice in Scotland.

> In regard to sugar being prejudicial to the teeth, this has long been known as a prudent old woman's bugbear, to frighten children; that they might not follow their natural inclination, by seizing opportunities, when they are not watched, of devouring all the sugar they can find.
>
> This story has had a good effect among the common people in Scotland. They are impressed with a notion that *sweeties* hurt the teeth; therefore they live contented without the article, not always within the compass of their finances.[2]

[1] From Edward's *History of the West India Islands*, quoted by Dundonald, *Treatise on Agriculture and Chemistry*, p. 217.

[2] B. Moseley, *A Treatise on Sugar*, p. 143.

Chapter XXIII
FERMENTATION INDUSTRIES

'There cannot be a doubt that beer is inferior in salubrity to plain water as a beverage.'

THOMAS THOMSON.

ALTHOUGH a single Scottish distillery contributed to the Exchequer a sum of money equivalent to the whole land tax of Scotland, the social technology of the fermentation industries has been singularly neglected by economic historians. Nevertheless, the interconnexions of brewing, distilling, and mechanical technology on the one hand, and their relation to agriculture on the other, merit consideration here. Although both brewing and fermenting may now be regarded as branches of technical chemistry, they were among the most highly developed eotechnic arts. For example, when Sir William Brereton visited Edinburgh in 1634, what he saw there caused him to comment on the scale of the capitalist brewery already in operation in that city:

> I took notice here of that common brewhouse which supplieth the whole city with beer and ale, and observed there the greatest, vastest leads, boiling keeres, cisterns and combs (wooden tubs), that ever I saw: the leads to cool the liquor in were as large as a whole house, which was as long as my court.[1]

In the era preceding industrial revolution steam economy, breweries were among the chief consumers of coal in making the vegetable extracts from which ale and beer are brewed. The industry too was one of the first to utilize mechanical power to expedite the grinding and mashing of the vast quantities of grain which are handled in the course of a twelvemonth, and it had been to the brewer's copper or the distiller's pot that steam-engine inventors turned when they were first faced with the problem of generating steam in large quantities. Throughout the eighteenth century the scale on which brewing and distilling was carried on progressively expanded, so that by the time Boulton and Watt, having exhausted the markets of Cornwall, were looking for a

[1] Wm. Brereton, *Travels, 1634–5*, p. 104.

further market for their engines, many brewers had sufficient capital to purchase them. Boulton and Watt's fifth engine went to a large London distillery.[1]

The acquisition of steam-engines by capitalist brewers enabled them further to expand their production at the expense of their smaller rivals, the brewing publicans.[2] This altered the organization of the brewing industry, leaving on one hand the domestic brewer, and on the other the vast commercial undertaking. As public breweries developed the brewing publican, of whom there had once been twenty-four thousand in the country, passed from the scene, but the continued co-existence of domestic craft and capitalist enterprise is one of the peculiarities of both brewing and, to a lesser extent, distilling.[3]

Of drink in the country at the middle of the eighteenth century it was said:

> Ale or beer brewed by every farmer at home from oats and heather—'so new that it was scarce cold when it was brought to table', was their chief beverage.[4]

But there is no doubt that capitalist brewing also made inroads into this private brewing, as we learn from the following citation from the *Session Papers*:

> James Rigg, of the Marischal Street Brewery, Aberdeen, the first common brewer there, 'was the first person who ever set up what is called a common or public brewery at Aberdeen (in 1765), from which the inhabitants, in place of brewing their own ale, are furnished with it both at a cheaper rate and of a better quality than they can possibly manufacture it for themselves'. When Rigg set up we are told that almost every inhabitant of the town brewed his own ale, but that after the lapse of fourteen years, 'there is hardly such a thing now practised, and five or six great breweries are set up upon the same plan with that of Rigg's'.[5]

Fermenting industries, for brewery and distillery are in principle the same, the latter simply proceeding one stage further than the former, have contributed considerably to the development of industrial techniques. The retort, that overworked symbol of chemical activity, was the laboratory analogue of the simple still, and as such has always been of prime importance in the

[1] J. Lord, *Capital and Steam Power*, p. 152; H. W. Dickinson, *James Watt*, p. 155; J. U. Nef, *Rise of the British Coal Industry*, Vol. 1, p. 215.
[2] A. Barnard, *Noted Breweries of Great Britain and Ireland*, Vol. 1, p. 209.
[3] S. Child, *A Practical Treatise on Brewing*, p. 23.
[4] H. G. Graham, citing Morer's *Short Account*, p. 179.
[5] *Session Papers*, Signet Library, 597.14.

preparation of pure substances. Both the pot-still and Aeneas Coffey's patent still (1831) used in the commercial production of grain spirit, were subsequently modified and adapted to the distillation of coal-tar and petroleum products as advancing palaeotechnics demanded new and more involved products. When the excise laws were altered so that the size of his stills became of significance to the distiller, Scottish inventive genius was turned to devise a still of maximum distillation rate. The results thus achieved were, in the words of a contemporary, 'beyond what had been practised in fact, or foreseen by theory'.

Only gradually did brewing and distilling come under even simple scientific control, but in the early years of the nineteenth century J. H. Baverstock saw fit to remark:

> The rapid advances which chemistry has made in the last forty years, and the use of accurately-constructed thermometers and hydrometers, have been the means of introducing a regular system in brewing, which has shown that the process is a science, depending for its success upon certain and invariable principles, and that it is not a mere mechanical operation, performed by any menial and illiterate person whom it may be convenient to employ in it. And it is, in consequence, beginning to rank as high among the arts and scientific manufactures, as the enormous duties which it pays entitles it to do among the revenues of the kingdom.[1]

Knowledge of the processes whereby a dilute solution of *ethyl alcohol* (C_2H_5OH) can be produced from farinaceous or saccharine material has constituted part of the craft knowledge of practically every culture. There is therefore no need to deal with the historical antecedents of the subject here, beyond recording the microscopic observation, which Leeuwenhoek (1632–1723) made in 1680, that the yeast which brings about the fermentation process is a mass of small cells—it was not till 1838 that Cagniard-Latour showed that it was a living organism which reproduced by budding.

In order to understand the relation of brewing and distilling to correlative mining and agrarian economy, we must consider in detail the operations whereby a potable liquor or spirit is produced, as well as define some of the terms used whose connotations have changed from time to time.

According to Thomas Thomson (1773–1852), who became Professor of Chemistry in Glasgow, and whose work on brewing may be said to have laid the basis of Scottish excise legislation:

[1] J. H. Baverstock, *Treatises on Brewing*, p. vi.

> The English word *ale* is obviously the same with the Swedish word *öl*, which is applied to the same kind of fermented liquor; while the word *beer* is synonymous with the German word *bier*. These two words in Great Britain are applied to two liquors obtained by fermentation from the malt of barley; but they differ from each other in several particulars. Ale is light-coloured, brisk, and sweetish, or, at least, free from bitter; while beer is dark-coloured, bitter, and much less brisk. What is called *porter* in England, is a species of beer, and the term porter at present signifies what was formerly called strong beer.[1]

Ale was the English drink in the fourteenth and early fifteenth centuries. Small quantities of beer initially came from Flanders, and beer-drinking had become general by the end of the fifteenth century. There was also a further change:

> The common drink of the people of old had been ale, which was brewed in every farmhouse and sold in every change-house or tavern; it was drunk everywhere by the peasantry; and, indeed, by all classes. But by the imposition of the malt tax, the production of it was greatly affected; and partly owing to the increasing cost and to the gradual influx of spirits, smuggled or legal, the use of ale diminished, while whisky—in 1780 only 10*d*. a quart, on which no excise was being paid—came more and more into use, and where formerly the workmen were regaled with 'twopenny', now they were presented with whisky.[2]

There is a little more to be said on the subject, however. In England before 1722, three species of malt liquor were kept by the publican: (*a*) ale, (*b*) beer, and (*c*) twopenny. These liquors were not consumed separately, and it was usual to call for a pint of *mixed*. Then a brewer of the name of Harwood introduced a liquor called *entire*, or entire butt beer, which gave the flavour of the above mixed liquors without the necessity of mixing. This single liquor was supposed to be particularly suitable for porters and other workers, and so it got the name *porter*. At a later date the publican got two beers from the brewer: *mild*, which was freshly brewed, and *old*, or *stale*, which was specially brewed for keeping. This was retailed in surreptitious admixture as *entire*.[3]

The raw material from which ale, beer, and whisky are made is grain of some sort, almost always barley. This greatly increased the demands on farmers for supplies of grain.

[1] *Encyclopaedia Britannica*, Supplement, IV–VI Editions, Article by Thomas Thomson, Vol. 2, p. 485. [2] H. G. Graham, *Social Life in Scotland*, p. 216.
[3] F. Accum, *Treatise on the Art of Brewing*, p. 6.

The distillery affords a great market for barley, which is the only sort of grain used for distillation in this country. This manufacture not only stimulates the production of as much barley as supplies the demand; but as this crop requires the land to be well cleaned, wrought, and manured, it excites the production of much greater quantities of other sorts of grain, than would otherwise be raised. Thus the distillery operates as a premium on the production not only of barley, but of all other sorts of grain, and causes a much greater surplus to be raised than is necessary for feeding the people. The same thing may be said of the manufacture of starch, and of all the other modes by which grain is consumed, that is not used for food. Were these manufactures abolished, the farmers would soon come to limit their exertions, and would raise no more grain than could be disposed of as food. In such circumstances should a deficient crop occur, unless the deficiency could be made up by importation, the consequences might be fatal. But in extreme cases such as we have supposed, government can stop these manufactures, and apply the surplus which they had occasioned to be raised, for feeding the people.[1]

The choice of barley depended on its high starch content and ease of germination. A contemporary analysis of barley by Einhoff gave the following constituents[2]:

Volatile matter	360 parts
Albumen	44
Saccharine matter	200
Mucilage	176
Phosphate of lime with mucilage	9
Gluten	135
Husk, with some gluten and starch	260
Starch not quite free from gluten	2,580
Loss	76
	3,840

In the eighteenth century there were two species of barley cultivated in Scotland; in the south *Hordeum vulgare*, with the seeds arranged in two rows; in the more exposed situations *Hordeum hexastichon*, in which the seeds appear to be in six rows. The latter was called *bear* in the south of Scotland, *big* in Aberdeenshire. Some authorities, however, use the term *bear* exclusively for a four-rowed variety.[3]

Scotland did not grow sufficient barley to meet all demands,

[1] James Headrick, *General View of Agriculture of Angus*, p. 573.

[2] Thomas Thomson, *op. cit.*, Vol. 2, p. 462.

[3] *Ibid.*, Vol. 2, p. 461.

and had recourse to import from England.[1] For example, in West Lothian:

> There may be about 11,000 bolls of barley consumed in the distilleries of the county, which in all probability is more than is actually raised in it. A considerable proportion of the barley used in the distilleries is brought from other parts of Scotland, and from England.[2]

The practice of brewing is fully described in an anonymous book published in 1738 called *The London and Country Brewer*, the object of which was to enlighten the public which had long suffered from 'unwholesome and unpleasant beers and ales, by the badness of malts, under-boiling of worts, mixing injurious ingredients and unskilfulness of the brewers'. The first stage in the brewing process is the conversion of starch contained in the grain, which is insoluble, into more soluble products. This is effected by the processes of *malting* and *mashing*.

MALTING: Having been thoroughly moistened by steeping in water for two to three days, and stacked for 24 hours, the grain is spread out on the malt-floor and allowed to begin germinating. This starts in three to four days, and to ensure even germination, the grain is continually turned over. In order to begin growing, the embryo of the grain has to draw upon the starch of the grain, and to render it available an enzymic ferment, *diastase*, develops which can convert large quantities of insoluble starch into soluble sugars. That germinating barley had the power of liquefying starch was first observed in the laboratory in 1814 by Kirchoff, and the enzyme preparation was first made by the French chemists Payen and Persoz in 1833. What is perhaps the first mention in English of diastase, a generic term in French for all ferments, is contained in *A Practical Treatise on Brewing* by William Black of Gilcomston Brewery, Aberdeen.

Barley at this stage is called *green malt*, and while retaining the shape of the original grain, has acquired a sweet taste, since, along with the diastase, a quantity of malt sugar (maltose, $C_{12}H_{22}O_{11}$) has been formed.

$$\underset{(C_6H_{10}O_5)_x}{\text{STARCH}} \xrightarrow{\text{diastase}} \underset{(C_{12}H_{22}O_{11})}{\text{MALTOSE}}$$

When sufficient enzyme, after perhaps 14–21 days, has developed, germination is stopped by drying the grains. The temperature of the drying kiln is gradually raised to 65–66° C. and ultimately to

[1] Thomas Thomson, *op. cit.*, Vol. 2, p. 486.

[2] J. Trotter, *General View of West Lothian*, p. 200.

93–94° C. The fuel used in the kiln has at various times been straw, coke, anthracite, and in the manufacture of Scotch whisky, peat. The original difference between ale and beer lay in the fact that ale malt was dried at a very low temperature, and in consequence retained its very pale colour, while beer or porter malt acquired a brown colour through being dried at a higher temperature. Along with the brown colour, beer malt acquired other agreeable qualities, in consequence of which it was made in large quantities, and beer became the drink of the lower classes.

The following figures for the year 1811 give some idea of the dependence of the farmer on the brewer for the sale of his barley. The crop of barley that year was 4,800,000 quarters, of which 3,200,000 quarters were malted, and 600,000 retained for seed. Only 1,000,000 quarters were used directly for food, although as we shall see later, much of that initially converted to malt was in the end used for fattening cattle and pigs. The malt was absorbed as follows: by common brewers 1,600,000 qrs., brewing publicans 800,000 qrs., private families, 800,000 qrs.

Malt was subject to direct taxation. A tax of 6*d.* per bushel was imposed on English malt in 1695, and on Scottish in 1725. The extension of the tax to Scotland was very unpopular, and serious rioting broke out in Glasgow in June of that year. General Wade, who was in Scotland, was forced to take over the city, and a considerable number of lives was lost. The next year, however, the tax was halved. The following table illustrates the fluctuations in the malt tax over a period of years. It is important to observe the different levies in England and Scotland.[1]

	England	*Scotland*	*Ireland*
	s. d.	*s. d.*	*s. d.*
1760	9¼	—	—
1780	1 4¼	8	—
1785	—	—	7
1795	—	—	1 3
1802	2 5	1 8¾	1 9½
1804	4 5¾	3 9½	2 3½
1813	—	—	3 3¾
1815	—	—	4 5
1816	2 5	1 8¾	1 4
1819	3 7¼	3 7¼	3 6¾
1822	2 7	2 7	2 7

In addition to a progressive increase in the tax, certain operational restrictions were also imposed during the course of the

[1] D. Bremner, *Industries of Scotland*, p. 437.

eighteenth century. For example, malt had to remain in the steeping cistern for at least forty hours, where it was gauged by the exciseman. Duty was levied on his best gauge. A further restriction was that the grain might not be moistened on the malting floor.[1]

One hundred parts of English barley gave 109 of malt, while Scottish barley and big only gave 103 and 100.6 respectively.[2] Owing to this inferiority Scottish duties were less than those levied in England in the following ratios:

	s. d. per bushel	i.e.
English malt . .	4 4	100 to
Scottish barley malt . .	3 8⅛	84·856
Scottish big malt . .	3 0½	69·472

With the advent of the agrarian revolution and consequently improved Scottish husbandry, English brewers and distillers complained to the Lords of the Treasury that the Scots could undersell them. The matter in consequence was referred to Dr. Thomas Charles Hope, Professor of Chemistry, Dr. Andrew Coventry, Professor of Agriculture, and Dr. Thomas Thomson, lecturer in Chemistry, at that time all of Edinburgh. The results of their deliberations appeared in 1806 as the *Scotch Report on Brewing* (*House of Commons*).

Some idea of the revenues raised by taxation of malt in the years 1813 and 1814 will be gained from the following figures:

England: Duty 4*s.* 4*d.* per bushel.
1813 £4,188,450 6*s.* 9*d.* from 2,416,384·8 qrs.
1814 £4,772,332 5*s.* 5½*d.* from 2,753,268·6 qrs.
Scotland: Duty 3*s.* 8⅛*d.* per bushel.
1813 £134,106 12*s.* 0¾*d.* from 91,436·32 qrs.
1814 £125,787 7*s.* 10¼*d.* from 85,521·18 qrs.

In 1816 the war duty of 2*s.* per bushel was taken off. In 1819 a new duty increased the total levy on malt to 28*s.* per quarter, but 3 Geo. IV, cap. 18, deducted 8*s.* leaving the malt duty at 20*s.* per quarter.

MASHING: Returning to the manufacturing processes, the next operation is one which needs power, whether derived from water or steam. The rootlets which developed during the process of germination are removed by screening and the malt is ground in order to facilitate subsequent solution of soluble products. At one time brewers were thirled to the barony mill, but with the

[1] Thomas Thomson, *op. cit.*, Vol. 2, p. 464. [2] *Ibid.*, p. 468.

introduction of steel mills, which seems to have taken place in the neighbourhood of Edinburgh about 1740, brewers invested in their own mills, in consequence of which they were involved in legal proceedings with the tacksman of the mill. In one case at least the brewers were adjudged liable for the thirlage, since if they abstracted their malt from the barony mill, the miller would be unable to pay his rent, the 'sixteen bolls multure malt'.[1]

The ground malt is then macerated with warm water for several hours in the *mash-tun*.

> A very strong iron screw, of the same height as the mash-tun, is fixed in the centre of this vessel, from which proceed two great arms or radii, also of iron, which beset with vertical teeth, a few inches asunder, in the manner of a double comb; by means of a steam-engine, or any other moving power, the iron arms, which at first rest on the false bottom, are made slowly to revolve upon the central screw, in consequence of which, in proportion as they revolve, they also ascend through the contents of the tun to the surface; then, inverting the circular motion, they descend again in the course of a few revolutions to the bottom. These alternate motions are continued till the grist and water are thoroughly incorporated.[2]

So familiar was the operation of *mashing* or *masking* to the common people of Scotland, that when tea displaced beer or *twopenny*, the terminology of brewing was used to describe the operations of its preparation. The addition of the hot water to the tea was called mashing or masking, and the resulting infusion a brew of tea.[3]

In the presence of water, diastase completely converts starch into a sugar called maltose. Diastase present in malt will also convert starch present in unmalted grain into maltose, but in brewery this was expressly forbidden by law.[4]

The aqueous extract goes by the name of wort, and contains roughly about four per cent. maltose. Its strength is important because it determines the ultimate strength of the liquor brewed from it. It may contain 75–85 lbs. of saccharine matter per barrel. Being a biological process, there is an optimum reaction temperature, but it was not till about 1762 that Michael Combrune introduced the use of the thermometer into this part of brewery.[5] Even then he did it with difficulty, and he records that he was forced to hide his thermometer from his father who objected to

[1] *Session Papers*, Signet Library, 42:5. [2] F. Accum, *op. cit.*, p. 47.
[3] H. G. Graham, *Social Life in Scotland*, p. 217. [4] 56 Geo. III, *c.* 58.
[5] M. Combrune, *The Theory and Practice of Brewing*, p. 30.

T. H. Croker, *Dictionary of Arts and Sciences*, 1764–6

100. Brewhouse at the middle of the eighteenth century.

101. Illicit distillation in the Highlands, showing the simple apparatus used. A lookout can be seen in the background and the Excisemen bearing down on the distillers, who will, no doubt, make off across the river and 'jink the gauger'.

102. Special fast stills. These were devised when the Excise was levied on still capacity and there was a premium on fast working (1788–99).

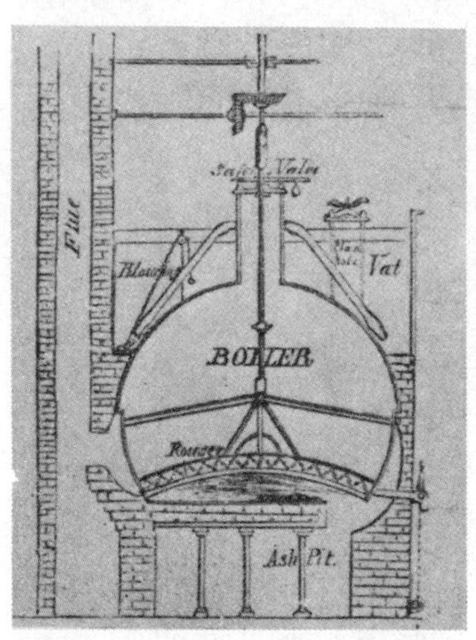

103. Brewer's copper.

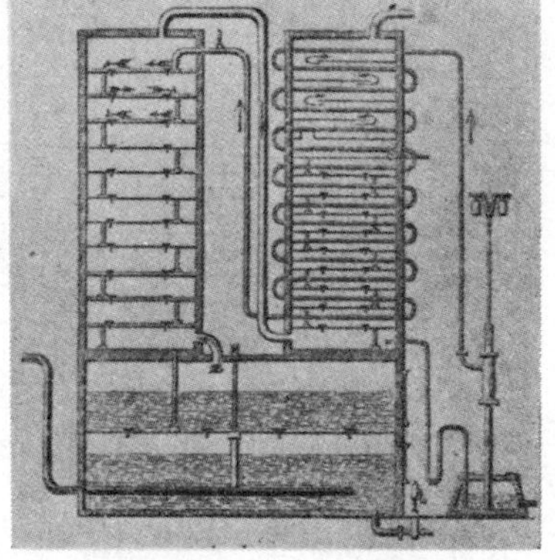

104. Coffey's Patent still (1831)

'experimental innovations'. Samuel Whitbread was as sceptical as Combrune's father about the use of another instrument—the hydrometer, used to determine the strength of the sweet liquor which is extracted from malt. John Richardson of Hull introduced what he called the saccharometer for determining the strength of wort in 1784, and James Baverstock (1741–1815) the hydrometer in 1768. By 1824 the hydrometer had been proved sufficiently reliable for an Act[1] to contemplate using it for excise purposes.

When the mashing and extraction process is complete, the solid residue is filtered off from the extract. The residue is considerable. According to Thomas Thomson, it amounts to the following:

From English barley . . .	50·63%
From Scottish barley . . .	50·78%
From Scottish big barley . .	52·69%[2]

This large percentage of the original grain is not wasted, however. It formed an important food for livestock; for example, of dairy cattle in Edinburgh it was written:

> Their food in summer is brewer's and distiller's grains and dreg, wheat shellings or small bran, grass and straw; and in the winter the same grain, dreg and bran, with turnips and potatoes, and hay instead of grass.[3]

In the manufacture of beer or ale the wort, freed from the residue of the malt, is next boiled with hops (*humulus lupulus*) which flavour the liquid as well as have a preservative effect on it. Larger quantities were used in English beer than in Scottish.[4] The wort is rapidly cooled and in both industries fermentation follows, in what is called the *gyle-tun* in the brewery. The wort was gauged both in the boiler and cooler as a check on the figure taken on the fermenting tuns. The object of this was that some of the wort might be abstracted between the boilers and coolers and fermented without the knowledge of the excise.

FERMENTATION: Yeast, which is a living organism capable of producing enzymes, is added to the wort. *Small beer* was disposed of by the brewer as soon as the yeast was added and before fermentation set in. One of the enzymes is called *invertase.* It causes the maltose or malt sugar to combine with a molecule of water (hydrolysis) forming two molecules of dextrose, another, but simpler, sugar.

[1] 1 and 2 Geo. IV, cap. 22. [2] Thomas Thomson, *op. cit.*, p. 471.

[3] *Encyclopaedia Britannica*, Supplement IV–VI Editions: Article *Dairy.*

[4] In the production of distilled spirits this process is omitted, since it is the distillate which is consumed and not the actual fermented liquor.

$$C_{12}H_{22}O_{11} + H_2O = 2C_6H_{12}O_6$$
Maltose + water = dextrose

But the dextrose is not in a propitious environment for its continued existence, since a second enzyme present in the yeast, *zymase*, breaks it down, this time with the formation of the gas, carbon dioxide, and alcohol.

$$C_6H_{12}O_6 = 2C_2H_5OH + 2CO_2$$
Dextrose = alcohol + carbon dioxide

Throughout the greater part of history the carbon dioxide has gone to waste, and it is only within the last decades that it has become available in large quantities for refrigeration. Yet of its collection Parkes says:

> It is a matter of surprise that no advantage has been taken of the vast quantities of carbonic acid gas, which are perpetually escaping from the vats of the large breweries in this metropolis. It might surely be collected and applied to many useful purposes, if the commissioners of excise would admit of it.[1]

In Edinburgh, however, much of the beer was bottled so that some of the carbon dioxide was prevented from escaping. This made it brisk and palatable.

The resulting aqueous solution formed along with the carbon dioxide is beer, ale, or porter, as the case may be, depending on the temperature at which the processes are carried out, the concentration of the raw materials, etc. In brewery, fermentation is not carried very far as it is not so much a maximum yield of alcohol which the brewer wishes to produce as a palatable drink. The above are the principal reactions in the process of fermentation, but side reactions produce by-products as well, and these have an important effect on the quality and flavour of the final product.

Among the substances produced in small quantities are acetic acid (CH_3COOH), ether ($(C_2H_5)_2O$), acetaldehyde (CH_3CHO), acetone ($(CH_3)_2CO$), amyl alcohol, the so-called fusel oil ($C_5H_{11}OH$), and glycerin ($C_3H_5(OH)_3$).

Starch can also be converted into a sugar by boiling with very dilute sulphuric acid. This was said to answer well for small beer, but was repressed by the Excise.

Since it was not till the time of Louis Pasteur that our knowledge of fermentation finally took shape, there is not much to say, over the period under discussion, of the theory of fermentation. The

[1] S. Parkes, *Chemical Catechism*, p. 279, *n*.

following citation does, however, illustrate the interest taken in the subject by the scientists of the time:

> Lavoisier was the first person who attempted to give anything like a theory of this intricate process. He attempted to determine the composition of common sugar, a substance which may be fermented just as well as the soluble part of malt, and which yields similar products. He endeavoured, likewise, to determine the constituents of alcohol, the substance formed by fermentation. With these data, and with a knowledge of the composition of water and carbonic acid, he formed a plausible theory, which was valuable as a first approximation, though there can be little doubt that it was erroneous in every particular. Since that time, several experiments on the subject have been made by Thénard. Gay-Lussac, and Thénard and Berzelius have determined the constituents of sugar with much care; and Theodore de Saussure has made very elaborate, and, we believe, accurate experiments on the composition of alcohol.[1]

The element of chance which entered into the brewing processes is commented on by Drummond and Wilbraham:

> It is possible by reading a book of the type of *The London and Country Brewer*, to get a clear idea of how great an element of chance there was in brewing at this date. The practice was carried out by rule of thumb methods, and the brewers were at a loss to explain why one brew kept quite well while another turned acid, cloudy, or 'ropy'. They were, of course, entirely ignorant that these changes were due to the growth of organisms, and all kinds of recipes were suggested for preventing and curing these 'diseased' beers. Some of them could scarcely be regarded as innocuous. 'An ounce of Alum very fine' mixed with 'two Handfuls of Horse-bean Flour' would 'cure a Butt of Fox'd or ropy Drink'. One addition suggested for 'curing' cloudy beer, 'Chemical Oil of Sulphur', is interesting, because the sulphites which would have been present in this preparation would have had a marked bactericidal action.[2]

We must now consider some of the effects of legislation on the technology of brewing.

Incidental reference has been made to the tax on malt. Towards the end of the eighteenth century, with the advent of the French wars, enormous duties were placed on malt which the brewers attempted to offset in different ways. For example, they observed that a stronger wort could be prepared from pale malt, so they

[1] Thomas Thomson, *op. cit.*, p. 479.
[2] J. C. Drummond and A. Wilbraham, *The Englishman's Food*, p. 238.

began to use a higher percentage of pale malt and attempted to simulate beer made entirely from dark malt by the addition, although forbidden, of foreign substances.[1] Some of the substances added, such as malt husks roasted like coffee, were harmless, and were permitted by the Excise. However, they did not stop at the addition of such innocuous materials. In an attempt to restore the original flavour of beer made from dark malt, they tried adding quassia, *cocculus indicus*, and even opium. Such was the deterioration which London porter underwent that at the end of the century an Act[2] cites the following as specially forbidden adulterants: molasses, honey, vitriol, quassia, grains of paradise, Guinea pepper, liquorice, capsicum, cocculus indicus, ginger, opium, and tobacco. According to Frederick Accum:

> The fraud of imparting to porter and ale an intoxicating quality by narcotic substances, appears to have flourished during the period of the French wars: for, if we examine the importation lists of drugs, it will be noticed that the quantities of cocculus indicus imported in a given time prior to that period, will bear no comparison with the quantity imported in the same space of time during the wars, although an additional duty was laid upon the commodity. Such has been the amount brought into this country in five years, that it far exceeds the quantity imported during twelve years anterior to the above epoch. The price of this drug has risen within these ten years from two shillings to seven shillings per pound.[3]

Although druggists were forbidden to sell 'illegal ingredients' to brewers under severe penalties, large-scale operation was engaged in, and it is perhaps worth recording some of the trade names under which adulterants were sold.[4] Their variety would stretch credulity were there not prosecutions to substantiate their having been used:

Black extract or hard multum	was	cocculus indicus
Multum	,,	gentian root, liquorice juice, etc.
Bittern	,,	the above with the addition of calcined ferrous sulphate
Beer heading	,,	alum and green vitriol
Flash (for addition to brandy)	,,	sugar and capsicum
Faba amara	,,	coriander seeds and nux vomica[5]

[1] 12 Anne, Sess. 1, c. 2. [2] 56 Geo. III, c. 58.
[3] F. Accum, *A Treatise on the Adulteration of Food*, p. 153.
[4] 56 Geo. III, c. 2. [5] F. Accum, *op. cit.*, *passim*.

It is also recorded that *salt* was added to beer in Edinburgh to make it 'brackish and intoxicating'.[1] This may also have been a trade designation.

It can hardly be wondered, perhaps, that quality fell and that adulterants were resorted to when we recollect that the price of barley rose by 100 per cent. between 1790 and 1807, and that the duty on malt was increased by 200 per cent. between 1803 and 1807.

Accum tells us that by 1820 the quality of London porter was on the upgrade again. He analysed samples from Barclay, Perkins and Company, Truman, Hanbury and Company, Henry Meux and Company, and other eminent London brewers, and found an average alcoholic content of 5·25 per cent.[2]

At the end of the eighteenth century while a warlike state Europe continued, there was a great scarcity of grain, which ultimately became so acute that legislation was introduced permitting the use of West Indian sugar as a supplement to malt in brewery. While barley normally cost 35*s*. per quarter, the price had risen to over £4. According to J. Baverstock, however, as a source of saccharine material for making beer, it was still 20 per cent. cheaper than sugar. He gives the relative costs of saccharine matter when 'equalised with a quarter of malt costing 76*s*.' as treacle 90*s*., sugar 107*s*., honey 161*s*. These were the three next cheapest materials, and in consequence of their considerably higher prices they were not adopted to any great extent.[3]

In addition to the tax on malt there was a direct tax on beer to be met as well. This was first introduced into England in the reign of Charles II (1660), when a levy of 2*s*. 6*d*. per barrel was imposed on *strong* beer, and 6*d*. a barrel on *small* beer. For many generations, strong and small beers were the only types of beer permitted to be brewed in England, as opposed to Scotland, where there was an intermediate variety, the *Scotch twopenny*.

Till 1684 the English tax was farmed. During the Revolution several additional dues were imposed, and in 1694 the excise was consolidated at 4*s*. 9*d*. per barrel on strong beer and 1*s*. 3*d*. on small.

In 1707, in the Treaty of Union, while English rates were imposed upon Scotland, there was a specific exception made in favour of the malt liquor of Scotland, *twopenny*. This variety was intermediate between the English beers and taxed at 2*s*. 4*d*. per

[1] E. Burt, *Letters from Scotland*, Vol. 1, p. 158 (5th Edition).
[2] F. Accum, *op. cit.*, p. 168. [3] J. Baverstock, *op. cit.*, pp. 110 and 124.

barrel.[1] From 1707–1737 while this state of affairs continued, some 400,000 to 500,000 barrels per annum was the Scottish production. In 1760 the Scottish duty was raised to 3*s.* 4¼*d.* per barrel, and production dropped rapidly to between 100,000 and 200,000 barrels. High duties encouraged the consumption of spirits.[2] The manufacture of twopenny ceased altogether in 1802. The difference in the type of liquor consumed in different parts of Great Britain has persisted into the present century. The following table, taken from Porter's *Progress of the Nation*, gives the annual consumption in England, Scotland, and Ireland for the year 1909.[3]

	Expenditure per head per annum		
Liquor	*England*	*Scotland*	*Ireland*
	£ *s.* *d.*	£ *s.* *d.*	£ *s.* *d.*
Spirits . . .	17 11¾	1 15 8	1 2 3½
Beer . . .	2 8 3	14 3½	1 16 0½
Wine . . .	4 10¾	4 2¾	2 4¼
Other Liquors .	9½	2½	2½

In the city of Edinburgh there was an additional levy of two pennies Scots per pint on all ale and beer either brewed in, or brought into, and sold within, the city. This was imposed in virtue of an Act of William and Mary, granted to the magistrates of Edinburgh:

> In consideration of the great debts due by the town of Edinburgh, which partly by their undertakings for, and service done to the Government upon several occasions, and partly through other causes, hath risen to that height, that the common good of the said town is not sufficient to pay the annual rent thereof, much less to defray and discharge so great burden.

The privilege continued into the nineteenth century, when it became the cause of a process at law through the contravention of the excise on the part of William Black and Company of the Devanha Brewery, Aberdeen, who were alleged to be sending beer to Leith 'to an extent ruinous to the Edinburgh brewers, as appears from the books of the Shore Dues Office kept at Leith'.[4]

[1] 5 Anne c. 8, Art. 17: confirmed 8 Anne c. 7, §§ 3 and 12; 12 Geo. I, c. 4 § 62. [2] *Edinburgh Review,* No. 98, Art. 4.

[3] G. R. Porter, *The Progress of the Nation*, p. 122.

[4] *Session Papers,* Signet Library, 267:3.

In England, taxation progressively increased throughout the eighteenth century and into the next:

	Strong	Small
1787	8*s*. 0*d*.	1*s*. 4*d*. per barrel
1802	9*s*. 5*d*.	
1804	10*s*. 0*d*.	

But the quantities brewed did not fall as they did in Scotland.

In 1830 beer duties were abolished, and anyone was allowed to open a public house on paying two guineas and executing a bond. This cut across the monopoly of the tied houses, and some thirty thousand houses licensed to sell beer and cider sprang up. In 1848 there was one public house for every 166 inhabitants of Birmingham.

DISTILLATION: Where the production of a distilled spirit is the object in view, the processes already described have to be carried one stage further. The objective now is to concentrate the alcohol which is produced in dilute solution in the fermentation of beer. This can be effected by making use of the different boiling-points of water (100° C.) and alcohol (78·5° C.). Reference must first be made, however, to certain specific alterations in the initial processes when the production of spirits is the end in view.

The original Scottish *Aqua vitae*, and it must be borne in mind that the first reference to *whisky* was made by Bailie John Stewart of Inverness so late as 1736, was made from malt alone.

> Little used in the Lowlands till 1750, it had long been much in vogue in the Highlands, where it was made in stills in the glens and drunk by persons of all classes. Best known of all was the 'Ferintosh' of Duncan Forbes of Culloden, which paid no duty, was sold cheap, and was so much drunk that 'Ferintosh' became a synonym for whisky.[1]

When the tax on malt became very heavy the Scottish distillers resorted to the use of larger and larger quantities of raw grain till as little as 20 per cent. of the grain mashed was previously malted. This could easily be done since the diastase in the malt was quite sufficient to convert the starch in the raw grain into saccharine material. According to F. Accum, the suggestion came from Dr. Irvine in 1785.[2] By doing this, Scottish distillers for a long time avoided paying malt tax, till the Government, with an apparent show of sound reason, put a tax on grain used in the distillery. It was reckoned, however, that the legislation was

[1] H. G. Graham, *op. cit.*, p. 529. [2] F. Accum, *op. cit.*, p. 21.

detrimental to agriculture, since distillers would only buy the best grain and farmers were left with inferior qualities on their hands.

There were also differences in the fermentation processes. The object of the brewer is not necessarily to produce the highest possible concentration of alcohol, but to produce a palatable beverage. On the other hand, the distiller wants as much alcohol as possible. To this end, distillers added to their wort much larger quantities of yeast, usually obtained from London porter brewers, and carried on the fermentation till the liquid in the fermenting tun was actually acid. This took some 9–13 days, depending on temperature, etc. The acid produced is acetic acid (CH_3COOH), and by combining with some of the alcohol to form the fruity ester, ethyl acetate, this may have contributed to the flavour of the final distillate.

$$\begin{array}{ccccccc} C_2H_5OH & + & CH_3COOH & = & CH_3COOC_2H_5 & + & H_2O \\ \text{Alcohol} & + & \text{acid} & = & \text{ester} & + & \text{water} \end{array}$$

Even more so than in the case of brewery, the development of the technology of spirit production was greatly influenced by the fiscal policy pursued by the Government, so perforce we must follow its intricacies over the period under discussion.

At the Union of England and Scotland in 1707, the 7th Article of the *Treaty of Union*[1] provided that the same excise should be levied on excisable liquors in both countries. This continued operative till 1736 when, with the idea of suppressing excessive consumption of spirits by the poor, the so-called *Gin Act* was introduced into England by Sir Joseph Jekyll.[2] The effect of the Act was (*a*) to impose an annual licence duty of £50 on all retailers of spirits in quantities less than two gallons, and (*b*) to impose a duty of 20*s*. per gallon on the spirits. The Gin Act did not apply to Scottish aqua vitae (§ 22), nor did the subsequent Act (16 Geo. II, c. 8), which modified the above Act. Thus a distinct difference between England and Scotland was established in 1736, and a duty on Scottish spirits imported into England was imposed, although the strength of the spirits was not specified, a clear indication that science had not developed quick and accurate methods of assessing the percentage of alcohol in an aqueous solution.

The control of the strength of alcoholic liquor was greatly simplified by the invention of a practical form of hydrometer by an instrument-maker called Clarke.[3] It was similar in form to the

[1] 5 Anne, c. 8. [2] 9 Geo. II, c. 39, §§ 1.4.3.
[3] *Phil. Trans. Roy. Soc.*, 1729–30, 36, 277.

hydrometer in use today but was made of copper. The stem bore three marks, one representing the depth to which the instrument would sink in 'proof' spirit and the other two one-tenth above and one-tenth under 'proof' respectively.[1]

In order to remind us of the close connexion of distilling with agriculture, it might be mentioned in passing that the crop failure of 1756 led the Government to stop distilling in Great Britain from March 1757 to December 1760.[2] Coincident with the dearth of 1756, a food-adulteration scare arose through the publication of an anonymous pamphlet (the author was probably Peter Markham) entitled *Poison Detected: or, Frightful Trusts: and alarming to the British Metropolis*. Other similar works followed, e.g., *The Nature of Bread, Honestly and Dishonestly Made*, by Joseph Manning, M.D. The millers and bakers not unnaturally took up the challenge, and there was much bitter controversy. However, out of it there came what Drummond considers the first reliable book on the chemical detection of food adulteration, H. Jackson's logical and rational *Essay on Bread* (1758).[3]

As the eighteenth century progressed, a number of large capitalist distilleries developed in Scotland, some of them of great size, but to find a market for their output they were forced to look to England, since the Scots themselves still showed a decided preference for illicit whisky. Thus in 1777 there were eight licensed stills in Edinburgh, against some four hundred which were operated outwith the law.[4] It is not impossible that a demand for illicit spirit came from England. It seems that whisky made outwith the law was undoubtedly of superior quality. The reasons given are that illicit whisky was still made entirely from malt, and that, with the simple equipment which of necessity had to be used, the spirit was obtained by slow distillation from a weak wort, two conditions consistent with a final product in a high state of purity. Of the general knowledge of the Highland distillers it was said:

> There are many that practise this Art who are ignorant of every other, and there are Distillers who boast that they make the best possible Whiskey, who cannot read or write, and who carry on their Manufacture in Parts of the Country where the Use of the Plough is unknown, and where the Face of an Exciseman was never seen.[5]

But the domestic production of spirit was by no means confined

[1] J. C. Drummond and A. Wilbraham, *op. cit.*, p. 242. [2] 30 Geo. II, c. 10.
[3] J. C. Drummond and A. Wilbraham, *op. cit.*, p. 225.
[4] H. Arnot, *History of Edinburgh*, p. 257.
[5] *Report concerning Distilleries in Scotland* (1799), p. 279.

to the illiterate, as the following description of the activities of the parish minister of Keith-Hall shows.

Experiments in the domestic production of spirits are described by Dr. George Skene Keith in his *Agricultural Survey*. In 1778 he planted a boll of potatoes.

> His object was merely to get a quantity of potatoes, not only for family use, but for making into spirits. It was necessary for him to lay in a stock of these (he was newly settled in the parish of Keith-Hall); and the state of his finances required that this should be done at as moderate an expense as possible. About a ton of the resulting crop was fermented with four hundred pounds of malt. The yield was about one hundred and sixty gallons of spirit at 10 per cent. above proof. They were distilled in three small stills, about ten gallons each, which private families were permitted to use without paying duty till 1779.

Of the quality of his production, he says:

> The potato spirit was not so pleasant as malt spirits, but superior in point of flavour to the corn spirits which were then distilled in the south country,

i.e., the product of the capitalist distillers.

Changes in the fiscal policy regarding excisable liquors appear to have put an end to the reverend doctor's experiments.

> The law having prohibited the distillation of spirits in private families, he could only plant potatoes in different modes and at different distances, marking their produce carefully, without attempting to distil them.[1]

In 1784 an Act which was known as the *Wash Act* was introduced.[2] This Act, having discontinued all the existing duties in both England and Scotland for a period of two years, imposed a duty of 5*d*. per gallon on the *wash*, i.e. the aqueous solution after fermentation. The Act was based on certain assumptions, one that distillers could produce twenty gallons of spirit from one hundred of wash (§ 4). This meant that a very high percentage of saccharine material had to be extracted from the malt in the preparation of the wort prior to fermentation, and to make sure that it did contain enough, the Scottish distillers *lobbed* their wort with a saccharine matter which they called *bub*. The composition of *bub* was kept a secret, but it is assumed that it consisted of treacle and sugar, practically the only readily available materials.

[1] G. S. Keith, *General View of the Agriculture of Aberdeenshire*, p. 268.
[2] 24 Geo. III, c. 46 § 1.

The use of *bub* was eventually declared illegal, and the amount of spirit required to be produced from a hundred gallons of wash reduced to fourteen, and later to thirteen. At twenty gallons, the tax of 5*d.* per gallon on wash was equivalent to 2*s.* 1*d.* per gallon on spirits, or 2*s.*3$\frac{7}{9}$*d.* if only 18 gallons could be produced, as the distillers maintained.

The same Act (§ 45) divided Scotland into two districts, Highland and Lowland, ostensibly with the idea of improving the agriculture in the Highlands, in which the ways of assessing the excise duty were quite different. It was basically an admission that the rule of Government at Westminster did not yet extend to the Highlands of Scotland, since there the levy was on the capacity of any still known to the law, rather than on the produce of that still. In the less lawless Lowlands the tax was based on quantity produced.

> As the restrictions on licensed distillers were increased, the proprietors of the unlicensed stills were encouraged to extend their operations, and to enter into competition with the legal manufacturers. Then began that system of smuggling which made a certain class of Highlander so notorious, and gave so much trouble to the Excise Department. The wild glens of the north afforded secure retreats for the working of the stills; and many ingenious modes of conveying the produce to market were devised. The great strongholds of the smugglers in the north were Glenlivet, Strathden, and the Glen of New Mill. The proprietor of the only distillery now in Glenlivet recollects seeing two hundred illicit stills at work in Glenlivet alone. Owing to the quality of the water and other causes, the whisky in the Glen became famous—indeed, smuggled whisky generally was preferred by customers, on account of its mildness and fine flavour.[1]

The line of demarcation, which was altered from time to time, was roughly by the east point of Loch Crinan, Loch Gilpin, Inveraray, Arrochar, Tarbet, the north side of Ben Lomond, Callendar, Crieff, Dunkeld, Fettercairn, Clatt, Huntly, Keith, Fochabers, Elgin, Forres, to the boat on the River Findhorn.[2] But the Highland Line was not without its absurdities, as the following citation shows.

> The line which bounds the privilege of small stills with moderated duties, prohibits the inhabitants on the north side of the street of Elgin, from the accommodation which is thereby permitted to their neighbours on the south side; other parts of

[1] D. Bremner, *Industries of Scotland*, p. 446.
[2] *Report: Distilleries* (*1799*), p. 28.

> the country in similar situations, are subjected to the same partiality: they must either drink of the smuggled distillation, or import their spirits at the vast additional expense of navigation, from the distant provinces of the southern quarters of the kingdom. This bounding line ought therefore to be laid down more suitably to the circumstances of the country.[1]

The same Act is interesting for the suggested, or rather, inferred use of the hydrometer as a means of determining the concentration of spirits. This indicated a very distinct advance in technical methods during the course of the century. A later Act[2] enjoined the use of a particular hydrometer.[3]

At the same time as the Wash Act was introduced, the privileged excise district of Ferintosh, near Inverness, was abolished. In 1689 the lands of Duncan Forbes of Ferintosh had been raided by Highland rebels, and in compensation an Act of Parliament was passed farming out to him and his successors for about £22 the yearly excise of the Lands of Ferintosh. In 1707 when the Board of Excise was formed, there were jealous complaints made against Forbes—his annual profit was reckoned at £18,000—but it was not till 1784 that he was bought out by the State for the sum of £21,580.[4]

About the time of the passing of the Wash Act (i.e. 1785), there was much discussion between the much-harassed capitalist distillers of the Scottish Lowlands and of London with a view to securing the Scottish home market to the legal distillers of Scotland. It will be remembered that the prevalence of smuggling from the Highland stills had forced the Lowland distillers to look for a new market for their products in England. Agreement was reached about 1785, James Haig signing on behalf of the Scottish distillers.[5]

In Scotland the Wash Act was only continued till 1786, when the first *Lowland Licensing or Scottish Distillery Act* was introduced.[6] By this Act, the London distillers hoped to secure the London market to themselves. The Act remained operative till 5th July, 1788. It imposed a levy of 30*s*. per gallon on the capacity of Lowland stills, and assumed that a still was run off once in 24 hours.

In 1786, when the Scottish Distillery Act was passed, an Equalizing Duty of 2*s*. on Scottish spirits exported to England was imposed, and it was also prohibited to export spirits to England *by land*. The Act seems to have been brought in on account of

[1] Wm. Leslie, *General View of the Agriculture of Nairn and Moray*, p. 429.
[2] 27 Geo. III, c. 31 § 17. [3] See *Phil Trans.* 1790, Pt. 2, Vol. 80.
[4] N. Gunn, *Whisky and Scotland*, p. 41.
[5] *Report: Distilleries* (1799), p. 291. [6] 26 Geo. III, c. 64.

pressure brought on Parliament by the English distillers, who were alarmed at the large imports of Scottish whisky. Bremner says that in 1787 upwards of 300,000 gallons crossed the Border without the knowledge of the Excise.[1]

The immediate effect of the Act was very detrimental to the lawful distillers of Scotland, many of whom went bankrupt, leaving the field more than ever in the hands of the illicit distillers.[2] The complicated repercussions of such alterations in the Excise is illustrated by the following citation from the *General View of the Agriculture of Nairn and Moray*:

> But the frequent alterations in the distillery law are, in a peculiar manner, felt distressing to the farmer; his whole system is embarrassed, and his rotation of cropping altered. Were distillers nearly in the same circumstances over all the kingdom, it would not be difficult to suggest alterations—(supposing the prohibition of distilling grain to be only temporary)—which would be advantageous to the country, and not injurious to the revenue; but by those who only see a small part of a whole, suggestions of amendment must always be presumptuous, and frequently erroneous. Since, however, there was no licensed still in the whole of the two counties of Aberdeen and Banff, and only two in Moray, when the distillation of grain was permitted, it must at once without dispute be admitted that the duties were too high for the circumstances of the country. For when these duties were lower in a former state of the law, the licensed small stills supplied in every quarter the whole consumption of these three counties, and fraudulent distillation, so extremely pernicious in every respect, was then almost totally and happily suppressed.[3]

The years 1786 and 1787 saw a veritable spate of pamphlets and the like dealing with distillery in Scotland. Walter Ross described *The Present State of the Distillery in Scotland* (Edinburgh, 1786), and the grievances of the distillers in Scotland were dealt with in the *Scots Magazine* (1786). Whether or not the facts presented there were true or not, a work, *Truths, in answer to the facts published respecting the Scottish Distillery*, was published in Edinburgh in the same year. The proprietors of land next had their say, in *Resolutions of the landed interest of Scotland respecting the Distillery* (Edinburgh, 1786). Then from London came the *Case for the capital corn distillers of Scotland* (London, 1787), and the *Case of the distillers of corn spirits in North Britain* (London, 1787). Thomas MacDonald reviewed the laws and regulations respecting the

[1] D. Bremner, *op. cit.*, p. 445.
[2] *Report*, p. 338.
[3] W. Leslie, *op. cit.*, p. 427.

distillery in Scotland (London, 1788), and the Scottish distillers presented to the Lords Commissioners of His Majesty's Treasury their *Answers to the memorial of the corn distillers in London* (1788).

In England the Wash Duty was progressively increased from 6*d.* to 10*d.* per gallon of wash, representing at 10*d.* a duty of from 4*s.* 2*d.* to 4*s.* 7*d.* per gallon on the spirits, depending on the quantity produced from a hundred gallons of wash.[1]

The next Act[2] passed in the year 1788, is one of great importance in the development of distillery technology. It introduced an entirely new principle, viz., a limiting of the size of the stills to be used in the different districts of Scotland. The effect of this was to make rapidity of working of prime importance to the distiller, and so we have a decade or so in which much ingenuity was exercised in still-designing, a subject to which we must now refer.

> Private families distilled whiskey for their own use; and the still they used was a large pot, globular, that for culinary purposes it might be capaceous; and to this pot, when they wished to distil, they luted an occasional head. Had distilling never become a distinct trade, and had not the Law stimulated distillers to invention, stills in Scotland would have been pots still.[3]

The still used in Scotland till after 1788 was a large, deep vessel which could only be brought to the boil slowly, and when boiling distilled off the alcohol slowly and gently. When the basis on which the excise levy was altered, it became of advantage to distillers to operate their stills as expeditiously as possible. The first alteration of which we hear is a modification in the shape of the still made, it is said at the suggestion of an Englishman, at the instance of John and William Sligo, rectifiers of Leith. Reference to the 'Englishman' will be made later. The modification was towards a shallower still, so that the rate of distillation was increased.[4]

As the spirit duty increased, distillers continually tried to keep at least one step ahead of the Government in the ingenuity of their distilling apparatus. In this they were so successful that, when a Parliamentary Committee was set up in 1799 to investigate the condition of Scottish distilleries, it is recorded that they found 'an increased rapidity of distillation beyond what had ever been practised in fact, or foreseen in theory'.

There is no indication in the *Scottish Report on Distillery* who was

[1] *Report*, p. 15. [2] 28 Geo. III, c. 46. [3] *Report*, p. 383.

[4] Details of stills will be found in the *Encyclopaedia Britannica*, Supplement IV–VI, Vol. 3, p. 51.

responsible for these remarkable advances in distillation technique, but a sidelight on some of them is obtained from Andrew Ure's *Chemical Dictionary*. We learn there that James Watt, having ascertained that liquids boil at lower temperatures under reduced pressure, applied this principle to the still.[1] Watt's still is said to have been able to pass 480 charges of 16 gallons in 24 hours.[2] This, however, does not by any means represent maximum output. In the *Report on Distillery*, there are mentioned eighty-gallon stills which could be distilled off, emptied, and ready for the next batch in three-and-a-half minutes, and stills of forty-gallon capacity in as little as two-and-a-half. It is not known whether reduced pressure was ever applied in practice, but it is unlikely.[3] Rapid distillation was not confined to Scotland. M. Chaptal in his *Elements of Chemistry* mentions shallow stills for distilling brandy at Languedoc.[4]

In order to keep stills working at such rates, large quantities of coal, some 60 tons per week in a '90-quarter house', were needed to fire the furnaces. Such consumption was regarded as a serious waste of fuel.[5] Just how great the consumption of coal in distillery was may be judged by Bald's estimates for consumption in 1812. They were 53,000 tons in distillery, compared with 25,000 in glass-making, and only 160,000 in the iron works.[6]

Difficulties were, as might be expected, brought in train with such rapid distillation. Wash, when it comes from the wash-backs, is saturated with carbon dioxide (CO_2), and on being heated is very liable to froth in the still. To get over this, soap was added, and Stephen Maxwell of Glasgow, a coppersmith, took out a patent for a still to obviate some of the difficulties.[7] Large distilleries used as much as £15 of soap per week.[8] Mechanical difficulties, such as blowing the head off the still, were not so easily dealt with.

One effect of accelerated distillation was a deterioration in the quality of the spirits produced. This came about from the lessened fractionation in the fast still as compared with the old slow-boiling pot-still. To get further information on this point the Committee set up to investigate the Scottish Distillery referred the matter to Dr. Joseph Black of Edinburgh, and to Dr. Ingenhousz. The substance of their findings was that the essential oils formed along with the alcohol were not so easily fractionated in the process of

[1] Article, *Distillation*. [2] Articles: *Distillation* and *Laboratory*.
[3] Report, p. 283. [4] M. Chaptal, *Elements of Chemistry*, Vol. 1, p. 6.
[5] *Report*, p. 342. [6] R. Bald, *A General View of the Coal Trade of Scotland*, p. 2.
[7] *Register of the Great Seal*, Edinburgh, Vol. 20, No. 324, 1787.
[8] *Report*, pp. 339 and 342.

distillation when it was fast as when it was slow, and that the presence of these oils in a fast-distilled spirit made it less palatable.[1] One of the difficulties was the communication of an *empyreuma* to the still contents, and to get over this, Henry Tritton patented a still in 1817 which communicated heat to the contents of the still by means of a water bath, and at the same time distilled under reduced pressure.[2] It was even suggested that a synthetic whisky might be made.

> A Method has been suggested, of joining to rapid distillation the advantage of that peculiarity in quality and flavour which belongs to the whisky made in the Highlands; a method which, as far as the Committee are capable of understanding it, seems somewhat analogous to the attempts so often made at the production of wine, mineral waters, and other liquors of a compound sort, by the artificial combination of the different elementary ingredients, which are found by chemical analysis, to be united in their original composition. But there is strong reason to believe that the best of those differ from, and fall short of, the compositions effected by what has been called the Chemistry of Nature.[3]

An interesting sidelight is the idea mooted by Dr. Jeffrey (presumably James Jeffrey (1759–1848), Professor of Anatomy at Glasgow), of adapting a steam-engine to a still.

> Mr. Cartwright's[4] Steam-engine is a Still and Steam-engine conjoined; and it may also be observed that a steam-engine fitted to a Scotch fast-going still would have great power; for the quantity of steam that rises from these stills in a given time is prodigious. There is no doubt that such an engine could grind the malt, turn the mash-stirring machine, work the pumps, etc., and it is, I think, likewise obvious, that under survey it would aid the Excise in detecting frauds; for it would show when the still was at work.[5]

Among the experimenters who tried to increase the variety of spirituous liquors was Dr. George Skene Keith, already referred to in connexion with domestic production. We find him carrying on experiments on behalf of the Scottish Distillery Committee, which he describes as follows in his *Report on the Agriculture of Aberdeenshire*:

> Ardent spirits, of an excellent quality, may be extracted from carrots. The Writer of this Report having mentioned to the

[1] *Report*, pp. 357 and 358.
[2] *Annals of Philosophy*, 1818, *11*, 445: with diagram. [3] *Report*, p. 21.
[4] Edmund Cartwright (1743–1823), inventor of Alcohol engine 1797.
[5] *Report*, p. 397.

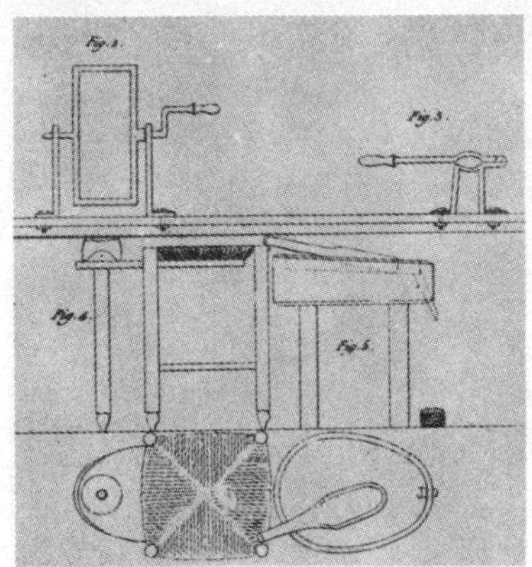

L'Art de Conserver (*English trans.* 1812)

105. Apparatus used by Appert in the preservation of food. Reel of wire, cork press, stands for holding bottles while being corked, etc.

106. One of Donkin, Hall and Gamble's tins of 1824.

107. John Moir's Preserving Works, Aberdeen, founded in 1822.

J. G. Bylaerd, 1763

108. Monastery where the University of Leyden was given its home from 1574.

109. Wedgwood portrait medallion of Hermann Boerhaave, by Flaxman.

110. Wedgwood portrait medallion of Dr. Joseph Black, by Tassie, 1786.

> Scotch Distilleries Committee, in 1798, that very good spirits could be extracted from both potatoes and carrots, was asked to undertake a series of experiments for that purpose; and a recommendation was given to the Scotch Commissioners in his favour. He made a number of experiments in 1799, an account of which he gave in the Scotch Distillery Committee of that year. And by other two sets of experiments, in 1802 and 1803, conducted under the authority of the Board of Excise, he found not only that a considerable quantity of spirits could be extracted from either of these roots, but that the flavour of the spirit is excellent. Indeed, he knows none equal to it, in respect of the pleasant taste of its essential or flavouring oils. The carrot, however, does not produce, from an equal weight, above two-thirds of the quantity of spirits that is produced by potatoes. But an acre of carrots, distilled into spirits, after paying all taxes, would be equal to at least four acres of wheat; and if the crop was good, to a much greater quantity. And when Britain attends more to its agricultural interest than to its colonial commerce, it may merit the attention of our Legislators, whether they ought not to encourage the distillation of roots raised in Great Britain and Ireland, rather than that of sugar from the West Indies, when corn is either scarce or dear in this country.[1]

He also experimented with liquors made from parsnips and scorzonera.

> A very good ardent spirit may be extracted from parsnips. This the Writer of the Report proved in 1802 when he distilled fifteen different kinds of spirits from various combinations of roots. In point of flavour, the spirit of parsnips is inferior to that from carrots, when distilled separately—but it may be of essential use if combined with other roots in the British Distillery.[2]

Of scorzonera he says:

> This root is much used on the continent, as a dish at the table; and when the Reporter distilled it, he found it produced a most agreeable flavoured spirit, and well adapted for making a *liqueur*. But its root is small, and the quantity raised on a given measure of ground is too little to render it an object to the agriculturist, though it might have its value in making *liqueurs for the ladies*.[3]

For his labours he was awarded £500 by Parliament.

Other interesting suggestions came from the Earl of Dundonald in his *Letters on Making Bread from Potatoes*. There he comments that although in the 'present state of chemical knowledge potato

[1] G. S. Keith, *op. cit.*, p. 318. [2] *Ibid.*, p. 319. [3] *Ibid.*, p. 321.

powder cannot be used to replace sugar, it might be used to very great advantage in making ardent spirits, by mixing it with the proportion of malted grain commonly used with unmalted grain at the distilleries'. On the quality of the product he hazards that 'the spirit will be much purer than what is got from a mixture of malted and unmalted grain'.[1]

We must now return to the problems which faced distillery in the concluding years of the eighteenth century. As Britain became involved in the wars of the late eighteenth century, the tax on excisable liquor increased. By 1796 it had been raised to £9 per gallon of still capacity, and in the years from 1797 onwards, advanced to the extraordinary amounts of £54 and £64 16*s*. 4*d*. per gallon, an equivalent of more than 9*d*. per gallon of spirit made.[2] The greatest increase in the rates of distillation coincided with the greatest increase in taxation. Still unable to raise sufficient revenue, a tax on raw grain used in distillery was imposed in 1798. At this time it was computed that the following quantity of spirits were consumed in Scotland:[3]

According to Distillery Report (1798) .	2,400,000 gallons
,, ,, Board of Excise . .	3,500,000 ,,
,, ,, Dr. Jeffrey, Glasgow . .	5,400,000 ,,
Average of several estimates . . .	3,686,136 ,,

Production in Scotland was still 'in a thousand hands'. By contrast there were only twelve distilleries in England.

As the skill of the Scottish still-designers increased, it became obvious to the Government that they were being defeated in their attempt to raise revenue by a tax on the still capacity. So, in the year 1799, this system was abandoned, and a duty of 4*s*. 10½*d*. was imposed on every gallon of spirit made for consumption at home. In the first year of the new duty, the 87 licensed distilleries of Scotland yielded £1,620,388 to the Excise. The change in the method of raising the revenue did not meet with approval from the distillers, and about a third of them disappeared or retired from the Trade. Revenue in consequence dropped to £775,750 in the next year, and in 1802 the duty was reduced to 3*s*. 10½*d*. This seems to have had a stimulating effect on production, for eight-eight distillers were registered in 1803, who contributed £2,022,409 to the Excise. In the next year, however, the duty was increased again, and as before a progressive decline in the number of registered distillers set in.

[1] Letters by the Earl of Dundonald on Making Bread from Potatoes. (Edinburgh, 1791: Signet Library, Pamphlet 42).

[2] *Report*, p. 340. [3] *Report*, p. 47.

As warlike conditions in Europe grew worse, grain became so scarce that in June 1808 distillation from grain was suspended and entirely prohibited. Despite the prohibition, the yield of revenue from the malt duty increased, which Baverstock took as a clear indication of the increased use of malt. To maintain supplies of liquor it was permissible for a time to make fermented liquors from sugar, but this could only be done at considerably increased cost, and the price of spirits increased by 2*s.* per gallon as a result of the prohibition of distilling from grain. At the same time the price of molasses increased from 39*s.* per hogshead to between 57*s.* and 59*s.*[1] As evidence of the use of sugar, we give the following quotation from the *General View of Renfrew*:

> Distilleries have been carried on in the town and neighbourhood of Paisley with great success, for more than twenty years, but they are now of less extent than formerly; some of the companies engaged in this business having removed their works to the banks of the Forth and Clyde Canal in the neighbourhood of Glasgow. The spirituous liquor prepared from malt is in great perfection; and the spirit produced from sugar, which is used during the suspension of the distillation from grain, though when first introduced not much esteemed, is now in very general request.[2]

The effect of the prohibition will be seen by the following table:

Table of Excise Duties charged for that Division of Renfrewshire, which comprehends Paisley, Renfrew, Neilston, and Johnstone.

	£	*s.*	*d.*
For the year from 6th July, 1807, to 5th July, 1808 (grain distilling this year) . .	103,854	18	2½
For the year from 6th July, 1808, to 5th July, 1809 (no distilling this year, or sugar distilling only)	43,963	5	6½
For the year from 6th July, 1809, to 5th July, 1810	79,072	11	8
For the year from 6th July, 1810, to 5th July, 1811	63,288	4	3¾

With the all-round increase in the cost of barley, and the increased taxes on both malt and spirits, the consumption of malt liquors began to decline from 1811–13.

[1] *Session Papers*, Signet Library, 428:45.
[2] J. Wilson, *General View of the Agriculture of Renfrew*, p. 264.

The prohibition of distillation from grain had a disturbing effect on agriculture. From the following long citations we get at one and the same time an idea of several of the great Lowland capitalist distilleries and of the effects of the prohibition upon agriculture. They are taken from the *Accounts of the Agriculture of Kinross and Clackmannan*, and of *Angus*. Speaking of Kinross and Clackmannan, the author says:

> This county has long been celebrated for the extensive scale on which its numerous distilleries have been conducted. The frequent instances which have lately occurred of the temporary suppression of distillation from grain, by acts of legislature, render it impossible to give an adequate account of the present condition of these establishments. As to the policy of these interruptions of this manufacture many doubts have been entertained and expressed. The discouragements given by this measure to the growing of barley, and the consequent derangement of some important agricultural rotations, but especially the diminished quantity of fattened cattle which these distilleries used to supply, have been considered as too great sacrifices to the inconsiderable addition of grain obtained by their suppression.
>
> Whilst distillation from grain was permitted, there were six considerable distilleries in this county, *viz*. Kennetpans, Kilbagie, Clackmannan, Carsebridge, Grange, and Cambus. The first two of these were by far the most extensive. Without entering into a particular detail, it may be sufficient to say that, previous to the year 1788, the quantity of grain used annually at Kilbagie alone was about 60,000 bolls,[1] the quantity of spirits produced, about 3,000 tons; the number of black cattle fed about 7,000; and of swine 2,000. The number of people employed was about 300. The buildings and utensils cost upwards of £40,000. The works occupy four acres of ground surrounded by a high wall. A small rivulet runs through the middle of the works, which drives the machinery of a thrashing-mill, and all the grinding mills necessary to the distillery. It besides supplies a canal of about a mile in length, which communicates with the Forth, and serves to convey the imports and exports of the establishment.
>
> The distillery at Kennetpans was in proportion to that at Kilbagie as three to five. These two paid to government an excise duty considerably greater than the land tax of Scotland. The Reporter finds in his notes, taken at the Manse of Clackmannan, on the respectable authority of Dr. M., that the Kilbagie distillery paid at one period a duty of half a million sterling to Government.[2]

[1] 45,000 quarters.

[2] Patrick Graham, *General View of Kinross and Clackmannan*, p. 360: See also *Statistical Account*, Vol. 14, p. 623.

The author of another Report, that on Angus, goes on to speak of the inter-relation of British with West Indian economy:

> The grain consumed in distillation is not wholly lost as human food. It has been ascertained that as much milk, beef, pork, or other animal food can be raised from the draff and dref after the spirit is extracted as could have been produced upon the land on which the barley grew, had it been sown with grass and thrown into pasture. From the dreg of sugar, after distillation, no food of any kind is produced. The beef and pork, fed in the distilleries, always supplied the markets, during the interval between the failure of the turnip-beef and the preparation of the grass-beef of the ensuing summer. In this way, animal food was always in regular supply, and never rose so high as it has done since the distilleries were prohibited from using grain, during several successive years. The consequence of this has been, that many farmers have thrown a greater proportion of their land into pasture, or have extended the cultivation of turnips, in order to supply that portion of animal food which used to be furnished by the distilleries. Many also sow wheat, where they were wont to sow barley. The latter agrees best with light friable soils, and used generally to succeed turnips or potatoes; whereas wheat thrives best on strong clays, or what are here esteemed to be well adapted for drilled beans. It is obvious, that if wheat, or any other grain, be too often repeated on soils that are ill-adapted for them, the productive powers of the land must suffer a rapid diminution. There is another circumstance attending distillation from grain that has no place in that of sugar. The straw on which the barley grew was all consumed in the farmer's yard, and returned to the land in the shape of putrescent manure, to raise wheat, potatoes, or other crops for feeding the people. The animals, too, that were fed on the draff and dreg, yielded great quantities of rich manure, in addition to what was produced from the straw. Thus, distillation from grain, not only produced a large supply of animal food for the use of the people, but the great addition it made to the stock of manure enabled it to replace the whole, or greatest part of the grain that was thus consumed. Distillation from sugar can have no more effect upon the land of this country than if the process were carried on in the West Indies.
>
> From these facts it seems evident that if the prohibition of grain-distillation be continued, the operations of the plough will soon be very much curtailed. There will be no occasion to prohibit this mode of distillation, because there will be no surplus of grain on which to operate. Nay, it follows from the above data that even the portion which was consumed as food by the people will be very much diminished.

With regard to the interest of the West India planters, it is acknowledged on all hands that they have long been in the habit of urging on the cultivation of sugar, and of other disposable produce, to a greater extent than all the markets of the world can take off their hands. All other cultivation but theirs feeds the people employed in it, in the first instance; and it is only the surplus that is sold. But they export sugar, etc., and import provisions. Now that the slave-trade is abolished, it behoves them to consider seriously whether it would not be more for their advantage to substitute bananas and other plants in place of part of their sugar canes, that so they may raise sufficient food among themselves for feeding the negroes engaged in the cultivation of these colonies. This may diminish the amount of sugar raised, which is by far too great at present; but it will make their negroes more comfortable and enable them not only to keep up, but to increase their numbers. At least, Government should be very cautious, lest, by holding out premiums for the excessive production of sugar, they may, unintentionally, occasion the destruction of the negro race in the West Indies; and by suppressing a manufacture, so obviously beneficial, they should inflict upon the agriculture of their country a mortal wound. Here we do not take into the account the damaged grain, which the distillers always took off the farmers' hands at a price equivalent to its value to them. When grain distillation is suppressed, damaged grain is wholly unsaleable.[1]

The above discussion has approached the subject entirely from one side, that of stock-feeding as an adjunct to distillery, but the reverse was also in operation. One John Finnie owned 'considerable farms in the neighbourhood of Edinburgh, and for some years he carried on a distillery, principally with a view of obtaining manure for the use of these farms'. Andrew Stein, whom he met in 1786, gave him advice on how to run it.[2]

Summarizing the economic data supplied by a number of distillers, Dr. Jeffray came to the following conclusions concerning the stock which could subsist on distillery spent grain:

One gallon of still will be able to manure more than two acres of land now; but taking it for granted that one gallon of still can furnish manure for two acres of land only, then it follows that if fifty stills of fifty gallons each, or 2,500 gallons in all, be about the number necessary for the Lowland and Middle Districts, the distillers in these districts should manure 5,000 acres annually; and if a six-year rotation be reckoned not improper

[1] James Headrick, *General View of the Agriculture of Angus*, p. 573, *et seq.*
[2] *Session Papers*, Signet Library, Bowbridge Distillery, 443:28.

for Scotland in general, the number of acres which these fifty distilleries could keep in a state of high improvement would be 30,000 acres, and that of course every distillery, of fifty gallons, could furnish manure for a farm of 600 acres.[1]

Following the Napoleonic Wars the duty on whisky amounted to 9*s.* 4½*d.* per gallon, but the yield in revenue had fallen to less than a million pounds. This, of course, does not of necessity mean that the actual production of whisky had fallen, but simply that it had been driven underground. At the end of the wars, when the duty was again lowered, the number of licensed distilleries increased, till in 1823 there were 243—ten years earlier it had been so low as twenty-four. As the number of licensed distilleries mounted, many lesser stills disappeared, without ever having been within the cognizance of the law. By 1830 the duty had been reduced to 3*s.* 4*d.* per gallon, less than it had been since duty per gallon produced had been levied, and the total yield of revenue was £5,988,556. The table on page 568 summarizes the various changes over the period considered, and gives figures for England and Ireland for comparison.

From 1821 onwards, vigorous steps were taken to suppress illicit distilling. In 1823 there were no less than 14,000 official detections of illicit operation. The attack on the illicit distiller made by the Government was of economic significance, since, in the north, distilling had been a common source of revenue to small land-owners. In one district of Argyll the majority of the labourers and cottagers supported large families on profits of smuggled illicit whisky.[2] In Glenlivet alone there were two hundred stills. Up to the end of the eighteenth century not more than one-fortieth of the whisky produced in Scotland had contributed to the revenue.[3] According to D. F. Macdonald, in his account of Scotland's shifting population, the suppression of illicit distillation was frequently, like the collapse of kelping, the cause of the migration of the inhabitants.[4]

In 1823 with the return of more normal conditions in Europe, the duty on spirits was reduced from 6*s.* 2*d.* to 2*s.* 4¾*d.*, and legal distillation encouraged. In that year George Smith started the first licensed distillery in Glenlivet. Others were tried in 1825 and 1826, but the smugglers got an upper hand, as they also did when they burned down the Banks o' Dee Distillery, Aberdeenshire, in 1825.

[1] *Report*, p. 451. [2] *New Statistical Account*, Vol. 7, p. 385.
[3] *Report*, p. 279. [4] D. F. Macdonald. *Scotland's Shifting Population*, p. 21.

SPIRIT DUTY, 1791–1830[1]

Years	England	Scotland		Ireland
		Lowland	*Highland*	
	Duty/Gallon	*Duty/Gallon of Still Content*		*Duty/Gallon*
	s. d.	*£ s. d.*	*£ s. d.*	*s. d.*
1791	3 4¾	3 12 0	1 4 0	1 1¼
1794	3 10¾	10 16 0½	1 16 0	1 1¼
1797	4 10¼	64 16 4	3 0 0	1 5¼
1800	5 4¼	64 16 4	7 16 0¼	2 4¼
				2 10¼
		Duty/Gallon Made		
1802	5 4¼	3 10½	3 4¼	
1804	8 0½	5 9¾	5 0½	4 1
1807	8 0½	5 8¾	4 11¼	4 1
1811	10 2¾	8 0¼	6 7½	2 6½
		s. d.		
1815	10 2¾	9 4½		6 1½
1817	10 2¾	6 2		5 7¼
1823	11 8¼	2 4¾		2 4¾
1825	11 8¼	2 4¾		2 4¾
1826	7 0	2 10		2 10
1830	7 6	3 4		3 4

The following table gives the gallonage on which duty was paid in the years, and gives an index of increased production and the decline of illegal distilling.

1822 . . .	2,225,124
1825 . . .	5,981,549
1840 . . .	9,032,353
1850 . . .	11,638,429[2]

Much of this production was already in the hands of a few large firms. J. Haig and J. Stein each produced more than 250,000 gallons in 1831, and six other firms south of the Highland line had outputs of over 100,000 gallons. On the other hand, the average daily output of the 239 distilleries known to the excise was only 120 gallons per distillery per working day, i.e. about 20,000 gallons per annum.

[1] From D. Bremner, *op. it.*, p. 450. [2] *Ordnance Gazeteer*, Vol. 6, p. 83.

Chapter XXIV
APPERTIZING

"Whereas, with the method at present practised of closing such vessels (tins) violence is requisite to open them. . . ."

Preamble to patent of Anglibert, 1833.

EXTENDING food supplies so as to obtain foodstuffs either out of season or in localities where they do not naturally occur has been a problem ever since man took to living under highly concentrated conditions, or since he took to exploring the globe and penetrated areas far removed from accessible supplies. This was particularly true of long, slow voyages, and of Arctic exploration in the early nineteenth century.

> As a considerable time often intervenes between the production of different articles of subsistence and their actual consumption, and as the wants and occupations of mankind frequently impose on them the necessity of storing up the superabundant produce of one period to meet the exigencies of another, it becomes likewise of importance to ascertain the best methods of *preserving* such articles, either as nearly as possible in their original state, or in some other in which, though their form and properties be altered, their nutritive powers may be retained. By such means, not only may the more perishable alimentary substances of one season be reserved for consumption at another, but the superfluous productions of distant countries be transported to others, where they are more needed. To mariners, in particular, every means of preserving articles of subsistence in a recent state, must be an object of great interest; and even though this should not be practicable to the extent of supplying daily food for a large crew, yet an occasional use of such food would be at all times a great luxury, and in many cases of sickness and disease, essential, perhaps, to the restoration of health.[1]

To meet this situation various expedients have been employed. To the rationale several scientists well known in other fields have contributed. Like vegetable physiology, the principles which underlie successful preservation of food took many generations to work out, and the whole subject is intimately connected with the

[1] *Edinburgh Review*. April 1814, p. 45, quoted by A. Bitting, *Appertizing*, or the *Art of Canning: its History and Development*, p. 25.

evolution of an understanding of putrefaction, fermentation, and the science of bacteriology, In fact, much of the technique of food preservation was adopted on the erroneous premise that oxygen, not bacterial action, was the cause of putrefaction, and the success of the methods employed resulted from attempts to exclude oxygen which at the same time inhibited bacterial action.

Prior to 1800, three methods of preserving food were in common use: (*a*) freezing, (*b*) dehydration, (*c*) methods employing chemical substances, all of which had changed little in the preceding hundred years. As far as we are concerned, the only use of the first was the extensive trade in Scottish salmon which from an early date were conveyed to London on ice.

> In the more temperate climate of this country, we seldom attempt to preserve animal substances for any length of time by a simple reduction of temperature, though the expedient is perhaps worthy of more attention than it has yet received. In one important instance, however, the Russian practice has been followed with great advantage. We allude to the mode of preserving fish, now adopted on all the eastern rivers and coasts of Scotland and, we believe, in some parts of Ireland, by which means salmon is conveyed fresh to the capital of the empire. The practice is said to have been first recommended by a public-spirited country gentleman in Scotland, Mr. Dempster of Dunichen; and its adoption has been to many a source of great private emolument, and productive of much national benefit. Every salmon fishery is now provided with an ice-house for laying in a stock of ice during the winter. The salmons are packed in large oblong wooden boxes with pounded ice interposed betwixt them; and in this manner they are conveyed to London as fresh as when they are taken out of the water. Till the introduction of the practice just mentioned, ice-houses were possessed chiefly by the opulent, and applied only to purposes of luxury. It is not unlikely that they will, ere long, be extended to the preservation of other necessities of life: for what advantage does the fish merchant derive from them, which the dealer in other animal substances might not equally obtain?[1]

This early nineteenth-century citation should be contrasted with Defoe's remark in his *Tour* (1724–27) on the 'stinking' fish brought to Billingsgate (founded 1699) from the east coast. It illustrates clearly improvement in certain directions in food supplies during the eighteenth century.

Of chemical methods more may be said. These are divisible into salting, pickling, smoking, fermenting, and candying. All

[1] *Edinburgh Review*, 1814, *23*, 104.

depend on the establishment of an environment unpropitious to bacterial development. This was achieved by sodium chloride (salting), acetic acid (pickling), antiseptic phenolic bodies in pyroligneous acid (smoking), or alcohol (fermenting). For comments on these, and for various patents covering the processes, Sir John Sinclair's *Code of Health and Longevity* (1807) may be consulted.

All these preservatives have, however, an objection in common. The process of preservation alters both the flavour and texture of the food. No one could confuse salt pork with the fresh article. Yet pyroligneous acid was not entirely objectionable, since it gave a distinctive character to, for example, smoked fish. Thus, in spite of the undoubted efficiency, and in some cases attractive distinctiveness, of the method of preservation, there was a pressing demand for a process for 'salting by which the freshness and flavour would not be impaired', to quote the *Royal Society of Arts*.

Such a process was indeed mentioned in a remarkable but little-known book published in London as early as 1680: *A Book of Receipts according to the Newest Method*. The author is unknown. Here is the *modus operandi*. It should be compared with the methods used domestically at the present day.

> Gather gooseberries at their full growth, but not ripe, top and tail them, and put into glass bottles, put corks on them but not too close, then set them on a gentle fire in a kettle of cold water up to the neck, but wet not the cork, let them stand till they turn white or begin to crack, and set them till cold, then beat in the corks hard and pitch them over.
>
> You may do them in the oven if you please, or cork them down hard and pitch them over, and they will keep without scalding.[1]

It appears that knowledge of this sort was quite general. Speaking of the claim made by Nicholas Appert (*vide infra*) that his method was entirely original, the *Edinburgh Review* says:

> We are not enough acquainted with the taste and knowledge of the French ladies, in matters of household economy, to know whether it will pass on the other side of the Channel; but on behalf of our fair countrywomen, we must observe, that, unless they must, in general, be more or less acquainted with the methods of Mr. Appert. We have, indeed, seen it stated in an excellent little compendium of the culinary art, composed, it is said, by a very respectable lady in the north of England, that 'there was a time when ladies knew nothing *beyond* their own family concerns; but in the present day there are many who know

[1] Quoted A. Bitting, *op. cit.*, p. 12.

> nothing *about* them'. The very extensive sale of the little work just alluded to, we are willing to accept as evidence of a desire, at least, to discredit such a censure: and we venture to predict that every young lady who shall ponder well on the 'miscellaneous observations' prefixed to that little volume, will find in them much useful instruction in many circumstances and situations of her future life, for which the 'Romance in the Forest', and the 'Sorrows of Werther' make no sort of preparation.[1]

It was not till a century after the *Book of Receipts* was published that any great advance took place in the large-scale preservation of food. During this period the mistaken idea that oxygen was the cause of putrefaction gained considerable ground. Nevertheless, this very idea, wrong though it was, led to the evolution of methods of preservation which were extremely effective. Tinned food was a valuable commodity contributing its share to increasing social amenity three-quarters of a century before the basis of the operations on which its production depended received a scientific explanation as a result of the researches of Louis Pasteur.

During the same period the fundamentals of the germ theory, upon which Pasteur's work rested, were laid by the Abbé Spallanzani, whom Lancelot Hogben characterizes as 'perhaps the most truly original experimental biologist of his age'. Spallanzani was born in 1729 and studied under his cousin, the gifted Laura Bassa, who combined the Professorship of Natural Philosophy at Bologna with the devoted motherhood of a dozen offspring. Her cousin and pupil, Spallanzani, held academic chairs at Reggio and Pavia, and till his death in 1789 contributed to widely diverse fields of learning. Of his work in connexion with putrefaction we cannot do better than cite the account given by A. Bitting:

> In judging any work of that period it must be taken into consideration that there was no knowledge of bacteria at the time, and that the only works that had a direct bearing on the subject were those of Needham, Spallanzani, and Scheele. Needham, an Englishman, published a paper in 1745 in which he gave supposed proof of spontaneous generation. The 'proof' consisted in boiling meat extract, then closing the flask airtight, after which it was left for several days. When the flask was opened the extract was swarming with 'infusoria', and as the author was positive the heat had destroyed the 'eggs' present in the extract and had precluded any entrance from outside, those present must have originated spontaneously. In 1765, Spallanzani, an Italian, claimed that in Needham's work there must have been entrance of air not subjected to heat, otherwise there would

[1] *Edinburgh Review*, quoted A. Bitting, *op. cit.*, p. 39.

be no life present, as his own experiments had demonstrated. Unfortunately, his own experiments were attended with a number of failures sufficient to throw doubt upon his conclusions, consequently they were not generally accepted. These *germs of the principle of sterilization* were not extended to any practical application except in the most meagre way by Scheele.[1]

Towards the end of the eighteenth century the problem of successful preservation of food was becoming pressing. No doubt a number of people in various parts of Europe were researching on it, though it is very doubtful if they were directly aware of the work of Spallanzani. In 1787 the Royal Society of Arts offered a prize of £52 10*s*. for the successful preservation of food with the fresh flavour retained. The following letter in the Boulton Collection, Assay Office, Birmingham, shows the interest taken in the problem by Matthew Boulton, James Watt, and Sir John Dalrymple of Edinburgh, in 1795.

> Please tell him (J. Watt) that Mr. Mellish's beef kept fresh six weeks, but at the end of that time the vessel broke by the heat of the weather. The beef had not the least taint. I conclude that it will not be safe to put in the Fixt Air in England and send it to the West Indies; but that it will be safe to put the Fixt Air in in the West Indies because it will receive no additional heat then. Ask (Mr. Watt) what he thinks.[2]

This letter is not very clear, but certain deductions can be made from it. Mellish was preserving beef in vessels, most likely of glass, and he was filling them with carbon dioxide (Fixt Air) to replace oxygen in the belief that the latter was the cause of putrefaction. In the same year, 1795, the French Government offered a prize for an improved method of preserving food. Social conditions in France were deplorable.

> France was suffering acutely from the shortage of certain kinds of foods and through unsuitable preparation of others. Scurvy took a tremendous toll in the navy, while malnutrition and dietary diseases were rife in the civilian population as well as in the army. The conditions were only slightly different from those in other European countries, but in France apparently were discerned earlier and the remedy sought. It was while endeavouring to assist the Government that Nicholas Appert developed his method of preserving food by heating in closed containers, and to which he gave unremitting effort until his death in 1841.[3]

[1] A. Bitting, *op. cit.*, p. 11.
[2] J. Dalrymple to M. Boulton, July 23rd, 1795.
[3] A. Bitting, *op. cit.*, p. 8.

In 1808 a partial award was made by the *Royal Society of Arts* to Thomas Saddington of Lower Thames Street for preserving fruit. Saddington had travelled on the Continent, and it is thought that he may have been aware of the work being carried on by Nicholas Appert. No immediate commercial development followed the reward to Saddington, however.[1]

Nicholas Appert (1750–1841) deserves the credit of being the real founder of the canning industry. Born at Chalons-sur-Marne, he spent the whole of his life experimenting with food. His most important observation was that food after being heated in closed vessels would keep indefinitely. From *c.* 1794 he was actively preserving a wide range of foodstuffs, including meat, vegetables, fruit, and even milk. The containers which he used were to begin with strong glass bottles which, before heating, he stoppered and luted with a mixture of cheese and lime. Although the conditions achieved by his method conform to the requirements of modern bacteriology, they were determined by the belief that oxygen was the cause of putrefaction. In 1804 his operations attracted the attention of the Minister of Marine: the *gourmands* of Paris in the following year. Bitting cites from the *Almanach des Gourmands* (1805):

> M. Appert has had such success that in each bottle and at slight cost is a bounteous *entremêt* that recalls the month of May in the heart of winter, and often deceives, when it is dressed by a skillful cook; it is not an exaggeration to say that the small peas particularly, thus prepared, in short are as green, as tender, as savoury, as those that are eaten in season.[2]

In 1809 Appert was awarded 12,000 francs by the *Bureau Consultatif des Arts et Manufactures* for his process, provided that he wrote an exact and detailed account of the process, printed it at his own expense, and presented two hundred copies to the Bureau.[3] In the next year, no doubt in conformation with these conditions, he published *Le Livre de tous les Ménages ou l'Art de Conserver pendant Plusieurs Années Toutes les Substances Animales et Végétales.* Almost immediately it was translated into other European languages, German in 1810, English and Swedish in 1811. The original went through many editions.

Knowledge of Appert's work was brought to the cognizance of the English-speaking world through Humphry Davy's lectures at the Royal Institution. According to Davy:

> Salt and alcohol appear to owe their powers of preserving animal and vegetable substances to their attraction for water, by

[1] *Trans. Soc. of Arts*, 3rd Editn., 1807, *26*, pp. 19 etc.

[2] *Almanach des Gourmands*, Vol. 3.

[3] A. Bitting, *op. cit.*, p. 10.

which they prevent its decomposing action, and likewise to their excluding air. The use of ice in preserving animal substances is owing to its keeping their temperature low. The efficacy of M. Appert's method of preserving animal and vegetable substances, an account of which has been lately published, entirely depends upon the exclusion of air. This method is by filling a vessel of tin-plate or glass with the meat or vegetables; soldering or cementing the top so as to render the vessel airtight; and then keeping it half immersed in a vessel of boiling water for a sufficient time to render the meat or vegetables proper for food. In this last process it is probable that the small quantity of oxygene remaining in the vessel is absorbed; for on opening a tinned iron canister which had been filled with raw beef and exposed to hot water the day before, I found that the minute quantity of elastic fluid which could be procured from it was a mixture of carbonic acid gas and azote.

Where meat or vegetable food is to be preserved on a large scale, for the use of the navy or army for instance, I am inclined to believe that by forcibly throwing a quantity of carbonic acid, hydrogene, or azote into the vessel by means of a compressing pump, similar to that used for making artificial Seltzer water, any change in the substance would be more effectually prevented. No elastic fluid in this case would have room to form by the decomposition of the meat; and the tightness and strength of the vessel would be proved by the process. No putrefaction can go on without the generation of elastic fluid; and pressure would probably act with as much efficacy as cold in the preservation of animal or vegetable food.[1]

We remark that Davy pays no attention to the fundamental work of the Abbé Spallanzani. The same is true of Gay-Lussac, who harboured the weird idea that oxygen 'becomes concrete by heat in the same way as albumen'.[2] The fact is Appert himself was nearer a sound theory of his operations than were eminent contemporary scientists. Here are his own words:

I owe to my experiments, and above all to great perseverence, to being convinced that the subject of heat has the essential quality in itself not only of changing the combination of the constituent parts of animal and vegetable products, but also that, if not destroying, at least arresting for many years the natural tendency of these same products to decomposition; second, that its application in a proper manner to all these products, after having deprived them in the most rigorous manner possible of contact with the air, effects preservation of these same products with all their natural qualities.[3]

[1] H. Davy, *Lectures*, pp. 242–3. [2] A. Bitting, *op. cit.*, p. 10.
[3] Quoted by A. Bitting, *op. cit.*, p. 11.

The House of Appert was formed in 1812, and continues to the present day, having been the recipient during its century of existence of many honours for the high quality of its products. In 1814 Appert came to London to demonstrate his methods in the hope of collecting orders. He received a disappointing reception. In England there already existed a tinplate industry, and some English manufacturers saw that they could develop a native canning industry on the basis of available tinplate. In the year 1810 two patents were taken out for the use of tinned iron containers for preserving food, those of Augustus de Heine and Peter Durand. Durand's patent is quoted in full in Bitting.[1] There is no evidence that either Heine or Durand exercised the privileges granted them, but Bryan Donkin and John Hall may have acquired them (*vide infra*) before setting up the first English canning factory.

Before going on to consider in detail the rise of the British industry, a comment might be made on Appert's relations with the scientists of his time. According to Bitting, he was 'known and respected for his achievements and for his rare and sterling character' by his contemporaries, in particular, Bardel, Gay-Lussac, Parmentier, Corbiere, and Careme. He also laid the foundations of the work which brought Louis Pasteur lasting fame. This was acknowledged by Pasteur himself in the following words, when speaking of his own researches on fermentation:

> This is the method that Appert had previously put in practice with so much success for the conservation of a multitude of alimentary substances, method the germ of which is likewise found in the experiments of Needham and Spallanzani on the subject of the generations called 'spontaneous', as I have remarked elsewhere.
>
> When I published the first results of my experiments on the possible conservation of wine by preliminary heating, it was evident that I only made a new application of the method of Appert, but I was absolutely ignorant that Appert had devised this same application a long time before me.
>
> It is, however, this expert manufacturer who, the first, has clearly indicated the possibility of conserving wine through the preliminary application of heat.[2]

The first British canning factories were in operation for several years before Appert's visit to London. It is not known if they were

[1] A. Bitting, *op. cit.*, p. 23.

[2] L. Pasteur, *Etudes sur le Vin*, second edition, 1873, p. 133. Translation by Bitting.

in any way connected with developments in France. The British industry starts with the setting up of a works by Bryan Donkin (1768–1855) and John Hall. Both had other important industrial connexions. Hall was the founder of the Dartford Iron Company, and it may have been with the idea of creating a market for iron sheet that Hall became interested in canning. Bryan Donkin, it will be remembered, was the engineer associated with J. Gamble in the introduction of the Fourdrinier paper machine.

In a history of the Hall concern[1] it is stated that Hall and his associates paid £1,000 for a patent registered by a French chemist called Appert. The same story is given by Bremner in his *Industries of Scotland*, but according to Drummond and Lewis, Appert took out no patent. Moreover, there is no record of the transaction in the archives of his firm, which is still in existence.[2] Be that as it may, in the next few years products from Donkin and Hall's factory at Blue Anchor Road, Bermondsey, began to attract considerable attention in naval and exploration circles. Among the first to record an opinion of British canned food was Admiral Cochrane, later tenth Earl Dundonald, who was then in charge of the West India Naval Station. There were doubtless others, for the firm brought out a pamphlet in 1817 containing, among other testimonials, a letter from Sir Joseph Banks, President of the Royal Society, of which, incidentally, Donkin was a Fellow.

As a result of their testimonies the firm got a Navy contract in 1818, and enjoyed a position of monopoly for several years. About the same time, J. H. Gamble became associated with Donkin and Hall, and ultimately he took the leading share in the conduct of the business, that is before it passed to the control of Crosse and Blackwell—some time before 1850.[3]

One of the immediate and important effects of canned food was to improve the diet of men at sea, particularly those engaged in the long voyages connected with the various explorations which were taking place in the second and third decades of the nineteenth century. Some of Donkin's tinned meat was taken to the Arctic by Ross in 1814, and by Otto von Kotzebue in his voyage in search of the North-West Passage in 1815. Similarly in 1819 Parry on his first voyage included tinned food in his stores on the *Heckla*

[1] E. Hesketh, *J. & E. Hall, Ltd.*, 1785–1935.

[2] J. C. Drummond and W. R. Lewis, *The Examination of Some Tinned Foods of Historic Interest*, Publications of the International Tin Research and Development Council, No. 85, p. 1.

[3] Some further information will be found in the 1841 edition of the *Encyclopaedia Britannica* (Seventh Edition, Article, *Food*).

and *Griper*. Still larger quantities were contracted for in 1824 on his second voyage.

In naval circles Donkin and Hall's preserved meat quickly gained a reputation as an antiscorbutic, a curiously mistaken impression, since men on scurvy-free voyages were being supplied with preserved lemon juice at the same time. Lemon juice was then preserved by the addition of 25 per cent. rum, with, on occasion, additional sugar. This method of preservation was not always satisfactory, due to oversight in the addition of the required quantity of rum. As a result of the reputation gained by preserved meats, a certain Captain Bagnold suggested to the Admiralty the preservation of lemon juice by a modification of Appert's method.

> He proposed to boil the strained juice for an hour and a half, to pour the hot fluid into bottles, leaving a very small space, just sufficient to hold the cork, and then to stopper tightly and cement over the outside of the cork. He was, of course, following the current idea that one must exclude air, but in actual fact, if properly carried out, his process was a simple sterilization. The only drawback would have been that so long boiling would have destroyed a great part of the vitamin C. He claimed that the bottled juice kept unchanged in a hot climate for two years, but it is not known whether it was ever used on any scale by H. M. ships.[1]

For several years Donkin, Hall and Gamble enjoyed a position without interference from competitors, but in the 'twenties a number of firms, some to attain great eminence during the course of the century for the quality of their products, were founded. The principal of these were (*a*) John Moir of Aberdeen, in 1822; (*b*) Hogarth and Dickson, also of Aberdeen, in 1824–5; (*c*) Aeneas Morrison of Glasgow, 1824; and (*d*) James Cooper of Clerkenwell, also in 1824.

The voyage which Parry made in 1824 was victualled by three of these: Gamble, Morrison, and Cooper. Whether Cooper actually engaged in preserving, or was merely a retailer, we do not know, but there is little doubt but that on occasion he was supplied with preserved foods by Moir of Aberdeen, as will be seen from the following letter:

> Aberdeen, 21st June, 1827
>
> Mr. James Cooper.
>
> Sir,
>
> Your letter of the 7th inst. arrived in due course, but having been in the country these last 14 days superintending the fishing, has prevented me from replying to it sooner.

[1] J. C. Drummond and A. Wilbraham, *The Englishman's Food*, p. 468.

The salmon still continues very scarce, but I hope to be able to send you part of your esteemed order early in July.

The lowest price that I can sell you Lobsters is 1*s*. 6*d*. for pint cases and 10*d*. per half pints.

Shall send you tomorrow per Triumph 1 pint canister and one half pint as samples. The meat is put into the 1 pint canister as whole as possible without sauce. The half pint contains meat cut into small pieces, and sauce made from fragments of the Lobsters —you can have them done in either way you choose.

I am,

Sir, Your obedient servant,

John Moir.

7 Johns Street,
Clarkenwell,
London.

Two cans from Parry's 1824 voyage were examined in 1938 by Drummond and Lewis,[1] and two cans from the 1826 voyage have their history recorded in the *Journal of the Royal Society of Arts* (1867).

A manuscript letter-book of Moir's has been preserved which throws some light on the early activities of the firm. The origin of John Moir (1766–1833) is obscure: it is thought that he came to Aberdeen from Leith. Nothing is known of the source from which he learned about preserving. One of the letters from the letter-book is subjoined:

18th December, 1826.

Mr. Berry.

Sir,

By the contents of your letter of the 16th inst. I apprehend that the 5 boxes tin plates being sent was occasioned by you not receiving my note which was left at the weighhouse for you and your not sending an invoice with the goods prevented me from knowing your terms or they would have been instantly returned.

To end the dispute I shall willingly pay you your own price for a box of each kind which are opened, and return the other three which are untouched.

Sir,

Your obedient servant,

John Moir.

In the letter-book are also some money transactions over the period September 1824 to February 1827. These are given for comparison with present-day prices and conditions.

[1] Drummond and Lewis, *op. cit.*, p. 4.

Mr. John McDonald, Aberdeen, Dr. to John Moir.

Date		Item	Rate	£	s.	d.
1824						
Sept.	8	To *Aberdeen Journal* charged but not paid		1	8	10
1825						
May	30	„ 1 month rent of John		1	6	8
July	15	„ 1 Grilse 6 lbs.	@ 6*d.*		3	0
Aug.	2	„ 1 do. 5 „	@ 6*d.*		2	6
	8	„ 1 do. 5 „	@ 6*d.*		2	6
July	9	„ 3 do. 3 „ heads . . .	@ 2*d.*			6
	18	„ 1 do. 3½ „	@ 6*d.*		1	9
	22	„ 1 do. 7 „	@ 7*d.*		4	1
Aug.	1	„ 2 do. 13 „	@ 7*d.*		7	7
1826						
April	26	„ 7 Pints preserved crabs . .	@ 1/-		7	0
Aug.	3	„ 1 Grilse 3½ lbs.	@ 6*d.*		1	9
	13	„ 1 Case preserved salmon . .	@ 1/6		6	0
		„ 1 Grilse 3½ lbs.	@ 6*d.*		1	9
	15	„ 1 Pint preserved crabs . . .	@ 1/-		1	0
	17	„ 1 Grilse 7½ lbs.	@ 7*d.*		4	2½
	28	„ 1 Cut salmon 6 lbs. . . .	@ 1/-		6	0
	22	„ 1 Case salmon 4 lbs. . . .	@ 1/6		6	0
	31	„ 1 Grilse 6½ lbs.	@ 7*d.*		3	9½
Sept.	6	„ 4 lbs. Trout	@ 3*d.*		1	0
Nov.	10	„ 1 Pint crabs	@ 1/-		1	0
1827						
Feb.	1	„ Cash		4	0	0
		„ Solder for closing the seams and putting on the rings of 149 doz. 4 lb. cases at 1 oz. per dozen is 9 lbs. 5 oz.	@ 9*d.*		6	11½
				10	3	10½
To solder closing seams and putting on the rings of 2,160 cases 2 lbs. allowing 1 oz. each 18 cases is 7 lbs. 9 oz.			@ 9*d.*		5	8
To interest on £58 19 9 paid 4 mons. before due					19	11
To do. £44 12 6 „ 1 „ do.					3	8
To difference of price of 19 boxes Tin plates between 2nd Nov. 1825 and 20th June 1826 to be deducted from Mr. McDonald and as per contract				4	15	0
				£16	8	1½

All we know as yet of the other Aberdeen firm, Dickson Hogarth and Company, is derived from evidence given by W. Hogarth before a Royal Commission (1852) appointed to enquire into the preservation of food. There Hogarth revealed that he had invented his own method of using high pressure steam in 1837. This invention of Hogarth's is important, since the ensuing period saw much preserved food retailed which was unfit for human consumption because the temperature of the preserving bath had not been sufficiently high to penetrate to the centre of the larger packs then coming into favour. With his pressure steam Hogarth got over this difficulty, and the firm's products were favourably commented upon by the Commander of the *North Star* and *Assistance*, which set out to find Franklin.[1] A tin preserved by Hogarth was among those examined by Drummond and Lewis.[2]

Hogarth's method was but one of the possible ways of obtaining higher sterilizing temperatures, and it was perhaps one of the more difficult to operate in industrial practice. Several patents attacking the same problem were registered in the same year. The most important of these were the patents of Stephen Goldner and John Wertheimer, both registered in 1841. Instead of making use of the increase in the boiling-point of water with increased pressure, they raised its boiling-point by the addition of calcium chloride ($CaCl_2$), thus attaining a temperature of 270–280° F., instead of 212°, the normal boiling-point at atmospheric pressure. These developments were on the right lines but it required the development of bacteriology and the passage of many years before the public became fully confident as to the safety and utility of preserved food.

[1] *Report on Preserved Meat* (*Navy*) 1852.
[2] Drummond and Lewis, *op. cit.*, p. 12.

Chapter XXV

SOCIAL PERSONNEL[1]

Tout homme qui reçoit une education liberale, compte aujourd' hui la chimie parmi les objects les plus indespensables de ses études.

FOURCROY.

IN the foregoing chapters there is ample evidence of the significant contribution which Scotland made to the development of social technology, especially during the eighteenth century, but in handling the masses of technical and economic data there was little opportunity to analyse and interpret the antecedents of these contributions. In this final chapter we shall address ourselves to an inquiry into the special conditions whereby Scotland emerged from a state of comparative obscurity to become one of the intellectual foci of Europe, with a world-wide reputation for fertility of invention and eminence of academic teaching. The analysis is limited to chemical and cognate subjects. Such limitation excludes medical and surgical advances, of which Scotland is so justly proud, but at the same time it has this advantage that it leaves a clearer picture of Scottish leadership in the English industrial revolution.

It is almost certain that it was Scotland's traditional poverty that planted the seeds of chemical experiment in the northern soil. Even Scotland, far removed from the flux of European experiment, had turned to alchemy as a possible solution of her coinage problems before the death of Van Helmont in 1644. Cambuskenneth Abbey and Stirling Castle housed alchemical laboratories where John Damien and King James IV pursued the usual alchemistic mumbo-jumbo of gold-making. The *Accounts of the Lord High Treasurer of Scotland* show that laboratory expenses included wood and coals for furnaces, and such chemicals, or chemical-like materials, as saltpetre, alum, salt, brint silver, fine tin, quicksilver, lithargyrum, aqua vitae, etc.[2] James V upheld the Scottish regal interest in precious metals; his successor James VI subsidized alchemy and alchemists at a time when there were

[1] One of the authors (A.C.) was awarded the *Blackwell Prize* for a thesis based on this chapter and entitled: *Scotland's Contribution to Industrial Development through the Application of Chemical Science since the 17th Century.* (Typescript: Aberdeen University Library.)

[2] J. D. Comrie, *History of Scottish Medicine*, p. 154.

sufficient active participants in the art to gain for Scotland an international reputation. Among those interested were Geo. Erskine of Innerteil, John Napier of Merchiston, David Lindsay, Earl of Balcarres, Pat. Ruthven, Alex. Seton of Edinburgh, and Peter Scott of Falkland. The alchemical books of Erskine and Ruthven are still preserved in Edinburgh. From these seeds the Scottish chemical efflorescence expanded.

Towards the end of the seventeenth century, by which time, as was related in Ch. I, a certain amount of industry was developing in Scotland, laboratories moved from church and castle to more secular surroundings in the halls of the Surgeons' Incorporation in Edinburgh. When, in the last decade of the century, the surgeons acquired a new hall, Alexander Monteith of Auldcathie equipped a chemical laboratory where apprentice apothecaries might receive training in the arts of chemistry. His laboratory consisted of a suite of three rooms, and was not unambitious in conception, since it contained four hundred *gally pigs* which were obtained at a cost of twenty pounds.[1] Not only did Monteith see to the furnishing of his laboratory, but he took steps to ensure that it was adequately supplied with funds by petitioning Parliament in 1700 'that the art discovered by him to draw Spirits from malt equal in goodness to true French Brandie, may be declared a manufactory with the same privileges and immunities as are granted to other manufactories'.[2] Two centuries earlier, Edinburgh barber-surgeons had been granted the sole right to manufacture and sell aqua vitae in Edinburgh, a privilege they neglected, to their great loss, to husband.

Following the establishment of Monteith's laboratory, it was not long before the teaching of chemistry in Edinburgh was noticed. In 1702 it was announced that a 'course of chemistrie' would be given in the surgeon apothecaries' hall.[3] Who gave the lectures is not known.

The next development is one of great importance from at least two points of view. In the first place, it established active chemical teaching in the University of Edinburgh, and in the second, the incumbent of the chair, Dr. James Crawford, cemented a connexion already formed between Edinburgh and Leyden University which was of the utmost significance to Scottish chemistry. This continental connexion is so important that it is desirable to digress for a moment to consider the University of Leyden and its teachers.

The University of Leyden was founded in 1574. At that date

[1] J. D. Comrie, *History of Scottish Medicine*, p. 253. [2] *Ibid.*, p. 151.
[3] *Ibid.*, p. 289.

Louvain was the only existing university in the Low Countries. Louvain was a Catholic foundation, but from the start Leyden, although strictly speaking a Protestant foundation, admitted students of all faiths. That alone was an attraction to the young Scot in search of learning on the Continent. Coupled with an outlook characterized by religious toleration, Leyden acquired an enviable reputation as a school of medicine on the basis of the teaching of Boerhaave, Albinus, Gaubius, Ostendyk, and Van Royen. While Hermann Boerhaave (1668–1738) lectured on chemistry there, his eminence was perhaps the chief attraction to Leyden.

> For a student, therefore, to possess every advantage for improving himself in his medical studies it was deemed necessary to repair to Leyden and attend Boerhaave's lectures on chemistry.[1]

It should be remembered, however, that Boerhaave was the pupil of Dr. Archibald Pitcairne (1652–1713), the greatest luminary in the Edinburgh medical world, till the foundation of the Faculty of Medicine in 1726, and one who worked to bring it into being. Incidentally, every founder member of the Edinburgh Faculty had studied at Leyden under the celebrated Boerhaave.

Thus Leyden became the focus of continental learning for Scotland during several generations. Son followed father in the lecture halls, returning to take up academic appointments in the universities of Scotland till, by the end of the eighteenth century, some forty holders of Scottish chairs had come under its influence. In Britain as a whole the influence of this great seat of learning was not confined to Scottish intellectual circles, but it was probably more intense there than elsewhere.

Archibald Pitcairne, professor of medicine at Leyden, was the son of a Fifeshire landed family, a man of great spirits and jovial habits. Intended by his father for the Church, he studied Arts and Divinity at Edinburgh, but excessive application to Law took him to Montpellier to recuperate. There he became interested in Medicine, which he decided to make his profession. He graduated at Rheims in 1675 and returned to Scotland to take up practice. He was a founder Fellow of the Royal College of Physicians. Such was his reputation that in 1692 he was invited to go to Leyden to occupy the chair of medicine. Thus was established a vital link between Leyden and Edinburgh which persisted throughout the greater part of the eighteenth century. Pitcairne did not stay long,

[1] A. Bower, *History of University of Edinburgh*, Vol. 2, p. 126.

but during his stay he had the good fortune to have Hermann Boerhaave as a student. Boerhaave followed Pitcairne as a professor at Leyden, and built up the reputation of the Leyden medical school. Pitcairne returned to Edinburgh and died there in 1713.[1]

The prodigious influence of Boerhaave is inestimable. Stahl, the propounder of the phlogiston theory, was one of his students. Others are found among the founders of celebrated faculties throughout Europe particularly at Vienna and Edinburgh. We can trace his influence on John Dalton, founder of the atomic theory, and on the founders of chemistry in Ireland, but perhaps nowhere was it of greater significance than in Scotland. The first two holders of the chair of chemistry at Edinburgh attended his lectures, as also did Francis Home, who, although professor of *materia medica*, took the first steps towards founding the palaeotechnic bleaching industry in this country by publicizing sulphuric acid as a souring agent. After Boerhaave's death in 1738, his influence was carried on at Leyden by his pupil Hieronymus David Gaubius (1705–80).

While we are primarily concerned with the personnel contributing directly to the chemical revolution, one cannot gainsay the multiple effect which the Leyden alumni had on technics as a whole, since it was the cumulative effect of their teaching and investigations that placed their pupils on the path to invention and discovery. The following examples are but a few instances of those occupying cardinal positions who are not otherwise mentioned elsewhere in this chapter. Of the holders of chairs in the Universities of Aberdeen there were Pat. Chalmers (*d.* 1727), James Gregory (1674–1733), James Gordon (*d.* 1755), and James Gregory (1707–55). To Glasgow went Thomas Brisbain (1684–1742), John Johnstoun (1685–1762), Robert Dick (1722–57), the friend of James Watt, and also Robert and Thomas Hamilton. Thomas Simson, who was first professor of medicine at St. Andrews, was also at Leyden, and farther afield we have the Percivals of Warrington and Manchester, Wm. Stephens and Brian Higgins of Dublin. Then there were the industrialist pupils of Plummer who continued their studies on the Continent, viz. Drs. Roebuck and Hutton, every one additional to the members of the faculty in Edinburgh.

In the year Pitcairne died the Town Council of Edinburgh decided to appoint a professor of chemistry, and at the end of the year Boerhaave's student, James Crawford, was appointed

[1] J. D. Comrie, *op. cit.*, p. 274.

professor of chemistry and physic. The professorship of chemistry at Cambridge dates from the same period, Oxford from much later. Crawford's appointment to Edinburgh is noteworthy since it gave professorial status to a teacher of chemistry before even the first text on the subject was printed, viz. Boerhaave's *Elementa Chemiae* (1732). Of Crawford's students or influence we know little, and there is some evidence that he did not attract many students. At any rate, in 1724 four Fellows of the Royal College of Physicians announced that they had purchased a house for a chemical laboratory, and indicated that they proposed to lecture extramurally on chemistry and *materia medica.* They were Drs. John Rutherford (1695–1779), Andrew Plummer (*d.* 1756), John Innes (*d.* 1733), and Andrew St. Clair. After a few years extramural teaching they insinuated themselves into the University, and for a time there were four 'professors' of chemistry. Of the four, Rutherford and Plummer are the most important from the chemical standpoint, though Innes also is thought to have assisted Rutherford in the teaching of chemistry.

Like his predecessor, Rutherford was a pupil of Boerhaave.

> Having studied under Boerhaave, he had formed a strong predilection for chemistry; and, in the plan which he and his associates had projected, that department of medical science was assigned to him as his province. The science of chemistry was then in its infancy, and possessed but few of the allurements which now accompany the study of it.[1]

Although lecturing on chemistry in what was an embryonic medical faculty, Rutherford recognized its right to independent existence, and laid stress in his lectures on 'useful processes'. At first he included pharmacy in his instruction, but towards the end of his life devoted himself wholly to the study of chemistry, while of course still teaching medicine.[2] None of the quartet made any revolutionary contribution to the advancement of chemistry, but Rutherford's son Daniel became professor of botany in the University, and is remembered as the discoverer of nitrogen (*vide infra*).

In spite of the fact that Plummer and Rutherford had not, so to speak, a research outlook, their place in social technology is assured as mentors of several founders of chemical industry and of academic chemistry as well. Oliver Goldsmith spoke highly of Plummer's teaching, even compared with that at Leyden to which

[1] A. Bower, *op. cit.*, Vol. 2, p. 216.
[2] J. D. Comrie, *op. cit.*, pp. 298 and 306.

he repaired from Edinburgh. Three prominent industrialists who studied under them come to mind at once. The first, Dr. John Roebuck (1718–94), has to his credit the founding of chemical works in both England and Scotland, and of probably even greater moment, the founding of Carron Iron Works, the initial step in opening up the carboniferous formations of central Scotland. The part he played in fostering James Watt at a critical period in the evolution of the steam engine has been referred to as also his attempted synthesis of alkali with Joseph Black. The second of Plummer's students, James Keir (1735–1820) was more successful in the solution of this pressing problem of eighteenth-century technology. Keir particularly was successful in establishing the manufacture of soda at Tipton. Both Keir and Roebuck had industrial connexions with the English Midlands, and the *élite* surrounding Boulton and Watt at Birmingham. The third of Plummer's students who calls for mention here is James Hutton (1726–97), best known as the founder of modern geology. At the same time it must be remembered that Hutton derived a considerable income from the manufacture of sal-ammoniac (NH_4Cl), and was thus not forced to earn a living by lecturing or teaching. Roebuck and Hutton completed their education at Leyden, most probably reading under H. D. Gaubius, who taught chemistry there till 1775. Thus a community of interest early sprang up between the scientists in the university and the germinal chemical industry, part of which took root near Edinburgh, as at Prestonpans, or for that matter at Carron. Keir, on the other hand, implanted his manufactory in an area served by no official intellectual focus at a time when the established universities of Oxford and Cambridge were lagging far behind Edinburgh in the encouragement of such subjects as chemistry.

On the academic side Plummer had two brilliant students, William Cullen and Joseph Black, who followed him successively in the chair of chemistry at Edinburgh.

Andrew Plummer, professor of chemistry 1726–55, continued to lecture on chemistry till 1755, when he was succeeded by William Cullen, the first of a series of men who came from the expanding industrial centre in the west to infuse the University of Edinburgh with a new vitality.

The advent of Plummer, Rutherford *et al.* in the University more or less constitutes the foundation of the Faculty of Medicine in Edinburgh. After 1726 students were examined by the Faculty and the degree in medicine was conferred by the Senatus, instead of the examination being carried out by the College of Physicians.

The submission of a thesis was a prerequisite. In the newly constituted Faculty, Plummer and Innes were professors of chemistry and medicine, St. Clair and Rutherford of the Institutes and practice of medicine. There was in addition a professor of botany. In 1738, however, it was decided to link botany with medicine, and yet another pupil of Boerhaave, Dr. Charles Alston (*d.* 1760), was appointed to the new chair. This arrangement continued for thirty years, when a chair of *materia medica* was founded and given to Dr. Francis Home (1719–1813), whose contributions to agriculture and technology we have had frequent occasion to refer to.[1]

After some twenty years the presence of the new Faculty began to make itself felt, and the number of students coming up for laureation increased rapidly, particularly while the chair of chemistry was occupied by William Cullen and later by Joseph Black.

It is a fact not without social significance that from Cullen's appointment in 1755, the chair of chemistry at Edinburgh was held for over a hundred years by chemists whose training and initial teaching experience was obtained in the newer industrial focus in the west of Scotland. If Edinburgh during the later eighteenth century is considered next to Paris as the intellectual centre of Europe, it should not be forgotten that in the chemical field at least its reputation was founded on imported personnel, and that one of the chief centres from which it imported personnel was Glasgow. So developed the intellectual-industrial nexus.

William Cullen (1710–90) studied Arts at Glasgow, and Medicine at Edinburgh under Plummer. He became attached to the University of Glasgow as a teacher of medicine in 1746. Stimulated by Plummer's teaching, he was vitally interested in chemistry, and in 1747 induced the University of Glasgow to establish the teaching of *Chemie*. Expenditure up to £52 was sanctioned, but before Cullen was satisfied a sum of £136 had been expended; £20 per annum was granted for the maintenance of the department. In 1747 Cullen and John Carrick were appointed lecturers in chemistry. Carrick died in 1750 and Cullen continued to lecture on his own. He lectured in English.[2] At the beginning of his second course of lectures he printed and distributed *The Plan of a Course of Chemical Lectures and Experiments directed chiefly to the improvement of Arts and Manufactures* to be given in the laboratory of the College of Glasgow during the session 1748. The objective of Cullen's lectures must be noted. MS. notes of his lectures are

[1] J. D. Comrie, *History of Scottish Medicine*, p. 303. [2] *Ibid.*, p. 313.

still extant in the Free Library at Paisley.[1] In 1751 he was appointed professor of medicine and lecturer in chemistry at Glasgow, which post he held till 1755 when he moved to Edinburgh and was appointed colleague and successor to Plummer, largely through the good offices of his friend Lord Kames. Plummer died within a year.[2]

Cullen was one of the first chemists to appreciate that chemistry would become a discipline standing on its own, subservient neither to medicine nor pharmacy. He recognised the importance of scientific chemistry and its application to manufactures and agriculture. While in Glasgow he directed a considerable part of his energy to the industrial chemistry of the day, particularly to salt-boiling, bleaching, and alkali supply, for which researches he received a premium from the Board of Trustees for Manufactures. This was the era of rapid expansion in the Scottish bleaching industry.

> He was a great master in the scientific branches of husbandry; a consummate botanist, and possessed a correct taste in the fine arts. In the year 1758, after finishing off chemistry, he delivered to a number of particular friends, and favourite pupils, nine lectures on the subject of agriculture. In these few lectures he, for the first time, laid open the true principle concerning the nature of soils and the operations of manures.[3]

Yet Adam Smith, who was Professor of Logic and Moral Philosophy at Glasgow from 1751 to 1763, and as such a colleague of Cullen, could say in the *Wealth of Nations* that universities were . . .

> . . . sanctuaries in which exploded systems and obsolete prejudices found shelter and protection after they had been hounded out of every other corner of the world.

Either Adam Smith was not referring to the Scottish universities, or he was unfamiliar with developments coextensive with his own.

On the then more academic side Cullen contributed, at Black's suggestion, to the early researches on heat which culminated in the latter's discovery of latent heat (1762) and James Watt's separate condenser engine (1769). Cullen's researches are contained in a pamphlet entitled *On the Cold produced by Evaporating Fluids*. To ensure the continued development of science in Scotland, he played an important part in the founding of the *Royal Society of Edinburgh* (1783).

For five years during his tenure of office in Glasgow Cullen had as pupil Joseph Black (1728–99), who subsequently succeeded

[1] J. D. Comrie, *History of Scottish Medicine*, p. 361.
[2] T. Thomson, *History of Chemistry*, Vol. 1, p. 204, *et seq.*
[3] A. Bower, *op. cit.*, Vol. 2, p. 392.

him in Glasgow and later in Edinburgh when Cullen transferred to a medical chair. These two men, fundamentally chemists, were the founders of the Glasgow medical school which fostered the development of chemistry in what became the industrial focus of Scotland. Of Black a great deal has been said, but he was by no means Cullen's only outstanding student. They included Dr. Withering of the Lunar Society, Dr. Geo. Fordyce of Harley Street, William Hunter, and, most important to us here, George French who established the teaching of chemistry in Aberdeen. As a student of Cullen's, the non-medical approach to chemistry which characterized French's teaching is significant (*vide infra*).

In 1756 Cullen was succeeded at Glasgow by Black. Black was born in Bordeaux to Scottish parents. In 1746 he matriculated as a student of medicine at Glasgow, and came under the influence of Robert Dick, professor of natural philosophy. He stayed for five years with Cullen before continuing his studies under Plummer at Edinburgh where his second cousin, James Russell, was professor of natural philosophy. In Edinburgh Black researched on a practical problem of current medical importance—the search for a solvent for urinary calculi. During the course of this research he elucidated the distinction between mild and caustic alkalis (carbonates and hydroxides) and was led on from that discovery to the isolation of *fixed air* (carbon dioxide, CO_2). His results were presented to the University as *Dissertatio de humore acido a Cibis Orto, et Magnesia Alba* in fulfilment of the regulations for the degree of Doctor of Medicine (1754). They were later incorporated in *Essays and Observations Literary and Physical* under the title *Experiments upon Magnesia Alba, Quicklime, and some other Alcaline Substances.* When he returned to Glasgow and succeeded Cullen he turned his attention to other problems which culminated in the discovery of latent heat. This discovery was announced to the Literary Society of the College of Glasgow in 1762. James Watt was then attached to Glasgow University, and was one of Black's friends. The practical outcome of Black's discovery was Watt's steam engine.

> Watt worked on the steam engine from 1765 on, applied for a patent in 1769, and between 1775 and 1800 erected 289 engines in England. His earlier steam engines were all pumps. Not until 1781 did Watt devote himself to inventing a rotary prime mover; and the answer to this problem was the great double-action fifty horse-power engine that his firm installed in the Albion Flour Mill in 1786, following the ten horse-power engine he first made for a brewery in London.[1]

[1] L. Mumford, *Technics and Civilization*, p. 160.

Of the association of these two great men we shall say more later. Adam Smith of course was another of Black's Glasgow associates.

Before leaving Glasgow Black made important provisions for improving the facilities for teaching chemistry there. In 1763 he persuaded the University to equip a new laboratory and lecture room. This they did at the very considerable expense of £500.[1]

In 1766 Black went to Edinburgh, again to succeed Cullen, who had been translated to another chair, and for thirty years he occupied the chair of chemistry there during one of the greatest formative periods through which chemistry both in theory and practice has gone. In everything Black laid great emphasis on accurate quantitative work, and so influenced a generation of men who were his students at Edinburgh. The academic magnitude of Black's discoveries makes one forget that he also kept in touch with contemporary industrial developments, especially in a consultative capacity. Of particular significance was his connexion with the initial stages of Lord Dundonald's Tar Works at Culross, and with Cort's process for the production of malleable iron. He advised on pottery problems. Specimens of ore and water from the lead mines were sent to him for analysis. His opinion was sought by the Committee investigating Scottish distillery. It is not such activities as these that find a permanent place in the annals of science, but they leave no doubt that Black was no cloistered academician and that his opinion must have been of great value to the expanding industry of Scotland based on accurate chemical observation.[2] Among those, apart from his relations, with whom he enjoyed intellectual companionship were David Hume, Alexander Carlisle, Sir Geo. Clark and his brother, Drs. Roebuck and Hutton.

Two of Black's Edinburgh colleagues though not chemists by profession made important contributions to chemistry. The first was Francis Home (1719–1813) professor of *materia medica* from 1768, the second Daniel Rutherford (1749–1819), appointed professor of botany in 1786.

When Home was on the Continent during the War of the Austrian Succession he took the opportunity of attending the lectures of Boerhaave. When in Leyden he possibly met an Edinburgh graduate and subsequent friend of Black's who was to play the leading role in the initiation of palaeotechnics in Scotland, John Roebuck (1718–94). Be that as it may, Roebuck deserted

[1] J. D. Comrie, *op. cit.*, pp. 317 and 361.

[2] T. Thomson, *History of Chemistry*, Vol. 1, p. 313 *et seq.*; W. Ramsay, *Life and Letters of Joseph Black*, *passim*.

medicine for chemistry and started sulphuric acid works at Birmingham (1746) and Prestonpans (1749). One of the principal markets for Roebuck's acid was the rapidly expanding bleaching industry of Scotland. Credit for publicizing it as a souring agent usually goes to Home, but Thomas Thomson, in his *History*, infers that Home's publication plagiarized work carried out by Cullen.[1] The results of his researches were published in 1756 and he was awarded a premium of £100 by the Board of Trustees who encouraged the researches.

About the same time Home published an important work in quite a different field, the *Principles of Vegetation.* This work is a pioneer in the field of vegetable physiology. It earned the author the gold medal of the recently founded Edinburgh Society for Encouragement of Arts, Manufacture and Commerce. In the history of the prevention of infection and the preservation of food, it is noteworthy that Home recommended boiling drinking water as a precautionary measure against epidemics, particularly for armies in the field.[2]

Daniel Rutherford's contribution was the discovery of nitrogen, which he announced in a thesis *De Aere Fixo Dicto, aut Mephitico* submitted to the University of Edinburgh in 1772 just before the great flux of pneumatic chemistry. Although professorially a botanist, Rutherford was stimulated to carry out his chemical researches by Black.

The interest that Cullen and Black took in problems appertaining to heat has been mentioned, so it is not surprising to find that Rutherford also made a contribution to the same subject. Problems of heat require instruments for measuring temperature, and Daniel Rutherford added to existing scientific instruments by the invention of a maximum and minimum thermometer.[3] The next step was a continuous recording thermometer. This was also devised by an Edinburgh man, Alexander Keith.[4]

In the same social context the basis of spectrum analysis was laid by Thomas Melvill, a young Edinburgh graduate.

> Melvill's pioneer experiments in spectrum analysis may be described very briefly in his own words. He mixed various salts with burning spirits, and 'having placed a paste-board with a circular hole in it between my eye and the flame of the spirits, in order to diminish and circumscribe my object, I examined the constitution of these different lights with a prism (holding the refracting angle upwards)'. He noted which colour predominated

[1] T. Thomson, *op. cit.*, Vol. 1, p. 93. [2] J. D. Comrie, *op. cit.*, p. 317.
[3] *Edin. Phil. Trans.* 1794, *3*, 247. [4] *Ibid.*, 1798, *4*, 203.

in each case, and observed that in the case of sea-salt a bright yellow light (now known to be characteristic of sodium) was predominant in a striking manner, and formed in the prism a sharp image of the aperture through which the flame was viewed. 'Because the hole appears thro' the prism quite circular and uniform in colour; the bright yellow . . . must be of one determined degree of refrangibility; and the transition from it to the fainter colour adjoining, not gradual, but immediate'.[1]

In the history of technology few associations have had such consequences as the association of Joseph Black, John Roebuck, and James Watt. Black left Glasgow for Edinburgh in 1766. In his new surroundings he must have become acquainted with John Roebuck, if indeed he had not already done so, because Roebuck was becoming known as one of the most enterprising industrialists in Scotland at that time. His success as a vitriol manufacturer at Prestonpans was assured and he was striving hard to get his other foundation, Carron Iron Works, on its feet. Alkali requirements of embryonic palaeotechnics were increasing, and Black was no doubt familiar with the pressing economic situation. There is evidence that he and James Watt, who showed considerable aptitude as a chemist, gave considerable thought to the problem and it is probable that Roebuck, as a successful industrial chemist, was co-opted. Their joint investigations did not end in commercial production, at least as far as is known, but they were evidently vitally interested in possible developments and hotly contested an attempt by Alex. and George Fordyce to acquire privileges through a Parliamentary petition. Watt, on behalf of himself and Black, and Garbett, Roebuck's partner, both lodged protests in the interests of their friend James Keir, who had already developed a process for alkali manufacture. Perhaps it is fortunate that Watt did not become an alkali manufacturer, since he might have been diverted from that problem, the solution of which was his main contribution to the advance of technology.

When Black announced his discovery of the principle of latent heat, he and Watt were together at the University of Glasgow. In latent heat was the clue to the uneconomical operation of the Newcomen engine. Watt meditated on the wasteful successive heating and cooling of the cylinder of Newcomen's engine, and to effect the much desired economy devised his separate condenser. But the subsequent development from the laboratory scale model to anything of a size of practical utility was to be long and

[1] A. Wolf, *History of Science*, etc., XVIII cent., p. 171.

arduous. The capital required was far in excess of the command of a university instrument maker. Black stepped in and supplied funds sufficient to develop Watt's project. By 1768 Watt was in his debt to the extent of £1,200. Roebuck heard of Watt's activities from Black, and appreciated that in the separate condenser engine lay the solution of the problem of draining certain coal mines that he had taken over to make Carron self-sufficient in raw materials. Roebuck paid up Watt's debts to Black and invited Watt to build a full-scale experimental engine at Kinneil, near Bo'ness. This Watt did successfully, and an arrangement was made to patent the invention. Roebuck was to have a two-thirds share in the patent in exchange for facilities and accommodation given to Watt, but when Roebuck himself failed in 1773 his share was bought by Matthew Boulton (1728–1809) of Birmingham. So was founded the famous partnership of Watt and Boulton, but we must not forget that the university professor, Joseph Black, and the industrial chemist, John Roebuck, were the foster-fathers of Watt's invention. Together they laid the corner stone of palaeotechnic economy.

Black's interests extended in many directions. During his professorship at Edinburgh, one of the first chemical societies in the world was founded. Indeed, Edinburgh and Glasgow both had *Chemical Societies*. In Ramsay's biography of Black there is mentioned a list of members of a *Chemical Society* existing in 1785. The personnel contained in the list has been analysed by James Kendall, of Edinburgh. Of Black's fifty-nine chemistry students, nineteen graduated in medicine at Edinburgh between 1784 and '90. Of these it is important to note that but three were Scottish, three were English, the rest Irish.[1] If the proportion of non-Scottish students attending lectures at Edinburgh University was normally as high as this it is a remarkable tribute to the fame of Black's teaching. In both Edinburgh and Glasgow chemical societies formed part of a student's facilities for gaining practical experience. Practical teaching was not as a rule provided by the University, and at least some of the societies enabled enthusiastic students to 'repeat the professor's experiments'. Also, most of the societies were provided with libraries of some magnitude, which rather suggests that there was a lack of adequate provision on the part of the parent institutions.

The 'eighties of the eighteenth century saw a spate of societies founded. Many were exclusively medical, as is natural in an expanding medical school, but non-medical interests were not far

[1] James Kendall, *Old Chemical Societies in Scotland, Chem. and Ind.* 1937, *15*, 141.

behind. In Edinburgh there was a *Wernerian Society* (*c.* 1786), the *Royal Physical Society*, a *Natural History Society*, as well as the above *Chemical Society*. Many of the small societies, having served their original purpose, were merged into the *Physical Society* which was incorporated by royal charter in 1788.

Perhaps the most outstanding name in the *Chemical Society* list is that of Dr. Thomas Beddoes (1760–1808), founder of the Pneumatic Institute, Bristol. Beddoes links up scientific schools in various centres in Britain. He studied in Edinburgh under Black and became lecturer in chemistry in Oxford. Impressed with the plethora of gases discovered in the second half of the eighteenth century, he conceived the idea of using some of them medicinally, and established a Pneumatic Institute for inhalation therapy. Boulton and Watt made apparatus for him.[1] Black, Darwin, Wedgwood, Kirwan, and Ingenhousz were subscribers to the venture. Gregory of Edinburgh declined.[2] Thus we have evidence of further cross connexions between Edinburgh, the industrial Midlands, and Bristol, where Beddoes established his Institute, and he strengthened his connexion with the Birmingham philosophers by marrying Anne, the daughter of Richard Lovell Edgeworth. But there is an even more important link. James Watt's business connexions with Cornwall, and the residence of his son Gregory there for health reasons, brought the Watts into contact with a boy who showed great enthusiasm for chemistry. To set him upon a chemical career the Watts recommended him to Beddoes as an assistant; he was Humphry Davy. At the Pneumatic Institute Davy, among other things, researched on nitrous oxide and investigated its anaesthetic properties.

We must, however, return to explore Black's influences through the appointment of his students to influential positions in the academic and industrial world. In the chemical world there is no doubt that Black was to Edinburgh what Boerhaave was to Leyden. Not only did his pupils, Robison, Irvine, Hope, Cleghorn, and Thomson follow him in his lectureship at Glasgow and chair at Edinburgh, but some of his pupils founded chemical schools throughout the world: Smithson Tennant developed the backward Cambridge school, Morgan and Rush founded the teaching of chemistry at Philadelphia. Ogilvie went to Aberdeen, Garnett to the Andersonian at Glasgow. There is in fact almost no end to the list, and those whom he must have influenced at second hand, that is were students of Black's pupils, are legion. From

[1] M. Boulton, 1st Dec., 1794, A.O.L., Birmingham.

[2] T. Beddoes to J. Watt, 9th and 20th Jan., 1794, B.R.L.

these men many present-day scientists may trace their scientific genealogy through Black to Boerhaave.

After Black left Glasgow for Edinburgh in 1766, the vacant lectureship in chemistry was filled by Dr. John Robison (1739–1805), one of his own students, and friend of the Wedgwoods. For three years the teaching of chemistry was in his hands, but he became better known subsequently as professor of natural philosophy at Edinburgh. Robison was succeeded in 1769 by William Irvine (1743–87). Irvine also was a pupil of Black and was associated with Black in the latter's researches on latent heat. Irvine died in 1787 and was succeeded by Thomas Charles Hope (1766–1844), son of the professor of botany at Edinburgh. Hope only occupied the chemistry lectureship for four years before transferring to the chair of medicine, but his interest in research maintained the reputation of the Glasgow school built up by Cullen and Black. Of particular interest was his continuance of researches on heat which characterized the school. While in Glasgow he discovered the maximum density of water. On his translation to medicine he was succeeded by Dr. Robert Cleghorn (1755–1821) and for a time, under pressure of extra-mural activities, original chemical work in Glasgow was discontinued.

Cleghorn had not the leisure to engage in original research, nor, for that matter, the right type of mind. Nevertheless he was a lucid and interesting lecturer. He continued to lecture with acceptance on chemistry till an independent chair was founded in 1818, but it was a great misfortune for Glasgow that she was unable to attract and hold a professor of chemistry who would have more ably followed the experimental tradition of Cullen and Black, especially as the social context was one to which a scientist interested in technology should have responded with enthusiasm.[1]

During either Irvine's or Hope's lectureship, some time *c.* 1786, a *Chemical Society* was founded in Glasgow. In the biographical note on Dr. John Thomson which prefaces his *Life, Lectures and Writings of William Cullen, M.D.*, there occurs the following passage:

> At the beginning of the winter session of 1788–9 Mr. John Thomson went to Glasgow to attend the medical classes. . . . He also joined a Chemical Society, which contained several members who afterwards attained great eminence as practical chemists.

[1] J. D. Comrie, *op. cit.*, pp. 361 and 363; Alex. Duncan, *Memorials of the Faculty of Physicians and Surgeons of Glasgow, passim.*

The *Society* is referred to in the Charles Macintosh *Memoir*.[1] It appears to have met in the University, and been subject to university discipline. We learn something of its activities in a letter from Wm. Couper to Charles Macintosh. Macintosh was then in Holland.

> Some convulsions in our Society, which are now happily over, induced me to put off writing till I saw their termination. Owing to a regulation of the College Faculty, every Society existing within the College was under the necessity of acquainting the Principal with the nature and intention of their meetings, and of sending him a list of the members. This, though it appeared to most of us a thing simple, and easy to be submitted to, had a different effect on others; and we lost Mr. Candlish, Mr. Tilloch, and Mr. Crawford.[2]

The Alexander Tilloch (1759–1825) referred to was founder of the *Philosophical Magazine* (1797), and editor of the *Star* newspaper from 1789–1821. Among the other members were John Wilson of Hurlet, Kirkman Finlay's brother John, and Dr. John McLean, subsequently professor of chemistry in the College of Princeton. It is possible that opposition from the University authorities may have led to this Glasgow society disbanding. Ramsay, in his *Life of Joseph Black*, mentions that his grandfather, also William Ramsay, was president of a *Chemical Society* founded in Glasgow in 1798. Extracts from the second minute-book of this Society found recently in the rooms of the *Royal Philosophical Society* were published by Forsyth J. Wilson.[3]

The fact that Glasgow became the industrial capital of Scotland is reflected in its extra-academical institutions. The *Glasgow Chamber of Commerce* was the pioneer in Great Britain. It was founded in January 1783, while Edinburgh's Chamber followed in December 1785. Glasgow's first president was Patrick Colquhoun, founder of the Dumbarton Glass Co. The following were among the members, many of whom are mentioned in other chapters:

> *Humphrey Barbour*, bleacher, etc. *George Boyle*, of the Glasgow Tan Works, the Eastern Sugar House, and the Cudbear Works. *James Brown* who invented glass bubbles (hence his soubriquet *Bubbly*) for testing the strength of spirits. *John Brown*, of Brown, Carrick and Co., bleachers. *John Buchanan*, agent in Glasgow for Richard Arkwright. *Richard Collins*, paper maker and bleacher.

[1] Geo. Macintosh, *Memoir*, p. 6.

[2] Wm. Couper to Charles Macintosh, 14th Dec., 1786, quoted by G. Macintosh, *op. cit.*, p. 6.

[3] *Annals of Science*, 1937, *2*, 451.

James Findlay of James Findlay and Co. *John Glassford* of the Glasgow Tan Works, Dyeing and Calico Printing Co., and the Cudbear Works. *John McGregor*, bleacher at Clober, whose daughter married James Watt. *Peter Murdoch*, sugar refiner. *John Semple*, bleacher. *Thos. Stewart*, bleacher. *John Tennent*, who sold his market garden ground to *Charles Tennant* for St. Rollox Chemical Works.[1]

Among the initial activities of the Chamber of Commerce were the organization and fostering of the cotton industry, the development of turkey red dyeing, calico-printing, and the improvement and extension of chemical manufactures.[2]

Although Glasgow had a Chamber of Commerce before Edinburgh, it was not till much later that a *Philosophical Society* was formed in Glasgow. The history of its first few years will be found in its *Transactions*,[3] in an article by Andrew Fergus. The *Society* was founded in 1802, and among the original subscribers were the Monteiths, Macintoshes, Birkbeck, Mushet, Syme, Tennant, etc. Also a Dr. James Watt, who was probably Watt of Boulton and Watt, Soho. The early discussions of the Society centred round problems of mechanics, coal-gas, etc.

There was no retiring age for Scottish professors in the eighteenth century, so when Black's health began to fail in 1795, Thomas Charles Hope, then holding the chair of medicine at Glasgow, was appointed joint professor. He became sole professor in 1799, the third Glasgow-trained chemist to hold the chair in the capital. While professor of medicine in Glasgow he studied minerals from the lead mines at Strontian in Argyll, and in 1793 sent to the Royal Society of Edinburgh a paper entitled *An Account of a Mineral from Strontian and of a Peculiar Species of Earth which it contains*. This was the culmination of work done in 1791–2 of which preliminary announcement was made in 1792 to the Glasgow College Literary Society. Thus Hope anticipated Klaproth's distinction of strontia from baryta. His contribution to the Scottish Report on Brewing will be referred to when we are dealing with Thomas Thomson.

The discovery of the new element strontium did not long remain in academic isolation. Lime ($Ca(OH)_2$) was largely used in the purification of sugar, and we find that before long the similar element strontium was experimented with in Glasgow by Dr. David Boswell Reid (1805–63), who was assistant to Hope at one time. Reid was one of the first to have a laboratory publicly

[1] Geo. Stewart, *Curiosities of Glasgow Citizenship*, p. 169.
[2] *Ibid.*, p. 140. [3] *Trans. Glas. Phil. Soc.*, Vol. 13, p. 1.

open to students for engaging in practical chemical operations.

Under Hope's influence the development of chemistry was rapid and of increasing economic importance. In some ways Hope, on account of his contacts, occupies a place of equal importance with Black, because an increasingly large proportion of industrialists were in a position to benefit by contact with the universities. The popularity of chemistry with all classes in Edinburgh became so great that Hope sometimes had five hundred students attending his lectures, while outside the University the interest was every bit as great. In 1826 Hope began giving popular lectures to the public. He continued to lecture till 1844, when he was succeeded by Dr. William Gregory, as independent professor of chemistry, fully quarter of a century later than the foundation of an independent chair in the more highly industrialized city of Glasgow.

Further evidence of the increasing interest in chemistry is the fact that, about the beginning of Hope's Edinburgh professorship, the existing *Natural History Society* was turned into a *Chemical Society* at the instigation of Dr. John Thomson (1765–1846), whose membership of the *Glasgow Chemical Society* has been referred to. John Thomson was an extra-mural teacher of Chemistry in Edinburgh in the last decade of the eighteenth century. Among the members of Thomson's society were Lords Brougham and Lauderdale. On one occasion the whole of the latter's service of plate was converted into a battery for a galvanic experiment performed before the society.

At the beginning of the nineteenth century, therefore, no less than four chemical societies had been founded in the scientific schools of Scotland. The next chemical society in the world, in point of time, was the *Philadelphia Chemical Society*. It was virtually the daughter of the Edinburgh society founded in 1785. Both John Morgan and Benjamin Rush, the first lecturers in chemistry at Philadelphia, were Edinburgh pupils of Black.[1]

All these factors contributed to build up a powerful chemical fraternity in Scotland, and it is little wonder that Sir John Sinclair wrote in 1814:

> At present there are a greater number of intelligent practical chemists in Scotland, in proportion to the population, than perhaps in any other country in the world.[2]

In the light of Sinclair's statement it is opportune to examine other centres of chemical research and teaching in Britain.

[1] Jas. Kendall, *Endeavour*, Vol. 1, No. 3, *passim*.

[2] Sir J. Sinclair, *General Report*, App. 2, p. 307.

Instead of correcting the perspective, as might be expected, it will be found, however, to throw Scottish achievement into even greater relief. As far as universities are concerned, there are only two to be considered, Cambridge and Oxford. As one would expect, Cambridge passed through the same alchemistic phase as did Scotland. Indeed, alchemical tradition persisted there longer than it did in Scotland, where rising interest in medicine and its adjuncts proved a fortunate corrective to the labours of gold-making. For example, Newton was still studying Boyle's method of gold-making from 1690–3. It was during that decade that he was appointed Warden of the Mint. In the early days of the eighteenth century, i.e. about the time that James Crawford became professor of chemistry at Edinburgh, the title of Honorary Professor of Chemistry at Cambridge was conferred on one J. F. Vigani (*c.* 1650–1713), a native of Verona. Of Vigani we know little, but he was probably the first chemist there to throw off the alchemical tradition. From records of purchases made to illustrate his lectures it is highly probable that they were biased towards pharmaceutical ends. It is interesting to note that one of Vigani's students was Stephen Hales (1677–1761), whose researches on the chemical reactions of plants laid the foundation on which Francis Home of Edinburgh was able to build his *Principles of Vegetation.*

Vigani was followed by John Waller, who lectured till 1718, and Waller in turn by John Mickleburgh, or Mickleborough, who brings us up to the middle of the century. If we compare the number of students reading chemistry at Edinburgh and at Cambridge at this period, it is likely that Edinburgh will be found to have the smaller number, but an expansion took place almost immediately. In the light of these figures we can sympathize with Davies, who wrote to Stephen Hales in 1759 lamenting of Cambridge that:

> Anatomy, botany, chemistry, and pharmacy have been but occasionally taught; when some person of superior talents has stayed up and has honoured the University by his first display of them, before his passage into the world.[1]

In Scotland at this date Cullen was teaching in Edinburgh and Black in Glasgow.

Mickleburgh gave way to John Hadley, and he to Richard Watson, subsequently Bishop of Llandaff, who was appointed professor of chemistry at Cambridge in 1764. At the time of his appointment it was said of Watson that 'he knew nothing at all

[1] R. Davies to S. Hales, 1759.

of chemistry, had never read a syllable on the subject, nor seen a single experiment in it'. In two years the illustrious Black was to succeed Cullen at Edinburgh. Small wonder that the evolution of chemical science in the two countries was so different. Yet Watson was no idle churchman. He took his new appointment seriously, and has related how at one period his conscience forced him to burn his chemical writings lest he be lost to the church altogether. His *Essays*, first published in 1781, contain useful pictures of various industries, particularly on coal, lead and zinc, and his researches on charcoal production for gunpowder by closed distillation of wood are known to have saved the Government large sums of money.

Although there is a great contrast between the evolution of chemical teaching in England and Scotland, it must not be overlooked that chemistry was then regarded not only as an adjunct of medicine, but of natural philosphy as well. Thus, it was I. Milner (1750–1820) who was Jacksonian professor of natural philosophy, and not a chemist as such, who, in 1788, observed that when ammonia is passed over heated manganese dioxide it is converted into red fumes which dissolve in water to form nitric acid. This oxidation of ammonia, now effected catalytically by atmospheric oxygen, is the basis of the modern commercial method of preparing nitric acid.

Both Milner and his successor F. J. H. Wollaston (1762–1823) taught chemistry and published plans of their courses.

Richard Watson was followed by Isaac Pennington, and he in 1793 by W. Farish, who, like Cullen in Glasgow, lectured on the *Application of Chemistry to the Arts and Manufactures of Britain*. In Farish's lectures we see a swing towards an appreciation of the important contribution that chemistry was making to the industrial revolution. They covered smelting metallic ores, the uses of coal, such industrial chemicals as sulphur, alum, salt, acids, and alkalis, the chemical arts of bleaching and preparing cloth, and the production of mordants, etc. This highly practical approach heralded the further break with tradition, namely the appointment of a chemist trained in the Scottish schools to the Cambridge chair. In 1813 Smithson Tennant (1761–1815), who had been in Cambridge since 1782, was appointed to the vacant chair of chemistry. With this insinuation of Black's student into Cambridge we can go on to the consideration of Oxford, after mentioning that it was Tennant's basic researches that made possible the invention of the miner's safety lamp.[1]

[1] R. H. Gunther, *Early Science in Cambridge, passim.*

On cursory examination it appears that, as a centre of chemical activity, Oxford has a better claim than Cambridge. This arises in the main from its association with the alchemistic Roger Bacon, nevertheless R. H. Gunther says that 'it is a moot point whether Roger Bacon really made much impression on his contemporaries; if any, it was evanescent; and in the succeeding centuries Oxford savants continued to wander in a maze of arbitrary figments and partial inductions, in which experimental science found no place'.[1] In the middle of the seventeenth century, we find that for a period Oxford did indeed give hospitality to one of the greatest of contemporary thinkers, Robert Boyle (1627–91). For fourteen years from 1654 Boyle was at Oxford. While there he became the centre of a small coterie of intellectuals who may have helped to bring about what J. U. Nef calls the first English industrial revolution. Boyle's influence was rather that of a patron experimentalist than a teacher, but he was responsible for introducing into Oxford the first regular teacher of practical chemistry. It was a long time, however, before the teaching of chemistry became continuous. In compensation for the paucity of chemical instruction, an important contribution to technics made by an Oxford B.C.L. may be mentioned.

> The mystery of salt-glazed stone ware was discovered by the ingenious John Dwight of Christ Church, who set up a manufacture at Fulham. . . . When and where John Dwight became acquainted with this use of salt is not known, but in 1671 he took out a patent for his process, and in the same year the first specimens of salt-glazed ware were being manufactured at Fulham. Soon after 1688 similar ware was being produced at Burslem by the Dutchman Elers, and in 1700 in Nottingham.[2]

Oxford suffered from its proximity to London, and the removal of Boyle (as well as many other intellectuals, following the more settled conditions of state established in 1660) did irreparable damage to its scientific life. The only scientist worthy of the name who remained was John Mayow. But Mayow too left Oxford in 1675, and died in 1678.

A succession of chemists followed, in a laboratory founded by Elias Ashmole (Robert Plot, Edward Hannes, John Freind, Richard Frewin), but none of them succeeded in establishing any sustained teaching or research school.

> The reason for this sterility was not far to seek. The Oxford contemporaries of Newton had not the enquiring mind; the

[1] R. H. Gunther, *Early Science in Oxford*, p. 7. [2] *Ibid.*, p. 27.

> most brilliant of her sons devoted their genius to other ends and developed their talents in other places; those who stayed behind were content to accept the statements of others without testing them for themselves, and to pass on to their students information acquired at second-hand. The business of teaching was set higher than the duty of research.[1]

Ashmole's inadequate foundation was, from the chemical point of view, a failure and chemistry continued to lag behind other expanding sciences. No university professor was appointed, with the result that students who wanted to acquire some familiarity with the science had no one better to instruct them than the college fireman.

The only interesting outcome of the Ashmolean period is the association of John Wall (1708–76) of Merton with the foundation of the Worcester Porcelain Company (1751), but such a connexion cannot be considered adequate to compensate for the new low level to which Oxford intellectual life sank in the earlier part of the eighteenth century. The Wall connexion with Oxford was continued when John's son Martin Wall (1747–1824) delivered a course of lectures there from 1781 in the capacity of Public Reader in chemistry.

The next development in the chemical history of Oxford is of greater interest, corresponding as it does with the appointment of Smithson Tennant to Cambridge: it was the appointment of Thomas Beddoes (1760–1808), also a student of Joseph Black's at Edinburgh, to be reader in chemistry. Beddoes was only at Oxford from 1788 to 1793, but for a time at least chemical interests there were stirred up by his enthusiasm, and it is on record that such was the interest he created that attendance at his lectures exceeded anything known in the university since the thirteenth century. Here in Oxford was the vivid effect of Black's infectious personality re-enacted. Beddoes' short readership in chemistry was but a phase in his life, and the Pneumatic Institute at Bristol which he founded and his nurturing of the young Humphry Davy have been referred to already.

It cannot be said that Beddoes' short sojourn at Oxford led to any permanent chemical efflorescence, but in 1803 a professorship of chemistry was endowed, and, with the establishment of regular teaching, something that had been in progress in Edinburgh for almost a century, we can, after a glance at the Continent, return to consider further development in the Scottish schools and activities associated with them.[2]

[1] R. H. Gunther, *Early Science in Oxford*, p. 53.

[2] *Ibid.*, p. 1, *et seq.*

The marked difference between England and Scotland makes it desirable to look for a moment at the situation on the Continent, with which Scotland still kept up a fruitful intercourse. On the Continent, as in Scotland, chemistry first emerged as an independent science from under the wing of a variety of disciplines; medicine, anatomy, botany, mineralogy, or physics. It was in France that it first achieved real independence. In 1770 the chair of medicine in the *Collège de France* was converted into a chair of chemistry, and five years later Jean d'Arcet was appointed professor. Like Cullen he broke with academic tradition by lecturing not in Latin but in the vulgar tongue, and was granted the additional privilege of being permitted to lecture without donning academic robes. D'Arcet was the pupil and son-in-law of G. F. Rouelle (1703–70), master also of the celebrated Lavoisier. In view of what has been said of the industrial connexions of the Scottish chemists, d'Arcet's connexion with industry is also worthy of note. He was technical director of the Sèvres porcelain works, and an inspector in the dye-house of the Gobelin tapestry factory.[1]

Scottish chemists figured conspicuously in the industrial application of the element chlorine discovered by the Swedish chemist Scheele in 1774. It will be remembered that industrial use in Scotland followed quickly upon information about it being introduced by Patrick Copland and James Watt. It was only towards the end of the century, when it was firmly established as a bleaching agent, that its chemical composition began to create interest. Much of the 'chlorine controversy' took place in Edinburgh between Dr. John Murray (*d.* 1820), lecturer on natural philosophy, chemistry, and *materia medica*, who adhered to the original view that chlorine was a compound of oxygen, and Dr. John Davy, brother of the illustrious Sir Humphry. Part of the confirmatory research was carried out in Hope's laboratory in the presence of Sir Humphry when he was on his tour of Scotland in 1812. The Davys' view that chlorine was an elemental substance was confirmed. The argument can be followed in *Nicholson's Philosophical Journal*.[2] It was John Davy who first observed the combination in the presence of sunlight of chlorine (Cl_2) and carbon monoxide (CO) with the formation of phosgene ($COCl_2$), a reaction of great importance to the synthetic dye industry and in the technology of war.

$$Cl_2 + CO = COCl_2$$

[1] *Ciba Review*, 1937, *1*, 132.

[2] *Nicholson's Phil. Journal*, 1811–13. Vols. 28, 32 and 34,

Chlorine was also tried as a disinfectant, particularly in the cholera epidemic which struck Paisley in 1832.

> Public health had not been studied much in those days, but something had to be done, and the bright idea developed in Paisley of fumigating the houses with diluted sulphurous acid and chloride of lime. No time was to be lost in carrying out the necessary measures, so six large tubs filled with the mixture were placed on barrows, and, preceded by torches, these were wheeled through the streets and lanes of Paisley. No wonder that this weird procession making its way through the narrow streets made a deep impression on the onlookers, who opened their windows to admit the pungent fumes of chlorine.[1]

There can be little doubt that extra-mural teaching and the like did much to stimulate the development of chemistry in the various Scottish centres. Perhaps the very existence of such activities outside the universities was symptomatic of need for continued evolution. That was certainly true in the particular case of the Andersonian Institution of Glasgow. Dr. John Anderson (1726–96) was Professor of Natural Philosophy in the University. He was a fervid supporter of the French revolutionary movement. Indeed he presented a fire-arm mechanism to the National Convention in 1791. His lectures in Glasgow were among the first to spread technical knowledge among the working classes. These facts alone suffice to demonstrate the dissimilarity of Anderson's outlook from that of the majority of his colleagues, with whom he frequently quarrelled over unduly delayed reforms. The final outcome of his differences with the University was that he conceived the idea of leaving his slender fortune to found an institution to rival the mediaeval University of Glasgow. This was the origin of the so-called Andersonian University, which evolved in time into the Royal Technical College. On paper Anderson's scheme far outstripped the provision that he made for bringing it into being. Nevertheless in 1796 Dr. Thomas Garnett (1766–1802) was appointed to deliver a course of 'Physical Lectures' covering chemistry and physics. Other chemists who made conspicuous contributions, both to the advance of their science, and to the popularization of its teaching, followed as the college expanded. They included George Birkbeck (1776–1841), Andrew Ure (1778–1857), Thomas Graham (1805–69), and William Gregory (1803–58).[2]

When Birkbeck left Glasgow he went to London, and founded

[1] E. Haldane, *The Scotland of our Fathers*, p. 271.
[2] J. D. Comrie, *op. cit.*, p. 516.

the institution which still bears his name. His lectures, which were free to working men, were of great popularity, and his audience often ran into many hundreds. His activities in association with Lord Brougham led to the foundation of Mechanics' Institutes. He was also associated with the foundation of University College, London, where the second professor of chemistry was Thomas Graham, referred to above.

One important industrial inventor who received his training in the Andersonian Institution was James Beaumont Neilson (1792–1865). From starting work in a colliery, Neilson became manager of the recently formed Glasgow Gas Works, a situation which he found very congenial.

> In the first place, it afforded him facilities for obtaining theoretical as well as practical knowledge in chemical science, of which he was a diligent student at the Andersonian University, as well as of natural philosophy and mathematics in their higher branches. In the next place it gave free scope for his ingenuity in introducing improvements in the manufacture of gas, then in its infancy. He was the first to employ clay retorts; and he introduced sulphate of iron as a self-acting purifier, passing the gas through beds of charcoal to remove its oily and tarry elements. The swallow-tail or union jet was also his invention.[1]

Taking up again the story of the development of teaching and research in the University of Glasgow during a period when an increasing number of Glasgow manufacturers came to depend on advancing chemical knowledge, we find Cleghorn in occupation of the lectureship till 1818. Cleghorn was not a researcher, but during his occupancy of the chair the general trend of chemistry was more and more away from medicine towards arts and manufactures. This is confirmed by the fact that a professorship independent of medicine was founded in 1818. Although the new chair went to Dr. Thomas Thomson (1773–1852) who was a pupil of Black's and as such perhaps a favoured candidate, there is evidence of an attempt to fill the chair with the English industrial chemist William Henry (of Henry's Law), son of Thomas Henry, secretary and president of the Manchester Literary and Philosophical Society. Dr. Brown of Miller Street, Glasgow, wrote Henry and informed him of a vacancy on the staff of the University and suggested that Henry should become a candidate. Here is Henry's reply, dated 18th August, 1817:

> Though I estimate most highly its value in this way, as well as conferring scientific rank, yet I am compelled, by a variety of

[1] S. Smiles, *Industrial Biography*, p. 152.

> circumstances which for some years to come must bind me to this spot, to forego all intentions of proposing myself for a situation which, in almost every respect, would have been more agreeable to my taste and habits than the sphere in which I am now moving. . . . As my family is a pretty large and increasing one, I have made great exertions to extend my chemical manufacture. These exertions have been attended with all the success I could have expected, and from one time to another I have been induced to lay out sums of money in buildings and utensils, the aggregate of which is to me of serious moment. I am of opinion, therefore, that it will be more consistent with my interest and happiness to persevere steadily in the course I am now pursuing than to yield even to so strong a temptation as that which you hold out.

He went on to make pertinent remarks on the change which had taken place in the position of chemistry in the universities during the earlier part of the industrial revolution. He continued:

> Now that the office is vacant, would it not be advisable to require that the future holder of it should not practise medicine. In the course of the lectures which I formerly gave, I found employment for my whole time as long as they lasted. . . . I cannot conceive even moderately extensive medical practice to be otherwise than incompatible with daily lectures on any experimental science.[1]

Incompatibility of medical practice with professorial duties was one of the principal shortcomings of Cleghorn's tenure of office. The University was evidently aware of this, and his successor, Thomas Thomson, was appointed to profess chemistry alone.

Thomas Thomson (1773–1852), like so many contemporary chemists, was a man of wide interests. He contributed to the second supplement of the *Encyclopaedia Britannica*. The periodic appearance of editions of this famous Encyclopaedia is an index of the association of science and letters in Scotland. The first edition in three volumes appeared from 1768–71 under the editorship of William Smellie (1740–95), Keeper of the Natural History Museum in Edinburgh. When the third edition was published in 1788–97, it had been expanded to eighteen volumes. Four years later a supplement appeared containing valuable contributions on scientific subjects by Dr. John Robison (1739–1805), formerly lecturer in chemistry at Glasgow and later Professor of Natural Philosophy at Edinburgh. In the same way the work was kept up to date by the publication of a supplement to the fourth, fifth,

[1] Quoted by K. Leowenfeld, *Manchester Lit. and Phil. Soc.*, 1913, 57, No. 19.

and sixth editions. This supplement contained elaborate articles on aerostatics by Sir John Leslie (1766–1832), gas lights by Henry Creighton, chemistry by Wm. Thos. Brande (1788–1866), professor of chemistry at the Royal Institution, and on arts and manufactures by Thomas Thomson. The eminently practical nature of the contributions of the Scottish professoriate should be remarked. Another of Thomson's interests was the history of Chemistry. In the more practical field he invented the oxy-hydrogen blowpipe, and followed up the Daltonian atomic system with the introduction of chemical symbols. The beneficial result of freedom from medical practice is very evident in Thomson's case. He had the opportunity of stimulating his students to engage in original research, and attracted large numbers to his laboratory. So great was the influx that a new department was required for them. It was opened in 1831 and is believed to have contained the first modern laboratory in the world devoted to chemical research.[1]

One of the most distinguished of Thomson's pupils was Thomas Graham (1805–69). Graham also studied under Hope at Edinburgh. For two years from 1828 he was an extra-mural teacher of chemistry in Glasgow before entering the Andersonian Institution for a period of seven years. He was one of Scotland's most illustrious chemists, who, during his lectureship at the Andersonian, researched on the diffusion of gases and discovered the law of gaseous diffusion which keeps his memory green. During the same period he extended his diffusion experiments to cover the passage of substances in solution through membranes, and so laid the foundation of colloid chemistry.

In 1837 Graham was appointed professor of chemistry in the recently founded University College of London, and subsequently taught there with great distinction. Clark of Aberdeen was an unsuccessful candidate for the post. Two of Graham's pupils merit passing mention, although their operative lives are outwith the period under discussion; the first was James (Paraffin) Young (1811–83), the founder of the Scottish shale industry; the second was Joseph Lister (1827–1912) who, after studying in London was advised by Graham to continue his studies with James Syme (1799–1870) at Edinburgh. Although a surgeon, Syme is of interest in the context of social technology. He was a member of the *Chemical Society* that flourished during Hope's professorship, and in his experiments at the age of eighteen discovered a solvent for india-rubber and the waterproofing properties of the solution.

[1] J. D. Comrie, *op. cit.*, p. 522.

His discovery was applied in industry by Charles Macintosh, who patented the process for rendering cloth waterproof with rubber solution.[1]

The only Scottish chemical teaching outside Edinburgh and Glasgow during the period of industrial revolution was in the Universities of Aberdeen. Although there was no chair of chemistry in either of the Universities there till 1793, some teaching was engaged in earlier, both extra-murally and at King's College. In the latter, William Ogilvie included chemistry in his course on natural history. He was a Glasgow student of Black. Aberdeen became conspicuous as a centre of interest in applied chemistry in 1787 through the activity not of a chemist, but of the professor of natural philosophy in Marischal College, Patrick Copland (1748–1822). Copland's contribution to bleaching is dealt with in Ch. IX. He was a pioneer social technologist, the recipient of grants from the Board of Trustees for an industrial museum, and may have included chemistry within his domain till the separate teaching of chemistry was established. Extra-mural teaching began in Aberdeen in 1783 and two years later the course given by the extra-mural lecturer, Dr. George French (1765–1833), was officially recognized by Marischal College. Of particular significance was the fact that French directed his lectures towards the application of chemistry to arts, manufactures, and agriculture. In this the institution of chemical teaching in Aberdeen is distinct from that in Edinburgh and Glasgow.

In 1793 Barbara, widow of Principal Thomas Blackwell, left an endowment of £40 per annum to provide a chair of chemistry in Marischal College. French, who was their nephew, became its first incumbent. According to J. D. Comrie, he 'so organized his lectures that the subject in his hands had nothing to do with medicine'. French held the chair till his death in 1833. He was succeeded by Dr. Thomas Clark.[2]

Thomas Clark (1801–60) continued the non-medical attitude of his predecessor. At the age of fifteen he became a clerk in the counting-house of the extensive industrial chemical concern, Charles Macintosh and Company, of Glasgow. But the actual chemical operations of the Macintosh firm must have interested him, for he next became chemist to Charles Tennant of St. Rollox, a position one would judge of considerable responsibility. Thus he was for ten years in contact with chemical manufacture in

[1] J. D. Comrie, *op. cit.*, p. 593; *Annals of Philosophy*, 1818, *12*, 112.

[2] *Ibid.*, p. 552; A. Findlay, *Teaching of Chemistry in the Universities of Aberdeen* p. 4.

Glasgow during the period of consolidation which took place between the efflorescence at the end of the eighteenth century, and the new flux associated with the development of alkali manufacture. From industrial chemistry Clark passed to teaching in the Mechanics' Institute, which occupation he apparently found congenial. He decided to make teaching his life's work. After formal study at Glasgow University he was appointed in 1833 to the chair at Aberdeen rendered vacant by the death of French. While a student at Glasgow, his mature understanding of chemistry was such that he produced a continuous flow of original contributions, particularly on applied chemistry. He was chemical consultant to a Staffordshire company, and an expert witness in the J. Beaumont Neilson hot-blast *cause célèbre* in the Court of Session. He is best remembered, however, for his contribution to water technology. According to Thomas Graham, his process for softening water (1841) was 'the most consummate example of applied science known in the whole circle of the arts'.

Although Clark was only able to lecture for eleven years, he occupied the Marischal College chair till 1860. In his active years, however, the number of students studying chemistry increased greatly, so as to necessitate improved teaching accommodation. The emoluments of the chair were also put on a more satisfactory basis during his professorship. In 1843 Clark became ill, and his lectures had to be given by a series of substitutes, John Shier from 1843–45, Andrew Fyfe (Professor of Chemistry, King's College) from 1845–47, John Smith from 1847–52, J. S. Brazier from 1852–60. Clark retired in 1860 and was succeeded in turn by Fyfe and Brazier.

The remaining Scottish professor of chemistry who calls for discussion is Dr. William Gregory, the last of the brilliant descendants of John Gregorie of Aberdeenshire. William Gregory (1803–58) followed the family tradition of graduating in medicine, but decided at an early date to make not medicine but chemistry his career. He was inspired to do so by hearing a lecture by T. C. Hope. Having studied medicine at Edinburgh, to extend his studies he went to Giessen where Justus von Liebig had established a celebrated continental school of chemistry. At Giessen he formed a lasting friendship with his master Liebig and helped to introduce his teaching into this country. Gregory returned to Edinburgh in 1832 and joined D. Boswell Reid as an extra-mural lecturer. During the ensuing decade he moved through the various scientific schools of the country: the Andersonian Institution in 1837,

Dublin soon after, King's College Aberdeen, 1839, and finally succeeded Hope in the chair at Edinburgh in 1844.

Gregory succeeded to the chair formerly occupied by Cullen, Black, and Hope at a time when a great unilateral expansion was taking place in chemistry. William Murdock's invention of gas lighting had been commercialized in the opening decades of the century by Boulton and Watt of Birmingham. Charles Macintosh patented a tar-burning furnace in 1826. Municipal gas lighting flooded the market with by-product tar, for which there was no demand, but in 1834 Runge discovered that it contained aniline: in 1842, Leigh similarly discovered benzene. In 1844 Hofmann began the systematic investigation of the chemical nature of coal tar. Chemistry was expanding rapidly and amassing vast stores of information about compounds of carbon. Organic chemistry, freed from the impediment of vitalism, was itself revitalized by exploration of the new vista which was opened up by the expansion of the new gas industry and the chemical examination of its by-products. This side of chemistry interested Gregory, and half his chemical lectures, no longer cluttered up with caloric and other imponderables, was devoted to it.

Gregory did another service to British science. At Giessen Liebig was a penetrating, if somewhat hasty, thinker and writer. By translations, Gregory made his ideas, techniques, and speculations known to the British public. Among the most important translations were: *Familiar Letters on Chemistry*, *Animal Chemistry*, *Researches on the Chemistry of Food*, *Chemistry in its Applications to Agriculture*, *Instructions for the Chemical Analysis of Organic Bodies*, *Agricultural Chemistry*.[1]

We must now, in conclusion, turn to the wider aspects of the influence of the scientists and technologists so far discussed. During the period of academic brilliance that coincided with Cullen's and Black's occupancy of the chair at Edinburgh, a group of men in England laid some of the cornerstones of palaeotechnic civilization. Some were English by birth and education; some were Scots, but the intellectual power behind their achievements was the direct offshoot of the cultural renaissance in eighteenth-century Edinburgh. Taken *en masse* this group probably represents the highest concentration of Fellows of the *Royal Societies* that has ever been associated at one time with any industrial undertaking. The social context of this industrial miscelle was Birmingham industrialism, and for convenience it may be referred to as the *Lunar Society*. For a period of years when the *Royal Society of London* and

[1] A. Findlay, *op. cit.*, p. 49.

the Universities of Oxford and Cambridge were characterized by their lethargy, and the *Royal Society of Edinburgh* was struggling with the millstone of polite learning that was attached to its neck at birth, the *Lunar Society*, a small, unofficial, private colloquium of philosophers, did much by mutual stimulation to bring to greater prominence and utility the application of chemistry to arts and manufactures. In H. C. Bolton's *Scientific Correspondence of Joseph Priestley*, the following are reckoned to have constituted the stable nucleus of the Society, to which from time to time, visiting savants were added: Matthew Boulton, Erasmus Darwin, William Small, Thomas Day, R. L. Edgeworth, William Withering, James Watt, James Keir, William Murdock, Joseph Priestley, Samuel Galton Senior and Junior, R. A. Johnson, Dr. Stoke, and John Baskerville. More than half of them had Scottish connections, and most were interested in the application of chemistry. In addition to meeting regularly, they kept up intellectual and social intercourse with Henry Moyes, Josiah Wedgwood, John Smeaton, Joseph Banks, Wm. Hershel, J. A. Solander, Joseph Black, and J. A. de Luc. Smiles tells us of the genesis of the *Lunar Society* in the following citation:

> Towards the close of the last century, there were many little clubs or coteries of scientific and literary men established in the provinces, the like of which do not now exist—probably because the communication with the metropolis is so much easier, and because London more than ever absorbs the active intelligence of England, especially in the higher departments of science, art, and literature. The provincial coteries of which we speak, were usually centres of the best and most intelligent society of their neighbourhoods, and were for the most part distinguished by an active and liberal spirit of inquiry. Leading minds attracted others of like tastes and pursuits, and social circles were formed which proved in many instances the source of great intellectual activity, as well as enjoyment. At Liverpool, Roscoe and Currie were the centres of one such group; at Warrington, Aikin, Enfield, and Priestley, of another; at Bristol, Dr. Beddoes and Humphry Davy of a third; and at Norwich, the Taylors and Martineaus of a fourth. But perhaps the most distinguished of these provincial societies was that at Birmingham, of which Boulton and Watt were among the most prominent members.
>
> From an early period the idea of a Society, meeting by turns at each other's houses, seems to have been entertained by Boulton. It was probably suggested in the first place by his friend Dr. Small. The object of the proposed Society was to be at the same time friendly and scientific. The members were to exchange views

> with each other on topics relating to literature, arts and science; each contributing his quota of entertainment and instruction. The meetings were appointed to be held monthly at the full of the moon, to enable distant members to drive home by moonlight; and this was the more necessary as some of them—such as Darwin and Wedgwood—lived at a considerable distance from Birmingham.[1]

The *Lunar Society* is of especial interest in that it provided both the necessary environment and the financial means for Joseph Priestley to pursue his researches in comfort. After passing through various vicissitudes Priestley settled in Birmingham and was introduced to the *Lunar Society*. His acquisition was no mere lion-hunting on the part of wealthy industrialists with money to spare. Priestley was a dissenter, and shared with some other members of the Society pronouncedly Jacobin views. He had no formal training in chemistry, and so was forced to develop his understanding of the subject on his own. Cut off by his views from emoluments necessary to create an atmosphere suitable for the pursuit of scientific research, the members of the *Lunar Society* gave him the shelter which in more liberal times he might have found in an academic institution. Not only was their friendship stimulating, but they tactfully undertook to subsidize his work with an annual grant of money. Although an informal arrangement, the contributions of Boulton and Wedgwood to Priestley's expenses constitute the pattern for the more eloquent foundations of Carnegie and Rockefeller. Unfortunately the flux of political controversy in the latter decades of the eighteenth century drew Priestley into the maelstrom. At the hands of the Birmingham rioters he suffered materially and mentally. For his safety he left the congenial Midland atmosphere, and after a time settled in America.

Priestley himself was the first to acknowledge the loss of his convivial friends, and lament the sting he felt at the neglect which the schizophrenic Cavendish meted out to him while in London. The important effect which the *Lunar Society* had in furthering the development of pneumatic chemistry at the hands of Priestley is illustrated by the following letter which he wrote to Withering.

> One of the things I regret most in being expelled from Birmingham is the loss of your company, and that of the rest of the *Lunar Society*. I feel that I want the spur to constant exertion which I had with you. My philosophical friends here are cold and distant. Mr. Cavendish never expressed the least concern on

[1] S. Smiles, *Boulton and Watt*, p. 367.

account of anything that I had suffered, though I joined a party with which he was, and talked with them some time. I do not expect to have much intercourse with any of them.[1]

He was right about the coldness with which he was to be received. The situation did not improve, and again he wrote:

> The *Lunar Society* was of irreplaceable advantage to me, and I am not able to replace it here.[2]

In another direction altogether, the *Lunar Society* was important —it formed the pattern for other more lasting societies, particularly the *Literary and Philosophical Society* at Newcastle-on-Tyne. In the Boulton Collection in the Assay Office Library, Birmingham, there is a letter from George Chalmers, Mayor of Newcastle, enclosing the plans of a *Literary Society* established there on the lines of the *Lunar Society*. It was Chalmers, too, who introduced Henry Cort to Matthew Boulton.[3]

Bowden discusses the part played by the *Manchester Literary and Philosophical Society* in the foundation of the Manchester College of Arts and Science (1783). Also, according to Wadsworth and Mann, papers by Wilson, Charles Taylor, and Henry, presented to the *Literary and Philosophical Society*, mark the beginning of the scientific method in the dyeing and finishing trades, and lectures on dyeing, bleaching, and calico-printing, held under the auspices of the Society, were the beginning of technical education in Manchester.[4]

Never before in history was there such an advantageous syncretism of pure science and advancing industry as in the *Lunar Society*. Boulton, the prince of industrial capitalists, wrote that he had 'annihilated Wm. Murdock's bedchamber, having taken away the floor, and made the chicken kitchen into one high room covered over with shelves, and these I have filled with chemical apparatus. . . . Next year I shall annex to these a laboratory with furnaces of all sorts, and all other utensils for dry chemistry'.

Phlogistians and anti-phlogistians dined at the same table: Watt and Priestley vied with Cavendish and Lavoisier over the discovery of the composition of water: Keir achieved the manufacture of alkali where Watt, Black, and Roebuck had failed: Murdock's gas-light illuminated the mills powered by Boulton and Watt's engines: Wedgwood supplied chemical utensils for

[1] Clapton, 2nd Oct., 1792, B.R.L. [2] Clapton, 15th April, 1793, B.R.L.

[3] G Chalmers to Matthew Boulton, 11th June, 1784; Henry Cort to Matthew Boulton, 3rd June, 1784.

[4] W. Bowden, *Industrial England*, p. 45 *et seq.*

Priestley's experiments: Priestley in turn analysed minerals of possible utility in pottery: a leaven of wit and philosophy was added by Darwin and Edgeworth. Indeed there was not an individual, institution, or industry with pretensions of contact with advancing technology throughout the length and breadth of the land, but some member of the *Lunar Society* group had connexions with it. Behind many of them was the inspiration of the environment in which they had passed the formative years of their adolescence, the Scotland of Cullen, Black, and Hope, faithful guardians of embryonic science who remained in the north while their pupils moved south to plant and nurture in English soil the seeds of the chemical revolution and reap their often rich reward.

APPENDIX TO CHAPTER XXV

TEACHERS IN THE MEDICAL AND CHEMICAL SCHOOLS OF EDINBURGH

	University	*Extra-Mural*
MEDICINE	1685 Archd. Pitcairne	
	1724 Wm. Porterfield	1724 Andrew St. Clair
	1726 And. St. Clair	John Rutherford
	John Rutherford	
	1747 Rob. Whytt	
	1766 John Gregory	1770 And. Duncan
	1773 William Cullen	Gregory Grant
	1790 James Gregory	1782 John Aitken
	1821 James Home	
CHEMISTRY AND MEDICINE		1697 A. Monteith
	1713 Jas. Crawford	
	1726 And. Plummer	1724 And. Plummer
	John Innes	John Innes
	1755 William Cullen	
	1766 Joseph Black	1776 And. Duncan
		Chas. Webster
		1780 John Brown
		1782 John Aitken
	1795 Thomas Chas. Hope	1790 Wm. Nelson
CHEMISTRY		1832 John Murray
		William Gregory
		1834 D. B. Reid
		1835 And. Fyfe
	1844 William Gregory	1845 William Reid
		George Wilson
		Thomas Anderson
	1858 Lyon Playfair	1856 Stevenson Macadam
MATERIA MEDICA	1768 Francis Home	
	1798 John Home	
	1821 And. Duncan (II)	
	1832 Rob. Christison	

(From J. D. Comrie, *History of Scottish Medicine.*)

TEACHERS IN THE UNIVERSITY AND ANDERSONIAN INSTITUTION, GLASGOW

	University	*Andersonian*
MEDICINE	1751 William Cullen	
	1756 Rob. Hamilton	
	1757 Jos. Black	
	1766 Alex. Stevenson	
	1789 Thos. C. Hope	
	1796 Robert Freer	
	1827 Chas. Badham	1828 Alex. Hannay
ANATOMY	1742 Rob. Hamilton	
	1756 Jos. Black	
	1757 Thos. Hamilton	
	1781 Wm. Hamilton	
	1790 Jas. Jeffray	1799 John Burns
	Lecturers	
CHEMISTRY	1747 William Cullen	
	1756 Joseph Black	
	1766 John Robison	
	1769 William Irvine	1796 Thomas Garnett
	1787 Thos. C. Hope	
	1791 Robert Cleghorn	*Physical Science*
		1796 George Birkbeck
	Professor	1804 Andrew Ure
	1818 Thomas Thomson	
		1830 Thomas Graham
		1837 William Gregory
		1839 Fred. Penny
MATERIA MEDICA	1766 William Irvine	
	1787 Thos. C. Hope	
	1788 Robert Cleghorn	
	1791 Rich. Millar	

(From *Memorials of the Faculty of Physicians and Surgeons of Glasgow, 1599–1850* by Alex. Duncan.)

GLOSSARY OF DEAD CHEMICAL LANGUAGE

IN the atmosphere of social change which characterized the end of the eighteenth century, the traditional alchemistic nomenclature which had been carried into chemistry was transformed. As is not surprising, the change was brought about through the collaboration of four French chemists.

During the development of eotechnics, names were given to materials used by the industries in the era, e.g. copperas, oil of vitriol, sal-ammoniac, etc.—traditional names which were quite unrelated to the chemical properties of the substance concerned. They often depended on accidents of discovery, or the names of the chemists who first isolated them, e.g. Epsom and Glauber's Salts.

In the second half of the eighteenth century, confusion arising from the use of such names attracted attention, and some simplified generic names were introduced. Thus similarity in chemical properties led to the adoption of group-names like calxes (oxides), pyrites (sulphides), vitriols (sulphates), and nitres (nitrates). In 1782, T. Bergman, a Swedish chemist, published a systematic nomenclature which, although confined to salts, and based on the Latin language, was an advance on the previous arbitrary nomenclature. A fundamental systematization dates from 1780, when Guyton de Morveau (1737–1816) took charge of the chemical articles in the *Encyclopédie Méthodique*. Editorial duties brought to his notice the confusion existing in chemical nomenclature, and in 1782 he published a contribution to reformed terminology. His ideas were based on the following principles:

> A chemical name should not be a phrase, it ought not to require circumlocutions to become definite; it should not be of the type 'Glauber's salt', which conveys nothing about the composition of the substance; it should recall the constituents of a compound; it should be non-committal if nothing is known about the substance; the names should preferably be coined from Latin or Greek, so that their meaning can be more widely and easily understood; the form of the words should be such that they fit easily into the language into which they are to be incorporated.

In the 'eighties the battle between the 'phlogistians' and their opponents was being fought out under the leadership of Antoine Laurent Lavoisier (1743–91). De Morveau was involved in the battle, and was won over by the reformers who were seeking to bring chemical experiment and theory into line. On 18th April, 1787, the reformers brought to the notice of the French Academy the fact that a reformed and perfected nomenclature for chemistry was overdue. On 2nd of May following, details of such a system were presented to the Academy by de Morveau. This system, with but few alterations, is still in use, and credit for its excellence is almost entirely due to the genius of de Morveau. The following are the salient points of the reformed nomenclature:

> All substances were divided into two classes: elements and compounds. The former, that is the materials that had not yet been divided into simpler substances, were divided into five groups: (1) those that approached nearest to a state of simplicity such as oxygen, hydrogen; (2) the acid-forming elements, such as sulphur, phosphorus, carbon; (3) the metals; (4) the earths; (5) the alkalis. The old names for the metals and some of the non-metals were retained; Lavoisier's names for the other non-metals, oxygen, hydrogen, azote (nitrogen) were introduced.
>
> Compounds were classed as binary or ternary. Binary compounds included the acids which were assumed to be made up of two elements, oxygen (the acid-former) and a non-metal. The names were composed of two words, one (acid) common to all, the other specific for each acid (carbonic acid, sulphuric acid, etc.).

The recommendations of de Morveau and his associates were published in book form (*Méthode de Nomenclature Chemique*) in Paris in 1787, and speedily translated into many languages. J. St. John's English translation appeared in 1788. They were quickly adopted by young chemists, though older chemists were almost universally prejudiced and opposed to them. This is revealed by a letter from M. Berthollet to James Watt, written from Paris on 9th November, 1788:

> Nous sommes bien flattés du jugement que vous portez de la nomenclature que nous avons proposée: un suffrage tel que le vôtre compense bien les plaisanteries et les mauvaises discussions qu'elle nous a attirées, quoique nous n'avons pu avoir en vue que de faciliter l'étude de la chimie, et de porter plus d'exactitude dans les idées par celle de langage. Les caractères dont vous vous servez sont réellement plus commodes que les nôtres. Que n'avons nous pu vous consulter! Nous l'aurions fait avec autant de confiance que d'empressement.[1]

[1] J. P. Muirhead, *Origin and Progress*, Vol. 2, p. 225.

The reformed language of chemistry proved of the utmost utility and remains to-day very much as it was originally formulated by de Morveau, Lavoisier, and their collaborators Antoine François Fourcroy (1755–1809) and Claude Louis Berthollet (1748–1822). From the historian's viewpoint, it has one disadvantage. Nomenclature in use to-day only dates from the French Revolution. Thus to interpret developments in the eighteenth century it is necessary to be familiar with a now dead scientific language, ignorance of which places a barrier in the way of tracing the development of social chemistry in the era of industrial revolution. To this end, the following short glossary is appended. The two columns give the terms by which the same thing was known in the eotechnic and palaeotechnic economies. Some of the eotechnic terms, it might be noted, are still in use in industry.

Eotechnic	*Palaeotechnic*
Air, An . . .	Substance in gaseous state
Air, dephlogisticated .	Oxygen (O_2)
Air, fixed . . .	Carbon dioxide (CO_2)
Air, hepatic . . .	Hydrogen sulphide (H_2S)
Air, inflammable . .	Hydrogen (H_2)
Air, marine acid . .	Hydrogen chloride (HCl)
Air, mephitic . .	Carbon dioxide (CO_2)
Air, phlogisticated .	Nitrogen (N_2)
Air, vital . . .	Oxygen (O_2)
Alkali, concrete volatile .	Ammonium carbonate ($(NH_4)_2CO_3$)
Alkali, common mineral .	Sodium carbonate ($Na_2CO_3.10H_2O$)
Alkalis, caustic . .	Hydroxides ($-OH$)
Alkali, fossil . . .	Sodium carbonate
Alkali, marine . .	Sodium carbonate
Alkali, mild . .	Carbonates ($-CO_3$)
Alkali, vegetable, fixed .	Potassium carbonate (K_2CO_3)
Alkali, vegetable, mild .	Potassium carbonate
Alkali, volatile . .	Ammonia (NH_3)
Alum	Double salt of Aluminium sulphate with Sodium, Potassium, or Ammonium Sulphate
Ash, black . . .	Impure Sodium carbonate
Ash, pot . . .	Potassium carbonate
Ash, pearl . . .	Potassium carbonate
Aqua fortis . . .	Nitric Acid (HNO_3)
Azote	Nitrogen (N_2)
Barilla . . .	Impure Soda extracted from soap-wort
Bittern . . .	Solution of Magnesium Salts
Bleaching powder . .	Calcium chloro-hypochlorite ($CaOCl_2$)
Brimstone . . .	Sulphur (S)

Eotechnic	*Palaeotechnic*
Calcination . . .	Oxidation
Calx	Oxide (—O)
Carbonic Acid . .	Carbon dioxide
Ceruse . . .	Lead carbonate ($PbCO_3$)
Chalk	Calcium carbonate ($CaCO_3$)
Colcothar . . .	Ferric oxide residue in sulphuric acid manufacture (Fe_2O_3)
Copperas . . .	Ferrous sulphate ($FeSO_4.7H_2O$)
Creech. . . .	Calcium sulphate ($CaSO_4$)
Dephlogisticated marine acid	Chlorine (Cl_2)
Earth, calcareous, caustic	Calcium hydroxide ($Ca(OH)_2$)
Earth, calcareous, mild .	Calcium carbonate
Earth, magnesian, mild .	Magnesium carbonate ($MgCO_3$)
Earth, silicious . .	Silica (SiO_2)
Galena . . .	Lead sulphide (PbS)
Glauber's salt . .	Sodium sulphate ($Na_2SO_4.10H_2O$)
Hepars . . .	Sulphides (—S)
Kelp	Impure soda from seaweed
Lime, carbonate of .	Calcium carbonate
Lime, slaked . .	Calcium hydroxide ($Ca(OH)_2$)
Lime, quick . . .	Calcium oxide (CaO)
Limestone . . .	Calcium carbonate
Litharge . . .	Lead oxide, yellow (PbO)
Lye	Aqueous extract of ashes
Manganese . . .	Manganese dioxide (MnO_2)
Marine Acid. . .	Hydrochloric Acid (Aqueous HCl)
Minium . . .	Lead Oxide, red (Pb_3O_4)
Muriatic Acid . .	Hydrochloric acid
Muriates . . .	Chlorides (—Cl)
Nitre	Potassium nitrate (KNO_3)
Oil of vitriol . . .	Sulphuric acid (H_2SO_4)
Oxymuriatic acid .	Chlorine (Cl_2)
Plaster of Paris . .	Calcium sulphate ($(CaSO_4)_2.H_2O$)
Pyrites . . .	Ferrous (or iron) sulphide (FeS_2)
Pyroligneous acid . .	Crude acetic acid from wood
Saccharum saturni .	Lead acetate (PbA_2)
Sal-ammoniac . .	Ammonium chloride (NH_4Cl)
Sal enixum . . .	Potassium sulphate (K_2SO_4)
Sal mirabili . . .	Sodium sulphate ($Na_2SO_4.10H_2O$)
Saltpetre . . .	Potassium nitrate (KNO_3)
Sandriver . . .	Scum on glass pot
Soda	Sodium carbonate
Spanish white . .	Bismuth oxychloride (BiOCl) or oxynitrate ($BiONO_3$)

Eotechnic	*Palaeotechnic*
Spirit of salt . . .	Hydrochloric acid
Spirit of wine . .	Ethyl alcohol (C_2H_5OH)
Sugar of lead . .	Lead Acetate
Sulphurets . . .	Sulphides (—S)
Trona	Natural sodium carbonate ($Na_2CO_3.NaHCO_3$)
Vitriol or Vitriolic acid .	Sulphuric acid (H_2SO_4)
Vitriols . . .	Sulphates ($-SO_4$)
Vitriol, blue . . .	Copper sulphate ($CuSO_4.5H_2O$)
Vitriol, green . .	Ferrous (or iron) sulphate ($FeSO_4.7H_2O$)
Vitriol, white . .	Zinc sulphate ($ZnSO_4.7H_2O$)
White lead	Lead carbonate ($2PbCO_3.Pb(OH)_2$)

CHEMICAL CHRONOLOGY

Seventeenth Century:

1610 Foundation of pneumatic chemistry (Van Helmont).
1613 Mining with gunpowder.
1616 First manufacture of window-glass at Newcastle (Mansell).
1624 Patent law protecting inventions, in England.
1635 Micro-organisms (Leeuwenhoek).
1650 Cobalt blue (Schürer).
Air-pump (O. von Guericke).
1660 Relation between gaseous volume and pressure (Boyle).
Royal Society founded in London. *cf.* 1783.
1661 Ban prohibiting importation of foreign dyes removed.
Chemista Scepticus (Boyle).
Methyl alcohol (Boyle).
1662 First publication on dyeing in English (Petty).
1663 Georgical Committee of Royal Society.
1664 Artificial silk (Hooke). *cf.* 1740.
1665 *Phil. Trans.* of Royal Society.
1667 Thomas Shirley observes gas given off by coal.
1668 Relation between combustion and respiration (Mayow).
Phosphorus discovered (Brand).
1669 Calico-printing (Hooke).
1670 Rock salt discovered in Cheshire (Marbury).
Tinplate at Pontypool (Capel Hanbury). *cf.* 1728.
1674 *Anatomy of Vegetables* (Grew).
1678 Coal substituted as fuel in alum manufacture.
1680 Salt-glazing in addition to lead-glazing.
Food preservation by bottling.
Yeast (Leeuwenhoek). *cf.* 1836.
1681 John Houghton's *Collections*.
1682 Patent for pumping brine (Marbury).
1683 Industrial Exhibition at Paris.
1688 Distillation of gas from coal (Clayton and Boyle).
1690 First textile printing factory in England at Richmond (Grillet).
1694 Patent for distilling coal (Ecle).
1695 Epsom salts isolated (Grew). *cf.* 1829.
1697 *Husbandry Anatomised* (Donaldson).
1699 *Mémoires de l'Académie des Sciences.*

Eighteenth Century:

1700 Importation of printed textiles prohibited.
1702 Boric acid (Homberg).

1704 Prussian blue discovered (Diesbach). *cf.* 1724.
1708 Wet sand iron casting (Abraham Darby I).
1709 Coke used in blast-furnaces (Darby).
Hard porcelain (Böttger).
1710 White dipped earthenware.
1714 Mercury thermometer (Fahrenheit).
1718 Paper pulp made with *Hollander*.
c. 1720 Start of Scottish kelp manufacture.
1720 Sulphuric acid made in England (Drebbel).
Vinegar from alcohol by Orleans or Boerhaave process. *cf.* 1823.
Calico Act prohibiting use of printed calicoes.
1723 Society of Improvers in Knowledge of Agriculture.
Flint added in earthenware manufacture (Astbury).
1724 Preparation of Prussian blue disclosed (Woodward).
1726 Gas from coal (S. Hales).
Flint grinding patent (Benson).
1727 Photosensitivity of silver nitrate (Schulze). *cf.* 1802, 1839.
Vegetable Staticks (Hales).
Board of Trustees for Manufactures, Fisheries, and Improvements in Scotland.
1728 Rolled tinplate (John Hanbury).
1729 Textile printing introduced into Scotland.
1730 Lead compounds introduced in glass-making.
1732 Bleaching with kelp introduced into Scotland (Holden).
1733 *Horse Hoeing Husbandry* (Tull).
1736 Commercial manufacture of sulphuric acid (Ward). *cf.* 1746 and 1749.
Potash distinguished from soda (Duhamel). *cf.* 1807.
India-rubber introduced into Europe. *cf.* 1791 and 1822.
Pale ale successfully bottled for India trade.
1740 Artificial silk (Bon).
Harlem Dye Company, Glasgow.
Cast steel (Huntsman).
1742 Gas patent (Betton).
c. 1744 Porcelain manufactured at Chelsea, Stratford-by-Bow, Derby, Worcester.
1744 Porcelain patent (Heylyn and Frye).
Definition of a *Salt* (Rouelle).
1745 Alleged proof of spontaneous generation (Needham).
1746 Gas patent (Haskin).
British Linen Company incorporated.
Sulphuric acid at Birmingham (Roebuck and Garbett). *cf.* 1749.
1747 Turkey Red dyeing at Rouen. *cf.* 1785.
Sugar in beet juice (Marggraf). *cf.* 1779.
1748 Copperas at Hartley (Delaval).

1749 Sulphuric acid at Prestonpans.

1750 Pure platinum prepared (Watson). *cf.* 1800.

Alizarin from European madder.

Papier mâché (Baskerville).

Vellum paper (Baskerville).

c. 1750 Pottery firing in two stages (Booth).

1752 Window-glass first manufactured in Scotland.

1753 Copperas at Hurlet by Liverpool company.

1754 Carbon dioxide (Black).

Alumina distinguished from lime (Marggraf). *cf.* 1827.

1755 Kaolin discovered (Cookworthy).

1756 Hydraulic cement manufacture (Smeaton).

Experiment on bleaching with sulphuric acid (Home).

1758 Cudbear manufactured (G. and C. Gordon). *cf.* 1771.

1759 Carron Iron Company founded (Roebuck, Garbett and Cadell).

Beginning of pottery research (Wedgwood).

1760 Gas from Coal (Dixon).

Basaltes ware (Wedgwood).

1761 Power-operated blast-furnace (Smeaton).

1762 Discovery of latent heat (Black).

Thermometer in brewing (Combrune).

1763 Queen's Ware (Wedgwood).

First Exhibition of the Industrial Arts, Paris.

Coke ovens.

Iron plough (Small).

1766 Specific gravity of hydrogen (Cavendish).

Black suggests experiments with Hydrogen-filled bladder.

Madder culture introduced into France (Altken).

Attempt to make alum in Scotland (Nicholson and Lightbody).

Reverberatory furnace (Brothers Cranage).

1767 Coal gas (Watson).

1768 Distillation of coal (Sultzbach).

Porcelain patent (Cookworthy).

1769 Separate condenser engine (Watt).

Water-frame (Arkwright).

Discovery of tartaric acid (Scheele).

1770 Attempted 'corner' in alum (Colebrook).

Sulphur dioxide (Priestley).

Spinning-jenny (Hargreaves).

1771 Cudbear at Glasgow (Macintosh).

Alkali manufacture established (Keir).

Sulphuric acid patent (Roebuck and Garbett).

Company of the Cast-plate Manufactory.

1772 Nitrogen isolated (Rutherford).

Manganese isolated (Scheele).

1773 Coke ovens at Carron Iron Works.
Urea isolated (Rouelle). *cf.* 1828.

1774 Chlorine discovered (Scheele). *cf.* 1785 and 1787.
Oxygen discovered (Priestley—Scheele).
Lime distinguished from baryta (Scheele).
Jasper ware (Wedgwood).

1775 *Experiments on Different Kinds of Air* (Priestley).
Phosphorus prepared from bone-ash (Scheele).
Académie des Sciences prize for manufacture of soda. *cf.* 1782 and 1790.

1776 Carbon monoxide (de Lassone).

1777 Theory of combustion (Lavoisier).
Research on glazes (Delaval).

1779 Alkali patent (Shannon).
Beet sugar manufactured (Achard).
Glycerine discovered (Scheele).
Spinning-mule invented (Crompton).

1780 Effect of alkalis on oil observed (Berthollet).
Lactic acid from milk (Scheele).

1781 Alkali patent (Higgins).
Phosphoric taper.
Experiments with hydrogen and soap bubbles (Cavallo).
Alcohol analysed (Lavoisier).

1782 Practicability of balloons demonstrated (J. M. and J. E. Montgolfier).
Alkali patent (Collison).
Soda works at Croisac: reduction of Na_2SO_4 by coal and iron (Morveau and Carney).
Duty on soap used by bleachers removed.
Absorption of carbon dioxide (Senebier).
Wedgwood orders Watt engines.
Systematic nomenclature (Bergman).

1783 Royal Society of Edinburgh founded.
Hot-air balloon (Montgolfiers).
Hydrogen balloon released from Champ de Mar, Paris.
Combining of hydrogen and oxygen to form water (Watt, *et al.*).
Alcohol and carbon dioxide by fermentation (Lavoisier).
Cod-liver oil used medicinally.
Iron manufacture patent (Onions).

1784 Balloon ascent in Great Britain (Tytler).
Experiment with coal-gas-filled balloon (Thysbaert). *cf.* 1821.
Balloons used for scientific observations (Jeffries).
Puddling process in reverberatory furnace (Cort).
Saccharometer in brewing (Richardson).

1785 Lecture theatre at Louvain lighted by gas.
Methane and ethylene (Berthollet).

Theory of combustion complete (Lavoisier).
Chlorine as bleaching agent (Berthollet).
Bleachers' Licence at £2 proposed (Pitt).
Cylinder-printing introduced (Bell).
Turkey-Red manufacture established (Macintosh).

1786 Details of Turkey-Red process published (Henry).
Sal ammoniac and lead acetate (G. Macintosh).
Briquet phosphorique.

c. 1786 Thermometric control of dye-bath (Gott).

1786–93 Monge, Berthollet, and Vandermonde on metallurgy.

1787 New chemical nomenclature (De Morveau, *et al.*).
Law of gaseous expansion (Charles). *cf.* 1802.
Society of Arts prize for food preservation. *cf.* 1808.
Loom (Cartwright).
Flax-spinning by machinery (Kendrew and Porthouse).
Bleaching by chlorine in Scotland (Copland and Watt).

1787 Tan bark substituted for dung in Dutch Process (Fishwick).

1788 Alteration in still design due to 28 Geo. III, *c.* 45. *cf.* 1817.

1789 Copperas manufacture at Walker (Barnes and Foster).
Annales de Chimie.

1790 Manufacture of soda from salt (Leblanc). (Projected 1787 and patented 1791.)
Zinc white (Courtois). *cf.* 1848.
Baryta distinguished from strontia (Hope). *cf.* 1808.
Voltaic electricity discovered (Galvani).

c. 1790 Phosphatic fertilizers (Dundonald).

1791 Patent for waterproofing with rubber dissolved in turpentine (Peel).
Gas engine (Barker).
Neutralization of acids and bases (Richter).

1792 Rags for paper bleached with chlorine (C. and G. Taylor).
Paper machine (Robert). *cf.* 1803.
Gas lighting (Murdock).

1793 Sulphuric acid made in America (Harrison).
Aluminium acetate manufactured (Macintosh).
Military use of balloons.

1795 Soda works established at Dalmuir.
French Government prize for food preservation.
Food canning (Appert).
Ecole Polytechnique.

1796 Soda works started at Walker (Losh). *cf.* 1808.
Lead carbonate patent (Dundonald).
Natural cement (Parker).
Carbon disulphide (Lampadius).

1797 St. Rollox Works started (Tennant).
Alum manufacture at Hurlet (Macintosh).
Alcohol engine (Cartwright).

Chromium discovered (Vauquelin).
Report on soap (d'Arcet, Lelièvre, and Pelletier).
1798 Bleach liquor patent (Tennant).
Experiments on hot blast (Sadler). *cf.* 1828.
1799 Bleaching powder manufactured (Tennant).
Enamelled hollow-ware (Hickling).
Prize for earthenware (French National Institute).
Anaesthetic properties of nitrous oxide demonstrated (Humphry Davy). *cf.* 1824.
Picric acid (Welter).
Royal Institution founded.

Nineteenth Century

1800 Blackband ironstone discovered (Mushet).
Lead chromates used in paint.
Galvanic cell (Volta).
Electrolysis of Water (Nicholson and Carlyle).
Loom (Jacquard).
Commercial production of platinum (Cock).
1801 House in Paris lit with gas (Lebon).
White lead (Thénard).
1802 Gas light (Boulton and Watt).
Spectroscopy (Wollaston).
Law of gaseous expansion (Gay Lussac).
Photography (Wedgwood and Davy). *cf.* 1839.
Discharge printing of dyed calico (Monteith).
1803 Paper-making machine (Foudrinier).
Glazes (Proust).
1805 Phillips and Lees' factory lighted with gas.
Oxymuriate or chemical matches (Chancel).
Electro-plating (Brugnatelli). *cf.* 1836 and 1840.
Chlorine liquefied (Northmore).
1806 Cartwright loom perfected.
Law of constant composition (Proust).
Sulphuric acid substituted for ferments in tinplate industry.
1807 Sodium and potassium isolated (Davy).
Law of Multiple Proportions: Partial Pressures: Atomic Theory (Dalton).
Patent for gas-driven automobile (Isaac de Rivaz).
1808 Royal Society of Arts award for food preservation (Saddington).
Calcium and barium isolated (Davy).
Second Alkali Works on Tyne (Doubleday and Easterby).
Alum manufactured at Campsie (Macintosh).
1809 Prize awarded to N. Appert for food preservation.

1810 Patents for tinned food (A. de Heine and P. Durand).
London and Westminster Chartered Gas Light Company.
Ammonia-soda reaction discovered (Fresnel).
Ceramic decoration in platinum and gold (Warburton).
Jahresberichte (Berzelius).
c. 1810 British canning industry founded.
1811 Molecular hypothesis (Avogadro).
Mineral charcoal filter for sugar (Figuier).
1812 Depression in Scottish kelp industry.
Iodine discovered (Courtois).
Bleaching works at Monaghan (Gamble).
Vacuum pan sugar-refining (Howard).
Locomotive (Matthew Murray). *cf.* 1829.
1814 Gum elastic catheters.
1815 Portable Gas Company, London.
Safety Lamp (Davy).
Optical activity of sugars (Biot).
Polytechnic (Vienna).
1816 Experiments on hot-blast furnace (Stirling). *cf.* 1828.
Chemical Manufactory at Dublin (Muspratt and Abbott). *cf.* 1823.
1817 Reduced pressure still (Tritton).
1818 Attempt to use pyrites in sulphuric acid manufacture (Hills).
Water-glass (Fuchs).
Navy contract for tinned meat.
1819 Contract with Glasgow Gas Works for tar liquor (C. Macintosh). *cf.* 1822.
Naphthalene discovered in coal products (Garden).
Isomorphism (Mitscherlich).
1820 Bichromate manufacture (J. and J. White).
Incandescent lamp (De la Rue).
Quinine isolated (Pelletier and Caventou).
Technical Institute (Berlin).
1821 Coal-gas balloon (Green).
Ferrous alloys (Faraday).
Caffeine isolated (Pelletier).
1822 Waterproof patent (C. Macintosh).
Tar distillery (Longstaff and Dalston).
Canning established in Aberdeen (Moir).
First Scientific Congress at Leipzig.
1823 Leblanc process started at Liverpool (Muspratt).
Researches on fats, soap, etc. published. Begun 1811 (Chevreul).
Quick or German vinegar process (Schützenback and Wagemann).
Liquefaction of Gases (Faraday).
Gun cotton prepared (Braconnot).

1824 Canning established in Glasgow (Morrison).
Nitrous oxide (Hickman).
Portland cement (Aspdin).
Volumetric analysis (Gay-Lussac).
Alcoholometry (Gay-Lussac).
1825 Soda-ash manufactured at St. Rollox.
Benzene discovered in oil gas (Faraday).
Plaited wicks in candles.
Annalen (Poggendorf).
1826 Tar-burning furnace (Macintosh).
Bromine discovered (Balard).
Aniline (Unverdorben).
1827 Gay-Lussac Tower at Chauny.
Friction matches (Walker).
Aluminium isolated by means of metallic potassium (Wöhler).
Tax on salt removed.
1828 Alkali works at St. Helens started (Gamble).
Promethean matches (Jones).
Hot blast in iron production (Neilson).
Manufacture of ultramarine (Guilmet and Gmelin).
Synthesis of alcohol (Hennell).
Synthesis of urea (Wöhler).
1829 Lucifer matches (Jones).
Magnesium isolated (Bussey)
Incandescence of rare earth oxides—basis of gas mantle (Berzelius).
Filtration plant for water (Chelsea Water Works, London).
The Rocket (George and Robert Stephenson).
1830 Congreve phosphorus matches (Sauria).
Sheet glass by mechanical process.
Analysis by combustion (Liebig).
1831 Raw coal in hot blast furnace (Dixon).
Patent still (Coffey).
Synthesis of sulphur trioxide (Phillips).
Chloroform isolated (Guthrie, Liebig, and Soubeiran).
1832 *Annalen der Chemie* (Liebig).
1833 Laws of Electrolysis (Faraday).
1834 Hydrocarbons from coal tar (Runge, Faraday, and Laurent).
Carbolic acid discovered (Runge).
Alkali Works at Newcastle (Attwood).
Contact theory of fermentation (Berzelius, Mitscherlich).
Lawes begins experiments on agricultural chemistry.
Journal für praktische Chemie.
1835 Alkali manufacture at Oldbury (Chance).
Constitution of bleaching powder.
Coloured sheet glass.
Guano imports begun.

Comptes Rendus de l'Académie française.
Wöhler's laboratory (Göttingen).
1836 Electro-plating (de la Rive). *cf.* 1840.
Ammonia-soda process tried at Camlachie (Thom).
Recovery of hydrochloric acid (Gossage).
Acetylene discovered (Edmond Davy).
Growth of yeast during fermentation (Cagniard de la Tour).
Duties on paper equalized.
1837 Galvanized iron (Crauford).
Fermentation due to organisms (Kützing).
High pressure steam in food-canning (Hogarth).
1838 Sulphur monopoly. *cf.* 1839.
Dyer and Hemming attempt ammonia-soda process.
Chromium oxide pigment (Pannetier and Binet).
Creosoting of wood (Bethell).
Agricultural Society.
1839 Pyrites used in sulphuric acid manufacture (Farmer).
Leadless enamelled hollow-ware (T. and C. Clark).
Manganese steel (Heath).
Hot vulcanization of rubber (Goodyear).
Callotype (Talbot).
Daguerrotype (Niépce and Daguerre).
1840 Silver-plating in Birmingham.
Incandescent lamp (Grove).
Patent plate glass.
Muspratt and Young attempt ammonia-soda process.
Bunsen's laboratory (Marburg).
1841 Glover Tower in sulphuric acid manufacture.
Iodine first manufactured in Glasgow.
Calcium chloride baths in canning (Goldner and Wertheimer).
Field experiments (Dumas and Boussingault).
Paper positives in photography (Talbot).
Chemical Society.
1842 Synthesis of cyanides (Farmes).
Seybel attempts ammonia-soda process.
Ether as anaesthetic (Long).
Benzene discovered in coal tar (Leigh).
Water gas (Selligue).
1843 Gutta-percha (Montgomery).
Manufacture of superphosphate established (Lawes).
Patent fertilizer (Liebig).
Agricultural Chemical Association.
Synthesis of acetic acid (Kolbe).
1844 Bunsen and Playfair on blast-furnace gases.
Caustic soda used in soap manufacture.
Wood pulp paper (Keller).
Hoffmann investigates coal-tar.

	Cork-and-rubber linoleum (Galloway).
	Nitrous oxide as anaesthetic (Wells).
1845	Alum made from coal shale at Manchester (Spence).
	Last use of Orkney kelp at Normanby's Works.
	Petroleum discovered in Pennsylvania.
	Red phosphorus discovered (Schrötter).
	'Phossy jaw' described (Loringer).
	Resin soaps.
	College of Chemistry (R.C.S. from 1890).
1846	Ether Anaesthesia (Warren and Morton).
	Nitroglycerine (Sobrero).
	Gun cotton (Schönbein).
1847	Chloroform anaesthesia (Simpson).
	Rolled plate glass.
	Collodion (Maynard).
	Synthesis of cyanides (Bunsen and Playfair).
	Collodion films (Niépce).
1848	Modern safety match (Böttger).
	Acetic acid from wood (Kestner and Halliday).
	John Miller and Company, Aberdeen.
	Zinc oxide as pigment on large scale.
	Journal of Chemical Society.
1848–56	Preservation of timber (Burt).
1849	Mansfield investigates coal-tar and manufactures benzene.
1850	Alkali works at Widnes (Gossage).
	Inversion of cane sugar (Wilhelmy).
c. 1850	Mechanical wood pulp.
1851	Collodion in photography (Scott Archer).
	Diffusion laws (Graham).
	Crystal Palace. International Exhibition of Industrial Arts (Joseph Paxton).
	Laboratory at Owens College, Manchester.
1853	Soda wood pulp (Wall and Burgess).
	Science Museum (London).
1854	Coke ovens at Saarbrucken.
	Roll films (Melhuish).
1855	Alkali works at Widnes (Gaskell-Deacon).
	Gossage takes up the manufacture of soap.
	Commercial production of aluminium (Deville).
	Dry-plate photography (J. M. Tampenot).
	Vitalistic theory of fermentation (Pasteur).
1856	Open hearth furnace (Siemens).
	Bessemer converter (Bessemer).
	Perkin discovers mauve. In production, 1857.
	Colour photography (Zenker).

BIBLIOGRAPHY

N.L.S. . . .	National Library of Scotland.
S.L. . . .	Library of Writers to H.M. Signet.
G.R.H.E. . .	H.M. General Register House, Edinburgh.
B.R.L. . . .	Birmingham Reference Library.
A.O.L.B. . .	Library of the Assay Office, Birmingham.
E.P.L. . . .	Edinburgh Public Library.
P.O. . . .	Patents Office, London.
B.M. . . .	British Museum.

DOCUMENTS

(Manuscript and Printed, including Theses in Typescript.)

Address and Proposals from Sir John Dalrymple, Bart., on the Subject of the Coal, Tar and Iron Branches of Trade. (Edinburgh, 1784) (N.L.S.).

Articles of Sugar Copartnery in Edinburgh. (n.d.) (Manchester Reference Library.)

Birmingham Reference Library:

(1) Soho MSS., 50 vols. of copies and about 8,000 original letters.

(2) Timmins MSS.

Birmingham Assay Office Library:

Tew MSS. (Papers of Matthew Boulton of Soho Foundry).

Board of Trustees for Fisheries, Manufactures, and Improvements in Scotland. Volume 96, Administration. (G.R.H.E.)

Board of Trustees, etc.

An Account of the Ashes imported into Scotland from 5th July, 1753 to 5th July, 1758.

Board of Trustees, etc.

Register of Theses on Technical Subjects: Memorial by Dr. William Cullen.

Customs House Returns: Prestonpans. (G.R.H.E.)

Papers of J. Moir and Sons, Ltd. (Founded in Aberdeen, 1822).

Thomas Cochrane's *MS. Notes of Black's Lectures*, in Library of Royal Technical College, Glasgow. *See Annals of Science*, 1936, *1*, 101.

Wm. Cullen's *MS. History of Chemistry*, Royal College of Physicians, Edinburgh. *See Annals of Science*, 1936, *1*, 138.

Documents (Printed) Legal and Other, relating to the suit instituted by the Carron Company versus Samuel Garbett, merchant of Birmingham. 1777–1779. (B.R.L.)

Documents concerning the Sulphur Monopoly, consisting of the Parliamentary Inquiry into the conduct of the Foreign Secretary. (London, 1841) (E.P.L.)

Papers of Archibald Cochrane, Ninth Earl of Dundonald, concerning *The British Tar Company*. (N.L.S.)

Edinburgh Sugar House Co. MS. Minute Book, April 27th, 1752–March 27th, 1763. Account Book, Feb. 18th, 1758–Nov. 22nd, 1759. (E.P.L.)

An Enquiry into the methods that are said to be now proposed in England to revive the Sugar Trade. (London, 1733.) (N.L.S.)

J. Gamble: *Collection of Documents respecting the Claims of L. Robert as the Original Inventor, and of J. Gamble as the First Introducer of the French Paper-machine.* (P.O.)

John Graham: *The Chemistry of Calico Printing, 1790–1835.* (Manchester Reference Library.)

Important Crisis in the Calico and Muslin Manufactory in Great Britain explained. (London, 1788.)

Mayer Collection. (Wedgwood and Bentley Papers.) (10 vols. of letters, 73 boxes and 22 portfolios of commercial documents.) (Liverpool Free Library.)

Memorandum from Orkney to the Lords of the Treasury, 13th Dec., 1822. (N.L.S.)

Memorial of the Glasswork Co. at Leith and the Glasswork Co. at Dumbarton, 1791. (N.L.S.)

Memorial of Manufacturers and Bleachers in Edinburgh stating the Articles on which they were subject to duty and from which the Irish were exempted. (N.L.S.)

Memorial relating to Ash-burning, etc. for the perusal of the Right Hon. the Lord Advocate for Scotland: by Ebenezer McCulloch. (N.L.S.)

Plan of the Literary and Philosophical Society of Newcastle on Tyne. 1793.

Prestonpans Pottery Co. MS. Ledger and Inventory, 1801–02. (E.P.L.)

Reasons for allowing English salt to be brought to Scotland at the Scotch Duty—1793. (N.L.S.)

Register of the Great Seal of Scotland. (G.R.H.E.)

Remarks (Anonymous) on the System of Laws in Scotland relative to Salt. (Edinburgh, 1794.) (G.R.H.E.)

Brief Remarks on the Existing Differences between the Gas Co. and the Inhabitants of Newcastle upon Tyne. (Newcastle, 1818.) (E.P.L.)

Session Papers. (S.L.)

Specification of Patents and Drawings 1767–87. (G.R.H.E.)

The Importance of the Sugar Colonies to Great Britain stated, and some Objections against the Sugar Colony Bill answered. (London, 1731.) (N.L.S.)

The Case of the Panmasters. (N.L.S.)

P. S. Bebbington: *Samuel Garbett (1717–1803).* (Thesis, B.R.L.)

Lysaght, J.: *Chemistry at the Royal Society (1687–1727).* (London University, 1937.)

Mackie, J. L.: *British Sugar Taxation since the Napoleonic Wars.* (London University, 1939.)

White, J. W.: *A Historv of the Phlogiston Theory.* (Thesis, London, 1932.)

HISTORIES OF CHEMISTRY, ETC.

Bauer, Hugo: *History of Chemistry*. (London, 1907.)
Brown, Jas. Campbell: *History of Chemistry*. (London, 1913.)
Comrie, J. D.: *History of Scottish Medicine*. (2 vols.) (London, 1932.)
Findlay, Alex.: *A Hundred Years of Chemistry*. (London, 1937.)
Findlay, Alex.: *The Teaching of Chemistry in the Universities of Aberdeen*. (Aberdeen, 1935.)
Gunther, R. T.: *Early Science in Cambridge*. (Oxford, 1937.)
Gunther, R. T.: *Early Science in Oxford*. (Oxford, 1923.)
Hilditch, T. P.: *Concise History of Chemistry*. (London, 1911.)
Holmyard, E. J.: *Chemistry to the Time of Dalton*. (London, 1925.)
Lowry, T. M.: *Historical Introduction to Chemistry*. (London, 1915.)
Mason, Irvine: *Three Centuries of Chemistry*. (London, 1925.)
Meyer, E. von: *History of Chemistry*. Translated by G. McGowan. (London, 1891.)
Miall, S.: *A History of British Chemical Industry, 1634–1928*. (London, 1931.)
Moore, F. J.: *History of Chemistry*. (New York, 1918.)
Partington, J. R.: *A Short History of Chemistry*. (London, 1937.)
Pledge, H. T.: *Science since 1500*. (London, 1939.)
Sarton, George: *Introduction to the History of Science*. (3 vols.) (Baltimore, 1927–31.)
Thomson, T.: *History of Chemistry*. (2 vols.) (London, 1830.)
Thorpe, T. E.: *Essays in Historical Chemistry*. (1st Edtn., London, 1894.)
Venables, Fr. P.: *History of Chemistry*. (London, 1923.)

SCIENTIFIC PERIODICALS AND PUBLICATIONS OF LEARNED SOCIETIES

Annals of Agriculture and other Useful Arts: collected and published by Arthur Young. (London, 1784–1815.)
Annales de Chimie. (Paris, 1789–1815.)
Annales de Chimie et de Physique. (Paris, 1816–)
Annals of Philosophy. (ed. Thomas Thomson.) 1st Series (London, 1813–20). 2nd Series (London, 1828–36).
Annals of Science. (London, 1936–)
Ciba Review.
Economic History. (Supplement to the *Economic Journal*.)
Economic History Review.
The Edinburgh Philosophical Journal. (Edinburgh, 1819–)
Farmer's Magazine. (With general index.) (Edinburgh, 1800–25.) Publication of same title started in London in 1830s.
A Journal of Natural Philosophy. (By Wm. Nicholson.) 1st series, 5 vols. (London, 1798–1802). 2nd series, (London, 1802–).
The Journal of Sciences and the Arts. (Published Quarterly.) (London, 1817–30.)
Journal of Chemical Education.

Chemistry and Industry. (Journal of the Society of Chemical Industry.)

Medical Essays and Observations by a Society in Edinburgh. (5 vols.) (Edinburgh, 1733–44.) Precursor of Trans. Royal Society of Edinburgh.

Memoirs of the Caledonian Horticultural Society.

Memoirs and Proceedings of the Manchester Literary and Philosophical Society. 1st series. (5 Vols.) (Warrington, 1785–1802.) 2nd series. (Manchester, 1805–)

Mechanics' Magazine.

Museum Rusticum et Commerciale: or, Select Papers on Agriculture, Commerce and Manufactures. By Members of the Society for the Encouragement of Arts, Manufactures, and Commerce. (London, 1764.)

Select Essays on Husbandry extracted from the Museum Rusticum, containing a variety of experiments all of which have been found to succeed in Scotland. (Edinburgh, 1767.)

The Philosophical Magazine. (ed. A. Tilloch.) (London, 1798–)

Philosophical Transactions of the Royal Society. (London, 1665–)

Proceedings of the Philosophical Society of Glasgow.

Scottish Historical Review.

Transactions of the Royal Society of Edinburgh. (Edinburgh, 1788–)

Transactions of the Newcomen Society for the Study of the History of Engineering and Technology.

Transactions of the Royal Scottish Society of Arts.

Transactions of the Society of Glass Technology.

Transactions (and Select Essays) of the Highland (and Agricultural) Society of Scotland. 1st series, Vols. 1–6 (1799–1824). 2nd series, Vols. 1–8 (1829–1843). 3rd series, Vols. 1–11 (1845–67).

Index to the 1st, 2nd and 3rd series of the *Transactions of the Highland and Agricultural Society of Scotland* for 1799–1865, compiled by F. N. Menzies. (2 parts.) (Edinburgh, 1869.)

Transactions of the Society for the Encouragement of Arts, Manufactures, and Commerce. 1st series (London, 1761–). 2nd series (London, 1783–1845).

Letters and Papers on Agriculture, etc., selected from the Correspondence-Book of the Society instituted at Bath for the Encouragement of Agriculture, Arts, Manufactures, and Commerce. (Bath, 1780.)

Reports of the British Association for the Advancement of Science. (London, 1831–)

PERIODICALS

Annual Register. (London, 1758) and various collected indices.

New Annual Register. (London, 1780.)

Edinburgh Review. (Edinburgh, 1755.)

Edinburgh Review or Critical Journal. (Edinburgh, 1802.)

European Magazine and London Review. (Published by Philosophical Society of London.) 1st series (London, 1782–1825), 2nd series (London, 1825–) United with *Monthly Magazine*.

Gentleman's Magazine. (London, 1731) and General Indices.
Historical Register. (London, 1714–38.) (Published by Sun Fire Office.)
London Magazine, or Gentleman's Monthly Intelligencer. (London, 1732–) (General Index.)
The Monthly Magazine. (London, 1796–1825.)
New London Magazine. (London, 1785–)
The Scots Magazine. (Edinburgh, 1739.) (For a time was continued as the *Edinburgh Magazine*.)
The Weekly Magazine. (Edinburgh, 1764.) (Continued as *Edinburgh Weekly Magazine*.)
Wonderful Magazine, or Marvellous Chronicle. (London, 1764.)

DICTIONARIES, ENCYCLOPAEDIAS, BIBLIOGRAPHIES, COLLECTED INDEXES, ETC.

Abridgements of all Specifications of Patented Inventions relating to:—
Manure; Preservation of Food; Iron and Steel, Parts 1, 2 and 3 (1621–1857); Paper, etc.; Bleaching, Dyeing, etc. (1617–1883); India Rubber and Waterproofing; Production and Applications of Gas; Metals and Alloys; Photography (1839–59); Pottery (1626–1866); Sugar (1663–1866); Acids, Alkalis, etc. (1622–1866).
Aikin, J., and Enfield, W.: *General Biography*. (10 vols.) (London, 1799–1815.)
Ashton, T. S.: *Bibliography of the Industrial Revolution*. Economic History Review, 1934–36, 5, 104.
Barclay, A.: *Handbook of the Collections illustrating Industrial Chemistry*. (Science Museum, South Kensington, London, 1929.)
Barrow, John: *A New and Universal Dictionary of Arts and Sciences*. (London, 1751.) Supplement (London, 1754).
Bolton, H. C.: *Bibliography of Chemistry*. Smithsonian Miscellaneous Collections, Vols. 36, 39, 41, 44.
Bolton, H. C.: *A Catalogue of Scientific and Technical Periodicals*. Smithsonian Miscellaneous Collections, Vols. 29 and 40.
Brewster, D.: *Edinburgh Encyclopaedia*. (18 vols.) (Edinburgh, 1830.)
Chambers, E.: *Cyclopaedia or Universal Dictionary of Arts and Sciences*. (London, 1738.)
Chambers, E., and Rees, A.: *Cyclopaedia: or, An Universal Dictionary of Arts and Sciences*. (4 vols.) (London, 1778.)
Catalogue of Books, Pamphlets, and Maps belonging to the Society for the Encouragement of Arts, Manufactures, and Commerce. (London, 1790.)
Cripps, F. S.: *The Earliest Works on Gas Lighting*. (London, 1907.)
Croker, T. H.: *Complete Dictionary of Arts and Sciences*. (3 vols.) (London, 1764–66.)
Curtis, T.: *The London Encyclopaedia, or Universal Dictionary of Science, etc*. (22 vols.) (London, 1829.)
Diderot, D., and d'Alembert, J. Le D.: *Encyclopédie* (17 vols.). (Paris 1751–65.)

Recueil de Planches, sur les Sciences, les Art Liberaux, et les Art Méchanique. (Supplement to Diderot and d'Alembert's *Encyclopédie.*) (Paris, 1762.)

Encyclopaedia Britannica :

1st Edition	3 Vols.	Edinburgh,	1768–1771.
2nd "	10 "	"	1777–1784.
3rd "	18 "	"	1788–1797.
4th "	20 "	"	1801–1810.
5th "	20 "	"	1815–1817.
6th "	20 "	"	1823–1824.

Encyclopaedia Britannica, Supplement to IV–VI Editions. (6 vols., with preliminary Dissertation on the History of the Sciences.) (Edinburgh, 1824.)

Encyclopédie Méthodique, par une Sociètè de Gens de Lettres. (301 vols.) (Paris, 1782–1832.)

Encyclopaedia Perthensis, or Universal Dictionary of Arts, Science and Literature. (Edinburgh, 1807.)

Feldhaus, Franz Maria: *Lexikon der Refindung und Entdeckungen auf den Gebieten der Naturwissenschaften und Technik.* (Heidelberg, 1904.)

Good, J. M.: *Pantologia, a New Cyclopaedia.* (12 vols.) (London, 1813.)

Gregory, Geo.: *A Dictionary of Arts and Sciences.* (2 vols.) (London, 1806.)

Groome, Francis H.: *Ordinance Gazetteer of Scotland.* (6 vols.) (Edinburgh, 1882.)

General Index to the Transactions of the Royal Society of Edinburgh. (Edinburgh, 1890.)

Guide to the Printed Materials for English Social and Economic History, 1750–1850. (2 vols.) (New York, 1926.)

Halkett, Sam., and Laing, John: *Dictionary of Anonymous and Pseudonymous Literature.* (4 vols.) (Edinburgh, 1882–8.)

Harland, J.: *Collectanea Relating to Manchester and its Neighbourhood.* Chetham Society, 1866, *68*; 1867, *72*.

Harris, J.: *Lexicon Technicum, or a Universal English Dictionary of all the Arts and Sciences.* (2 vols.) (London, 1704–10.)

Harris, W.: *Oxford Encyclopaedia and Dictionary of Arts, Sciences, etc.* (7 vols.) (London, 1828–31.)

Hough, W.: *Collection of Heating and Lighting Utensils in the U.S. National Museum.* Smithsonian Bulletin 141.

Howard, G. S.: *A New Royal Encyclopaedia.* (3 vols.) (London, 1788.)

Humphreys, A. L.: *Handbook of County Bibliography, Being a Bibliography of Bibliographies Relating to the Counties and Towns of Great Britain and Ireland.* (London, 1917.)

Index to Abstracts of Specifications. (In 1879 Index Society published an Index to 100 volumes of Abstracts of Specifications as Appendix 2 to the Report of its Second Annual Meeting.)

Jameson, Alex. (Editor): *A Dictionary of Mechanical Science, Arts, Manufactures, and Miscellaneous Knowledge*. (London, 1827.)

K(eir), J(ames): The first part of a *Dictionary of Chemistry*. (Birmingham, 1789.)

Lardner, D.: *Cabinet Cyclopaedia*. (133 vols.) (London, 1830–49.)

Loudon, J. C.: *An Encyclopaedia of Agriculture*. (London, 1831.)

McCulloch, J. R.: *A Dictionary, Practical, Theoretical, and Historical, of Commerce and Commercial Navigation*. (New Edition, with Supplement.) (London, 1838.)

Macquer, P. J.: *Dictionnaire de Chymie*. (2 vols.) (Paris, 1766.)

Marwick, W. H.: *Bibliography of Scottish Economic History*. (Economic History Review, 1931–2, *3*, 117.)

Morison, W. M.: *Dictionary of Decisions*. (42 vols.) (Edinburgh, 1811.)

New and Complete Dictionary of Arts and Sciences. (4 vols.) (London, 1754–5.)

New Royal and Universal Dictionary of Arts and Sciences. (London, 1772.) (*Chemistry* by M. Hinde.)

Nicholson, Wm.: *A Dictionary of Chemistry*. (2 vols.) (London, 1795.)

Nicholson, Wm.: *A Dictionary of Theoretical and Practical Chemistry*. (London, 1808.)

Nicholson, Wm.: *The British Encyclopaedia or Dictionary of Arts and Sciences*. (London, 1809.)

Passmore, J. B., and Spencer, A. J.: *Agricultural Implements and Machinery*. Science Museum, South Kensington, London, 1930.

Peacock, E.: *Index of English Speaking Students at Leyden University*. (London, 1883.)

Power, D'A., and Thompson, C. J. S.: *Chronologia Medica*. (London, 1923.)

Rees, Abraham: *The New Cyclopedia or Universal Dictionary*. (39 vols.) (London, 1819.)

Roth, H. L.: *Guide to the Literature of Sugar*. (London, 1890.)

Savary, —: *Postlethwayte's Universal Dictionary of Trade and Commerce*. (2 Vols.) (London, 1757.)

Stephen, L.: *Dictionary of National Biography*. (63 volumes and supplements.) (London, 1885.)

Tomlinson: *Encyclopaedia of the Useful Arts*. (2 vols.) (London, 1854.)

Ure, Andrew: *A Dictionary of Chemistry on the Basis of Mr. Nicholson's*. (London, 1821.)

Ure, Andrew: *Recent Improvements in Arts, Manufactures, and Mines*. (London, 1844.)

Watt, Rob.: *Bibliotheca Britannica*. (4 vols.) (Edinburgh, 1824.)

Wilkes, J.: *Encyclopaedia Londinensis*. (24 vols.) (London, 1810–29.)

Woodcroft, B.: *Subject matter Index of Patents of Invention from March 2nd, 1617, 14 James I, to October 1st, 1852, 16 Victoria*. (2 vols.) (London, 1854.)

Woodcroft, B.: *Titles of Patents of Invention Chronologically arranged from March 2nd, 1617, to October 1st, 1852.* (2 vols.) (London, 1860.)

BOOKS AND PAMPHLETS

(* Denotes an article in a periodical publication.)

A. B.: *Some Account of the rise, progress, and present state of the brewery.* (London, 1757.)

Accum, Frederick: *Description of the Process of Manufacturing Coal Gas, etc.* (London, 1819.)

Accum, F.: *A Practical Treatise on Gas-Light.* (London, 1815.)

Accum, F.: *Treatise on the Adulteration of Food.* (London, 1820.)

Accum, F.: *Treatise on the Art of Brewing.* (London, 1820.)

Achard, Franz C.: *Traité Complet sur le Sucre Européen de Betteraves traduction abrégé M. D. Augar, precédé d'une introduction et accompagné de notes et observations par M. Ch. Derosne.* (Paris, 1812.)

Acts of the Parliament of Scotland 1124–1707. (11 vols.) (London, 1814–44.)

Agricola, Georgius: *De Re Metallica.* (1546.) Translated from the 1556 edition by H. C. Hoover and L. H. Hoover. (London, 1912.)

Agriculture: *General Views of the Agriculture and Rural Economy of the County of* . . . 1st and 2nd series. Second series:—

Aberdeenshire	Keith, G. S.	1811
Angus and Forfar	Headrich, J.	1813
Argyll	Smith, J.	1813
Ayr	Aiton, W.	1811
Banff	Souter, D.	1812
Berwick	Kerr, R.	1809
Bute	Aiton, W.	1816
Caithness	Henderson, J.	1812
Clydesdale	Naismith, J.	1813
Dumbarton	Whyte, A., and Macfarlan, D.	1811
Dumfries	Singer, Dr.	1812
Fife	Thomson, J.	1800
Galloway	Smith, S.	1810
Hebrides	Macdonald, J.	1811
Inverness	Robertson, J.	1808
Kincardine	Robertson, G.	1813
Kinross and Clackmannan	Graham, P.	1814
East Lothian	Somerville, R.	1805
Mid-Lothian	Robertson, G.	1795
West Lothian	Trotter, J.	1811

Nairn and Moray	Leslie, W.	1813
Orkney Islands	Shirreff, J.	1814
Peebles	Findlater, C.	1802
Perth	Robertson, J.	1799
Renfrew	Wilson, J.	1812
Ross and Cromarty	Mackenzie, G. S.	1810
Roxburgh and Selkirk	Douglas, R.	1798
Shetland Islands	Shirreff, J.	1813
Stirling	Graham, P.	1812
Sutherland	Henderson, J.	1812

Aikin, J.: *Description of the Country from Thirty to Forty Miles round Manchester.* (London, 1795.)

Aitken, W. C.: *Glass Manufacturers of Birmingham and Stourbridge.* Reprinted privately from the Birmingham Journal, 1851, May 31st.

Alderson, J.: *Essay on the Improvement of poor soils.* (2nd Edtn. London, 1807.)

Allen, G. C.: *The Industrial Development of Birmingham and the Black Country. 1860–1927.* (London, 1929.)

Allen, J. Fenwick: *Some Founders of the Chemical Industry.* (London and Manchester, 1906.)

Anderson, A.: *History of the Origin of Commerce.* (London, 1764.)

Anderson, James: *Essays Relating to Agriculture and Rural Affairs.* (Edin., 1775.)

Anderson, James: *Practical Treatise on Chimneys.* (Edin. 1776.)

Anderson, James: *Observations on Means of Exciting a Spirit of National Industry.* (Edinburgh, 1777.)

Anon.: *A Book of Receipts according to the Newest Method.* (London, 1680.)

Anon.: *The London and Country Brewer.* (1738.)

Anon.: *An Essay on Means of Enclosing and Fallowing.* (1729.)

Appert, N.: *Le Livre de tous les Ménages ou l'Art de Conserver pendant Plusieurs Années Toutes les Substances Animales et Végétales.* (Paris, 1810.)

Armstrong, Sir W. G. (and others): *The Industrial Resources of the District of the Three Northern Rivers, the Tyne, Wear, and Tees.* (London, 1864.)

Arnot, Hugo: *The History of Edinburgh from the Earliest Accounts to the Year 1780.* (4th Edtn. Edinburgh, 1818.)

Ashley, Wm.: *The Bread of Our Forefathers.* (Oxford, 1928.)

Ashton, T. S.: *Iron and Steel in the Industrial Revolution.* (Manchester Univ. Press, 1924.)

Ashton, T. S., and Sykes, J. L.: *The Coal Industry of the 18th Century.* (Manchester Univ. Press, 1929.)

Babbage, Ch.: *Reflections on the Decline of Science in England.* (London, 1830.)

Babbage, Ch.: *On the Economy of Machines and Manufactures.* (London, 1832.)

Babbage, Ch.: *Exposition of 1851, or, Views of the Industry, the Science, and the Government of England.* (London, 1851.)

Bailey, A. M.: *One Hundred and Six Copper Plates of Mechanical Machines and Implements of Husbandry Approved and Adopted by the Society for the Encouragement of Arts, Manufactures and Commerce.* (2 vols.) (London, 1782.)

Bailey, W.: *Advancement of Arts, Manufactures and Commerce.* (London, 1772.)

Baines, E.: *History of the County Palatine and Duchy of Lancaster.* (4 vols.) (London, 1836.)

Baines, E., Jr.: *History of the Cotton Manufacture in Great Britain.* (London, 1835.)

Baines, T.: *History of the Commerce and Town of Liverpool, and of the Rise of Manufacturing Industry in the adjoining Counties.* (London and Liverpool, 1852.)

Baines, T., and Fairburn, W.: *Lancashire and Cheshire, Past and Present.* (2 vols.) (London, 1868–9.)

Baker, Julian L.: *The Brewing Industry.* (London, 1905.)

Bald, R.: *General View of the Coal Trade of Scotland.* (Edin., 1812.)

Baldwin, T.: *Airopaidia or Aerial Recreation.* (Chester, 1786.)

Bancroft, Ed.: *The Philosophy of Permanent Colours.* (2 vols.) (London, 1813.)

Barnard, Alfred: *Whiskey Distilleries of the United Kingdom.* (London, 1887.)

Barnard, Alfred: *Noted Breweries of Great Britain and Ireland.* (4 vols.) (London, 1889–91.)

Baverstock, J. H.: *Treatises on Brewing by the late James Baverstock.* (London, 1824.)

*Beaton, Angus: *On the Art of Making Kelp.* (Trans. High. Soc. 1799, *1* (1st series), 32.

Beckmann, J.: *History of Inventions, Discoveries and Origins.* Trans. from the German by W. Johnson. (London, 1797.)

Belhaven, Lord: *The Countryman's Rudiments.* (Edinburgh, 1723.)

Berthelot, P. E. M.: *La Révolution Chimique Lavoisier.* (Paris, 1890.)

*Berthollet, C. L.: *Description du blanchîment des toiles et des fils par l'acide muriatique oxigéné, etc.* (Annales de Chimie, 1789, *2*.)

Berthollet, C. L.: *Essay on the New Method of Bleaching.* (English trans. by Robert Kerr for the Dublin Trustees for the Linen Manufacture.) (Dublin, 1790.)

Berthollet, C. L.: *Eléments de l'Art de la Teinture.* (Paris, 1791.) (English Trans. 2 vols., by W. Hamilton, London, 1791.)

Berthollet, C. L. and A. B.: *Elements of the Art of Dyeing with a Description of the Art of Bleaching by Oxymuriatic Acid.* Translated by Andrew Ure. (2 vols.) (London, 1824.)

Birch, T.: *History of the Royal Society.* (4 vols.) (London, 1756.)

Birnie, A.: *An Economic History of the British Isles.* (London, 1935.)

Bischoff, J.: *Comprehensive History of the Woollen and Worsted Manufactures.* (2 vols.) (London, 1842.)

Bitting, A. W.: *Appertizing, or the Art of Canning: Its History and Development.* (San Francisco, 1937.)

Black, J.: *Dissertatio medica inauguralis, de humore acido a cibis orbo, et magnesia alba.* (Edinburgh, 1754.)

Black, J.: *Lectures on the Elements of Chemistry.* (2 vols.) (Edin., 1809.)

Black, Wm.: *A Practical Treatise on Brewing.* (London, 1831.)

Blackwell, A.: *A New Method of Improving cold, wet, and barren land.* (1741.)

*Bladen, V. W.: *The Potteries in the Industrial Revolution.* Economic History (Supplement to Economic Journal) 1920, *1*, 117.

Bland, A. E., Brown, P. A., and Tawney, R. H.: *English Economic History: Select Documents.* (London, 1914.)

Boerhaave, H.: *Elementa Chemiae.* (2 vols.) (Lugduni Batavorum, 1732.) (English Trans. by T. Dallowe.) (2 vols.) (London, 1735.)

*Bolton, H. C.: *The Lunar Society.* Trans. New York Acad. of Science, 1888, 7, No. 8. Similar to essay in Smithsonian Misc. by same author and to *Priestley Letters.* Copy in Birmingham Reference Library has added press cuttings '*Dr. Priestley's Laboratory*' (reprinted from *Birmingham Weekly Post*, 15th, 22nd, 29th March and 5th April, 1890, by S. Timmins). Also *Priestley Commemoration* (*Birmingham Daily Gazette*, 4th April, 1867) and portraits of the members of the Lunar Society.

Bolton, H. C.: *Scientific Correspondence of Joseph Priestley.* (New York, 1892.) (Privately printed. Appendix on the Lunar Society of Birmingham.)

*Bolton, H. C.: *Early American Chemical Societies.* Journal of the American Chemical Society, 1897.

Bontemps, Georges: *Le Guide du Verrier.* (Paris, 1868.)

Bourné, H. R. F.: *Romance of Trade.* (London, 1871.)

Boussingault, J. B., and Dumas, J.: *Essai de statique chimique desêtres organisés.* (Paris, 1841.)

Bowden, Witt: *Industrial Society in England towards the End of the Eighteenth Century.* (New York, 1925.)

Bower, A.: *History of the University of Edinburgh.* (2 vols.) (Edin., 1817.)

Boyle, R.: *The Sceptical Chemist.* (London, 1661.)

Brand, Henderson W.: *The Modern Method for the Preservation of all Alimentary Substances.* (London, 1838.)

Bremner, D.: *The Industries of Scotland, their Rise, Progress, and Present Position.* (Edinburgh, 1869.)

Brereton, Sir W.: *Travels in Holland . . . Scotland, etc., 1634–5.* (Chetham Society Publication, Vol. 1.)

British Association: *See* Day, St. John V.; Gossage, Wm.; Richardson, T.; McLean, A.; Paton, A. W.; Thomson, T.

Brooke, R.: *Liverpool as it was during the Last Quarter of the Eighteenth Century.* (Liverpool, 1853.)

Brougham, Henry, Lord: *Lives of Men of Letters and Science in time of George III.* (2 vols.) (London, 1845–6.)

Brown, A.: *History of Glasgow and of Paisley, Greenock and Port Glasgow.* (2 vols.) (Glasgow, 1795.)

*Browne, C. A.: *The Life and Chemical Services of Frederick Accum.* Journal of Chemical Education, 1925, *2*, Nos. 10, 11, 12.

*Brownlie, David: *The Early History of the Coal Gas Process.* Trans. Newcomen Soc., 1923, *3*, 57.

Brownrigg, W.: *The Art of Making Common Salt.* (London, 1749.)

Brydall, R.: *Art in Scotland, its Origin and Progress.* (Edinburgh and London, 1889.)

Buchanan, Robertson: *Practical and Descriptive Essays on the Economy of Fuel, and Management of Heat.* (Glasgow, 1810.)

*Buckle, William: *On the Inventions and the Life of William Murdock.* Proc. Inst. Mech. Eng., 1850.

*Buckley, F.: *Notes on the Glasshouses of Stourbridge 1700–1830.* Trans. Soc. Glass Tech., 1927, *11*, 106.

Burt, Ed.: *Letters from a Gentleman in the North of Scotland.* (2 vols.) (n.p. 1754.)

Burton, W.: *Josiah Wedgwood and his Pottery.* (London, 1922.)

Byrn, Edward W.: *The Progress of Invention in the Nineteenth Century.* (New York, 1900.)

*Cadell, Henry M.: *Sketch of the History of the Iron and Steel Industry.* Trans. Royal Scottish Soc. of Arts, 1833, *10*, 506.

Cadell, H. M.: *The Story of the Forth.* (Glasgow, 1913.)

Callico Printers' Assistant, including a history of Calico Printing. (London, 1789.)

Calvert, A. F.: *Salt in Cheshire.* (London, 1915.)

Campbell, J.: *Political Survey of Britain.* (2 vols.) (London, 1774.)

Cavallo, T.: *The History and Practice of Aerostation.* (London, 1785.)

Cavendish, H.: *Scientific Papers*, ed. Sir E. Thorpe. (2 vols.) (Camb., 1921.)

Chalmers, G.: *Caledonia, An Account of North Britain.* (3 vols.) (London, 1807–24.)

Chambers, Rob.: *Domestic Annals of Scotland.* (3 vols.) (Edinburgh, Vol. 1, 1858; Vol. 2, 1858; Vol. 3, 1861.)

Chance, J. F.: *History of the Firm of Chance Brothers & Co., Glass and Alkali Manufacturers.* (London, 1919.)

Chandler, Dean: *Outline of History of Lighting by Gas.* (London, 1936.)

Chapman, S. J.: *The Lancashire Cotton Industry. A Study in Economic Development.* (Manchester, 1904.)

Chapoy, –.: *L'Invention des Allumettes Chimiques.* (n.p. 1894.)

Chaptal, J. A. C.: *Eléments de Chimie.* (Montpellier, 1790.)

Chaptal, J. A. C.: *Traité de la Teinture du Cotton en Rouge.* (1805.)

Chaptal, J. A. C.: *Chimie appliqué aux arts.* (Paris, 1807.)

Chaptal, J. A. C.: *Chemistry applied to Arts and Manufactures.* (4 vols.) (London, 1807.)

Chaptal, J. A. C.: *Sur l'Industrie Française.* (2 vols.) (Paris, 1819.)

Chaptal, J. A. C.: *Chimie appliqué à l'agriculture.* (Paris, 1829.) (English translation by W. P. Page, New York, 1840.)

Charmes, Pajot des: *The Art of Bleaching.* (English translation by Nicholson.) (London, 1799.)

Chase, Stuart: *Men and Machines.* (New York, 1929.)

Cheyney, E. P.: *Industrial and Social History of England.* (Revised edition, New York, 1921.)

Chevreul, M. E.: *Recherches Chimiques sur les Corps gras d'Origine Animale.* (Paris, 1823.)

Child, S.: *Every Man his own Brewer, A practical treatise on brewing.* (London, 1790?.)

Chitty, J.: *A Treatise on the Laws of Commerce and Manufactures.* (4 vols.) (London, 1820–24.)

Christy, M.: *The Bryant and May Museum of fire-making Appliances.* (London, 1926.)

Church, Arthur H.: *English Earthenware.* (London, 1904.)

Church, A. H.: *English Porcelain.* (London, 1904.)

Church, A. H.: *Josiah Wedgwood.* (London, 1903.)

Clapham, J. H.: *Economic History of Modern Britain.* (3 vols.) (London, 1932.)

Clapperton, R. H.: *Modern Paper-making.* (Oxford, 1941.)

Clark, G. N.: *Science and Social Welfare in the Age of Newton.* (Oxford, 1937.)

Clark, Victor S.: *History of Manufactures in the United States, 1607–1928.* (3 vols.) (New York, 1929.)

*Clayton, E. G.: *The Character and Chemical Composition of some early Matches.* Chemical News, 1911, *104*, 223; and Proc. Chem. Soc., 1911, *27*, 229.

*Clegg, Samuel (Sr.): *Apparatus for making Carburated Hydrogen Gas from Pit Coal, and Lighting Factories therewith.* Trans. Soc. for the Encouragement of Arts, Manufactures, and Commerce, 1808, *26*, 202.

Clegg, Samuel (Jr.): *A Practical Treatise on the Manufacture and Distribution of Coal-Gas.* (London, 1841.)

Cleland, J.: *Annals of Glasgow.* (2 vols.) (Glasgow, 1816.)

Cleland, J.: *Rise and Progress of Glasgow.* (Glasgow, 1820.)

Cleland, J.: *Description of the City of Glasgow.* (Glasgow, 1843.)

Cochran-Patrick, R. W.: *Early Records relating to Mining in Scotland.* (Edinburgh, 1878.)

Cochran-Patrick, R. W.: *Mediaeval Scotland.* (Glasgow, 1892.)

Collet, C. D.: *History of Taxes on Knowledge.* (London, 1899.)

Collins, John: *Salt and Fishery, a discourse thereon.* (London, 1682.)

Collins, J. H.: *The Story of Canned Food.* (New York, 1924.)

Comber, N. M.: *An Introduction to the Scientific Study of the Soil.* (London, 1927.)

Combrune, Michael: *The Theory and Practice of Brewing.* (London, 1762.)

*Cook, B.: *Methods of Producing Heat, Light, and various useful Articles from Pit-coal.* Trans. Soc. for the Encouragement of Arts, Manufactures, and Commerce, 1810, *28*, 73.

Court, W. H. B.: *The Rise of the Midland Industries, 1600–1838.* (Oxford, 1938.)

Coutts, J.: *A History of the University of Glasgow, 1451–1909.* (Glasgow, 1909.)

Coventry, Andrew: *Discourses Explanatory of the Object and Plan of the course of Lectures on Agriculture and Rural Economy.* (Edin., 1808.)

Coventry, Andrew: *Notes on the culture and cropping of arable land.* (Edinburgh, 1811.)

Croome, H. M. and Hammond, R. J.: *The Economy of Britain—A History.* (London, 1938.)

Crowther, J. G.: *Social Relations of Science.* (New York, 1941.)

Crump, W. B.: *The Leeds Woollen Industry, 1780–1820.* (Thoresby Society, 1931.)

Cullen, W.: *The Plan of a Course of Chemical Lectures and Experiments directed chiefly to the improvement of Arts and Manufactures.* (1748.)

Cullen, W.: *Lectures on the Materia Medica.* (London, 1773.)

Cullen, W.: *On the Cold produced by Evaporating Fluids.* (n.p., 1777.)

Cullen, W.: *The Substance of nine lectures on vegetation and agriculture, with notes by* G. Pearson. (n.p., 1796.)

Cunningham, W.: *The Growth of English Industry and Commerce.* (Cambridge, 1885.)

Dacre, Rev. B.: *Testimonies in favour of Salt.* (Manchester, 1825.)

Daniels, G. W.: *Early English Cotton Industry.* (Manchester, 1920.)

Darwin, Erasmus: *Phytologia, or the Philosophy of Agriculture and Gardening.* (London, 1800.)

Davy, Humphry: *Elements of Agricultural Chemistry (in a Course of Lectures for the Board of Agriculture).* (London, 1813.)

Day, St. John V., Mayer, J., and Paton, J.: *Notices of Some of the Principal Manufactures in the West of Scotland.* (Glasgow, 1876.)

Defoe, D.: *A Tour through the whole Island of Great Britain, divided into Circuits or Journeys.* (London, 1724–27.)

Desaguliers, J. T.: *A Course of Mechanical and Experimental Philosophy.* (2 vols.) (London, 1729–44.)

Devereux, Roy: *John Loudon McAdam.* (London, 1936.)

Dickinson, H. W.: *John Wilkinson, Iron Master.* (Ulverston, 1914.)

Dickinson, H. W.: *James Watt.* (Cambridge, 1936.)

Dickinson, H. W.: *Matthew Boulton* (Cambridge, 1937.)

Dickinson, H. W.: *A Short History of the Steam Engine.* (Camb., 1939.)

*Dickinson, H. W.: *Manufacture of Sulphuric Acid.* Trans. Newcomen Soc., 1937, *18*, 43.

Dickson, Adam: *A Treatise of Agriculture.* (Edinburgh, 1762.)

Dircks, H.: *Life, Times, and Scientific Labours of the Second Marquis of Worcester.* Contains the *Century of Inventions, 1663.* (London, 1865.)

*Distillery: *History of British Distillery.* (Scots Magazine, 1743, *5*, 443.)

*Distillery: *Laws regulating Distillers.* (Scots Magazine, 1786, *48*, 358.)

Distillery: *The Present State of Distillery in Scotland.* (Edinburgh, 1786.) Author: Walter Ross.

Distillery: *Truths, in answer to the facts published respecting the Scottish Distillery.* (Edinburgh, 1786.)

Distillery: *Resolutions of the landed interest in Scotland respecting the Distillery.* (Edinburgh, 1786.)

Distillery: *Case of the capital corn distillers of Scotland.* (London, 1787.)

Distillery: *Case of the distillers of corn spirits in North Britain.* (London, 1787.)

Distillery: *Answers to the Memorial of the Corn Distillers in London.* (1788.)

Dixon, Wm.: *The Match Industry: its origin and Development.* (1925.)

Dobbin, L.: *Collected Papers of C. W. Scheele.* (London, 1931.)

Dodd, G.: *Days at the Factories.* Series I: London. (London, 1843.)

Dodd, G.: *British Manufacturers.* (London, 1844.)

Dodds, A. E.: *Education and Social Movements, 1700–1850.* (London, 1919.)

Donaldson, Jas.: *Husbandry Anatomised, or an Enquiry into the present manner of tilling and manuring the ground in Scotland.* (Edin., 1697.)

Dossie, Robert: *The Elaboratory laid open.* (London, 1758.)

Dossie, R.: *Observations on the Pot-Ash brought from America. Processes for making Pot-Ash and Barilla in North America.* (London, 1767.)

Dossie, R.: *Memoirs of Agriculture and Other Economical Arts.* (3 vols.) (London, 1768–82.)

Douglas, Francis: *General Description of the East Coast of Scotland.* (Paisley, 1782.)

Drummond, J. C. and Wilbraham, Anne: *The Englishman's Food.* (London, 1939.)

Dudley, Dud.: *Metallum Martis, or Iron made with Pit Coale, Sea Coal, &c., and with the same Fire to melt and fire imperfect Metalls, and refine perfect Metalls.* (London, 1665.)

Dumas, J., and Boussingault, J. B.: *The Chemical and Physiological Balance of Organic Nature.* (London, 1841.)

Dunbar, James: *Smegmatologia, or The Art of Making Potashes, Soap, and Bleaching of Linen.* (Edin., 1736.)

Duncan, Alex.: *Memorials of the Faculty of Physicians and Surgeons of Glasgow, 1599–1850.* (Glas., 1896.)

Dundonald, Archibald Cochrane, 9th Earl of,: *Account of the Quality and Uses of Coal Tar and Coal Varnish.* (London, 1785.)

Dundonald, Archibald Cochrane, 9th Earl of: *The Present State of the Manufacture of Salt Explained.* (London, 1785.)

Dundonald, Archibald Cochrane, 9th Earl of: *Letters by the Earl of Dundonald on Making Bread from Potatoes.* (Edin., 1791.)

Dundonald, Archibald Cochrane, 9th Earl of: *A Treatise shewing the Intimate Connection that subsists between Agriculture and Chemistry.* (London, 1795.)

Dundonald, Archibald Cochrane, 9th Earl of: *Directions by Lord Dundonald for extracting Gum from the Lichen or Tree Moss.* (Glas., 1801.)

Dundonald, Thomas Cochrane, 10th Earl of: *Autobiography of a Seaman.* (2 vols.) (London, 1860.)

Espinas, Alfred: *Les Origines de la Technologie.* (Paris, 1899.)

Fairbairn, W.: *Iron, its History, Properties and Processes of Manufacture.* (Edinburgh, 1869.)

Fairrie, G.: *Sugar.* (Liverpool, 1925.)

Falkner, F.: *The Wood Family of Burslem.* (London, 1912.)

Farrer, K. E.: *Correspondence of Josiah Wedgwood.* (London, 1906.)

Faujas de Saint-Fond, Barthelemy: *Description des Expériences de la Machine Aërostatique de MM. de Montgolfier.* (2 vols.) (Paris, 1783–4.)

Faujas de Saint-Fond, Barthelemy: *Essai sur le Goudron du charbon de terre.* (Paris, 1790.)

Faujas de Saint-Fond, Barthelemy: *Voyage en Angleterre en Ecosse et aux Iles Hebrides, 1784.* (2 vols.) (Paris, 1797.) (English trans. ed. by A. Geikie, 2 vols., Glasgow, 1907.)

Fay, C. R.: *From Adam Smith to the Present Time.* (London, 1928.)

Figuier, Louis: *Les Merveilles de l'Industrie.* (3 vols.) (1781–4.)

Filby, F. A.: *A History of Food Adulteration and Analysis.* (London, 1934.)

Findlay, A.: *The Teaching of Chemistry in the Universities of Aberdeen.* (Aberdeen, 1935.)

Fleming, J. Arnold: *Scottish Pottery.* (*Glasgow, 1923.*)

Fleming, J. A.: *Scottish and Jacobite Glass.* (Glasgow, 1938.)

*Forsyth, Chas.: *Mines, Minerals and Geology of West Lothian.* Trans. Highland Soc., 1847, *2*, 229. (3rd series.)

Forsyth, R.: *The Beauties of Scotland.* (5 vols.) (Edinburgh, 1805–8.)

Foster, W.: *The Romance of Chemistry.* (London, 1927.)

Fourcroy, Antoine François de: *The Philosophy of Chemistry, or fundamental truths of modern science, arranged in a new order.* (English translation, London, 1795.)

French, G. J.: *Life and Times of Samuel Compton.* (Manchester, 1859.)

Franklin, Benj.: *Complete Works.* (New York, 1888.)

*Fyfe, A.: *Essay on the Comparative Value of Kelp and Barilla.* Trans. Highland Soc. 1820, *5* (1st series), 10.

Fyfe, J. G.: *Scottish Diaries and Memoirs, 1746–1843.* (Stirling, 1942.)

Galt, John: *The Annals of the Parish.* (1821.)

Gardner, W. M. (Editor): *The British Coal-Tar Industry.* (London, 1915.)

The Gas Light and Coke Company, 1812–1912. (London, 1912.)

Gaskell, P.: *The Manufacturing Population of England, its Moral, Social and Physical Conditions, and the Changes that have arisen from the Use of Steam Machinery*. (London, 1833.)

Geer, William C.: *The Reign of Rubber*. (New York, 1922.)

Gerspach, Edouard: *History of the Art of Glass Making*. (Paris, *c*. 1890.)

*Gibbs, F. W.: *The History of the Manufacture of Soap*. Annals of Science, 1939, *4*, 169.

Gibson, John: *The History of Glasgow*. (Glasgow, 1777.)

Gill, Conrad: *The Rise of the Irish Linen Industry*. (Oxford, 1925.)

Girardin, J. P. L.: *Leçons de Chimie Elémentaire appliquées aux Arts industriels*. (Paris and Rouen, 1846.)

Girvin, John: *The Impolicy of prohibiting the exportation of rock-salt from England to Scotland, to be refined there*. (London, 1799.)

Goodyear, Charles: *Gum Elastic and its Varieties*. (New Haven, 1855.)

Gordon, J. (Secretary to the Society for B.S.D.C.): *New Statistical Account of Scotland*. (15 vols.) (Edinburgh, 1845.)

Gordon, J. F. S.: *Glasghu Facies. See under* J. M'Ure.

*Gossage, Wm.: *A History of the Alkali Manufacture*. Report of the British Association, Manchester, 1861, p. 80.

Graham, H. G.: *Scottish Men of Letters in the Eighteenth Century*. (Edinburgh, 1901.)

Graham, H. G.: *Social Life in Scotland in the 18th Century*. (London, 1909.)

Grant, A.: *The Story of the University of Edinburgh*. (Edinburgh, 1884.)

Grant, I. F.: *The Economic History of Scotland*. (London, 1934.)

Grew, Nehemiah: *The Anatomy of Vegetables Begun*. (London, 1674.)

Grew, Nehemiah: *The Anatomy of Plants*. (London, 1682.)

Gunn, N. M.: *Whisky and Scotland*. (London, 1935.)

Haldane, E.: *The Scotland of our Fathers*. (London, 1933.)

Hales, S.: *Vegetable Staticks*. (London, 1727.)

*Hamilton, H.: *The Founding of Carron Iron Works*. Scottish Historical Review, 1928, *25*, 185.

Hamilton, H.: *The Industrial Revolution in Scotland*. (Oxford, 1932.)

Hands, W.: *Law and Practice of Patents for Inventions*. (London, 1808.)

Harris, H. G.: *The Principles of Glass Making, with Treatise on Crown and Sheet Glass by H. Chance, and Plate Glass by H.G.H.* (London, 1883.)

Hassall, A. H.: *Food and its Adulteration*. (London, 1855.)

Hawkes, S. M.: *Historical Notes on Beer and Brewing*. (London, 1850.)

Haydon, F. S.: *Aeronautics in the Union and Confederate Armies, with a Survey of Military Aeronautics prior to 1861*. (Baltimore, 1941.)

Heaton, H.: *Yorkshire Woollen and Worsted Industries from the Earliest Times up to the Industrial Revolution*. (Oxford, 1922.)

Heavisides, Mich.: *History of the Invention of the Lucifer Match*. (Stockton-on-Tees, 1909.)

Hellot, J.: *L'Art de la Teinture des Laines*. (Paris, 1750.)

Henry, W.: *Elements of Experimental Chemistry*. (6th Edition.) (London, 1810.)
Henry, W.: *Experiments on the gas from coal*. (Manchester, 1819.)
Henry, W.: *A general view of the nature and objects of Chemistry and of its application to Arts and Manufactures*. (Manchester, 1799.)
Henry, W. C.: *Memoirs of the Life and Scientific Writings of John Dalton*. (London, 1854.)
Herring, R.: *Paper and Paper-making*. (London, 1856.)
*Hertz, G. B.: *The English Silk Industry in the Eighteenth Century*. English Historical Review, 1909, *24*, 721.
Hesketh, E.: *J. & E. Hall, Ltd., 1785–1935*. (Glasgow, 1935.)
Higgins, Samuel H.: *A History of Bleaching*. (London, 1924.)
Higgins, Wm.: *An Essay on the Theory and Practice of Bleaching wherein the Sulphuret of Lime is recommended as a Substitute for pot-ash*. (Dublin, 1799.)
Histoire Centennale du Sucre de Betterave, published by Syndicat des Fabricants de Sucre de France. (1912.)
Hodgson, J. E.: *History of Aeronautics in Great Britain*. (London, 1924.)
Hogben, L.: *Science for the Citizen*. (London, 1938.)
Home, Francis: *Experiments on Bleaching*. (Edinburgh, 1754.)
Home, Francis: *The Principles of Agriculture and Vegetation*. (Edinburgh, 1757.)
Home, Henry (Lord Kames): *The Gentleman Farmer*. (Edinburgh, 1776.)
Honey, W. B.: *Old English Porcelain*. (London, 1928).
Horne, Henry: *Essay concerning Iron and Steel*. (London, 1773.)
*Hough, Walter: *Fire as an Agent in Human Culture*. (Smithsonian Institution, Bulletin 139.) (Washington, 1926.)
Houghton, J.: *A Collection of Letters for the Improvement of Husbandry and Trade*. (1st series, 1681–3, monthly: 2nd series, 1692–1702, weekly.)
Hughes, Ed.: *Studies in Administration and Finance, 1558–1825*. (Manchester University Press, 1934.)
Hunt, Chas.: *A History of the Introduction of Gas Lighting*. (London, 1907.)
*Hunter, J. R. S.: *The Silurian Districts of Leadhills and Wanlockhead, and their Early and Recent Mining History*. Trans. Geol. Soc. Glasgow, 1884, *7*, 373.
Hutcheson, John M.: *Notes on the Sugar Industry of the United Kingdom*. (Greenock, 1901.)
Hutton, W.: *History of Birmingham*. (Birmingham, 1st Edition, 1781.)
Ingenhousz, Jan: *Essay on the food of Plants and the Renovation of Soils*. (London, 1796.)
Ingenhousz, Jan: *Experiments upon Vegetables*. (London, 1799.)
Irving, Geo. Vere, and Murray, Alex.: *The Upper Ward of Lanarkshire*. (3 vols.) (Glasgow, 1864.)
Jackson, H.: *An Essay on Bread*. (London, 1758.)

Jaegear, F. M.: *Cornelius Drebbel and his Contemporaries.* (Gröningen, 1922.)

Jameson, Rob.: *Outline of the Mineralogy of the Shetlands, etc., with an Appendix on Peat, Kelp and Coal.* (Edinburgh, 1789.)

Jameson, Rob.: *Mineralogy.* (3 vols.) (Edinburgh, 1816.)

*Jameson, Rob.: *Observations on Kelp.* Trans. High. Soc. 1799, *1* (1st series), 43.

Jameson, Rob.: *Letters from the North of Scotland, I and II* (edited by R.J.) *with an Appendix on the State of the Highlands in the beginning of the 17th century.* (London, 1822.)

*Jardine, R.: *An Account of John Roebuck, M.D., F.R.S.* Trans. Roy. Soc. Edin., 1796, *4*, 65.

Jars, G.: *Voyages Métallurgiques, ou Recherches et Observations sur les Mines et Forges de Fer, la Fabrication de l'Acier, etc.* (3 vols.) (Lyons, 1774.)

*Jenkins, Rhys: *Early Attempts at Papermaking in England.* Library Association Record, 1900, *2*, 479.

Jewitt, Llewellyn: *The Wedgwoods, being a life of Josiah Wedgwood.* (London, 1865.)

Jewitt, Llewellyn: *The Ceramic Art of Great Britain from Prehistoric Times down to the Present Day.* (2 vols.) (London, 1878.)

Johnson, J. E.: *Principles, Operation, and Products of the Blast Furnace.* (New York, 1918.)

*Johnson, William: *The early History of Gas Lighting in Glasgow.* Proc. Glasgow Architectural Soc., 1867.

Kay, John: *Original Portraits and Caricature Etchings.* (Edinburgh, 1842.)

Kay, John: *Paper, its History.* (London, 1893.)

Keith, Theodora: *Commercial Relations between England and Scotland, 1603–1707.* (Cambridge, 1910.)

*Kemp, D. W.: *An Unwritten Chapter on the Early History of the Iron Industry—Smelting in Sutherlandshire and the North of Scotland.* Trans. Royal Scottish Soc. of Arts, 1886, *11*, 286.

*Kendall, J.: *Old Chemical Societies in Scotland.* Chemistry and Industry, 1937, *15*, 141. *Endeavour*, Vol. 1.

*Kennedy, J.: *On the Rise and Progress of the Cotton Trade.* Mem. Manch. Lit. and Phil. Soc., 1819, *8*, 115.

King, W.: *Treatise on the Science and Practice of the Manufacture and Distribution of Coal Gas.* (3 vols.) (London, 1878–1882.)

Kingzett, C. T.: *History of the Alkali Trade.* (London, 1877.)

Kirwan, R.: *On the Manures most advantageously applicable to various Soils, and the Causes of their Beneficial Influence.* (London, 1796.)

Knott, C. G. (Editor): *Edinburgh's Place in Scientific Progress.* (Edinburgh, 1921.)

Knox, John: *Tour through the Highlands of Scotland and the Hebride Isles.* (London, 1787.)

Koops, M.: *Historical Account of Paper, and the Substances used prior to its Invention.* (London, 1800.)

Langford, J. A.: *Century of Birmingham Life*. (2 vols.) (Birmingham, 1868.)

Lauder, Sir T. D.: *The Morayshire Floods*. (Elgin, 1873.)

Lavoisier, A., Berthollet, C. L., and Fourcroy, A. F.: *Méthode de Nomenclature Chemique*. (Paris, 1787.) (English trans. by J. St. John, 1788.)

Lefèvre, N.: *A Compleat Body of Chymistry*. (Eng. trans. by P.D.C., Esq.) (London, 1664).

Le Normand, L. S.: *Essai sur l'Art de la Distillation*. (Paris, 1811.)

Lewis, G. R.: *The Stanneries, A Study of the English Tin Miner*. (Boston, 1908.)

Liebig, J.: *Instructions for the Analysis of Organic Bodies*. (Trans. by W. Gregory.) (Glasgow, 1839.)

Liebig J.: *Chemistry in its Application to Agriculture and Physiology*. (2nd Edition, ed. L. Playfair.) (London, 1842.)

Liebig, J.: *Animal Chemistry, or Organic Chemistry in its application to Physiology and Pathology*. (Edited by W. Gregory.) (London, 1842.)

Liebig, J.: *Familiar Letters on Chemistry*. (Edited by J. Gardner.) (London, 1844.)

Liebig, J.: *Researches on the Chemistry of Food*. Edited from the MS. of the Author by W. Gregory. (London, 1847.)

Liebig, J.: *The Principles of Agricultural Chemistry:* with special reference to late researches made in England; Trans. by W. Gregory. (London, 1855.)

Liebig, J.: *The Relations of Chemistry to Agriculture and the Agricultural Experiments of Mr. J. B. Lawes*. Trans. by S. W. Johnson. (Albany, N.Y., 1855.)

Lindsay, Patrick: *The Interest of Scotland considered*. (Edin., 1733.)

Lock, C. G. W., Wigner, G. W., and Harland, A.: *On Sugar Growing and Refining*. (London, 1882.)

*Loewenfeld, K. L.: *Contributions to the History of Science*. (Period of Priestley-Lavoisier-Dalton.) Mem. Manch. Lit. and Phil. Soc., 1913, *52*, No. 19.

Lord, J.: *Capital and Steam Power, 1750–1800*. (London, 1923.)

Lunge, G.: *The Manufacture of Sulphuric Acid and Alkalis*. (Revised Edition, 1923.)

Lunardi, V.: *Account of Five Aerial Voyagers in Scotland*. (London, 1786.)

*Macadam, W. I.: *The Ancient Iron Industry of Scotland*. Proc. Soc. of Antiquaries, 1887, *21*, 130.

McArthur, J.: *Short Sketches of things old and new in and around Wanlockhead*. Collected from Dumfries and Galloway Herald and Register, 1852.

MacCulloch, John: *Description of the Western Isles of Scotland*. (London, 1819.)

Macdonald, D. F.: *Scotland's Shifting Population, 1770–1850*. (Glasgow, 1937.)

MacGregor, Geo.: *History of Glasgow*. (Glasgow, 1881.)

MacGregor, M., *et. al.*: *Memoirs of the Geological Survey of Scotland: Special Reports on the Mineral Resources of Great Britain: Vol. 11, The Iron Ores of Scotland.* (Edinburgh, 1920.)

Macintosh, Geo.: *A Memoir of Charles Macintosh, F.R.S., of Campsie and Dunchattan.* (Glasgow, 1847.)

Mackenzie, Thos. B.: *Life of James Beaumont Neilson, F.R.S.* (West of Scotland Iron and Steel Institute, 1929.)

Mackenzie, W. C.: *A Short History of the Scottish Highlands and Isles.* 3rd Edition. (Paisley, 1908.)

*McKie, D.: *Daniel Rutherford.* Science Progress, 1935, *29*, 650.

Mackinnon, J.: *The Social and Industrial History of Scotland from the Union to the Present Day.* (London, 1921.)

McLean, A. (Editor): *Local Industries of Glasgow and the West of Scotland.* (Glasgow, 1901.)

Macleod, Donald: *Dumbarton, Vale of Leven, Etc.* (n.d.)

M'Ure, John: *Glasghu Facies. A View of the City of Glasgow*, originally published 1736: several editions with varying titles. (Edited by J. F. S. Gordon.) (4 vols.) (Glasgow, 1873.)

Maiben, J. & Co.: *A Statement of the Advantages to be derived from the Introduction of Coal Gas into factories and dwelling houses.* (Perth, 1813.)

Manning, Jas.: *An Essay on Bread.* (1757.) (Reprinted, London, 1825.)

Manning, Jas.: *The Nature of bread honestly and dishonestly made.*

Mantoux, Paul: *The Industrial Revolution in the 18th Century.* (London, 1927.)

March, R.: *A Treatise on Silk, Wool, Worsted, and Cotton, with Instructions to clean the Manufactures in the Hosiery Branch.* (London, 1779.)

Markham, Peter: *Poison Detected or Frightful Truths: and alarming to the British Metropolis.* (1757.)

*Marshall, T. H.: *Jethro Tull and the New Husbandry of the 18th Century.* Econ. Hist. Rev., 1929, *2*, 4.

Marshall, W.: *Review and complete Abstract of the Reports of the Board of Agriculture from the Northern (1818), Western (1810), Eastern (1812), Midland (1815), Southern (1817) Departments of England.*

*Marwick, H. W.: *The Limited Company in Scottish Economic Development.* Econ. Hist. Rev., 1934–7, *3*, 415.

*Marwick, H. W.: *The Cotton Industry and the Industrial Revolution in Scotland.* Scottish Historical Review, 1924, *21*, 207.

Mathieson, Wm. L.: *Scotland and the Union: A History of Scotland from 1695–1747.* (Glasgow, 1905.)

Matthews, J. M.: *Bleaching and Related Processes.* (New York, 1921.)

Mawe, J.: *The Mineralogy of Derbyshire, with a Description of the most Interesting Mines in the North of England, in Scotland, and in Wales.* (London, 1802.)

Maxwell, Rob.: *Select Transactions of the Honourable the Society of Improvers in the Knowledge of Agriculture.* (Edinburgh, 1743.)

Maxwell, R.: *The Practical Husbandman.* (Edinburgh, 1757.)

Mayne, W.: *Nouveau manuel complet du fabricant de Briquettes et d'Allumettes*. (Paris, 1903.)

Meikle, H. W.: *Scotland and the French Revolution*. (Glasgow, 1912.)

Menzies, G. K.: *The Story of the Royal Society of Arts*. (London, 1935.)

Meteyard, E.: *A Group of Englishmen (1795–1815): Records of the Younger Wedgwoods*. (London, 1871.)

Meteyard, E.: *The Life of Josiah Wedgwood from his private correspondence and family papers*. (2 vols.) (London, 1865–6.)

Miller, And.: *The Rise of Coatbridge and the Surrounding Neighbourhood*. (Glasgow, 1864.)

Mitchell, John: *Treatise on the Falsification of Food and the Chemical Means of Detecting them*. (London, 1848.)

Moilliet, A.: *Sketch of the Life of J. Keir, F.R.S.* (London, 1868.)

Morgan, G. T. and Pratt, D. D.: *British Chemical Industry*. (London, 1938.)

Morgan, Pat.: *Annals of Woodside and Newhills*. (Aberdeen, 1886.)

Morland, W. H.: *From Akbar to Aurangzeb*. (London, 1923.)

Moseley, Benj.: *The Origin, Rise, and Progress of Sugar Refining and Raw Sugar Manufacture*. (London, 1800.)

Muir, R.: *History of Liverpool*. (Liverpool, 1907.)

Muirhead, Jas. P.: *Correspondence of the late James Watt on his discovery of the composition of water*. (London, 1846.)

Muirhead, Jas. P.: *Origin and Progress of the Mechanical Inventions of James Watt*. (3 vols.) (London, 1854.)

Muirhead, Jas. P.: *Life of James Watt, with Selections from his Correspondence*. (London, 1858.)

Mumford, L.: *Technics and Civilization*. (New York, 1934.)

Mumford, L.: *The Culture of Cities*. (London, 1938.)

M(urdoch), A(lexander): *Light without a Wick. A Century of Gas-Lighting, 1792–1892. A Sketch of William Murdoch, the Inventor*. (Glasgow, 1892.)

Murdock, Wm.: *An Account of the application of the Gas from Coal to economical purposes, read before the Royal Society on 25th Feb., 1808*. (London, 1808.)

Murray, David: *The York Buildings Company: A Chapter in Scotch History*. (Glasgow, 1883.)

Murray, James: *Life in Scotland a Hundred Years Ago*. (Paisley, 1905.)

Muschamp, Robert: *History of the Township of Ainsworth*. (Bury, 1930.)

Mushet, D.: *Papers on Iron and Steel, Practical and Experimental*. (London, 1840.)

Muspratt, Sheridan: *Chemistry, Theoretical, Practical, and Analytical, as applied and relating to the Arts and Manufactures*. (2 vols.) (London, 1860.)

Naismith, John: *Thoughts on various Objects of Industry pursued in Scotland*. (2 vols.) (Edin., 1790.)

Nef, J. U.: *The Rise of the British Coal Industry*. (2 vols.) (London, 1932.)

Neill, Patrick: *A Tour through some of the Islands of Orkney and Shetland.* (Edinburgh, 1806.)

Neri, Antonio: *The Art of Glass.* (London, 1662.) Trans. by Dr. Reid Messet, who was appointed English representative of the Leith Glass Works.

Neuburger, Albert: *The Technical Arts and Sciences of the Ancients.* (New York, 1930.)

Newbigging, Thos.: *A Hundred Years of Gas Enterprize.* (London, 1901.)

*Newbury, N. F.: *History of the Common Salt Industry on Merseyside.* Annals of Science, 1938, *3*, 138.

Niccol, Rob.: *Essay on Sugar and General Treatise on Sugar Refining, as practised in the Clyde Refineries.* (Greenock, 1864.)

*Nicholson, F.: *Notes on the Wilkinsons, Iron Masters.* Mem. Lit. and Phil. Soc. Manch. 1905.

Oakley, C. A.: *Scottish Industry Today.* (Edinburgh, 1937.)

Offor, R.: *The Papers of Benjamin Gott in the Library of the University of Leeds.* Reprinted from the *Leeds Woollen Industry, 1780–1820.* Thoresby Society, 1931, *32*, 167–253.

Owen, Hugh: *Two Centuries of Ceramic Art in Bristol.* (London, 1873.)

Owen, Robert: *Life of Robert Owen by Himself.* (2 vols.) (London, 1857.)

Palmer, A. N.: *John Wilkinson and the Old Bersham Ironworks.* (London, 1899.)

Paris, J. A.: *The Life of Sir Humphry Davy.* (2 vols.) (London, 1831.)

Parkes, Samuel: *Chemical Essays.* (4 vols.) (London, 1815.)

Parkes, Samuel: *Chemical Catechism.* (8th Edition.) (London, 1818.)

*Parkes, Samuel: *A Descriptive Account of the Several Processes which are usually pursued in the Manufacture of the Article known in commerce by the name of Tinplate.* Mem. Manch. Lit. and Phil. Soc., 1819, *8*, 347.

*Parkes, Samuel: *Essay on the Comparative Value of Kelp and Barilla.* Trans. Highland Soc., 1820, *5* (1st series), 65.

Partington, J. R.: *The Alkali Industry.* (London, 1918.)

Pasteur, Louis: *Etudes sur le Vin.* (Paris, 1866.)

Paton, A. W. and Miller, A. H.: *British Association Handbook.* (Dundee, 1912.)

Paton, T. S.: *Reports of Cases decided in the House of Lords upon Appeal from Scotland.* (n.p., 1849.)

Peckston, T. S.: *The Theory and Practice of Gas-Lighting.* (London, 1819.)

Péligot, Eugène M.: *Le Verre, Son Histoire, sa Fabrication.* (Paris, 1877.)

Pellatt, Apsley: *Curiosities of Glass Making.* (London, 1849.)

Pennant, Thos.: *Tour in Scotland in 1769.* (Chester 1771.) The King's College, Aberdeen, copy contains MS. notes by Ty. H. of a tour made *c.* 1765.

Pennant, Thos.: *A Tour of London, 1780.* (In *The British Tourists,* by W. Mavor, 1798–1800.)

Percival, T.: *Essays, Medical, and Experimental.* (London, 1777.)

Percy, J.: *The Art of Extracting Iron and Steel.* (London, 1864.)

Percy, J.: *The Metallurgy of Lead.* (London, 1870.)

Phillips, Richard: *The Book of English Trades.* (1823.)

Pilcher, R. B., and Butler-Jones, F.: *What Industry owes to Chemical Science.* (London, 1918.)

Pilkington, J.: *View of the Present State of Derbyshire.* (2 vols.) (Derby, 1789.)

*Pitt, Wm.: *Dundonald's Coke Ovens nr. Birmingham.* Trans. Soc. Encourag. of Arts, 1791, *9*, 131–40.

Playfair, John: *Illustrations of the Huttonian Theory of the Earth.* (Edinburgh, 1802.)

Plot, R.: *Natural History of Staffordshire.* (Oxford, 1686.)

Pococke, Richard: *Tours in Scotland, 1747, 1750, and 1760.* Scottish Historical Society. (Edinburgh. 1887.)

Pococke, Richard: *Travels through England, 1751.* (Camden Society, 1888 and 1889.)

Porteous, Alex.: *History of Crieff.* (Edinburgh, 1912.)

Porteous, J. Moir: *God's Treasure House in Scotland. A History of Times, Mines, and Lands in the Southern Highlands.* (London, 1876.)

Porter, G. R.: *The Progress of the Nation in its various Social and Economic Relations.* (London, 1851.) (Revised Edition, 1912.)

Powell, H. J.: *Glass Making in England.* (Cambridge, 1923.)

Prentice, A.: *Historical Sketches and Personal Recollections of Manchester.* (2nd Edition.) (London, 1851.)

Priestley, J.: *Experiments and Observations on Different Kinds of Airs.* (2 vols.) (London, 1774–5.)

Prosser, Richard B.: *Birmingham Inventors and Inventions.* (Birmingham, 1881.)

Prothero, R. E. (Rowland Ernle, 1st Baron): *The Pioneers and Progress of English Farming* (London, 1888.) Edition under title, *English Farming, past and present.* (London, 1936.)

Ramsay, John: *Scotland and Scotsmen in the Eighteenth Century.* (Edinburgh and London, 1888.)

Ramsay, Sir W.: *Life and Letters of Joseph Black, M.D.* (London, 1918.)

Ray, John, F.R.S.: *Itinery* in P. Hume Brown's *Early Travellers in Scotland.*

Reid, Robert, (Senex): *Glasgow Past and Present.* (3 vols.) (Glasgow, 1874.)

*Report (Second) of the *Committee of the Highland Society* upon the *Manufacture of Kelp.* Trans. Highland Soc. 1820, *5* (1st series), 1.

Reports (Parliamentary): *Distilleries of Scotland* (1799). *Brewing* (1806). *Preserved Meat* (1852).

Richardson, John: *Theoretic Hints on an improved practice of brewing Malt Liquors.* (2nd Edition.) (London, 1777.)

Richardson, John: *Statical Estimates of the Materials of Brewing, showing the Use of the Saccharometer.* (London, 1784.)

*Richardson, T., Stevenson, J. C., and Chapman, R. C.: *On Chemical Manufactures in the Northern Districts.* (Newcastle.) Report of the British Association, 1863, p. 701.

Richardson, W.: *The Chemical Principles of the Metallic Arts: designed chiefly for the use of Manufacturers.* (Birmingham, 1790.)

Rickard, Thomas A.: *Man and Metals: A History of Mining in Relation to the Development of Civilization.* (2 vols.) (New York, 1932.)

Rogers, J. E. T.: *Six Centuries of Work and Wages. The History of English Labour.* (2 vols.) (London, 1894.)

Roll, Erick: *An Early Experiment in Industrial Organisation, being a History of the Firm of Boulton and Watt, 1775–1805.* (London, 1930.)

Routledge, Robert: *Discoveries and Inventions of the Nineteenth Century.* (London, 1896.)

Russell, E. J.: *Soil Conditions and Plant Growth.* (6th Edition.) (London, 1932.)

Salmon, T. S.: *Borrowstoneness and District, c. 1550–1850.* (Edinburgh, 1913.)

Salzmann, L. F.: *English Industries of the Middle Ages.* (London, 1913.)

Saussure, Nicolas Theodore de: *Recherches chimiques sur la végétation.* (Paris, 1804.)

Scott, W. R.: *Joint Stock Companies to 1700.* (3 vols.) (Cambridge, Vol. 1, 1912; Vol. 2, 1910; and Vol. 3, 1911.)

Scott, W. R. *et al.*: *An Industrial Survey of the South-west of Scotland.* (London, Board of Trade, 1932.)

Scrivenor, H.: *A Comprehensive History of the Iron Trade from the earliest Records to the Present Period.* (London, 1854.)

Senex: *see* Reid, Robert.

Shaw, Peter: *Chemical Lectures for the Improvement of Arts, Trades, and Natural Philosophy.* (London, *c.* 1731.)

Shaw, S.: *History of the Staffordshire Potteries and the Rise and Progress of the Manufacture of Pottery and Porcelain.* (Hanley, 1829.)

Shaw, S.: *The Chemistry of compounds used in the Manufacture of Porcelain, Glass, and Pottery.* (London, 1837.)

Simmonds, P. L.: *Science and Commerce, their influence on our Manufactures.* (London, 1872.)

Sinclair, Sir John (Editor): *The Statistical Account of Scotland drawn from the Communications of the Ministers of the different Parishes* (21 vols.) (Edinburgh, 1791–1799.)

Sinclair, Sir J.: *Account of the Origin of the Board of Agriculture and its Progress for Three Years after its Establishment.* (London, 1793.) Contained in *Correspondence to the Board of Agriculture.* (1797.)

Sinclair, Sir J.: *Communications to the Board of Agriculture.* (2 vols.) (London, 1797.)

Sinclair, Sir J.: *Essays on Miscellaneous Subjects.* (London, 1802.)

Sinclair, Sir J.: *Code of Health and Longevity.* (4 vols.) (Edinburgh, 1807.)

Sinclair, Sir J.: *An Account of the Systems of Husbandry of Scotland.* (2 vols.) (Edinburgh, 1814.)

Sinclair, Sir J.: *General Report of the Agricultural State and Political Circumstances of Scotland.* (3 vols. and Appendix, 2 vols.) (Edinburgh, 1814.)

Sinclair, Sir J.: *The Agricultural State of the Netherlands compared with Great Britain.* (London, 1815.)

Sinclair, Sir J.: *Correspondence of the Right Honourable Sir John Sinclair, Bart., with Reminiscences.* (2 vols.) (London, 1831.)

Sinclair, Sir J.: *Analysis of the Statistical Account of Scotland.* (In two parts.) (Edinburgh, 1831.)

Sindall, R. W.: *The Manufacture of Paper.* (London, 1919.)

Smiles, Samuel: *Lives of the Engineers.* (3 vols.) (London, 1861–62.)

Smiles, Samuel: *Industrial Biography.* (London, 1863.)

Smiles, Samuel: *Lives of Boulton and Watt.* (London, 1865.)

Smiles, Samuel: *Men of Invention and Industry.* (London, 1884.)

Smiles, Samuel: *Josiah Wedgwood, F.R.S., his Personal History.* (London, 1894.)

Smith, Adam: *An Inquiry into the Nature and Causes of the Wealth of Nations.* (2 vols.) (London, 1776.)

Smith, Jas.: *Remarks on Thorough Draining and Deep Ploughing.* (n.p., 1833.)

*Smith, R. A.: *A Centenary of Science in Manchester.* Mems. Manch. Lit. and Phil. Soc., vol. 9, 2nd series. (1883.)

Smith, R. W. I.: *Students of Medicine at Leyden.* (Edinburgh, 1932.)

Smollett, T.: *Tour in England and Scotland in 1785.* (London, 1788.)

Solon, M. L.: *History of Old English Porcelain.* (London and Derby, 1903.)

Somerville, Thos.: *My Own Life and Times, 1741–1814.* (Edinburgh, 1861.)

Southern, J.: *A Treatise upon Aerostatic Machines, containing Rules for Calculating their Powers of Ascension.* (Birmingham, 1785.)

Sprat, Thos.: *The History of the Royal Society of London.* (London, 1667.)

Stewart, Geo.: *Curiosities of Glasgow Citizenship.* (Glasgow, 1881.)

Stock, J. E.: *Memoirs of the Life of Thomas Beddoes.* (London, 1811.)

Straker, Ernest: *Wealden Iron.* (London, 1931.)

Straker, Ernest: *Wealden Glass: The Surrey-Sussex Glass Industry (1226–1615).* (London, 1933.)

Surface, G. T.: *The Story of Sugar.* (London and New York, 1910.)

Sykes, Sir Alan J.: *Concerning the Bleaching Industry.* (Bleacher's Association, 1925.)

Tennant, E. W. D.: *One Hundred and Forty Years of the Tennant Companies.* (London, 1937.)

*Thomas, John: *The Pottery Industry and the Industrial Revolution.* Economic History (Supp. to Econ. J.), 1937, *3*, 399.

*Thomas, P. J.: *The Beginnings of Calico Printing in England.* Eng. Hist. Rev., 1924, *39*, 206.

Thomson, John: *Life of William Cullen, M.D.* (2 vols.) (London, and Edinburgh, 1832 and 1859.)

*Thomson, T.: *On the most important Chemical Manufactures carried on in Glasgow and the Neighbourhood.* (B.A. Report, 1840, p. 58.)

Thorpe, T. E. (Editor): Green, Miall, and Others: *Coal, its History and Uses.* (London, 1878.)

Tilden, W. A.: *Chemical Discovery and Invention in the Twentieth Century.* Edited by S. Glasstone. (London, 1936.)

Timmins, S.: *The Resources of the Birmingham Hardware District.* (London, 1866.)

*Timmins, S.: *James Keir, F.R.S. (1735–1820).* Trans. Birmingham and Midland Institute, 1891, *24*, 1.

Tomlinson, C. T.: *Illustrations of Manufacturers and Trades.* (S.P.C.K.)

Tomlinson, C. T.: *Useful Arts and Manufactures.* (Several series published by S.P.C.K. from *c.* 1840.)

Tournée faite en 1788 dans la Grande-Bretagne par un Français parlant la langue anglaise. (Paris, 1790.)

Toynbee, A.: *Lectures on the Industrial Revolution in England.* (London, 1884.)

*Trade: *On the Neglect of Trade and Manufactures.* Scots Magazine, 1740, *2*, 475.

Traill, H. D. and Mann, J. S.: *Social England.* (6 vols.) (London, 1904.)

Troughton, T.: *History of Liverpool.* (Liverpool, 1810.)

Tull, Jethro: *The New Horse Hoeing Husbandry, or an Essay on the Principles of Tillage and Vegetation.* (London, 1731.)

Turner, W. (Editor): *Williams Adams, an Old English Potter.* (London, 1904.)

Unwin, George: *Industrial Organization in the 16th and 17th Centuries.* (Oxford, 1904.)

Ure, Andrew: *The Philosophy of Manufactures, or an Exposition of the Scientific, Moral, and Commercial Economy of the Factory System of Great Britain.* (London, 1835.)

Ure, Andrew: *The Cotton Manufacture of Great Britain Systematically Investigated.* (2 vols.) (London, 1836.)

Usher, Abbot P.: *Industrial History of England.* (Boston, 1920.)

Usher, Abbot P.: *A History of Mechanical Inventions* (New York, 1929.).

Vitalis, J. B.: *Cours Elémentaire de Teinture.* (Paris, 1823.)

Vogel, Max: *On Beer, a Statistical Sketch.* (London, 1874.)

Wadsworth, A. P. and Mann, J. de L.: *The Cotton Trade and Industrial Lancashire, 1600–1780.* (Manchester, 1931.)

Walford, Thos.: *The Scientific Tourist through England, Wales and Scotland.* (2 vols.) (London, 1818.)

*Walker, J.: *An Essay on Kelp containing the Rise and Progress of that Manufacture in the North of Scotland; its present State; and the means of carrying it to a greater extent.* Trans. High. Soc., 1799, *1* (1st Series), 1.

*Walker, J.: *An Essay on the Natural, Commercial, and Economic History of the Herring.* Trans. High. Soc., 1803, *2* (1st Series), 297.

Walker, John: *An Oeconomical History of the Hebrides and Highlands of Scotland.* (Edinburgh, 1808.)

Walker, W.: *Memoir of Distinguished Men of Science of Great Britain living in 1807–8.* (London, 1862.)

Ward, John: *History of Stoke-on-Trent.* (London, 1843.)

Warden, A. J.: *The Linen Trade.* (London, 1864.)

Warington, R.: *Chemistry of the Farm: Morton's Handbooks of the Farm, No. 1.* (8th Edition.) (London, n.d.)

*Watson, J. A. S.: *The Agrarian Revolution in Scotland, 1750–1810.* Trans. High. and Agri. Soc., 1929, *41* (5th Series), 1.

Watson, J. A. S., and Hobbs, M. E.: *Great Farmers.* (London, 1937.)

Watson, Richard, (Bishop of Llandaff): *Chemical Essays.* (5 vols.) (London, 1782–7.)

*Webster, Thos.: *Memoir of Henry Cort.* Mechanics' Magazine, 1859, *2* (2nd Series), 3, 36, 52, etc.

Wedgwood, Julia: *Personal Life of Josiah Wedgwood.* (London, 1915.)

Weir, Dan.: *History of the Town of Greenock.* (Greenock, 1829.)

Weld, C. R.: *History of the Royal Society.* (London, 1848.)

Welford, R.: *History of Newcastle.* (3 vols.) (London, 1884–7.)

Williamson, G.: *Memorials of the Lineage, Early Life, Education and Development of the Genius of James Watt.* (Greenock, 1856.)

Wilson, D.: *Archaeology and Prehistoric Annals of Scotland.* (Edinburgh, 1851.)

Wilson, G.: *A Compleat Course of Chymistry.* (London, 1699.)

Wilson, G.: *The Life of the Hon. H. Cavendish.* (London, 1851.)

Wilson, G. V. and Flett, J. S.: *Memoirs of the Geological Survey of Scotland: Special Reports on the Mineral Resources of Great Britain,* Vol. 17—*The Lead etc., Ores of Scotland.* (Edinburgh, 1921.)

Wilson, Rob.: *Historical Account and Delineation of Aberdeen.* (Aberdeen, 1822.)

Winsor, F. A.: Various *Pamphlets on Gas Lighting.*

Winter, George: *A New and Compendious System of Husbandry, containing the Mechanical, Chemical, and Philosophical Elements of Agriculture.* (Bristol, 1787.)

Wolf, A.: *History of Science, Technology, and Philosophy, 17th Century.* (London, 1935.)

Wolf, A.: *History of Science, Technology, and Philosophy, 18th Century.* (London, 1938.)

Wood, H. T.: *Industrial England in the Middle of the Eighteenth Century.* (London, 1910.)

Wood, H. T.: *History of the Royal Society of Arts.* (London, 1913.)

Woodcock, F. H. and Lewis, W. R.: *Canned Food and the Canning Industry.* (London, 1938.)

Wordsworth, D.: *Recollections of a Tour made in Scotland, A.D. 1803.* (Edinburgh, 1874.)

Wright, G. H.: *Chronicle of the Birmingham Chamber of Commerce, A.D. 1813–1913, and of the Birmingham Commercial Society, A.D. 1783–1812.* (1913.)

Young, And.: *History of Burntisland.* (Kirkcaldy, 1913.)

Young, Arthur: *The Farmer's Letters to the People of England.* (London, 1767.)

Young, Arthur: *Rural Oeconomy.* (London, 1770.)

Young, Arthur: *Political Arithmetic, containing Observations on the Present State of Great Britain and the Principles of her Policy in the Encouragement of Agriculture.* (London, 1st part, 1774; 2nd part, 1779.)

INDEX

C

D

E

F

I

J

N

O